THE MOLECULAR BIOLOGY OF PLANT DEVELOPMENT

BOTANICAL MONOGRAPHS · VOLUME 18

THE MOLECULAR BIOLOGY OF PLANT DEVELOPMENT

EDITED BY

H. SMITH
PhD, DSc, FIBiol
Professor of Botany
University of Leicester

D. GRIERSON
BSc, PhD
Reader in Plant Physiology
University of Nottingham

UNIVERSITY OF CALIFORNIA PRESS
BERKELEY AND LOS ANGELES

UNIVERSITY OF
CALIFORNIA PRESS
Berkeley and Los Angeles, California

Library of Congress Cataloging in Publication Data

Main entry under title:

The Molecular biology of plant development.

(Botanical monographs; v. 18)
Includes bibliographical references.
1. Plants-Development. 2. Molecular biology.
I. Smith, Harry, 1921– . II. Grierson, D.
III. Series.
QK731.M63 581.3 81–43687

ISBN 0-520-04675-7 AACR2

Printed in Great Britain

CONTENTS

CONTRIBUTORS

P. W. BARLOW ARC Letcombe Laboratory, Wantage, Oxon, OX12 9JT, UK.

J. A. CALLOW Department of Plant Sciences, The University of Leeds, Baines Wing, Leeds, LS2 9JT, UK.

M. E. CALLOW Department of Plant Sciences, The University of Leeds, Baines Wing, Leeds LS2 9JT, UK.

E. EARLE Department of Plant Breeding and Biometry, Cornell University, 252 Emerson Hall, Ithaca, NY 14853, USA.

L. V. EVANS Department of Plant Sciences, The University of Leeds, Baines Wing, Leeds LS2 9JT, UK.

J. GRESSEL Department of Plant Genetics, The Weizmann Institute of Science, Rehovot, Israel.

D. GRIERSON Department of Physiology & Environmental Studies, University of Nottingham, School of Agriculture, Sutton Bonington, Loughborough, Leics LE12 5RD, UK.

T. J. V. HIGGINS CSIRO, Division of Plant Industry, PO Box 1600, Canberra City, A.C.T. 2601, Australia.

B. HORWITZ Department of Plant Genetics, The Weizmann Institute of Science, Rehovot, Israel.

C. B. JOHNSON Department of Botany, University of Reading, Whiteknights, Reading, Berks RG6 2AH, UK.

K. KLOPPSTECH Institut für Botanik, Universität Hannover, Herrenhäuser Strasse 2, 3000 Hannover 21, W. Germany.

J. S. LOVETT Department of Biological Sciences, Purdue University, Lilly Hall of Life Sciences, West Lafayette, Indiana 47907, USA.

A. MARCUS The Institute for Cancer Research, 7701 Burholme Avenue, Fox Chase, Philadelphia, PA 19111 USA.

T. P. O'BRIEN Botany Department, Monash University, Clayton, Victoria, Australia 3168.

J. V. POSSINGHAM CSIRO, Division of Horticultural Research, Box 350, G.P.O. Adelaide, South Australia 5001.

J. A. RAVEN Department of Biological Sciences, University of Dundee, Dundee DD1 4HN, UK.

S. RODAWAY Institute for Cancer Research, 7701 Burholme Avenue, Fox Chase, Philadelphia PA 19111 USA.

P. H. RUBERY Department of Biochemistry, Tennis Court Road, Cambridge CB2 1QW, UK.

J. SCHELL Max-Planck-Institut für Züchtungsforschung D-500 Köln 30, W. Germany, *and* Laboratorium voor Genetica, Riksuniversiteit-Gent B-9000 Gent, Belgium.

N. STEELE SCOTT CSIRO, Division of Horticultural Research, Box 350, G.P.O. Adelaide, South Australia 5001.

H. SMITH Botany Department, Adrian Building, University of Leicester, University Road, Leicester LE1 7RH, UK.

D. SPENCER CSIRO, Division of Plant Industry, PO Box 1600, Canberra City, A.C.T. 2601, Australia.

A. TREWAVAS Department of Botany, The King's Buildings, Mayfield Road, Edinburgh EH9 3JH, UK.

D. P. S. VERMA Department of Biology, McGill University, 1205 Avenue Docteur Penfield, Montreal, PQ, Canada H3A 1B1.

P. F. WAREING School of Biological Sciences, Department of Botany, University College of Wales, Aberystwyth, SY23 3DA, UK.

H. W. WOOLHOUSE John Innes Institute, Colney Lane, Norwich NR4 7UH, UK.

PREFACE

The Preface to a text-book usually begins with an encapsulated definition of the title. This book is an exception, for it would be a brave editor indeed who would attempt to provide simultaneous definitions of 'molecular biology' and 'plant development' to satisfy all potential readers. Molecular biology, once the sole territory of those investigating the macromolecules responsible for the inheritance and expression of genetic characteristics, now, through familiarity of usage, emcompasses virtually all quantitative investigations of biological processes carried out at the molecular and cellular levels. The erstwhile concentration on nucleic acids and proteins has been replaced by a more catholic enquiry into the relationships and interactions of all biological molecules. Perhaps the most important central theme of this general approach to biological problems is the primary concern with biological organization, and with the storage, replication, transmission, selection and expression of the information necessary to provide and maintain that organization. Thus, the orientation and deposition of cellulose microfibrils in the primary wall of a plant cell is as much a part of molecular biology as is the nature of the controlling element on the *E. coli* chromosome. Development, quintessentially, is a function of organization; the creation of a giant redwood from a fertilized ovum, or, on a rather smaller scale, the differentiation of heterocysts at regular intervals along a cyanobacterial filament, is each a marvel of controlled organization in time and space. Our natural hope is ultimately to appreciate the infinite beauty of such phenomena from the standpoint of intellectual understanding, rather than merely to experience them as mysterious wonders. The role of molecular biology in the attainment of such an objective will obviously be dependent upon the degree to which molecular biological theory will be found to account for developmental phenomena—and the extent to which it will be found wanting.

This book sets out to provide a balanced treatment of the application of the molecular biological approach to the study of plant development, as exemplified by current knowledge. It does not seek to offer the up-to-the-minute detail found in reports of research symposia; rather, it attempts to identify those selective aspects of plant development which have proved to be particularly amenable to molecular biological techniques. The book was originally conceived as a complement to 'The Molecular Biology of Plant Cells', edited by one of us (H.S.) and published in 1977. Although the present volume concentrates on the developmental biology of multi-cellular plants and depends, to some extent at least, on the coverage of plant cell biology in that volume, nevertheless it has been found necessary to include here certain cellular aspects of the development of plant cells. Following the brief Introduction, the first section of the book comprises a treatment of the principal fundamental concepts of plant development. Here there is an exposition of the scope of the book, with a realistic and, in some senses, quite sceptical consideration of the benefits of the molecular approach to problems of development which appear most intractable at the level of tissues, organs or whole organisms. The authors of these chapters (Chapters 2, 3 & 4) address themselves to the extremely difficult and, to a large extent, philosophical question of whether phenomena expressed at a high level of organization can be understood in terms of mechanisms operating at a much lower level of complexity. Following this fundamental introduction, three case histories of development in lower plants are presented (Chapters 5, 6 & 7). The authors of these chapters concentrate on fundamental questions, such as stored messenger-RNA and the molecular basis of polarity, and provide a framework within which to consider more complex developmental phenomena in multicellular plants. The majority of the rest of the book is concerned with realistic assessments of the progress made in using the molecular approach to understand the phenomena of vegetative development (Chapters 8, 9 & 10), reproductive development (Chapters 11, 12 & 13), the environmental control of development (Chapters 14 & 15) and developmental interactions between plants and other organisms (Chapters 16, 17 & 18). Finally, prospects for the application of the molecular approach towards the understanding of some of the most intractable problems of plant development, including determination, competence and phase changes, are speculated upon in Chapter 19.

This book should not be seen as a research monograph; it is intended as a text-book for advanced undergraduates and post-graduate students, and the individual chapter authors were strongly urged to keep the aspiring student in mind. Putting the book together has taken nearly three years to-date, and therefore some of the chapters may already, in certain respects at least, be outstripped by the rapid march of discovery in this field. The prime consideration of the editors, however, has been to provide a balanced coverage; not merely reflecting the enthusiasm for everything that is 'molecular', but giving the reader the necessary biological framework into which the molecular fragments of the jigsaw are fitted.

March 1982

Harry Smith
Donald Grierson

CHAPTER 1

MOLECULAR BIOLOGY OF PLANT DEVELOPMENT: PROSPECTS AND CONSTRAINTS

Within the last 30 years we have progressed from the first insights into the chemical basis of heredity and gene expression to a position where the purification, sequencing and manipulation of identified genes from bacteria, animals and plants is an everyday occurrence. This exciting revolution in molecular biology has continued at an ever-increasing pace, providing new information relating to gene structure, evolution and function and pointing the way to possible biological manipulations of benefit to humans. Thus an area which was once the preserve of a small group of biologists eager to understand, for its own sake, the basis of cell function and heredity, now attracts agriculturalists, technologists and industrialists in large numbers.

All the indications are that the methods of molecular analysis and manipulation developed by molecular biologists are generally applicable to the solution of cellular problems in plant biology. Achievements which testify that this is so include the isolation, mapping and sequence analysis of chloroplast DNA, the unravelling of the inter-relationships between genome and plastome during chloroplast biogenesis and the elucidation of the molecular basis of plant tumour induction. Other accomplishments seem just over the horizon, such as the determination of the complete structure and mode of expression of plant nuclear genes, understanding the control of cell division by tumour genes from Ti plasmids and learning how instructions for intracellular transport of proteins are genetically specified. However, although it is important to recognize how plant biology has benefited from the introduction of the molecular approach, it would be unrealistic to expect that the same methods will solve all botanical problems.

TRENDS AND FASHIONS

Science, as other human endeavours, tends to proceed in fashions. New discoveries, particularly those of a spectacular nature, set trends for all connected subject areas, leading often to an exciting explosion of ideas and information, but carrying the related danger of an uncritical transfer of hypothesis to problems as yet unready for solution. The too-facile application of molecular biological theory to major questions of plant development may already have been a contributory cause of the reluctance of many botanists to recognize the revolutionary power of the molecular approach. This unfortunate tendency may be illustrated effectively by a brief consideration of the fashions in auxin and phytochrome research.

The discovery and chemical identification of auxin almost half a century ago led to an enormous proliferation of what would now be regarded as 'phenomenological' experimentation, through which the significance of auxin as a regulator and integrator of growth and development in the whole plant became evident. During this period, ideas on the mechanism of action of auxin, although primitive, centred on the mechanics of cell enlargement—e.g. effects on Ca^{2+} cross-linking of cell wall polysaccharides, on water pumping or on vacuolar osmotic potential. When the molecular biological revolution broke in the early 1960s such concepts were swept away, to be replaced by the more fashionable hypothesis that auxin operates via the regulation of gene transcription. In a sense, this enthusiasm for molecular biological ideas, although providing the impetus and techniques for much-needed analysis, impeded overall progress, due to a reluctance to accept that certain actions of auxin could not be explained through differential control of gene expression. It is, ironically, probably true that the more balanced view which now prevails, in which it is realized that auxin almost certainly acts both on membranes and, although perhaps indirectly, on gene expression, is a direct result of the more recent fashionable interest in membranes and membrane-associated receptors.

A remarkably similar sequence of ideas can be discerned with regard to phytochrome action. Although evidence that phytochrome could act extremely rapidly, and was probably associated with membranes, already existed, it was possible in 1964, as a direct result of discoveries in the regulation of enzyme synthesis in bacteria, for the general hypothesis to be proposed that phytochrome operates through the selective regulation of gene transcription (See Chapter 14). Although the poverty of this concept was quickly evident, it has taken many years for a more balanced view, again based on the 'membrane fashion', to emerge. Even this fashion, however, has been taken to extremes resulting in the proposal, based on minimal data, of the existence of specific membrane-associated receptors for phytochrome.

Such wholesale and uncritical transfer of hypothesis is clearly to be avoided. On the other hand, and as discussed in many of the chapters in this book, the application of the *techniques* of molecular biology to problems of plant development is of great value.

THE MOLECULAR APPROACH

Molecular biology can be viewed as a basic tool for the appreciation of biological complexity: it has shown us that an understanding of structure leads to an understanding of function. Although the emphasis often seems to be placed on nucleic acids, molecular biology is proving just as valuable in understanding the organization and function of proteins and pigments involved in light harvesting and electron transport in thylakoid membranes as it has been in providing details about nucleosomes and ribosomes.

More recently, the genetic aspects of molecular biology have received great publicity. The purification of restriction enzymes for the molecular dissection of DNA sequences and the development of methods for ligating DNA segments together has made it possible to clone specific DNA sequences in bacteria and yeast. Gene segments isolated and multiplied in this way can now be sequenced and various functions ascribed to particular regions. Furthermore, mRNA molecules can also be copied into cDNA, using reverse transcriptase, and these molecules cloned and sequenced. This has led to the discovery that genes sometimes contain transcribed 'leader sequences' which may be important for recognition or control. Sometimes parts of the leader sequence are translated to generate precursor amino acid sequences at the N-terminus of mature proteins that provide a signal code for the destination of the proteins. In addition at least some genes are 'split' so that coding regions (exons) are interrupted by non-coding regions (introns). Although present in genomic DNA, introns are not present in mRNA. The pathway leading to mRNA production is therefore quite complex. It begins with the binding of the appropriate RNA polymerase to the gene and the production of an Hn-RNA (heterogeneous nuclear RNA) transcipt substantially larger than the final mRNA. Subsequently, sequences may be removed from the 5′ and 3′ ends of the molecule and specific additions may be made, such as the 'capping' of the 5′ end or the addition of poly A to the 3′ end. Finally, nucleolytic processing enzymes remove the transcribed introns from the mRNA precursor and ligate the mRNA coding sequences to produce an mRNA molecule colinear with its protein product. Clearly the details of this pathway are of special interest to those studying the control of gene expression. Perhaps more importantly the methods which have been devised to study these processes are now finding general application in plant biology.

The key to some recent successes has generally been the identification and purification of a specific protein known to be important at a particular stage of the life of a plant cell. Work has generally concentrated on proteins present in large quantities, because the purification is easier. Examples are the large and small subunits of the chloroplast CO_2-fixing enzyme ribulose bisphosphate carboxylase (Chapter 9), cereal storage proteins (Chapter 12), leghaemoglobin in root nodules of legumes (Chapter 16) and glyoxysomal enzymes. Once the protein has been purified the fingerprint pattern or a specific antibody can be used to show the subcellular location and measure the amount of translatable mRNA coding for that protein. For example, the mRNA for the small subunit of ribulose bisphosphate carboxylase, the product of a nuclear gene, is located in the cytoplasm and the large subunit, coded for by the plastid DNA, is located in the chloroplast. Further, gibberellic acid has been shown to stimulate the accumulation of α-amylase mRNA in barley aleurone layers (Chapter 13). The identification and purification of these mRNAs has led to the production of DNA copies of the mRNA and in some instances to the identification and sequencing of the parent genes. Thus, the genes for *Phaseolus* storage proteins have introns (Chapter 12) and the gene for the small subunit of ribulose-bisphosphate carboxylase codes for 'extra' N-terminal amino acids which function in the uptake of the small subunit precursor protein by chloroplasts.

In addition to providing this type of information there are other advantages of cloning genes. A cloned gene sequence can be used for the quantitative measurement of the amount of mRNA coding for a specific protein present in a given cell type. Clones can be used to detect Hn-RNA molecules and processing intermediates, thus helping to work out the processing sequence. They can also be used to determine the nucleotide sequence of regulatory and coding regions in a gene, and as defined genes whose transcription and translation can be studied *in vitro*.

This technology is capable of giving answers to very precise questions about gene expression: it can in principle determine whether a hormone or other signal induces the synthesis or accumulation of a specific mRNA, whether the gene was previously being transcribed to Hn-RNA or not and whether specific post-transcriptional controls operate in a given situation. It also offers the possibility ultimately of testing aspects of gene-based theories of control in a test tube.

Students of animal embryology have recognized for a long time the importance of the interaction between nucleus and cytoplasm during development. This concept seems especially appropriate for plants since they frequently respond to external stimuli, either from the environment or from neighbouring cells. Thus hormone gradients, light quality, photoperiod and temperature can all have profound effects on plant development and it is clear that the perception of at least some of these stimuli occurs outside the nucleus sometimes in association with membranes. Furthermore, there is much evidence to suggest that other membrane properties and functions can play an important part in controlling development (Chapters 2 and 3). There is undoubtedly considerable complexity involved in this area of research which is discussed at some length in later chapters. Some argue that the problems of plant development are too complicated for solution by the molecular reductionist approach (Chapter 4). Nevertheless the conviction of others is that wherever our ideas about a particular process have been sufficiently well worked out to allow the identification of the appropriate cellular components involved, a detailed study of their structure and properties will be necessary before the perception and transduction mechanisms can be properly understood. Here also, application of the techniques of molecular biology as applied to protein synthesis, structure and function should pay handsome dividends.

FUNDAMENTALS OF PLANT DEVELOPMENT

CHAPTER 2

POSSIBLE CONTROL POINTS IN PLANT DEVELOPMENT

2.1 INTRODUCTION

It is a common assumption by plant biologists that differentiation and development result from changes in gene expression; the visible differences in structure of various cell types and their specific and significant metabolic characteristics result from the possession of unique patterns of proteins and enzymes. Furthermore, it is also commonly assumed that the origin of this process lies in a continuing interaction between nucleus and cytoplasm. Controlling factors in the cytoplasm are unequally partitioned during the first cell division of the fertilized embryo. The interaction of these with an initially constant genome and continuing cell proliferation eventually induces the selective (and reversible) gene expression associated with specific cell and organ types.

These 'epigenetic' theories of development, perhaps most clearly stated in plant terms by Brown (1963) and Bonner (1965), concentrated attention on events at the level of transcription. Both assumed that the production of new types of messenger RNA were the critical controlling features. At the time these ideas were produced, there was an almost total dearth of experimental molecular data involving plant systems. These theories therefore leant heavily on the well-established concepts of animal development, particularly those derived from embryological research.

A decade and a half later we can afford to be more wary of any such simple parallels between plants and animals. In evolutionary terms these two major groups of organisms separated as single cells. Developmental interactions and controls which obviously involve groups of cells have therefore evolved in isolation from each other. It is now perhaps clearer just how radical those developmental differences actually are, answering as they do the specific needs of the characteristic life style of each; sessile and autotrophic in plants but motile and heterotrophic in animals.

This chapter provides not only some critical assessment of the epigenetic/transcription theory but also outlines a molecular basis for some of the novel developmental properties of plants. This indicates the growing importance which membrane systems and electrical controls are occupying in the minds of plant developmentalists. It may also help to counteract the all-too-common 'plant is a green animal' syndrome with its obsession for gene-centred control.

2.2 PROTEIN AND MESSENGER RNA DIFFERENTIATION ACCOMPANY DEVELOPMENT

2.2.1 PROTEIN OR ENZYME DIFFERENTIATION DURING DEVELOPMENT

Research in the last decade has revealed that many aspects of plant development are accompanied by specific changes in enzyme activity. Table 2.1 lists a number of these and includes the developmental stage or specific tissue or cell stage involved, with an indication of the metabolic change induced by the accumulation of the enzyme. Those listed in Table 2.1 have been included, either because their possession by discrete tissues confers a significant and specific metabolic property or because the appearance or increased activity of the enzyme induces a visible developmental event.

A number of these enzymes (marked by asterisk) seem only to be detectable within the tissue indicated, although their apparent absence elsewhere may simply represent a limitation in sensitivity of the assay procedure. For those enzymes which are not uniquely located, selection has been limited to enzymes whose activity increases by a factor of 10-fold or more; in the majority of examples the increase is upwards of 100-fold. For nearly all the enzymes or proteins listed in Table 2.1 the factors controlling their appearance are unknown. In the remainder, light or growth substances appear to be the regulating stimulus.

The list covers enzymes concerned not only with major aspects of specific organ development and differentiation of specific tissues but also with such general phenomena as embryogenesis and germination. The evidence in Table 2.1 should leave little doubt that enzyme changes accompany, and indeed are responsible for some aspects of plant development, but it would be a mistake to extrapolate that notion to all. With the present state of knowledge it is rather easy to identify enzymes responsible for visible chemical changes (e.g. chlorophyll synthesis or starch breakdown) and the enzymes listed in Table 2.1 have been mainly identified by this means. Many other aspects of development have no such easy enzyme candidate to test the hypothesis. What enzymes for example, are responsible for determining the plane of, or the initiation of cell division, the sphericality of many fruits or the shape of a leaf? Are these processes a function of enzyme changes at all? Those who believe implicitly that all development is the result of protein or enzyme changes should be aware that while the initial evidence supports this contention, when much fuller information is available protein changes may be seen to be much more trivial than might at first have been supposed.The morphological changes induced by naked, apparently untranslated RNA molecules such as the potato spindle tuber viroid indicate one exciting alternative (Hall *et al.*, 1974; Davies *et al.*, 1974).

2.2.2 SPECIFIC MESSENGER RNA ACCUMULATION ACCOMPANIES DEVELOPMENT. THE DIFFICULTY OF DEMONSTRATING TRANSCRIPTIONAL CONTROL

It is perhaps a measure of the times that publication of Table 2.1 a decade ago would have been immediately followed by a consideration of whether such changes represented de-novo synthesis. Technical advances seem largely to have made that problem redundant. Now that methods for assaying mRNA are so widely used, more pertinent questions can be asked concerning the mechanisms of enzyme accumulation. Table 2.2 summarizes some of this information. It lists the enzymes whose accumulation is likewise accompanied by an accumulation of the appropriate mRNA. In only one case, cellulase, has it been established that there is a distinct lag between appearance of mRNA and the enzyme. In the others, where time courses are available, there is a close correspondence between mRNA and enzyme/protein accumulation. Such results are often loosely (and incorrectly) referred to as illustrating specific transcriptional control. While it is certainly true that an increased rate of transcription of a particular gene can give rise to an accumulation of mRNA a brief reiteration of the pathway of mRNA synthesis will illustrate the difficulties in interpretation.

As far as is known at present, genes are not represented by contiguous nucleotide sequences on DNA, but have interspersed sections. The initial RNA transcript (Heterogeneous nuclear RNA or HnRNA) has these additional interspersed pieces removed during a stage of nuclear processing. A capping sequence and a stretch of poly A up to 200 nucleotides in length is added and the mRNA may then combine with proteins before passing through the nuclear membrane to the cytoplasm. In the cytoplasm, mRNAs compete with each other for the available ribosomes, ATP, initiation factors etc. Messenger RNAs are subject to degradation both by nucleoplasmic and cytoplasmic ribonucleases (details of the process of mRNA synthesis may be found in Jordan *et al.*, 1979 and Grierson, 1981).

Each of these steps in this rather lengthy synthesis sequence represents a possible control point at which accumulation of the final mRNA could be controlled. For example, convincing evidence indicates that many single copy nuclear transcripts (i.e. RNA transcripts of structural genes) are degraded in the nucleus and never reach the cytoplasm (Ford, 1976). Polyadenylation or combination with proteins confers stability upon the mRNA and could thus increase the chances of movement to the cytoplasm, with subsequent translation (Harris & Dure, 1978; Ford, 1976). Again there is evidence that selective degradation of HnRNAs occurs in the nucleus (Ford, 1976). This could either reflect variation in secondary structure of HnRNAs and thus resistance to ribonuclease digestion or some selective gating mechanism which only permits the transit of specific mRNAs to the cytoplasm at a particular time. There is, however, an alternative simpler mechanism which follows directly from these.

Table 2.1 Enzyme differentiation in plants. This table includes only the enzyme changes which are (a) characteristic of a particular cell or tissue type or (b) responsible for visible developmental changes

Developmental stage	Tissue or cell type involved in change	Metabolic change induced	Enzymes or proteins responsible or involved	Controlling factor (if known)	Ref.
Embryogenesis	Cotyledon or endosperm	Starch accumulation	Starch synthase Sucrose synthase		1
		Storage protein accumulation	*Zeins, *Hordeins, *Glutellins		2
Seed germination	Cotyledon and aleurone	Metabolism of fats by glyxoylate cycle	*Isocitrate lyase *Malate synthetase		3
	Aleurone	Starch and protein hydrolysis	Amylase, Protease	Gibberellin, osmotic pressure	4
Shoot development	Most etiolated shoots	Chlorophyll synthesis and breakdown	δAla synthetase Chlorophyllase	Light	5
	Stem and leaf (absent in mesophyll cells of C_4 plants)	Major enzymes for Calvin cycle	*Ribulose bisphosphate carboxylase *Phosphoribulokinase		6, 7
	Stem and leaf	Plastid replication	Chloroplast membrane and structural proteins		6
	Etiolated stem and leaf	Photosynthetic light collection	*Chlorophyll a/b harvesting protein	Light	8
	Leaf, stem and some cotyledons	Photorespiration	Peroxisomal enzymes	Light	3
Leaf development	Mesophyll cells of C_4 plants	Hatch-Slack pathway	*Pyruvate phosphate dikinase		9
	Abscission zone	Abscission	Cellulase	Ethylene	10
	Gland cells in insectivorous plants	Digestion of prey	Proteases	Uric acid	11, 12

	Leaf hairs in *Urtica*	Stinging principle	Choline transacetylase		13
Adventitious root development	Cortex	Cavity formation	Cellulase	Auxin	14
Root nodule formation	Cortex	Anaerobic atmosphere for nitrogen fixation	Leghaemoglobin	Rhizobial infection	22
Fruit development	Ripening fruit	Softening and pH changes	Amylase, Malate decarboxylase, Pectinase, Polygalacturonase	Ethylene	10, 23
Flower development	Petals (some leaf and fruit epidermal cells)	Flavonoid production	*Flavone synthetase, *Chalcone isomerase *Chalcone methyl transferase	Light	15
Cell differentiation	Epidermal cells	Cutin formation	*Palmitate hydroxylase activity		16
	Laticifers	Rubber formation	*Rubber transferase		17
	Phloem	Callose synthesis	*Callose synthetase		
		Phloem movement	*P-Protein		18
	Xylem	Lignin formation	Phenylalanine ammonia lyase, cinnamic acid 4-hydroxylase, Coumarate Co A ligase		15
	Root cap	Slime formation	*Slime synthetase		19
Death	Most cell types	Cell lysis	RNase, DNase β1-3 Glucan hydrolyase		20, 21

* Enzyme unique to organ, tissue or cell type.

References: 1. Perez *et al.* (1975) 2. Boulter (1973) 3. Beevers (1979) 4. Yomo and Varner (1971) 5. Bogorad (1976) 6. Bradbeer *et al.* (1974) 7. Link *et al.* (1979) 8. Apel and Kloppstech (1978) 9. Hatch and Slack (1968) 10. Abeles (1973) 11. Robins (1976) 12. Heslop-Harrison (1975) 13. Barlow and Dixon (1973) 14. Fan and McClachlan (1967) 15. Hahlbrock and Grisebach (1974) 16. Soliday and Kolattukudy (1977) 17. Archer and Cockban (1969) 18. Cook and Stoddart (1973) 19. Harris and Northcote (1970) 20. Mathile and Winkenbach (1971) 21. Moore and Stone (1972) 22. Verma *et al.* (1974) 23. Tucker *et al.* (1980).

Table 2.2 Developmental enzyme and protein changes which are accompanied by accumulation of the appropriate mRNA

Developmental event concerned	Enzyme or protein mRNA which accumulates	Ref.
Embryogenesis	Seed storage proteins including zeins, hordeins and globulins	1–4, 13–17
Germination	Amylase	6
	Malate synthetase	7
	Isocitrate lyase	7
Shoot development	Chlorophyll a/b light harvesting protein	8
	RuBP carboxylase	5, 8, 18, 19
	Differential accumulation of RuBP carboxylase between mesophyll and bundle sheath cells of C_4 plants	9
Vascular tissue and flower formation	Phenylalanine ammonia lyase	10
	Flavonone synthetase, UDP-Apiose synthetase	
Adventitious root formation	Cellulase	11
Root nodule formation	Leghaemoglobin	12

References: 1. Larkins *et al.* (1976) 2. Miflin *et al.* (1979) 3. Hall *et al.* (1978) 4. Evans *et al.* (1979) 5. Cashmore (1979) 6. Higgins *et al.* (1976) 7. Weir *et al.* (1980) 8. Apel and Kloppstech (1978) 9. Link *et al.* (1979) 10. Schroder *et al.* (1979) 11. Verma *et al.* (1975) 12. Verma *et al.* (1974) 13. Jones *et al.* (1977) 14. Sun *et al.* (1978) 15. Fox *et al.* (1977) 16. Beachy *et al.* (1978) 17. Luthe & Peterson (1977) 18. Cashmore *et al.* (1978) 19. Ellis (1979).

That mRNAs vary in their stability to hydrolytic enzymes is a property they share in common with proteins, but it does mean that non-specific stimuli to mRNA synthesis are likely to give rise to specific accumulation of selected mRNA species. The details of the mechanism are considered at length in Kafatos (1972) and in an abbreviated version for proteins in Trewavas (1976a). The interested reader is referred to both of these for more information. The conclusions of this argument may be simply stated. *In any population of macromolecules of varying stability a non-specific stimulus to synthesis results in vastly different rates of accumulation for the individual members of the population. Those macromolecules with the highest turnover rates accumulate more quickly than those with lower turnover.* Apparently-specific mRNA accumulation could arise from a very non-specific stimulus to overall mRNA synthesis. The importance for cytoplasmic mRNA levels may be even more dramatic since in this case the effective rate of accumulation is equal to the rate of release from the nucleus. Factors non-specifically stabilizing mRNAs in the nucleus (such as polyadenylation or protein combination) or which lead to a greater rate of release from the nucleus could lead to an 'apparent' increased rate of cytoplasmic mRNA synthesis and thus to specific modulation of cytoplasmic mRNA levels.

Conversely, if mRNA synthesis were non-specifically inhibited (or degradation increased) the differential stability amongst the population would ensure the relative survival of the most stable. As a proportion of the mRNA population in fact the stable mRNAs would appear to increase. Furthermore, with competition among mRNAs for available ribosomes the more stable mRNAs would be presented with a greater chance of translation. The super-induction of a number of enzymes, produced anachronistically with the transcription inhibitor actinomycin D, appears to occur by just such a mechanism (Palmiter & Schimke, 1973); the enzymes which are super-induced are translated from the more stable messengers. The possibility of competition amongst mRNAs for ribosomes raises some interesting points since it is clear from the data of Palmiter and Schimke that the most long-lived mRNA species compete less effectively. Perhaps really stable mRNAs can only be effectively translated when other shorter-lived cellular mRNAs have decreased substantially in amount. In this respect they would then appear to possess the characteristics of exceptionally long-lived or even stored mRNA and only be translated at specific times. It must be added that the mRNAs of most terminally-differentiating cells seem to be very stable (Kafatos, 1972; Ford, 1976).

Competition amongst mRNAs for ribosomes may not only occur *in vivo*. The standard method used in fact for assaying most of the mRNAs listed in Table 2.2 is *in vitro* translation and this is likely to suffer from the same defect. A recent and careful analysis by Schwinghamer and Symons (1977) using purified mRNAs and four different *in vitro* translation systems illustrates the immense difficulties investigators face in interpretation. Consistent results between the different translation systems and the different RNAs were hardly obtained and what was translatable in one system was either not translatable in another or yielded different products.

The simple equation, 'apparent' mRNA accumulation equals transcriptional control, should be seen by now to be not sustainable. Direct demonstrations of transcriptional control are in fact extremely difficult and alternative methods of mRNA assay (such as complementary or copy DNA) would seem now to be essential if a critical demonstration of even mRNA accumulation is to be accomplished. As a first step however, the indications are promising for the molecular basis for enzyme variations to be understood. Certainly with the wealth of enzyme differentiation available (Table 2.1) there is no shortage of systems for investigation.

2.2.3 ARE PROTEIN OR ENZYME LEVELS CONTROLLED AT THE LEVEL OF TRANSLATION AND PERHAPS POST-TRANSLATION?

To prove that an enzyme is being controlled at the translational level it is essential to show that specific mRNAs for proteins which are not being synthesized exist concurrently in cells with other mRNAs which are being translated. Using this criterion the control of cellulase levels by auxin, in which there is an approximate 24 hour lag between mRNA accumulation and subsequent cellulase synthesis, may qualify as a translational control (Verma *et al.*, 1975). It should be borne in mind that the estimates of cellulase mRNA were a very early use of *in vitro* translation assays and caution should be exercised concerning their exact quantitation as a consequence.

An enzyme often discussed in the context of translational control is the protease of cotton cotyledons (Ihle & Dure, 1972). This protease is synthesized *de novo* upon the onset of germination and is partly responsible for the mobilization of storage proteins.Enzyme accumulation is, however, unimpeded even if the seeds are germinated in the presence of high concentration of actinomycin D. This implies that the protease is translated from a stable mRNA which was transcribed prior to germination, i.e. during embryogenesis. Using precocious germination as an investigative tool, Ihle and Dure (1972) were able to pinpoint a time in embryogenesis at which they considered protease mRNA transcription commenced.

The problem here is that the evidence for translational control really hinges on the notion that actinomycin D will inhibit protease mRNA transcription. However, it has been clearly shown that the transcription of mRNAs for a number of enzymes can escape inhibition by this substance (Ford, 1976) and this must throw the evidence for protease mRNA into considerable doubt. Clearly more exacting evidence is going to have to be produced before translational controls can be considered to operate in this case; some form of direct mRNA assay would seem to be the minimum requirement.

The question of stored mRNA is often brought up in the context of germination studies. There is no doubt that RNA can be extracted from ungerminated seeds and translated *in vitro* (see Chapter 13). Furthermore, some of the labelled products are at least identical (on a two-dimensional gel basis) with cytoplasmic proteins synthesized *in vivo* during normal germination (Trewavas, 1979d). Whether these 'stored' mRNAs are capable of directing proteins unique to germination is, however, the crucial question since they may merely represent relatively stable mRNAs directing the synthesis of 'housekeeping' proteins in both embryogenesis and germination. There is at present no published evidence which convincingly discriminates between these two possibilities.

Critical and exacting evidence for translational control of an enzyme in plants is in fact lacking at present, but this may simply result from the paucity of work in these areas rather than the absence of such control mechanisms. One line of investigation which may show promise for the future is the indication that the synthesis of the rough endoplasmic reticulum be a necessary precedent to the synthesis and secretion of enzymes such as cellulase and amylase (Yomo & Varner, 1971; Trewavas, 1976a). If confirmed, such results would suggest that a wide variety of enzymes may be subject to a similar type of control.

Although post-translational control is a well-established phenomenon in animals it has proved to be far more difficult to discover to what extent it occurs in plants. The critical criterion is the appearance or disappearance of biological activity in the absence of synthesis or degradation of the protein or enzyme. An unusual form of such control was demonstrated in germinating lettuce cotyledons (Mayer & Poljakoff-Mayber, 1975) in which the apparent appearance of protease activity was shown to result from the degradation of an inhibiting protein. Although many plant proteins are phosphorylated (Trewavas, 1976b) or glucosylated, evidence that this form of post-translational modification is used to control biological activity is almost non-existent. Pyruvate dehydrogenase is the only known plant enzyme at present in which activity can be modified by phosphorylation (Rubin & Randall, 1977) whilst histone I phosphorylation may play an essential role in chromatin condensation during mitosis in artichoke much as it does in other tissues (Trewavas & Stratton, 1977).

The general features of allosteric control were outlined in Volume 14 of this series (Davies, 1977), but from the developmental point of view the interaction of growth substances with their putative protein receptors should represent an important example of this phenomenon. Again the plant scientist faces a dearth of experimental fact (see Chapter 3). Only one binding protein has been convincingly isolated (Venis, 1977) and this is specific for auxin. Unfortunately, this protein has no known function and is located in the rough endoplasmic reticulum and it would be premature in that case to refer to it as a receptor. Even the existence of binding proteins for other growth substances is at present in an uncertain state and the problem will probably not be resolved until a satisfactory molecular function can be ascribed to these putative receptor proteins and suitable *in vitro* systems developed for their detection and assay (Trewavas, 1979a).

2.3 PLANTS POSSESS NOVEL DEVELOPMENTAL PROPERTIES. THE CONCEPT OF PLASTIC DEVELOPMENT

The evidence quoted in the previous sections is certainly consistent with the epigenetic transcription theory of development as described in the introduction. This leads to a central dilemma which must now be considered.

This theory of development originated and has derived its main supporting evidence from animal embryological studies. Here, development is a highly-controlled and exactly-specified phenomenon. Once a developmental choice is initiated, animal cells and tissues remain irreversibly fixed in their chosen pathway, a process aptly described as canalization. Individual animal cells are totipotent only as far as the four-celled embryo and regional distributions of cytoplasmic-located genome-controlling factors in the unfertilized egg have been commonly observed. Furthermore, it is obvious from the movement of many animal embryonic cells and their exact positioning after such movement that an elaborate surface-sensing mechanism is an essential part of the process (Ebert & Sussex, 1970).

Developmental control features of this sort are essential if the production of the specialized, highly-organized animal is to be reproducibly reiterated. Most higher animals have well over 100 cell types (Calow, 1979) and the majority of these have individually-recognizable cytoplasmic characteristics. The presence of numerous cell-type specific proteins such as keratin, silk fibroin, haemaglobin etc., leads naturally to developmental models in which precise transcriptional control and a continuous monitoring of gene expression by nucleo-cytoplasmic interactions form the essential framework. The poor, almost non-existent, regenerative capacity of most animal tissues, organs and cells suggests that gene expression in development is an essentially irreversible process with progressive repression as development proceeds (Caplan & Ordahl, 1978).*

In contrast to the above, plant development and organization is extraordinarily plastic. The properties which give rise to this notion are listed in Table 2.3 and described in greater detail elsewhere (Trewavas, 1976a; Trewavas, 1979a–c). The open-ended nature of plant development and the lack of commitment of many cells and tissue which thereby remain labile become the essential features. Totipotency and truly-remarkable regenerative capabilities are commonplace whilst cell surface contact and cell movement are almost non-existent. Some of these unusual plastic properties may derive from the very simplicity of plant organisms. At the most there are only some 15–20 cell types (Bloch, 1965) and the majority of these are only distinguishable by virtue of an unusual cell wall structure rather than specific cytoplasmic differentiation. Secretion is therefore a critical event in plant cell differentiation. Plastic development is an evolutionary response to the sessile form of life, in which the environment and predation have to be accommodated rather than avoided; it enables the plant to avoid irreversible commitments in any situation. As pointed out in the introduction, such properties have evolved in isolation; their molecular basis may therefore be entirely novel.

This brings us to the basic dilemma. If the epigenetic/transcription theory is invoked to account for the very characteristic specific features of animal development can it be used to explain phenomena in plants which, whilst bearing the same name, may be fundamentally different? Plant scientists have yet to face the unpalatable fact that in attempting to understand plant development, they may have elected to examine only those features of plant development which bear similarity to those in animals (the production of vascular tissue, for example) but have ignored much more interesting and novel characteristics. The evidence quoted in the previous sections as agreeing with transcription theory may therefore represent no more than a pronouncedly unbalanced view of the mecha-

* Nuclear transplantation experiments are often quoted as evidence against irreversibility in gene expression in animals. Caplan and Ordahl (1978) contain a reasoned and correct assessment of the very limited conclusions which can be drawn from the data on this point.

Table 2.3 Properties of plants which indicate that development and organization are plastic

1. *Regeneration*
Whole plants can be regenerated from sections of shoot or root, from leaves, isolated buds, shoot meristems, bulb scales, from pollen and even single cells or protoplasts in culture.

2. *Callus formation and meristem reorganization*
By appropriate nutrient and hormone manipulation most plant tissues can be induced to produce callus. Both shoot and root meristems can be reorganized from this dividing mass by modification of hormone balance.

3. *Pattern formation*
Complexity of vascular tissue patterns can be dependent on shoot or root size and may be altered by environmental manipulation. Complexity of leaf patterns may be dependent on shoot size.

4. *Use of 'nutrients' as developmental agents*
Mineral nutrients, water or sucrose can modify apical dominance in roots and shoots and thus control branching by means of regulating cell division. Sucrose or water drastically modify leaf shape, stomatal frequency and can control embryogenesis and vascular tissue formation.

5. *Meristem plasticity*
Shoot meristems can be surgically divided into six parts and each of them becomes an autonomous meristem in its own right. Meristems can be rendered dormant by dehydration or by growth substances and be simply reactivated.

6. *Genome plasticity*
New genotypes can be induced in flax and tobacco by treatment with inorganic ions. Phenomenon of callus habituation.

7. *Polarity plasticity*
Although axial polarity is normally constant and irreversible, in a small number of root and shoot cuttings it may be reversed by simply turning the cutting upside down (e.g. Massart, 1922).

8. *Cell plasticity*
Moss buds can be induced by cytokinin and dedifferentiate if the stimulus is removed. Mesophyll cells of *Monstera* can dedifferentiate and redifferentiate to epidermal cells.

Data summarized from Trewavas (1976a); Trewavas (1979a–c).

nisms of plant development and the true picture when it emerges may be very different.

Some indications of the underlying molecular mechanisms involved in plastic development can be deduced from their properties. Totipotency and the phenomenon of meristem regeneration suggests that the plant genome must be reversibly controlled rather like that of bacteria (the influence of nutrients on growth and development may be a further similarity) and not irreversibly as in animals (Caplan & Ordahl, 1978). To initiate regeneration, it is usually necessary to excise a part of the plant thus removing adjacent or surrounding tissue, and interrupting cell-to-cell communication (Carr, 1976). Treatment of whole tissues, with hormones or nutrients, is rarely sufficient in itself. Such results imply that the developmental state of many cells in a plant is continuously maintained by a set of signals from the surrounding tissues. When these signals are interrupted developmental reiteration results. Clearly the permeability characteristics of the cell membrane become important in this context and will now be considered. As will be seen, a number of plastic developmental properties become explicable when the results of plant membrane studies are considered.

2.4 MOLECULAR BASES FOR DEVELOPMENTAL PLASTICITY

2.4.1 THE RELATIONSHIP OF MEMBRANES AND MICROTUBULES

The basic matrix of the membrane is considered to be the lipid bilayer in which hydrophobic proteins are found to be dissolved. In recent years it has been found that membranes exhibit fluid characteristics; the structure is thought to be fluid in the sense that individual molecules are able to diffuse readily in the plane of the membrane to form regions of distinctive affinity or character.

Perhaps in an analogous way to proteins in aqueous solution, individual membrane proteins can be made to aggregate (precipitate) by appropriate manipulation of experimental conditions, e.g. antibody treatment, pH changes etc. Such aggregates, now termed patches, were first convincingly demonstrated by Pinto da Silva (1972) but have been observed in membrane structures as far apart as angiosperm chloroplast (Torres-Perrera *et al.*, 1974) and *Fucus* (Peng & Jaffe, 1976). Under extreme conditions (lectin treatment, for example) all patches may aggregate to one single membrane region to form a cap (Edelman, 1976). An important aspect of patch formation is that it may exhibit threshold characteristics (Gershon, 1978). Patches below a certain critical size are unstable and tend to disaggregate; those above a critical size are stable and grow by further precipitation of proteins onto the patch. Stabilization of the patch permits other more fluid membrane components to flow directionally away from the regions in which the patch is formed. Patch formation is prevented by respiratory inhibitors, low temperature, cytochalasin B etc. (Quinn, 1976).

These results in themselves would not be particularly interesting were it not for the evidence that some membrane proteins are directly associated with actin, myosin and tubulin located in the cytoplasm (Bourguignon & Singer, 1977; Ash *et al.*, 1977; Albertini & Anderson, 1977; Edelman, 1976 and references in this paper). Once a certain patch size is reached the associated actin, myosin and tubulin act as nucleation centres for the growth of cytoplasmic microfilaments and microtubules and this in its turn effectively anchors the patch within the membrane, preventing further movement. It is this ability of patches to act as nucleation centres which is most exciting. Microtubules for example seem responsible not only for the direction of secretion of cell wall precursors, but form the skeletal basis of the spindle and phragmoplast during cell division and through the preprophase band determine the plane of cell division (Hart & Sabnis, 1977) (see Chapter 4). Positioning of this critically-important latter structure by membrane proteins contains within it the seeds of understanding of a molecular basis for plant morphogenesis.

The anchorage of membrane proteins by microfilaments and microtubules helps to explain the paradox of an apparently fluid membrane with properties which indicate strict localization of certain membrane functions. Thus, for example, in an elongating coleoptile cell, cell wall precursor vesicles are clearly secreted through the side membranes and not through the top or bottom. The auxin transport system on the other hand seems to be located only in the bottom cell membrane (see Chapter 3).

Anchorage mechanisms may help to explain one of the basic plant/animal differences. Recognition between adjacent surface membranes of animal cells occurs by a lock-and-key type of mechanism. Lectin-like proteins on one membrane have a covalently-attached sugar which is specifically recognized by a protein (possessing perhaps a recognition centre similar to the active site of glycosidase) on the other membrane. Proteins of this latter group are termed surface receptors and can be induced to form patches by added lectins with subsequent anchorage of the patch by microtubule formation (Edelman, 1976). Interaction and recognition of adjacent animal membranes occurs by means of surface receptors and surface membrane-located lectins on both membranes (Hughes, 1976). The resulting dual anchorage provides stability to the interaction perhaps helping to maintain cells in a particular developmental state or pathway. In contrast to this, [with a few exceptions, such as interaction between egg and sperm in *Fucus* (see Chapter 7),] the surface membranes of plant cells are prevented from interaction by the cell wall.* As will be indicated later, patch formation and thus localization and anchorage of critical sets of microtubules and membrane proteins will be dependent on a continuous supply of transmissible signals. Interruption of these could lead to alterations in the plane of cell division as occurs in callus formation or regeneration.

2.4.2 THE IMPORTANCE OF ELECTRICAL FIELDS IN DEVELOPMENT

Most of the early work on production of surface membrane patches used antibodies or lectins to control formation. Poo and Robinson (1977) have demonstrated an intriguing alternative of direct relevance to plant developmental studies. They have shown that cell surface membrane proteins of embryonic cells can be localized and induced to form patches by applying very low potential electrical fields.

The study of electrical fields in plants has had a chequered history. A substantial body of measurements made mainly by Lund and his associates (see Rosene & Lund, 1953) in the early part of this century established that many plant tissues and organs maintained electrical potential gradients across them-

* If interaction could occur by the plasmodesmata, which seems unlikely, this would still involve an area considerably less than 1% of the total surface membrane (Robards, 1976).

selves and that these potentials varied significantly with the structure and shape of the organ. More recent work by Etherington and Dedolph (1972) has established the same conclusion for individual cells; the top and bottom of single coleoptile cells differ in potential by several millivolts. The significance of much of Lund's early data could hardly be appreciated at the time it was collected and interest in this area of work which lapsed some 30 years ago is only just recommencing. Two important groups of experimental observations form the basis for this recovery of interest.

(1) The polarity of growth of single cells or spores of *Acetabularia, Griffithsia, Fucus, Equisetum, Funaria* and even single pollen grains can be determined by an applied electrical field (see Chapter 7). The fields required are of the same size as those used to localize surface proteins and growth is usually to the positive pole (Jaffe & Nuticcelli, 1977).

(2) Many isolated plant cells or hyphae have been found to drive small electrical currents themselves during their initial development and growth. Such currents have been observed in *Pithophora, Acetabularia, Pelvetia, Lilium* pollen and germinating spores of *Funaria* and *Equisetum*. (See the discussion of electric currents and regeneration in *Acetabularia*, Chapter 6). Growth in all cases takes place at the positive pole of the cell (Jaffe & Nuticcelli, 1977). Again the size of the current which is driven between the two ends of the cell is of the same order as that used artificially to establish polarity. To this list can also be added growing roots of a number of higher plants which drive a current through their terminal centimetre (Weisenseel *et al.*, 1979).

The relevance of these natural currents is best seen by considering in brief their importance in the developing eggs of *Fucus* or *Pelvetia* (see Chapter 7). Much of this work has been summarized in Jaffe (1969a and b) and Jaffe *et al.*, (1974) and concerns the development of polarity.

The unfertilized *Fucus* egg is spherical and apolar. It lacks a cell wall, is low in turgor pressure and exhibits poor ion selectivity. Several hours after fertilization, wall synthesis commences and a turgor pressure develops commensurate with the uptake of potassium ions. Some sixteen to twenty hours after fertilization the first visible expression of polarity can be detected as a protruberance or wall outgrowth on one side of the embryo which thus becomes pear-shaped. This is followed about four hours later by an unequal cell division. The smaller daughter cell containing the protuberance develops into the rhizoid whilst the larger becomes the thallus. The positioning of this rhizoid outgrowth under natural conditions is specified by a light gradient across the egg, but in the laboratory it can be determined by a whole variety of physical and chemical gradients (see Table 2.4 later).

Some five to ten hours prior to the rhizoid outgrowth a marked electrical asymmetry develops across the embryo. The future rhizoid pole becomes positive with respect to the future thallus pole and this seems to result from the development of a calcium leak at the rhizoid end (Jaffe *et al.*, 1974) resulting probably from the accumulation of calcium transporting proteins in the surface membrane. Free calcium ions are relatively-immobile in cytoplasm and thus the rhizoid pole acquires a higher concentration of positively-charged ions establishing an electrical field. The development of a calcium pump at the thallus pole completes the electrical circuit through the egg and surrounding sea water.

The electrical field developed (several mV/cm) is certainly sufficient in size to localize appropriately-charged, freely-mobile membrane proteins at one or other of the two poles. Such a mechanism could be used to help establish the electrical polarity in the first place. If the membrane-located calcium-transporting proteins have a net negative charge, small initial electrical differences between the two poles would be greatly amplified by a process of positive feedback. Once sufficient numbers of membrane proteins are localized, patch formation or membrane protein aggregation can occur. Peng and Jaffe (1976) using freeze-fracture techniques have demonstrated the accumulation of surface membrane protein aggregates at the rhizoid pole before wall outgrowth commences. Anchorage of these proteins by microtubule formation will then direct the secretion of vesicles containing cell wall polysaccharides and proteins.

It is probable that cytoplasmic constituents carrying a net charge will also move in the electrical field, a phenomenon termed self-electrophoresis. Calculations indicate that an approximate three-fold concentration of certain protein species or nucleic acids could be expected at the appropriate pole. Microscopic observations bear this conclusion out; negatively-charged vesicles and RNA accumulate at the positive pole (Jaffe, 1969a). The electrical field thus serves to polarize the cytoplasm and the consequence of this will be seen after the first mitosis in which the polarization is rendered permanent (see Chapter 7). Thus, the two initially-identical daughter nuclei are separated into cytoplasmic environments which differ both in their spectrum and content of proteins and nucleic acids. It can then be supposed that gene expression might be modified accordingly. Unequal cell divisions, such as that in *Fucus*, occur also during the development of root hairs, stomata, higher plant embryos, phloem and pollen and are often quoted as evidence for the epigenetic theory of development since it is assumed that cytoplasmic controlling factors will be unequally partitioned during this event. However, self-electrophoresis would on its own ensure an unequal cytoplasmic distribution during an apparently equal cell division.

The presence of electrical fields in higher plants is well-established. These fields vary with their position in the appropriate organ and at any point are the summation of surrounding individual cell potentials (Rosene & Lund, 1953). It should be clear from the previous discussion that these observations lead to several important conclusions.

Dividing cells in meristematic regions synthesize new membranes and membrane proteins and undergo mitosis in established electrical fields. Similar conclusions to those described for *Fucus* should apply; the two daughter cells of any mitotic division should have different proportions of cytoplasmic constituents, particularly proteins and nucleic acids, and a different spectrum of membrane proteins. Each division will, therefore, represent a polarized cell division with the ultimate fate of each daughter cell dependent on its position within the field. The electrical currents in the *Fucus* egg also determine the plane of cell division and it is reasonable to suppose that this results from localizing membrane proteins attached to tubulin which form either the preprophase band of microtubules or the mitotic spindle itself. In meristems therefore it could be expected that the plane of cell division would reflect the particular magnitude and direction of the electrical field in which each cell finds itself.

Electrical controls may not be the only factor however. Rhizoid outgrowth and thus polarity in *Fucus* can be specified by gradients of potassium and calcium salts as well as gradients of auxin, pH, osmotic and mechanical pressure and light (Jaffe 1969b). Do these physical and chemical factors all control polarity by localizing calcium currents or can they alternatively induce patch formation in surface protein/tubulin complexes without the necessary intervention of the calcium leak/pump system? If the latter is the case then an alternative simple mechanism for establishing polarity and the plane of cell division in higher plants follows directly. The notion that pattern formation, and morphogenesis is specified by physical and chemical gradients of ions and growth substances is of course well-established, see for example, the auxin flow/gradient proposal of Sachs (1978) for vascular tissue patterns, but clearly a testable molecular mechanism is now available to explain it.

As higher plant cells leave the dividing zone and mature, two consequences follow. Membrane patches, stabilized solely by electrical forces will change as the electrical field changes. Dramatic changes in a stable electrical field, which could result from wounding, hormone treatment etc., should substantially alter the distribution of membrane proteins as the fluid characteristics of the membrane reasserts itself. It would be expected then that reprogramming of mature cells into division could follow such a stimulus. On the other hand protein patches stabilized by microfilament and microtubule anchorage would resist modification as the surrounding electrical field alters unless the tubule or filament itself is disaggregated. One mechanism to initiate this disaggregation is by modulation of cytoplasmic calcium levels and this will now be considered.

2.4.3 CYTOPLASMIC CALCIUM; A BIOCHEMICAL SWITCH MECHANISM FOR TERMINATING THE RESTING STATE?

The production and termination of the resting stage are essential phases in plant development and an integral part of its plastic nature (Table 2.3). There are many obvious examples such as dormant buds and root pericycle cells and the various resting organs, tubers, rhizomes etc. However, cortical cells, which may form up to 80% of the total cell population in young plants, may also be said to be in a state of rest, since they can be reactivated and induced to divide by various environmental manipulations. These developmental features are almost certainly an evolutionary response to predation, environmental damage and stress and the necessity for adjusting the morphology of the plant to its immediate surroundings.

The metabolic responses which accompany reactivation or the breaking of dormancy have been examined in several diverse examples but they do show remarkable uniformity. Generally, there are increases in respiration and potassium uptake, increases in RNA and protein synthesis, accompanied by transcriptional modification and usually initiation of DNA synthesis and cell division. This set of metabolic changes has been observed in storage tissue discs incubated in water or nutrients (Kahl, 1973), in lateral buds freed from apical dominance (Usciati *et al.*, 1972 and references therein), in cortical cells induced to division by hormones (Trewavas, 1976a; Trewavas, 1979a), in dormant buds reactivated by ethylene chlorohydrin (Rappaport & Wolf, 1969) and in embryo germination (Trewavas, 1979d). This general similarity suggests that the underlying metabolic mechanism involved may be the same in all cases. A general and unusual feature of the termination of rest of dormant buds is the large number of agents which initiate it (Samish, 1955). If we assume this to be a model for the reactivation switch in other systems then this provides a clue to the molecular mechanism involved.

The sea urchin egg is an isolated, resting animal cell which is activated by sperm entry. The metabolic changes which occur subsequent to fertilization are similar to those described above; they involve changes in membrane potential, increased ion uptake and respiration, increased nucleic acid and protein synthesis accompanied by transcriptional modification and DNA synthesis and cell division (Berridge, 1976; Balinsky, 1970). Remarkably, sperm entry can be mimicked by a wide variety of physical and chemical stimuli which thus induce parthenogenesis. The primary event in echinoderm egg fertilization is a temporary increase in cytoplasmic calcium ion concentration (Berridge, 1976); the other physical and chemical factors induce parthenogenesis by a similar effect on the permeability of membranes to calcium ions.

Table 2.4 provides a comparison of the factors which induce parthenogenesis in the echinoderm egg and those which terminate rest (and usually induce cell division) in dormant buds. A brief perusal indicates the remarkable similarity between the two lists both in the breadth of the treatments involved and even down to single chemicals. If the comparison is valid this suggests that the *primary molecular event in terminating the rest of plant cells is likewise an increase in cytoplasmic calcium ion concentration.* To emphasize the point Table 2.4 includes the factors known to polarize the *Fucus* egg (see Chapter 7) and to alter membrane potentials in plants. Again there is remarkable similarity in all four cases and as has been discussed previously the polarity of the *Fucus* egg is determined by alteration of membrane permeability to calcium ions. The inclusion of growth substances in the list of factors terminating rest has obvious implications for their molecular mode of action and the ability of these substances to modify permeability is well known (see Trewavas, 1976a; Marre,

Table 2.4 A comparison of the factors inducing parthogenesis in sea urchin, inducing polarity in *Fucus*, terminating rest in buds and modifying potential differences across plant membranes

Factors inducing parthenogenesis in sea urchin eggs	Factors promoting cell division in shoot buds	Factors inducing polarity in *Fucus* egg	Factors affecting potential differences across plant membrane
Temperature shock	Temperature shock	Temperature gradient	Temperature
Mechanical and osmotic shock	Excision or wounding (mechanical or osmotic shock)	Osmotic potential gradient. Flow gradient	Mechanical force
U.V. light	Photoperiod or low light	Light gradient or U.V. light gradient	Light
Inorganic ions	Inorganic ions	Inorganic ions	Inorganic ions
Uncoupling agents Fatty acids such as butyric acid and oleic acid, ethanol, ether, urea, sucrose, acetone, chloroform	Dinitrophenol Ethanol, ether, thiourea, sucrose acetone, chloroform	Dinitrophenol gradient	Dinitrophenol ether, chloroform
	Gibberellin, cytokinin, ethylene, auxin	Auxin gradient	Auxin
pH change	HCl or ammonia (citric, malic and succinic acids)	pH gradient	pH dependent
Calcium ionophore	Not known but since it inhibits extension, effects would be difficult to detect	Calcium gradient	Not known in plants but affects membrane potential in animals
Sperm entry		Sperm entry	

Data from Phillips (1969); Samish (1955); Hemberg (1965); Jaffe (1969); Balinsky (1970); Berridge (1976); Rosene and Lund (1953); Waddington (1956); Doorenbos (1953). Evocators in animal embryology seem likely to come in the same category since their effects can be mimicked by pH, thiocyanate, CO_2, NH_3, sterols, nucleotides, fatty acids, glycerol etc. Balinsky (1970); Waddington (1956).

1979). Location of an auxin-binding protein in the microsomal fraction (Venis, 1977) may be significant since this cell compartment represents a major cellular store of calcium ions. Likewise, since phytochrome acts as the mediator of phenomena which can terminate rest, its initial mode of action may also be directly related to calcium permeability. There is already strong evidence that phytochrome may control a light-sensitive potassium gate and the possible relationship to cytoplasmic calcium has been outlined by Haupt and Weisenseel (1976).

How then could an increase in cytoplasmic calcium concentration function to terminate rest? There are at present two possibilities which can be considered.

The first of these supposes that the membranes of resting cells exhibit low ion selectivity and may be said to be leaky. As a consequence membrane potentials are low and the ability of resting cells to synthesize ATP is severely impaired. All metabolic reactions directly requiring ATP will therefore operate at a reduced rate. The effect of an increase in cytoplasmic calcium ion concentration is to reimpose stringent ion discrimination, to re-establish full membrane semipermeability thus increasing the flow of energy via ATP into synthetic reactions. The selectivity of response is built into the cell type responding. The evidence which supports this view may be briefly considered.

Calcium is an important component of plant membranes. In its absence membranes show increased permeability to ions such as K^+ and become leaky. Addition of Ca^{2+} ions to such membranes increases the membrane potential and makes them relatively insensitive to external ion concentration, the so-called clamping effect (Higinbotham, 1973). The membranes of many unfertilized, resting plant eggs exhibit low selectivity to many ions and are relatively-speaking leaky, (see extensive data, summarized in Allen *et al.*, 1972). After fertilization the membrane potential increases, accompanied by a 'tightening' of the ion selectivity and an accumulation of K^+ ions providing the necessary turgor for growth. In an identical way dormant root cell tissue, when it is first cut, is extremely leaky and this leakiness can be maintained if the tissue is incubated at low temperatures (Laties, 1959). The well known phenomenon of potassium accumulation which develops when such tissue is aged is primarily caused by a 10–20-fold drop in efflux (van Stevenninck, 1964; Maklon, 1976); rates of K^+ influx only increase a few fold on ageing. Abscisic acid induces the dormant state in many plant systems. In common with this it appears to open channels in membranes promoting two-way ion flux but preventing net accumulation (Lea & Collins, 1979; Andres & Smith, 1976; Glinka & Reinhold, 1971; Marre, 1979). The possibility of the dormant state in plants being produced by impaired ATP accumulation has been dealt with by Jones (1972).

The second possibility points to a control of microtubule aggregation by calcium ions as a means of explaining the termination of the resting state. This may happen in the following way. Cytoplasmic microtubules seem to have their polymerization directed by nucleation or organizing centres [abbreviated to

MTOC (Pickett-Heaps, 1969)] (see also Chapter 4). Low concentrations of calcium ions not only inhibit microtubule formation from tubulin but will disaggregate microtubules already present. The inhibition may be modulated by a calcium-binding protein (Marcum *et al.*, 1978) located in the nucleation or organizing centre itself. These MTOCs are sometimes obvious structures (e.g. centrioles) but in plants are often amorphous, electron-dense bodies. As outlined by Lloyd (1979) a cell may have a number of MTOCs (their structure significantly may involve RNA, perhaps explaining the morphological changes induced by the potato spindle tuber viroid) and these compete with each other for the available tubulin. Thus in a dividing cell the MTOCs for cell wall growth have to compete for the available tubulin with the MTOCs of the phragmoplast, spindle and pre-prophase band. In a growing system this leads to effective competition between cell growth and cell division; the rate of cell growth is low in a dividing cell, but increases markedly when division ceases (Brown & Broadbent, 1950; Wright, 1963). The hinge point between cell division and cell expansion may be the requirement for the replication of an MTOC at the G1/S boundary (Lloyd, 1979).

It is not difficult to envisage that both cell growth and cell division together could be limited by the level of available tubulin and that the necessity for an increase in cytoplasmic calcium is to disaggregate effectively-inactive or nominally-stable microtubules in the resting cell. As pointed out by Hart and Sabnis (1977) plant microtubules seem to be especially resistant to disaggregation. Intriguingly, low temperature treatment is an alternative method of disaggregating microtubules. That the chilling requirements for the breaking of the dormant state in some plants may reflect a requirement for disaggregating microtubules is in this case well within the bounds of possibility.

2.5 CONCLUSIONS

A molecular attack on the problems of control points in development requires a multiplicity of approaches. That at least should be obvious from the information described in this chapter. Understanding the molecular basis of development will come neither from studies of transcriptional control nor from membrane systems in isolation, but requires an integration of molecular information from many fields.

This chapter is deliberately provocative and critical of the epigenetic/transcription theory in an attempt to rebalance some of the current weight of research emphasis. The attention this theory receives originates (in the writer's view) from the mistaken assumption that plant development proceeds in a linear direction as a result of a closely-controlled continuous two-way transfer of information between nucleus and cytoplasm. Such a theory, based as it is on a purely animal parallel, takes little or no cognizance of the plastic and uncommitted nature of much of plant development with its frequent resting states and its easy ability totally to reiterate developmental programmes. Such development might perhaps be better conceived as essentially circular in nature with a continuous outflow of cells to terminal differentiation and death. As was discussed, knowledge of membrane properties provides a more accurate understanding of these plastic phenomena.

FURTHER READING

TREWAVAS A. J. (1976) Plant Growth Substances. In *Molecular Aspects of Gene Expression in Plants.* (ed. Bryant J. A.) pp. 249–99. Academic Press, London and New York.

LEAVER C. J. (1980) *Genome Organisation and Expression in Plants.* Plenum Press, New York and London.

EDELMAN G. M. (1976) Surface Modulation in cell recognition and cell growth *Science,* **192,** 218–26.

JORDAN G., TIMMIS J. & TREWAVAS A. J. (1980) The Nucleus. In *A Comprehensive Treatise of Plant Biochemistry,* Vol. 1. (ed. Tolbert E.) pp. 387–487, Academic Press, London and New York.

CHAPTER 3

COORDINATION OF DEVELOPMENT: HORMONE RECEPTORS, HORMONE ACTION AND HORMONE TRANSPORT

3.1 INTRODUCTION

The emphasis in this chapter is on the role of growth substances in the spatial organization and differentiation of plant cells. The experimental investigation of morphogenesis encompasses systems which range in complexity from the study of organogenesis by pre-existing meristems to the formation of embryoids, and ultimately whole plants, from single cells in suspension culture. Although we do not understand pattern *inception* in plants, there exist theoretical frameworks (Turing, 1952) and model chemical reactions (Winfree, 1973) allowing relatively stable regularities to arise from an initially homogenous condition. While these may be useful paradigms, the molecular organization of cells themselves is not uniform; for example, the polarity of carrot embryoids may first become visibly established by an unequal division of the originating single somatic cell (Backs-Husemann & Reinert, 1970). Also, tissues such as meristems or even callus cultures will inevitably be non-unifrom due to the physical and chemical gradients inherent in bulk associations and to the cells' responses at different positions in such gradients (see Chapter 2). Cells communicate with one another using specific chemical messengers (including hormones) and by electrical (ionic) messages—circulating currents on which may be superimposed action potentials (Bentrup, 1977; Jaffe, 1979) (see Chapter 2). Given a degree of initial organization, or even transient heterogeneity, epigenetic mechanisms for the propagation and elaboration of pattern can be formulated with an ease inversely related to their specificity and testability. However, the urge to generalize must be balanced against the enormously detailed variety of form shown in the plant kingdom—indeed it is the basis of taxonomy—whose genetic specification we acknowledge but do not understand.

Differentiation in eukaryotes is thought to involve differential gene expression in the various cells and tissues of the organism. However, if we wish to understand development, which is essentially hierarchical, although we must ultimately look to the regulation of transcription, RNA processing and translation, it should also be recognized that developmental processes can be regulated by control of the activity and intracellular localization of 'pre-existing' gene products. These will include enzymes, membrane-associated transport proteins and receptors for hormones and other regulatory compounds. A further consideration is the communication between cells and interactions with the environment which influence or coordinate the patterns of cell division, enlargement and specialization in the developing organism.

For some purposes, a useful distinction can be made between spatial and temporal coordination during the life cycle of a higher plant (Wareing, 1977) although, since pattern (i.e. spatial) development occurs in a time continuum, the distinction between the two concepts is not absolute. For instance, the development of shoot phyllotaxy and root branching patterns depend largely on autonomic spatial coordination of the activities of meristematic regions whereas the matching of a plant's activity to a climatic cycle depends on the perception of 'temporal' environmental signals such as day length and seasonal temperature changes. An important feature of coordinated spatial responses is likely to be gradient(s) of hormone(s)* which can differentially activate the receptors and target reactions in one part of the organism relative to another. In principle this might occur even if the receptors are uniformly distributed among 'pre-differentiated' cells. However, a cell's position in the organism due to its developmental (temporal) history is also likely to influence its response to its hormonal milieu. For example, auxin-stimulated lateral root initiation starts with periclinical cell divisions specifically in the pericycle. In contrast, 'temporal' hormones are envisaged as conveying information directing an organism or tissue to change its behaviour in response to some exogenous or endogenous signal. For these hormones it is possible that the receptors may already be spatially arranged (e.g. at a high concentration in a vegetative apex which is potentially reproductive) so the notion of hormone gradients is less relevant.

The growth substances which may be involved include the five plant hormones, or groups of hormones, that have been characterized so far (Fig. 3.1) (Cherry, 1977; Letham *et al.*, 1978a,b). Of these, two (IAA† and cytokinins) may be produced in part by dying cells such as differentiating xylem elements—IAA from tryptophane released by protein breakdown and cytokinins from tRNA (Sheldrake, 1973). A number of pathways from tryptophane to IAA are possible (Schneider & Wightman, 1974). Recent evidence (Burrows, 1978) suggests that not all cytokinins are derived from tRNA breakdown, since some free cytokinins have been identified in tissues in which the corresponding adenine substituents are absent from tRNA. Two others, abscisic acid and the gibberellins, are terpenoids. There are still ambiguities in our understanding of abscisic acid biosynthesis (Walton, 1980); there are also lacunae in our comprehension of the pathway of synthesis of gibberellins (Hedden, *et al.*, 1979). It is likely that

* Wareing (1977) has suggested that in the plant context the term 'hormone' is often inappropriate and prefers the more general designation 'growth-regulating substance'. In this chapter plant growth substances are often referred to as 'hormones' without necessarily implying separation of sites of production and action.

† Abbreviations: A^- = Generalized anion; AH = Generalized weak acid; ABA = Abscisic acid (neutral + anion); ABA^- = Abscisate anion; ABAH = Neutral abscisic acid; C^+ = Generalized cation; GA = Gibberellic acid (neutral + anion); GA^- = Gibberellate anion; GAH = Neutral gibberellic acid; IAA = Indoleacetic acid (neutral + anion); IAA^- = Indoleacetate anion; IAHH = Neutral indoleacetic acid; K_D = Ether-water partition coefficient; P = Permeability coefficient; ψ = Electrical potential difference.

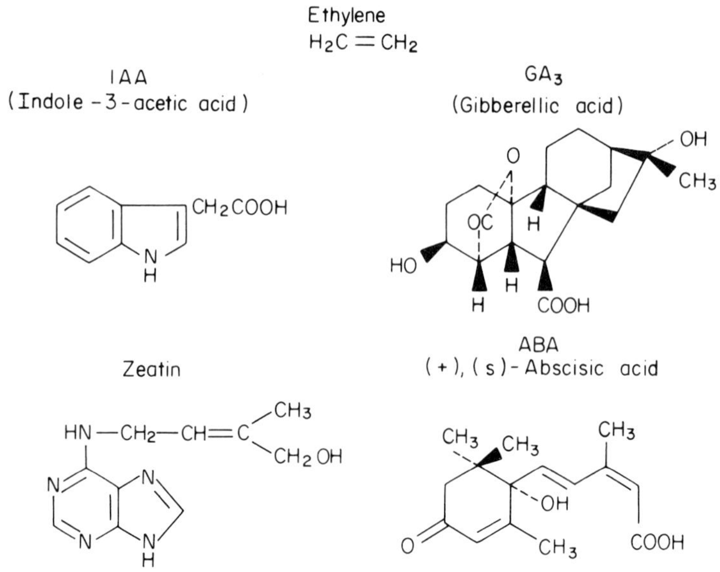

Fig. 3.1 The five widely-recognized plant growth substances.

abscisic acid and the gibberellins can be synthesized in plastids, although abscisic acid degradation is cytoplasmic (Walton, 1980; Hedden, *et al.*, 1979; Hartung *et al.*, 1980). Ethylene is probably synthesized from methionine via 1-aminocyclopropane-1-carboxylic acid (Adams & Yang, 1979; Lieberman, 1979). At the whole plant level, it appears that the shoot supplies most of the IAA and ABA, while the roots produce most of the cytokinin complement; gibberellic acid is synthesized in both roots and shoots, as is ethylene (Torrey, 1976; Wareing, 1977; Goodwin, *et al.*, 1978). The cellular levels and overall distribution of these growth substances depends on the controlled balance of transport and the rates and localization of their formation, removal and possible reversible sequestration (see van Staden & Davey, 1979).

As Wareing (1977) has pointed out, there is evidence that spatial coordination primarily depends on IAA which can move preferentially in a morphologically defined direction due to its polar transport system and, in the case of cambium, appears to be active in continuously stimulating cell division and differentiation along the path of transport itself. Wareing (1977) also suggested that other hormones (cytokinins, gibberellins, abscisic acid and ethylene) have a major role in temporal organization. We may note that the mammalian hormones are all essentially temporal regulators, with spatially well-defined target organs. Of course, animals must also have 'spatial organizers', but in determinate organisms like most metazoans these substances are presumably only active in early organogenesis. The indeterminate nature of the growth of most plants means that spatial and temporal organizers coexist throughout the life cycle.

This chapter deals with the chemical nature, action and movement of plant hormones and presents a speculative model based on IAA action and transport, which is testable and encompasses the minimal mechanisms thought to be involved in spatial differentiation in plants. A feature of this model is a feed-forward mechanism to maintain and strengthen polarity during growth, and which can maintain polarity during non-growth periods (e.g. seed dormancy, when the embryo is dehydrated).

3.2.1 STRUCTURE AND PROPERTIES OF PLANT HORMONES RELATED TO MEMBRANE TRANSPORT

Examples of the five types of plant hormone are shown in Figure 3.1. An important chemical aspect of these plant hormones is that, with the exception of ethylene, they are all weak electrolytes whose uncharged species are relatively hydrophobic and lipid soluble. This is shown in Table 3.1 as K_D, the ether/water partition coefficient for the neutral molecules of the hormones. Comparison of the pK_a values (Table 3.1) with the typical pH values found in the major compartments of terrestrial vascular plants (Table 3.2) shows that at least one major compartment could have a significant concentration of both neutral and charged forms of the growth substances present.

Their high lipid solubility means that these substances can passively cross 'fluid mosaic' plant membranes at physiologically significant rates by a 'lipid solution' mechanism that is not dependent on carrier mediation (Collander, 1959). Permeability coefficients (the proportionality constants relating flux to driving force) for the neutral forms of the hormones are also shown in Table 3.1. Except for the cytokinins, where some delocalization of the charge on the protonated ring N atom is possible (Panjipan, 1979), the permeability coefficients for non-mediated 'lipid solution' movement of the neutral species are likely to be much greater than for the charged forms (Le Blanc, 1971). Most data are available for IAA (Albaum *et al.*, 1937; Sutter, 1944; Collander, 1959; Rubery & Sheldrake, 1973, 1974; Higgins & Jacobsen, 1978; Raven, 1975, 1979; Rubery, 1979 a,b,c), the permeability of whose neutral form is quoted in Table 3.1. This P_{IAAH} is considerably higher than that for most plant metabolites, but is lower than those for relatively small molecules such as H_2O, O_2, CO_2, N_2 and NH_3 irrespective of their low lipid solubility (Raven, 1980a).

Very significant data on IAA permeation through lipid bilayers *in vitro* have recently been produced by Gutknecht and Walter (1980). They found a P_{IAAH} for phospholipid bilayers of $3.4 \times 10^{-5} m\,s^{-1}$ (corrected for diffusion through the unstirred layers on either side of the membrane); this value is very close to what might be expected from the membrane thickness and the partition coefficient for IAA between water and decane. Measurements of membrane conductance showed that P_{IAA^-} could not be greater than $10^{-11} m\,s^{-1}$, so that P_{IAAH}/P_{IAA^-} was at least 10^6. This reinforces the likelihood that the permeability of plant cell membrane to IAA^- reported by Rubery (1979a,c) will

Table 3.1 Values of molecular weight, pKa, ether-water partition coefficient and permeability coefficient for some plant growth substances

Growth Substance	rel. mol. mass	pK_a(1)	K_D(ether/water partition coefficient)	P(Permeability coefficient) for neutral molecule
IAA	175	4.60–4.75	20	$6.10^{-6} m\,s^{-1}$
ABA	264	4.8–4.9	3.3–5.0	$10^{-6} m\,s^{-1}$
GA	346 (GA_3)	3.8–4.0 (GA_3)	0.32 (GA_3)	$3.10^{-8} m\,s^{-1}$ (GA_1)
kinetin	215	$pK_{a1} = 3.8$ $pK_{a2} = 10.0$	0.81	$1.10^{-8} m\,s^{-1}$
zeatin	219	$pK_{a1} = 4.4$ $pK_{a2} = 9.8$	0.032	?

Table 3.1—notes

(1) pK_a values are from Rubery and Sheldrake (1973) for IAA; Demmig & Gimmler (1979) and Milborrow (quoted by Lea & Collins, 1978) for ABA; Cross (1954), Tidd (1964) and Durley & Pharis (1972) for GA_3; and Miller *et al.*, (1956) and Letham *et al.*, (1967) for zeatin. pK_a refers to the reaction:

$$R{-}COOH \rightleftharpoons R{-}COO^- + H^+$$

for IAA, ABA and GA_3. For kinetin and zeatin, pK_{a_1} is for the reaction:

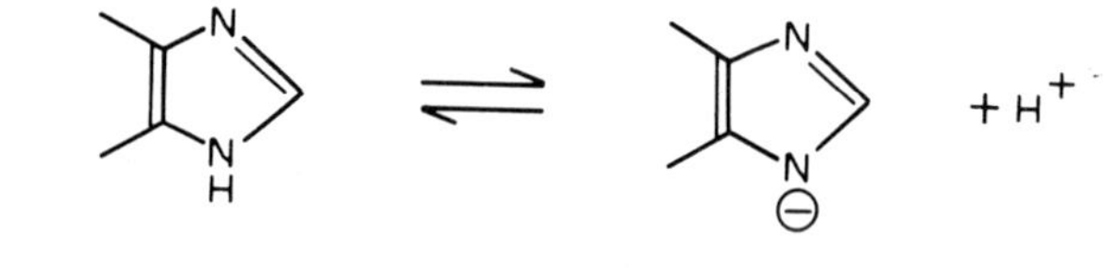

pK_{a_2} is for the reaction:

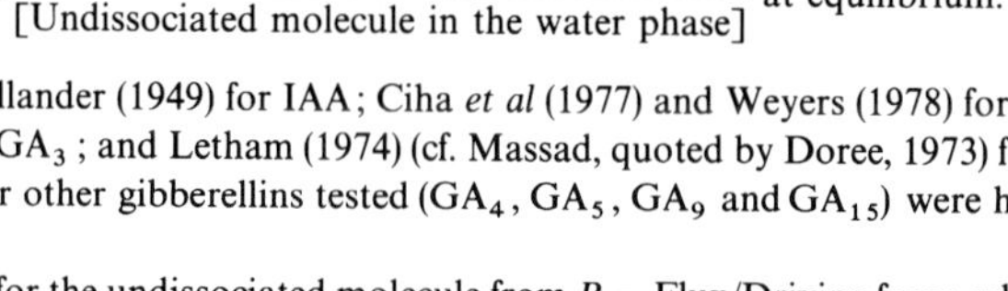

(2) K_D refers to the distribution of the undissociated molecule, i.e.

$$K_D = \frac{[\text{Undissociated molecule in the ether phase}]}{[\text{Undissociated molecule in the water phase}]} \text{ at equilibrium.}$$

Values are from Collander (1949) for IAA; Ciha *et al* (1977) and Weyers (1978) for ABA; Durley & Pharis (1972) for GA_3; and Letham (1974) (cf. Massad, quoted by Doree, 1973) for kinetin and zeatin. K_D values for other gibberellins tested (GA_4, GA_5, GA_9 and GA_{15}) were higher than for GA_3.

(3) P is computed for the undissociated molecule from P = Flux/Driving force, where: P = Permeability Coefficient, $m\,s^{-1}$; Flux = Net flux of undissociated molecule, mol. $m^{-2}\,s^{-1}$; and Driv-

involve specific uniporters for IAA^-. Gutknecht and Walter (1980) also showed that the 'anomalous' relationship between external pH and IAA uptake (with more IAA entry at high external pH values than can be explained in terms of bulk phase IAAH concentrations: Rubery & Sheldrake, 1973; Raven, 1975) can be explained quantitatively in terms of IAA^- diffusion at high external pH values ($> pK$) applied with rapid protonation to IAAH reducing the limitation on IAAH uptake due to IAAH diffusion through the unstirred layers outside the cells.

Data on other plant growth substances are more limited. The high P_{ABAH} for chloroplast permeation (Table 3.1) is consistent with the ability of ABA to partition into lipid bilayers to a sufficient extent to permit the formation of ion-uniport channels (Lea & Collins, 1979), although some data are better explained by a lower P_{ABAH} of the chloroplast envelope, at least in plants not subjected to dehydration stress (Milborrow, 1979). Also P_{ABAH} for crown gall suspension cells may be closer to 10^{-7} than $10^{-6} m\,s^{-1}$ (Rubery, 1979c). For GA_1H, the permeability coefficient is derived from estimates of GA_1 fluxes at

---

ing force = Concentration difference ($mol.m^{-3}$) for the undissociated molecule between the 'source' and 'sink' sides of the membrane.

(4) P_{IAAH} is from Raven (1975).

(5) P_{ABAH} was computed from the data of Demmig and Gimmler (1979), who showed that the $t_{1/2}$ for equilibration of ABAH between the stroma of isolated intact *Spinacia oleracea* chloroplasts and the bathing medium was about 60s at 25°C. P_{ABAH} was deduced from this $t_{1/2}$, assuming $pH_{stroma} = pH_{medium} = pH\ 7.6$; $pK_a = 4.8$; and that the surface area of an isolated, intact *Spinacea oleracea* chloroplast is $3.8\ km^2\ m^{-3}$ (Heber, 1974; Heldt *et al*, 1973). A similar value is consistent with the work of Wellburn and Hampp (1976) (cf. Hampp & Wellburn, 1976) on 48h greened *Avena sativa* etiochloroplasts, although these workers only measured transport over one time interval (60s) and gave no direct evidence for passive equilibration of IAAH (cf. Demmig & Gimmler, 1979).

(6) P_{GA_1H} was computed from the work of Drake (1979), who carried out flux analysis of $^3H - GA_1H$ distribution in *Avena sativa* coleoptile segments. From the estimated plasmalemma influx in pmol (g fresh wt)$^{-1}$h^{-1}, the influx in $nmol\,m^{-2}s^{-1}$ was derived assuming a plasmalemma area/wet weight value for cylindrical cells 150μm long and 20μm in radius (Drake & Carr, 1978). P_{GA_1H} was calculated from this flux and the observed $pH_{medium} = 6.0$, $pK_a = 4.09$ (Tidd, 1967) and an external concentration of dissociated plus undissociated GA_1 of $116\mu mol\,m^{-3}$. Drake (1979) did not directly test the passive nature of GA_1 distribution, although such a distribution is compatible with his data (3.2.1). K_D (ether/water) is probably similar to that for GA_3H, based on the similarity of K_D (ethyl acetate/water) for GA_1H and GA_3H (Durley & Pharis, 1972). P_{GA_3H} for *Avena sativa* plastid envelope membranes (subject to the same provisos as were mentioned for P_{ABAH} under (5) above) is at least $10^{-7}m\,s^{-1}$ (Wellburn & Hampp, 1976).

(7) $P_{kinetin}$ was computed from the data of Doree (1973) who showed the linearity of the initial influx of ^{14}C-kinetin (pmol min^{-1}(10^6 *Acer pseudoplatanus* cultured cells)$^{-1}$) with external concentration of undissociated kinetin in the range 0.2–100mmol m^{-3}; the fluxes were converted to an area basis using a cell surface area of $0.005m^2$ per 10^6 cells (King & Street, 1977). Doree (1973) points out that while the initial influx from a given external concentration of 6-N-substituted adenines increased with their K_D (ether/water) values, the effects of 2,4-dinitrophenol on the influx (Doree, 1973) and the non-linear relationship between the steady-state intracellular concentration of free kinetin and the external kinetin concentration (Doree & Guern, 1973) suggests that other processes may be superimposed on passive equilibration.

Table 3.2 Quantitative aspects of compartments and transport paths in terrestrial vascular plants

Parameter	Gas spaces	Apoplast (cell walls)	Cell vacuoles	Symplast	'Super-apoplast' (xylem conduits)	'Super-symplast' (phloem conduits)
pH (1)	—	5–7	4–6	7–8	5–6	7–8
Fraction of typical herb volume occupied by the compartment (2)	0.05	0.05	0.84	0.05	0.05	0.01
Velocity of movement of solutes (3)	$3 mm\,s^{-1}$	$3 \mu m\,s^{-1}$	$3 \mu m\,s^{-1}$	$3 \mu m\,s^{-1}$	$3 mm\,s^{-1}$	$300 \mu m\,s^{-1}$
Mechanism of movement (4)	Diffusion	Diffusion (+ mass flow)	Diffusion	Diffusion (+ cyclosis?)	mass flow	mass flow
Flux of a mean hormone concentration of 1mmol m^{-3}($nmol\,s^{-1}\,m^{-2}$ of transport path) (2, 3)	3000	3	3	3	3000	300
Flux of a mean hormone concentration of 1mmol m^{-3}($nmol\,s^{-1}\,m^{-2}$ of plant cross-section (2, 3)	150	0.15	2.5	0.15	150	3

Table 3.2—notes

(1) pH values are taken from Raven (1977a) and Raven and Smith (1976).

(2) Data abstracted from that compiled byRaven (1977b) and Geiger (1975); values are illustrative rather than referring to a particular plant.

(3) 'Mean' concentration of hormone is cited because there is generally a *decrease* in concentration between 'source' and 'sink'. For mass flow, lateral leakage of solute *can* lead to a decrement of concentration with passage from source to sink. For a process driven by passive diffusive forces alone, there is an *obligatory* decrement of concentration according to Fick's law [see Note (3) for Table 3.1]. Note that, for the IAA polar transport system discussed in 3.2.3, the polar application of the electrochemical driving force leads to an increment of concentration in the direction of net transport.

(4) Cell vacuole and symplast are quoted together here for the case of polar IAA transport, where vacuole and cytoplasm may form parallel paths for transcellular IAA flux (if P_{IAAH} at the tonoplast is high) in the alternating cell-apoplast-cell pathway (see 3.2.3).

only a single external [GA_1] and pH value, so that there is no direct evidence for passive permeation based on proportionality of net flux with the concentration difference for the undissociated neutral species. The situation improves somewhat for cytokinins, where there is reasonable evidence (Table 3.1, notes) that P_{kinetin} does apply to a lipid solution transfer. The totality of results on permeability summarized in Table 3.1 shows that $P_{\text{neutral hormone species}}$ is relatively high and is explicable in terms of K_D and relative molecular mass (see Collander, 1959, for empirical relationships between K_D, rel. mol. mass and permeability).

Since the neutral form of the hormone is likely to have a higher intrinsic membrane permeability than the charged species (Le Blanc, 1971; Raven, 1975), the transmembrane distribution of the growth substance as passive flux equilibrium is approached often appears to reflect primarily the equilibration of the neutral species across the cell membranes, with the relative concentrations of undissociated and dissociated forms in each compartment being determined by the pK_a of the solute and the prevailing pH value. Provided that energy is supplied to ensure that the compartments are adequately buffered against proton release or uptake due to growth substance ionization, this means that acidic hormones will tend to be accumulated by relatively alkaline compartments which act as 'anion traps'. The opposite would be the case for the cytokinins: the cationic form would accumulate in relatively acidic compartments, but only to a marked extent when the pH value is equal to or below pK_{a1} (see Table 3.1). The distribution of ([IAAH] + [IAA^-]) between the vacuole, cytoplasm, and bathing medium of *Hydrodictyon africanum* can be largely explained by the permeation of undissociated IAAH (Raven, 1975).

For ABA, the distribution between isolated intact *Spinacia* chloroplasts and their incubation medium is accounted for by ABAH permeation (Demmig & Gimmler, 1979), as is ABA compartmentation in leaf cells (Heilman *et al.*, 1980). It is tempting to explain some findings by Hoad (1978) and Milborrow (1979) in this way. Dehydration stress increased ABA levels in both leaves and phloem sap of *Lupinus albus*, the concentration in the phloem sap being up to 30 times that in the leaf (Hoad, 1978). The phloem and leaf cytoplasm have a much higher pH than the leaf cell vacuoles (Table 3.2) and it is possible that the phloem acts as an alkaline trap for ABA^- thus largely explaining the distribution data (c.f. Collander, 1959). The effects noted by Setter, Brun and Brenner (1980a,b) on leaf ABA content under conditions of restricted translocation (sink removal or petiole girdling) in *Glycine max* could also be interpreted in 'alkaline trap' terms (see also Section 3.2.3).

Milborrow (1979) found much higher total ABA in a fraction enriched with cell walls and protoplasm from *Citrus* fruit juice-vesicles than in a vacuolar fraction. Since the low pH of Citrus fruit is a function of their large acidic vacuoles (Table 3.2), the distribution of ABA can again be related to an alkaline trap—in this case the cytoplasm. Raven, Hartburg, Cowan & Farquhar (in preparation; cf. Raven, 1981) have suggested that the weak acid nature of ABA may partly account for parallelism between stomatal conductance and the activity of photosynthetic biochemistry in leaves under a variety of conditions (Wong, Cowan & Farquhar, 1979). The light-induced alkalinization of the chloroplast stroma would act as an alkaline trap for ABA^- (Demmig & Gimmler, 1979; Heilman *et al.*, 1980); thus the increased rate of photosynthesis at high irradiances would mean more alkaline trap activity on the part of the stroma, which would accordingly sequester a greater fraction of the total leaf ABA. Cowan *et al.* (1982) have shown that the change from darkness to a photosynthesis-saturating irradiance could lead to a halving (at least) of the cell wall ABA, which might account for at least part of the change in stomatal conductance. This mechanism could explain at least a part of the parallelism of stomatal conductance and the biochemical activity of the mesophyll cells when incident irradiance is varied; however, the irradiance response of stromal pH shows an earlier light saturation than is most readily accommodated by this model (Enser & Heber, 1980).

Finally, a compartmental analysis for GA_1 in *Avena* coleoptile segments (Drake, 1979) showed that ($[GA_1^-] + [GA_1H]$) for the compartments corresponding to the 'slow' and 'fast' phases were similar when expressed as pmol (g fresh wt)$^{-1}$, although the vacuolar and cytoplasmic volumes were in the ratio of about 19 : 1. Again, the cytoplasm could act as an alkaline trap for GA^- since the 19 : 1 concentration ratio implied by the data is explicable if pK_a for GA_1 were 4 (Table 3.1), $pH_{vacuole}$ were 6.1 (Bonner, 1934), $pH_{cytoplasm}$ were 7.3 and only GA_1H could cross the tonoplast. Thus, the slowly exchanging compartment could still be equated with the vacuole and the faster exchanging compartment could be the cytoplasm. However, some inhibitor data (Drake, 1979) do not entirely accord with this interpretation. The conclusions of Ohlrogge *et al.*, (1980) that passive diffusion of GA_1H across membranes does not make a significant contribution to $[^3H]GA_1$ uptake by cowpea leaf section may be premature in view of the length of the washing period to which the segments were subjected before measurement of their radioactivity.

The high permeability of the neutral hormone species does not rule out the possibility of transport by membrane-associated proteins (porters in the terminology of Mitchell, 1970). Evidence consistent with an IAA^-/H^+ symport has been presented for several higher plant tissues (Rubery, 1977; Rubery, 1978; Davies & Rubery, 1978) and this may under certain conditions mimic an increase in P_{IAAH} (Rubery, 1978, 1979c). Also, both carrier-mediated and diffusive transport contribute to the uptake of abscisic acid by segments from the apical region of runner bean roots (Astle & Rubery, 1980). The interpretation of hormone distribution between compartments of different pH values is complicated by membrane permeation of the charged species which introduces the transmembrane electrical potential difference as a component of the driving force for hormone movement. If the alkaline compartment is electrically negative, then net anion efflux can occur down the gradient of electrochemical potential, reducing the distribution ratio and effectively acting as a partial leak on accumulation, (for equations and further detail, see Rubery, 1979c). The intrinsic anion permeability may be enhanced by carriers such as the IAA^- uniport suggested to occur in addition to IAA^-/H^+ symport (Rubery, 1979a,c). The far-reaching consequences of a potentially variable auxin anion flux component are discussed below (3.2.3). However, it is worth noting here that when the intrinsic permeability of lipid membranes to solutes is high, it is difficult qualitatively to confine the solute to the apoplast, the symplast, or intracellular compartments, even granted the indirect input of metabolic energy (via active H^+ uniport; Fig. 3.2) to any hormone carriers. The significance of ready, uncatalysed entry of the hormone into 'target cells' is discussed below in connection with the location of 'receptors' (Oostrom *et al.*, 1975; Higgins & Jacobsen, 1978).

3.2.2 MODE OF ACTION AND RECEPTORS

Plant hormones can modulate both the activities and the cellular levels of transporters and enzymes. For example, the much-studied stimulation of cell extension by auxin encompasses early wall-loosening events associated with activation of H^+ secretion from the protoplast *and* the enhanced synthesis of macromolecules necessary for cell enlargement to be sustained (Rayle & Cleland, 1977; Penny, 1971). Although in certain systems at least (e.g. gibberellin action on cereal grain aleurone layers—Varner, 1978) our knowledge of *what* hormones do is becoming increasingly detailed and sophisticated at the molecular level, we are still ignorant of *how* they work.

Investigation of the mechanism of hormone action requires detailed molecular knowledge of the target process so that a direct test *in vitro* of the influence of the hormone can be made. Whenever a likely target has been examined, no consistent evidence has been found for the direct interaction of hormones with enzymes or membrane-associated porters (e.g. Jacobsen, 1977) or with the genetic apparatus (Venis, 1977). Rather, the hormone action is thought to be initiated by combination with a receptor. Subsequent interaction of the hormone-receptor complex with an enzyme or transporter, or with the genome, could be direct or could depend on a further 'coupling' reaction sequence that could also amplify the response to the initial stimulus. The differentiation of a particular cell may be seen as reflecting that cell's complement of receptors and receptor-mediated events. The high intrinsic permeability of plant hormones (3.2.1) means that they are not restricted to surface receptors which need 'second messengers' to communicate their occupancy to the intracellular machinery. In this respect, plant hormones resemble vertebrate steroid hormones (Bischoff *et al.*, 1954; Peck *et al.*, 1973; Gorski & Gannon, 1976; Jonkers *et al.*, 1980) rather than the hydrophilic 'non-penetrant' animal hormones, many of which use cyclic AMP as second messenger (Sands & Rickenberg, 1978).

Receptors in plant extracts are recognized *inter alia* by having a saturable binding site for the relevant hormone, with a dissociation constant in the range of concentrations at which the hormone exerts its *in vivo* effects, and with a

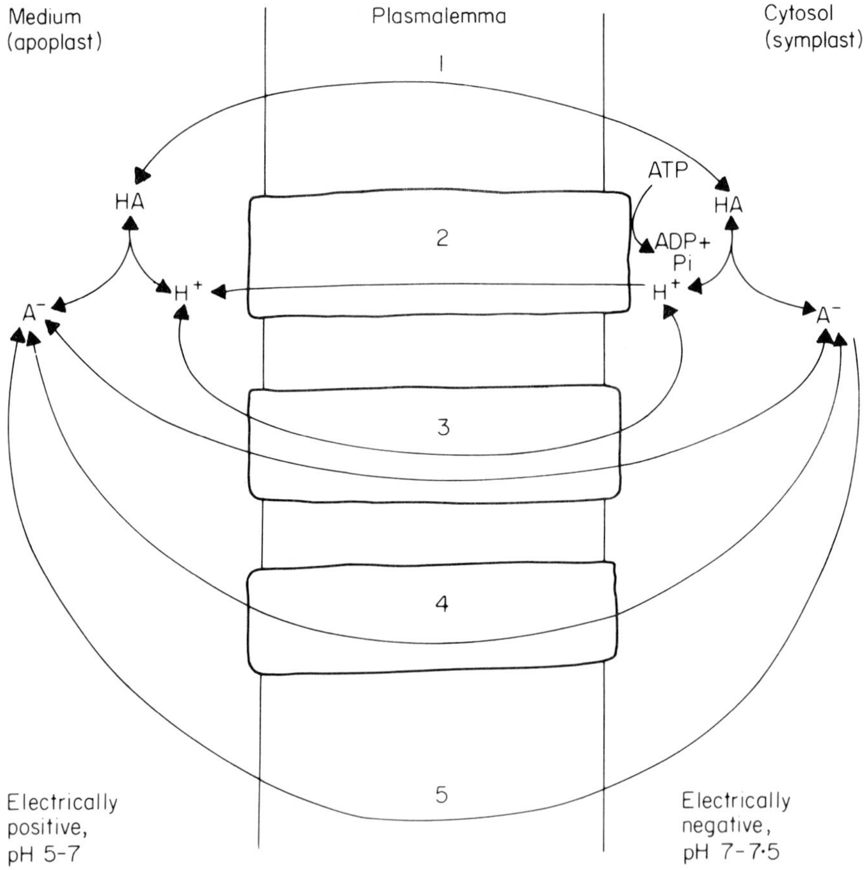

Fig. 3.2 Processes by which a weak acid growth substance (HA) and its anion (A^-) can cross the plant cell plasmalemma (adapted from Rubery, 1979c). The principles are of general application; the scheme refers specifically to IAAH (neutral indoleacetic acid) and IAA^- (indoleacetate anion).

(1) 'Lipid Solution' permeation of IAAH (P_{IAAH} $6.10^{-6}\mathrm{m\,s^{-1}}$).

(2) Active, electrogenic H^+ efflux (H^+ active uniport); an essential part of the mechanism which keeps the cytosol electrically negative with respect to the medium, and at a higher pH than the medium. Capacity 100–200nmol $\mathrm{m^{-2}s^{-1}}$. In the real plasmalemma there are many more processes returning H^+ across the membrane than (3) and (4) shown here, which have a low capacity compared to (2).

(3) Catalyzed H^+/IAA^- symport; if 1 : 1 coupled, and without kinetic constraints, behaves as an additional component to (1).

(4) Catalyzed IAA^- uniport; with the prevailing $\Delta\psi$ acts mainly in catalyzing IAA^- efflux. Supplements (often with a much larger capacity) process (5).

(5) Intrinsic IAA^- uniport; depends on lipid solubility of IAA^-; $P_{\mathrm{IAA}} - < P_{\mathrm{IAAH}}$, where P = permeability coefficient.

specificity (for agonists and antagonists) similar to that of the in-vivo responses. Such specific high affinity binding is most likely to be an attribute of a protein. However, caution (and confirmatory evidence) is necessary in view of the ability of non-biological materials such as talc to show both saturable and selective binding of, for example, adrenalin (Cuatrecasas *et al.*, 1975) and cytokinins (Gardner *et al.*, 1978). The molecular features responsible for such binding may contribute extensively to the often large 'non-specific' low affinity binding capacity shown by tissue extracts. Receptors detected by binding assays could, in principle, include enzymes, action receptors and transport proteins; the recent demonstration that bovine serum albumin would be classed as an auxin receptor if it had been extracted from plants (Murphy, 1979) shows that apparently biologically irrelevant protein-binding may further complicate interpretation. Also, the possibility that binding activity may be retained by partially degraded proteins such as IAA-metabolizing enzymes should not be overlooked. It is to be hoped that the use of mutants allowing correlations of lost or modified hormone responses and binding components (Gardner *et al.*, 1978; Ashton *et al.*, 1979) and of reconstituted systems using purified components will in the future allow more confident attribution of function to 'receptors'.

The majority of data on receptors concerns auxin binding (e.g. reviews by Kende & Gardner, 1976; Jacobsen, 1977; Venis. 1977; Letham *et al.*, 1978a,b; Rubery, 1981). At least three such receptors (in terms of auxin binding) have been recognized on membranes (Dohrmann *et al.*, 1978; Jacobs & Hertel, 1978) and there are also rather less well-characterized reports of soluble receptors (e.g. Oostrom *et al.*, 1975; Ihl, 1976), one of which (Wardrop & Polya, 1977) has since been shown to be ribulose bis-phosphate carboxylase (Wardrop & Polya, 1980a,b). The auxin binding sites are predominantly intracellular with 'site I' occurring on the endoplasmic reticulum, 'site II' tentatively assigned to the tonoplast (Dohrmann *et al.*, 1978) and 'site III', which may be a transport protein, apparently associated with the plasmamembrane (Jacobs & Hertel, 1978). Recently auxin receptors have been solubilized from microsomes, partially purified, and shown to be proteins but to be devoid of ATPase activity (Cross & Briggs, 1978; Cross *et al.*, 1978).

For cytokinins, there is evidence for a proteinaceous receptor that can be made soluble of unknown function associated with plant ribosomes (Fox & Erion, 1975; Chung *et al.*, 1979). Ethylene has been shown to be associated *in vivo* with binding protein, present to different extents in different tissues (Jerie *et al.*, 1978, 1979). Ethylene binding by cell-free preparations from bean cotyledons has recently been characterized (Bengochea *et al.*, 1980a; Bengochea *et al.*, 1980b).

There is thus some evidence for receptors. Difficulties remain nevertheless: one is the large concentration range (up to six orders of magnitude) over which plant hormones usually exert their effects (Kende & Gardner, 1976). This wide dose-response range contrasts with sensitivity over two orders of magnitude for *some* responses to penetrant and non-penetrant vertebrate hormones

(Katzenellenbogen & Gorski, 1972; Cuatrecasas & Hollenberg, 1976) and resembles the span of signal intensities perceived by many animal sensor receptor systems or the range of light energies capable of photomorphogenic effects in plants (Smith, 1975). However, the further that receptor binding is removed from both the site of hormone application and from the ultimate biological response monitored, the more complex the relationship between applied hormone concentration, receptor occupancy, and magnitude of response is likely to be.

The ready uncatalysed transport of plant hormones across membranes means (cf. Higgins & Jacobsen, 1978) that there need be no *spatial* restriction on receptor location within cells such as is the case for the 'impermeant' vertebrate hormones which have receptors on the outer surface of the plasmalemma (Cotte & Duffau, 1977). Even the most rapid plant responses to the application of hormone can probably be explained by an influx of hormone across the plasmalemma (and other membranes) to a receptor, which subsequently influences the 'target' catalyst (Penny & Penny, 1978; Demmig & Gimmler, 1979; Rubery & Sheldrake, 1973).

A further 'spatial' consideration relates to the tissue distribution of receptors. Clearly, in the light of the above discussion of the difficulty of *recognizing* receptors, it is not possible to present detailed data on this topic, although tissue specificity of response to hormones (e.g. cell enlargement) could reflect receptor distribution. Nevertheless, one theoretical point is worth emphasizing. The great tissue specificity of vertebrate hormone receptors ('target cells') is symptomatic of a role for the hormones in what Wareing (1977) calls 'temporal regulation'; the spatial information has already been provided to determine which cells have the receptors. In the earliest phases of the provision of 'spatial information', however, we must by definition have a minimum of anticipatory spatial differentiation. As will be shown later (3.2.3), the simplest model requires inhomogeneity of hormone concentration but not (initially) inhomogeneity of hormone receptors concentration.

In discussing the events that follow the binding of hormone to the receptor, it is perhaps worth distinguishing the following mechanisms. (a) The coupling mechanism leading to initial transporter or enzyme modulation; this could depend *inter alia* on protein–protein interaction, perhaps within a fluid mosaic membrane. The large body of work on a stimulus–response coupling and adenylylate cyclase in vertebrates shows the complexity that can be reached (Insel, 1978). (b) The possibility that what might be called a 'second messenger' could be generated due to activation of the initial catalyst. However, since this term has a particular connotation for non-penetrant animal hormones, we suggest that 'coupling messenger' is a preferable alternative. (c) The further chain of consequences of a discrete hormonal effect could also be generated. This clearly overlaps somewhat with (b). As an illustration of (c), consider the auxin stimulation of H^+ secretion alluded to earlier which, according to the acid growth theory of auxin action (Rayle & Cleland, 1977), leads to plasticization of the cell wall and so to turgor-driven cell extension. Irrespective of the site of the primary proton pumping event (at the plasmalemma or into the lumen of the endoplasmic reticulum cisternae followed by vesiculation as Ray (1977) suggests); the acidification of the wall space is accompanied by hyperpolarization of the plasmalemma and alkalinization of the cytoplasm thus stimulating auxin uptake (Rubery, 1980). This positive feed-forward increase in intracellular IAA levels has obvious implications for other auxin-dependent processes, including gene expression. Marre (1979) has discussed other possible ion transport and metabolic responses to these changes. He also points out that comparison of the overall effect of fusicoccin and auxin on plant tissues may be fruitful since the fungal toxin stimulates proton pumping, but does not have other auxin attributes, such as target tissues specificity (fusicoccin effects can act on auxin-insensitive tissues such as mature leaves) and gene expression effects (fusicoccin effects are primarily on the plasmalemma). Until more is known about hormone receptors, it is only possible to speculate about the identity and mechanism of action of 'coupling messengers'. Acetyl choline and cyclic AMP seem unlikely candidates in higher plants (Amrhein, 1977; Letham, 1978; but see also Brown *et al.*, 1979). Ca^{2+} does seem a possibility (see Chapter 2). Indirect evidence suggests a low free [Ca^{2+}] in plant cytoplasm (as in other eukaryotes: Raven, 1977a), and recently calmodulin (otherwise known as calcium-dependent regulator protein or CDR) has been demonstrated in higher plants (Anderson & Cormier, 1978; Waisman *et al.*, 1978; Van Eldik *et al.*, 1980). This protein shows activity in activating enzymes (e.g. NAD kinase: Anderson *et al.*, 1980) when free [Ca^{2+}] is raised from a basal (100μmol m^{-3}) to a higher (1–10mmol m^{-3}) value. The immediate cause, in the hormonal context, of the higher free [Ca^{2+}] in the cytosol might be 'hormone gating' of a Ca^{2+} channel. Gradients of cytosol free Ca^{2+} are certainly correlated with polar cell development in higher plants (Reiss & Herth, 1978); these may be related to electric current circuits in developing organs (Jaffe, 1979).

A possible objection to Ca^{2+} as a messenger is that free Ca^{2+}-calmodulin would make a poor intermediate for more than one hormone action. This has been overcome by Lowenstein and Rose (1978); they argue that the rapidity with which additional free Ca^{2+} reacts with endogenous chelators in the cytosol provides a very great spatial and temporal restriction on the spread of the Ca^{2+} message. This sort of buffering could underlie the spatial restriction of the phytochrome-mediated chloroplast movement in the alge *Mougeotia*. Here (Haupt & Weisenseel, 1976) the use of microbeams and polarized light has shown that active phytochrome is associated with the plasmalemma, and that only that portion of the chloroplast adjacent to P_{fr}-containing plasmalemma is acted upon. Dreyer and Weisenseel (1979) have recently shown directly that P_{fr} increases $^{45}Ca^{2+}$ influx in *Mougeotia*.

The rapid buffering of Ca^{2+} could then give different hormone receptors 'private lines' (Loewenstein & Rose, 1978) to target reactions. However, we do not have enough plant data at the moment to produce a directory of such lines, nor of any similar lines that might result from H^+ fluxes (cf. Raven & Smith, 1977; Polevoy & Salmatova, 1977).

A more specifically 'Plant' coupling messenger has been proposed in ethylene (see Wareing, 1977); again, the mobility of this compound is relatively restricted due to the presence of ethylene binding proteins which cause a chromatographic effect (Jerie *et al.*, 1978, 1979). Generally a great deal of study has been devoted to the question of the interacting effects of plant hormones (Wareing, 1977) in the control of development. Molecular understanding of this important topic is negligible.

3.2.3 HORMONE TRANSPORT

The higher plant has three main intra-organism transport systems (Table 3.2): the intercellular gas spaces; the wet cell walls (apoplast) plus the water-filled dead cells (xylem = super-apoplast); and the cell cytoplasm connected by plasmodesmata (symplast) plus the phloem (super-symplast) (Raven, 1977b; Letham *et al.*, 1978c). Transport of solutes in the gas phase, and in the symplast and the apoplast, is (sometimes) by diffusion with mass flow of the solvent. In the xylem and phloem, mass flow predominates.

We have already noted (3.2.2.) that the gaseous hormone ethylene is less mobile in the gas phase of plants than might have been anticipated due to the 'chromatography' effect. The non-volatile growth substances have all been reported as moving in both the xylem and the phloem as the free hormone (see Letham *et al.*, 1978c), although recent work (Allen *et al.*, 1979) suggests that IAA in the xylem may have originated from microbial contamination subsequent to detopping the plant for collection of xylem exudate. However, the evidence for cytokinin movement from root to shoot in the xylem is overwhelming (see Wareing, 1977; van Staden & Button, 1978).

Turning to the phloem transport, all of the major growth substances (together with various growth inhibitors) seem to be phloem-mobile. A problem with long-distance transport of growth substances, with their generally high membrane permeabilities, is that of reversible interchange between the apoplastic xylem and the symplastic phloem (Bowen & Wareing, 1969). This problem has been discussed elsewhere for boric acid and for xenobiotics (Oertli & Richardson, 1970; Tyree et al., 1979; Raven, 1980b); for the highly membrane-permeable growth substances one might predict accumulation in transpirational termini, i.e. mature leaves.

One method by which counter-current exchange between xylem and phloem might be reduced is the occurrence of continuous, energy-dependent 'reaccumulation' all along the sieve tubes (Raven, 1980b) based on (e.g.) H^+/A^- antiport (for weak acids) or C^+ uniport (for cytokinins). Another possibility would be to transport the hormone in the phloem as some conjugate with a hydrophilic compound [e.g. aspartyl-IAA; glucose ester of ABA (Milborrow, 1978) glucosylated GA and cytokinin (Wareing, 1977; van Staden & Button, 1978)] which cannot cross a membrane unless some specific porter is present. Conjugated cytokinin is certainly present in phloem (van Staden & Button,

1978). However, such a mechanism requires (if, as seems likely, the non-conjugated hormone is the growth-active form) enzymes for 'deconjugation' at the target end of the pathway and (if the enzymes are apoplastic) specific permeases to unload the conjugate. This implies a rather complex 'receiving apparatus' in the target organ. Such complexity seems compatible with a role for xylem and/or phloem transport in 'temporal regulation' (Wareing, 1977), but is only with difficulty reconciled with such hormone transport as conveying 'positional information'—the target tissue is required to know too much about where it is, and what information to expect.

At a more fundamental level, and regardless of problems of retaining the hormone in the transport pathway *en route*, there are other objections to the use of xylem and phloem for the transmission of *primary* positional information. The presence of functional xylem and phloem presupposes considerable development, so information transferred in the xylem and phloem can only help to *confirm* the pre-existing polarity of the plant. Furthermore, these impose severe constraints as to where informational signals are able to go in the xylem and phloem. The xylem conducts dissolved solutes from the roots to transpirational termini, while the phloem conducts them from sources (photosynthetic organs; emptying storage organs) to sinks (filling storage organs; meristems) of carbohydrate. Indeed, once the basic polarity of the plant and its position (if terrestrial) at the soil-air interface has been established, the gross features of shoot–root balance (as influenced by light, temperature, and the supply of water and nutrients) can be adequately modelled by considering the transport of root-acquired resources to the shoot and vice-versa, without the explicit involvement of hormones (Thornley, 1972; 1977). This is one point at which the 'plumbing' arrangement of plants, with apoplastic and symplastic elements, leads to a regulatory possibility not readily encompassed in the vertebrate's blood system (cf. Gunning & Robards, 1976, p.2).

Thus the long-distance transport systems of plants seem to be optimized for the transport of materials (including energy sources) and of 'temporal information'; for the transmission of 'positional information' we must turn to the other two pathways mentioned in Table 3.2, the symplast and the apoplast. All hormones seem to be able to negotiate apoplastic, symplastic or an alternation of apoplastic and symplastic pathways (Goldsmith, 1977; Letham, *et al.*, 1978c; Cande & Ray, 1976; Sheldrake, 1979; Drake & Carr, 1978, 1979; Drake, 1979) in parenchymatous tissue, but only one hormone transport system seems to have the attributes which one might expect of a system for initiating and perpetuating polarity in an undifferentiated system. This is 'polar auxin transport', whose major properties (see Goldsmith, 1977; Rubery, 1979c; Raven 1979) are outlined below.

(1) The transport is specific for IAA and related compounds. This implies (Rubery, 1979c) that polarity is a function of specific porters rather than of a driving force, which would not be expected to exert so selective an influence.

(2) An uphill concentration gradient of auxin can be established along the

tissue in the direction of polar transport.

(3) The transport occurs through parenchymatous tissues associated with the vascular system (Wangermann, 1974). It is initiated early in the embryonic development of angiosperms, and functions throughout the life of herbaceous annuals such as *Pisum sativum*. In a mature *P. sativum* plant, exogenous IAA (and probably endogenous IAA) is transported from the shoot apex down to at least the mature regions of the root; what happens in immature portions of roots is more controversial, but polar transport in opposite directions in stele and cortex is a possibility (Goldsmith, 1977; Morris *et al.*, 1973; Morris & Thomas, 1974; 1978). The polar transport is readily distinguishable from transport in the phloem (Morris & Thomas, 1974; 1978; Morris *et al.*, 1973).

(4) The transport seems to involve cell–apoplast–cell movement rather than symplastic transport (Cande & Ray, 1976; Drake & Carr, 1978; Sheldrake, 1979).

(5) The transport occurs at apparent velocities which are consistent with overall diffusive movement (Goldsmith, 1977).

Recently a mechanism has been proposed for polar IAA transport (Rubery & Sheldrake, 1974; Raven, 1975; 1979; Rubery, 1980) which has been termed the 'chemi-osmotic polar diffusion model' by Goldsmith (1977). The basis for this proposal is the difference in H^+ electrochemical potential across the plasmalemma (cytoplasm electrically negative, and alkaline, relative to the apoplast); the observed ready permeation of undissociated IAA; and the occurrence of IAA^- uniporters at a higher density (mol. uniporter per area of plasmalemma) at the lower (basal) end of the cells in shoots and mature portions of roots. Only the final postulate (the asymmetry of IAA^- uniporter distribution) lacks direct experimental support.

The proposed operation of the mechanism is as follows: auxin (IAAH + IAA^-) in the apoplast encounters the first cell in the sequence. It enters the cell as IAAH, and dissociates inside to give IAA^-. At equilibrium, the influx of *total* auxin species are equal but the cytoplasmic concentration of IAA^- is greater than the external concentration of IAA^- because the inside is more alkaline (3.2.2). Also important in determining the direction of net IAA^- flux is the inside-negative value of the electrical potential difference ($\Delta\psi$). The net IAA^- *efflux* catalyzed by the uniporters is, *ex hypothesis* preferentially at the basal end of the cell. The overall result is a higher concentration of (IAAH + IAA^-) in the apoplast at the basal end than at the apical end of the cell (cf. characteristic 2 and 4 above). This process is repeated for successive cells in the transporting parenchyma, bringing about polar transport. The uphill concentration gradients developed, depend on the expenditure of metabolic energy to maintain the pH and potential differences which might otherwise be dissipated by the proton movements accompanying transmembrane auxin transport itself.

Recently, a mathematical and modelling and computer simulation approach, based on the assumptions of the chemi-osmotic theory (Rubery, 1980), has been adopted in an attempt to account for the dynamic aspects of polar auxin transport quantitatively, and not simply to account for its directional and energetic characteristics (Mitchison, 1980). The experimentally observed features of both auxin pulse propagation (Goldsmith, 1967) and of hormone transport in tissue sections exposed to a continuous auxin source—distribution profiles (Wangermann, 1974) and arrival curves (Hertel & Leopold, 1963)—can be accommodated by the chemi-osmotic model with the following proviso: the permeability coefficients for AH and A^- movement across the plasmalemma (and perhaps the tonoplast if the vacuole is a major transport route through the cell) must have values in the same range as the velocity of pulse propagation or 'front' movement—about $1 cm h^{-1}$ or $2.8 \times 10^{-6} m s^{-1}$ (Goldsmith, 1977). For auxin anions at least this is considerably larger than the likely intrinsic permeability of 10^{-7}–$10^{-8} m s^{-1}$ (Raven, 1975). It suggests that porters (A^- uniport and perhaps A^-/H^+ symport) may be essential not only to give direction and specificity to polar transport, but perhaps also to account for its high rates. These considerations would also predict substantial carrier involvement in any polar transport of the less intrinsically permeant growth substances (Table 3.1), such as may occur in roots for abscisic acid and gibberellin (Wareing, 1977).

The local carrier densities required are not likely to be excessive: transmembrane fluxes of about $10 nmol\, m^{-2} s^{-1}$ should be adequate for most polar transport needs, and these could be satisfied by about 1 μmol carrier m^{-2}, granted a specific reaction rate of $10 s^{-1}$ which is limited by the diffusive supply of substrate. With these assumptions, a 100 000 rel. mol. mass carrier would occupy about 1% of the membrane area. In contrast, the major plasmalemma porter—the proton translocating ATPase (Raven, 1980a)—may be estimated to be present at $10 nmol\, m^{-2}$ to account for its transport capacity, since it probably has a specific reaction rate nearer $10^2 s^{-1}$ permitted by its better substrate supply.

As outlined above, the chemi-osmotic hypothesis lacks both direct experimental evidence for the asymmetric distribution of IAA^- uniporters, and a mechanism whereby such a distribution could be brought about. Raven (1979) has invoked membrane electrophoresis to account for the origin and perpetuation of this asymmetric distribution. It is now clear (Jaffe, 1979; cf Lund, 1947; Scott, 1967) that substantial electric current flows are associated with the development of polar cells and tissues (see Chapter 2). Current (as electrically unbalanced ion flux) occurs *outward* across the plasmalemma in one portion of the tissue, through the external solution, and *into* the tissue at a site of current entry. In higher plants, the outward current is mainly active H^+ uniport, while the inward current may have an H^+ (passive) contribution. Ca^{2+} is also involved. The circuit is completed by symplastic current flow which is carried overall by low molecular weight ions (since protein and nucleic acids cannot generally pass through plasmodesmata); the apoplastic current is also carried by low molecular weight ions. The resistances of the apoplastic and symplastic pathways are generally not identical, so to maintain the symplastic current equal to the apoplastic current, the longitudinal $\Delta\psi$ must be different in the two compartments (with $\Delta\psi$ larger in the more resistive compartment). For coleoptiles, positive current leaves the symplast in the elongating zone and enters more apically, so that there is an apoplastic $\Delta\psi$, base positive to apex, and a symplastic $\Delta\psi$, base negative to apex. These potential differences will tend to move any charged entity. Although (see below) there will not be a net cell-to-cell movement of soluble proteins and nucleic acids, the symplastic potential gradient will tend to polarize them. In this instance, proteins which bear a net positive charge at cytoplasmic pH will move basipetally and those with a negative charge will move acropetally in individual cells. The same will happen to plasma membrane (integral) proteins; those with a net positive charge exposed at the cytoplasmic face of the membrane will tend to migrate basipetally in response to the symplastic $\Delta\psi$, while negatively charged ones could move acropetally. The situation is complicated by the fact that many integral proteins (especially transport catalysts) span the membrane and are exposed to the apoplastic as well as the symplastic solution.

The result of these various interactions is proposed to be that some proteins congregate apically in the plasmalemma, while others concentrate basally (see Poo, *et al.*, 1979). In the case of the IAA^- uniporters, which our hypothesis in its simplest form requires to congregate basally, it is necessary to propose a net internal positive charge or a net external negative charge to account for basal accumulation. An apical uniporter localization (though of lower density than the basal distribution), necessary to permit high transport velocities, could reflect the presence of 'isoporters' with different isoelectric points. Freeze fracture evidence for plasmamembrane inhomogeneity with respect to the distribution of membrane proteins has been presented for algal cells (Weiss *et al.*, 1977; Miller & Miller, 1978); the mechanism of maintenance of spatial inhomogeneity of porters in membranes is discussed by Bourguignon and Singer (1977), Anderson (1979), Spertell (1978) and Hewitt (1978).

While there is no direct evidence for our proposal in IAA-transporting plant cells, and indeed lateral protein movement in plant plasmamembranes remains conjectural, it has the advantage of incorporting positive feedback; the elongating tissue to which IAA is transported will have its H^+ efflux pumps stimulated (the acid growth hypothesis) which, under some circumstances, could perpetuate the symplastic and apoplastic $\Delta\psi$s and hence the IAA^- porter asymmetry. It is known that auxin can stimulate and maintain its own polar transport (Goldsmith, 1977). Also, the development or reinforcement of electrical and pH gradients in tissues due to polar auxin transport in conjunction with stimulated proton extrusion could be an important aspect of auxin-mediated physiological and developmental coordination. Transport systems that involve electrochemical H^+ gradients such as movement of weak electrolytes (including other growth substances) or sucrose loading into phloem (Giaquinta, 1979) may be regarded as potential secondary auxin targets.

Raven (1979) has speculated that early asymmetry in the embryo could result from maternal influences which determine polarity and hence current flow—embryoid formation in suspension tissue cultures however, remains a problem (see Section 3.1). Gradients of fixed negative charge in the cell wall could maintain polarity in dormant tissues such as seeds (Raven, 1979).

We feel that the scheme proposed above for polar IAA transport, and its initiation and perpetuation, has the advantage of requiring minimal 'receiving apparatus' for the signal; all cells in the pathway are required to make the IAA^- uniporter and (at least while expanding) to respond to IAA by increasing the rate of H^+ extrusion and thus hyperpolarizing. Granted this, the system is self-perpetuating after the initial polarity has been determined in the early stages of embryogenesis. Polarity of receptors for IAA, or even of IAA synthesis and metabolism, are not crucial to the scheme. Sheldrake, (Sheldrake & Northcote, 1968) has also presented xylem differentiation as an autocatalytic process initiated by auxin.

Polar auxin transport of the type we have discussed seems to fit quite closely to a 'design specification' for a transmitter of positional information which relies on a minimum of additional *ad hoc* assumptions about a receiving apparatus. Perhaps it is because we know more about auxin than other growth substances, but nevertheless it still seems to merit the title 'master hormone' awarded over a quarter of a century ago (Thimann, 1952).

3.3 CONCLUSIONS AND PROSPECTS

Of all the plant growth substances which have been well characterized, IAA (and possibly cytokinin) best fulfill the criteria for transmitting positional information; other mobile plant growth substances are more concerned with temporal coordination, although they may interact with auxin to bring about spatial effects.

Hormones probably act via proteinaceous receptors, distinct from the genes or catalysts (enzymes or porters) which they ultimately affect. Analysis of these receptors and *a fortiori* of 'messengers', which may act between hormone-receptor complex and the ultimate target, is in a rather primitive stage.

The ability of IAA to transmit positional information is due to the fact that more information resides in the IAA gradient generated by polar transport than need be put in to accomplish polar transport. Granted the presence of IAA receptors, the catalyst on which the IAA-receptor complex acts, IAA^- porters, and also the ability of these IAA^- porters to undergo membrane electrophoresis, an initial environmental (e.g. maternal) polarizing event can lead to an established polarity reinforced by positive feedback.

The circulating currents which may underpin polar transport could well have informational roles aside from initiating and perpetuating polar hormone transport; in plants the best evidence for a causal relationship between these currents (and related action potentials) and differentiation comes from the coenocytic alga *Acetabularia* (Bentrup, 1977; Goodwin & Pateromichelakis, 1979).

Further investigation is required for these points.

FURTHER READING

CHERRY J. H. (1977) Hormone Action. *The Molecular Biology of Plant Cells*, (ed. Smith H.) pp. 329–64. Blackwell Scientific Publications, Oxford.

GOLDSMITH M. H. M. (1977) The Polar Transport of auxin. *Ann. Rev. Plant Physiol.*, **27**, 439–78.

GRAHAM C. F. & WAREING P. F. (1976) *The Developmental Biology of Plants and Animals*, pp. 393. Blackwell Scientific Publications, Oxford.

JACOBSEN J. V. (1977) Regulation of ribonucleic acid metabolism by plant hormones. *Ann. Rev. Plant Physiol.*, **28**, 537–64.

KENDE H. & GARDNER G. (1976). Hormone binding in plants. *Ann. Rev. Plant Physiol.*, **26**, 267–90.

LETHAM D. S., GOODWIN P. B. & HIGGINS T. J. V. (1978) *Phytohormones and Related Compounds—A Comprehensive Treatise. Vol. 1*, pp. 641, *The biochemistry of phytohormones and related compounds.* Vol. 2, pp. 648, *Phytohormones and the development of higher plants.* Elsevier/North Holland, Amsterdam.

QUANTRANO R. S. (1978) Development of Cell polarity. *Ann. Rev. Plant Physiol.*, **29**, 487–510.

TORREY J. G. (1976) Root hormones and plant growth. *Ann. Rev. Plant Physiol.*, **27**, 435–59.

WAREING P. F. (1977) Growth substances and integration in the whole plant. *Symp. Soc. Exp. Biol.*, **31**, 337–45.

CHAPTER 4

CELL GROWTH AND DIVISION

4.1 INTRODUCTION

Many research workers in biology hold to the view that the phenomena of plant development which are recognized at our level of scale (production of roots, buds, leaves, stems, rots, flowers, new plants) can be understood if enough is known about the structure and function of the molecules which make up the developing organs or plants. This line of thought seeks to explain plant physiology in terms of biochemistry and cell biology and to understand the phenomena discovered at these levels of scale by what has come to be known as 'molecular biology'. Put another way, the proposition is that the phenomena of plant development, occurring on a scale of a few tenths of a millimetre up to a few metres, can be understood in terms of events occurring at a scale of nanometres, 10^5–10^9 times finer. This pattern of thinking may not be the most useful way to approach developmental problems. It is possible that living organisms may be closer to probabilistic, rather than deterministic, systems (see Beer, 1975; Ackoff, 1980). It is characteristic of probabilistic systems that one cannot predict their behaviour with any degree of confidence by analysis of their components, however detailed or sophisticated that analysis may be. This fact, which at first sight seems quite disturbing, may explain why the problems of plant development have been so hard to solve. The classical reductionist methods clearly will not be adequate if the developmental phenomena which we wish to understand flow from the activities of sets of interconnected probabilistic systems, connected at various levels of scale in both space and time.

When a living system is killed with a fixative or when it is snap-frozen, the system is transformed into particular states about which a great deal can be learned. If the freezing is done carefully, the process may be reversible, and living material 'stopped in its tracks' in that manner can be processed to provide information at many levels of scale about a *moment* in the life of a probabilistic system. Unfortunately, information cannot be obtained from the same specimen at any other moment in its life. When living tissues are killed by chemical fixation, the living systems are transformed into particular states that depend in detail on the nature of the living system, the fixative, and the conditions surrounding its use. Careful studies of the actions of fixatives on living plant cells have been few (see Mersey & McCully, 1978 for references) but they all agree about two important facts: all presently known techniques of chemical fixation alter the visible structure as seen by light microscopy and, even in thin hair-cells, fixation is far from instantaneous. Two to ten minutes elapse between the first detectable arrival of fixative in hair cells and cessation of Brownian motion of cytoplasmic particles visible by light microscopy and this time is probably longer in multicellular organs such as roots and leaves. The image of chemically fixed material is therefore not even a 'snapshot' of the living state as with snap-frozen material, but the image of a state that has been slowly transformed over many minutes, and which is known to contain visible artefacts.

It is also easy to forget that most microscopic information about structure comes as images created by the interaction of electromagnetic radiation with the reality. The images created by modern light microscopes are so good it is easy to fall into the trap of believing that one is 'looking at the reality made big enough to see'. This is never the case. The image is always the product of the interaction of the viewing radiation with the real structure. Even under the most ideal circumstances, the information that can be gathered from images of living systems is very incomplete and needs the utmost care in interpretation.

None of this denies the progress that has been made in the study of plant structure in the last three decades. Frey-Wyssling's masterpiece 'Submicroscopic Morphology of Protoplasm' (1953) was a most remarkable synthesis of knowledge about structure of biological materials using information obtained largely prior to the introduction of electron microscopy. His work still contains ideas, observations and conclusions to challenge modern students but there is no doubt that, if one compares what was known about subcellular organization 30 years ago with what is in Gunning and Steer's equally superb book of 1975, the field has undergone a quantum jump. The introduction of glutaraldehyde as a non-coagulant fixative (Sabatini *et al.*, 1963), of epoxy resins as embedding media (Glauert & Glauert 1958; Spurr, 1959), of the bypass ultramicrotome, of glass and diamond knives (see Wachtel *et al.*, 1966) and of the technology of freeze etching and freeze fracturing (Branton *et al.*, 1975), all combined with steadily improved electron-optical systems to produce this quantum jump. The detail available today about the structure of killed cells or about replicas of the freeze-fractured state of living cells is inspiring and about as daunting and overwhelming as a first exposure to Frey-Wyssling's book 30 years ago—but to what extent is that detail representative of reality?

4.1.1 THE ABSENCE OF CELL MIGRATION

The two terms with which this chapter begins—cell growth and division—are names applied to major phenomena which accompany plant development. Students of animal development have also to consider cell migration. With the exception of motile gametes, and motile vegetative and sexual spores, cell motility, in the sense of cell migration, plays little role in plant development and will not be treated in this chapter. Before setting aside the topic it is worthwhile asking why cell migration, one of the most important morphogenetic events in animal development, is so restricted in plants. In part the answer lies in the physical chemistry of the extracellular matrices—the cell walls—that invest plant cells. These walls are constructed from polymers (chiefly polysaccharides, some proteins, and the complex aromatic derivative lignin, among others) which form strong, largely water-insoluble gels, in which the dry matter content may lie between 10 and 60%. These gels resist deformation and separate the cell membranes of adjacent cells from one another everywhere except at plasmodesmata. Even unicellular and filamentous organisms are coated with similar constituents and cannot make cellular contact with one another unless these coats (rarely less than 0.1μm, and in many cases up to 10μm thick) are digested or unless naked cells are generated by some other method (as for example, in the formation of certain algal gametes). The presence of thick walls is only part of the story because even when the walls are removed to produce living protoplasts, the protoplasts show little evidence of the vigorous cell-migration behaviour that is so characteristic of many types of animal cell *in vitro* or *in vivo*.

4.1.2 THE INTERRELATIONSHIPS BETWEEN GROWTH, DIVISION, DIFFERENTIATION AND DEVELOPMENT

Three basic phenomena—cell growth, cell division, and cell differentiation—accompany development wherever it occurs. How are these three phenomena related to development? The relationships are not simple and depend on how one views 'development' (see Fig. 4.1). One tends to think first of development in terms of the production of something new; buds, flowers, fruits, leaves, roots, new plants and so on. In most cases, anything new is produced by growth, defined as an irreversible increase in volume (the production and maturation of some spores and pollen grains is sometimes accompanied by negative growth). Furthermore, under natural conditions the production of new organs of all

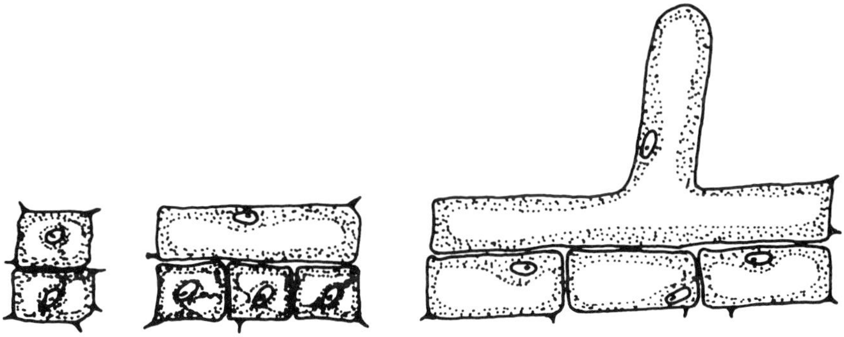

Fig. 4.1 A diagram to illustrate (from left to right) the relationships between division, growth (here represented by extension) and differentiation as these processes may affect the progenies of two adjacent cells in a meristem. The progeny of the upper cell undergo extension and differentiation (formation of a root hair) while in the lower sequence, division first accompanies growth in such a way that cell size stays constant. This is followed by a period of further growth in which overt differentiation does not ensue.

kinds involves cell division and cell differentiation (the interesting exception posed by multinucleate single-celled organisms will be mentioned later). Indeed, it is the characteristic patterns of division and differentiation that are used by anatomists to classify organs into types. However, if 'development' is extended to include *all* changes in state that occur naturally in plants, one meets events such as abscission, bark shedding, root loss, senescence, the development of symbiotic states and processes such as the dispersal of spores or seeds by passive motor organs. Some of these events involve cell and tissue death, sometimes on a grand scale (measurable in tonnes per hectare). Some of them involve no growth, or negative growth (senescence, abscissions of various kinds). The passive discharge processes seem to involve none of the three basic phenomena, though it is true that the passive mechanisms could not occur if growth and differentiation did not produce the various structural anisotropies (i.e. inequalities) that make them work (see Haberlandt, 1965).

For our purposes, development may be regarded as *any permanent change in state*. If death and destruction of tissues (either partial or total) are included as aspects of cellular differentiation, and if it is agreed that growth can be negative when organs or tissues are shed, then plant development is always accompanied by at least two, and often by all three of the phenomena, cell growth, division and differentiation.

What then is the relationship between these three phenomena? Despite misleading statements in many elementary texts, there appears to be no *necessary* connection between them. It is undeniable that whenever new organs or tissues are being generated, either in apical meristems or in the lateral meristems of the cambia and phellogen, the three basic phenomena are closely associated with one another both in space and in time. That they are indeed separable phenomena is now very well documented. Cortical parenchyma cells in appropriately wounded internodes of *Coleus* will differentiate directly into tracheary elements without prior growth or division (O'Brien & McCully, 1969). Isolated callus cells can also be provoked into carrying out the same conversion (see Torrey, 1975; Roberts, 1976; Fukuda and Komamine, 1980). The study of the development of γ-irradiated grass seeds confirms that essentially normal coleoptile and leaf development can occur in a plant totally deprived of the capacity for cell division (see Haber 1962 and references therein).

Division and growth are processes that often, but not necessarily, accompany one another (O'Brien & McCully, 1969). Green (1976), in a long overdue critique of concepts such as 'growth with and without cell division' has introduced the concepts of extension rate and partitioning rate, and analyses six modes in which they may be coupled (see Fig. 4.2 and its caption for details). To the six modes set out by Green in Fig. 4.2, one other set should be added namely, the modifications that arise when the plane of division is parallel to, or at some intermediate angle to, the plane of extension. Such situations are commonly encountered in the development of provascular tissue (Esau, 1977).

It is helpful to think about the interrelationship between these phenomena of growth, division and differentiation as follows. In the development of new organs, growth produces new volume. The shape of new organs is determined by the manner in which this new volume is distributed in space. The shape of any new organ can usually be derived by imagining it to have developed as the result of various rates of expansion along three axes at right angles to one another (for the case of tip growth, see Green, 1969; 1976). If the growth increment is equal along all axes, the organ will tend to be isodiametric; if growth in one axis is restricted, flat plates may develop. The conditions that are needed to generate something like a palm frond, a fern crozier, or a leaf full of holes (as in *Monstera deliciosa*) provide interesting challenges. Some reservations are necessary here to allow for the difference between the way the tissue or organ would expand if left to its own devices, and how it behaves if it is subject to external constraints (e.g. the development of buds in the axils of hard, woody leaf-sheaths). Within any expanding mass of tissue, cell division will generate new cells. The shape of the cells that are generated depends almost entirely on the plane(s) of the divisions that generate them and cell size is determined by the balance struck between extension rate and partitioning as set out in Fig. 4.2. Having been produced, the cells may now follow any of the pathways of differentiation that are possible for that tissue. Because they adopt different structures and functions, we speak of this event as *cellular differentiation.*

Some qualifications are necessary before applying this simple way of viewing these interrelationships to what one sees in real plant organs. Firstly, the production of internal gas spaces, the death of some cells and tissues, the osmotic properties of the protoplasts and the elasticity of certain wall-types may all interact such that the real shape of mature cells in turgid tissues may be rather different from the simple shapes they may have when they are first produced (see Esau, 1977, Ch. 5 for details). Secondly, many cells have the capacity for tip growth (root hairs, pollen tubes, fungal and some algal filaments, fibres, fusiform initials in the cambia and young secondary xylem tracheary elements [Wardrop, 1964] are all good examples). Here, the cell apex is the site of extension (in some cases, two or more growing apical regions can form). Such cells have the capacity to force their way between other cells, an activity that may seriously complicate the interpretation of the mature structure. Thus, in the medulla of the brown alga *Fucus vesiculosus*, the fibres lie in a matrix which they have invaded, but not synthesized (O'Brien & McCully, 1969, Fig. 4.2). Thirdly, in some species, certain of the fusiform initials in the cambium apparently grow in spring at a radial rate of growth that exceeds that of their neighbours. In the mature wood, several files of cells are often missing near the tracheary elements that these initials produce, and adjacent files often show distinct signs of compression (see Esau, 1977, Fig. 10.1).

These qualifications (and others which could be introduced, e.g. the production of complex shapes by single cells such as desmids) do not seriously detract from the usefulness of viewing growth, division and differentiation in the manner treated above. Rather they simply underscore the existence of natural

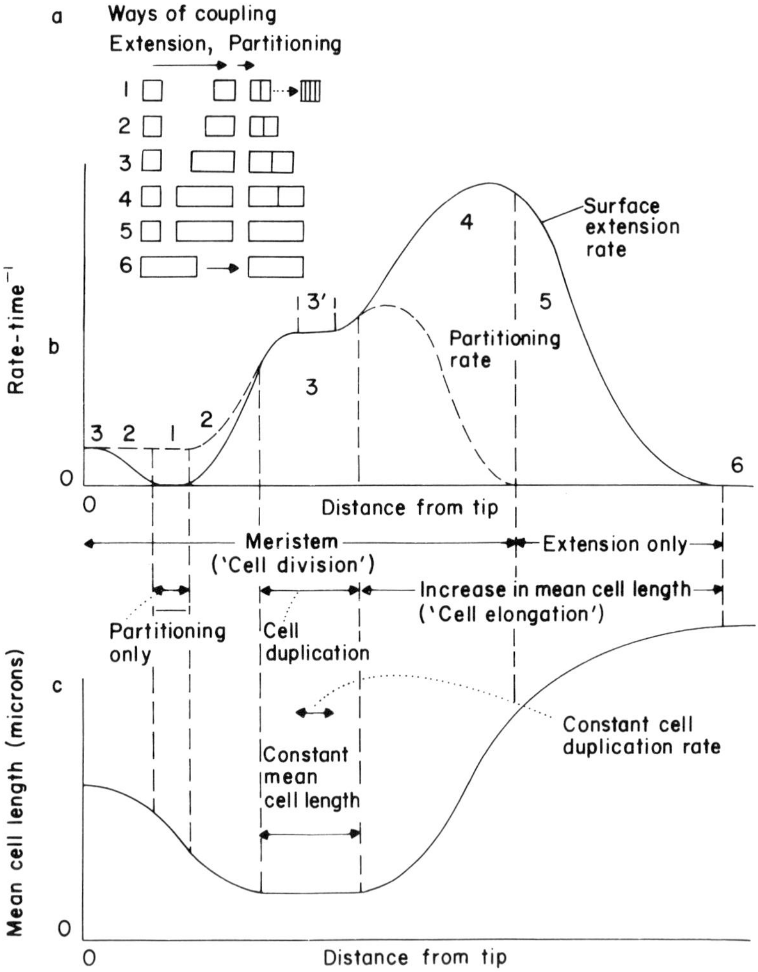

Fig. 4.2 An analysis of a hypothetical growing axis (tip is at left) which is characterized by curves of extension rate and partitioning rate. The six qualitative types of histogenetic behaviour given in *a* are all-inclusive for cell pattern down a line. Each type of behavior brings on a change (or constancy) in mean cell length. Cross-reference between the curves in *b* and the histogenetic images in *a* is given by the numbers. Cross-reference between the curves of *b* and the change in mean cell length, *c*, is given by the vertical dashed lines. *a*, 1: Cell partitioning only. Here the extension rate is zero and the partitioning rate is finite. This behavior is typical for xylem regeneration from parenchyma and is not typical for apices. It is given here for completeness of illustration. Cells in such a zone would be pushed to the right by growth of cells to the left and hence would resume expansion. 2: Both rates finite, partitioning having the higher value. Mean cell length decreases but less rapidly than before if the partitioning rate is unchanged. This type of behavior is seen in roots. 3: Here the rates are equal to each other but not necessarily constant with position or time. There is no change in mean cell length; hence such a zone would be one of cell duplication. In corn this condition exists only at the point where the curves cross. 3′: Here the rates are not only equal but are constant with position and hence, for a while,

diversity. Nature has had at least 400 million years in which to experiment with ways of building plant bodies; it would be disappointing if all of the richness of those experiments could be encompassed by the simple views given above.

4.1.3 MULTICELLULARITY AND THE CELL AS THE UNIT OF LIFE

Before treating cell growth and division in more detail, there are a few more general points to discuss. Nearly a century has elapsed since a number of great minds wrestled with the problems posed by multicellularity and the status of the cell as a unit of life. Two major views competed for support at that time. One held that multicellular organisms arose from colonial organisms which in turn arose from unicells by aggregation. Volvox, with its beautifully regulated cilia, was often cited as an example of the kind of colonial organism which, with further integration of its units and increased differentiation, would be an effective precursor for multicellular green plants. Anton de Bary was a firm believer in the other view, namely that multicellular organisms and colonial ones arose from unicellular ones by partitioning and growth and differentiation. Here, the emphasis is on the organism as the unit of life and the cell is seen in a derivative sense. De Bary (1862) (cited in Sharp, 1926) put it very firmly: 'die Pflanze bildet Zellen, nicht die Zelle bildet Pflanzen'. This old argument is no longer a central issue of enquiry, and it is possible to dismiss it casually. Unicellular, colonial and multicellular organisms exist and any of the three types could have given rise to the other two. And yet, in a curious way, the stand one takes on this old question affects the way one thinks about development and its regulation. For if the basic unit of life is the cell, then the life of organisms will

---

constant with time. This zone of constant cell-duplication rate is the only one where the presumed characteristics of cell-colony growth obtain: constant exponential production of cells of constant mean dimensions. In the corn root, case 3′ does not exist as the curves are not horizontal while identical. The colony-growth situation, from which traditional concepts relating to axis differentiation have evolved, is clearly a special case in the general frame of analysis being reviewed here. 4: Both partitioning and extension rates are finite, the extension rate predominating. Here mean cell length must increase even though cell partitioning is occurring. 5: Partitioning rate is zero; extension rate is finite. Mean cell length increases. 6: the case in which both parameters are zero—the 'maturation' zone. The major generalization is that change in mean cell length in a region is proportional to the difference between the curves in the region. If partitioning predominates, cell length decreases; if extension predominates, cell length increases. If the values are identical, cell length is constant. The generality has application in the definition of zones along the axis. Apparent old names are given in quotes. Zones are all defined by the relative values of the two curves and hence have unambiguous meaning and well-defined boundaries. The names are chosen to imply the correct histogenetic characteristics. To have constant histology at the tip, an axis would have to begin with cells showing no growth (case 6) or cell duplication (case 3). This need be true only over an infinitesimal distance, as shown (from Green P. B. 1976. *Bot. Gaz.*, **137**, 187–202. Reproduced with permission from Chicago University Press).

ultimately be 'understandable' in terms of cell biology provided cybernetically-based ways of investigating the problems and of thinking about them are developed. In this case it means that the 'life units' range in size from small spheres about 0.2μm in diameter to the giant fibre cells of about 1m that are found in the leaves of *Boehmeria*. If the basic unit of life is the organism (which means they range in size from submicroscopic single cells up to whales and once included *Brontosaurus*), then the multicellular state might be viewed profitably as a response by the organism to problems it might encounter if it tried to live its life as a single cell. What are the problems that might limit the development of *Sequoiadendron*-sized single cells?

The most important problems are those created by diffusive limitations. While diffusion is remarkably rapid over distances up to a few tenths of a millimetre, it is extremely slow once one exceeds a centimetre or so. It is no accident that even the largest cells have at least one dimension that is reduced to a few millimetres, and usually to less than 500μm. Diffusion could not meet the needs of a mouse-sized cell, let alone a *Sequoiadendron*-sized one. Perhaps of equal importance are the problems created by the surface area available for exchange between the organism and its environment. It is interesting to look at cell division in this context. With very few exceptions, cell division does three things simultaneously. New cells are generated; additional cell membrane surface is created and the new surface is stabilized by the development of additional apoplast, in contact with the original apoplast (see Fig. 4.3). These last two

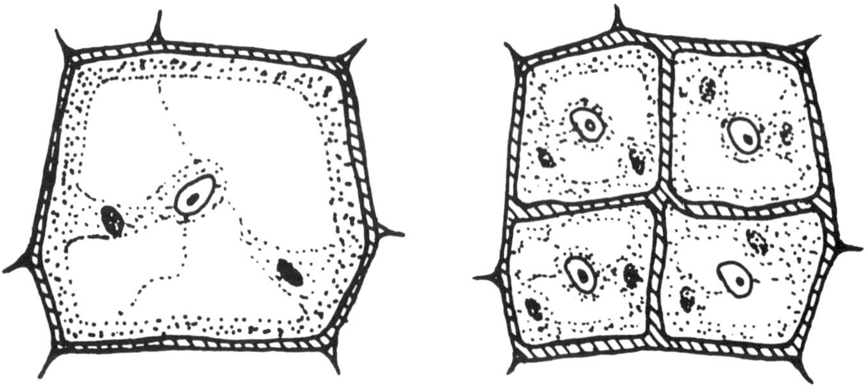

Fig. 4.3 Division, even when unaccompanied by growth, increases the cell membrane surface area in contact with the apoplast.

aspects of cell division are not often discussed and yet it seems likely that these are two solutions to one of the most difficult problems created by increasing organism size.

The events surrounding the cellularization of the central cell as it forms endosperm in cereals provide one with an ideal opportunity to witness the problem developing and the way it is solved. In most cereals, the central cell of the embryo sac is a giant, highly vacuolate cell containing one or more polar nuclei. Immediately after double fertilization, the endosperm nucleus undergoes an extremely quick set of divisions, producing a set of nuclei that line the periphery of the cell. This phase is complete in about 36 hours. During this process, the central cell has already commenced a phase of expansion, an expansion which will go on rapidly for the next three weeks. Cellularization begins by a process of wall invagination that begins to carve up the large cell into a series of bays. Cellularization is completed by a mixture of normal phragmoplast activity, and further wall ingrowths (see Mares *et al.*, 1977; Morrison & O'Brien, 1976). Three to five days after fertilization the whole system is completely cellularized (Briarty *et al.*, 1979). In the next 30–40 days, the structure will both enlarge and add many more cells at its periphery (to give 10^5 overall), the outermost of which become aleurone cells. At the same time the cells will become absolutely crammed full of reserves of starch, protein and lipid, all of which have been manufactured from precursors that had to cross a cell membrane surface to enter this system. Internal cell walls increase from 0 at anthesis, through $6 \times 10^6 \mu m^3$ at 2 days post anthesis, to $8 \times 10^8 \mu m^3$ at 36 days post anthesis (Briarty *et al.*, 1979). The boundary of the original central cell and the outer surfaces of all of the cells which finally surround the endosperm lack plasmodesmata, so that a symplastic route of entry is not possible. Apart from the aleurone layer, the rest of the endosperm (about 90% by weight) has no morphogenetic future. Only the aleurone layer will live; the rest will be digested after germination. The cell walls and the cell membranes that formed during grain development will all be destroyed. Why make them in the first place? It seems likely that the answer lies in producing from this central cell a large internal cell-membrane surface with an apoplastic contact, a surface across which an adequate rate of nutrient uptake can be sustained to fill the grain. (For a discussion of the problems of diffusive transfer in such embedded wall/membrane interfaces, see the parallel discussion for transfer cells in Gunning & Pate, 1974).

A further problem that might hinder the evolution of giant single-celled organisms might be viewed as a problem in 'nuclear administration of the cytoplasm'. Shortage of DNA could, in principle at least, be dealt with by processes like those which give rise to polytene chromosomes, or produce endopolyploidy (a widespread phenomenon in plants: see d'Amato, 1964; 1977; Datta, 1971; Oksala & Therman 1977). However, a great deal of information flows in the form of RNA from the DNA into the cytoplasm, creating a special case of diffusive limitation which can be stated as follows: what is the volume of cytoplasm that can be administered by a nucleus (or its polytene or endopolyploid equivalent)? The answer will undoubtedly depend upon the rate of streaming in the cell. This problem can be dealt with very effectively by 'going multinuclear' so that no area of cytoplasm is ever very far from an administratively competent nucleus.

It is quite clear that by combining nuclear division (karyokinesis) and cell partitioning (cytokinesis), the cell (or organism) can combine the solutions to both of the last two problems.

However, there is at least one further major problem, that of the highly non-random distribution of functions in even unicellular organisms (see Atkinson *et al.*, 1974 for a beautiful 3-D reconstruction of *Chlorella*). Even here no clear picture exists of how spatial control of function is exercised. Multicellularity may have the further function of allowing certain regions of an organism to have particular structures and functions. This is entirely speculative since even bacterial cells have considerable spatial differentiation. It could even be that spatial differentiation is not directly regulated at all and that spatially-structured states are examples of stable configurations that highly interconnected probabilistic systems adopt in the presence of certain constraints.

4.1.4 CELLULAR OR ORGANISMAL LIFE-UNITS?

There is no easy way to choose between a cellular and organismal view of life. At present, most studies of development appear to be based, at least tacitly, on the assumption that the unit of structure and function, the 'life-unit', is the cell. Certainly, structure and function often change sharply at cell boundaries, a fact which supports this view. But other boundaries also show sharp discontinuities in structure and function (e.g. all sub-cellular compartment boundaries). On the other hand, it is possible, using coleoptiles or sporangiophores, to so arrange the conditions under which phototropic stimuli are given that all that happens is a re-distribution of the same amount of growth about the axis (see Gamov & Bottger, 1977 and references therein and Thimann & Curry, 1960). Under these conditions, the organ seems to behave as unit, and the behaviour of the separate cells (in the case of the coleoptile) or of different regions of the same cell (in the case of the fungal sporangiophore) is somehow regulated by their position within the whole.

There can be no doubt that a single cell is enough to reconstruct the organism. In all non-apogamous plants with alternation of generations, the organism is cyclically reduced to a single cell (spore and zygote) twice every life-cycle. Studies of carrot phloem-explants showed that individual cells of a callus culture, derived from mature, secondary-phloem parenchyma cells, could regenerate whole embryos and plants. This extended the concept of totipotency of plant tissues (well known from cases of nucellar embryos and various types of vegetative reproduction) to the single cell. These facts are not disputable. What affects the way one thinks about development is the impact of the interconnectedness of cells upon events that take place in a particular cell or group of cells. Though the plant body lacks a nervous system, all of its living protoplasm is electrotonically coupled via plasmodesmata, except for spores and zygotes, the two unicellular gateways that connect the generations (see Carr, 1976 for an excellent review and Drake *et al.*, 1978 for more recent work). It has been known for a long time that plant protoplasm is irritable and that a great deal of electrical activity takes place associated with stimuli of various kinds (Bose, 1913). Unfortunately, there is no good theory of the significance of this electrical activity at any level of analysis, be it whole organ, cellular or sub-cellular behaviour that is examined. What information, if any, is exchanged around the plant body by this electrical activity is unknown. Whatever the significance of electrical activity may be to the plant body, there is no doubt that chemical information is exchanged between different parts and that the exchange has effects at various levels of scale.

These considerations suggest that one should be cautious in assuming that the unit of stimulus reception or response in the plant body is always the cell. It is possible that many developmental responses result from signals that are perceived by large numbers of cells acting in concert and that they lead to localized (or diffuse) responses in the same (or distant) cells as a result of concerted activity that does not respect cell boundaries. The fact that a highly-controlled growth response can take place in a single cell (tropistic curvature in a *Phycomyces* sporangiophore or conversion from one to two-dimensional growth in fern gametrophytes responding to light quality: see Wada & O'Brien, 1975) should warn us not to ignore the possibility that large volumes of interconnected protoplasm may be capable of carrying out the same phenomena, whether they are cellularized or not (for a fascinating set of essays on this topic see Bonner, 1955).

4.2 CELL GROWTH

The measurement of plant growth, the concept of growth zones and the existence of circumnutation and tropisms were elegantly described by Darwin (1880). One wonders what his reaction would have been to the superb time-lapse movies of growing plants and of growing and streaming cells now available. However, it is still easy to overlook the remarkable complexity of plant growth.

Consider the problem posed to the engineer and tradesman of trying to build a factory with the capacity to produce 3000 compounds. The production lines of all these syntheses have to be speedily repairable and their capacity must increase with time. The structure has to be built to resist high internal pressures (several atmospheres) and yet be able to extend continuously as the factory expands. New construction materials must be added without destroying the integrity of the original structure. Supplies of key intermediates used on the production lines must be compartmented and the activities of any single production line must be regulated so that the functioning of the whole is not jeopardized. It must be possible to shut some of them down without interrupting the work of others. No process requires a temperature of more than about 25°C. A daunting requirement and yet this is only a very simplified view of the specifications met by any growing plant cell. Moreover, in a multicellular system, each factory maintains links with all the other factories in the area and adjusts its own activities accordingly.

4.2.1 TYPES OF GROWTH AND DEFINITIONS

Thimann long ago pointed out the merits of defining plant growth as irreversible increase in volume (Went & Thimann, 1937). Clearly, for any plant cell to grow, its extracellular jacket must expand, either locally (as in tip growth: Fig. 4.4) or more generally. Single cells and filaments can also grow over their whole surface (surface growth, Fig. 4.5: see also the lovely marking experiments on a desmid by Hogetsu & Shibaoka, 1978a; 1978b). When cells grow as part of a tissue, they do not, in general, sever their plasmodesmatal connections with one another. This form of growth is termed *symplastic growth* (Fig. 4.6) and carries with it the implication that, if one were able to measure the rate of extension at any point on an imaginary plane taken through the tissue at right angles to the axis of extension, the growth rate would be constant (cf. Figs 4.5, 4.6). Nearby planes might well have very different growth rates (Fig. 4.2) (Green, 1976). The existence of circumnutation suggests that this condition is not quite met in practice since the circumnutatory movements must arise from local departures in growth rate about the axis of extension, so that, at least for short periods, some points on the planes are extending faster than others. Even when an organ responds to a tropic stimulus, the circumnutatory movements persist but the movements become elliptical with the long axis of the ellipse lying in the plane

Fig. 4.4 Tip growth. Markers originally equidistant from one another remain the same distance apart except at the extreme apex where new wall synthesis leads to their displacement to the flanks of the dome.

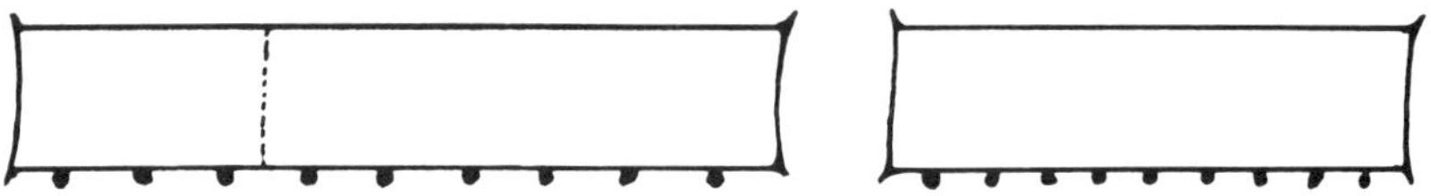

Fig. 4.5 Surface growth. All elements of the surface expand at a constant rate as the cell grows from its former length (shown dotted) to its new length. Marker particles behave as shown.

Fig. 4.6 Symplastic growth. Two cells separated by a common wall perforated by pits. As the cells extend, pits maintain their relative positions.

of curvature (Darwin, 1880).

Water accounts for 70–95% of the fresh weight in hydrated tissues with growing tissues tending to lie in the upper part of this range. Growth therefore involves water uptake, much of which is stored in the vacuole(s) (see Burström, 1979, for a different way of viewing this fact). There was a time when doubts were expressed about whether the volume of cytoplasm actually increased or not during expansion. Inspection of sections shows how this doubt arose. If one examines cells shortly after they are produced in meristems (especially in the root: it is less obvious in the shoot, and totally untrue for cells of the cambia and phellogen) and compares their appearance with that of the same cells after a period of growth, the former appear to be what has come to be called 'densely cytoplasmic'. The latter appear to be almost all vacuole, with a very thin peripheral lining of cytoplasm around the wall. It turns out that appearances are deceptive. A peripheral layer of cytoplasm 1μm thick in a cell 500μm long and radius 20μm has a volume at least twice that of a cell that appears to be 90% cytoplasm and is 10μm long, 6μm radius (see Fig. 4.7). Wright (1961) showed convincingly that coleoptile cells increase cytoplasmic constituents several fold during expansion so on both counts we can assume that cytoplasm increases during growth. The amount of extracellular matrix increases also since, with the exception of a few special cases (e.g. *Oedogonium*; Pickett-Heaps, 1975b), cell walls do not get thinner as cells expand.

New walls form in a variety of situations (see O'Brien, 1972). The term *cell plate* is used for any newly formed wall that is partly completed or just been completed (the latter is difficult to judge). *Primary wall* is the term used for the material deposited by cells displaying surface expansion at the site where the wall is being deposited; new walls formed by tip growth are therefore considered to be primary walls. *Secondary wall* is the term used for the situation where the surface has completely finished expanding at the site where the wall is being deposited. Different regions of the same cell may be depositing primary or secondary walls at the same time (root hairs [Newcomb & Bonnett, 1965], perhaps fibres [Robards & Kidwai, 1972] and protoxylem tracheary elements [O'Brien, 1970]). Since concern here is with growth, secondary walls will not be discussed except when they are deposited in a growing cell.

So, in growing cells, the processes that produce new cytoplasm, new wall material and which enlarge the vacuole(s) are all active and if the growth rate is rapid (1mm/h^{-1}) they must be proceeding apace. However, in cells undergoing surface growth or symplastic growth, these processes are not highly localized in space. The increment of 1mm/h^{-1} shown by the organ as a whole is the result of growth taking place throughout the whole length of the extension zone (refer to Fig. 4.2). At any plane within the zone of extension the growth rate will vary. If the growth rate were to be measured carefully on any particular organ, one could plot it in the same manner as the theoretical curve for surface extension rate as a function of distance shown in Fig. 4.2. The details of the curve will vary with organ and growing conditions.

4.2.2 BIOPHYSICS OF CELL GROWTH AND CORTICAL MICROTUBULES

The review by Green (1969) is an excellent starting point in any discussion of the mechanics of cell growth and its control. Note that he does not attempt to cover the difficult field of symplastic growth where tissue tensions, arising from adjacent tissues, can profoundly influence the environment in which any cell is expanding (but see Green, 1980). Green's earlier review concentrates on situations where the stresses within the wall are generated primarily by the expansion of the cell itself.

Fig. 4.7 The volume of cytoplasm in the highly vacuolated cell is double that of its 'densely cytoplasmic' precursor cell. Lower cell and vacuolated cell to same scale but vertical axis of upper cell × 1/4.

The first case to consider is growth of a cylindrical cell, growing by highly directed surface-expansion along its entire length. This mode of growth has been shown to be accompanied by a typical wall texture (Roelofsen, 1965). The cell wall microfibrils are deposited with a predominantly transverse orientation at the cytoplasmic surface (Fig. 4.8) and are gradually realigned until they have a

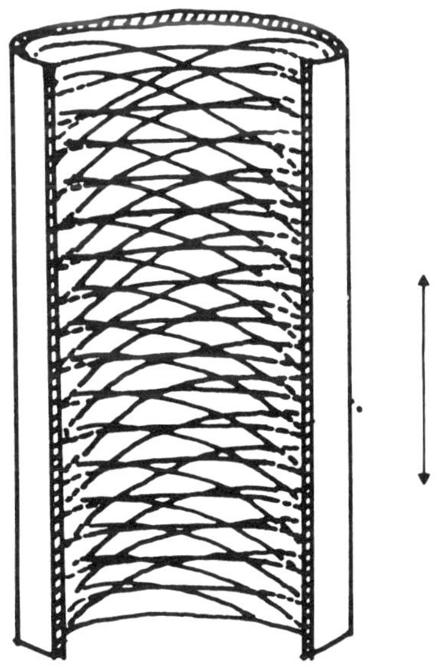

Fig. 4.8 Approximately transverse texture of cellulose microfibrils on the inner face of an element of a cylindrical cell expanding in the direction arrowed, by surface growth.

nearly longitudinal orientation at the outer surface of the wall (but see Roland & Vian, 1979 for a re-evaluation of this model of wall texture). This transverse orientation so strongly reinforces the wall against radial expansion that the cell shows predominantly axial expansion. It seems plain that it is the orientation of the cellulose microfibrils in the most recently deposited layer of wall that determines the direction of growth in the growing cell. The discovery of arrays of cortical microtubules lying just beneath the cell membrane and with an orientation more or less parallel to the orientation of cell wall microfibrils (Ledbetter & Porter, 1963) generated a great deal of interest for it seemed that here was a cytoplasmic structure that might be the mediator between genes and morphogenesis (Fig. 4.9). To the extent that cell expansion is regulated by cell-wall microfibril orientation in the layer immediately adjacent to the cell surface in expanding cells, here was a cytoplasmic candidate that had the right orientation. Since that time, a large number of investigations have been carried out on systems ranging from unicells to hypocotyls (see O'Brien, 1972; Stebbings & Bennett, 1975; Hepler, 1976; Srivastava *et al.*, 1977; Hogetsu & Shibaoka, 1978a, for references) in attempts to test the correlation between the presence of

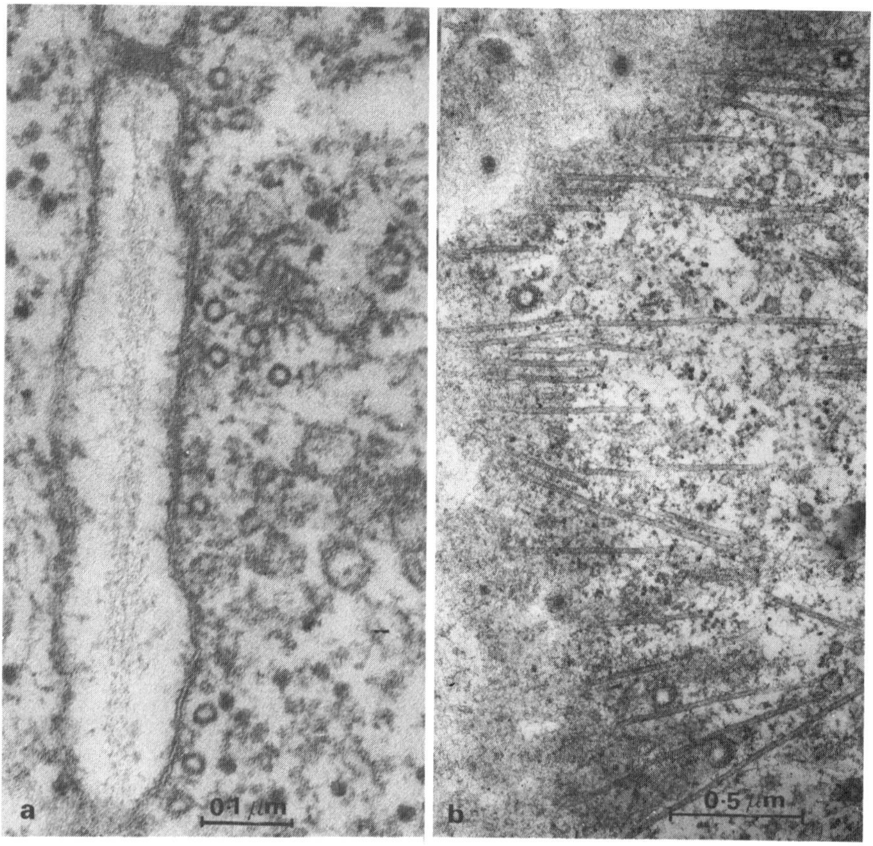

Fig. 4.9 Cortical microtubules seen in TS (a) and in face view (b) in parenchyma cells of the root tip of *Azolla pinnata*. (Micrographs courtesy of Dr. A. R. Hardham.)

ordered transverse arrays of cortical microtubules and transversely deposited wall microfibrils. Although many objections can be raised to the details of these experiments and observations (see O'Brien, 1972 for some of them) the consensus of opinion is undoubtedly that, in cells showing surface growth, the correlation is strong. The effects of treating cells with the microtubule-depolymerizing drug colchicine are central to much of this consensus. In general, expanding cells or tissues so treated tend to swell up, microtubules disappear (at least the majority) from the cell cortex, and the cell carries on depositing wall microfibrils but with a disorientated arrangement. In some systems removal of the drug leads to reinstatement of the regular arrays of microtubules, a return of orientated cellulose deposition and the re-establishment of normal growth. The interpretation of these experiments has always suffered slightly from the fact that colchicine (and probably others among the anti-mitotic drugs that have been studied: see Tippit & Pickett-Heaps, 1974; Hepler, 1976) is known to have effects other than those it has on microtubules. It is to be hoped that the availability of lumicolchicine (Keith & Srivastava, 1978) which mimics the effects of colchicine on respiration, but which lacks the capacity to depolymerize microtubules or to cause swelling of hypocotyl tissues, may allow this problem to be laid to rest.

To conclude with Green's (1969) words:

> 'In summary, the directed expansion of cylindrical cell surfaces appears based on continuous oriented wall synthesis which is related to transversely aligned microtubules'.

He went on to pose two questions: how is the microtubule array set up in the first place and how is it maintained during the expansion of the surface which accompanies growth? These questions will be considered after tip growth has been examined.

> 'Tip growth designates the activity in a dome-shaped growth zone which typically generates a cylindrical cell surface at its base. The best studied examples are pollen tubes, hyphae, and root hairs. Ordinarily the cylindrical part of a tip growing cell does not change shape except to initiate branches . . . Assuming the growth is radially symmetrical, there are three interacting parameters: shape of the growth zone, the gradient in local rate of area expansion (at different distances down the dome) and directionality of this expansion'. (Green, 1969).

Marking experiments suggest that in tip-growing cells, the expansion rate is maximal at the extreme tip and approximately isotropic, and sometimes transverse at the base. The microfibrillar texture on the inner surface of such growth zones is isotropic or slightly transverse and Green concludes that the maintenance of a steep descending gradient in relative rate of expansion within the apical dome appears to be the essential basis of generation of cylindrical form in tip-growing cells. How is the gradient in rate of expansion maintained?

Microtubules are either absent from the tip of tip-growing cells or, if present as in some root hairs, their orientation is not correlated with the orientation of wall microfibrils (see Fig. 4.10 and Pickett-Heaps, 1975b, p. 480). Green suggests that while, in principle, a wall could yield without any active loosening by the cell, a concurrent lytic action by the protoplast on the wall is highly probable.

It is instructive to compare the characteristics of tip growth and surface growth. In surface growth the cell surface contains a highly ordered array of microtubules underlying the cell membrane which in turn underlies a highly-ordered array of wall microfibrils. Marker displacement shows that every element of the outer surface is expanding smoothly and at the same rate. In tip growth, there is no microtubular array or, if a few are present, their orientation is not well correlated with the microfibrils in the wall. The wall has a random microfibril texture and the surface shows a strong local gradient in rate of expansion from the tip to the base of the dome.

In both cases it is important not to forget that new wall material is being

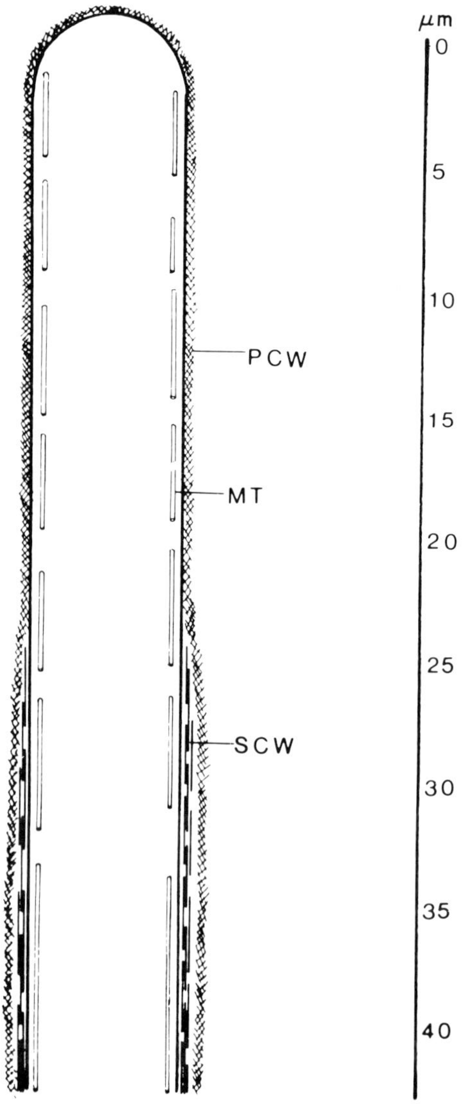

Fig. 4.10 Diagram illustrating the randomly arranged wall microfibrils in the primary wall [PCW], the axially oriented wall microfibrils in the secondary wall [SCW], and the disposition of microtubules [MT] in the terminal 40μm of a root-hair cell. Primary wall synthesis is believed to be confined to the apical dome which is free of microtubules which maintain an orderly disposition beneath the primary wall for the next 20μm until secondary wall synthesis is initiated. (Diagram courtesy of Dr. R. Seagull.)

deposited and new cell membrane surface is being generated. Cytoplasmic volume is also increasing though some tip growing cells 'cheat' a little here, by withdrawing their cytoplasm from older parts of the cell (or organism) and sealing off the old wall with a plug. This is common in pollen tubes in the style and in some fungi (see Robinow, 1963; Tippett *et al.*, 1976). In the case of symplastic growth we have the added complication that parts of the cell surface (the primary pit fields) maintain their length during extension though they separate from one another (Wardrop, 1955; Wilson, 1957). Presumably they maintain their plasmodesmatal connections through the pit field regions in spite of the extension that occurs near them (see Fig. 4.6).

Two questions suggest themselves. What are the sources of the cell-wall polymers and of the new cell membrane? How might microtubule arrays act to regulate microfibril deposition?

4.2.3 ORIGINS OF WALL POLYMERS IN GROWING CELLS

There are two classes of wall polymer to consider: the microfibrillar components (usually, but not always cellulose; see Robinson, 1977; Preston, 1979) and the matrix materials which include an array of non-cellulosic polysaccharides, glycoproteins (Roberts, 1974; Preston, 1979) and ester-linked phenolics (see references in Smith & O'Brien, 1979; Smart & O'Brien, 1979). Additionally, as Chrispeels (1976) and Preston (1979) point out, the cell wall is a metabolic compartment and contains a variety of enzymes (see review by Lamport, 1970; Knox *et al.*, 1975; and references in Smith & O'Brien, 1979). Some aspects of the cell-wall phenolics and cell-wall enzymes are treated in a later section, but as far as the polysaccharides are concerned, one could postulate for each of these two classes of polymer—fibril and matrix—two major compartments of synthesis. They could be made within the cytoplasm and discharged outside the cell or they could be made at or close to the cell membrane surface by synthetases located in, on, or near to it. Note that this does not preclude the possibility that both components are synthesized in both compartments, nor does it preclude the possibility that once new wall-polymers have been intercalated that they are further modified by extracellular enzyme activity.

4.2.4 THE EVIDENCE FROM MORPHOLOGICAL STUDIES

Eight years ago the literature on this topic was reviewed critically (O'Brien, 1972) and the main conclusions were as follows:

(1) Electron microscopy rarely provides useful information about dynamic processes that involve either exchange between cellular compartments or short-term, time-dependent changes within a compartment. The only exception to this rule is when the pathway between the compartments is very specifically marked, as is the case for the assembly within the Golgi apparatus of certain flagellar and body scales that bedeck some algae (see e.g. Manton, 1966; Brown & Romanovicz, 1976).

(2) Identity of electron contrast is not evidence of chemical similarity between cellular constituents unless it is backed up with good histochemical evidence.

(3) Electron microscopy can provide detailed information about the structure of the fixed state only if attempts are made to establish three-dimensional structure, either by the use of serial sections or stereoscopic examination of sections (now made easier by the availability of high-voltage microscopes).

(4) There has been a strong tendency to assign a wall-forming role to any evident differentiation of the cytoplasm (or cell membrane) in a cell known to be undergoing wall synthesis, ignoring other possible functions for the differentiated state observed.

It is easiest to appreciate the problems that are involved by considering specific instances.

4.2.4.1 *The lily pollen tube tip in vitro*

The paper by van der Woude *et al.*, (1971) is one of the most comprehensive on tip-growing cells. They show that the tip 25μm (about the length of the hemispherical dome) contains a population of vesicles 150–300nm in diameter, a large amount of smooth and semi-smooth ER, and many dictyosomes (Fig. 4.11). The ground substance contains ribosomes [earlier reported absent from the extreme tip in lily pollen tubes by Rosen *et al.*, (1964) and Dashek and Rosen (1966)]. Less well documented is a smaller 'vesicle', 90nm diameter, which may in fact be the result of sectioning tubules since no three-dimensional study was carried out. Cell wall preparations were carried out from pollen tubes of mean length 8mm and analyses of them show that 90% of the wall is non-cellulosic polysaccharides. The extraction used was Ray's (1963) sequential procedure (hot water, hot 0.05 N H_2SO_4, 4N KOH, 72% H_2SO_4): the material that finally dissolves in the last step is taken to be cellulose and amounted to about 6% of wall weight. Up to seven sugars were identified in the hydrolysates of each fraction and, although the recoveries were low, the sugar spectra of these hydrolysates of wall fractions were compared to the spectrum obtained from a population of isolated large vesicles. These vesicles were also shown to stain by the periodic acid, silver-hexamine reaction (PASH) both *in situ* and isolated. Since the periodic acid control failed to stain (not illustrated), this is presumptive evidence for the existence of polysaccharide(s) in the larger vesicles. The authors state that:

> 'the sugar distribution of the vesicle fraction resembled most closely that of the hot-water-soluble cell wall polysaccharide'.

While this statement is not untrue, it is misleading. It would have been better to say that the sugar spectrum in the isolated vesicle fraction does not resemble closely the spectra either of the whole wall or any of its fractions. The vesicles have about the same glucose content as the hot water fraction, but two and a half times as much galactose, nine times as much mannose, four times as much

Fig. 4.11 Longitudinal section (approximately median) of the pollen tube tip showing the presence and distribution of cytoplasmic components. Secretory vesicles (v) are concentrated at the tube apex, the region of cell wall formation, (d) dictyosome, (1) lipid body, (m) mitochondrion. (From van der Woude *et al.*, 1971, *J. Cell Sci.* **8**, 331–51. Reproduced with permission from Cambridge University Press.)

xylose, more than twice as much rhamnose, 25% of the arabinose, and a trace of fucose which is absent from the hot water extract of the cell walls. The authors' conclusion in the abstract that: 'the secretory vesicles contained polysaccharides which were similar in composition to those of the hot water soluble fraction of the pollen tube cell wall' does not seem to be justified.

On re-analysis, the evidence from this paper actually suggests that either the polysaccharides of the secretory vesicles do not remain in the wall as a structural component of it, or their contents are modified after they are intercalated. The fact that the analyses are of total wall, 8mm long, while the vesicles may be incorporated largely in the tip 25μm introduces further uncertainty. One cannot obtain evidence on this point by electron microscopy because the vesicle fusion process may proceed at different rates along the cell surface. The profusion of secretory vesicles at the tip of the cell might arise either because the process of incorporation is restricted to the tip or because it is slower at the tip than everywhere else; since ER membranes and contents are also PASH positive, they might also contribute polysaccharide and/or membrane to the cell surface.

This study is a good illustration of the dangers of assigning a wall-forming role to evident differentiations of the cytoplasm. It might be better to propose that the polysaccharide content of the vesicles is part of the secretion of a non-wall mucilage, one which, like the root cap slime in grass roots, has its origin in a hypertrophied Golgi system (see O'Brien, 1972 for early references, and Paull *et al.*, 1975). The chemical composition of maize root-cap slime has some features in common with that of the lily vesicles described, including the unusual sugar fucose, and high contents of glucose, galactose and galacturonic acid (see Jones & Morré, 1967; Paull *et al.*, 1975). As in the root-cap, this material is added to the cell wall probably by reverse micropinocytosis, though a way of proving this has not yet been devised (O'Brien, 1972).

As in the cap, the externalized mucilage finally swells and disperses (Fig. 4.12). Some of this material is almost certainly present as the compartmented cap in the work of Dashek and Rosen (1966) on this same lily system, while the rest must have passed into the liquid culture medium during tube growth, or was lost from the tube surface during fixation. As in the root-cap cell, what we have is a process of active secretion, involving the release of 2000 vesicles per min (van der Woude & Morré, 1968) occurring probably in the same region where cell wall formation is taking place. Until very specific cytochemical probes become available that will allow analysis of the chemical content of the ER, the secretory vesicles, the ground substance and the smaller 'vesicles', it will not be possible to analyse wall formation in this system.

Before leaving this system, it is worth noting the potential impact of vesicle incorporation on this tip zone. If one accepts the figure of 2000 vesicles per min of average radius 0.1μm and if one further assumes that these are incorporated more or less uniformly in an apical dome 5μm in radius, one may calculate the area of the dome at $50\pi\ \mu m^2$, and the area of vesicle membrane incorporated at $80\pi\ \mu m^2$/min. The vesicle membrane is more than adequate to replace the cell

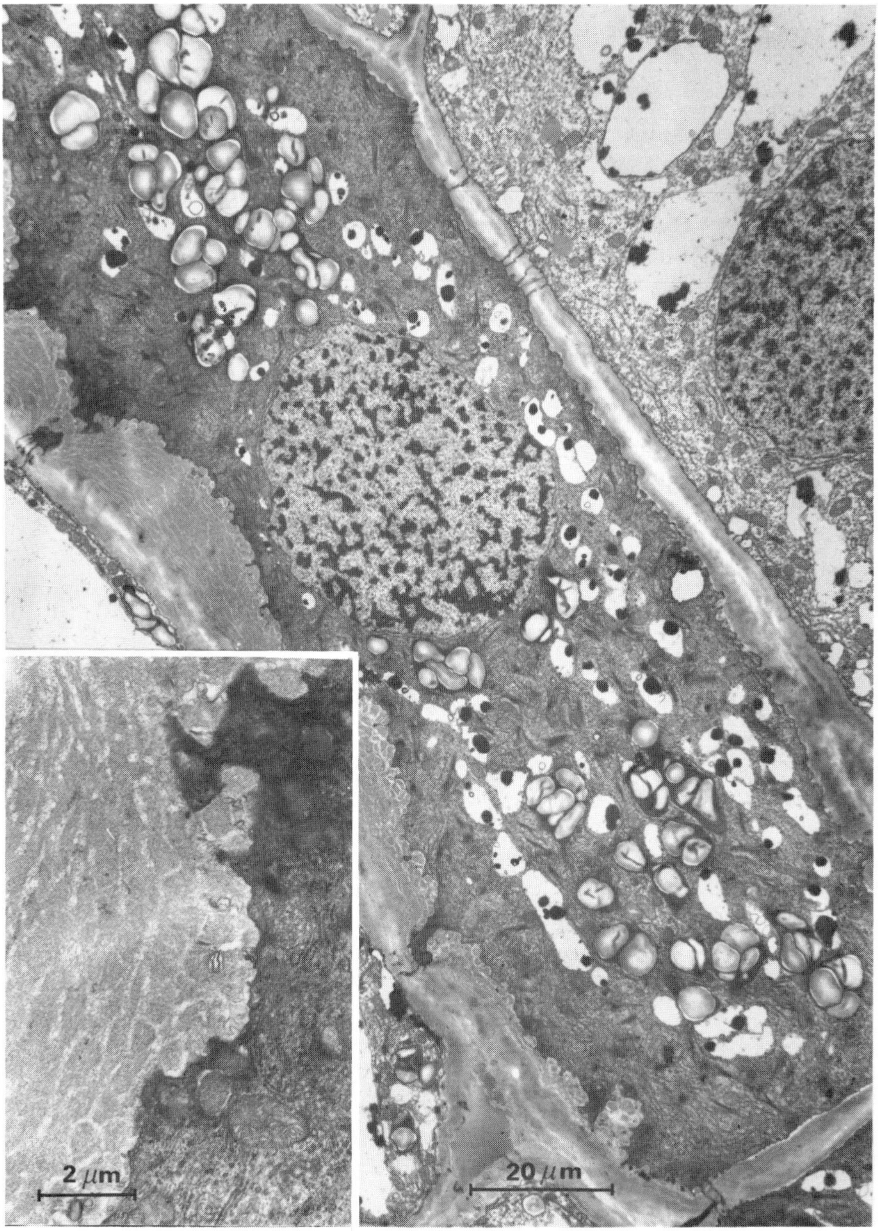

Fig. 4.12 Electron micrograph of outer root cap cell of wheat, showing extensive accumulations of mucilage beneath the outer wall. Inset: detail of mucilage deposit. Glutaraldehyde/OsO4 fixation.

membrane surface of the apical dome every minute. Even the volume relationships are significant. The wall volume over the apical 5μm dome is about 32μm^3 if one takes a figure of 0.4μm for its thickness. 2000 vesicles would add a volume of approximately 8μm^3/min. It is easy to see why it is important to know what components other than the polysaccharides described by van der Woude *et al.*, (1971) are contained within the vesicles. If they contained as little as 5% structural or enzymatically active protein, their potential for contribution to the wall is enormous. If any other membrane-bound cytoplasmic organelle also adds its contents to the wall in this manner (e.g. the smaller 'vesicle' or the ER) the area of cell membrane that is incorporated must be correspondingly increased.

4.2.4.2 *Cell-plate formation*

> 'To what, then, does our core of hard fact reduce in the case of cell-plate cytology? In many cells, but not in all, the cell plate forms within phragmoplast fibres that lie at right angles to the axis of plate formation. The fibres consist of an overlapped set of microtubules, encased within the region of overlap in an electron-dense differentiated region of the ground substance. In living endosperm cells, cytoplasmic droplets appear to slide down and coalesce with nodules that form on the fibres. Electron microscopy of the forming plate reveals that its fine structure is much more complex than the images revealed by light microscopy, and that the forming plate is encased in a lattice work of spatially differentiated ER. It is possible, but not proven, that this ER network and/or dictyosomes contribute to plate formation . . . in the face of such complexity of structure and with no knowledge about the temporal stability or lability of the associations revealed in these micrographs, speculations about the contributions of these systems to plate formation seem to me to be idle. I do not understand how anyone can examine those micrographs carefully and remain confident that they understand the structures or processes that form the plate'. (O'Brien, 1972).

The illustrations in Hepler and Newcomb (1967) and Figs 4.13 and 4.14 show the complexity of the forming plate. Since 1972, some interesting additional systems have been investigated.

Cellularization of endosperms can be classified into three types: nuclear, cellular and helobial (see Newcomb, 1978 for references). In wheat (nuclear; Mares *et al.*, 1975; 1977; Morrison & O'Brien, 1976), and in *Haemanthus katherinae* (helobial; Newcomb, 1978) early cellularization is accomplished by wall ingrowths which at least in the early stages lack a normal phragmoplast. These infurrowings can remain as embayments or fuse to form a peripheral layer of cells. In either case, further cellularization is by transverse or oblique divisions accompanied by normal phragmoplast development and the formation of a cell plate. Later stages of wall ingrowths in both wheat and *Haemanthus* sometimes contain a phragmoplast. This has remarkable parallels to that reported

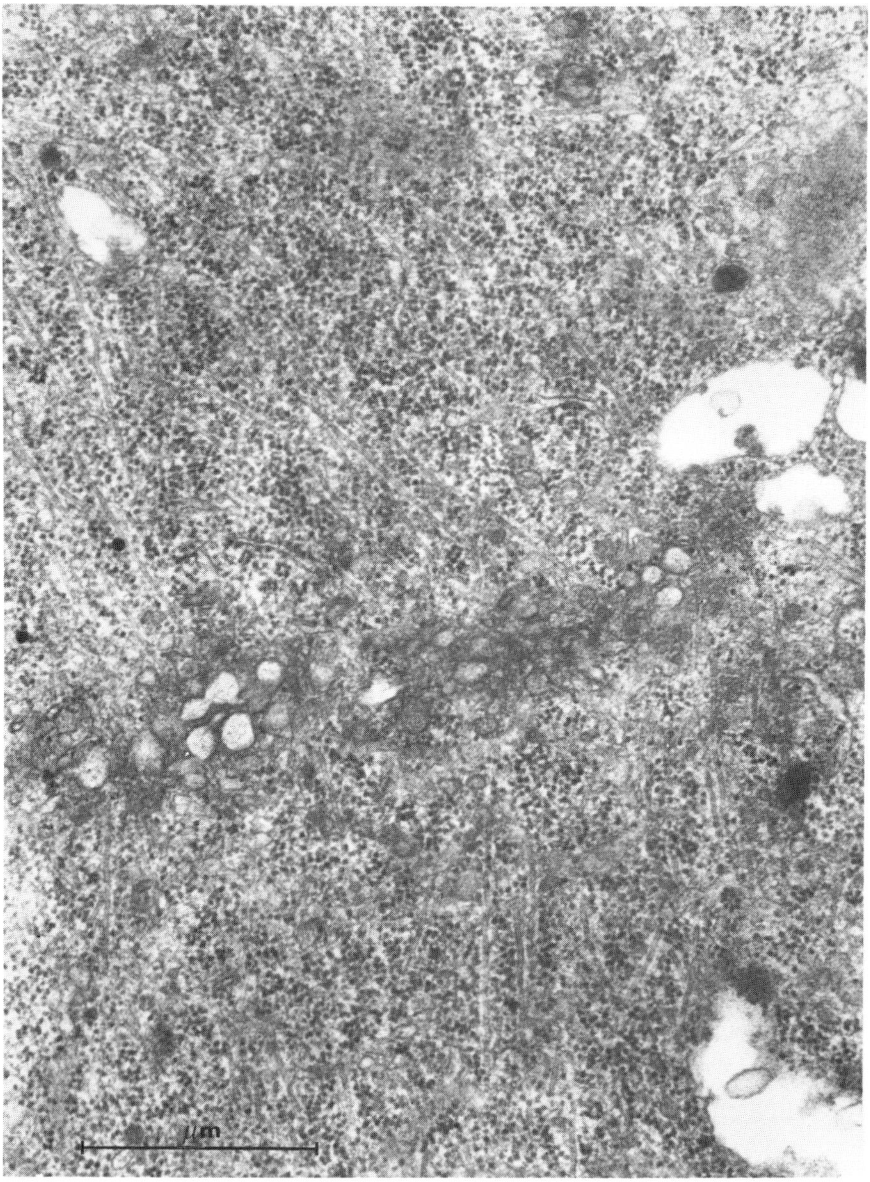

Fig. 4.13 Longitudinal section of margin of developing cell plate in dividing epidermal cell of wheat root. Numerous phragmoplast microtubules may be seen converging on developing cell plate material. Glutaraldehyde/OsO4 fixation.

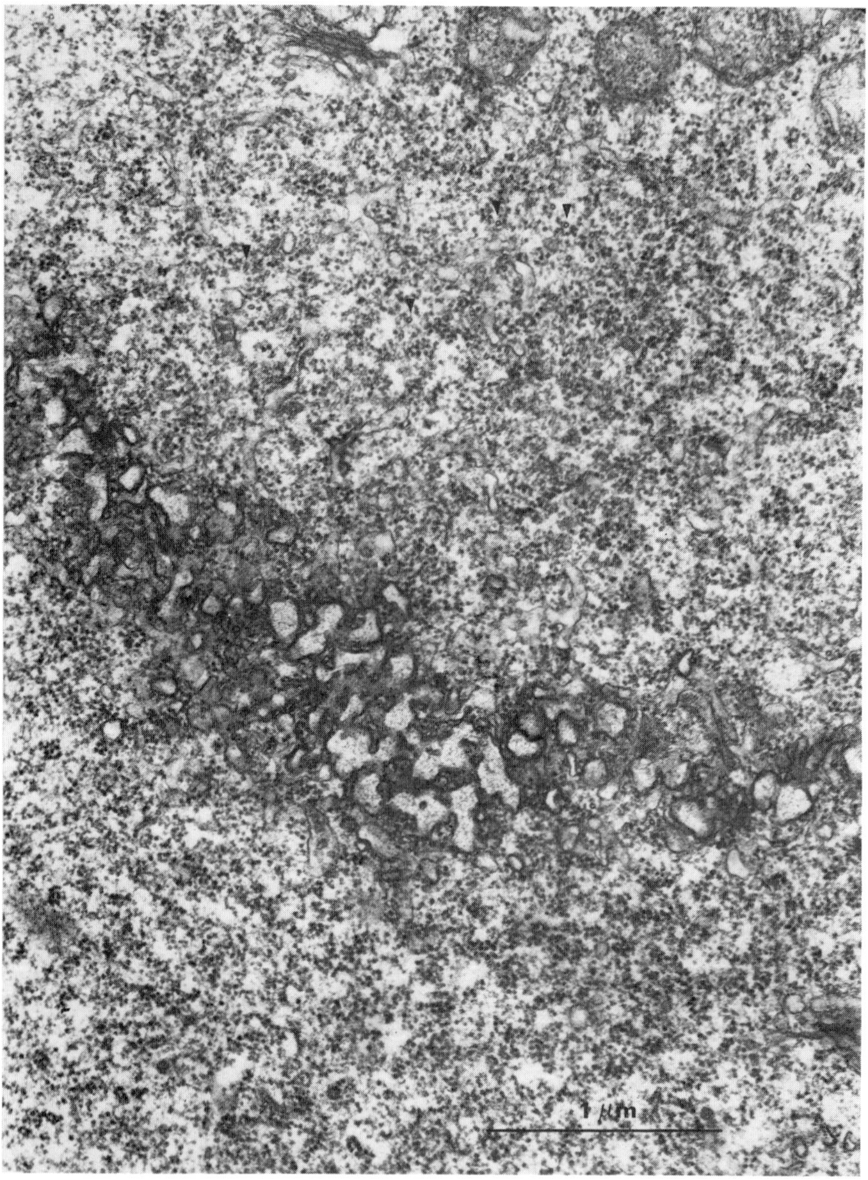

Fig. 4.14 Face view of part of developing cell plate in a cortical cell of a wheat root. Note the transected phragmoplast microtubules (arrows) and the irregularly disposed and complex nature of the material of the surface of the forming cell plate. Glutaraldehyde/OsO4 fixation.

in *Spirogyra* where the filaments are cellularized at cytokinesis by an infurrowing septum. When this has closed off about 70% of the cell diameter, cellularization is completed by the activity of a phragmoplast with cell plate formation (Pickett-Heaps, 1975b). Although dictyosomes and ER are present in the cytoplasm near these infurrowing walls, neither Newcomb (1978) nor Morrison and O'Brien (1976) were convinced that these systems contributed any material to the developing ingrowths.

Pickett-Heaps and his colleagues (see Pickett-Heaps, 1975b) have studied cytokinesis in many algal species, especially in the green algae. Many of these divide by a cleavage furrow. In some species (e.g. *Kirchneriella, Tetraedron*), cleavage takes place within a set of phycoplast microtubules but no evidence was obtained that the cleavage furrows which grew through these sets of phycoplast microtubules did so by the addition of vesicles to them (Pickett-Heaps, 1972). In others (e.g. *Chlorella*) opinion is divided as to whether the cell membrane furrows inwards or whether dictyosome vesicles form it (Pickett-Heaps, 1975b). Perhaps the most remarkable case is the formation of a new cross wall in *Oedogonium*. Here a typical phycoplast appears between the daughter nuclei which collapse back on one another after telophase (see Figs 5.15 and 5.16 in Pickett-Heaps 1975b). Within this phycoplast a septum forms, composed in thin section, of an array of tubular and cisternal elements which resemble smooth, semi-smooth and rough ER. After ring breakage, the septum slides up the cell, retaining its coherence, and attaches itself to the lower lip of the expanding ring (see Pickett-Heaps & Fowke 1969 for details). The septum is then transformed into a structure which closely resembles a young cell plate in a higher plant cell (see Figs 4.15a and b).

> 'There is no clear evidence that these [cell plates] could be derived from golgi bodies nor do they contain polysaccharide demonstrable by the PASH procedure'. (Pickett-Heaps, 1975b).

This situation changes rapidly as the plate matures and it quickly acquires PASH positivity.

This observation raises the issue of the nature of the materials that are enclosed by the forming membrane of developing cell plates. Fulcher *et al.*, (1976) show that aniline-blue-positive material, probably β-1, 3-linked glucans, is present in cell plates during their formation. Whether such 'callose' is present *in vivo* is always a difficult problem to determine since it is known to form so rapidly in cells in response to wounding (Hughes & Gunning, 1980). Whether it is present *in vivo* or not, there is not much doubt that it is responsible for at least some of the relatively electron-lucent material visible in electron micrographs of developing cell plates. That other components are also present is very likely from the fact that even quite young cell plates are PASH positive (see Figs 17–20 in Pickett-Heaps, 1968), but in view of the lack of specificity of this reaction, it is not safe to conclude that this is due to polysaccharides. In the same paper, Pickett-Heaps shows that sieve-plate callose deposits may be either PASH positive or PASH negative. Since such deposits are negative in the periodic acid

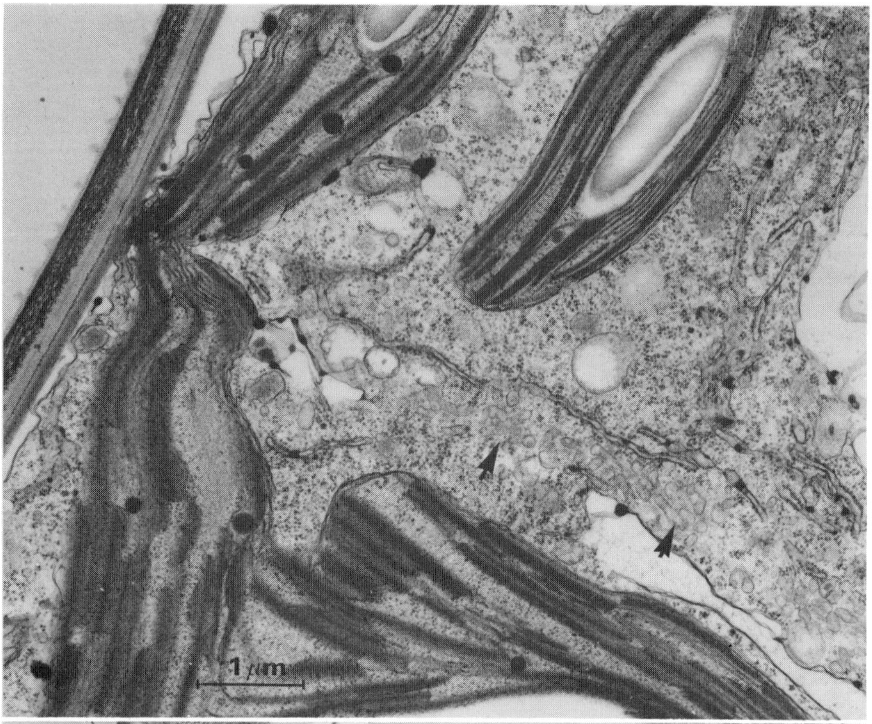

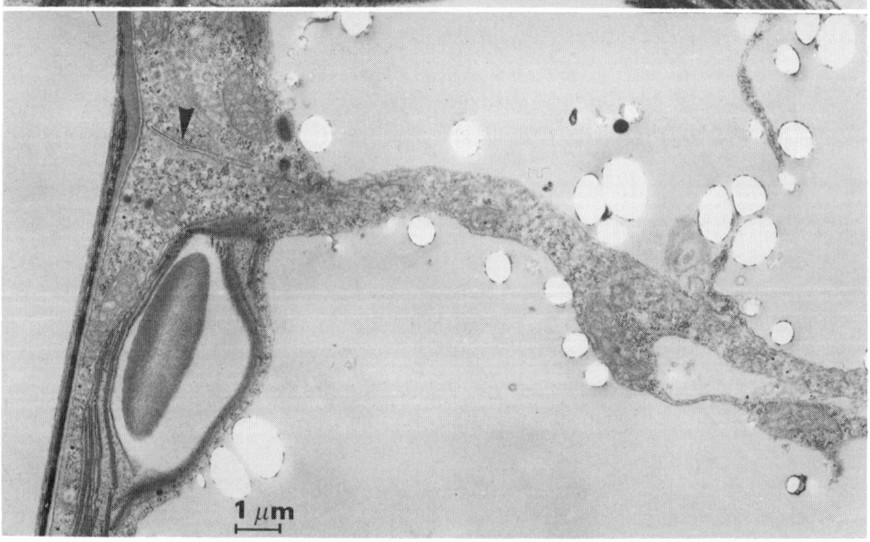

Fig. 4.15a Longitudinal section of the septum forming within a phycoplast of *Oedogonium*. The tubular elements of the septum are very conspicuous (arrows). The phycoplast is about to cut through the chloroplast at the left hand side of the picture.

Fig. 4.15b The beginning of cell wall formation within a phycoplast *Bulbochaete* (both figures courtesy of Professor J. D. Pickett-Heaps).

Schiff's reaction, and since pure β-1, 3-linked polysaccharides should be negative in both tests because of their linkage, the results are not easy to interpret. Perhaps a mixed-link glucan is present or has become trapped by the structure of the callose (see Preston, 1979).

In summary, study of cell plates does not reveal much about the origins of either the wall polymers or the cell membrane. In cleaving or infurrowing systems, microtubules may be present (phycoplasts) or absent (endosperms); vesicles may be entirely absent from the developing furrow (*Kirchneriella*) or present and perhaps contributing to it (*Chlorella*). In a system that forms the plate as a separate entity (the *Oedogonium* septum) the plate shows all of the features seen in cell plates forming in the presence of phragmoplasts. In none of these cases is there clear evidence for a contribution of membrane or vesicle contents to the plate, though it certainly could happen in many systems (e.g. *Spirogyra* phragmoplasts are 'bathed' in vesicles, and in many higher plant cells vesicles are abundant; O'Brien, 1972 for references). In fixed cells callose-like polysaccharides are present in many plates and slowly seem to give way to normal wall constituents but the origin of these substances is obscure.

4.2.4.3 *The cytoplasm/wall interface in cells undergoing surface or symplastic growth*

In 1963, Newcomb stated that 'despite much speculation and experimentation on the mechanism of cell wall growth, little is known about the relationship of the cytoplasm to cell wall formation'.

The view was put (O'Brien, 1972) that the only evident differentiation of the cytoplasm that looked as if it might be involved in primary cell-wall formation was a cortical network of ER. It was also pointed out that there was no really good evidence to support the idea that this ER network had a wall-forming role (for an opposed view, see Chrispeels, 1976). It is not known how stable the ER network is with time or even whether it moves in cytoplasmic streaming. Radioautography has been inconclusive basically because of its low resolution (see Pickett-Heaps, 1966 and discussion in O'Brien, 1972).

The number of published papers that include micrographs of expanding cell surfaces has increased enormously since 1972 and includes major studies of the algae (Pickett-Heaps, 1975b) as well as higher plants. To these we can add an impressive literature on wall formation in primary xylem tracheary elements (see Esau & Charvat 1978 for references) where wall thickenings are deposited in zones of relative stasis in a cell otherwise expanding its surface. A few papers also exist on fibre differentiation and on the development of cambial cells (Robards & Kidwai, 1972). With few exceptions the story that emerges from a consideration of these micrographs is always the same. Except in cells known to be secreting mucilage (see Chrispeels, 1976) or manufacturing scales, no clear-cut *morphological* evidence exists that tells us where wall polysaccharides are formed or where the cell membrane is synthesized. One can underline the problems by examining Fig. 4.16. These are all taken from a plane about 5μm wide

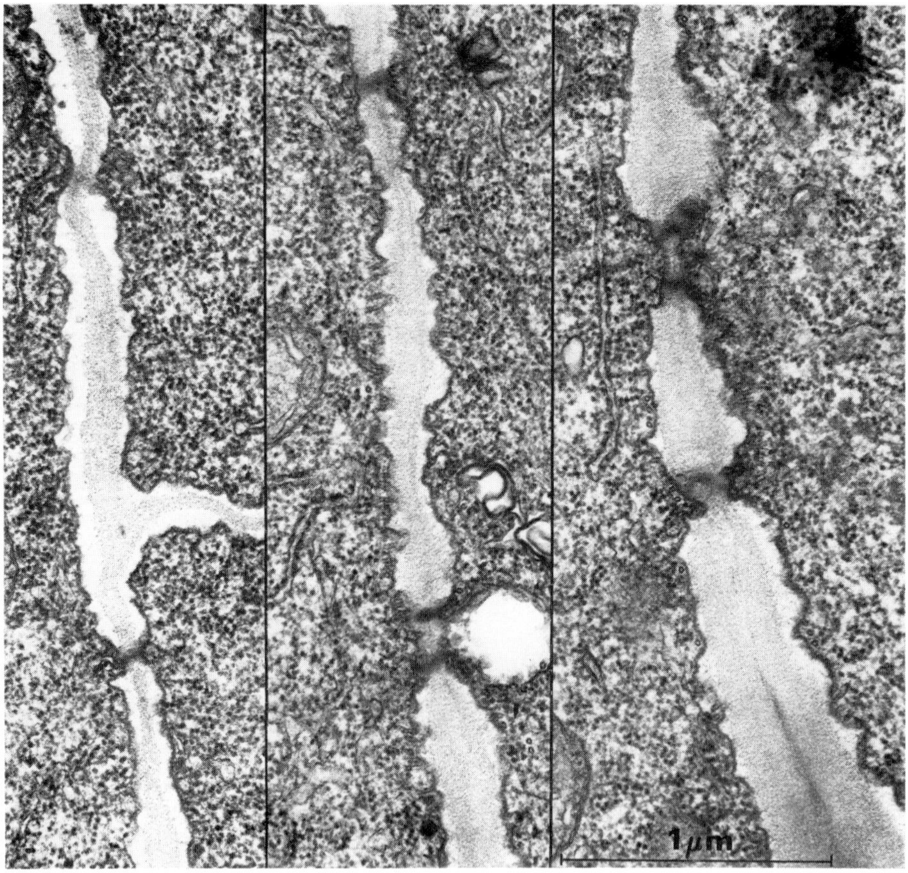

Fig. 4.16 Longitudinal sections taken from three neighbouring cells in a zone about 5μm wide in a wheat root. At the time of fixation, all of these cells would presumably have been growing at close to the same rate. Nevertheless, the cytoplasm-cell wall interface shows a good deal of local variability in the disposition of ER, microtubules and vesicles, and the microtopology is not interpretable in terms of the growth process.

that extended across several cell files in a root cut in LS. There is much small-scale variability in the disposition of reticular ER, microtubules, cell membrane bulges, presence of pit fields and local wall thickness in the cytoplasm on opposite sides of the same cell or on either side of a common wall, although all of these areas are presumably growing at approximately the same rate. The events that must occur wherever a transverse wall (which does not extend in the expanding axis?) abuts a longitudinal wall are also interesting to contemplate. Wall material, cytoplasmic contents, and cell membrane are all growing but the microtopography is highly variable and not interpretable in terms of the growth processes.

4.2.4.4 *Scale formation in Pleurochrysis (Brown & Romanovicz, 1976; Romanovicz & Brown, 1976)*

In sharp contrast to what has gone before, this elegant work shows very clearly the endomembrane system's contribution to the synthesis and transport of the scales which bedeck this beautiful micro-organism. They show that the scale is assembled in the *Golgi cisternae* and externalized by exocytosis. This naturally occurring, specifically marked pathway points up the direction in which research must move if comparable information for other systems is to be obtained. Electron dense markers that are specific for particular wall polymers and for the components that make up cell membranes must be developed. EM cytochemistry that allows one to localize enzymes by products that are electron dense and specific for cell membrane enzymes will also be a help. Until research moves more firmly in these directions, progress cannot be rapid (see Hall, 1978).

4.2.4.5 *Wall formation in Oocystis apiculata (Brown & Montezinos 1976)*

> 'Cellulose microfibril biosynthesis, assembly and orientation in the unicellular green alga, *Oocystis*, is visualized in association with a linear enzyme complex embedded in the B face of the plasma membrane'.

Strictly speaking this paper is not concerned with growth since the observations were made on cells engaged in secondary wall formation. It is included here to allow a discussion of certain attitudes and the limitations of freeze-fracture technique. The authors state that 'this study gives visual confirmation of cellulose synthesis in a eukaryotic cell.' This claim, and the ones above, are made on the basis of the distribution of particles, grooves and microfibrils demonstrated very elegantly in metal replicas of fracture faces of the cell membrane. One cannot measure the composition and activity of particles seen in metal replicas and to label such particles as enzyme complexes engaged in cellulose synthesis is premature. The authors (Brown & Montezinos, 1976) know that the wall of *Oocystis* consists of polymers other than cellulose (their Figs 1 & 2), and that cells growing in liquid culture must have enzyme complexes in their cell membrane that are active in functions other than cellulose synthesis. Chrispeels (1976) appears to criticise this attitude in these matters on the grounds that later work sometimes shows that earlier hypotheses are correct. The objection is not to the formulation of hypotheses, or to reasonable speculation, but rather to unwarranted speculations and the process which leads to metamorphosis of these *untried* speculations into 'facts'. As Williams (1967) put it so well in another context 'Any work of man—whether it be a work of art, a tool, or a scientific discovery—is diminished in stature if it is made the subject of exaggerated claims.' Whatever the future may prove, it is important to realize that one cannot deduce the composition or activity of particles seen in metal replicas by looking at them (see also Giddings *et al*, 1980; Mueller & Brown, 1980). What is needed are new approaches in the hope that the relationship may be discovered between the entirely morphological image seen in metal replicas of a fracture plane and living functions.

4.2.5 THE EVIDENCE FROM BIOCHEMICAL STUDIES

This subject is very controversial, and the existence of molecular secretion in plant cells complicates the interpretation of the production of cell-wall constituents. The latter point is especially troublesome where, as is the case for developing root-cap cells and root-epidermal cells, the processes of macromolecular secretion, molecular secretion and absorption, and cell-wall deposition and growth are almost certainly going on in the one cell at the same time.

It is not even possible reliably to distinguish whether or not a polymer should be regarded as a wall precursor. To a cell of *Acetobacter*, cellulose is a secretion and not considered to be part of the wall, whereas in plant cells it is a load-bearing polymer, plays an integral role in wall dynamics, and it is not helpful to think of it merely as a secretion. However, if one considers matrix materials, one meets a continuum from water-soluble secretions to constituents normally regarded as cell-wall components. Similarly, how should the sporopollenin layer in *Chlorella* walls be regarded (Atkinson *et al.*, 1972), or the range of 'walls' built up of scales in certain algal species (Gunning & Steer, 1975)? Equally difficult are the cell-wall enzymes; esterases in pollen walls could easily be treated as secretions associated with recognition or tube penetration but in the root epidermis they probably have a role in regulating ester linkages between phenolics and polysaccharides (Smith & O'Brien 1979; see also Preston, 1979). Perhaps the best to be hoped for at present is to localize the cellular fractions that carry the synthetases that make the fibrillar and matrix components and the glycoproteins and mucilages. The easiest systems to study will in many cases involve those engaged in the synthesis of a massive amount of secretion (e.g., root-cap cells). Unfortunately, the production of biochemical amounts of isolated root-caps is a formidable task and so work is done with root tips (0.1–0.5cm). This introduces an inevitable heterogeneity into the analysis which is very evident in the work of Paull and Jones (1976). Despite the specifity of the fucose label they used, their EM autoradiographs show less than half of the dictyosomes labelled in any one preparation. Presumably this arises in part because at least some of the dictyosomes are from cells with low rates of incorporation of fucose into dictyosome products. Thus, dictyosomes in different cells, and possibly even within the one cell, do not carry the same synthetases.

In an excellent review, Quail (1979) has assessed the evidence from cell fractionation studies that implicates dictyosomes, ER, and cell membranes as sites of synthetase activity which *may* be related to synthesis of wall polymers. Several glucosyl transferases have been detected in dictyosome-rich fractions. Of these, two β-glucan synthetases (I and II) have received most study. There is comprehensive *in vivo* and *in vitro* autoradiographic and cell fractionation evidence that these *in vivo* labelled polysaccharide products are predominantly localized in the dictyosomes and serve as markers for them. However, Quail stresses that, despite these facts, the products of the glucan synthetase I assayed *in vitro* indicate that this enzyme is *not* primarily responsible for the major incorporation products detected in dictyosomes following labelling *in vivo*. There is also now convincing evidence that neither the *in vivo* products nor the glucan synthetase I are ER localized (see discussion of ER labelling following ^{3}H-glucose feeding in O'Brien, 1972; cf., Northcote, 1972). The in-vivo role of the synthetase I is a matter of debate. Despite its capacity to form β 1–4 linkages, it seems not to be involved in cellulose synthesis in dictyosomes. The demonstrated synthesis of cellulose in the dictyosomes of *Pleurochrysis* must either be an exception or else it uses a different synthetase.

Gardiner and Chrispeels (1975) have data which indicate that at least the glycosylation step of extensin synthesis takes place in the dictyosome (see also Lamport, 1977). On the question of the much-proposed role of the dictyosome as a locus of 'dynamic, steady-state differentiation where ER-like membranes are progressively transformed into PM-like (cell-membrane-like) membranes by insertion or modification of constituents during migration across the dictyosome stack', Quail states that 'few studies have directly addressed this question to date', but that Morré and his colleagues have assembled considerable evidence which is consistent with this view (references in Quail, 1979).

On the question of cell-membrane synthetase activity, there is substantial evidence that cellulose synthesis occurs directly at the cell surface. This suggests that isolated cell membrane will show cellulose synthetase activity. 'Despite much effort this expectation is yet to be realized' (Quail, 1979). the glucan synthetase II activity, which may be chiefly located in the cell membrane, synthesizes largely β 1–3 linkages, and may well be involved in callose synthesis.

It has been known for some time that coumarin inhibits cellulose synthesis without affecting the incorporation of glucose into pectins and hemicellulose (see discussion in Elbein, 1979, p. 261). In the alga *Prototheca zopfii* coumarin inhibits transfer of an oligosaccharide chain (formed from UDP-glucose) from its lipid carrier to a protein receptor which is in turn believed to be a precursor of the cellulose chain (see Hopp *et al.*, 1978).

It is interesting that, despite so much work, cellulose synthetase activity in isolated cell membranes is still not realized. Perhaps the reason lies in the fact that in eukaryotic cells growing in nature, cellulose synthesis takes place in a surface which is usually under pressure from within. As the intriguing hypothesis put forward by Zimmermann (1978) stresses, the compressive and shearing effects produced in the cell membrane by turgor pressure may well be very important in maintaining cellulose synthetase activity. Of course, turgor pressure is not a *sine qua non* for cellulose synthesis *in vivo* since protoplasts and plasmolysed cells engage in cellulose synthesis (see Robenek & Peveling, 1977). Even in these systems, the membrane at the surface is still in a more ordered form than is likely in the isolated state after cell fractionation. This aspect perhaps needs more attention.

If the biochemical and morphological information is put together, it is plain that there is still a great deal of uncertainty about the sub-cellular origins of the fibrillar and matrix components of cell walls and of the cell membrane. The working hypotheses clearly are that matrix components (including extensin) are dictyosome-derived and that fibrillar components, especially cellulose, are synthesized at, on or near to the cell membrane. Ironically, in the one case where there is good morphological evidence for the sub-cellular site of synthesis (the scale of *Pleurochrysis*), both cellulose and glycoprotein appear to be packaged, and perhaps synthesized, in the dictyosome. The implication that the ER is directly involved in synthetase activity is less strong than it was in 1972, though it is a very likely source of membrane-lipid precursors (Quail, 1979). What is needed are very specific cytochemical probes to monitor in-vivo systems and to monitor fractionated cells *in vitro*.

Two comments will close this section. Firstly Chambers and Chambers, (1961) microsurgical experiments with amoebae, in which more than 50% of the cell surface was removed without lysis, show that a significant fraction of the surface can be regenerated almost instantaneously. Perhaps this is a special case since such cells can form pseudopodial outgrowths very quickly. However, it indicates that cell membranes may be able to form rapidly from precursors which may simply be dispersed in the ground substance of the cytoplasm, a *de novo* self-assembly, without the immediate use of cisternae or vesicles.

Secondly, hypotheses about the involvement of the dictyosome do not pay enough attention to its three-dimensional structure. Most hypotheses involve a flow of membrane from the ER face to the secreting face and invoke clusters of 'secretory vesicles' at the margin of the dictyosome, especially at the 'secreting face' (see discussion in Gunning & Steer, 1975). What many investigators call 'free vesicles' are in fact sections of perforated or tubular cisternae. Brownian motion, either in living or fixing cells, would rapidly disperse free vesicles from the neighbourhood of the dictyosomes in just a few seconds, unless cytoplasmic viscosity is very different in that region. Remember that fixation takes 10 minutes in single cells to stop Brownian motion of particles ten times the diameter of secretory vesicles (Mersey & McCully, 1978).

4.2.6 CORTICAL MICROTUBULE ARRAYS: THEIR STRUCTURE, INITIATION AND MAINTENANCE

Originally it was postulated that the cortical arrays of microtubules were perhaps disposed as hoops, running circumferentially around the cell at right angles to the axis of extension of the cell (Ledbetter and Porter 1963; see discussion in Hardham and Gunning 1978). There is little evidence to support this view. Microtubule maps of *Sphagnum* leaflet cells (Schnepf *et al.*, 1976) and sections of the protonema of the fern *Adiantum capillus-veneris* (Wada &

O'Brien, 1975) both produced evidence which did not support this view. In root-tip cells microtubules are disposed as overlapping arcs during interphase. Each arc has an average length of 2–4μm, corresponding to about one-eighth of the cell circumference. Usually, microtubule terminations are random, but terminations in lines parallel to the long axis of the cell have been detected (Hardham & Gunning 1977, 1978 and Fig. 4.17). Overlapping elements have also been detected in *Avena* and *Phleum* (Palevitz, 1978) and in radish root-hairs

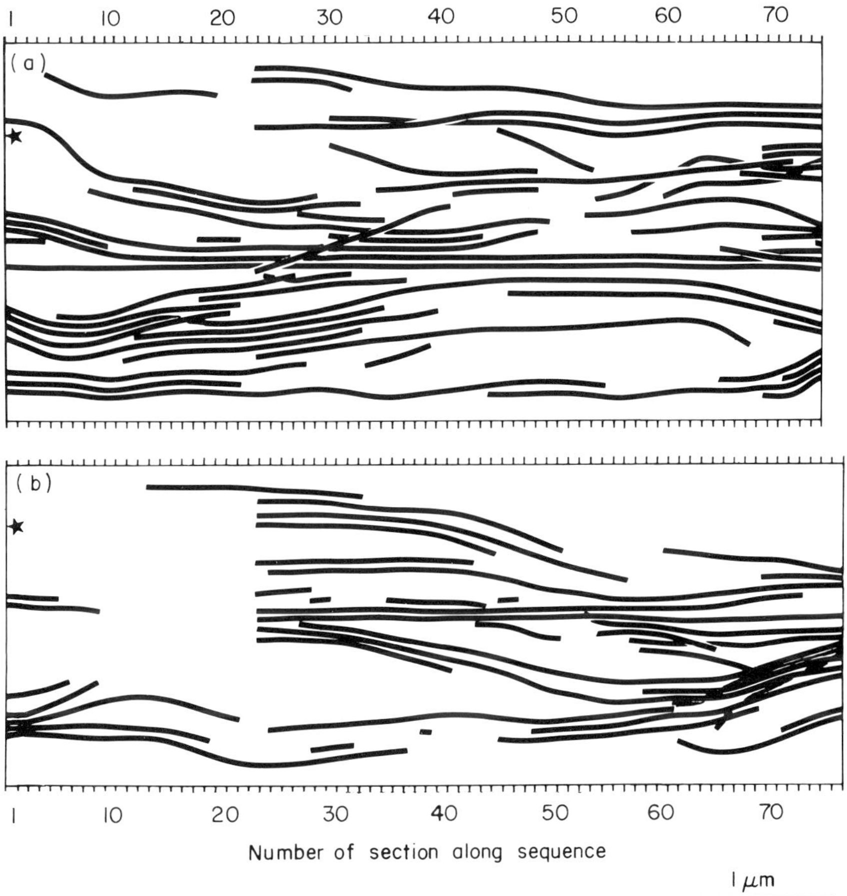

Fig. 4.17 Scale representation of two arrays of cortical microtubules against the longitudinal cell walls at interphase. The view in a is as from the centre of the cell, looking outwards. b represents the cortex of the neighbouring cell, viewed by looking on through A and the intervening wall. The two arrays can be superimposed in register by using the left-hand frame and the asterisks as reference marks. The axis in which the cells were elongating runs up and down the page. *Azolla pinnata.* (From Hardham A. R. & Gunning B. E. S., (1978) *J. Cell Biol.*, **77**, 14–34. Reproduced with permission from the Rockefeller Press.)

(Seagull, 1978). Hardham and Gunnings' analysis of these arrays concentrated initially on the structure of the arrays as seen in serial longitudinal sections of cells which, for simplicity, were treated as if they had an approximately circular cross-section. In a later series of remarkable papers (Gunning *et al.*, 1978a; 1978b; 1978c: 'this study is a landmark in the area of organ extension and cell differentiation'; Green, 1980), Gunning and his co-workers have made use of the small roots, each of which shows determinate growth, of the aquatic fern *Azolla pinnata*. In this root, the apical cell shows a precise pattern of segmentation producing units (merophytes) whose subsequent patterns of division and expansion are remarkably precise (Gunning *et al.*, 1978a: see Figs 4.18 & 4.19).

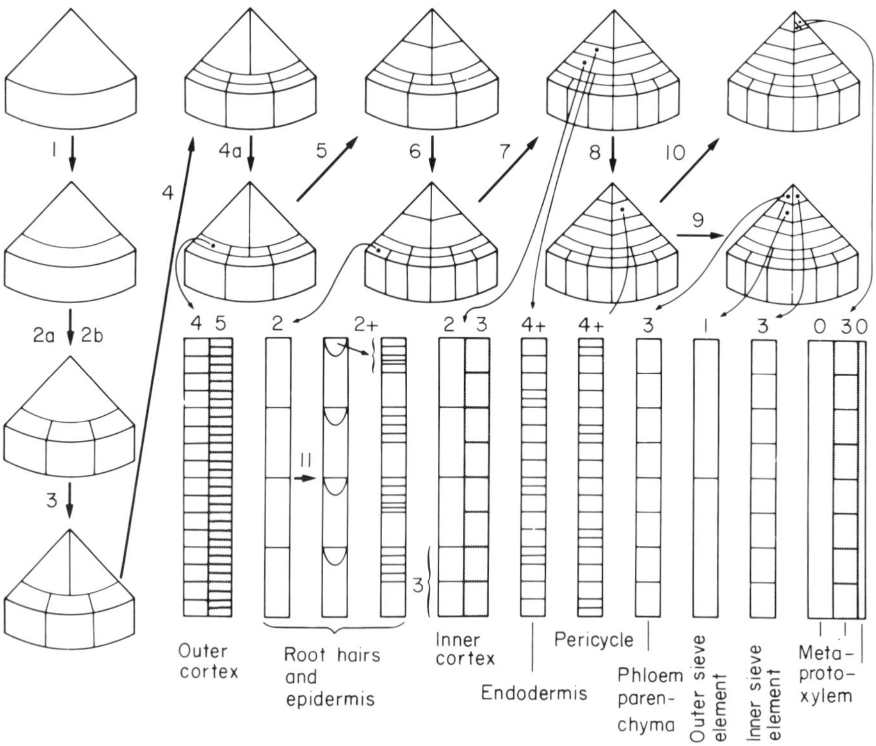

Fig. 4.18 Diagram summarizing the formative division stages and the final outcome of the proliferative divisions in the root of *Azolla*. Segmentation of merophytes by the cell lineage divisions is shown at the left and along the top, including division 4a in the outer cortex of *A. pinnata*, and division 10, in the xylem of *A. filiculoides*. The columns represent the final state of the cell files within merophytes after cell proliferation. The small numbers indicate the numbers of rounds of division in each case. Not all differences between *A. pinnata* (open columns) and *A. filiculoides* (shaded columns) are shown. (From Gunning *et al.*, (1978a) *Planta*, **143**, 121–144. Reproduced with permission of Springer-Verlag.)

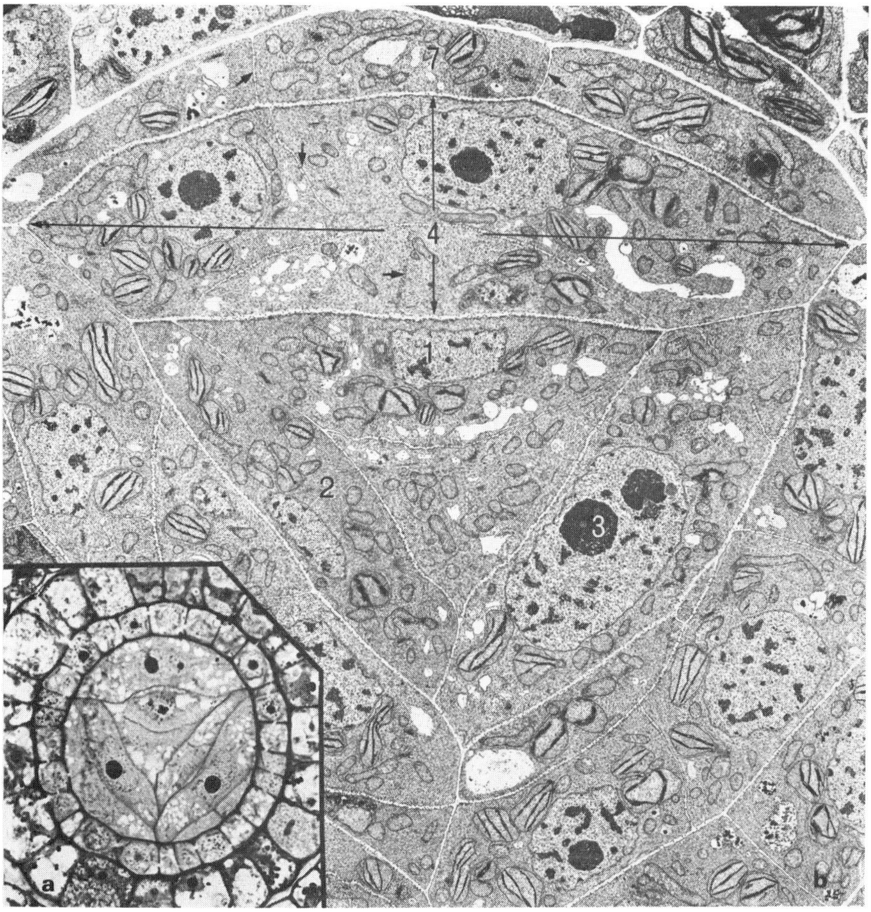

Fig. 4.19 *Azolla filiculoides*. Transverse sections of a root cut about half-way down the sloping sides of the apical cell (a, telophase in apical cell, clockwise) and closer to its apex (b, counter-clockwise). Successive merophytes are numbered in b. The first 3 are undivided. No. 4 has completed division 2a (vertical arrow) and is in late telophase of division 3 (horizontal arrow). All that is seen in No. 7 is part of the outermost region with divisions 2a and 2b completed (sloping arrows). (From Gunning *et al.*, (1978a), *Planta*, **143**, 121–144. Reproduced with permission of Springer-Verlag.)

To understand fully the significance of their findings, it is important to have a clear grasp of the 'microtubule cycle' as it occurs in higher plant cells going through interphase, karyokinesis, cytokinesis and back to interphase (see p. 95 for more detail, and Fig. 4.20). Briefly, in interphase in root-tip meristem cells, microtubules lie in the cortex; at pre-prophase they appear as a band (the pre-prophase band), whose position has now been shown to mirror faithfully the

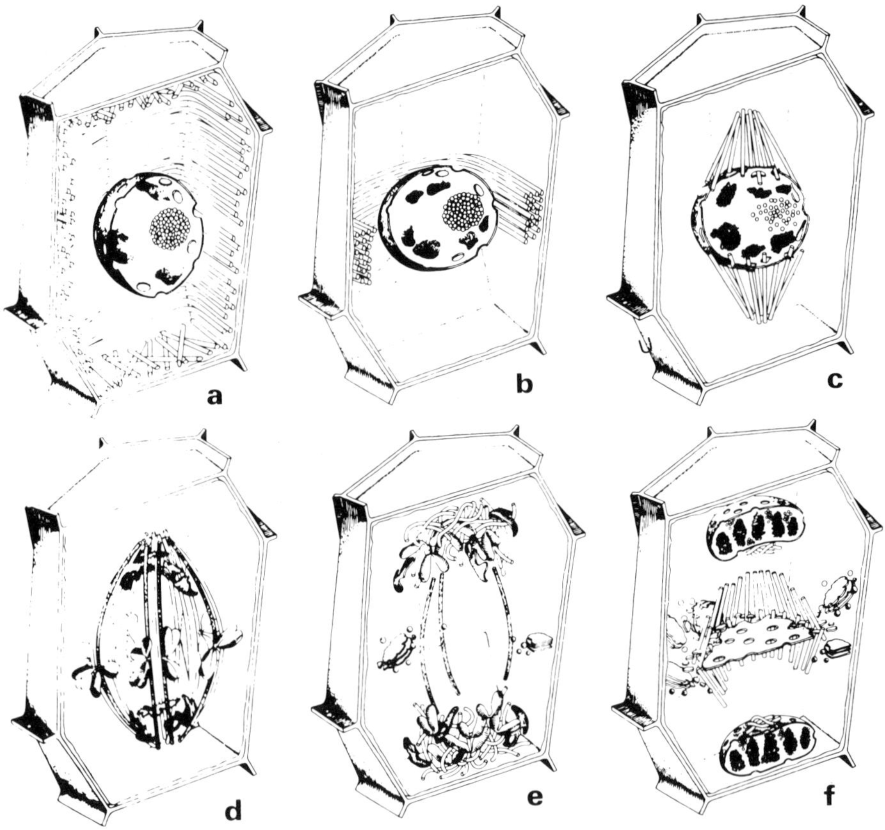

Fig. 4.20 Diagram of changes in fine structure during mitosis. (a) shows interphase disposition of cortical microtubules (the tubules are now known to be arranged as overlapping arcs: see Fig. 4.18); (b) shows the formation of the preprophase band (see Fig. 4.24). During prophase (c) the chromatin condenses to form chromosomes and the microtubules of the spindle make their appearance. The metaphase arrangement is shown in (d), while at anaphase (e) the chromosomes have moved apart. The phragmoplast forms the cell plate while the daughter nuclei are reconstituted during telophase (f). (From Ledbetter M. C. & Porter K. R. Introduction to the Fine Structure of Plant Cells, 1970. Reproduced with permission of Springer-Verlag.)

plane of the future cell plate that will divide the daughter cells (see Busby & Gunning, 1980 for references). During karyokinesis, microtubules are absent from the cortex but present as components of the spindle and at cytokinesis they are present in the phragmoplast. At the conclusion of cytokinesis, cortical arrays are re-established in the daughter cells.

Capitalizing to the full on the small size and precise developmental sequences of the merophytes, Gunning and his co-workers have identified a specialized complex of microtubules closely associated with vesicles 50–60nm in diameter, the whole embedded in an electron-dense enrichment of the ground substance of the cell (see Figs 4.21a–d). These complexes are found in the angles formed where new walls have recently joined the parent walls in all of the division types examined in the root. In some views, microtubules seem to fan out from these complexes such that tubules run initially in a variety of directions from the cell edges (see Fig. 4.22). These complexes are envisaged as microtubule organizing centres (MTOCS: see also p. 104) involved in the establishment of a cortical array after cytokinesis, an array whose final form may be determined in part by growth of the cell surface containing the array, by the establishment of intermicrotubule crosslinks, and by microtubule-plasmalemma cross-links. Evidence for the existence of such cross links is well-known from other studies (Roberts & Hyams, 1979). Interestingly, microtubule/vesicle/matrix clusters have also been seen in these cells along edges in dividing cells where the edge in question was not the immediate product of a division.

Now that it is known where to expect such complexes, no doubt a thorough search will be made for them in other systems. The fact that microtubules are seen to run from these edge-complexes initially in directions that are transverse, longitudinal and intermediate to the axis of extension may turn out to have great significance in the current controversy about the texture of the primary wall and the disposition of cortical microtubules beneath the primary wall. It has recently been proposed that polylamellate, crossed-fibrillar patterns of cellulose microfibrils, long known from collenchyma (Chafe & Wardrop, 1970) and the outer epidermal wall of coleoptiles (Bayley *et al.*, 1957) are common in other primary walls (Roland & Vian, 1979). Together with the corner ribs of thickening in parenchyma cells (Böhmer, 1958), which have microfibrils aligned parallel to the growth axis, such wall textures have always stood as difficult exceptions for those who seek a 1 : 1 relationship between the orientations of microtubule and microfibril. However, if the cortical MTOCS have the capacity to generate transverse, vertical and intermediate arrays of microtubules, perhaps at different times or at the same time in different parts of the same cell, an explanation for these exceptions which is in keeping with the rest of the evidence on the microtubule/microfibril orientation, may not be far away.

Analysing different cell files in the same roots, Hardham and Gunning (1979) have also gained some insight into the maintenance of the arrays during extension. Briefly, cells augment their population of microtubules as they extend and divide, showing a net increase of 0.6μm of microtubule per min. When a longitudinal wall is first formed, it has a slightly higher number of microtubules than it will have later when expanded, i.e., the initial quota is diluted. However, the cells interpolate new microtubules at a rate which is characteristic for the cell type and in the case of the endodermal cell, for the face of the cell in question, as the cell elongates. A most intriguing observation is that the initial set, and the set added during extension, behave differently when challenged with microtubule-perturbing treatments (colchicine, high pressure, D_2O). Studying

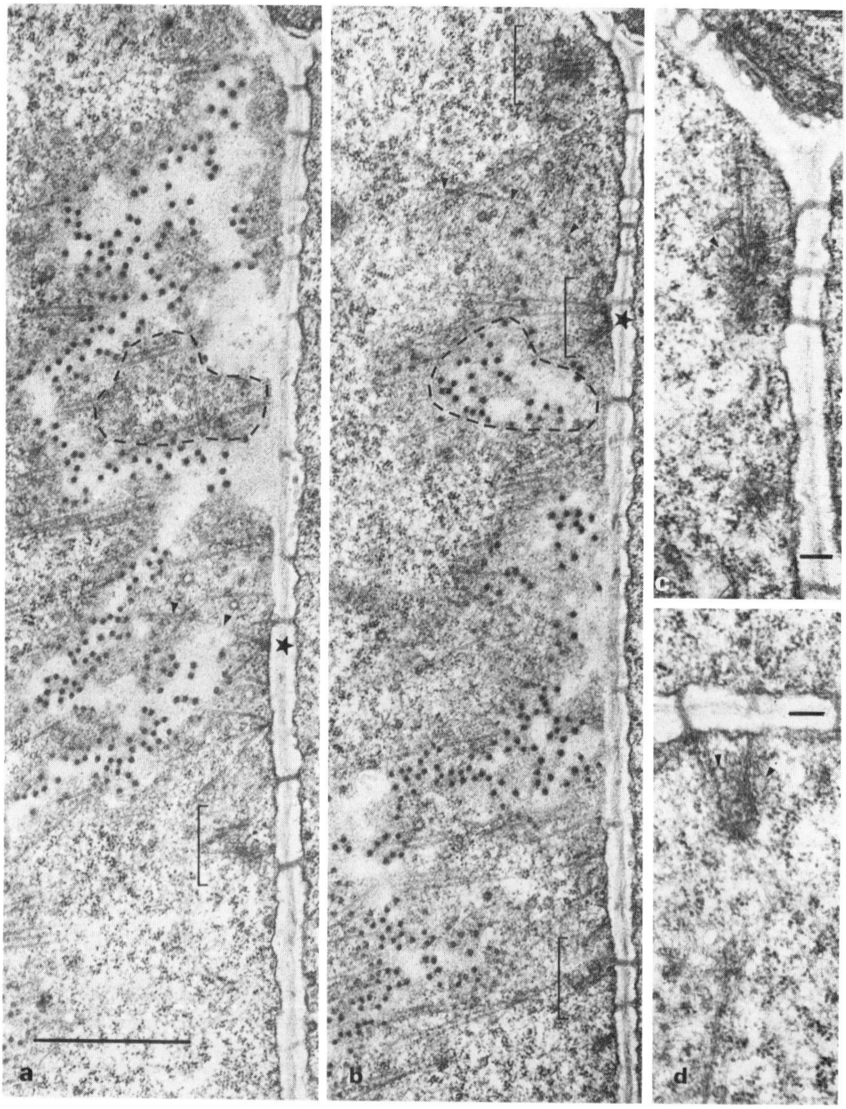

Fig. 4.21a–d Two adjacent sections of the junction of a radial longitudinal intermerophyte wall (vertical, right-hand side) and the first representative of a tangential longitudinal wall between pericycle and endodermal layers (wall in grazing section) in merophyte No. 18 of an *Azolla pinnata* root. The portions of wall in the grazing sections fit together (e.g., dashed line). The microtubule-vesicle-matric complexes (brackets and stars) are both above and below the new wall. Those indicated by the star and bracket in (a) lie in one daughter cell, as does the cluster (bracket) at the lowest bracket in (b); those above the dashed line in (b) lie in the other daughter cell. Microtubules radiate away from certain complexes sited very close to the plasma membrane (e.g. stars), some of them (arrowheads) curving towards the predominantly transverse alignment. The plasmodes-

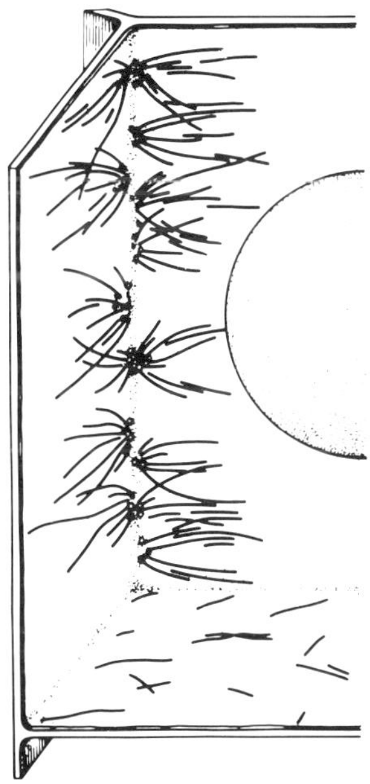

Fig. 4.22 Diagram to illustrate the emergence of microtubules from complexes lying at intervals along a vertical cell edge in an *Azolla* root cell. (From Gunning *et al.* (1978c), *Plant*, **143**, 161–79. Reproduced with permission from Springer-Verlag.)

the much more complex meristems of the roots of wheat, oats, barley and ryegrass, Withers and O'Brien (unpublished) have also established the existence of cell-specific 'base numbers' of microtubules, and cell-specific (i.e., epidermis different from cortex different from stele) substitution rates during elongation. In addition, it has been found that root-cap cells show, essentially, dilution of the base set without significant additions during expansion and that some cell files show two different rates of intercalation during expansion, one about half the other. This shift to a lower rate of substitution may be related to the time when cells in the file stop dividing. Some evidence for a parallel phenomenon in *A. pinnata* roots is presented by Hardham and Gunning (1979, their Fig. 4 and discussion).

The net result is that most cells maintain a set density of microtubules in the cortex during the phase in which elongation is accompanied by division.

---

mata are not in rows, nor can any predominating orientation of microfibrils in the new wall be detected. Scale marker 1 μm.

In (c) & (d) details of complexes from sections close to those used for (a) & (b). Arrowheads indicate vesicles with distinctly angular outlines. Scale markers 0.1 μm. (From Gunning *et al.*, (1978c), *Planta*, **143**, 161–79. Reproduced with permission from Springer-Verlag.)

4.2.7 THE ACTION(S) OF CORTICAL ARRAYS OF MICROTUBULES UPON WALL DEPOSITION

Most of the easily imaginable and more-or-less reasonable functions for cortical arrays of microtubules in wall deposition have by now appeared in print (for reviews see Newcomb, 1969; O'Brien, 1972; Heath, 1974; Hepler & Palevitz, 1974; Pickett-Heaps, 1974; Gunning & Steer, 1975; Hepler, 1976; Gunning *et al.*, 1978c; Hardham & Gunning, 1978; Schnepf *et al.*, 1976). Many of the early propositions were put forward before it was known that the arrays consist of a population of relatively short microtubules. New speculations and hypotheses that the array is involved in ordering cellulose microfibrils need to bear this fact in mind, but there are other, much more cogent reasons for restraint upon speculation in this area.

First, despite 40 years of relatively intensive work, it is still not clear how the cellulose microfibril is assembled. The basic biosynthetic chain that leads from glucose to high-energy intermediates is clear enough but the role and identity of the lipoprotein intermediaries between phosphorylated sugars, oligosaccharides and cellulose molecules are still very unclear (see Elbein, 1979; Preston, 1979; Quail, 1979). The structural role of glycoproteins, considered in their own right (Clarke *et al.* 1979) or as molecules with a demonstrated capacity for interaction with cellulose microfibrils in algal scales (Romanovicz & Brown, 1976) is equally unclear in most primary walls in extending organs. Whether the microfibril is synthesized as a complete unit by tip growth, or whether cellulose molecules are produced which then somehow 'self-assemble' into microfibrils, is also not known (Wardrop *et al.*, 1979; Mueller & Brown, 1980; Giddings *et al.*, 1980). If, as appears likely, the chains of cellulose molecules within the microfibril lie in anti-parallel arrays, two different complexes of synthetases must be active at any growing end of a microfibril (whether they would grow at one or both ends is not known, nor is it known whether they can branch). The problems of fitting 50–100 synthetases on the growing end are formidable and it is not certain that a growing end has ever been seen in an image, despite such claims from examination of freeze-etched material (see Brown & Montezinos, 1976 and the more cautious interpretations given by Mueller & Brown, 1980; Giddings *et al.*, 1980). If the microfibril does not grow at its tip as a unit, and if the chains are anti-parallel, how is this arranged and how does interweaving of microfibrils between adjacent lamellae come about?

Another important matter is the length of the cellulose molecule relative to the length of the cellulose microfibril. Some microfibrils can be traced in digested wall preparations for many micrometres and degrees of polymerization measured in extracted cellulose suggest a minimum cellulose-molecule length of about 2μm. Curiously, except at cut or digested edges, it is rare to find an image of an end that can be unequivocally designated as the tip of a microfibril. In short there are so many gaps in knowledge about the details of microfibril formation that it is premature to adopt a rigid stance on the way cortical arrays of microtubules might affect cellulose-microfibril assembly.

A further point which needs to be stressed is that despite the consensus of opinion that there is a 1 : 1 agreement between microtubule orientation and wall microfibril orientation, the evidence upon which this consensus has been reached is not very strong. As discussed earlier, there are important exceptions, of which the case of the root-hair tip is still unanswered. Here (see Fig. 4.10) the cell grows by tip growth and the extreme apical dome has no microtubules (Newcomb & Bonnett, 1965). The wall over this area has randomly-arranged fibrils. For the next 25μm, the wall is underlain by an axially-arranged layer of microtubules, but it is not known in detail if there is any wall deposition in that zone. Finally, the cell develops a secondary wall with an inner layer of axially-arranged fibrils over the layer of microtubules. Correspondence between the orientations of microfibrils and microtubules in the region of secondary wall deposition is fair, but some tubules depart very considerably from a strict axial alignment.

Roland and Vian (1979) correctly emphasized that the arrangement of microfibrils in growing walls is very imperfectly known. Many of the older data were based upon very drastic preparation procedures which may have disrupted the structure that existed *in vivo* (Deshpande, 1976a,b). More use of the shadowed-section technique (Maser *et al.*, 1967), in which microfibril orientation is visualized without removal of the non-cellulosic components of the wall, might help with this problem. However, it will not help with another major problem, namely the contraction that walls undergo when turgor is released. This problem, known for a very long time (see Heyn, 1933), is usually not acknowledged by workers who attempt to deduce *in vivo* microfibril orientations from killed cells. As Figs 4.23a–c show, very pronounced passive re-arrangement of microfibril orientations can take place in a wall without the fact being obvious in the plane of section examined. In walls with the capacity for elastic contraction upon turgor loss, this fact must be born in mind. A good deal of the 'herring-bone patterns' so often seen in walls is likely to originate from the effects of contraction upon lamellate walls (see Fig. 4.23a & b).

There is the further problem of attempting to deduce the orientation of cellulose microfibrils in stained EM sections. What evidence there is suggests that cellulose microfibrils are *not* positively stained by heavy metal stains used traditionally to enhance section contrast (Heyn, 1966). It is not known whether the PASH and the allied Thiery stains (Roland & Vian, 1979) react with cellulose microfibrils in native walls or not, but the low level of reactivity of cellulose towards the periodic acid-Schiff reaction suggests that the PASH and Thiery stains would not stain cellulose microfibrils positively (see O'Brien & Thimann, 1967; O'Brien, 1970). It is therefore unwise to deduce cellulose microfibril orientation from the appearance of positively stained fibrils in EM sections, especially since cell wall phenolics (see references in Smith & O'Brien, 1979; Smart & O'Brien, 1979) are known to be widespread in primary walls and to react with such stains. Further as Roland and Vian (1979) stress, wall components other than cellulose can be fibrillar.

The notion that at least some components of the walls of plant cells are

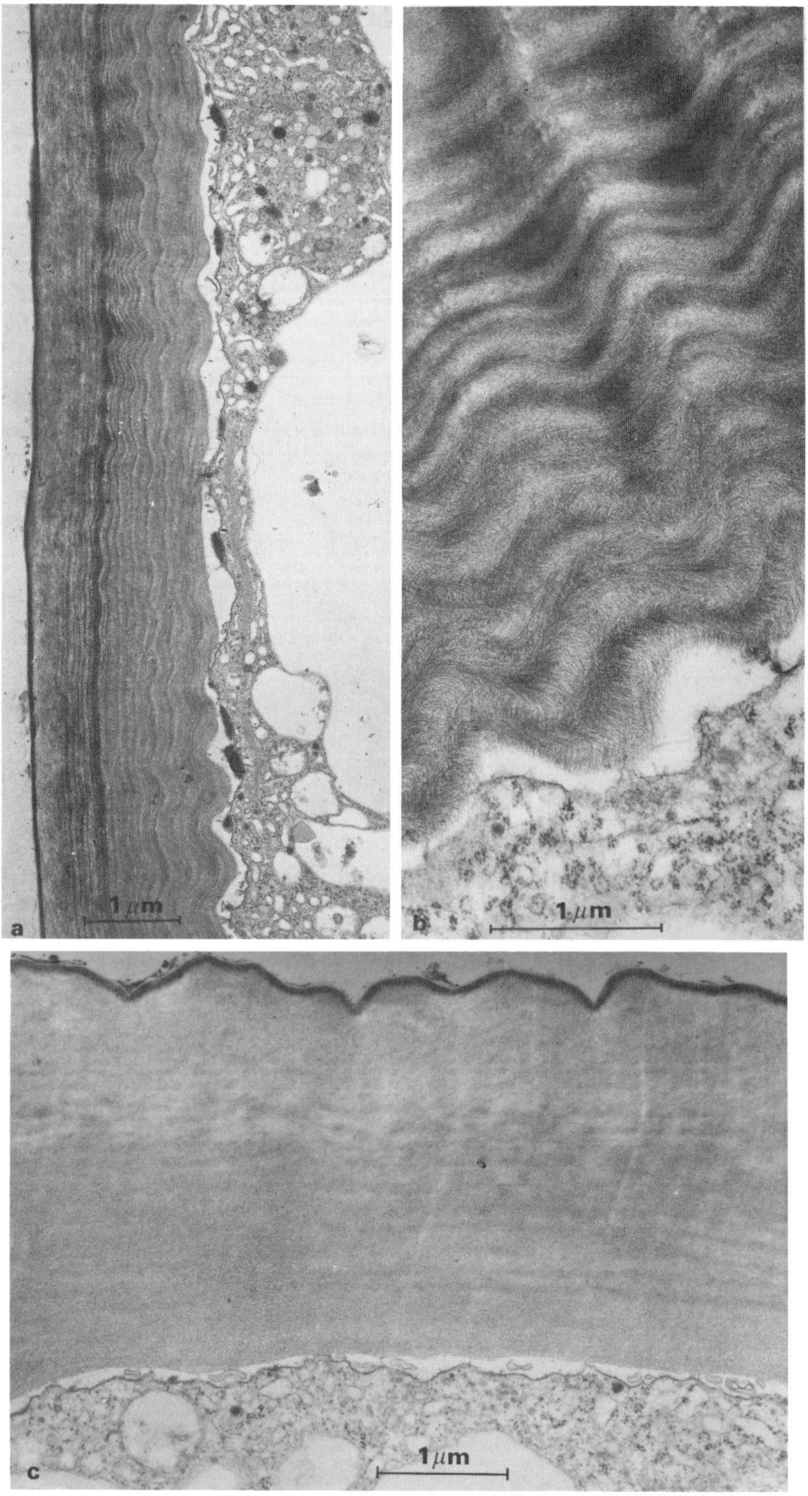
1 μm
a
1 μm
b
1 μm
c

synthesized within the endo-membrane system of the cell (ER, Golgi bodies, secretory vesicles), delivered from there to the cell surface within a membrane-bound compartment, and then incorporated into the cell surface is a deeply-rooted working hypothesis of the majority of workers in this field (see Gunning & Steer, 1975; Chrispeels, 1976; Preston, 1979; Quail, 1979; Northcote, 1979 *inter alia*). The evidence that supports the generality of this view is inconclusive, largely due to the technical difficulties involved.

A role for microtubules in 'channelling' vesicles charged with supposed wall-precursors for the developing cell plate was an early speculation (see O'Brien, 1972 for history) and it is natural that the parallel speculation that cortical arrays carry out a similar function beneath primary walls should have been put. The speculation has been put in two basic forms (see Gunning & Steer, 1975). In one form, the cortical microtubules are thought to collect vesicles and channel them to the cell surface (or to form cell plate). Viewed historically, it is curious that this idea should have become so popular. First, in nearly mature root-cap cells (where the evidence for vesicle incorporation into the wall is best: see Fig. 4.12) cortical microtubules are so scarce it is hard to find them (see Sawhney & Srivastava, 1975). Later work has shown that their sparse distribution in these cells arises because new tubules are not added to the cortical arrays during expansion of the cell (Withers & O'Brien unpublished). Secondly, the early images of cortical arrays (Ledbetter & Porter, 1963) showed that the tubules were so closely packed that it would be difficult to get a vesicle between them. Thirdly, in the two tip-growing systems studied earliest (pollen tubes: Dashek & Rosen, 1966; root-hair tip: Newcomb & Bonnett, 1965) microtubules were absent from the extreme tip in both cases, the zone where wall deposition was believed to be taking place.

Perhaps partly because of the second point, a second form of speculation arose in which it was suggested that areas of cell surface covered by cortical arrays were regions to which vesicle access was denied. Proponents of this view had probably already begun to suspect that the cortical array of tubules might not be erected over all of the surface at any one time during interphase.

Both of these speculations were probably not warranted by the evidence at the time and are not necessary now. Firstly, for vesicles up to at least 0.1μm diameter, diffusion is quite adequate to permit vesicle/cell membrane contact. The distances from dictyosomes and other parts of the endomembrane systems

---

Fig. 4.23a Longitudinal section of outer epidermal wall of the oat coleoptile showing compression and folding of the inner wall lamellae due to elastic contraction of outer layers when cell was killed in fixative.

b Detail of inner layers of outer wall in LS, to show the development of so-called 'herring-bone pattern'. The degree of herring-bone pattern appears to be more marked in the innermost layers of the wall where wall deformation has been greatest.

c The same as for a, but in transverse section. The distortion in wall structure is much more difficult to detect in this view.

All glutaraldehyde/OsO4 fixation, with uranyl acetate/lead citrate staining.

to the cell surface are never more than a few micrometres and over these distances, there is no need to postulate the existence of a special transport or guidance system. Secondly, membrane–membrane recognition may well be sufficient to allow vesicles to 'find' the 'right place' on the cell surface with which to fuse, if indeed they do so fuse. Finally, the discovery that the arrays are made up chiefly of short tubules largely invalidates the 'restricted access' speculations, since in most cases (but *not* for the regions underlying pre-prophase bands; see Fig. 4.24) there would be adequate opportunity for the vesicles to 'slip between the microtubules'.

The two questions to which one would most like an answer are:

(1) What is the spatial and temporal stability of the array and its component tubules?

(2) What regulates the distribution, activation and activity of MTOCS?

It is extremely difficult to devise ways to study the first in unsynchronized cell

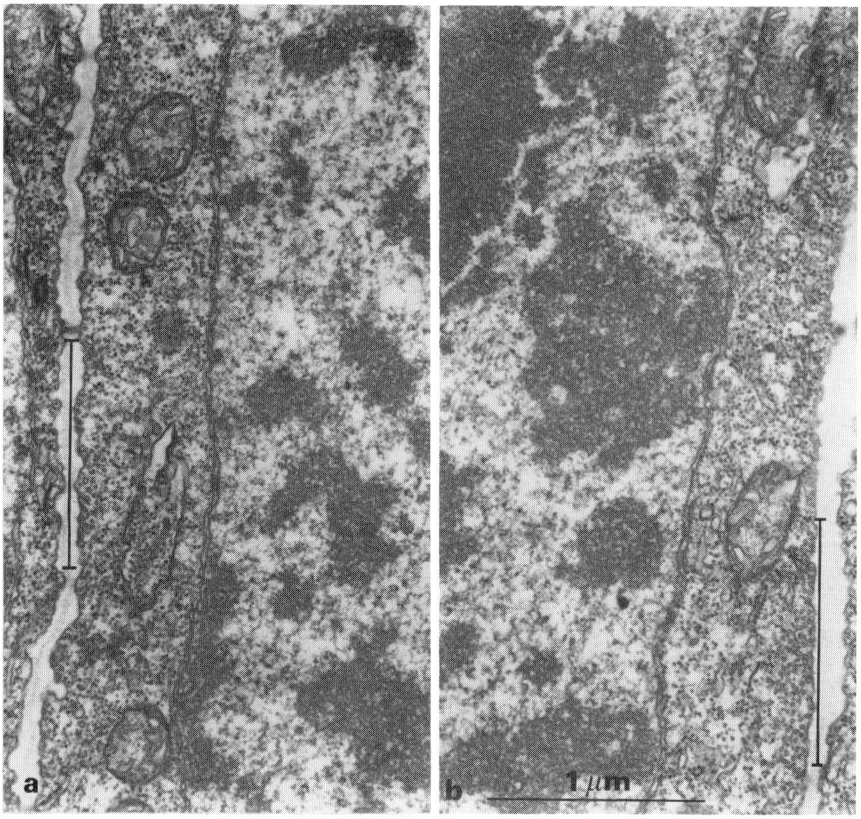

Fig. 4.24 Preprophase band of microtubules (cut in transverse section) from a longitudinal section of a provascular cell in a wheat root tip. a and b show opposite faces of the same cell. Glutaraldehyde/OsO4 fixation.

populations (most organs of higher plants). The second is one of the key questions to our understanding of cellular differentiation.

4.3 CELL DIVISION

Division in prokaryotic cells is beyond the scope of this chapter. In eukaryotic cells, the process is considered traditionally as the sum of two events: *karyokinesis* (division of the nuclear material or nucleus) and *cytokinesis* (partitioning of the cytoplasm). These two events usually occur in concert such that the former is almost or totally completed before the other begins. However, like the connection between growth and division, there is no obligatory coupling between karyokinesis and cytokinesis. Replication of chromatin (the DNA/histone/non-histone protein complex of the nucleoplasm in eukaryotes) usually precedes but is not always followed by karyokinesis. Polyteny (chromatin replication without the production of new sets of chromosomes) is known to occur in specialized plant cells as well as in a variety of animal tissues (Nagl, 1976a; 1976b; D'Amato, 1977) and polyploidy, the replication of chromatin accompanied by increase in chromosome number, is widespread in plant tissues (D'Amato, 1952; 1964; 1977). The most common kind is endopolyploidy in which chromatin replication occurs with a concomitant increase in nucleoprotein content; such nuclei often fail to divide unless they are provoked by manipulating tissue nutrition (D'Amato, 1977).

Polyploid cells can also be induced by manipulation of tissues with antimitotic drugs of which colchicine has enjoyed a long use. These drugs inhibit spindle formation and karyokinesis stops with the production of a characteristic colchicine- or c-metaphase (D'Amato, 1959). The daughter chromosomes separate from one another but do not migrate in the absence of a spindle. Provided the concentration of drug used is not too high, it can be washed out, the nuclear envelope will reform around the mass of chromosomes, reforming an interphase nucleus with a doubled chromosome number.

Binucleate cells occur naturally in anther tapeta (see Oksala & Therman, 1977; D'Amato, 1977 and references therein) where they arise because cytokinesis does not follow karyokinesis. Guard cells of some grasses are also functionally binucleate due to incomplete cytokinesis (Brown & Johnson, 1962). The multinucleate condition is widespread in fungi where it is accompanied by heterokaryosis in some groups. Polysiphonous algae afford other examples, as do early stages of division in endosperm in many angiosperms (D'Amato, 1977).

These examples, show clearly that the events of chromatin replication and chromosome separation, the key events of karyokinesis, and the partitioning of the parent protoplast, which normally occur in concert, are not coupled in an obligatory manner. Endopolyploidy, polyteny, and even the bi- and multinucleate states are often interpreted as various strategies that bring about gene amplification. Many developmental events in plants seem to be associated with prior or concurrent, partial or full, genome amplification (D'Amato, 1977). However, it is far from clear in most cases whether the amplification is full or partial, how it is regulated or whether it is a cause or effect of the developmental events with which it is associated (Evans & Van't Hoff, 1975; Nougarède & Rembur, 1980). More recent studies (see Nagl, 1976b; Ciba Foundation 1975; Li & Eckhardt, 1977) show that the views on chromatin replication current in the 1960s and early 1970s were naïve and simplistic and that the process is much more complex and fascinating than anyone had dared to imagine in public at that time.

There is an important fact to be grasped here. In examining the fascinating events of karyokinesis which follow, bear in mind that almost all of the events associated with chromatin replication and packaging have been completed before there is any structural evidence that such a nucleus is going to divide. It seems likely that the detailed structural basis of chromatin replication and packaging will elude the microscopist for some time to come (see Nagl, 1979).

In what follows the term 'mitosis' is not used because it is ambiguous. Though useful in speech as a short-hand for 'karyokinesis and cytokinesis', it is often used interchangeably for either or both of these processes and without specifying just what is meant.

4.3.1 KARYOKINESES IN 'TYPICAL' HIGHER PLANT CELLS (most observations are based on root tips or endosperm cells)

Traditionally, four main stages were recognized: prophase, metaphase, anaphase and telophase. It is now fairly widely accepted that it is useful to add two more, one on either side of prophase: pre-prophase and prometaphase. What follows is based upon a synthesis of descriptions from different systems, and some events are added from the study of dividing, living endosperm cells of the African blood lily, *Haemanthus katherinae* (see O'Brien, 1972; Gunning & Steer, 1975; Newcomb, 1978 for references). Students should note that the illustrations and accompanying discussions of karyokinesis in many elementary text books are often inaccurate and superseded by more recent studies. More accurate descriptions were published long ago by Haberlandt (1909), from which Fig. 4.25 is taken.

Karyokinesis in these systems is accompanied by changes in both the nucleoplasm and cytoplasm and these changes are summarized in Table 4.1.

The descriptions given in Table 4.1 differ in three major respects from the events seen in most cells of animals. Karyokinesis in plant root-tips is *acentric* (no centrioles are present) and *anastral* (no asters are formed, the star-shaped arrays of fibres that radiate out from spindle poles in such classroom-classics as dividing sea-urchin eggs). Thirdly, pre-prophase bands of cortical microtubules have not been reported in animal cells, though it will not be surprising if they are discovered in animal cells which divide inside rigid jackets that produce cell lineages with the same precise patterns that are found in plants.

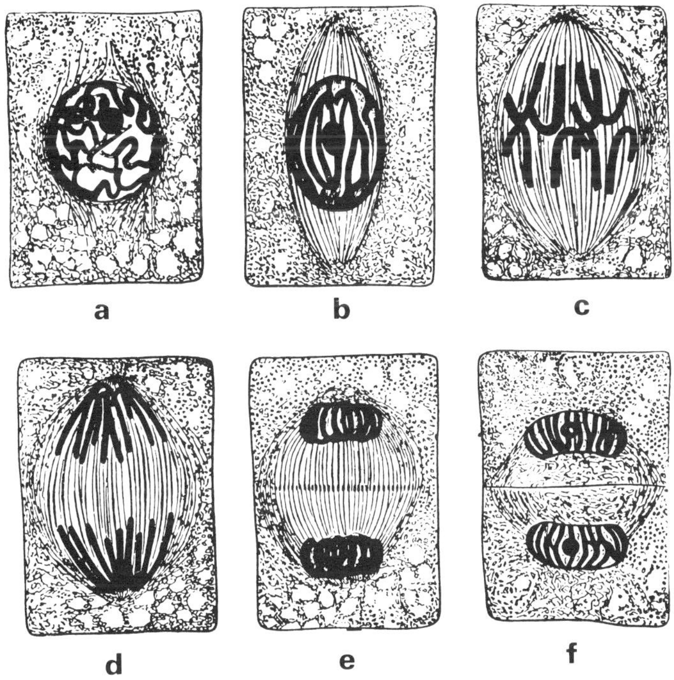

Fig. 4.25 The stages of mitosis and the structure of the spindle and phragmoplast as drawn by Haberlandt in 1909. Compare with Fig. 4.20. (From Haberlandt G. (1965) *Physiological Plant Anatomy*, Reprint Edn. Reproduced with permission from The MacMillan Co., London).

4.3.2 CYTOKINESIS IN 'TYPICAL' HIGHER PLANT CELLS

The operation is accomplished apparently by the action of the *phragmoplast*, yet another fibrous entity which consists again of microtubules, in this case arranged as a set of oppositely-directed tubules which appear to overlap in the plane of the metaphase plate in electron-dense 'fuzzy stuff' (Fig. 4.13). In living endosperm cells the phragmoplast appears as a set of fibres, each of which sticks out from a swollen central nodule. These nodules contain the regions of overlap of the microtubules which make up the fibres seen by light microscopy and presumably contain the 'fuzzy stuff'. (Scientific language is sometimes amusingly helpless when confronted with biological reality; nonetheless 'fuzzy stuff' is preferable to the pseudoscientific 'fibrillogranular material'. 'Fuzzy stuff' is somehow a more honest statement of the state of knowledge). A similar nodule plus fibre image can occasionally be produced in root-tip cells (O'Brien & McCully, 1969).

The cell plate forms within the phragmoplast. In living endosperm cells, tiny droplets (0.2μm in diameter) appear to invade the phragmoplast from the surrounding cytoplasm. Some influence causes these droplets to slide down the

Table 4.1 Karyokinetic events in 'typical' cells of higher plants

Stage*	State of nucleoplasm	State of cytoplasm
Preprophase	Nucleolus begins to disperse in many but not in all cells (Busby & Gunning, 1980).	Preprophase band of microtubules marks future plane of cytokinesis. Clear zones begin to form outside nuclear envelope at sites of future spindle poles.
Prophase	Chromatin begins to condense. Chromosomes (attached to nuclear envelope) are seen to be doubled, with arms of chromatids twisted about each other.	Microtubules run from the clear zones and encase the nuclear envelope as in a basket. Preprophase band usually disappears.
Prometaphase	Chromosomes separate from nuclear envelope, which fragments into cisternae of ER. Nucleolar material disperses. Kinetochore microtubules grow out from each of the two chromatids of each prophase chromosome. Non-random chromosome movement begins.	Bundles of spindle fibres run between the chromosomes. Each fibre consists of several (2–20) microtubules. These fibres stretch from pole to pole and are the continuous spindle fibres.
Metaphase	Each chromosome lies with its kinetochores aligned in an imaginary plane. If the chromosomes are short the arms lie in the same plane. If large, the arms lie at right angles to the axes of the kinetochores. Chromosomes 'rock' about their kinetochores (seen in living endosperm cells). Chromosome contraction is usually close to its maximum.	The spindle is fully developed: continuous fibres stretch from pole to pole, kinetochore fibres lie between them, one fibre from each kinetochore pair of any one chromosome being directed towards each pole. This fully-developed spindle, or sometimes some components of it, is referred to in older literature as the achromatic figure (did not stain with chromatin stains).
Anaphase	Kinetochores 'stretch' (living endosperm cells) chromatids separate from one another as they unwind, and they move to opposite poles as daughter chromosomes.	Kinetochore fibres shorten as daughter chromosomes move to opposite poles. Continuous fibres either elongate or stay the same length.
Telophase	Chromosomes gradually lose their distinctness and return to the interphase state. ER cisternae become applied to their surface and slowly reform an intact nuclear envelope Nucleoli (1-several) start to reform.	The spindle fibres disappear (interzonal fibres, whose relationship to continuous spindle fibres is not entirely clear, may persist: data from living endosperm cells). If cytokinesis is to follow, a phragmoplast starts to form between the two sets of daughter chromosomes, sometimes as early as mid-anaphase.

* The stages are arbitrary, but very useful and time-honoured classifications of a continuous process and must not be regarded as if they had sharp boundaries. Overlap of events, described here as if they 'belonged' to a particular stage, is common.

fibres towards the nodules where they *appear* to fuse to form the cell plate. There has been some debate about the nature of these particles and their contribution to the forming plate (reviewed by O'Brien, 1972). Some authors believe the particles are dictyosomes or their secretion vesicles, while others believe they are

ER cisternae. It is important to realize that the apparent size of these particles (0.2μm diameter) may not reflect their real size. As McLaren *et al.*, (1970) have shown so elegantly, the apparent size of structures that are not resolved by the numerical aperture of an optical system is a function of the numerical aperture (NA). For NA = 1.3, the size of visible objects below the resolution limit is about 0.5μm (image details that appear to be less than 0.5μm in diameter are likely to be optical artefacts: see discussion in McLaren *et al.*, 1970). Thus, any small vesicles whose real diameter is less than 0.5μm will appear in the image of an optical system as an image detail about 0.5μm in diameter at NA = 1.3, provided the vesicle has sufficient contrast to be visible. Phase contrast, Nomarski and fluorescence optics often render objects visible that are well below the resolution limit (e.g., Bajer & Jensen, 1969; for a theoretical discussion of this question see Françon, 1961). One must distinguish, then, between two possibilities. The particles seen in the living phragmoplast might be objects close to the resolution limit (0.25μm; segments of ER, whole dictyosomes) or they might be structures well below the resolution limit and rendered visible by their high contrast in the particular optical system used. Correlated light and EM has not yet resolved these questions adequately.

Equally difficult is the question of fusion of these particles with one another or the forming cell plate. EM of forming cell plates shows that their substructure is enormously more complex than the best images of living cells would lead one to expect, especially when viewed in tangential sections of the plate (see illustrations in Hepler & Newcomb, 1967 and Figs 4.13 and 4.14). The plate material that is bounded by a membrane has an irregular structure and the growing plate is encased in a subtly-differentiated lattice-work of rough, semi-smooth and smooth ER (see O'Brien, 1972 for references). Since the temporal or spatial stability of these components is unknown, it is impossible to dissect their contribution(s) to plate formation just by examining micrographs. The apparent fusion between the cytoplasmic particles and the forming plate is also not direct evidence of a real fusion. If the particle image really is an unresolved image and if particle visibility in the image is due solely to its contrast, then the movement of the particle into a region close to the plate where the particle contrast drops relative to the background will create the impression that the particle has fused with the plate. Such particles may well have fused with any of the components in the region of the plate (any of the forms of the ER or plate material). Since there is as yet no unequivocal evidence for what these particles are in a resolved image, it is not possible to be sure even whether they are membrane-bound.

Two approaches might resolve these questions. In systems such as isolated endosperm cells, it might be possible to kill dividing cells fast enough to 'trap' a particle, seen in the living cell, on its way to the plate, and, by correlated light and EM, to establish the structure of the particle beyond doubt. Alternatively, EM histochemistry might allow location of enzyme function(s) previously shown by LM histochemistry to be a feature of these particles (for methods see Hall, 1978).

4.3.3 KARYOKINESIS IN OTHER PLANT CELLS

During the last 15 years karyokinesis has been studied in a wide range of plant cells including representatives of many algal and fungal groups. Study of this literature yields a complex array of karyokinetic systems which differ in several respects from the situation described in Table 4.1 for the 'typical' plant cell (see Pickett-Heaps, 1969; 1971; 1975a; 1975b; Ris, 1975; Kubai, 1975). Differences in cytological behaviour may reflect evolutionary trends (Pickett-Heaps, 1975a). It is suggested that closed spindles (ones in which the nuclear envelope remains closed throughout karyokinesis) are more primitive than open ones since the closed condition is widespread in primitive fungi, algae and certain protozoans. Similarly, direct attachment of the chromosomes to the nuclear envelope (widespread in dinoflagellates: see Leadbeter & Dodge, 1967; Oakley & Dodge, 1974; Kubai, 1975) is regarded as more primitive than the situation described in Table 4.1 in which the kinetochores are attached to the chromosomes which let go from the nuclear envelope at prophase. Recent studies of the dinoflagellates and of other flagellates show that the kinetochore may have originated from a modified region of the nuclear envelope (see Kubai, 1975; Ritter *et al.*, 1978; Inoué & Ritter, 1978).

The function of the kinetochore is among the least clear aspects of karyokinesis. The studies on flagellates reviewed by Kubai (1975) suggest that microtubules are often attached to kinetochores even when the kinetochore is clearly embedded in (or perhaps sticking through) the nuclear envelope. Whether these microtubules originate from MTOC activity associated with these kinetochores does not seem to have been determined. Similarly, it is not possible to assess whether the 'astral' fibres that apparently 'engage' the intranuclear kinetochores in the flagellate *Barbulanympha* (Ritter *et al.*, 1978) have really originated solely in the aster, or whether they have trapped tubules growing from the kinetochores*. These difficulties are made worse by the fact that there are now a large number of examples in which there is little evidence to suggest that there are any kinetochores or chromosomally-derived microtubules involved in the karyokinesis (see Kubai, 1975). It is also hard to reconcile the presence and absence of kinetochores, or the degree of specialization of their structure, with 'primitiveness'. Thus, kinetochores are very clear by light microscopy in endosperm cells and cells of root tips but their EM image is disappointingly diffuse compared to the complex kinetochores found in certain green algae (cf. Gunning & Steer, 1975 with Pickett-Heaps, 1975b). Study of metakinesis (chromosome movement during prometaphase) in living cells of endosperm shows that each chromosome behaves independently of its neighbours, that chromosomes which lack kinetochores are not partitioned reliably during karyokinesis, and that the kinetochore regions of the chromosomes become aligned in the metaphase plate, with each chromatid's kinetochore directed to opposite poles of the

* The same query can be raised about the origin of the kinetochore microtubules in *Syndinium* (Kubai, 1975).

spindle (see Gunning & Steer, 1975). It would not yet seem safe to speculate how this system may have evolved from those systems found in lower forms, where metakinetic movement may be absent, metaphase alignment may not occur, and where microtubule attachments to the chromosomes are not always detectable. Similarly, it is necessary to remember that no clearly defined MTOCS have been detected at the spindle poles, on the nuclear envelope or even in the phragmoplast in angiosperm cells.

Five other features need to be discussed: the preprophase band, the behaviour of the nucleolus, the presence of ER in the spindle, the role(s) of the centriole, and the mechanism(s) of chromosome movement. The preprophase band has only been found in cells which perform cytokinesis with a phragmoplast (Fig. 4.25). This band probably has nothing to do with karyokinesis *per se* but reflects the activity of a cortical MTOC whose function is associated with future cytokinetic and early post-cytokinetic events. The absence of preprophase bands from cells that divide in whole or in part by invagination does not support a role for this band in karyokinetic events. Rather it seems likely that it has a role in determining that the phragmoplast, initiated in the interzonal region between departing anaphase chromosomes, will impinge finally at the cell margin, in the appropriate place, even after highly asymmetric divisions (see Busby & Gunning, 1980; Gunning *et al.*, 1978b). It is also possible that it is present at the time when future edge-complexes of MTOCS are put in place prior to their controlled reactivation after cytokinesis to build the cortical arrays (Gunning *et al.*, 1978c).

The nucleolus shows divergent behaviour in different systems. Persistent nucleoli during karyokinesis are a feature of the Psilotaceae and the phenomenon also occurs in a variety of animal cells. It seems likely that the different nucleolar behaviour seen in different cells reflects differences in the timing of discharge of nucleolar products to the cytoplasm (Gunning & Steer, 1975) and that it is incidental to, not critical to, the events of karyokinesis.

The presence of ER in the spindle is a widespread phenomenon, though some dividing cells show it better than others (see Hepler, 1980 and references therein). It is particularly striking in wheat as Fig. 4.26a & b shows. In part, these ER profiles seen in the spindle may represent the process whereby new nuclear envelope cisternae are transported into position for reapplication to the chromatin at telophase. There is a suggestion (Wick & Hepler, 1980) that they are the means of regulating the Ca^{2+} level in the spindle.

Transport of structures other than chromosomes by spindles, asters and phragmoplasts is well documented in the older literature based on microsurgurgy (see Chambers & Chambers, 1961 for references). The observation of 'chains of vesicles' in the phragmoplast of *Haemanthus* (Hepler & Jackson, 1968) might perhaps be interpreted better as transport of perforated cisternae by this structure also. It would be of interest to know whether nuclear envelope was still being reapplied to telophase nuclei in that study or whether the study was made on cells which had definitely completed all karyokinetic events. If the

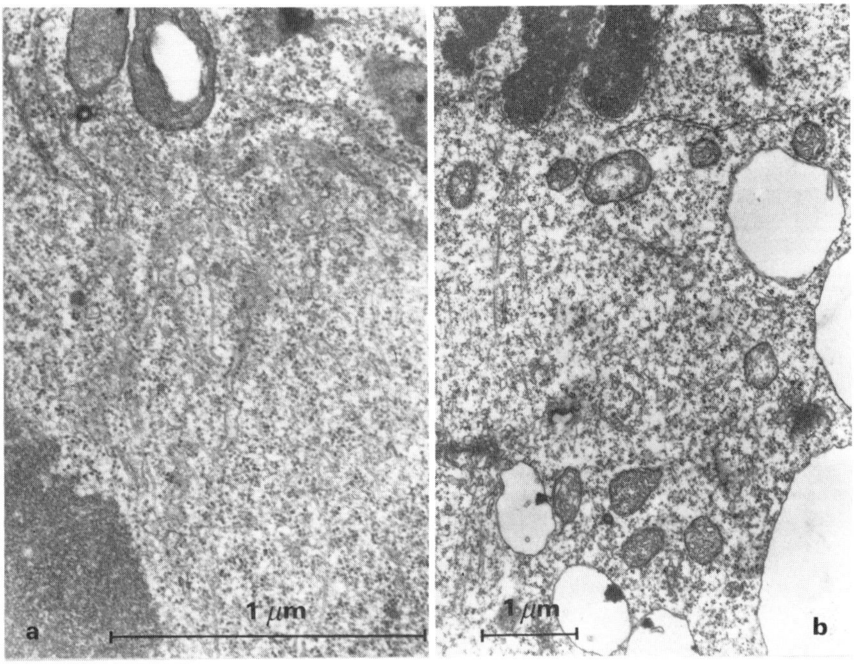

Fig. 4.26 ER in the spindle (a) and phragmoplast (b) in what root tip cells. Glutaraldehyde/OsO4 fixation.

stages studied were slightly different, it might help to explain the extraordinary divergence of opinions expressed by Bajer (1968) and Hepler and Jackson (1968), about the origins of the cell plates in the same system, a divergence commented on in some detail previously (O'Brien, 1972).

Centrioles are not found in the angiosperms though they are common enough in cycads, ferns and many groups of lower plants (Fig. 4.27). Centrioles are associated universally with the capacity to form cilia or flagella (Gunning & Steer, 1975). In addition to the production of flagellated male gametes, ciliated cells are very common in the somatic tissues of animals (see Fawcett, 1966). However, outside of the ciliates and flagellates, the presence of centrioles is restricted in plants to tissues which will form motile gametes or motile spores of various kinds. In some cases, the cells which will produce motile forms have no centrioles and the latter form *de novo* (e.g., in *Marsilea*: Hepler, 1976). In others, centrioles are always present in the cell lineages that lead to production of the motile phase (see Pickett-Heaps, 1975b for references).

The common, but not obligatory relationship between centrioles and spindle poles in cells which have centrioles during division led in the past to an over-emphasis on a spindle-forming role for the centriole. Pickett-Heaps (1969; 1971; 1975a) has soundly criticised this view, pointing out that karyokinetic structure and events are often identical in cells with and without centrioles. He proposed

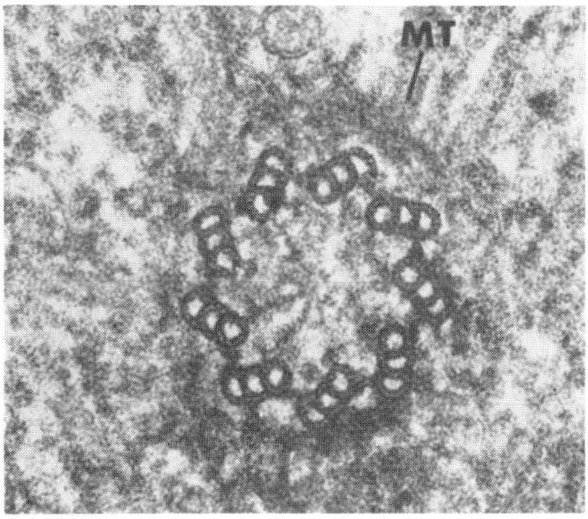

Fig. 4.27 Cross section of a centriole, showing the 9 triplet tubules embedded in a ring of pericentriolar material into which microtubules also penetrate. (From McNitt R. (1974) *Protoplasma*, **80**, 91–108. Reproduced with permission from Springer-Verlag.)

that the close association of centrioles with the spindle is a device that ensures equal partitioning of centrioles to the daughter cells. He stressed the case of *Chlamydomonas* in which the centrioles become attached to each side of the cleavage furrow, guaranteeing proper partitioning into the daughter cells. Despite these comments, there is still a case that must be put. Chambers and Chambers (1961) show, beyond any doubt, that in many invertebrates, an MTOC is present near to the centriole of the male gamete once it enters the egg. A monaster forms around it, and is associated both with the movement of the male pronucleus to the site of fusion with the egg nucleus, and with directed movement of 'hyaloplasm' (thin, less viscous cytoplasm) along the astral rays (see Harris *et al.*, 1980). Astral rays are now known to consist of microtubules and the important question is: are astral microtubules associated with the triplet microtubules of the centriole in any causative way, or are they formed by a diffuse MTOC that lies outside the centriolar triplets (Hyams & Borisy, 1978)?

This question has yet to be resolved. Chambers and Chambers (1961) state that the astral centre is non-fibrous, yet it is now known that it contains centriolar triplet fibres. That microtubules penetrate very close to centriolar triplet fibres has been known for many years (Fawcett, 1966) and confirmed more recently in a very elegant analysis (McNitt, 1974). However, these observations do not deny the possibility that some astral microtubules are generated by polar MTOCS with which centriolar triplets also associate. Telzer and Rosenbaum (1979) have shown that isolated pericentriolar complexes are competent as MTOCS only if isolated from dividing cells and that the microtubules they initiate assemble predominantly upon the pericentriolar material, but also on to the centriolar triplets. Other evidence implicating the pericentriolar material as an important MTOC in animal cells is given in Berns *et al.*, (1977) and Brenner *et al.*, (1977).

If it be postulated that an MTOC is always present in each pole of the spindle and that in cells that can form monasters or diasters, another MTOC is associated with each centriole, it is possible to rationalize a great many observations in both plant and animal cells, provided it is assumed that the degree of expression of these MTOCS is different in different cells and in the same cell at different times. Consider the following. The spindle pole MTOCS clearly have the capacity to produce spindle fibres in all cells that have them. In lower plants, these MTOCS are sometimes visible as distinct structures, (e.g. spindle pole bodies in certain fungi: Kubai, 1975; Lu, 1978; Peabody & Motta, 1979), but in higher plants their existence can be inferred only from their capacity to produce spindle microtubules. In the monaster, the pericentriolar MTOC generates a set of astral microtubules which acts as a cytoskeleton, upon which nuclear migration can take place. It is likely that an analogous structure is present around all nuclei that show directed migration not just simply due to the nuclei being transported in a bulk flow of cytoplasm. Specifically, the polar microtubules around migrating nuclei in fern gametophytes (Wada & O'Brien, 1975), and the set of microtubules associated with nuclear migration in desmids (Pickett-Heaps, 1975b), may all be examples of such structures. Furthermore, it is possible that the beautiful sets of microtubules that form the manchette in many flagellated gametes originate from a comparable MTOC, one which is usually closely associated with the site of a centriole that may have formed a flagellum or not (see Heidemann & Kirschner, 1975). Manchette microtubules are often arranged as some form of hollow basket and are clearly involved in determining cell shape (see Pickett-Heaps, 1975b, p. 498).

It is interesting to speculate that following gamete fusion in cells which have centrioles, the spindle that forms during karyokinesis might in some systems contain microtubules from three different sets of MTOCS. One set of microtubules may have been produced by the MTOCS that surround the centrioles and can also form asters; perhaps it is the spindle component of these MTOCS upon which centrioles normally migrate, ensuring partitioning of the daughter centrioles. *Chlamydomonas* would then stand as an example which shows the essential independence of this system from the second component of the spindle, namely microtubules produced from the spindle-pole MTOCS *sensu stricto*. The third spindle component, the kinetochore fibres, probably take their origin from chromatid-related MTOCS in higher cells but their evolutionary origins are obscure. A further example of the independence of centriolar and spindle MTOCS is given by the case of those fungi in which centrioles remain always external to the nucleus, separated by an intact nuclear envelope from the spindle which separates the chromosomes inside the nuclear envelope (Kubai, 1975; Heath, 1980).

The working hypothesis here is that the different forms of karyokinetic system may represent different balances between the presence (or activity) of three classes of MTOCS; centriolar (capable of generating asters and manchettes), spindle pole, and kinetochore MTOCS. As a corollary, the association of the centriolar triplet fibres with the spindle and aster (when present) would be coincident, rather than causative. Whether pericentriolar MTOCS have the capacity to replicate centriolar triplet fibres on their inner face and to form aster and/or manchette microtubules on their outer face, or whether there are really two classes of MTOC present, is not yet clear, since centriolar MTOCS raise another problem. The triplet system, and the tubules it produces in cilia and flagella, have microtubules with properties different from those of the central doublet (when present) in cilia and flagella, and also different from spindle and many other microtubules (Gunning & Steer, 1975). Whether centriolar MTOCS should be regarded as having the capacity to manufacture different types of microtubule, or whether there is really more than one type of MTOC present in such sites, is a matter for the future.

The preceding discussion on spindle structure in different organisms shows that one may no longer regard the simple question 'what causes chromosome movement?' as meaningful, for the mechanism of chromosome transport may well be different at different times in the same cell (metakinesis v. anaphase movements) and is very likely to be different in different types of plant cell. As Chambers and Chambers (1961) put it so perceptively, anaphase movement of chromosomes takes place upon a spindle which is itself enlarging by elongation. Several studies by Pickett-Heaps (1975b) and especially the analysis of spindle microtubule distribution at varying stages of karyokinesis in the colonial diatom *Fragilaria* (Tippit *et al.*, 1978) show that the central spindle elongates during anaphase movement and that the degree of overlap of its constituent microtubules decreases. They propose the simple explanation (which is by no means the only one that could be offered) that the microtubules slide past one another. However, these authors do not believe that it is this elongation of the spindle that causes the chromosomes to move. In a very different organism (*Syndinium*: Kubai, 1975) the kinetochore microtubules apparently remain at a fixed length while the spindle between the two centrioles elongates. In *Barbulanympha*, the kinetochores are moved towards the opposite ends of the spindle apparently by astral fibres that act as chromosome fibres by engaging the kinetochores (Ritter *et al.*, 1978). The elongation of the spindle in that organism is also a clearly separate phenomenon. In most dinoflagellates, it is not known whether the microtubules that attach to the kinetochores on the nuclear envelope actually shorten during chromosome movement. As Kubai (1975) discusses in some detail, whether microtubules play any active role (as distinct from a purely cytoskeletal role) in the many organisms in which connection between chromatin and microtubule is not well demonstrated, remains problematical.

As with so much of the literature that has ascribed functions to microtubules, it appears that the data are still inadequate to paint a clear picture in which one can have much confidence. The metakinetic movement of chromosomes

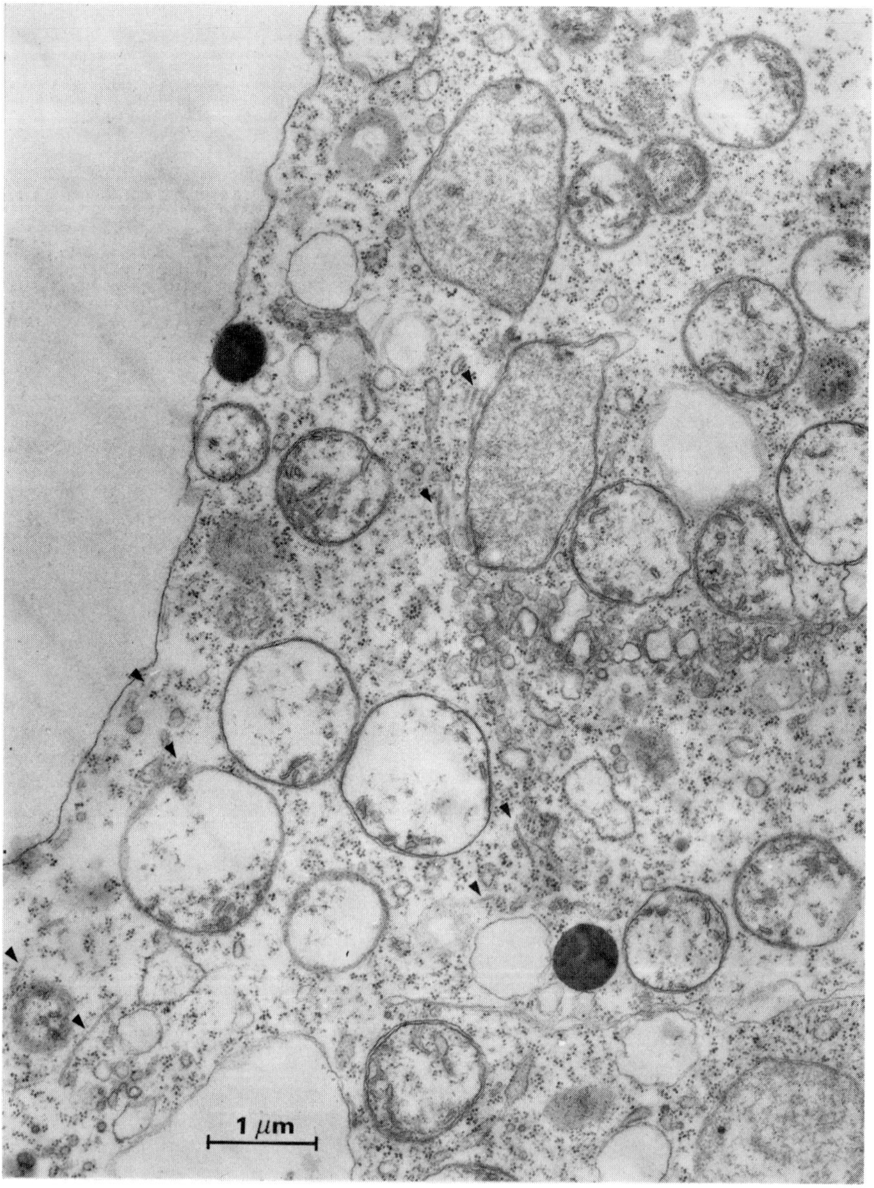

Fig. 4.28 Margin of an advancing wall ingrowth during cellularization of wheat endosperm. Though microtubules (arrows) are sometimes associated with these ingrowths, they are not disposed as in normal phragmoplasts [cf., Figs 4.13 and 4.14]. Glutaraldehyde/OsO4 fixation.

which ends with the production of the metaphase plate is equally mysterious though Pickett-Heaps (1967c) has proposed an interesting model.

Assuming that microtubules play some sort of 'active' role in chromosome movements, there are several theories about how this movement could come about. Bajer (1973) proposed the 'zipper' theory, and a number of sliding tubule theories are currently fashionable (see Tippit *et al.*, 1978; Hyams & Roberts, 1979). The latest appears to be a rotating microtubule theory (Schultz & Jarosch, 1980), a paper which gives many references to the other models.

4.3.4 CYTOKINESIS IN OTHER PLANT CELLS

As already discussed in connection with cell-wall formation, cytokinetic systems have now been studied in some detail. Where phragmoplast activity is not involved, most systems appear to make use of a cleavage apparatus which produces infurrowing of the cell membrane. In some algae, the infurrowing is associated with a phycoplast, a set of microtubules which lie in the plane of the furrow, but in others no phycoplast is present (see Pickett-Heaps, 1975b). The special cases of *Oedogonium* and *Spirogyra* have already been discussed (see p. 77).

In endosperm cells there is now widespread evidence for the existence of freely-growing walls (see Mares *et al.*, 1975; Newcomb, 1978 for references). The structure at the margin of these freely-growing walls is a matter of some dispute. Morrison and O'Brien (1976) failed to find any evidence for phragmoplast activity at the growing margin in the early stages of the division, but Mares *et al.*, (1975) show some ingrowths with phragmoplasts. More recently, a few microtubules have been found associated with these ingrowths of freely-growing walls, but the microtubule orientation and general disposition is not what one would expect of a phragmoplast (see Fig. 4.28). In view of the uncertainty that exists about the origin of endosperm in a phylogenetic sense, this system needs much more detailed study.

As Bailey stressed so long ago (Bailey, 1920), the problem of guidance of the phragmoplast in the cambial initials of the pine where the cell divides over a distance of 4000μm stands as a challenge to ingenuity and as a marvel of nature. While recent studies on the preprophase band (Gunning *et al.*, 1978b; Busby & Gunning, 1980) show the precision with which phragmoplast activity forms a cell plate which bisects the former site of the pre-prophase band, the guidance system which underlies and programmes this precise activity is unknown.

SUMMARY

Development, defined as a permanent change in shape, is accompanied by at least two and sometimes all three of the basic phenomena of cell growth, division and differentiation. Although these processes are often temporally and spatially linked, careful analysis shows that there is no necessary connection between them. In general, growth leads to an increase in volume. The frequency and planes of division and the extent to which growth accompanies division determine cell shape and differentiation regulates cell function.

Plant cells are physically interconnected by plasmodesmata and are also in communication via the gas spaces and by movement of small molecules in the phloem and xylem. The overall behaviour of the network thus derives from the integrated effects of numerous stimuli that may pass between responsive elements across many orders of magnitude of scale. To what extent is any one cell autonomous and to what extent is the behaviour of a cell imposed on it by its position in the organism? It is argued that if this problem is approached analytically at a particular scale, the network's behaviour tends to be 'explained' in terms of the approach that is followed. Thus, a physiologist interested in flowering may deduce that the process is controlled by light and temperature, but a biochemist studying the response in the same plant may uncover changes in enzyme levels or hormone outputs associated with the response. A cell biologist will find dramatic changes in the patterns of growth and differentiation, while a molecular biologist may find alterations in the pattern of gene expression that are associated with the response. None of these observations of what happens are wrong: it is simply that development is accompanied by changes in the overall behaviour of the network and that what one uncovers depends on where in the network one looks. The complexity of the network precludes simple cause and effect analysis of the response, for the fact is that a large number of responses are occurring at different rates and at different levels of scale. Many of the responding elements can generate signals that may affect what other elements are doing at other levels of scale.

At the cellular level, the nucleus, with its enormous information store, clearly acts both as a powerful regulator of the possible states of the cytoplasm it administers, and as a clearing house for signals going up and down the scale. It is always responsive to at least some of the messages that arrive from other cells, and from the physical environment (light quality, temperature, etc.). When cells become isolated from other cells, the pattern of stimuli to which they were contributing and responding changes. This can often lead to a new pathway of development; the classic alternation of generations in Pteridophytes appears to be a naturally-occurring case of just such a change in the cellular environment. Tissue culture experiments are also informative in this connection. It seems that the majority of mature tissues require a considerable chemical or physical 'jolt' to get them to embark upon a new pathway of development, even though it is perfectly clear that all blueprints for all possible pathways for that species reside in the nucleus of each cell in the tissue.

Good models are now available for the generation of form in growing cells and tissues and tip, surface and symplastic growth modes have been clearly identified. The activity of zones growing by one or more of these modes has been analysed in some detail and the role of the structural anisotropy of the cell wall carefully described. It is clear that the disposition of arrays of cortical microtubules is important, at least during and shortly after the time when cell partitioning accompanies growth, but the cytoplasmic control of the structural anisotropy of the wall has not been elucidated. There is very little evidence to implicate microtubules in this process once cells pass out of this phase and it should be borne in mind that the great bulk of tissue growth takes place after this phase is completed.

The events associated with cell division have been intensively studied in the last decade and represent one of the great contributions of the electron microscope to plant biology. An outline of the evolution of chromosome and cell partitioning is beginning to emerge. Broadly, one can identify two trends: a shift from chromatin division within an intact nuclear envelope to the completely open spindles of higher plant and animal cells and an increasing development of chromosome-associated spindle fibres. The role(s) of the centriole have been re-investigated and its significance as a source of spindle microtubules needs to be re-evaluated. It seems that a variety of cytoplasmic structures, some with a very regular sub-structure, some almost amorphous, are associated with microtubule initiation. In the search for the factors regulating these microtubule organizing centres of the cell (MTOCs) it may be necessary to depart from the limitations of the analytical mode of cause and effect thinking.

FURTHER READING

GREEN P. B. (1969) Cell morphogenesis. *Ann. Rev. Plant Physiol.*, **20**, 365–94.

GREEN P. B. (1976) Growth and pattern formation on an axis: critique of concepts, terminology, and modes of study. *Botan. Gaz.*, **137**, 187–202

GREEN P. B. 1980) Organogenesis: a biophysical overview. *Ann. Rev. Plant Physiol.*, **31**, 51–82.

GUNNING B. E. S. & STEER M. W. (1975) *Ultrastructure and the Biology of Plant Cells.* Edward Arnold. London.

KUBAI D. F. (1975) The evolution of the mitotic spindle. *Int. Rev. Cytol.*, **43**, 167–227.

PICKETT-HEAPS J. D. (1975) *Green Algae.* Sinauer Associates Inc. Massachusetts.

MOLECULAR BASIS OF DEVELOPMENT IN LOWER PLANTS

CHAPTER 5

BLASTOCLADIELLA

5.1 INTRODUCTION

Blastocladiella emersonii is a simple nonfilamentous water mould that grows as a saprobe in freshwater habitats. Although of little known significance in nature, it has been studied extensively since its developmental potential was first thoroughly described by Dr. E. C. Cantino in 1961 (Cantino, 1961). The zoospore of this microscopic fungus represents a highly differentiated cell, 7 × 9μm in size, propelled by a single posterior whiplash flagellum (Fig. 5.1). The kinetosome (basal body) of the flagellum lies adjacent to the nucleus and is enveloped by a single, large acentrically positioned mitochondrion. A group of lipid droplets and a large fused microbody complex lie between the mitochondrion and plasma membrane forming the so called 'sidebody'. Flagellar rootlets extend laterally from the kinetosome and nine sets of three microtubules extend forward around the nucleus. The anterior two-thirds of the nucleus is enclosed within a closely packed aggregate of the cellular ribosomes. This unusual organelle, called the nuclear cap, is surrounded by a double-layered cytoplasmic membrane. The cap membrane layers and the inner and outer nuclear envelope membranes are fused, thus compartmentalizing the cell into the cap area, the nucleus, and the remaining cytoplasm (nuclear pores only occur within the cap region). Finally, several 0·5μm membrane-enclosed electron dense 'gamma' particles lie in the cytoplasmic compartment. There is no rigid cell wall at this stage.

Zoospores can swim for considerable periods but, when triggered to germinate, undergo a very rapid developmental sequence that in a few minutes reverses the highly-ordered intracellular organization of the spore (Fig. 5.2). The spores first encyst by rounding up, retracting the flagellar axoneme, and secreting a thin polysaccharide cell wall. Simultaneously the intracellular anchoring microtubules disappear, the nuclear cap membrane begins to disperse releasing the previously 'packaged' ribosomes, and the gamma particles produce membranous vesicles (multivesicular bodies) that fuse with the plasma membrane. The latter process is part of wall formation and is proposed to release the enzyme chitin synthase at the surface (Mills & Cantino, 1978). After 20–25 minutes the retracted flagellar axoneme has disappeared and the germlings produce a primary rhizoid as a tubular outgrowth of the cell wall. The tubulins released by disassembly of the cytoplasmic microtubules return to a soluble pool, but the flagellar axoneme proteins appear to be degraded (Abe & Lovett, unpublished). Formation of the primary rhizoid is followed in the subsequent

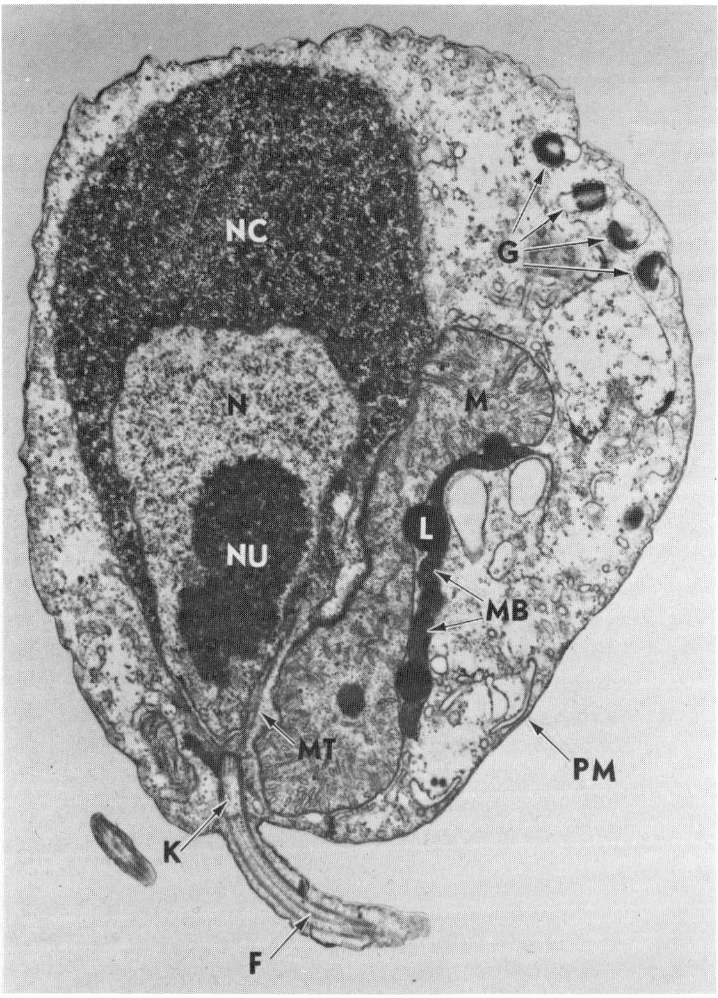

Fig. 5.1 The internal organization of a *Blastocladiella* zoospore. An electron micrograph of a longitudinal thin section through a motile zoospore. F, flagellum; G, gamma particles; K, kinetosome; L, lipid droplet; M, mitochondrion; MB, fused microbody; MT, anchoring microtubule; N, nucleus; NC, ribosomal nuclear cap; NU, nucleolus; PM, plasma-membrane. (× 14 500) (Courtesy of Dr. W. E. Barstow.)

growth phase by steady enlargement of the cyst and concomitant rhizoid elaboration and branching. The result is the typical multinucleate (coenocytic) chytridiacious thallus with a basal tuft of branched and tapering rhizoids.

In batch cultures grown with a complete medium the vegetative thalli grow rapidly with a doubling time of 2.1h at 24°C. They eventually convert to zoosporangia and 'cleave out' numerous uninucleate zoospores which are then released by breakdown of the discharge papilla. Growing cells exposed to a variety of stress conditions such as crowding and high bicarbonate or KCl concentrations, develop much more slowly to form dormant, thick-walled, pigmented, resistant sporangia. When mature these can be induced to release zoospores if immersed in fresh water. Despite uncertainty about the cytological events in resistant sporangia, the zoospores they release appear similar to those produced by the zoosporangia and give rise to the latter under growth-sufficient conditions.

It is of considerable interest that synchronous development of zoosporangia and zoospore differentiation can be induced at any time by shifting the growing cells to starvation conditions (e.g., if they are washed and suspended in dilute phosphate buffer). Even tiny uninucleate cells grown for as little as 30 minutes can be induced to sporulate in this manner and yield minature sporangia with a single zoospore (Hennessy & Cantino, 1972). This is similar to the well characterized process of microcyclic sporulation in bacteria. A minicycle can by produced in a similar manner by inducing sporulation at the binucleate stage (140min). In this case the entire vegetative growth and developmental cycle can be completed with good synchrony in seven hours which provides a very convenient experimental system (Barstow & Lovett, 1978).

Because *Blastocladiella* can be cultured using standard microbiological techniques, it grows and develops rapidly, and can be induced to develop synchronously without special treatments, it has been used extensively as a model system. Almost all stages have been examined and many different experimental approaches have been used including electron microscopy, enzyme studies, amino acid and nucleotide pool changes, cyclic nucleotide metabolism, chitin biosynthesis, polysaccharide and lipid composition and metabolism, tubulin pool changes, and the role of ionic currents as well as RNA and protein synthesis. In recent years the structure and function of the zoospores and their germination have received particular attention. RNA and protein synthesis and their regulation in zoospores and early development will be emphasized in this chapter. It should be kept in mind, however, that a number of other metabolic activities and structures also undergo significant changes during zoospore germination. As examples: the respiratory rate increases (Suberkropp & Cantino, 1972), the levels of cyclic AMP and cyclic GMP drop sharply (Silverman & Epstein, 1975), the lipid content decreases by some 60–70% (Smith & Silverman, 1973), the large stores of glycogen may be reduced by 93% (Suberkropp & Cantino, 1972) and a chitinous wall is layed down (Selitrennikoff, *et al.*, 1976). Thus, the changes in RNA and protein metabolism to be discussed below are accompanied by many other metabolic shifts. The causal relationships between these several activities and the regulation of development still remain largely unexplored.

Superficially, motile zoospores would not seem to have much in common with dormant *Neurospora* ascospores, dormant seeds or animal eggs. But, in a functional sense all of these must be able to transmit a functional system for protein synthesis to their 'progeny.' In this they face the common need to

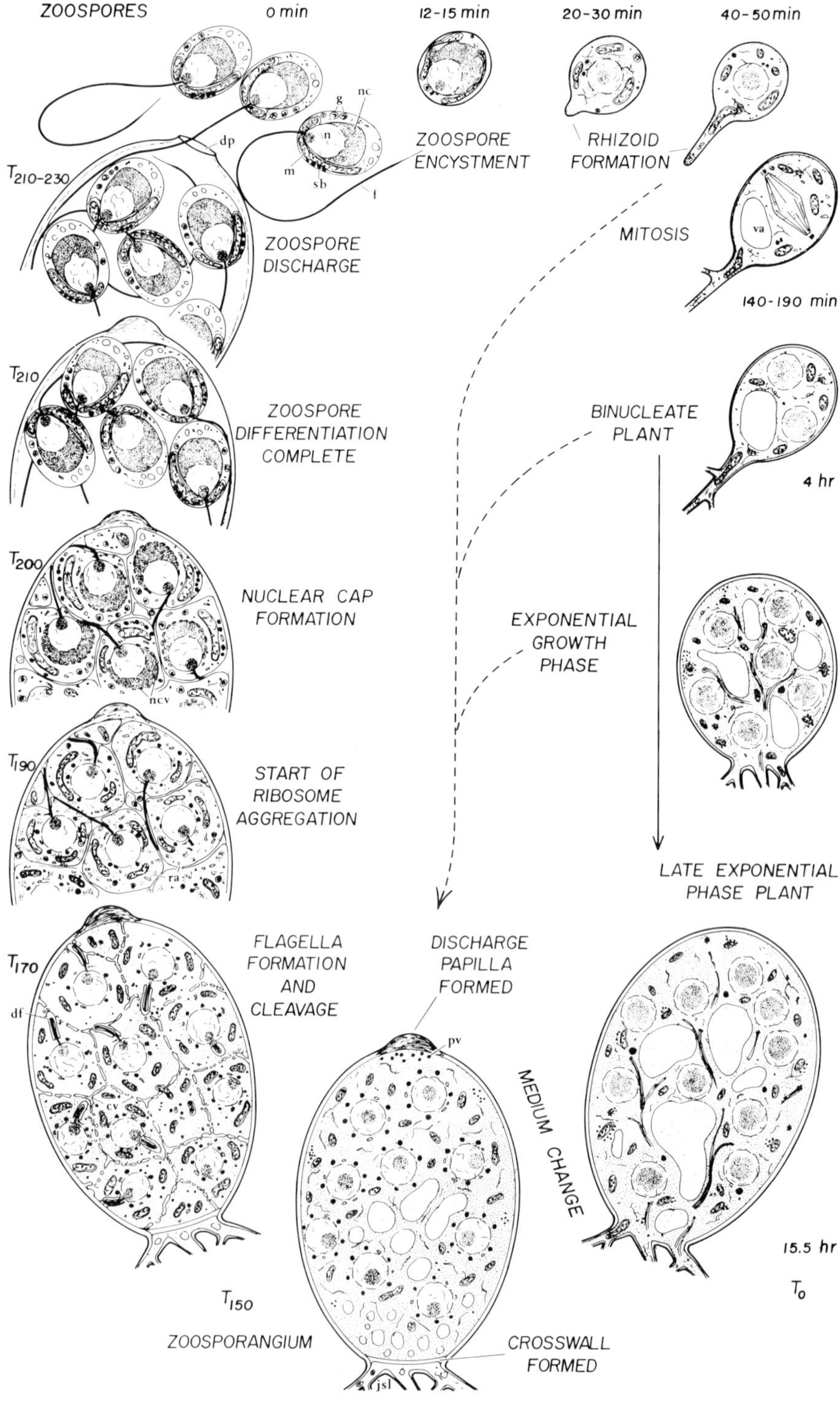

ZOOSPORES
0 min
12-15 min
20-30 min
40-50 min
ZOOSPORE ENCYSTMENT
RHIZOID FORMATION
MITOSIS
140-190 min
$T_{210-230}$
ZOOSPORE DISCHARGE
T_{210}
ZOOSPORE DIFFERENTIATION COMPLETE
BINUCLEATE PLANT
4 hr
T_{200}
NUCLEAR CAP FORMATION
EXPONENTIAL GROWTH PHASE
T_{190}
START OF RIBOSOME AGGREGATION
LATE EXPONENTIAL PHASE PLANT
T_{170}
FLAGELLA FORMATION AND CLEAVAGE
DISCHARGE PAPILLA FORMED
MEDIUM CHANGE
15.5 hr
T_0
T_{150}
ZOOSPORANGIUM
CROSSWALL FORMED
dp
g
nc
n
m
sb
f
va
ncv
ra
df
cv
pv
jsl

suppress or severely restrict biosynthetic activity, sometimes for many years, yet maintain the synthetic system in a functional state. In the case of protein synthesis more than 50 large and small molecules and macromolecular aggregates are required to make a single polypeptide (Boulter, 1977). At least minimal levels of all the proteins essential to the synthesis of new proteins must be conserved or the system could not be reactivated. A simple organism like *Blastocladiella emersonii* seems particularly useful for the analysis of how such a complex system is regulated during development.

5.2 ACTIVATION OF BIOSYNTHESIS DURING ZOOSPORE GERMINATION

Rather simple incorporation experiments with radioactively-labelled precursors ([^{14}C]-uracil and [^{3}H]-uridine for RNA; [^{14}C]-amino acids for protein) were initially used to assess the biosynthetic activity of *Blastocladiella* zoospores before and during their germination. The results (Figs. 5.3, 5.4) showed RNA and protein synthesis to be absent in zoospores and that both systems were rapidly reactivated at an early stage in the developmental process (Lovett, 1968). Neither inhibition of RNA synthesis with actinomycin D, nor blocking protein synthesis with cycloheximide, prevented the encystment of zoospores (Lovett, 1968; Soll & Sonneborn, 1971), establishing that neither biosynthetic activity was required for this initial event. For further development, including disassembly of the retracted axoneme and rhizoid emergence, protein synthesis was essential. On the other hand, cells could develop up to the 45–50 minute stage (Fig. 5.2) in actinomycin before becoming arrested. Thus, new RNA synthesis was not crucial up to this stage even though it began as early as did protein synthesis. From these original experiments it seemed probable: (1) that zoospores, despite their respiratory activity, contained a preformed and functional but inactive system for protein synthesis; (2) that one or more new proteins had

Fig. 5.2 Zoosporangial cycle of *Blastocladiella emersonii*. The principal intracellular changes during growth and differentiation are shown, as observed by transmission electron microscopy. The early stages are disproportionately enlarged to show internal details, and the rhizoid systems are not shown in their entirety. The timing from 0–4h indicates germination and early growth in synthetic medium; the timing from T_0 to T_{230} represents the time in minutes after the inducing medium shift starting at 15.5h. The broken lines indicate that cells shifted to inducing medium at the respective stages will begin zoosporagium and zoospore differentiation but in proportionately smaller sporangia and with proportionately fewer zoospores produced, depending on the number of nuclei present (i.e., 1–256). cv, cleavage vesicles and furrows; df, developing flagella; dp, discharge pore; f, flagella; m, mitochondrion; n, nucleus; nc, nuclear cap; ncv, nuclear cap membrane vesicle; pv, papilla vesicles; ra, ribosome aggregations; sb, side body granules of lipid and complex microbody; va, vacuole. (From Lovett (1975) courtesy of *Bacteriological Reviews*.)

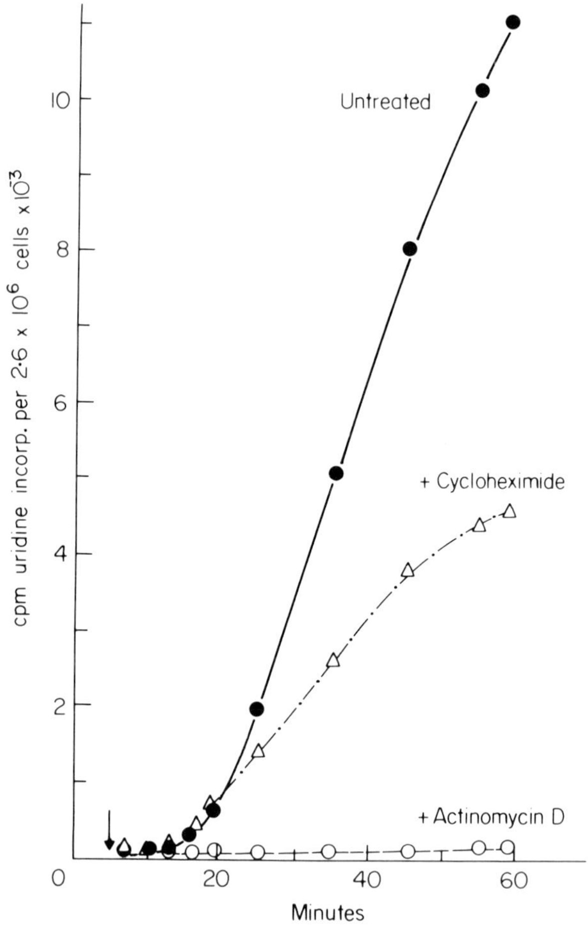

Fig. 5.3 Uridine incorporation during germination of *Blastocladiella* zoospores. Zoospores were inoculated at 2.6×10^6 cells/ml in complete synthetic medium with and without inhibitors at zero time. 5μCi of 2-^{14}C-uridine were added at 5 min and 1ml samples removed at the times indicated to measure the cold trichloroacetic acid insoluble radioactivity. Cycloheximide was present at 5μg/ml and Actinomycin D at 20μg/ml. (Reprinted from Leaver & Lovett (1974) by courtesy of *Cell Differentiation*.)

to be made at the earliest stages for normal development to occur; and (3) that the messenger-RNA for this essential new translation was already present before the spores were triggered to germinate; that is, existed in some cryptic form. However, because inhibitors can sometimes either fail to block the intended target activity or cause nonspecific inhibition, a more detailed analysis was obviously necessary to evaluate these conclusions.

To test directly whether any RNA synthesis at all occurs in zoospores prior to encystment, spores were inoculated in growth medium containing the RNA

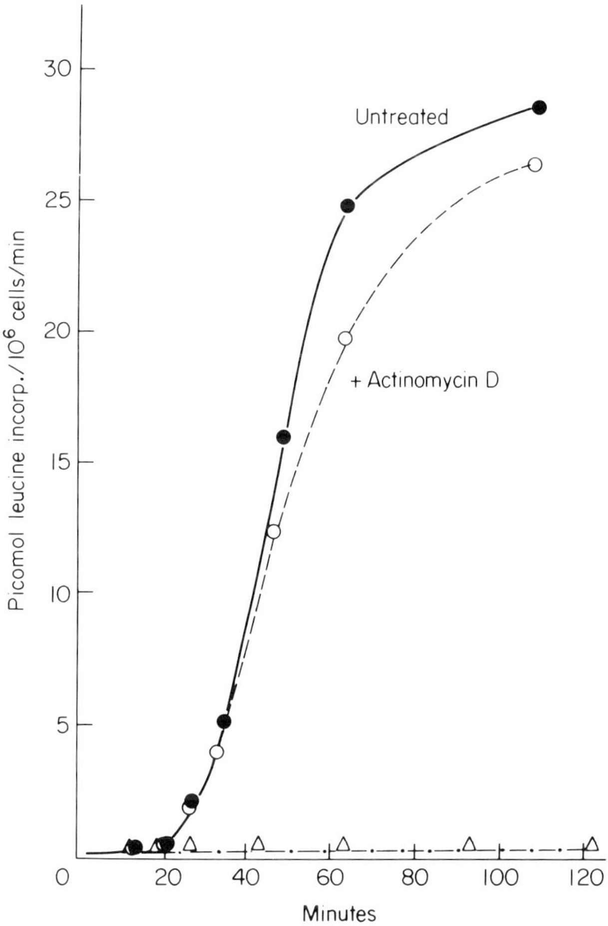

Fig. 5.4 Leucine incorporation during germination of *Blastocladiella* zoospores. Zoospores were inoculated at 1×10^6 cells/ml in complete synthetic medium at zero time. For each sample time 5ml aliquots were removed and preincubated for 5min with or without 50μg Actinomycin D per ml, and pulsed with 0.5μCi 1-^{14}C-leucine for 5min. The hot trichloroacetic acid insoluble radioactivity was then determined. (Reprinted from Lovett (1968) by courtesy of the *Journal of Bacteriology*.)

precursor [^{3}H]-adenosine. Samples were removed at close intervals and the specific activities of the ATP pool and total cellular RNA were determined. The results were unequivocal (Fig. 5.5). Label in ATP could be detected within one minute and its specific activity increased rapidly, while on the other hand label in RNA could only be found at 12 minutes, exactly the point at which the majority of zoospores had encysted. RNA synthesis subsequently increased linearly. Thus, all RNA synthesis appears to be suppressed in motile zoospores, but can be quickly restarted upon encystment. Fractionation experiments have

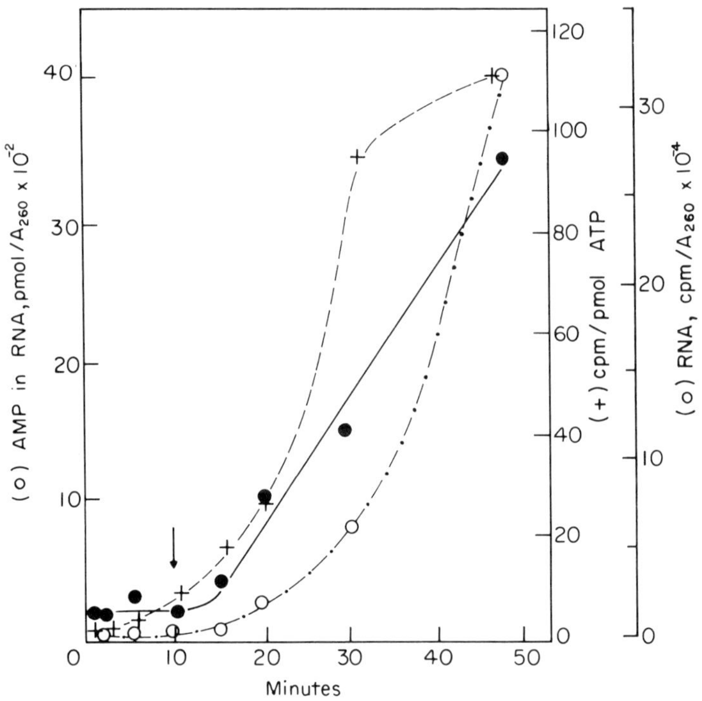

Fig. 5.5 RNA synthesis during early germination. Zoospores were inoculated at 2.67×10^6 cells/ml into complete synthetic medium containing 1μCi/ml of 2-^{3}H-uridine. Aliquots of cells were removed at the times indicated to measure the ATP pools and incorporation into RNA. (+ --- +) ATP pool specific activity, cpm/pmol ATP; (○ –·–·– ○) incorporation into RNA, cpm/A_{260} unit RNA; (● —— ●) molar accumulation of new RNA, pmoles AMP incorporated/A_{260} unit RNA. (Reprinted from Johnson *et al.* (1977) by courtesy of *Developmental Biology*.)

further shown that all major classes of RNA are transcribed from the start. The evidence does suggest that mRNA is made preferentially during the first 20 minutes of this developmental sequence (Johnson *et al.*, 1977).

Other labelling and cell fractionation experiments have shown that amino acids are taken up slowly by zoospores prior to encystment (Soll & Sonneborn, 1971). However, the rate of incorporation observed was very low and no labelled amino acids were found as polypeptides in the ribosome region of sucrose density gradients (Gong & Lovett, 1977). This small amount of whole-cell incorporation probably represents turnover of amino acids in the charged aminoacyl-tRNA pool. Just as in the case of RNA, encystment triggers the sudden start of new protein synthesis when measured by amino acid incorporation (Fig. 5.4) or the appearance of polysomes (Fig. 5.6). By 20 minutes at least 50% of the ribosomes are found in polysomal complexes compared to 10% in

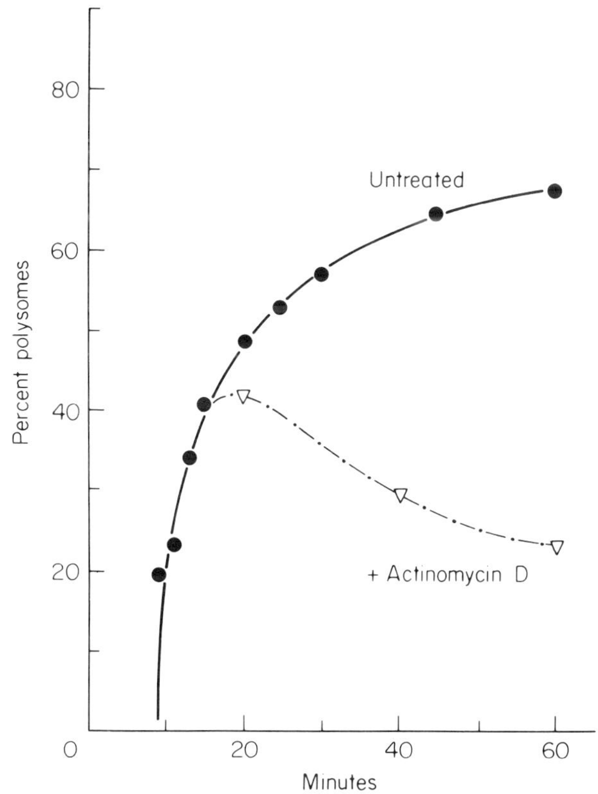

Fig. 5.6 The accumulation of ribosomes in the polysome fraction during zoospore germination and early development. Zoospores were inoculated into complete synthetic medium with or without 20μg/ml of actinomycin D. At the times indicated aliquots of cells were removed and the fraction of ribosomes in polysomes was analyzed on 10–35% exponential sucrose gradients. (Reprinted from Leaver & Lovett (1974) by courtesy of *Cell Differentiation*.)

zoospores (see later). It is significant that this initial very rapid formation of polysomes occurs even when all new transcription to produce RNA is blocked by the antibiotic actinomycin D. After 20 minutes the level of polysomes in the control cells continues to increase but at a declining rate, reaching a growth phase level at 70–80 minutes. When new RNA synthesis is blocked after 20 minutes the percent of ribosomes in polysomes decreases exponentially with a half-life of 45 minutes. This is the expected result if the original polysomes are formed by addition of ribosomes to pre-existing mRNA, and the latter are turned over during translation without replacement by new mRNA synthesis.

The compound 2-(4-methyl-2,6-dinitroanalino)-N-methylpropionamide (D-MDMP) selectively inhibits the initiation step of polypeptide formation in eukaryotic cells. When added to *B. emersonii* zoospores, D-MDMP did not inhibit encystment but did block germ tube formation by the encysted cells. It also inhibited amino acid incorporation 85–90% as compared to the non-inhibitory L-isomer (Leaver & Lovett, 1974). This complemented the incorporation experiments by showing that the appearance of polysomes required significant initiation activity to recruit the stored mRNA into this fraction for translation. Although it appears not to be essential for rhizoid emergence, at least some newly synthesized mRNA begins to appear in polysomes as early as 25 minutes, and perhaps sooner. Labelling experiments with [^{3}H]-adenosine further indicate that considerable new synthesis and/or turnover of the 3′-poly(A) end group sequences of the mRNA takes place during these early stages (Jaworski *et al.*, 1976; Johnson *et al.*, 1977). The actual quantitative contribution of newly made mRNA to the polysome fraction at this stage is uncertain, but it appears to be negligible prior to 20 minutes. After that stage it may replace and augment the preformed mRNA as it turns over, though this has not been examined.

Despite the rapid initiation of new synthesis that can be detected with radiolabelled precursors, a measurable accumulation of total RNA (mainly tRNA and rRNA) only begins at 40 minutes. Total protein starts to accumulate at a measurable rate even later at about 70–80 minutes. This has the effect of returning the RNA : protein ratio to 0.41, a value typical for growing cells (it is decreased to 0.32 through RNA degradation during sporulation). This presumably provides more ribosomes to support the increasing rate of protein synthesis. The ratio does not increase if spores germinate in buffered salts where the rate of protein synthesis and the polysome levels are depressed.

5.3 REGULATION OF BIOSYNTHESIS IN ZOOSPORES

The foregoing has presented a picture of the *Blastocladiella* zoospore as a cell with a high degree of intracellular differentiation and a stringently suppressed biosynthetic system. Both of these are quickly reversed when zoospores are triggered to germinate and enter the growth phase. How the biosynthetic functions are selectively suspended in motile cells that have an active respiratory energy generation system is an intriguing question. It seems highly probable that the ATP produced by mitochondrial respiration must be channelled in some way, or otherwise isolated from the pathways for macromolecule assembly. From a developmental perspective the way in which this selective regulation is achieved during sporulation and reversed in germination is also of considerable interest. There are three immediately obvious mechanisms that could control protein synthesis in the zoospore phase: (1) compartmentation to isolate one or more components of the synthetic system from the remainder (e.g., the ribosomes in the nuclear cap); (2) a deficiency of one or more of the essential proteins, nucleic acids or precursors; and (3) the production of an inhibitor for one or more of the synthetic steps (e.g., aminoacylation, polypeptide initiation, chain elongation, etc.).

Table 5.1 Distribution of components for protein synthesis in the nuclear caps and extra-cap fractions of zoospores

Component	Percentage of total	
	Cap	Extra-cap
Soluble protein	14	86
Ribosomes	100	0
Elongation factor-1	1.2 (8.6)*	98.8
Elongation factor-2	4.1 (29.3)	95.9
Transfer RNA	61	39
5S RNA	100	0
Messenger-RNA [Poly(A)RNA]	80.2	19.8
Met-tRNA synthetase (purified)	5.8 (41.4)	94.2
Phenylalanine synthetase (purified)	8.1 (57.9)	91.9
Group synthetase (assayed with protein hydrolysate)	29.6	70.4

* Numbers in parentheses indicate the percentage of the expected activity found, based on the assumption that 14% of each soluble component should be present in the cap fraction as found for the soluble protein. (Reprinted from Lovett *et al.*, (1977) by courtesy of Marcel Dekker, Inc.)

It had seemed possible early on that selective control could be related to the rigid segregation of the cap ribosomes from the remainder of the cytoplasm. The close association of the mitochondrion with the rootlets and kinetosome of the flagellum might further ensure the efficient transfer of ATP to that energy using system. At the same time ATP and other nucleotides could be excluded from the cap and nuclear regions by the continuous fused membrane system. The specific question of ATP channelling or compartmentation has not yet been addressed. It is possible to isolate the nuclear caps as intact organelles (Lovett, 1963) and such experiments appear feasible. On the other hand advantage has been taken of the cap isolation procedure to establish whether all of the macromolecular components for protein system are present in zoospores and their possible localization in, or exclusion from, the cap compartment. The results are summarized in Table 5.1, from which two conclusions can immediately be drawn. First, all of the macromolecular species needed to synthesize proteins appear to be present in zoospore extracts. Second, there is little evidence to suggest that activity is blocked by completely excluding any required proteins (enzymes), nucleic acids, or charged aminoacyl-tRNA molecules from the ribosomes in the cap compartment. The evidence against compartmentation as a regulatory mechanism does, nevertheless, require some qualification because all the factors have not been measured individually. For example, the initiation factors have not been directly measured, the elongation factors are present at quite low levels (particularly EF-1), and the twenty aminoacyl-tRNA's and aminoacyl-tRNA synthetases have not been assayed individually. Also the data do not yet rule out the deficiency of a single amino acid although this seems unlikely. On the other hand, even low levels of elongation factors should support *some* polypeptide formation and none is observed. Also indirect *in vitro* amino acid incorporation assays using zoospore extracts with and without specific initiation inhibitors have shown initiation activity to be present (Table 5.2). The same kinds of experiments have shown that aminoacylation and polypeptide synthesis can be achieved with dialyzed extracts when only labelled amino acids and an energy source are provided. Finally, in this regard, no evidence for a deficiency in synthetases or tRNAs has been found in spores of other fungi (Lovett, 1976). *In toto* the evidence suggests that no major factor needed for protein synthesis is missing from spores, or excluded from the ribosomal cap. Except for possible channelling of ATP and/or GTP, these conclusions have led us tentatively to reject two of the control mechanisms proposed above: compartmentation and a deficiency of a required component.

The third regulatory possibility put forth was the presence of a specific inhibitor and the evidence for this kind of regulation is strongly suggestive. As a result of experiments to measure the functional activity of isolated nuclear cap ribosomes, it was found that they would only support polypeptide formation *in vitro* if first washed with high-salt buffers (e.g., containing 0.25–0.5M KCl) (Schmoyer & Lovett, 1969). Furthermore the KCl-wash from cap ribosomes inhibited active ribosomes from growth phase cells and could reinactivate activated zoospore ribosomes. Although attempts to isolate the inhibitory material have been unsuccessful, considerable evidence supports its presence. The activity binds to ribosomes reversibly. It appears during sporulation and disappears again during germination of the zoospores (measured as ribosome-bound inhibitor and by the rate of polypeptide formation per ribosome *in vivo*). And finally, whole-zoospore lysates (23 000*g* supernatants) will support amino acid incorporation *in vitro* if first dialyzed against high salt buffer to remove the inhibitory material (Table 5.2). The mechanism of action remains uncertain, but based on the evidence it seems probable that the major effect is to block polypeptide chain elongation.

At present the ribosome-associated inhibitor is considered to be the major control factor to modulate protein synthesis in zoospores, but the channelling of ATP could serve as an additional factor. Certainly the function of the nuclear cap membrane, if it does not serve to exclude small molecules, must remain a subject for speculation. The only other reasonable function to suggest would be protection of the ribosomes from degradation while they are not involved in translation. Such degradation does often occur when growing bacterial and fungal cells are subjected to starvation for a required amino acid. The aggregation of ribosomes to form a nuclear cap is found in zoospores of several simple aquatic fungi, but the totally enclosed condition seems restricted to the *Blastocladiaceae.*

Table 5.2 Amino acid incorporation by whole-zoospore cell-free extracts

Assay system	Sp act (counts/min mg of protein)	Percentage of control
Complete system*	35 843	100
Complete − S-23 fraction	0	0
Complete − GTP	14 105	39
Complete + aurin-tricarboxylic acid (0.1mM)	5147	14
Complete + pactamycin (0.1mM)	8961	25

* The S-23 zoospore fraction was prepared by 15-s sonic treatment (Branson Sonifier model S125, microprobe, setting 1) of 2ml of packed zoospores suspended in 5ml of extraction buffer (20mM Tris-hydrochloride (pH 7.5), 500mM KCl, 1mM MgCl2, 0.1mM ethylenediaminetetraacetic acid, 5mM dithiothreitol), followed by 15-min centrifugation at 23 000*g* and dialysis of the supernatant (S-23) against the same buffer with 100mM KCl. Incorporation activity was measured by 30-min incubation at 30°C in a 0.5-ml volume containing S-23 (455μg of protein); Tris-hydrochloride (pH 7.8), 30μmol; MgCl2, 2μmol; KCl, 25μmol; ATP, 0.5μmol; GTP, 0.025μmol; phosphoenol pyruvate, 1.25μmol; pyruvate kinase, 20μg; yeast tRNA, 50μg; reconstituted ^{3}H-labelled yeast protein hydrolysate, 2μCi; amino acids not in the hydrolysate, 10mmol each; and dithiothreitol, 2.5μmol. Radioactivity was measured as hot trichloroacetic acid-insoluble material. (Reprinted from Lovett (1975) by courtesy of *Bacteriological Reviews*.)

5.4 INDUCTION OF ZOOSPORE GERMINATION

Setting aside the still uncertain role played by the nuclear cap, what we do now know is that zoospores have an apparently intact but stringently suppressed protein synthetic system. Virtually nothing is known about how starvation triggers the differentiation of zoospores and leads to this state. It is clear that extensive turnover and degradation of RNA and protein accompany the process (Murphy & Lovett, 1966). Although it is better studied, not much more is understood about the induction of zoospore encystment and germination. Calcium ions (10^{-3}M) help maintain extended motility and the addition of the appropriate concentration of potassium ions (50mM) induces rapid and synchronous encystment (Soll & Sonneborn, 1972). Studies of zoospore induction with potassium have shown that K^+ is very actively taken up and that it depolarizes the plasma membrane (van Brunt & Harold, 1980). The calcium ionophore A23187 was also found to be an effective inducer by the same workers. Although such inducing conditions are not those operative in nature, the results do suggest that the initial event may be a depolarization of the membrane, and that the subsequent membrane rearrangements involved in encystment could result from a critical shift in the amount of bound calcium. Similar events are of considerable interest in other developmental systems, as for example the wave of calcium release elicited by fertilization of *Medaca* eggs (Gilkey *et al.*, 1978). Localized currents are associated with several tip growing systems such as pollen tubes (Weisenseel *et al.*, 1975) and rootlets (Weisenseel *et al.*, 1979); and the calcium ionophore A23187 also alters the tip growth of *Neurospora* hyphae (Reissig, 1977). Thus, ion fluxes could be involved in the localization of rhizoid emergence in *Blastocladiella* as well as in the primary encystment event.

5.5 INFORMATION CONTENT OF STORED MESSENGER RNA

Once triggered, germination of zoospores is an all or none response; each spore undergoes encystment and produces a germ tube in an irreversible sequence (Lovett, 1968; Truesdell & Cantino, 1971). This occurs even under conditions that are unsuitable for growth such as germination in water, in organic salt solutions, or in the presence of actinomycin D added to block all new RNA metabolism. Since protein synthesis is required for the cells to proceed beyond encystment, the ability to germinate in actinomycin shows that the mRNAs for all needed proteins are already present. The existence of this putative mRNA has been mentioned earlier as well as its primary location in the cap organelle. The independence of germination from RNA synthesis suggests that, at a minimum, the stored messenger pool includes a set of mRNA's that serve as a 'germination program'. These would-be messenger templates enter the polysome fraction during the first 20–40 minutes following encystment and code for the essential, germination-specific proteins. The analysis of the stored mRNA pool has provided data that are quite consistent with this working hypothesis.

5.5.1 CHARACTERIZATION OF STORED MESSENGER RNA

Experiments to characterize the stored messenger fraction were based on the assumption that the polyadenylate-containing RNA (poly(A)RNA) represents the major messenger population, as it does in other eukaryotes. The poly(A)RNA fraction used in these experiments was isolated from total spore RNA by chromatography on oligo(dT) cellulose. The isolated poly(A)RNA was found to represent 2.5% of the total zoospore RNA, to carry poly(A) base sequences 32 bases long at the 3′-end (Fig. 5.7), and to have a modal molecular weight of

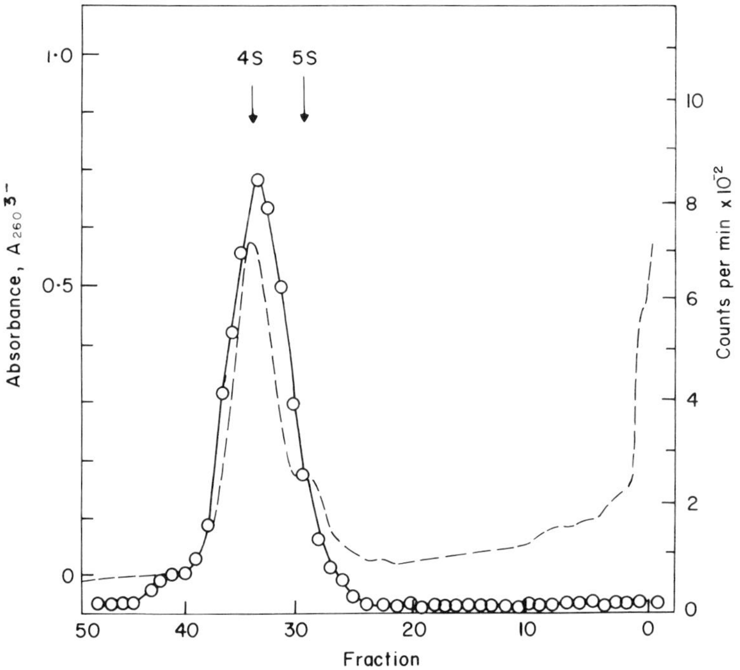

Fig. 5.7 Gel electrophoresis of poly(A) isostichs from zoospore poly(A)RNA. The poly(A) was isolated as the nuclease-resistant fraction of purified poly(A)RNA that had been uniformly labelled with ^{3}H-adenosine. After electrophoresis on 5% polyacrylamide gels, slices were cut for radioactivity measurements. The absorbance scan shows the position of 4S transfer RNA and 5S ribosomal RNA run on a parallel gel. (Reprinted from Johnson *et al.* (1977) by courtesy of *Developmental Biology*.)

425 000 daltons (1288 nucleotides) (Fig. 5.8). It also appeared from preliminary analyses that 7-methylguanosine groups 'capped' the 5′-end of the molecules, as is typical for eukaryotic mRNA. Poly(A)RNA isolated by similar procedures from growing cells at two hours was found to be present at the same concentration. The 3′-poly(A) sequences of this poly(A)RNA averaged slightly larger at 38 bases (Fig. 5.7) while the modal molecular weight was somewhat lower at 360 000 daltons (Fig. 5.8). The significance of these small variations, if any, is not known. These data did not provide any obvious indication that structural differences could easily explain the lack of translation in zoospores. The spore poly(A)RNA has since been translated in a standard cell free wheat germ system which demonstrated that at least some of it was functional after isolation (Fig. 5.9). All these results are consistent with the presence of a ribosome-associated translation inhibitor.

It was surprising to find that biosynthetically 'dormant' zoospores contained as large a percentage of mRNA as did actively-synthetic growing cells at two hours. This indicated that in addition to the messages for the germination

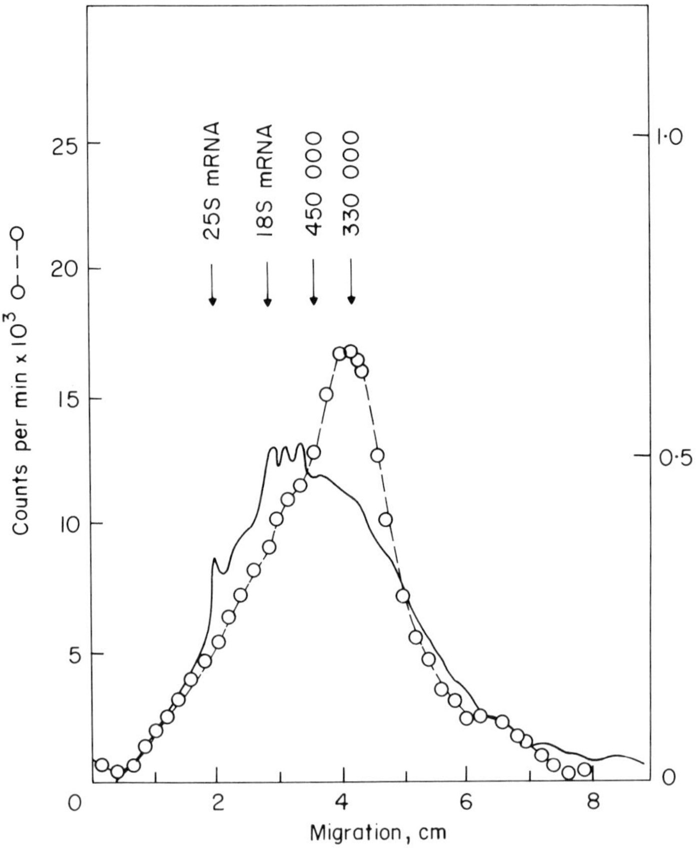

Fig. 5.8 Gel electrophoresis of poly(A)RNA purified from zoospores and growth phase cells at 2h. The poly(A)RNA from each cell type was isolated by oligo(dT)-cellulose chromatography from a total RNA preparation. 50μg of zoospore poly(A)RNA and 200 000 cpm of uniformly labelled growth phase cell poly(A)RNA were run on 2.4% polyacrylamide gels for 2h at 6ma/gel. The labelled gel was sliced and counted to locate radioactivity; the zoospore RNA gel was scanned at 260nm. The arrows indicate molecular weight based on the linear relationship between migration distance and log molecular weight. (Reprinted from Johnson *et al.*, 1977, by courtesy of *Developmental Biology*)

'developmental program' the spores could harbour a significant variety of other messages. It also suggested the interesting possibility that the spores could store specific messages for a wide spectrum of proteins to meet varying nutritional or other environmental demands on their synthetic capacity (e.g., environments that demand specific extracellular enzymes and/or special metabolic pathways to obtain carbon, nitrogen and energy for growth). If so, any of the needed subsets of messages could be mobilized and translated quickly to give the small cells a good start. Those subsets not needed might then be degraded. An alter-

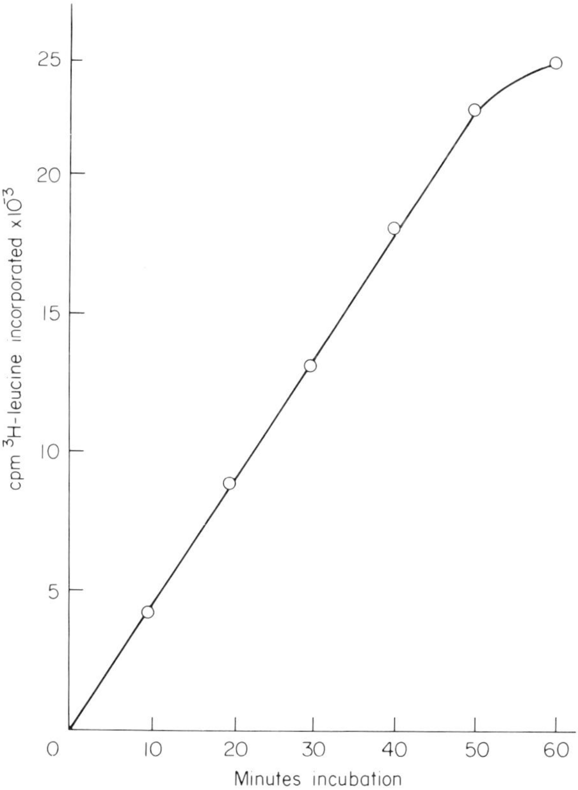

Fig. 5.9 *In vitro* translation of zoospore poly(A)RNA. Poly(A)RNA purified from *Blastocladiella* zoospores by oligo(dT)-cellulose chromatography was incubated in a standard cell-free wheat germ assay mixture containing [^{3}H]-leucine. At the indicated times aliquots were removed and the incorporation of [^{3}H]-leucine into hot trichloroacetic acid insoluble polypeptides was determined. Each point represents the radioactivity of a 50μl sample containing 1.5μg of poly(A)RNA. (Lovett, unpublished data.)

native possibility, of course, was that the large mRNA pool in zoospores merely represented a significant degree of redundancy with fewer kinds of messages present as many duplicate copies.

It was of obvious interest to know the form in which the mRNA was stored in zoospores while remaining untranslated, since this could aid understanding of how the system was controlled. It also seemed of value to examine the spectrum of different mRNA molecules stored by zoospores and their ultimate expression through translation, or other fate, during germination and early growth. It has been assumed by analogy with other organisms that the mRNA exists as RNA-protein (mRNP) complexes, but is not yet clear how it is actually 'stored' in zoospores. At least 80% of it is found in the nuclear cap along with the ribosomes (Table 5.3), but only about 7% can be accounted for by the

Table 5.3 Intracellular distribution of poly(A) in zoospores

Fraction	RNA (μg)	% of total RNA	% of total Poly(A)	DNA (μg)	% of total DNA
Zoospore homogenate	15 808	100	100	280.8	100
Low speed supernatant (extra cap)	2860	18.1	19.8	237.2	84.5
Nuclear cap pellet	10 680	67.6	80.2	14.6	5.2
Nuclear cap lysate supernatant	10 680	67.6	61.8	7.7	2.7
Nuclear cap lysate pellet	604	3.8	15.8	4.7	1.7

Washed zoospores (3.68×10^9 cells) were fractionated into nuclear cap pellet and extra-cap supernatant fractions. The nuclear cap pellet was lysed in 20% sucrose, 10mM Tris-HCl (pH 8.5), followed by centrifugation for 10min at 12 000***g*** to yield the nuclear cap lysate and cap lysate pellet (residue). RNA was isolated from an aliquot of each fraction and the A_{260} determined (μg RNA = $A_{260} \times 40$); poly(A) was measured by hybridization to ^{3}H-Poly(U) and DNA by the fluorometric assay of Hinegardner (1971). (Reprinted from Johnson *et al.* (1977) by courtesy of *Developmental Biology*.)

inactive polysome fraction found in zoospores (Fig. 5.10). Approximately 70% of the nuclear cap poly(A)RNA sediments with the 80S monoribosomes on sucrose gradients (Fig. 5.11), but it is not known whether this is due to actual association with the 80S ribosomes or results from co-sedimentation of the two types of ribonucleoprotein in gradients due to overlapping size and density ranges. Two observations would suggest an arrested but functional association of mRNP with the 80S ribosomes, e.g., as initiation complexes. One is the evidence for a ribosome bound inhibitor and the other is the extremely rapid formation of polysomes following zoospore encystment. In the latter case the level increases from 10–50% in 10 minutes (Fig. 5.6). The stored mRNA of sea urchins, on the other hand, exists as mRNP particles that overlap the monosomes on gradients (Young & Raff, 1979). In urchins these particles have a higher protein : RNA ratio than the newly synthesized RNPs found at early developmental stages. It is an obvious but unproven possibility that the extra protein on these mRNPs could in some way block translation. A very similar situation may exist in *Blastocladiella* zoospores, a possibility of considerable current interest.

5.5.1.1 *Hybridization Studies*

In another approach to the function of the stored mRNA, RNA/DNA hydridization experiments are being used to assess the amount of genetic

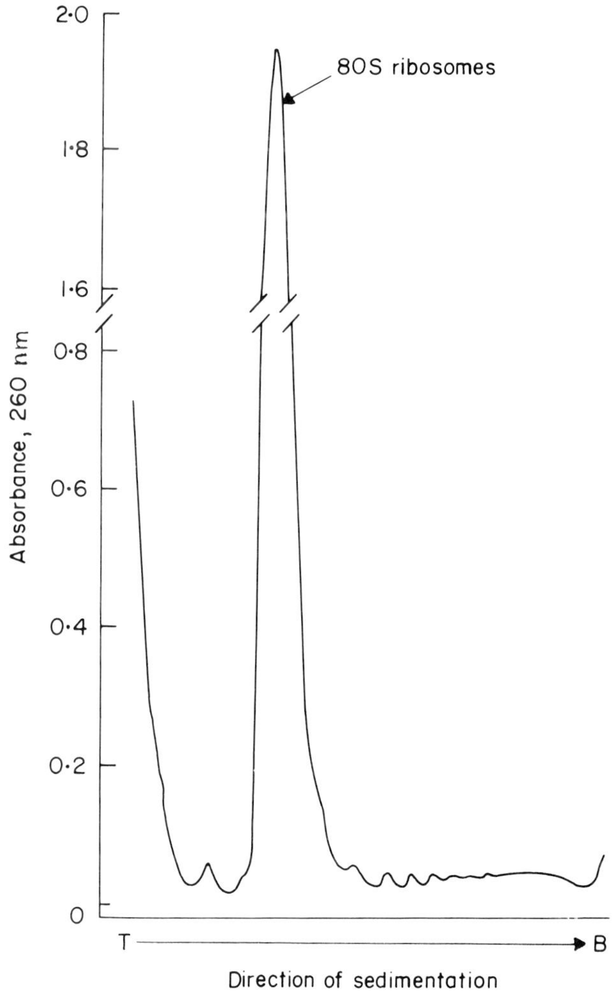

Fig. 5.10 Sedimentation analysis of zoospore nuclear cap ribosomes. Nuclear caps were isolated from zoospores and lysed in buffer containing 25mM EGTA to inhibit ribonuclease activity (Gong & Lovett, 1977). After a low speed centrifugation to remove insoluble material the lysate was layered over a 5ml, 0–32.5% sucrose linear-log gradient for 30min at 189 000*g*, then scanned at 254mn to measure the absorbance. (Lovett, unpublished data.)

'information' stored in the cryptic mRNA of zoospores. Based on DNA/DNA reassociation kinetics the genome of *Blastocladiella emersonii* contains about 2.67×10^{10} daltons of DNA and has a genetic complexity of 3.41×10^7 nucleotide pairs [using *Escherichia coli* DNA as a complexity standard (Davidson, 1976)]. The zoospore messenger fraction when hybridized against the total unique sequence DNA (i.e., total DNA after removal of the repetitive fraction) is

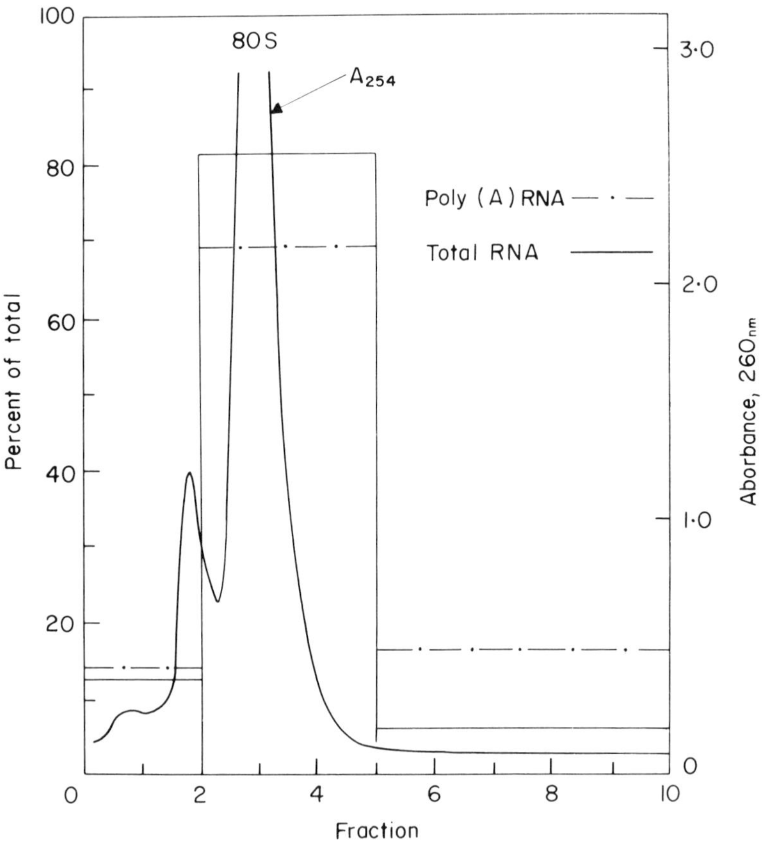

Fig. 5.11 Gradient analysis of messenger ribonuclear proteins from zoospore nuclear caps. Nuclear caps were isolated and lysed as described in Fig. 5.10 and the lysate centrifuged over a 10–35% linear sucrose gradient at 49 250*g* to place the monoribosomes and polysomes near the bottom of the tube. The monosome plus polysome region was collected and pelleted, then recentrifuged over a 0–32.5% linear-log sucrose gradient for 90min at 189 000*g*. The gradient was scanned at 254mn and the RNA isolated from pooled fractions, as indicated by the histogram; the percent of the RNA present as total RNA (——) and poly(A)RNA (–·–·–) was determined by oligo(dT)-cellulose chromatography. (Reprinted from Johnson *et al.* (1977) by courtesy of *Developmental Biology*.)

found to represent 1.45×10^7 nucleotide pairs of DNA base sequence complexity. This is equivalent to 42% of the complexity of the unique sequence DNA, a surprisingly large fraction. Using this figure and the modal value of 4.25×10^5 daltons (1288 nucleotides) for the molecular weight of the zoospore poly(A)RNA, one can estimate that the unique sequences stored in this mRNA pool represent approximately 11 600 different messages. The kinetics of the hybridization reactions further show that 95% of the DNA complexity represented is hybridized by only 4.8% of the spore poly(A)RNA. The remaining

95.2% of the poly(A)RNA hybridizes with only 5% of the unique sequence complexity. This result suggests that the zoospores store very many copies of a small number of mRNAs and very few copies each of a much larger number of different mRNAs. The presence of a redundant low-complexity, high-frequency class(es) of stored messengers does not alter the fact that a large number of different messages are stored. Rather, it means that a significant fraction of the total stored mRNA molecules is made up of such multiple copy sequences.

5.5.2 EXPRESSION OF STORED MESSENGER RNA

The amount of the stored mRNA-encoded information that is actually used during early germination has been estimated by labelling experiments. In this case zoospores were germinated in two cultures. One contained Actinomycin D to prevent new mRNA synthesis and one did not. At intervals during germination aliquots of cells from each culture were removed and pulse-labelled for 15 minutes with [^{14}C]-leucine to label any proteins being synthesized in the pulse interval. The soluble proteins were then extracted and separated by two-dimensional gel electrophoresis to visualize the individual labelled polypeptides (O'Farrell, 1975). Radioautogrammes were then prepared to locate the labelled proteins and these were counted. At 30 minutes (labelled during the 0–30 minute interval) 132 radioactive polypeptides were detectable in extracts from the control culture without added antibiotic. Most interestingly, all of the same proteins except two minor ones were also produced (labelled) by cells germinating in the presence of actinomycin. The total number of proteins detected by this method represents a minimal estimate of the number actually produced, for two reasons. First, because the fractionation does not recover and separate all proteins (particularly basic ones), and second, because proteins produced in very small amounts might fail to incorporate sufficient radioactivity to register on the film with the pulse and film exposure periods chosen. However, with this qualification it is evident that all but two of the 132 proteins must have been translated from stored messengers. Moreover the number of sequences expressed is quite small. The 130 proteins would require only 1% of the 11 600 or so different mRNAs estimated to be stored in the zoospores. The most probable explanation is that the 130 proteins are ones that are translated from messengers representing the predominant low-complexity, high-frequency class mentioned earlier. It thus seems reasonable to propose that two groups of proteins are represented, a set specific for germination (the 'developmental program'), and another set required to amplify the basic cellular processes of metabolism and biosynthesis. The latter suggestion would be consistent with the high levels of mRNA for glycolytic enzymes that have been found in yeast cells (Holland & Holland, 1978) and the increased levels of such enzymes that have been reported to appear during germination of other fungal spores (Ohmori & Gottlieb, 1965).

Until the early proteins are identified it will be impossible to assess the validity of the above interpretation, or, the relative numbers of developmentally-important proteins that are produced in the first 30 minutes. Since some protein synthesisis is essential for development beyond the encystment stage while mRNA synthesis is not, the two proteins missing in actinomycin-D-treated cells must not have an essential *early* function. There is virtually no information that would indicate the number or kinds of proteins that might be developmentally essential. Because so many proteins are translated from the stored mRNA the approach with general inhibitors used to date seems clearly inadequate to identify the proteins of interest. The identification of such proteins will require the much more difficult analysis of individual polypeptides with known functions, work that has hardly begun. Chitin synthesis is essential for cyst wall formation and germ tube growth, but all of the necessary enzymes pre-exist in zoospores (Selitrennikoff *et al.*, 1976). The tubulin proteins have also been examined on the assumption that microtubules might be required for rhizoid elongation. Again, however, the results do not show new microtubular protein synthesis (α and β tubulin subunits) to be required as an early event in germination. In fact, the tubulin pool is actually expanded by disassembly of the intracellular microtubules during zoospore encystment and new synthesis could not be detected until 30 minutes. Hybridization experiments using mRNA from the first-formed polysomes (i.e., stored mRNA in the process of translation) might provide a more accurate estimate of the contribution of stored mRNA complexity to new translation. However, this approach also would not serve to identify specific proteins or their functions, although it could be used to test the hypothesis that zoospores store a spectrum of mRNA molecules from which different subsets are selected under different environmental conditions.

5.6 CONCLUSIONS

A considerable amount of information is now available concerning *Blastocladiella* zoospores and the developmental process of germination. Much remains to be discovered, but a better understanding of zoospore function has emerged. Despite this progress we still know far too little about how both stages are regulated. These propagative cells, though actively motile to allow dispersal, have apparently complete and functional, but stringently-arrested, systems for macromolecule synthesis. The latter is a characteristic zoospores share with other dormant and semidormant systems such as seeds and animal eggs. The tiny zoospores, again like eggs, are preprogrammed with all of the 'information' (i.e., mRNA) that is required for early development; what is here called the 'developmental program'. It seems probable that they also carry a wide variety of additional preprogrammed messenger molecules that can be used to increase the general levels of metabolic and biosynthetic enzymes. More speculatively they may be preprogrammed to meet a range of environmental demands. In this regard they may be rather different from seeds and animal eggs, since, unlike these closed systems, zoospores do not carry a large reservoir of reserve materials to support extensive development. This makes the prompt establishment of a nutritional base a critical aspect of survival. One obvious 'value' of a preprogrammed developmental sequence for germination should be the capacity it provides for the very rapid production of a germ tube. That is, to penetrate quickly the organic substrate upon which it will grow and for which it must compete with other microorganisms. This stage in *Blastocladiella* germination correlates with the depletion of glycogen reserves (Suberkropp & Cantino, 1972) and the ability to undergo induced microcyclic sporulation (for which competence is achieved only after 30 minutes of growth; Hennessy & Cantino, 1972). The correspondence between these events suggests that the 40–50 minute stage could be a balance point. That is to say, the cells at this stage may in some way assess the sufficiency of the environment to support further growth. They could then proceed to grow if adequate nutrients are available, or switch to the alternate sporulation 'developmental program' if they are not. In this regard an analogy might be drawn to the arrest of yeast cells at the G_1 phase of the cell cycle when nutrients are depleted (Unger & Hartwell, 1976). In *Blastocladiella* the postulated assessment mechanism could well be related to the maximal rates of intracellular RNA and protein synthesis that can be sustained by the available nutrient supply. These rates are very sensitive to the supply of external precursors (Lovett, 1975). The useful ability to decamp via zoospore production when nutrients become limiting exists throughout the growth phase of *Blastocladiella* and probably most other aquatic fungi as well.

FURTHER READING

Brambl R., Dunkle L. D. & van Etten J. L. (1978) Nucleic acid and protein synthesis during fungal spore germination. *The Filamentous Fungi*, vol. 3, ch. 6, pp. 94–118, (eds Smith J. E. & Berry D. R.) John Wiley & Sons, New York.

Lovett J. S. (1975) Growth and differentiation of the water mold *Blastocladiella emersonii*: Cytodifferentiation and the role of ribonucleic acid and protein syntheses. *Bacteriol. Rev.*, **39**, 345–404.

O'Day D. H. & Horgen P. A., (eds) (1977) *Eukaryotic Microbes as Model Developmental Systems*, 440pp. Marcel Dekker, New York, Basel.

Smith J. E. & Berry D. R. (1974) *An Introduction to Biochemistry of Fungal Development*, 326pp. Academic Press, New York.

Weber J. & Hess W. M. (1976) *The Fungal Spore : Form and Function*, 895pp. John Wiley & Sons, New York.

CHAPTER 6

ACETABULARIA

6.1 INTRODUCTION

One of the landmark discoveries in cell biology was made around 1930 when J. Hämmerling observed that the green alga *Acetabularia* is not only unicellular, but is also uninuclear during the vegetative phase of the life cycle. *Acetabularia* cells may attain enormous size (up to 200mm in length), exhibit polarity with respect to distribution of organelles and macromolecules, and, perhaps most importantly, readily survive various merotomy experiments.

Merotomy experiments that divide a single cell into several fragments have shown that the basal fragment containing the nucleus is able to regenerate a complete cell and, in addition, that the anucleate apical fragment is also capable of significant morphogenesis. Individual nuclei from several easily morphologically-distinguished species may be readily isolated from contaminating cytoplasm and have been shown to survive implantation into homologous and heterologous cytoplasms. Thus, *Acetabularia* represents an eminently suitable system for the study of the relationship between nucleus and cytoplasm.

The study of nuclear-cytoplasmic relationships was continued in experiments on regulation of enzyme activities in nucleate and anucleate cells, as well as in heterologous combinations. Additional impetus was gained around 1960, when evidence was accumulated that chloroplasts and mitochondria possess their own genetic information, distinct from that of the nucleus.The mutual interaction between nuclear and organellar genomes has attracted the attention of a number of groups.

6.2 THE LIFE CYCLE

The life cycle of an individual *Acetabularia* cell begins with the fusion of two morphologically identical, but physiologically distinct gametes (Fig. 6.1.I). At the moment of fusion, the structural polarity of the cell is established; the outgrowth of the apical tip of the zygote occurs towards the light and that of the rhizoid takes place closest to the nucleus. The rhizoid anchors the cell to the substrate. Since the zygote has only two chloroplasts and mitochondria, the initial rate of growth is very low.

At about the 10mm stage (Fig. 6.1.IV), the cell forms rings of branched hairs, the sterile whorls. Development of whorls and renewed growth of the stalk cycle occur repeatedly, in succession, until the species-specific cap is developed

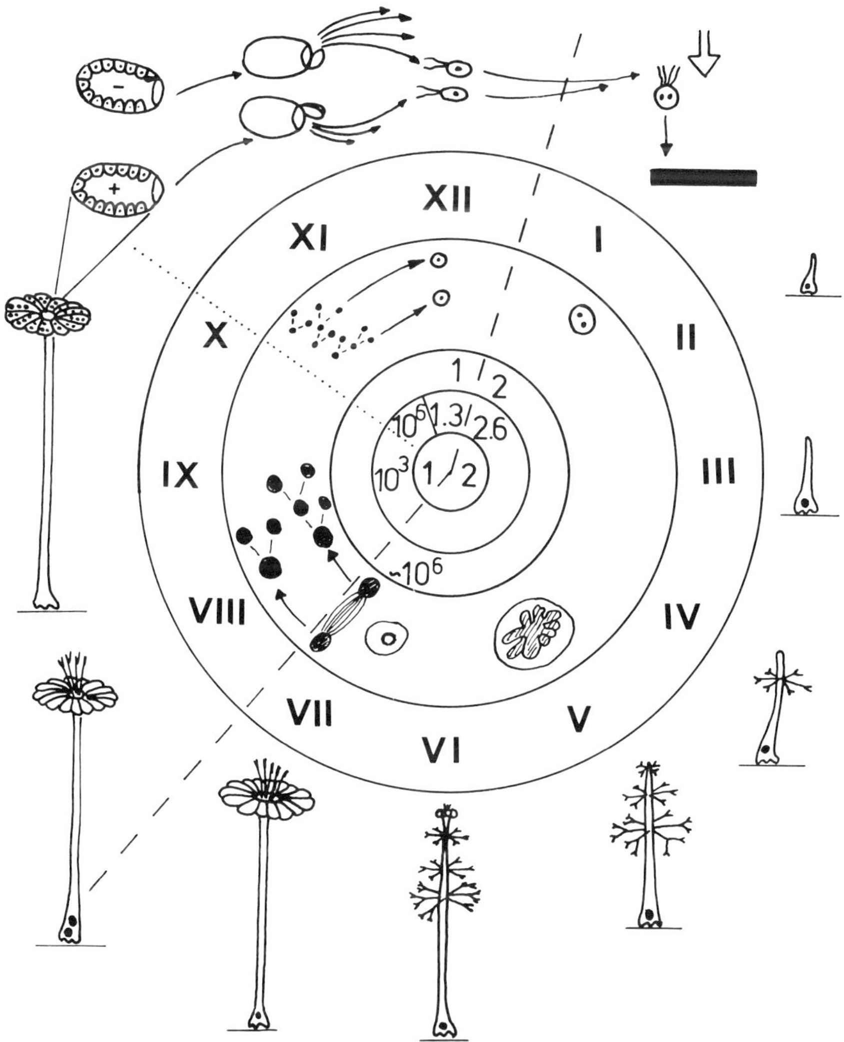

Fig. 6.1 Life cycle of *Acetabularia.* The concentric circles show, starting from the centre: 1. The relative amount of DNA per nucleus. 2. The absolute amount of nuclear DNA per cell in pg. 3. The number of chloroplasts and mitochondria per cell. 4. The nuclear cycle, in particular the increase in nuclear volume and nucleolar mass from stage I to V, the reduction of nuclear mass prior to stage VII, the first nuclear division and reduction of DNA content per nucleus at stage VII, and the division of secondary nuclei (stages VIII to XI, the different size of these nuclei does not reflect the natural situation). 5. The specification of stages used in the text. 6. Schematic drawings of the life cycle. The open arrow indicates the direction of light, the dark bar represents the substrate. The dashed line separates the haploid and diploid phases. Further explanations are given in section 6.2.

(Fig. 6.1.VI, VII). The cap is built up of a number of chambers or rays connected with each other and the stalk via small channels. The number of rays and the form and shape of the corona superior and inferior are important taxonomic criteria. Figures 6.2a and 6.2b show the species-specific characteristics of *A. mediterranea* and *A. cliftonii.*

Shortly after the cap reaches its maximum diameter, the primary nucleus divides (stage VII). Each of the secondary nuclei may then undergo approximately 10 additional divisions such that about 10^3 secondary nuclei are formed. These secondary nuclei migrate through the stalk into the cap rays, where they become anchored by microtubules (Woodcock, 1971) and associated with a number of organelles. They will later be segregated from the rest of the cytoplasm by a membrane and a cyst wall (gametangium, Fig. 6.1.IX and X). The enclosed nuclei divide again and isogametes, which require a period of maturation before they are released from the cyst, are formed. The gametes of one cyst are of the same mating type and may be induced to form haploid monster cells. Normally, the gametes of two different mating types fuse to form a zygote, which becomes negatively phototactic and adheres to the substrate. The cycle is completed and up to 10^6 zygotes may be obtained from a single cell.

The life cycle normally requires three to four months in laboratory conditions. Laboratory culturing of *Acetabularia* is relatively easy and has been extensively described (Berger & Schweiger, 1980). For further details of the life cycle and references, see Koop (1979).

6.3 THE BASIC EXPERIMENTS

These experiments, originally designed by J. Hämmerling about 50 years ago, still form the basis for biochemical and cell physiology experiments with these algae.

6.3.1 REMOVAL OF THE NUCLEUS

Since the nucleus lies within the rhizoid, it may be removed simply by amputating the rhizoid with a pair of scissors: thus, a nucleate and an anucleate cell fragment are obtained (Fig. 6.3a). Both fragments develop quite normally until stage VII in Figure 6.1, when development in the anucleate cell stops, although it may remain alive for another four to eight weeks. This simple experiment supports the following conclusions. Anucleate cells may function, grow and develop over a long period of time. Their development is temporally and spatially controlled and is very similar to that of the nucleate cell, even though no nuclear DNA is present. (But, as we shall see, the information for the morphogenetic events that occur during development originally and largely stems from nuclear DNA). Transcriptional control of gene expression plays a minor role in the development of these algae, at least up to the stage of cap formation.

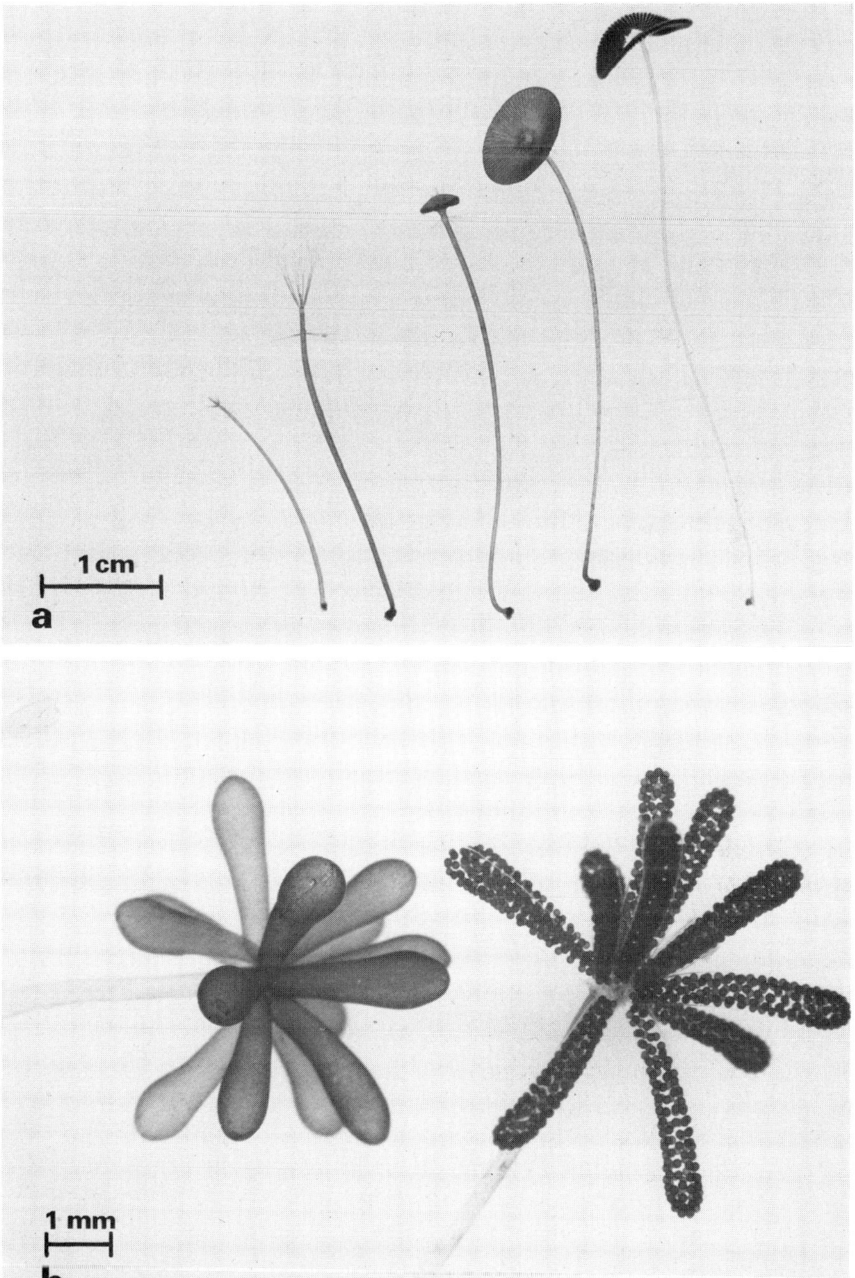

Fig. 6.2 Species-specific characteristics of a. *Acetabularia mediterranea* (photographs of stages IV to IX) and b. *Acetabularia* (Polyphysa) *cliftonii* (details of caps corresponding to stages VII and IX). (Photographs by courtesy of Dr. S. Berger, Ladenburg.)

6.3.2 FRAGMENTATION

The anucleate fragments, which in some species may be 20cm long, can be subdivided into segments that exhibit a developmental potential that reflects their original position in the stalk (Fig. 6.3b). The most apical fragment is capable of forming a cap, whereas the most basal fragment exhibits no growth and development. The failure of the basal fragments to grow and develop is not the result of a general absence of cellular activities, as these fragments synthesize organellar RNA and photosynthesis and chloroplast division have been observed (Clauss *et al.*, 1970).

6.3.3 TRANSPLANTATION AND IMPLANTATION EXPERIMENTS

The mechanical stability of the fragments permits the isolation and recombination of various cell parts. It is possible to prepare homologous and heterologous combinations. In the heterologous combinations where the rhizoid, containing the nucleus, is from *A. mediterranea* and the anucleate stalk is from *A. cliftonii* and vice versa, the cap formed is that of the species of the nucleus (Fig. 6.3.c, middle), whereas the anucleate fragment alone will form a typical, species-specific cap (Fig. 6.3c, left and right).

From these experiments, Hämmerling concluded that morphogenetic substances (MS) originally present in the anucleate cell become neutralized and supplanted by the morphogenetic substances produced by the transplanted nucleus. As can be concluded from figure 6.3a, MS are species-specific, nuclear dependent and normally released weeks before they become manifest. The possibility still existed that the small amount of cytoplasm of the rhizoid, together with the nucleus, could have influenced morphogenesis. In order to exclude this possibility, the nucleus was removed from the rhizoid, purified from the adhering cytoplasm and reimplanted into an anucleate fragment (Fig. 6.3c, lower part, implantation). The results from the implantation experiments were identical with and confirm the conclusions drawn from the transplantation experiments.

The implantation experiments are especially important in the light of the fact that chloroplasts and mitochondria possess their own DNA. Thus, it can be seen that heterologous combinations of nuclear and organellar genomes are compatible and functional and this type of experiment thereby provides the first evidence that the information of the organelles must be relatively constant between different species.

6.3.4 THE NUCLEATE BASAL FRAGMENT

As mentioned in 6.3b, the basal fragment normally does not grow and differentiate if it has no nucleus. If, however, the cell is fragmented such that the nucleus remains in the basal fragment for a few days before the rhizoid is removed, the

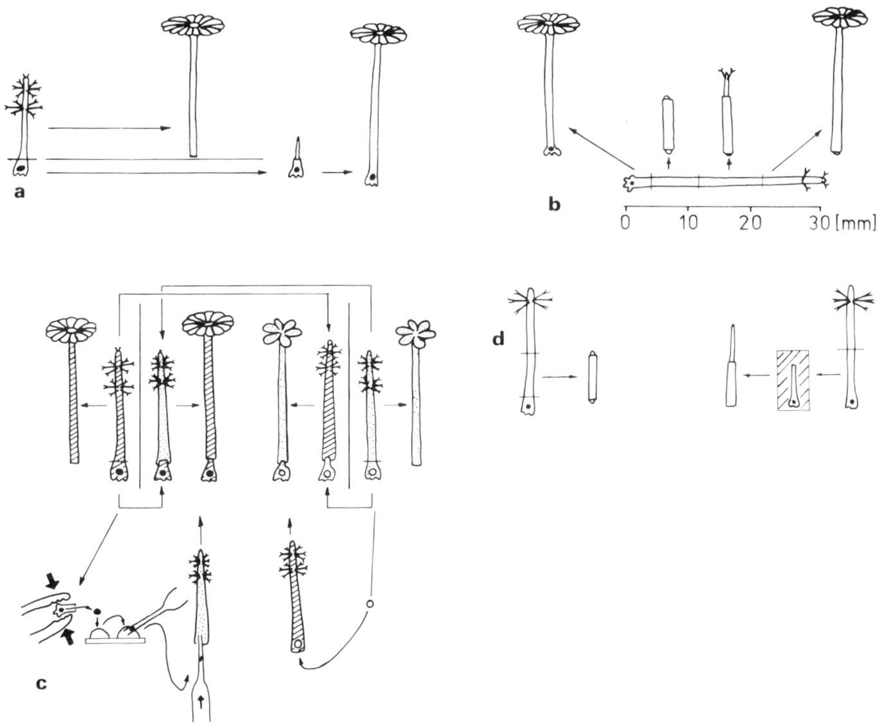

Fig. 6.3 Schematic drawings of the basic merotomy, transplantation and implantation experiments. The experimental details are described in Section 6.3.

now anucleate basal fragment will grow (Fig. 6.3d). Therefore, this system can be used to describe the kinetics of transfer of MS from the nucleus to the cytoplasm. This process also occurs in darkness, during which the nucleus is diminished in size, presumably because of export of material to the cytoplasm. These basal fragments are relatively stable and have also been used in electrophysiology experiments.

6.3.5 RECONSTITUTION EXPERIMENTS

In these experiments, performed by Sandakhchiev and his group (1972), the localization of the MS in the cortex of the cell has been demonstrated. The cortex is a plasma layer adjacent to the plasma membrane and remains *in situ* when the cell contents are centrifuged into the rhizoid at low centrifugal forces. If the centrifuged cytoplasm is reinjected into the empty stalk, regeneration will only occur when the cytoplasm is introduced into apical fragments, regardless of the site of origin of the cytoplasm. The centrifuged apical fragments, devoid of chloroplasts and mitochondria, are not capable of growth and differentiation.

6.4 A WORKING HYPOTHESIS: MORPHOGENETIC SUBSTANCES ARE IDENTICAL WITH MESSENGER RNA

Different groups (Brachet, 1960; Hämmerling, 1963) originally made suggestions as to the molecular basis of the morphogenetic substances based on a number of individually inconclusive experimental hints, such as sensitivity of MS to ribonuclease and UV irradiation. They concluded that the MS might be identical with mRNA that possesses extraordinary stability to account for the long-term persistence of information in the anucleate cell. This was a very unusual suggestion to make in the light of the ideas about mRNA at that time.

The long-term stability was postulated to be achieved by compartmentalization, such as might be brought about by the combination of the mRNA with basic proteins and thus forming messenger-ribonucleoprotein particles. The mRNA was predicted to be arrayed in an apico-basal gradient, as postulated for MS.

6.5 PRESENT KNOWLEDGE: CELLULAR COMPARTMENTS

The following list is an attempt to describe the morphology and activity of the organelles and cellular compartments during growth and development. This list, of course, will be incomplete since some organelles have attracted more interest than others.

6.5.1 THE NUCLEUS

It is apparent that the single nucleus must show considerable alterations in its activity and structure while the volume of the cell is changing from $10^2 \mu m^3$ to $10^{10} \mu m^3$, as it does in *A. mediterranea*. The morphology of the nucleus is altered as the volume of the nucleus changes by a factor of 4×10^5 between stage I and VI and the number of nuclear pores increases from 2.5×10^3 to more than 2×10^6 (Zerban *et al.*, 1973; Franke *et al.*, 1974), indicating a tremendous increase in exchange between nucleus and cytoplasm. The nucleus is surrounded by a cytoplasmic layer, the perinuclear cytoplasm, which is free of chloroplasts and mitochondria and is clearly separated from the rest of the cytoplasm. The perinuclear cytoplasm is connected with the cytoplasm by secondary pores. These pores are fewer in number than the nuclear pores and probably also function in regulation of transport between the nucleus and cytoplasm. The perinuclear cytoplasmic layer increases in the thickness during development (Fig. 6.4).

The amount of DNA in the zygote nucleus has been determined to be about 2pg, which is somewhat higher than that of other algae, such as *Chlamydomonas* (Spring *et al.*, 1978). The huge *Acetabularia* nucleus is therefore diploid or, at most, oligoploid until the first division of the nucleus. Because of a tremendous increase in nucleolar mass, it is not possible to determine if minor changes in the amount of DNA occur during development, such as amplification of DNA coding for rRNA (rDNA). This increase in nucleolar mass is due to the expression of the huge number of rRNA genes, which have been estimated by different methods to be on the order of 25 000 copies in *A. mediterranea* (Trendelenburg *et al.*, 1974) and 40 000 copies in *A. major* (Berger & Schweiger, 1975a). Figures 6.5a and b show examples of rDNA. It is unclear whether these genes are exclusively redundant or whether they are amplified during development. By hybridization methods, Spring *et al.* (1978) found 1900 copies of rDNA in the gamete genome. They postulate that this number of genes is sufficient to account for the rate of synthesis of 80S-type rRNA, which is at least 1.7ng per cell per day (Kloppstech & Schweiger, 1975a).

The rRNA genes have attracted great interest because of their extreme variability between species. Not only does the length of the cistrons and spacers vary, but also the mode of transcription may be unidirectional (Fig. 6.5a) or bidirectional (Fig. 6.5b; Berger *et al.*, 1978). A close correlation has been found between the length of the cistrons and the primary transcripts of rRNA (Kloppstech *et al.*, 1977).

From its morphology and biochemical data, it can be concluded that the nucleus remains active until stage VII. Thereafter, dramatic changes occur in its morphology. The first nuclear division is preceded by an intensive reduction of nucleolar mass and export of material to the cytoplasm. The first division is presumed to be reductive, the mode of division is intranuclear (Koop *et al.*, 1979) and a residual body is eliminated. The number of chromosomes per haploid genome is apparently ten.

The replication of DNA parallels the nuclear divisions, as indicated by the fact that no increase in DNA is detected during the vegetative phase, after which the amount of DNA per cell is multiplied by a factor of 10^6 within a few days. The coincidence of DNA replication and nuclear division is supported by the increases in phosphorylation of thymidine and activity of one of the phosphorylating enzymes, which parallel the formation of the secondary nuclei (Bannwarth & Schweiger, 1975).

6.5.2 CHLOROPLASTS

The number of chloroplasts increases exponentially from two in the zygote to 10^7 or more in some species. They are found to be arranged in an apico-basal gradient. Small chloroplasts, rich in lamellae, reside in the extreme apex and in the neighbourhood of the nucleus, indicating an interaction between the nucleus and the developing chloroplast. On the other hand, the size of the chloroplasts and the amount of starch increase from the apical to the basal end. The nucleus is therefore surrounded by a population of very old and very young chloroplasts (Puiseux-Dao, 1970).

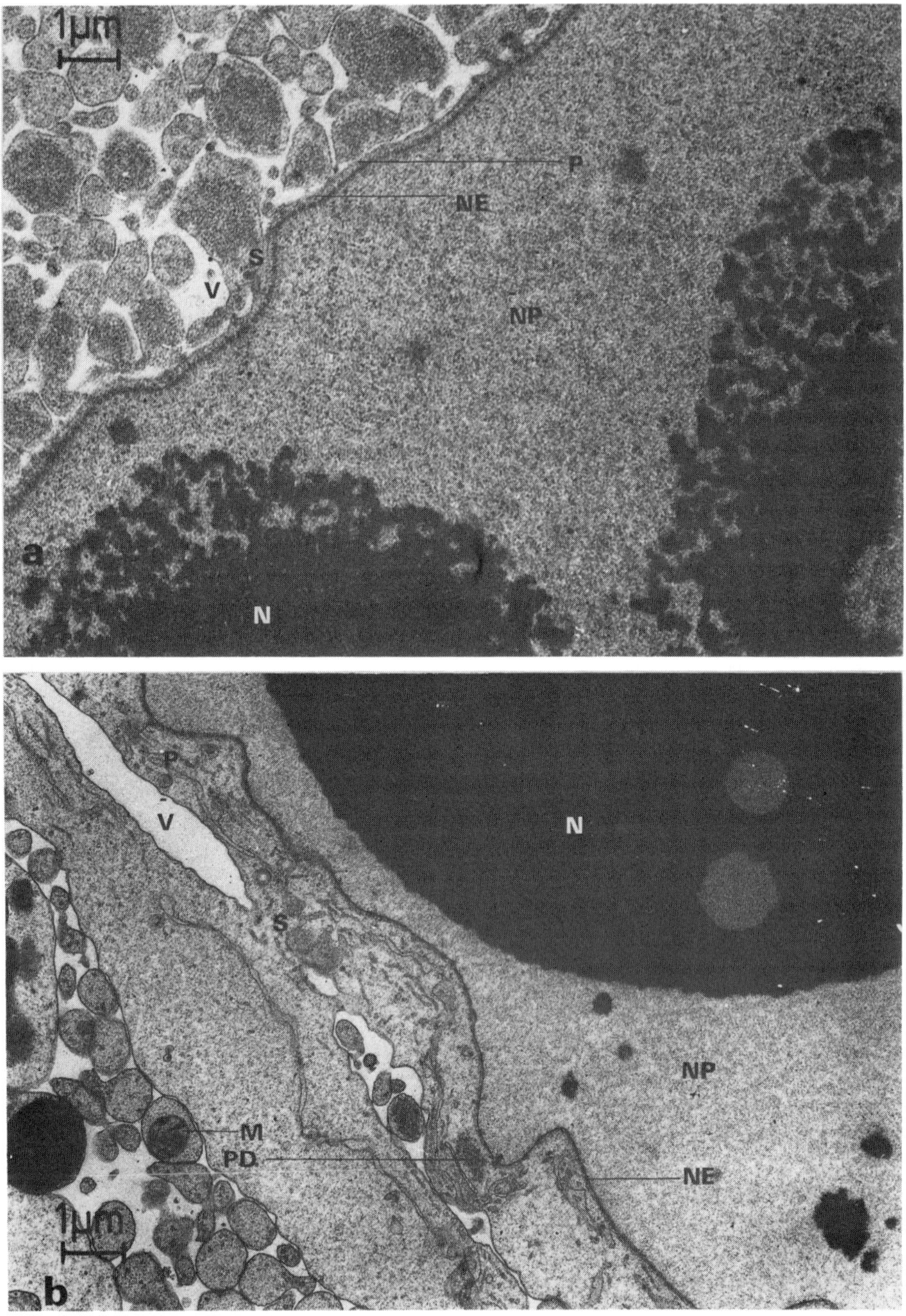

Fig. 6.4 Electron micrographs of nuclei from a, a young and b, an old cell of *Acetabularia mediterranea*. Note the differences in the thickness of the perinuclear cytoplasmic layer and in the structure of chromatin. N, nucleolar mass; NP, nucleoplasm; NE, nuclear envelope with pore complexes; P, perinuclear cytoplasmic layer; S, secondary pore; V, vacuole; PD, perinuclear dense body; M, mitochondrion. (Photographs by courtesy of Dr. S. Berger, Ladenburg.)

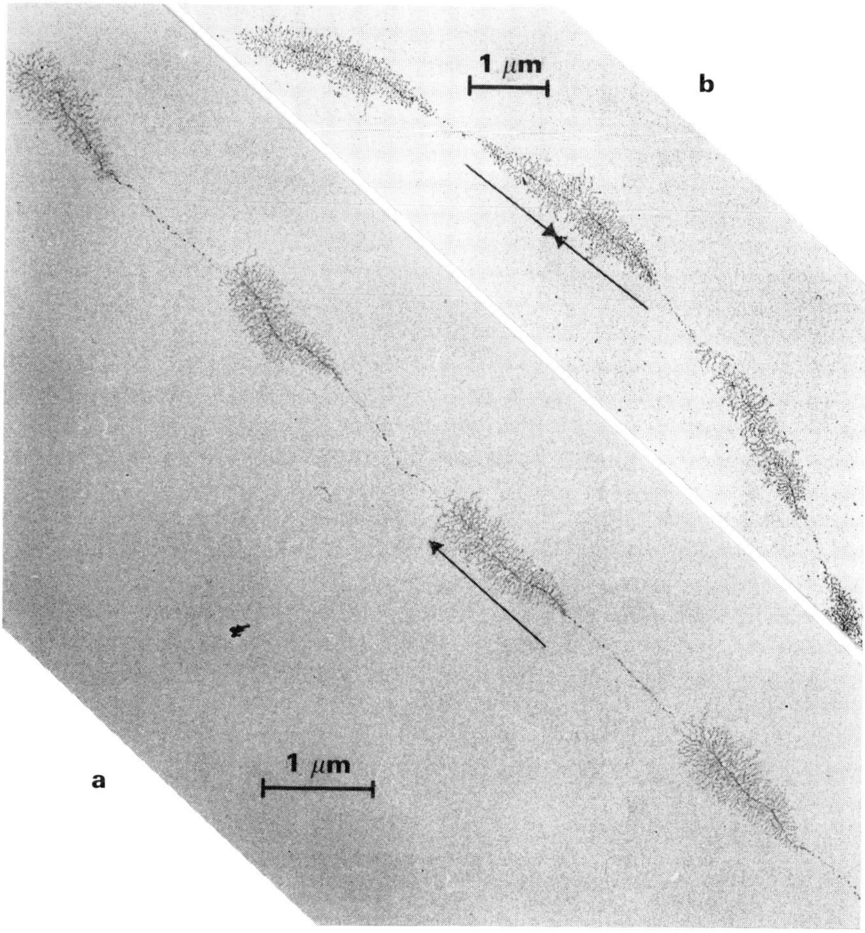

Fig. 6.5 Genes for ribosomal RNA from *Acetabularia crenulata* (a) and *Acetabularia exigua* (b). Nucleolar material was spread according to the technique of Miller. Note the unidirectional mode of transcription in *A. crenulata* and the bidirectional mode of transcription in *A. exigua* (arrows). (Photographs by courtesy of Dr. S. Berger, Ladenburg.)

Other important observations include circadian rhythmicity in chloroplast transport processes (Koop *et al.*, 1978) and structural changes in chloroplasts during the day, as observed by electron microscopy (Vanden Driessche, 1966).

Isolated chloroplasts have been shown to synthesize RNA and protein *de novo*. This is further support for the existence of DNA in the *Acetabularia* chloroplasts. Thus, they are similar to other chloroplasts, with two important exceptions. First, not all chloroplasts contain the same amount of DNA during the vegetative phase; however, prior to the formation of secondary nuclei, all chloroplasts possess detectable amounts of DNA. This is as expected, since only one chloroplast is present in a gamete, but no plausible explanation of this phenomenon has been offered. Second, the kinetic complexity of chloroplast DNA has been determined to be 1.5×10^9 daltons (Padmanabhan & Green, 1978), which is several times higher than that for all other algal and plant chloroplasts investigated. This value may be overestimated by a factor of two due to methodological problems and the presence of mitochondrial DNA in the chloroplast preparations has not been excluded. Re-examination might eventually yield smaller complexities.

The only identified gene product of chloroplast DNA in *Acetabularia* is ribosomal RNA, which has been determined to be 100ng per cell, which corresponds to 10^4 molecules per chloroplast (see Chapter 9 for a discussion of the role of chloroplast DNA in higher plants).

Species-specific differences have been demonstrated in the electrophoretic mobility of the thylakoid membrane proteins (Apel & Schweiger, 1972) and ribosomal proteins of the large subunit of the 70S chloroplast ribosomes (Kloppstech & Schweiger, 1973).

The composition of the chloroplast membranes has been intensively investigated by Apel (1976). An EDTA-insoluble membrane fraction binds most of the chlorophyll and photosystem I- and II-chlorophyll complexes have been isolated. The major chlorophyll a- and b-binding complex of photosystem II may be isolated in a complex with an apparent molecular weight of 67 000 and can be dissociated into two proteins of 21 500 and 23 000 daltons. The smaller protein binds the chlorophyll (light harvesting chlorophyll a/b protein, LHCP) and is an integral part of the thylakoid membrane, whereas the larger protein is accessible to pronase digestion. Both proteins are very similar in their amino acid composition, yield similar peptide maps, and react with the antibody raised against the 21 500 dalton protein species.

6.5.3 CYTOSOL

The cytosol will be considered separately, since it is segregated from the other organelles discussed so far by a lipid bilayer or unit membrane. It is the compartment where protein synthesis on 80S ribosomes takes place and which mediates communication between the nucleus and organelles (Section 6.6.1) and among the organelles themselves (Section 6.6.2). As will be shown, the cytosol in these algae is polar in its organization. Cytoplasmic streaming has been observed to occur at two different velocities: chloroplasts are transported at 1–$2 \mu m\ s^{-1}$ and secondary nuclei move in 'head-streamed bands' at 3–$11 \mu m\ s^{-1}$. The streamings also differ in the mechanisms of force generation (Koop & Kiermayer, 1980).

6.5.4 CELL MEMBRANE AND CELL WALL

Together with the cortex, the cell membrane and cell wall may be considered to be a functional entity. This is supported by studies on cell wall formation and from the developmental changes of stalk to cap to cyst formation, as well as from experiments on electrophysiology (Section 6.6.2.3). The composition of the cell wall changes during the transition from the uninuclear to the multinuclear stage. The wall of the stalk is composed mostly of mannans, whereas the cyst wall contains up to 80% cellulose. This alteration in composition is reflected in the activities of mannan synthetase and GDP-mannose-pyrophosphorylase. Both enzymes have their highest specific activities in the apical region and their distribution parallels that of incorporation of precursors into the cell wall. Furthermore, the activities of both enzymes decline drastically when cysts are formed (Bachmann & Zetsche, 1979). It is as this time that the activity of UDPG-pyrophosphorylase increases (Zetsche, 1968).

A prominent protein of 14 000 daltons has been found to be closely associated with the polysaccharides of the cell wall in *A. cliftonii* and may influence cap morphogenesis, although its biological function has not been clearly defined (Paradies *et al.*, 1977).

6.6 PRESENT KNOWLEDGE: INTERRELATIONSHIPS BETWEEN THE ORGANELLES

Although not yet fully documented, it is quite apparent that all these interrelationships are bidirectional.

6.6.1 NUCLEUS-CYTOPLASM

The importance of this interaction has been generally discussed in the Introduction.

6.6.1.1 *Morphology of the nucleus and developmental stage of the cytoplasm*

In his merotomy and transplantation experiments, Hämmerling showed that the cytoplasm may influence the nucleus. Repeated amputation of regenerated caps results in a practically indefinite delay of the division of the primary nucleus. Transplantation of rhizoids with young nuclei (stage IV) into cap-bearing anucleate fragments leads to the precocious division of the nuclei, as compared to the controls. On the other hand, combinations of nuclei from cap-bearing cells and young anucleate fragments resulted in delays in nuclear division, indicating a rejuvenation of the nucleus.

Recently, further experiments have shown that changes in the ultrastructure of the perinuclear cytoplasm, described in Section 6.5.1, as well as changes in the structure of chromatin, occur prematurely in the presence of old cytoplasm and are delayed in the presence of young cytoplasm (Berger & Schweiger, 1975b).

Removal of the greater part of the cytoplasm of the cell leads to a transient tenfold increase in the rate of transcription of 80S-type rRNA. These experiments clearly indicate that the cytoplasm influences or controls the activities of the nucleus. The molecular nature of the controlling factor or factors is unknown.

6.6.1.2 *Nuclear activity : 80S ribosomes*

The influence of the nucleus on the cytoplasm has been studied in great detail, as, for example, in the case of the aforementioned transplantation experiments.

A mature cell of *A. major* contains approximately 200ng of 80S ribosomes, which are synthesized during a two month period. This corresponds to a total of 3×10^{10} ribosomes, which are synthesized at a rate of $6000 s^{-1}$. Considering the number of nuclear pores, the transport of ribosomes from the nucleus to the cytoplasm must be a very slow event and require about six minutes per ribosome. Afterwards, they are transported to the apex of the cell at a speed faster than cell growth, but definitely slower than the previously described cytoplasmic streamings. Thus, a novel transport mechanism for ribosomes exists or ribosomes bind to membranes during the process of transport (Kloppstech & Schweiger, 1975a). These results suggest that the cytoplasm is highly ordered and organized, as otherwise the pulse-labelled ribosomes would appear evenly distributed in the cell after some time.

6.6.1.3 *Nuclear activity : messenger RNA*

As discussed in Section 6.4, it has been postulated that the morphogenetic substances are actually mRNA with high stability, arrayed in an apico-basal gradient and coded for by the nuclear genome. A large percentage of eukaryotic mRNA is polyadenylated at the 3′ end and may be readily isolated by affinity chromatography. Polyadenylated mRNA (poly(A)mRNA) has been isolated from *Acetabularia.* It is synthesized in the nucleus, is heterogeneous in size, is of relatively high molecular mass with an average of 1.5×10^6 daltons and its rate of synthesis declines as development proceeds.

If these high molecular weight mRNAs are assumed to be monocistronic, it can be calculated that about 8000 genes are transcribed in young cells and about 2500 genes are transcribed in stage V cells. This mRNA has been found to be highly stable up to stage VI; thereafter, the pulse-labelled mRNA synthesized at stage IV has a half-life of ten days.

Fully grown cells (stage VI) yield 15ng of poly(A)RNA or 6×10^9 molecules per cell. Recalling the estimate of the number of ribosomes per cell, it follows that each molecule of mRNA could be occupied by an average of four ribosomes. Non-polyadenylated mRNA has not been considered in this calculation, since there are no estimates available on the amount of such mRNA. The average number of ribosomes per mRNA molecule is surprisingly low, in light of the high molecular weight of this mRNA. This also lends credence to speculation that part of this mRNA may be stored in an inactive form, as postulated in the hypothesis of identity between MS and poly(A)mRNA. This identity would predict a distribution gradient of poly(A)mRNA in the cell. No biochemical evidence has been found for the existence of a concentration gradient of poly(A)mRNA. Functional gradients have been postulated, but experimental evidence is lacking.

The transport of poly(A)mRNA from the nucleus through the cytoplasm occurs at a higher rate than that of the 80S ribosomes, showing that both are transported separately and not in the form of polysomal complexes (Kloppstech & Schweiger, 1975b).

The messenger properties of the poly(A)mRNA have been confirmed by recent experiments in *in vitro* translation systems. Using the antibody against the light-harvesting chlorophyll a/b protein complex, four bands have been precipitated from the mixture of proteins obtained in the *in vitro* experiment (Fig. 6.6). Two of the bands have molecular weights of 22 000 and 24 000 and correspond to the two mature LHCP species described in Section 6.5.2. The other two bands have molecular weights of 26 000 and 28 000 and, on the basis of their immunoreactivity, are considered to be precursors to the two mature proteins. These results confirm that there are two messengers for the two mature proteins, which have been found otherwise to be very similar with respect to

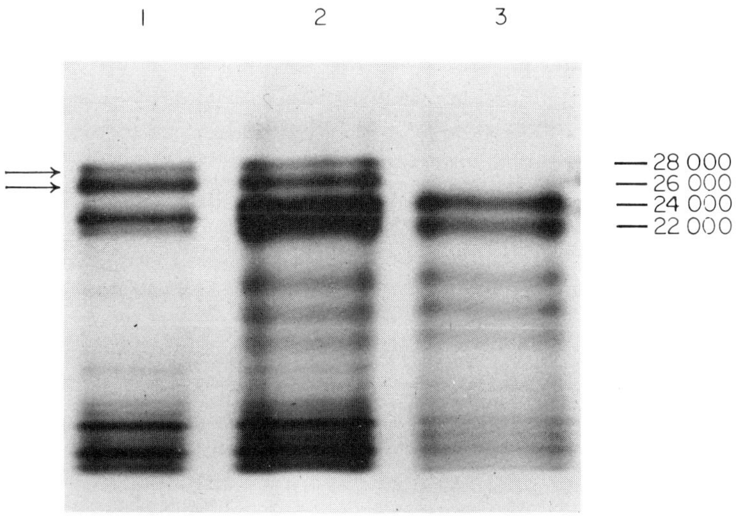

Fig. 6.6 *In vitro* translation products of poly(A)mRNA of *Acetabularia mediterranea* after immunoprecipitation. Poly(A)RNA was translated in the wheat germ cell-free system in the presence of labelled methionine and the *in vitro* products were treated with the antibody raised against LHCP and separated by polyacrylamide gel electrophoresis (track 1). The two light harvesting protein species of 22 000 and 24 000 daltons were labelled *in vivo*, isolated and separated solitarily (track 3) and in a mixture with the *in vitro* products. Note the two precursor bands in tracks 1 and 2 (arrows). The picture is a photograph taken from an radioautograph. (Viro *et al.*, unpublished.)

their primary structure. The mRNA for the LHCP is thought to be transcribed from nuclear DNA and translated in the cytoplasm on 80S ribosomes to give a precursor protein which is subsequently transported into the chloroplasts, processed and inserted into the thylakoid membranes.

6.6.1.4 *Nuclear activity : ribonucleoprotein complexes*

Different groups have attempted to isolate ribonucleoprotein particles (RNP), which might be responsible for the stability of mRNA, in accordance with the informosome theory of Spirin. Alexeev *et al.* (1974) have claimed the induction of morphogenesis by injection of RNP into non-growing basal fragments. Unfortunately, these results have not been confirmed by other laboratories. RNP of the same sedimentation coefficient and buoyant density have been found by others, but these RNP are labelled with radioactive precursors of RNA in the absence of the nucleus and therefore are not identical with the MS. These RNP have some of the properties of mitochondrial ribosomes (Hamilton & O'Brien, 1974).

6.6.1.5 *Organelle proteins coded for by the nuclear genome*

As mentioned in Section 6.3c, when there exist species-specific differences in morphogenetic or molecular properties of some gene product, the origin of the gene is shown to be in the nuclear genome when these properties vary according to the species of the grafted nucleus in heterologous combinations. Such species-specific differences have been found in electrophoretic mobilities of enzymes like phosphatases, malate dehydrogenases of different organellar origin, chloroplast membrane proteins and chloroplast ribosomal proteins (see Schweiger, 1976, for references). In these cases, the grafted nucleus influences the electrophoretic mobilities of the proteins. These results are especially interesting with respect to the question of chloroplast autonomy. It is generally agreed that the chloroplast genome is not sufficiently large to encode for all the chloroplast proteins. Consistent with this view is the observation that the differences in chloroplast ribosomal proteins in two species of *Acetabularia* are coded for by the nuclear genome (Fig. 6.7).

Chloramphenicol, an inhibitor of protein synthesis on 70S ribosomes, does not inhibit the appearance of or affect the sedimentation coefficient of the chloroplast ribosomes. This indicates that most, if not all, of the chloroplast ribosome proteins are coded for by nuclear DNA and synthesized on 80S ribosomes. To date, no chloroplast ribosome proteins have been shown to be coded for by the chloroplast genome.

On theoretical grounds, these results must be judiciously interpreted, as it is not possible to distinguish differences in primary structure and secondary modification of proteins, such as phosphorylation, by merely comparing electrophoretic mobilities. Although not very plausible, differences in mobilities

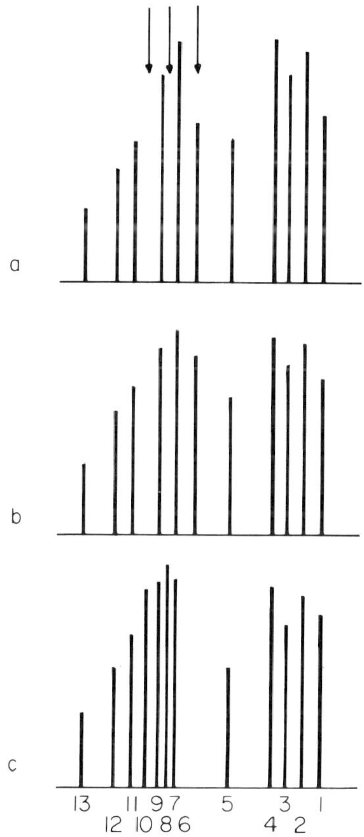

Fig. 6.7 Proteins of the large subunit of chloroplast ribosomes of *Acetabularia mediterranea* (a), *A. cliftonii* (c), and the F_1-generation of anucleate *A. cliftonii* cells to which the nucleus of *A. mediterranea* had been implanted (b). Note the species-specific differences between the two species (arrows) and the conformity between the heterologous cells and *A. mediterranea*. Semi-schematic drawings of the low molecular weight regions of gels, vertical bars indicate the position of protein bands. Migration was right to left. (Data taken from Kloppstech and Schweiger, 1973.)

could result from enzymatic modifications of otherwise identical proteins. Such enzymes would be, presumably, gene products of nuclear DNA. The only known protein-modifying enzyme in *Acetabularia* is a protein kinase (Pai *et al.*, 1975). This enzyme has not yet been completely characterized with respect to species-specific differences in substrate specificity and electrophoretic mobility.

Since the nuclear genome codes for the chloroplast proteins and they are synthesized on 80S ribosomes in the cytosol, the chloroplast proteins must be transported to the organelles. This must also be true for other proteins, such as the two LHCP species obtained in the cell-free translation system (Section

6.6.1.3). The extra peptide of the LHCP precursor protein is thought to be involved in recognition of the chloroplast membrane after transport. The exact mechanism, however, remains to be elucidated in other plant systems which yield more material for analysis.

An interesting observation concerning this intracellular transport is that part of the proteins of the thylakoid membrane that are translated in the cytosol are made on 80S ribosomes far-removed from the chloroplasts, whereas other proteins are made on 80S ribosomes which show a very close structural or functional relationship to the chloroplasts. The latter class of ribosomes remain present and active in a preparation of isolated chloroplasts, whereas the former class does not (Apel & Schweiger, 1973). The proteins made by the two classes of ribosomes are clearly distinguished from each other and one might speculate that the soluble proteins are made on free ribosomes and the membrane-bound proteins are made on ribosomes closely associated with the chloroplasts.

6.6.1.6 *Gene products coded for by the organellar genomes*

In vitro incorporation experiments with isolated chloroplasts from *Acetabularia* incubated in the presence of radioactive precursors of RNA, have established that the chloroplast genome codes for rRNA and, presumably, tRNA. Most of the knowledge of structure and information content of the chloroplast genome derives from the studies of a number of groups with different goals and is summarized in a review by Bedbrook and Kolodner (1979) (see Chapter 9).

Additional *in vitro* experiments have shown that isolated chloroplasts also synthesize proteins. Apel and Schweiger (1972) showed that part of the thylakoid membrane proteins are synthesized on 70S ribosomes. It is believed that most or all of them are coded for by the chloroplast genome. Further analysis will be facilitated by the use of more sensitive techniques, such as the coupled transcription and translation of chloroplast DNA (Bottomley & Whitfield, 1979).

Complex organellar structures, such as ribosomes and thylakoid membranes, are composite structures coded for by both the nuclear and the chloroplast genomes. This raises questions concerning the regulation and synchronization of gene expression in the nucleus and organelles, as it is obvious that functional ribosomes and membranes are obtained only if all their components are synthesized and combined in stoichiometric ratios. Experiments with inhibitors of chloroplast and cytosol protein synthesis indicate that the coupling may not be as strict as the demand for stoichiometric ratios might imply, as it may well be that small pools of gene products of both systems allow for some flexibility. As in other systems, it is possible to obtain with isolated *Acetabularia* chloroplasts not only the synthesis of membrane proteins, but also their incorporation into the thylakoid membrane.

In experiments with isolated chloroplasts, Green (1980) found that the apoprotein of the photosystem I complex (CP I) is synthesized within the organelle, thereby confirming results obtained from other systems.

Unfortunately, it has not been possible to detect differences in the proteins of the organelles which are coded for by the organellar genome in the various species of *Acetabularia*. This is in agreement with the postulated constancy of the information of the chloroplast genome already described. Heterologous combinations have failed to produce such differences. In addition to the CP I apoprotein, thymidine kinase, mentioned previously, might be coded for by the chloroplast genome and both proteins as well as the large subunit of ribulose-1,5-bisphosphate carboxylase may be likely candidates for further investigations on species-specific differences originating in the organellar genome. The increase in thymidine kinase activity that normally occurs in anucleate cells can be blocked by rifampicin, which prevents the synthesis of RNA in chloroplasts. However, as in all studies with inhibitors, these results must be cautiously interpreted.

6.6.1.7 *Acetabularia as a tool for in vivo studies of gene expression*

The oocyte of the African clawed frog, *Xenopus*, has been used almost exclusively and with great success for *in vivo* studies on the regulation of translation of mRNA. An equivalent system of plant origin has been lacking until recently. Different viral RNAs and DNAs have been introduced into the cytoplasm of *Acetabularia* by first injecting these nucleic acids into an isolated nucleus and then implanting the injected nucleus into an anucleate cell fragment (Cairns *et al*., 1978). The nucleus protects the exogenous nucleic acid from degradation.

These experiments have shown that it is possible not only to translate exogenous RNA but also to transcribe animal virus DNA to RNA, thereby using the *Acetabularia* cell as a coupled transcription-translation system. Finally, it has been shown that RNA isolated from the cytoplasm of eukaryotes, after injection into and implantation of a nucleus, is transported from the nucleus to the cytoplasm a second time. This indicates that the specificity of mRNA transport is not determined by precursor molecules, but rather should be looked for in the polarity of the nuclear pore complexes.

6.6.2 EVENTS OCCURRING PREDOMINANTLY IN THE CYTOPLASM

6.6.2.1 *Studies of gene expression in the anucleate cell*

Early experiments by Spencer and Harris (1964) showed that the activities of three phosphatases are regulated during cap formation in nucleate and anucleate cells of *Acetabularia*. The activity of UDPG-pyrophosphorylase increases during cap formation in both nucleate and anucleate cells. This increase in activity is probably correlated with the developmental changes in the composition of cell and cyst wall polysaccharides (Zetsche, 1968). Recent investigations show that the activity of UDPG-pyrophosphorylase per unit of protein does not increase during cap formation (Dillard *et al*., unpublished). Studies with inhibitors of RNA and protein synthesis have shown that the mRNA for this protein is coded for by the nuclear genome. This mRNA is exported to the cytoplasm at stage IV, whereas the increase in activity does not occur until cap formation. Expression of this gene is therefore not controlled at the level of transcription.

Furthermore, UDPG-pyrophosphorylase and mannan synthetase are synthesized predominantly in the apex of the cell. This indicates not only temporal but also spatial regulation of gene expression in the cytoplasm.

6.6.2.2 *Intracellular transport and communication*

The transfer of information and material between organelles may, in the simplest case, be achieved by diffusion. Diffusion alone may be adequate for cells that are of the order of a few hundred μm, but would hardly suffice for the giant *Acetabularia* cells. To transport materials over 20cm of cytoplasm in *A. major*, which resembles a nerve cell in its length and in its electrophysiology, the cell requires systems more specialized than simple diffusion, especially since it has been observed that transport does not occur at random, but is rather highly ordered and specific.

Biochemical analysis has shown that the transport mechanisms for ribosomes and mRNA run at rates of 0.1 and 0.2μm s^{-1}, respectively. Two additional systems have been observed via time lapse photography and were found to run at velocities of 1–2 and 2–11μm s^{-1}. The slower process, which is responsible for the transport of chloroplasts, seems to depend on the action of microfilaments, since it can be inhibited by cytochalasin B. The faster process transports secondary nuclei to the rays of the cap and is inhibited by cytochalasin B and colchicine. Thus, two types of filaments, microfilaments and microtubules, are thought to be involved in the faster type of cytoplasmic streaming (Koop & Kiermayer, 1980).

6.6.2.3 *Electrophysiology*

Another system for intracellular communication has been revealed by investigations of the electrophysiology of *Acetabularia.* As discussed in Section 6.3.4, basal fragments may regenerate if the nucleus remains in the fragment for five days in the dark before removal of the nucleus by amputation of the rhizoid. The first signs of regeneration appear about 50 hours after amputation of the rhizoid and transfer of the anucleate fragments to light. About 80% of the fragments will regenerate at their formal apical end.

If these fragments are transferred immediately after amputation to a subdivided and illuminated vessel, in which the chambers are insulated with silicone grease (Fig. 6.8), it is possible to measure a potential difference between chambers A and C. After 30 hours, one end becomes constantly positive, presaging the future site of regeneration (Novak & Bentrup, 1972). The current flows from the non-regenerating pole to the regenerating pole, which is always positive. In addition, spontaneous depolarizations occur at the regenerating end and, in mirror image, reflect the action potentials measured by Gradmann (1976). These spikes have their greatest magnitude at the regenerating end and

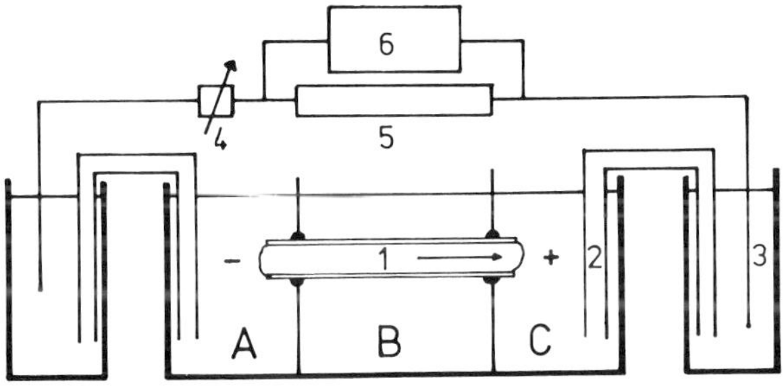

Fig. 6.8 Apparatus for the continuous measurement of extracellular potential differences: 1 = basal fragment of an *Acetabularia* cell in sea water, compartments A to C insulated against each other, 2 = KCl-bridge connecting the cultivation chamber with the electrode vessels, 3 = electrode, 4 = voltage compensation to account for potential differences between the electrodes, 5 = electrometer, 6 = recorder. The arrow indicates the direction of current. The cell will regenerate at the positive end in compartment C. (Redrawn from Novak & Bentrup, 1972.)

are light-dependent. If compartments A and C, which are normally connected via the measuring electrode circuit, are completely isolated, allowing no current to flow, regeneration does not take place (Novak & Sironval, 1975). Additionally, it was shown that the site of regeneration may be experimentally manipulated by application of an exogenous current to neutralize or reverse the normal current established by the cell. When the potential difference between compartments A and C is maintained at zero, no regeneration occurs. If the normal potential is reversed, a high percentage of cells will regenerate at the 'wrong' end, which has been experimentally made positive. Unfortunately, these results have not been confirmed by other laboratories. It is not clear how tightly-coupled morphogenesis or structural polarity in the cell is to potential differences, since application of the ionophore A23187 suppresses cap formation, but not the potential gradient or longitudinal growth (Goodwin & Pateromichelakis, 1979).

6.6.2.4 *Circadian rhythms*

Circadian rhythms are maintained even under constant conditions, have a period of approximately 24 hours and are temperature compensated, which means that they respond to long term temperature shifts with a temperature coefficient of approximately one. Finally, certain phase-shifting stimuli such as light, temperature or drugs may advance or delay the oscillations, depending on the time of the circadian day during which the phase shifting pulse is applied.

Evolution of oxygen under conditions of constant light is the most thoroughly investigated circadian rhythm observed in *Acetabularia.* Although the nucleus may exert a phase shifting influence on the rhythm, presumably via transcription as revealed by transplantation and inhibitor experiments, the rhythm is maintained in the anucleate cell over a period of weeks and therefore does not depend on continuous activity of the nucleus. The cytosol obviously plays an important role in synchronizing the 10^6 chloroplasts responsible for this rhythm.

The circadian rhythm in evolution of oxygen under constant conditions was first observed by Sweeney and Haxo (1961). More refined methods for the measurement of oxygen have been developed to observe single cells (Mergenhagen & Schweiger, 1973). This is a major improvement, since free running rhythms in individual cells have periods from 23–27 hours and the rhythms quickly overlap and damp out when measurements are made on cell populations.

Measurements of single cells have shown that the maintenance of the rhythm under constant conditions:

(1) is not dependent on continuous nuclear transcription, since anucleate cells retain the rhythm over many weeks;

(2) is not dependent on continuous transcription of the chloroplast genome, since rifampicin, an inhibitor of RNA synthesis in this organelle, does not affect the rhythm;

(3) is not dependent on protein synthesis on 70S ribosomes, since chloramphenicol has no effect on the rhythm;

(4) is dependent on protein synthesis on 80S ribosomes of the cytosol.

Cycloheximide at very low concentrations leads to a suppression of the rhythm. The inhibitory effect of cycloheximide is reversible; the rhythm is re-established upon removal of the drug. When applied in eight hour pulses at different times in the circadian day, it has been observed that there are cycloheximide-sensitive and -insensitive phases. Of special interest is the observation that the culture influences the period of cycloheximide sensitivity. At 25°C, the sensitive phase is from 3^{00} to 9^{00}, while at 20°C, the sensitive phase is from 11^{00} to 23^{00} (Table 6.1; Karakashian & Schweiger, 1976).

Observed circadian changes in the ultrastructure of chloroplasts suggest that the composition of thylakoid membranes may undergo diurnal alterations. There is at least one protein which appears periodically in these membranes (Leong & Schweiger, 1979). This protein has a molecular weight of 39 000 and appears complementary to the described cycloheximide-sensitive phases of the circadian rhythm, that is, during those periods of maximal sensitivity to cycloheximide. Incorporation of labelled amino acids into this protein is unaffected or stimulated by cycloheximide during the non-sensitive phases, suggesting that this protein is synthesized on 70S ribosomes and is probably a gene product of the chloroplast genome.

The 39 000 dalton protein is a major constituent of the thylakoid membrane. It seems unlikely, although not impossible, that such fluctuation of a membrane protein might be achieved by daily synthesis and degradation. An alternative explanation might be that the incorporation into and removal of the protein from the membrane is regulated by another protein which is synthesized in small amounts on 80S ribosomes. Cycloheximide exerts its effect at concentrations insufficient for complete inhibition of protein synthesis. Therefore, it

Table 6.1 The effect of cycloheximide pulses on the phase of oxygen evolution (A), the appearance of the labelled 39 000 dalton protein within the thylakoid membranes (B), and a hypothetical regulatory protein (C) as affected by the temperature of cultivation. The light and dark bars on top represent the circadian day and night. (Data from Karakashian & Schweiger, 1976 and Leong, 1978.)

	Phase response of oxygen evolution to pulse treatments with cycloheximide		
A	delay		25°C
		delay	20°C
	Appearance of labelled 39 000 dalton protein within the thylakoid membranes		
B	stimulation	inhibition	25°C
	inhibition	stimulation	20°C
	Synthesis of the hypothetical regulatory protein		
C	inhibition		25°C
		inhibition	20°C

may be that the rate of synthesis, rather than the synthesis *per se*, controls the circadian rhythm. Above a threshold concentration of the proposed regulator protein, the 39 000 dalton protein would be removed from the membrane, whereas below this concentration, the 39 000 dalton protein would reappear in the membrane (Table 6.1).

These observations have fostered a remarkable understanding of the molecular mechanism of the circadian clock. They suggest that the circadian rhythm, phase-dependent protein synthesis and phase-dependent alterations in membrane composition are all interrelated. Schweiger and Schweiger (1977) have developed the coupled translation-membrane model of the circadian rhythm in an attempt to unify these observations.

The investigation of the molecular mechanism of circadian rhythms has been hampered by the measurement of only one parameter, oxygen evolution. Two additional circadian rhythms have recently been described. Under conditions of constant light, the cell-generated current described in Section 6.6.2.3 shows rhythmic variations which are presumably in phase with oxygen evolution and persist in anucleate fragments (Novak & Sironval, 1976). Interesting features of this rhythm that warrant further study are that the amplitude of the current is light-dependent and thus connected with photophosphorylation, whereas the generation of the current results from the electrogenic chloride pump situated in the plasma membrane. The second rhythm was first observed via time lapse photography and is measured by recording changes in the optical density of the cytoplasm at a wavelength that favours the absorption of light by chloroplasts. It was found that the chloroplasts migrate with a circadian rhythm. Migration to the rhizoid occurs at the beginning of the circadian night and migration to the apical end occurs at the beginning of the circadian day (Koop *et al.*, 1978). The measurement and elaboration of these three independent circadian rhythms should provide new insight into the basic mechanisms of rhythmic phenomena.

6.7 FUTURE EXPERIMENTAL APPROACHES

Lines of research on *Acetabularia* that are likely to be extended have been mentioned in the previous sections. The size and polarity of the *Acetabularia* cell make it an ideal subject in which to study the topography of protein synthesis. This subject could be studied by *in vivo* incorporation studies, *in vitro* translation of mRNA isolated from different regions or compartments in the cell and finally, by comparisons of the kinetic complexity of these mRNAs. Proteins of the cell membranes and those associated with the polysaccharides of the cell wall will be of special interest in this respect.

FURTHER READING

Berger S. & Schweiger H. G. (1980) Acetabularia: Techniques for study of nucleo-cytoplasmic interrelationships. *Handbook of Phycological Methods*, pp. 47–57. E. Gaunt (ed.) Cambridge University Press.

Bonotto S., Lurquin P., & Mazza A. (1976) Recent advances in the research on the marine alga Acetabularia. *Adv. Mar. Biol.*, **14**, 123–250.

Brachet J. (1971) Nucleocytoplasmic interactions in morphogenesis. *Proc. Roy. Soc. B.* **178**, 227–43.

Hämmerling J. (1963) Nucleocytoplasmic interactions in Acetabularia and other cells. *Ann. Rev. Plant Physiol.*, **14**, 65–92.

Koop H. U. (1979) The life cycle of Acetabularia (Dasycladales, Chlorophyceae: a compilation of evidence for meiosis in the primary nucleus. *Protoplasma* **100**, 353–66.

Schweiger H. G. (1969) Cell Biology of Acetabularia. *Curr. Top. Microbiol. Immunol.* **50**, 1–36.

Schweiger H. G. & Berger S. (1979) Nucleocytoplasmic interrelationships in Acetabularia and some other Dasycladaceae. *Int. Rev. Cytol.*, Supplement 9, pp. 11–44.

Schweiger H. G. & Schweiger M. (1977) Circadian rhythms in unicellular organisms: an endeavor to explain the molecular mechanism. *Int. Rev. Cytol.*, **51**, 315–42.

Werz G. (1974) Fine structural aspects of morphogenesis in Acetabularia. *Int. Rev. Cytol.*, **38**, 319–67.

CHAPTER 7

FUCUS

7.1 INTRODUCTION

Species of the brown algal genus *Fucus* occupy the major part of the intertidal belt on rocky shores of the northern hemisphere. The fertile tips of the thallus (receptacles) bear many flask-shaped cavities (conceptacles) opening to the surface through a pore or ostiole. Conceptacles of dioecious spp. (*F. serratus*, *F. vesiculosus*) contain either antheridia or oogonia whilst monoecious spp. (*F. spiralis*, *F. distichus*) contain both. Eggs and sperm are released from oogonia and antheridia respectively (Fig. 7.1), when the plants are rewetted by the incoming tide.

Eggs of *Fucus serratus* are spherical, 60–100μm in diameter and contain a single haploid nucleus surrounded by chloroplasts and polysaccharide- and polyphenol-containing vesicles (Matthews *et al.*, 1976). The former type of vesicle is largely responsible for the rough appearance of eggs in the scanning electron microscope since they protrude beneath the plasma membrane (Fig. 7.6). Newly-released eggs are surrounded by oogonial mucilage which can be readily washed away to leave a naked plasma membrane. The biflagellate sperm of *F. serratus* are some 5μm long and bear a long posterior flagellum and a much shorter, Flimmer-bearing anterior flagellum through which the sperm initially attaches to the egg surface. The anterior end of the sperm body is prolonged into a flattened structure, the proboscis, which possesses an anterior framework of microtubules derived from a modified flagellar root. The cytoplasm of the sperm contains a nucleus, chloroplast (with a large eyespot) and several large mitochondria (Manton & Clarke, 1955).

Eggs and sperm of *Fucus* spp. have long provided an ideal model system for the investigation of many aspects of cellular development.

(1) Male and female plants of dioecious spp. may be readily distinguished. Plants can be transferred to the laboratory and under careful storage will liberate large quantities of viable gametes for a number of days.

(2) The oogamous form of fertilization presented, with a large, non-motile egg cell and motile sperm proves to be useful in chemoattraction studies (Section 7.2.2).

(3) The naked gametes provide one of the few plant systems in which surface fertilization receptors can be studied without intervening cell walls (Section 7.2.3).

(4) The large egg cell allows the insertion of microelectrodes to study the electrophysiology of developmental processes (Section 7.3.1.3).

(5) Studies on the control and determination of polarity in the initially

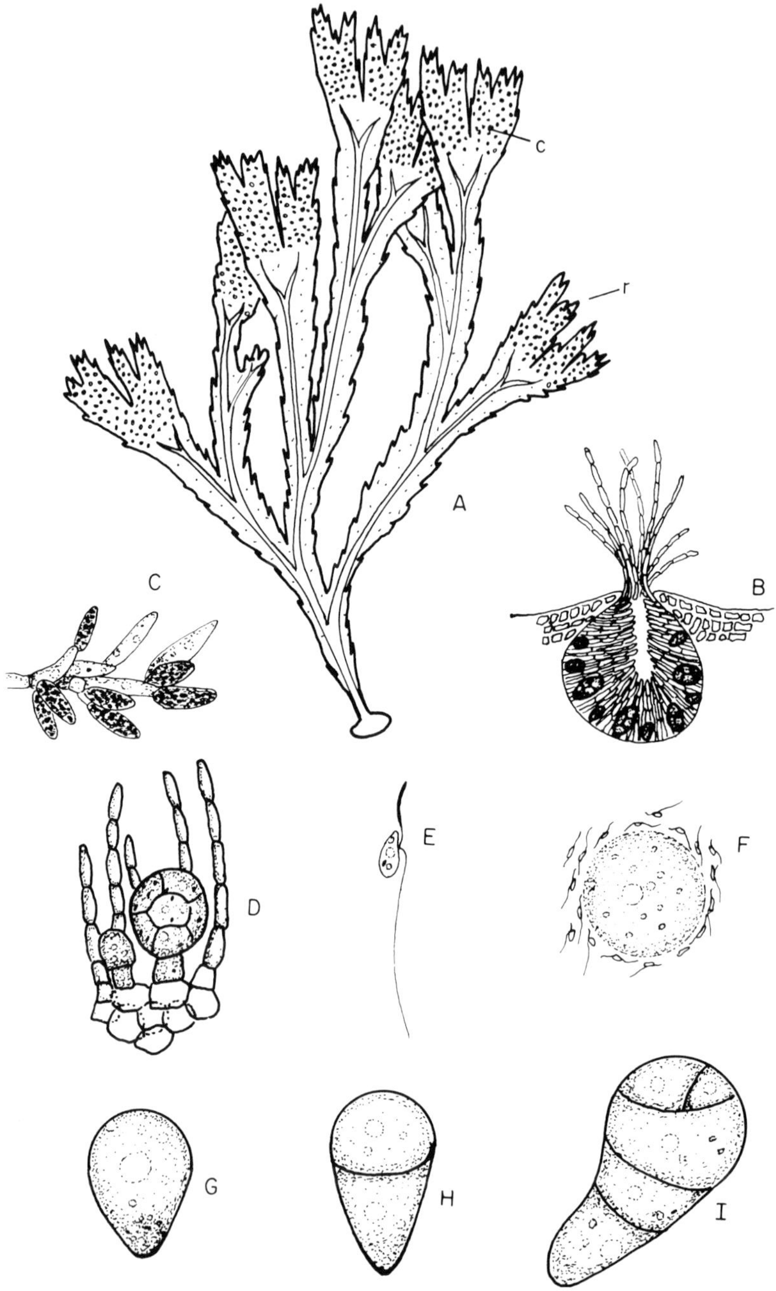

Fig. 7.1 Diagram showing the main features of the life-cycle of *Fucus serratus*. The mature plant (A), some 60–180cm long, bears fertile tips, the receptacles (r) which bear many conceptacles (c). Each conceptacle opens onto the surface by a pore from which sterile filaments protrude (B, transverse section of female conceptacle, × 20). Conceptacles contain either antheridia (C, × 400) in which male gametes (spermatozoids) develop, or oogonia (D, × 100) containing 8 egg cells. Spermatozoids (E, × 1000) are attracted to eggs (F, × 250), the result being a fertilized zygote which germinates as shown in G–I).

apolar egg cell form a model for studies on angiosperm polar embryo development. Many higher plant cells also undergo a pattern of differentiation similar to that of the first two cells of the *Fucus* embryo (i.e. polarity followed by unequal cell division). Synchronous single-cell cultures of *Fucus* can be grown aseptically in defined media, free from accessory cells and thus provide a less complex model system for the investigation of factors potentiating higher plant cell development.

(6) The release of cell wall material following the fertilization signal permits detailed studies of the temporal aspects of cell wall differentiation (Section 7.4.4).

7.2 GAMETE RECOGNITION

7.2.1 SPECIFICITY

Despite early reports of inter-specific crosses between gametes of different *Fucus* spp., experimental studies in the laboratory have failed to demonstrate any substantial degree of cross-fertilization (Bolwell *et al.*, 1977). In addition to ecological barriers such as non-coincidence of fruiting periods on the shore, there exist therefore, strong physiological barriers preventing the merging of the different spp. into one polymorphic complex.

7.2.2 CHEMOATTRACTION

Simple microscopical observation of the behaviour of gametes on mixing shows that sperm are attracted by egg cells (Fig. 7.2). Early studies on the nature of the

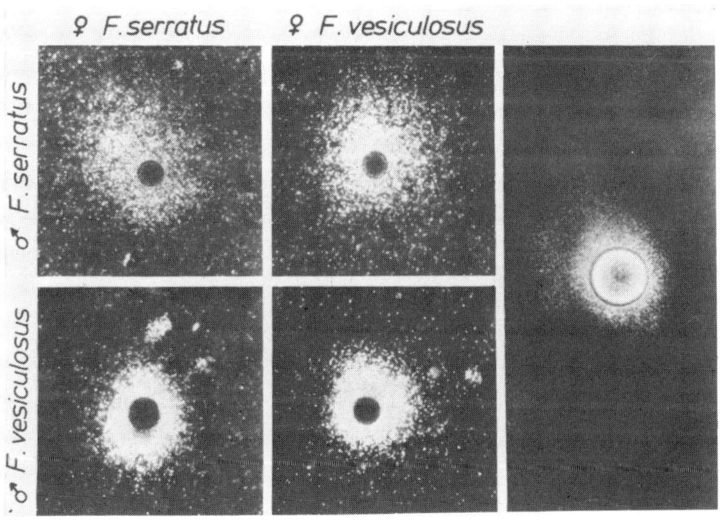

Fig. 7.2 Intra- and interspecific combinations of gametes of *F. serratus* and *F. vesiculosus* (left) and attraction of *F. serratus* sperm by an artificial globule containing 10^{-4}M fucoserraten in a fluorinated solvent. (Reproduced from Müller & Seferiadis, 1977, with permission.)

chemoattractant demonstrated its volatility and the fact that a number of simple hydrocarbons, ethers and esters could mimic its activity in low concentration. The exemplary studies of Müller & Jaenicke (1973) and Jaenicke & Seferiadis (1975) resulted in the isolation, characterization and synthesis of a conjugated hydrocarbon, 1,3-trans,5-cis-octatriene, given the trivial name *fucoserraten* (Fig. 7.3). Subsequent studies (Müller & Seferiadis, 1977) revealed the

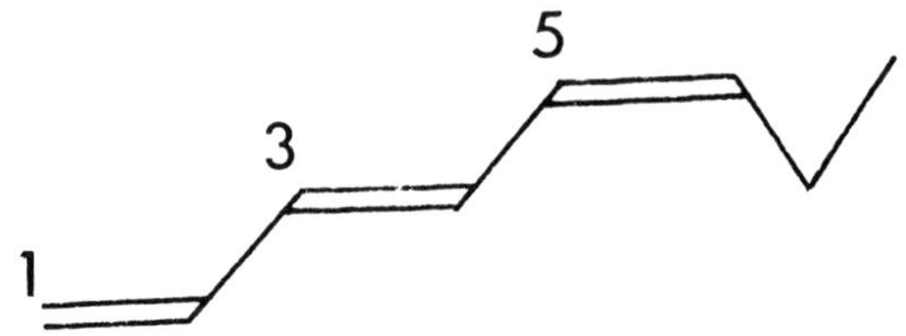

Fig. 7.3 Fucoserraten(e), (1,3-trans, 5-cis-octatriene) the sperm attractant secreted by eggs of *F. serratus*.

threshold concentration for attraction to be 10^{-6}M but no species-specificity was detected, i.e. sperm of *F. vesiculosus* reacted just as sensitively to fucoserraten as *F. serratus* sperm. This is in accordance with the absence of specific chemoattraction of sperm by eggs (Fig. 7.2).

The absence of specificity at this level clearly indicates that the barriers preventing cross-fertilization lie at a level other than gametic approach, presumably in the binding of sperm to egg and surface membrane fusion.

7.2.3 SURFACE RECEPTORS AND GAMETE RECOGNITION

The recognition and binding of sperm to egg surfaces and the fusion of their membranes represents the initial event in a chain of interactions culminating in fertilization. The capacity of all living cells and organisms to recognize and discriminate between other cells, organisms or exogenous molecules is fundamental, and in general, recognition systems are mediated by the association of surface-localized complementary molecular systems (Fig. 7.4).

In many cases the biochemical counterparts of the surface moieties appear to be specific saccharide structures ('ligands') that form part of membrane glycoproteins or glycolipids (Fig. 7.5) and complementary carbohydrate-binding proteins ('lectins') associating in a 'lock-key' relationship. The necessary degree of structural variability to account for the high degree of biological specificity in recognition systems is provided by the vast level of diversity known to exist in heterosaccharide chains through variation in sugar composition, anomeric configuration and branching patterns on the one hand, and on the other hand, the high degree of specificity shown by the binding site of enzymes and other proteins.

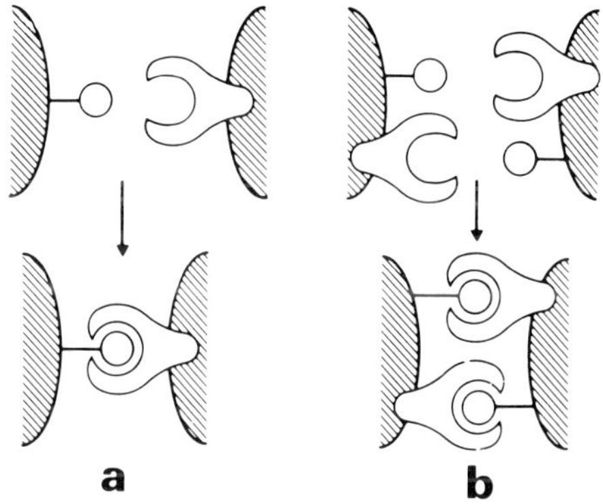

Fig. 7.4 A model illustrating specific contact adhesion of two cell surfaces assuming (a) unipolar complementarity, and (b) bipolar complementarity. (From Heslop-Harrison, 1978, with permission).

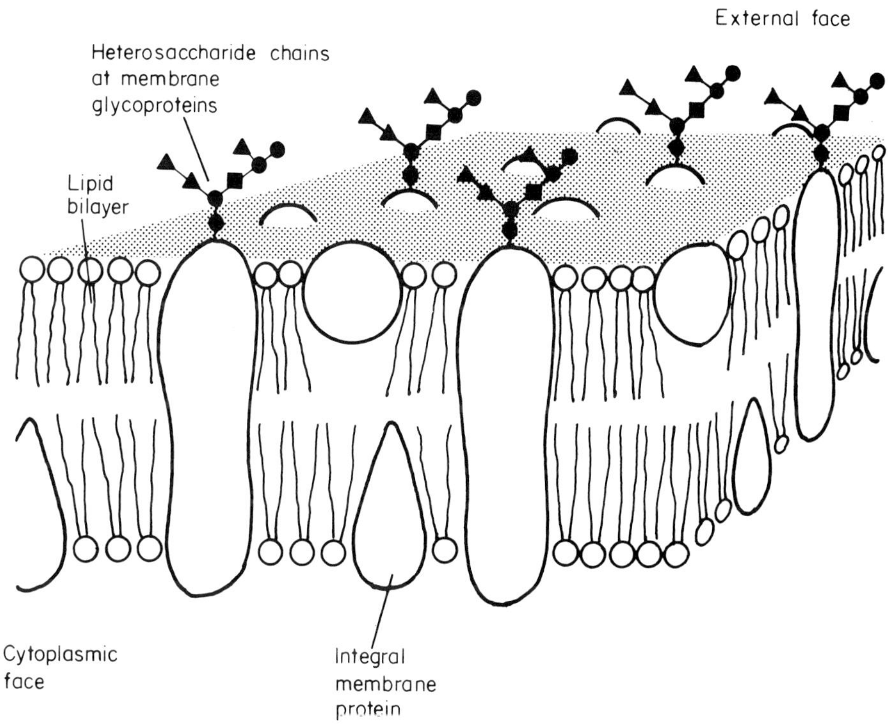

Fig. 7.5 3-D view of a generalized surface membrane illustrating the distribution of phospholipids, integral membrane proteins and glycoproteins, the latter with branched, complex heterosaccharide chains forming the surface 'glycocalyx' ('sugar-coating'). (After Heslop-Harrison, 1978 and Callow, 1977.)

In lower plants the biochemical basis of gamete recognition has been mainly studied in *Chlamydomonas* (reviewed by van den Ende, 1976), and in the brown alga *Fucus serratus*. In both cases substantial evidence indicates the presence of a surface-localized complementary mechanism involving specific saccharide binding by lectin-like proteins. The evidence for *F. serratus* will now be reviewed.

7.2.3.1 *Fertilization assay*

To permit detailed studies on the molecular basis of egg-sperm recognition a rapid, quantitative assay for fertilization was devised (Bolwell *et al.*, 1977; Callow *et al.*, 1978a). Within seconds of sperm entry the egg cell releases the β-linked polyuronide, alginic acid, held within vesicles just below the plasma membrane of the unfertilized egg (Fig. 7.6). The released material thus forms an incipient cell wall to the developing zygote and presumably serves as a barrier to polyspermy. The rapid fertilization assay is based on the staining of this cell wall with the fluorescent brightener Calcofluor White ST. The proportion of fluorescing (fertilized) eggs can then be estimated by direct counting under a UV

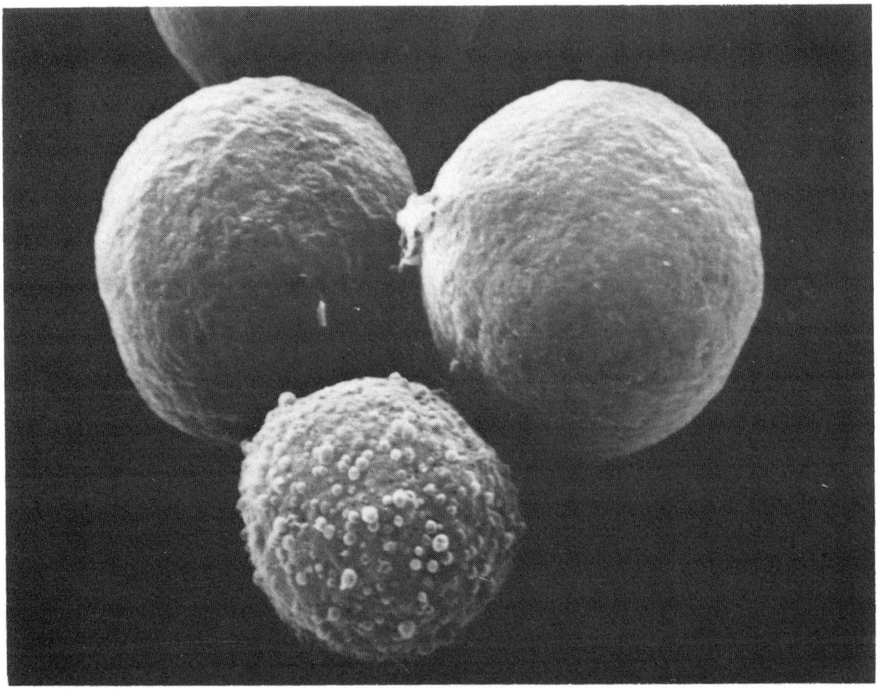

Fig. 7.6 Group of *F. serratus* eggs taken 10min after mixing with sperm as seen in the SEM. The surface of the unfertilized egg is rough (bottom) whilst surfaces of fertilized eggs are smooth due to the release of a cell wall. (From Callow *et al.*, 1978a.)

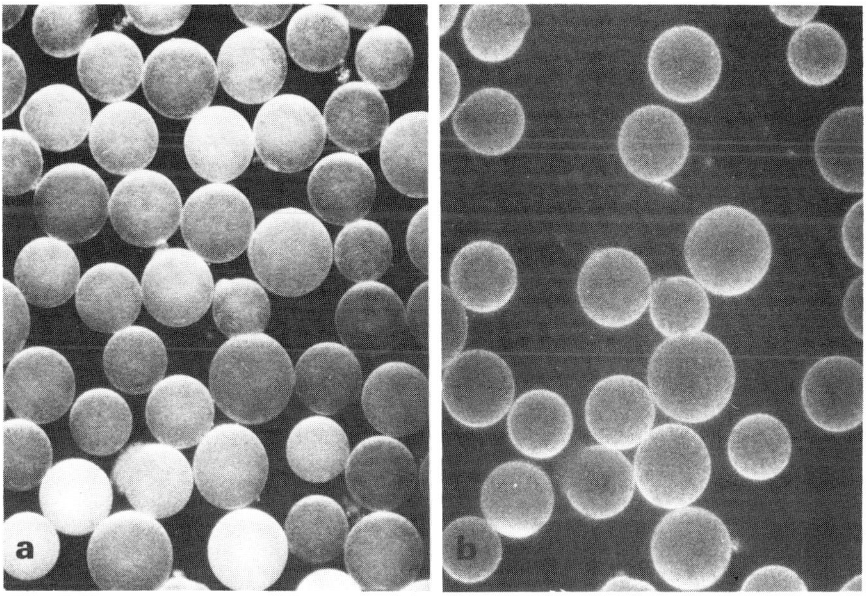

Fig. 7.7 Eggs of *F. serratus* stained with Calcofluor White ST under a fluorescence microscope, with barrier filters GG13 + W2B only (a), and (b) same field of view with same barrier filters plus exciter filters BG38 + UG1. In b, only fertilized cells which are stained by Calcofluor fluoresce ($\times$ 95). (From Callow *et al.*, 1978a.)

microscope (Fig. 7.7). Although this method is indirect, careful checks have revealed it to be a reliable estimate of the degree of fertilization. Nuclear fusion takes place 8–10min after cell wall release.

7.2.3.2 *Indirect evidence for receptors*

Indirect information on fertilization receptors has been obtained by examining the effects of a number of agents chosen to potentially perturb the processes of recognition, viz. plant lectins (purified specific carbohydrate-binding proteins isolated from plants, Callow, 1977), proteases and carbohydrases, sugars and various polysaccharides. In careful kinetic studies, Bolwell *et al.*, (1979) showed that the following lectins did not have inhibitory effects on fertilization at low concentration (high concentrations of many lectins exhibit non-specific, cytotoxic effects): Wheat Germ Agglutinin (specifically binds N-acetyl-glucosamine residues), Phytohaemagglutinin and Soyabean Agglutinin (both N-acetyl-galactosamine specific), Ricin_{120} (galactose-specific). On the other hand, two lectins, Concanavalin A (Con A, binds α-D-mannose and α-D-glucose residues) and Fucose-Binding-Protein (FBP, binds α-L-fucose residues) both inhibited fertilization strongly at low concentration (10^{-6}–10^{-8}g cm^{-3}). Furthermore, the effects of the lectins appeared to be directed specifically towards the eggs.

Thus, if eggs were lectin-treated, then washed to remove excess lectin before adding untreated sperm, fertilization was inhibited (Fig. 7.8a). However, the reciprocal experiment, pretreating sperm with lectin, was ineffective. The inhibition by both Con A and FBP at low concentration could be fully reversed by washing lectin-treated eggs with the appropriate hapten (competitive) sugars. By analogy with classical enzyme kinetics (i.e. treating the reaction between egg and sperm in the same way as the reaction between an enzyme and its substrate) the inhibitory effects of low concentrations of both lectins were shown to be of a competitive character, the lectins in effect competing with sperm for binding sites on the egg surface (Fig. 7.8b). However, high inhibiting concentrations of Con A or FBP (10^{-4}–10^{-5}M) were not specific for egg or sperm and in these inhibition was of a 'mixed' type (competitive plus uncompetitive), could not be reversed totally by the hapten sugars methyl-mannoside and fucose, and microscopic examination of gametes revealed membrane damage.

The majority of the biological effects of plant lectins on cells result from the binding of lectins to surface oligosaccharides. The specific, competitive effects of Con A and FBP on *Fucus serratus* suggest that gamete recognition is based on some aspect of saccharide binding and that α-mannose-like and α-fucose-like residues on the egg surface may be important. The lectins, by binding to specific saccharide residues that form part of putative sperm receptors, effectively block sperm binding. Further evidence consistent with this view was obtained by treating gametes with carbohydrases, proteases and purified polysaccharides. Fertilization was inhibited by polysaccharides containing either mannose or

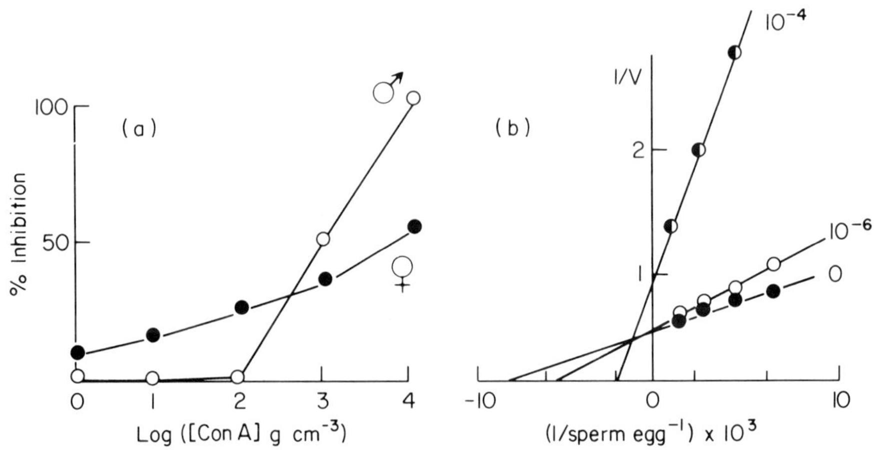

Fig. 7.8 The effects of Con A on fertilization. In (a) the inhibitory effect of Con A is shown for either egg (●—●) or sperm (○—○) pretreatments. (b), double-reciprocal plot of Con A inhibition at high (10^{-4}g cm^{-3} Con A ◐—◐), and low (10^{-6}g cm^{-3} Con A ○—○) concentration. Control (●—●) is without Con A. Abscissa is the reciprocal of sperm concentration relative to the number of eggs; ordinate, the reciprocal of the degree of fertilization. (From Bolwell *et al.*, 1979.)

Table 7.1 Method of isolation and some properties of putative gamete receptors of *Fucus serratus* (Bolwell *et al.*, 1980)

	Sperm receptor from eggs	Egg receptor from sperm
Source	Crude egg membranes	Sperm lysates
Method of purification	Affinity chromatography on Con A-agarose	Affinity chromatography on fucan-Sepharose
Species-specificity in inhibiting fertilization	Species-specific by binding to sperm	Species-specific by binding to eggs
Inhibitory activity destroyed by	α-L-fucosidase	Heat, trypsin
Inhibition reversed by		Fucoidan
Chemical form	Glycoprotein	Protein
Apparent mol. wt.	30 000	60 000

fucose residues, and in pretreatments of each gamete separately the evidence indicated that these polysaccharides were specifically binding to sperm. Furthermore, a number of glycan hydrolases were tested and at low concentrations, enzyme pretreatment of eggs with either α-L-mannosidase or α-L-fucosidase was particularly effective. Trypsin inhibited fertilization in a specific manner in sperm pretreatments but other proteases exhibited non-specific inhibition.

7.2.3.3 *Direct evidence for receptors*

Direct evidence for receptors has recently been obtained through the isolation of factors with the ability to inhibit fertilization in a gamete- and species-specific manner (Table 7.1). The properties of these putative receptors are consistent with those predicted from the studies using lectins etc. Thus, the glycoprotein sperm receptor from eggs binds to Con A, indicating the presence of glucose and/or mannose moieties in the glycan moiety of the molecule. The protein egg receptor from sperm binds to the fucan affinity matrix as predicted, and in effect therefore, constitutes a plant lectin. However, whilst the properties of the factors described in Table 7.1 are consistent with a role in gamete recognition, it remains to be demonstrated that these factors are located at the gamete surfaces and do interact with each other in the manner anticipated.

7.3 POLARITY

Although not rigorously conducted, the work of Knapp (1931) on *Cystoseira* suggests that in the absence of other factors rhizoid formation occurs at the site of entry of the sperm into the apolar fucoid egg. Although not necessary for inducing polarity, application of one of a range of external factors will reorientate the early established plane, since this remains labile until a few hours before rhizoid formation begins. Light, chemical gradients and electrical fields are the chief polarity-inducing factors (vectors), and each is perceived by the embryo at different (but overlapping) periods after sperm entry. These findings are summarized in reviews by Jaffe (1968), Quatrano (1974; 1978) and Jaffe & Nuccitelli (1977). In addition, Weisenseel (1979) comprehensively reviews polarity in a wider context. The existence of a range of vectors ensures that polarity, and thus, normal development can occur in any environmental situation.

In cell polarity, the interaction of external factors with genetic mechanisms leads to unequal distribution of nuclei and other cell components, and the resulting nucleocytoplasmic interactions are the basis for a divergence of gene expression in different regions of the cell, and later the developing embryo (Bloch, 1965). Much recent work has aimed to improve our understanding of the processes of perception and transduction of these external vectors, and how the labile polarity is stabilized and amplified into the asymmetry of the 2-celled embryo.

7.3.1 POLARITY-INDUCING FACTORS

7.3.1.1 *Light*

The light-induced polar axis is always parallel to the largest gradient of absorption. Strong unilateral illumination induces formation of a protrusion (the presumptive rhizoid) from the shaded hemisphere (i.e. the region of least absorption) (Fig. 7.9a). In unilateral illumination by light of moderate intensity

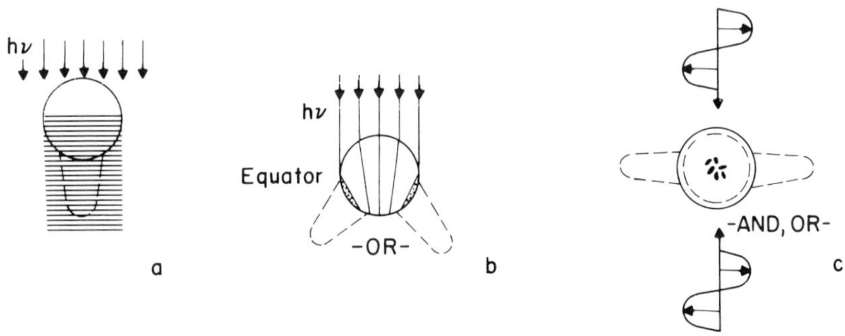

Fig. 7.9 (a) Light-induced orientation of rhizoidal outgrowth (broken line) of a *Fucus* zygote, as a result of strong unilateral light (shaded areas are hatched). (b) The pattern of lit and shaded regions (dotted) in unilateral light of medium intensity; the outgrowth (broken line) will appear at one of the 2 sites indicated. (c) The inferred orientation of the photoreceptor molecules (peripheral and lying parallel to the surface—dashes within egg) from the direction of the outgrowth(s) which arise in plane-polarized light shining on the zygote from opposite sides. (From Weisenseel, 1979, with permission.)

(10^{-15}–10^{-16} quanta cm^{-2}) the protrusion arises 'subequatorially' (from 90°–135° away from the source) (Fig. 7.9b), and in unidirectional illumination with plane-polarized light the zygotes also germinate subequatorially, but in the plane of vibration (Fig. 7.9c), and many of the embryos so produced may be bipolar.

Growth tends to occur where certain photoreceptor molecules absorb the least light. These molecules are believed to be located at the cell surface, and under unpolarized illumination the subequatorial and peripheral receptors absorb least light because they are bypassed by the focusing action of the cell (Fig. 7.9b). Experiments with plane-polarized light indicate that the photoreceptors are dichroic and orientated periclinally, i.e. so that their axes of maximum absorption are arranged parallel to the cell surface. Absorption by these molecules at the germinating regions are thus minimal (Fig. 7.9c) while under polarized illumination.

Only the shorter wavelengths of the blue end of the spectrum (approximately 400–560nm) and ultraviolet light have been found to be as effective as white light of equivalent intensity in establishing polarity in *Fucus* zygotes. A low-intensity receptor and a high-intensity receptor are believed to be involved in absorption of polarity-inducing light, but in both cases the action spectra show that again blue light and near ultraviolet light are the most effective. However, small differences between the action spectra suggest a protein-bound carotenoid low-intensity receptor and a riboflavin high-intensity receptor. The plasma membrane is a likely site for the preferentially-orientated photoreceptor molecules.

7.3.1.2 *Chemical Gradients*

Polarity can also be induced by self-generated or imposed gradients of inorganic or organic substances. There is circumstantial evidence that developing zygotes emit substances which affect rhizoid formation, e.g. zygotes separated by around four zygote diameters tend to form rhizoids towards each other (the positive group effect), whilst at a distance of approximately one zygote diameter apart they form rhizoids away from each other (the negative group effect). Again, in a laminar flow of 0.01–10μm s^{-1} at pH 6.5 zygotes germinate downstream (the direction in which a diffusible substance would be carried). Weisenseel (1979) suggests that the rhizoid promoting substance secreted by zygotes is part of a positive feedback mechanism which stabilizes a labile axis by amplifying its own release or production at the site of growth. There has been considerable experimentation and speculation as to the molecular basis for these observations. In his 1968 review paper Jaffe proposed that *Fucus* zygotes secrete two factors, a small unstable rhizoid-promoting molecule (rhizin) and a stable rhizoid-inhibiting molecule (antirhizin). Jaffe proposed that rhizin was indoleacetic acid. Environmentally-induced apolarity in rhizin secretion or distribution could presumably control the site of rhizin initiation. Polarization of rhizoid development in *Fucus* by blue light appears to have many similarities to phototropic responses in higher plants, but whether the processes are analogous is not yet known. The chemical nature of the inducing substances in *Fucus* is also unknown, but H^+, CO_2, O_2, IAA, AMP and c-AMP have been excluded.

Imposed gradients of H^+, K^+, Ca^{2+} and the uncoupler 2,4-dinitrophenol (DNP) also induce polarity in *Fucus* zygotes. (Rhizoid formation occurs at the side where the H^+ or DNP concentration is highest.) These are all well-known to affect membrane permeability, and the inducing effect may result from a gradient of one or more of these across cells causing pole to pole differences in membrane permeability and potential, and in the concentration of internal ions. For example, in *Pelvetia* (a genus closely related to *Fucus*), more Ca^{2+} enters the zygotes at a lower external concentration (1.5mM) than at a higher one (10mM, i.e. the Ca^{2+} concentration of natural seawater), so that Ca^{2+} permeability increases at lower external Ca^{2+} concentration. It has also been shown that in a Ca^{2+} gradient rhizoids form at the side of increased Ca^{2+} permeability. This is presumably the side of higher internal Ca^{2+} concentration, since it has been found that Ca^{2+} enters the rhizoid hemisphere preferentially during axis fixation.

7.3.1.3 *Electrical Fields*

Electrical currents flowing through cells will bring about redistribution of charged subcellular components ('self-electrophoresis'—see Section 7.3.4) and induce asymmetry in membrane potential, viz. a hyperpolarization at the side towards the positive electrode where the current is driven inward through the membrane, and a depolarization at the opposite side (see Chapter 2). A field of 3–4mV/cell measurably polarizes *Fucus serratus* zygotes. *Fucus* and *Pelvetia* zygotes orientate their rhizoids parallel to the applied field, the percentage doing so increasing the greater the magnitude of the field. However, varying responses are recorded with respect to the site of growth, i.e. whether rhizoids are initiated towards the positive pole (positive galvanotropism), the negative pole (negative galvanotropism) or whether the response is mixed. In positively galvanotropic zygotes, Ca^{2+} entry is believed to be faster at the positive pole because the membrane potential, and hence the force driving Ca^{2+} inwards, would be greatest there. In negatively galvanotropic zygotes Ca^{2+} entry is suggested to be greatest at the negative pole because in these, depolarization 'excites' the membrane and opens extra calcium gates to cause faster entry, despite the lower driving force. It is not yet known why these differences in membrane properties should occur. AC fields in the range 1–100Hz also induce polarity. Orientation is parallel to the field and zygotes produce a rhizoid which grows towards one electrode or the other (and rarely, towards both). Changes in membrane potential appear to be involved, but cytoplasmic changes may also occur. Induced currents, sufficient to polarize cells, are similar in magnitude to natural currents driven through themselves by polarizing cells (see Section 7.3.3).

7.3.2 PERCEPTION OF VECTORS

As stated earlier, a vector can be perceived and amplified into a stable fixed polarity over a limited period during development, and during this period the receptivity for a factor gradually increases to a maximum and then declines to zero. The receptive periods for different vectors occur at different times and may almost coincide. However, in practice, the timing and duration of the stages of receptivity for a particular factor do vary with different experimental conditions. For example, a maximum receptivity for light at 11–13h after fertilization has been shown in *Fucus serratus*, but other investigators reported two such maxima in this species, at 12 and 15h. An induction may be counterinduced by a second induction if the polarity axis resulting from the first was not stably fixed. However, integration of the two inductions will occur if the first induction cannot be cancelled by the second, resulting in a polarity axis orientated somewhere in between the two directions of induction. Endogenous processes therefore appear to underlie polarization, and these processes are not set in motion by induction. However, polarity can be induced by one of a range of vectors and these appear to constitute a hierarchy, the most important member of which is light (Weisenseel, 1979).

7.3.3 ENDOGENOUS CHANGES

In *Fucus* (and *Pelvetia*) eggs, the use of microelectrodes has shown that natural hyperpolarization of the membrane begins at fertilization, bringing the plasma membrane potential from -20 to -30mV (unfertilized) to -70 to -80mV (7–9h after fertilization). This hyperpolarization appears to result from KCl accumulation by the embryos and from the embryo membranes developing an increased permeability to K^+. KCl accumulation is mediated by activation of a membrane pump which causes Cl^- influx against its electrochemical gradient, and by sealing off Cl^- leaks in the membrane.

Using an ultrasensitive extracellular vibrating electrode to record extracellular voltage differences in the medium near the cell, a cell-generated electrical current of up to ca. $1\mu A\ cm^{-2}$ has been found to enter one or more small regions of the zygote membrane of *Pelvetia*; germination occurs from the final site of current entry and the current will generate a transcellular electrical field which could pull negatively charged cytoplasmic components to the growth region (see Section 7.4.4.3). Ca^{2+} ions have been shown to be major current-carriers in *Pelvetia* zygotes, and the use of the nickel screen method (Fig. 7.10) by Robinson & Jaffe (1975) for determining pole to pole differences in membrane permeability to calcium ions showed that a current of $0.03\mu A\ cm^{-2}$ traverses the

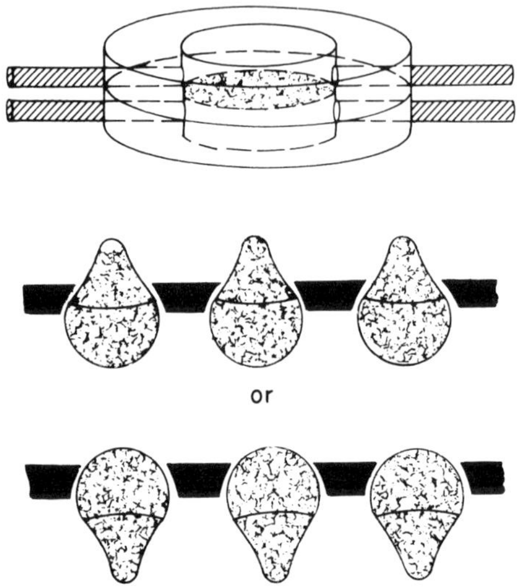

Fig. 7.10 (a) The chamber used for measuring ion fluxes across either half of the plasma membrane of a developing *Pelvetia* zygote. The nickel screen (25mm wide, 25μm thick and with 25 000 round holes) is held in a Plexiglass chamber filled with sea water. (b) Cross-sectional view of a screen to show the funnel-shaped holes with developing zygotes. Zygote suspensions were injected into the lower part of the chamber, forced upwards through the screen (placed with its wider apertures downwards) and kept in place initially by a slight upward flow of seawater. $^{45}Ca^{2+}$-labelled seawater was passed through the upper part of two chambers, one chamber being illuminated from above and the other from below, for calcium influx determinations. (From Robinson & Jaffe, 1975, with permission.)

cells as they are polarized by unilateral light. The current is greatest in the hours immediately following fertilization as a result of calcium influx at the dark (rhizoidal) pole occurring faster than at the opposite (thallus) end, and efflux at the latter being faster than at the rhizoidal pole (Fig. 7.11). A fixed number of channels and pumps are thought to be redistributed within the membrane to enable this to occur. (Later, polarity in the movement of Na^+ and Cl^- ions also occurs.)

7.3.4 POLARITY AMPLIFICATION

Zygotes clearly possess a mechanism for amplifying and stabilizing the feeble and labile axis induced by external vectors into the morphological asymmetry of the young embryo.

Polarization in unilateral light is not prevented by cycloheximide (an inhibitor of protein synthesis), colchicine (an inhibitor of microtubule assembly) or by sucrose (which lowers the potential of the external medium and thus prevents water uptake). However, addition of cytochalasin B (known to disrupt micro-

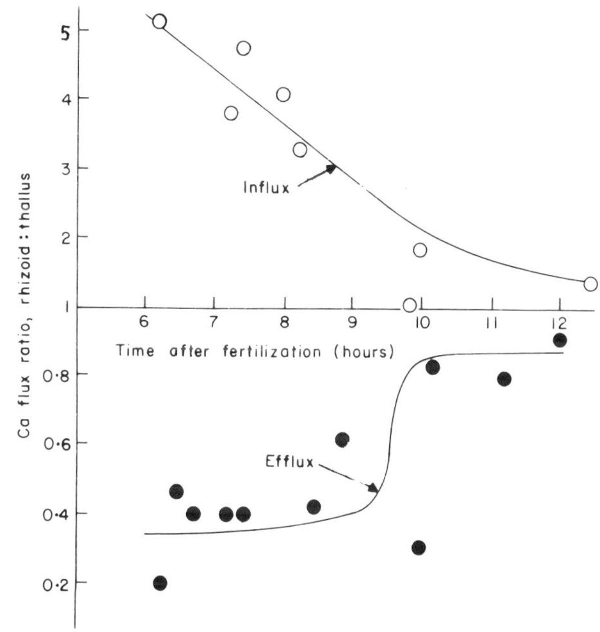

Fig. 7.11 Calcium influx and efflux ratios in developing *Pelvetia* zygotes, determined using the nickel screen method as in Fig. 7.10. Six hours after fertilization five times as much calcium enters the prospective rhizoid end as the prospective thallus end of zygotes. Initially, three times as much calcium left the prospective thallus end as the prospective rhizoid end of zygotes. (From Robinson & Jaffe, 1975, with permission.)

filaments and affect membrane function) at the same time prevents polarization by light (i.e. the polar axis remains labile) although it does not disrupt orientation of a previously induced axis. Amplification appears therefore to be independent of several important metabolic functions.

The amplification and stabilization of a labile polarity has been suggested to occur by self-electrophoresis (see Section 7.3.1.3). In this hypothesis, natural electrical fields across a cell cause charged cytoplasmic and membrane components to form gradients and these then act back to amplify and stabilize local changes induced by the polarizing factor. An alternative hypothesis is that the site of growth is that at which the membrane potential is changed most from its normal value. The two hypotheses appear to emphasize different aspects of what is likely to be a common electrical mechanism, and it is not yet known which of the two is the more important in the establishment of a stable polarization.

The foregoing, representing the present state of knowledge of polarity induction in *Fucus*, is summarized elegantly in a diagram by Weisenseel (1979), (reproduced in Fig. 7.12). As can be seen, there are still many gaps in our knowledge, which can only be filled by further research, in particular on vector perception, stabilization and amplification and the site of polarity.

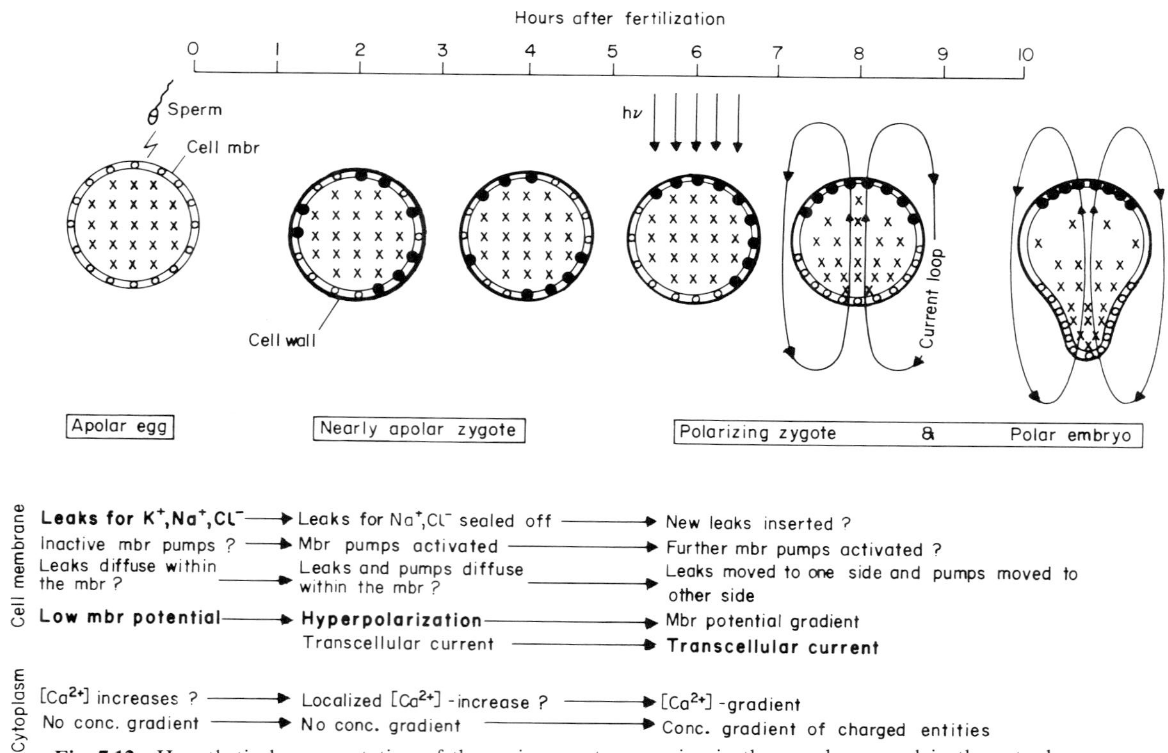

Fig. 7.12 Hypothetical representation of the major events occurring in the membrane and in the cytoplasm during the early development of a fucoid zygote polarized by unilateral light. Properties and events that have already been established experimentally are written in bold letters; those needing more experimental support are in normal letters; speculative points are labelled with a question mark. mbr, membrane; conc. concentration; ○ membrane leaky site; ● active membrane pump. (From Weisenseel, 1979, with permission.)

7.4 POST-FERTILIZATION CHANGES AND ZYGOTE DEVELOPMENT

7.4.1 EARLY CHANGES IN ULTRASTRUCTURE

Following sperm entry there are rapid changes in the ultrastructure of the egg. Alginic acid, contained in peripheral vesicles in the unfertilized egg, is rapidly released to form an incipient cell wall (Fig. 7.6 and Section 7.2.3.1). The envelope surrounding the sperm nucleus disintegrates as the latter migrates through the cytoplasm. The sperm eyespot and mitochondria migrate behind the sperm nucleus and prior to fusion the membranes of the egg nucleus become convoluted along the surface nearest to the sperm nucleus. Many egg mitochondria aggregate producing an asymmetry that may determine the orientation of the polar axis since it reflects the site of sperm penetration (Brawley *et al.*, 1976). In the absence of external gradients the asymmetrically-distributed stimulus of sperm entry is sufficient to orientate the polar axis of the developing embryo (Section 7.3). There is also marked hypertrophy of Golgi bodies and the accumulation of various types of cytoplasmic vesicle containing cell wall materials.

7.4.2 METABOLIC CHANGES FOLLOWING FERTILIZATION

Little attention has been given to changes in nucleic acid or protein synthesis following fertilization and there are no studies which have used modern techniques. In an investigation of *Fucus vesiculosus* zygotes treated with inhibitors of RNA synthesis (actinomycin D) and protein synthesis (cycloheximide and puromycin), Quatrano (1968) concluded that specific periods of RNA synthesis (1–5h) and protein synthesis (9–13h) were essential for the final expression of polarity. However, the effects on the individual components required for polarity to be expressed have not been investigated and the work with inhibitors requires substantiation by alternative techniques.

Increased rates of respiration and protein synthesis have been observed following fertilization. The former is likely to arise from the rapid breakdown of the storage glucan laminaran which decreases during the first 7h after fertilization. The enzyme mediating this breakdown appears to be a membrane-bound β-1, 3-exoglucanase (Quatrano & Stevens, 1976). The released glucose may provide building blocks for synthesis of cellulose or other polymers, as well as a substrate for ATP formation. The events following fertilization will take place in the absence of light and any external carbon source. However, an endogenous energy source is required for cellulose synthesis at least, since this process is blocked by an uncoupler of oxidative phosphorylation (p-trifluoromethoxyphenylhydrazone) (Quatrano & Stevens, 1976). The ability of eggs to fix CO_2 in the dark (unpublished results from this laboratory) may also result in a significant supply of carbon substrates for protein synthesis and wall formation.

7.4.3 LATER CHANGES. DEVELOPMENT OF A POLAR EMBRYO

At approximately 12h after fertilization a marked change in the distribution of cytoplasmic components occurs, prior to the first visible appearance of polarity. In *Fucus vesiculosus* the zygote nucleus becomes polarized with finger-like projections pointing towards the presumptive rhizoidal pole of the cell, and in *Fucus serratus* the rhizoidal protruberance is formed at approximately 16h. This pole of the cell will eventually develop into the multicellular rhizoid, the root-like structure which anchors the plant to the substratum. The opposite pole develops into the branched thallus.

Chloroplasts accumulate at the thallus pole and mitochondria, osmiophilic bodies and Golgi bodies at the rhizoidal pole (Quatrano, 1974). This pattern of distribution of Golgi bodies is related to asymmetry in the development of cell wall components, particularly sulphated fucans (Callow *et al.*, 1978b; Fig. 7.13 and 7.17a–e).

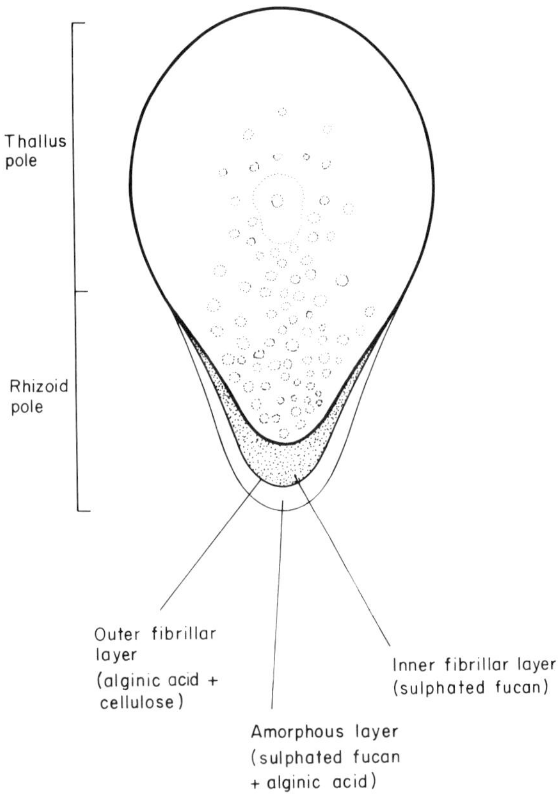

Fig. 7.13 Zygote of *F. serratus* 20h after fertilization showing relative thickness and composition of cell wall layers. The distribution of sulphated fucan-containing vesicles is also shown.

7.4.4 CELL WALL DEVELOPMENT

7.4.4.1 *Cell wall polymers*

It is pertinent here to outline briefly the elementary structures of the polysaccharide components of the *Fucus* cell wall. Cellulose occurs only in relatively small quantities in the mature plant and the chief structural role is taken on by alginic acid, which is composed of D-mannuronic and L-guluronic acids (Fig. 7.14a and b). These occur within the molecule in three types of block subunit, viz. M–M blocks, G–G blocks or M–G blocks (Fig. 7.14c). Within the plant alginate occurs as a mixed salt, the amount of calcium being mainly responsible for its gel strength (Rees, 1969).

Fucoid seaweeds also contain sulphated polysaccharides (sulphated fucans) typified by the predominance of fucose and sulphate, although other constituents such as galactose, xylose, glucuronic acid and protein are usually also present. Fucoidan is the most well-known of a family of glucuronoxylofucans, and a possible representation of its structure is shown in Fig. 7.15 (Percival & Mian, 1973). Although α1–2 linked fucopyranose units are the most common (with sulphate on C_4), α1–4 and α1–3 linkages (with sulphate on C_3 and C_4 respectively) are also present.

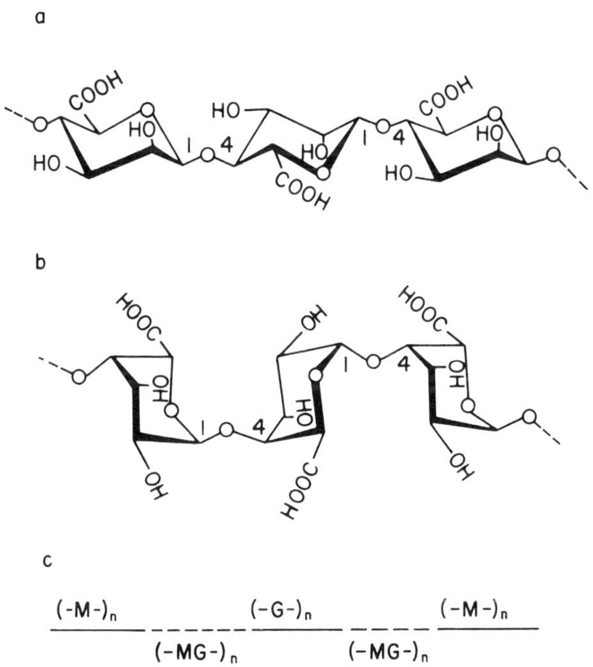

Fig. 7.14 Structure of alginic acid. Segment of polymannuronic acid chain (a); segment of polyguluronic acid chain (b); possible block structure (c).

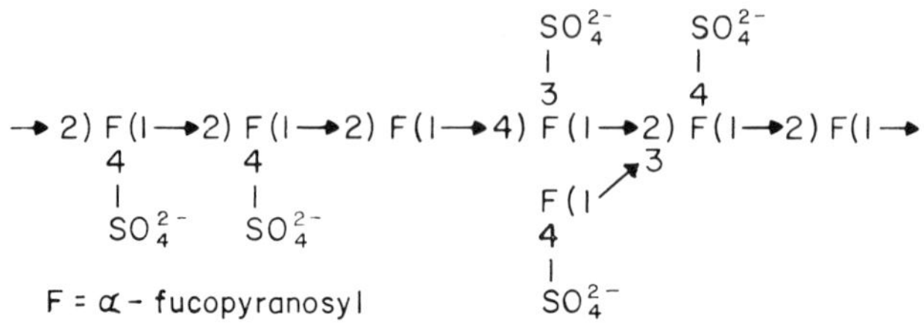

Fig. 7.15 Structure of fucoidan.

The rapid release of alginic acid following sperm entry has already been referred to (7.2.3.1). The composition of zygote walls has been studied in three *Fucus* spp. viz. *F. vesiculosus*, *F. distichus* and *F. inflatus* (Quatrano & Stevens, 1976). Although initially the wall is composed of equal amounts of alginic acid and cellulose, by 4h a fucan (F1) is also present, characterized by a low fucose : xylose ratio and containing little sulphate. From 6–24h the proportions of the three major wall constituents are constant, viz. alginic acid (60%), cellulose (20%) and fucan (20%). However, at 12h, purified cell walls contain another fucan (F2) which is electrophoretically distinct from fucan F1, being highly sulphated and having a high fucose : xylose ratio. Fucan F2 is responsible for the metachromatic staining reaction of toluidine blue, characteristic of the rhizoidal region of zygote walls, which is the first observable manifestation of the establishment of polarity (Fig. 7.17a–c).

Changes in the plasma membrane associated with microfibril synthesis following fertilization in *Pelvetia* have been examined by freeze-fracture techniques (Peng & Jaffe, 1976). The distribution of particles (and associated microfibrils and vesicle fusions) is initially uniform over the whole zygote plasma membrane surface. Prior to rhizoid formation, incorporation of new membranes and secretory activity becomes highly concentrated in this region, thus morphological differentiation may be closely related to the physiological changes accompanying polarization.

7.4.4.2 *Localization of sulphated fucan synthesis*

After uptake, sulphate is first incorporated into the two universal high energy intermediates APS (adenosine 5′-phosphosulphate) and PAPS (adenosine 3′-phosphate-5′phosphosulphate) in the cytoplasm (Fig. 7.16 a & b). Sulphate is

a $$SO_4 + ATP \xrightleftharpoons{\text{ATP sulphurylase}} APS + PP$$

b $$APS + ATP \xrightarrow{\text{APS kinase}} PAPS + ADP$$

c $$PAPS + \text{fucan} + \text{acceptor} \xrightarrow{\text{sulphotransferase}} \text{fucan-}SO_4^{2-} + ATP$$

Fig. 7.16 Steps in the biosynthesis of sulphated fucan.

then donated from PAPS, probably to a fucan acceptor, as outlined in Fig. 7.16 c.

Autoradiography of zygotes (22h) of *F. serratus* pulsed with $^{35}SO_4^{2-}$ shows that fucan sulphation occurs in the Golgi bodies (Fig. 7.17d), the fucans then being transported in vesicles and preferentially secreted at the rhizoid tip (Fig. 7.17e). Two sulphated fucans have been recovered from Golgi-rich fractions isolated from zygotes treated as above. One is highly sulphated, of low molecu-

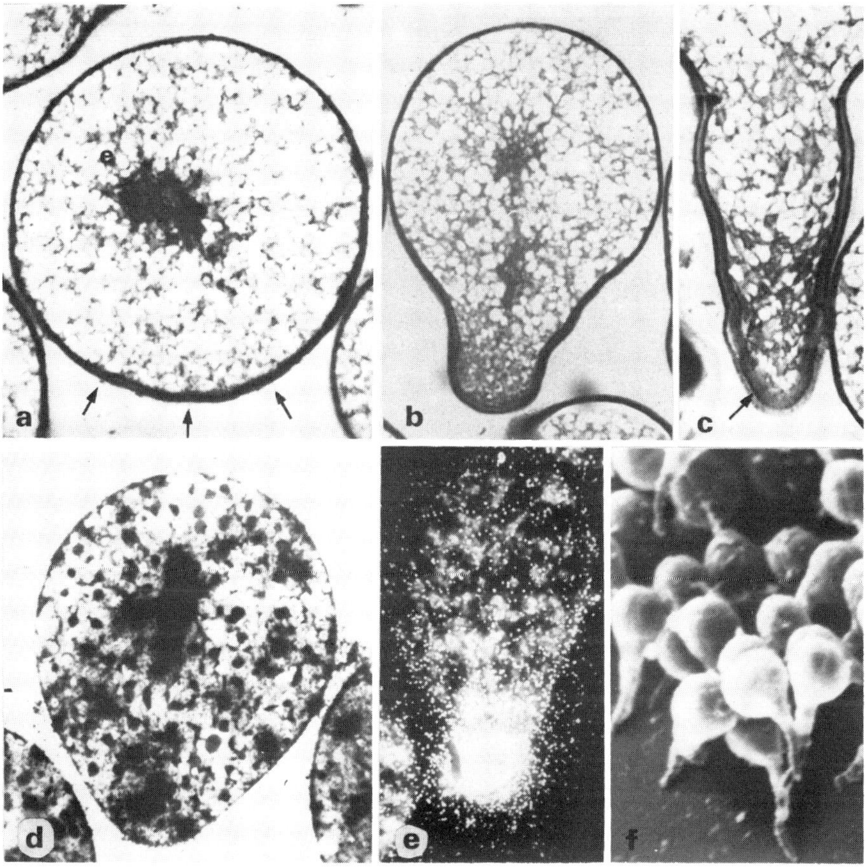

Fig. 7.17 a–c, L.S. through *F. serratus* zygotes at different stages of development, 1μm glycol methacrylate sections stained with toluidine blue, all showing metachromatic staining reaction of wall at rhizoid pole. (a) 16h zygote showing thickening of wall at presumptive rhizoid pole, ×600. (b) 24h zygote, ×750; (c) 48h zygote; three wall layers at rhizoid pole can be distinguished. The darker-staining middle region is the outer fibrillar layer (arrow); inside this is the inner fibrillar layer, but the outermost amorphous layer has been largely lost during the fixation procedure (×675). (d) autoradiograph of a 22h zygote labelled for 15min with $^{35}SO_4^{2-}$ showing dense patches of silver grains (black dots) in the cytoplasm, ×600. (e) autoradiograph treated as in (d) but with a 3.5h 'chase' period in unlabelled sulphate. Silver grains (white dots here due to dark ground illumination) are preferentially associated with the rhizoid pole, ×400. (f) SEM of 48h zygotes, ×120. (From Callow *et al.*, 1978b).

lar weight and may represent the initial sulphate acceptor molecule, whilst the second is less highly sulphated and of high molecular weight. Both Golgi-fucans are different from the fucans isolated from the cell wall and it has been proposed that further polymerization and elaboration of these may occur, either during transport through the cytoplasm in vesicles or after secretion into the cell wall (Callow *et al.*, 1978b).

7.4.4.3 *Expression of polarity*

Since the establishment of a stable polar axis is invariably associated with the deposition of sulphated fucan in the area of cell wall which will develop into the rhizoidal protruberance, one may ask whether these events are related. There is some evidence that such deposition of fucans preferentially at the rhizoid pole is not the event which establishes polarity, but one of the earliest expressions of a stable polar axis. Fucan sulphation is completely inhibited by cycloheximide and is not affected by cytochalasin B. However, as described in Section 7.3.4 these inhibitors have the opposite effects on polar axis fixation. In 16h embryos treated with cytochalasin B, sulphated polysaccharides are not localized at the rhizoid pole but are distributed evenly throughout the whole zygote wall (Novotny & Forman, 1975). Thus cytochalasin B does not affect sulphation or incorporation into the cell wall but apparently interferes with the directed deposition of fucan F2. Furthermore, Quatrano *et al.* (1978) have shown that the enzymatic sulphation of fucan is necessary for its transport to and/or incorporation into the rhizoid wall. Two possible mechanisms could control this type of intracellular molecular distribution, (a) a cytoplasmic electrical potential gradient or field which could separate molecules or particles (vesicles) on the basis of their net charge (electrophoresis) or, (b) a contractile mechanism involving actin and microfilaments (contractile). Although *Fucus* zygotes contain actin, and sulphated fucan deposition at the rhizoid pole is sensitive to cytochalasin B, there is no direct evidence in favour of the latter mechanism.The electric current which *Fucus* zygotes can drive through their cytoplasm (see Section 7.3.4) is sufficient to localize large macromolecules and small particles. However, although the net negative charge on the sulphated fucans is sufficient to account for its intracellular electrophoretic transport, the charge carried by the Golgi-derived vesicles within which it is contained is unknown (secretory vesicles from other systems do appear to be negatively charged). Furthermore, the results outlined above indicate that the sulphated fucans contained within vesicles are more highly sulphated than those isolated from the walls. Determination of the extent to which the electrical fields generated in *Fucus* zygotes control localization of sulphated fucans in the rhizoid awaits further research.

FURTHER READING

BONEY A. D. (1966) *A Biology of Marine Algae.* Hutchinson Educational, London.
ENDE VAN DEN H. (1976) *Sexual Interactions in Plants.* Academic Press, London.

HESLOP-HARRISON H. (1978) *Cellular Recognition Systems in Plants.* Studies in Biology, no. 100. Edward Arnold, London.

QUATRANO R. S. (1974) Developmental Biology: development in marine organisms. In Experimental Marine Biology (ed. Mariscal R. N.) pp. 303–46. Academic Press, New York.

WEISENSEEL M. H. (1979) Growth movements. In *Encyclopaedia of Plant Physiology*, New Series, vol. 7, (eds. Haupt W. & Feinleib M. E.) pp. 485–505, Springer-Verlag, Berlin.

MOLECULAR ASPECTS OF VEGETATIVE DEVELOPMENT

CHAPTER 8

ROOT DEVELOPMENT

8.1 INTRODUCTION

The development, growth and internal structure of roots have all been topics of scientific enquiry for well over a century. The observation by Sachs (1873), and before him by Duhamel du Monceau (1758), of the way in which marks on the root surface were displaced by growth, led to the familiar concept that roots have a meristematic zone, an elongation zone and a zone of maturity. The development, in the last century, of methods of cutting and staining tissue sections, together with improvements to the light microscope, allowed notable botanists such as Hanstein (1868) and Janczewski (1874) to lay much of the foundation of our present knowledge of the anatomy of roots. The invention of the electron microscope placed in our hands, 30 or so years ago, a tool that still continues to open up new vistas in the study of cell structure. More recently, the increasing sophistication of biochemical and biophysical techniques has made feasible attempts to explain the mechanisms whereby the language of the genetic material is translated into the cellular and developmental processes that cause organisms to manifest themselves as living entities.

The root and the root system are complex living structures and theoretically might be considered 'explicable' within the frame of reference of molecular biology. However, the essential characteristics that define a root and its cells, and which contrast them with, say, a shoot, are unknown. Therefore, at the present time, the best that can be done is to discuss the properties possessed by the cells that constitute the root in terms which, even if they are not explicitly molecular, at least hint at an underlying molecular basis for these properties. The subject of this chapter on the molecular biology of root development is the two intertwined levels of tissue organization that exist in every living organism. The first is concerned with the nature of the individual cells. The second is concerned with the way in which the cells interact, both amongst themselves and with the external environment, for it is through these interactions that cell differentiation comes about during growth. Most of what follows applies to actively growing roots some time after seed germination. But root development is also an integral part of embryogenesis, and so it is at this point that the discussion of the molecular biology of root development begins.

8.2 THE EMBRYONIC ROOT AND ROOT NEOFORMATION

The first cleavage division of the zygote results in a proembryo that is asymmetric with respect to both cellular contents and shape. This asymmetry contributes to the differentiation of the suspensor/basal cell complex from the cells that give rise to the embryo proper. Only when the embryo reaches the globular stage do the first traces of root organization appear. This is indicated by a feeble staining reaction for RNA and proteins in a group of cells that abut the suspensor complex. These cells are the forerunners of the quiescent centre which, within actively growing roots, occupies a zone at the pole of the stele and cortex and whose cells have low rates of synthesis (protein, RNA) and cell proliferation. The genesis of this quiescent zone may play a role in the further structuring of the root since its presence causes the surrounding proliferative cells to be separated into two groups: one group giving rise to the initials from which the root cap is formed, the other group contributing to the meristem of the remainder of the root. The molecular mechanism at work in the development of the quiescent zone is unknown. Quiescence may be due to some influence from within the embryo, or from the suspensor, or could even by programmed to operate within certain cells at a particular stage in embryogenesis.

The embryo is a difficult system with which to perform experiments that could throw light on the essential question of how root development is determined. A more amenable cell system which seemingly has the potential for elucidating the conditions specific to root development is an in-vitro cell culture system. One of the most useful of these is one in which a few layers of non-dividing cells are placed in contact with a medium, the composition of which can be varied, that causes the cells to divide and execute different morphogenetic programmes. In this way cells of the epidermal and sub-epidermal layers from flowering stems of *Nicotiana tabacum* can be induced to form either roots, shoots or flowers (these are called neoformations) depending on the auxin : cytokinin ratio in the medium (Tran Thanh Van, 1977). Before being placed in culture the epidermal layers contain terminally differentiated cells and it is supposed that they contribute relatively little information, in the form of endogenous growth regulators, to the subsequent development of the neoformation. The molecular requirement for the evocation of root formation is a high molar ratio of auxin to cytokinin; other ratios favour the evocation of other tissues (Table 8.1). Since the molecular requirements for the various neoformations are quite distinct, the question arises as to how these particular chemical conditions divert the epidermal cells into one or other of the morphogenetic programmes possible for them. Differences in the orientation of cell divisions are ultimately responsible for the distinctive morphologies of the tissues, but there are also changes to the cytoplasmic organelles during the early stages of neoformation (Tran Thanh Van & Chlyah, 1976). Although relatively modest, these changes must reflect subtle modifications of the cells' metabolism in the

Table 8.1 Optimal conditions for different types of morphogenetic programmes, the starting point being small explants of floral branches of tobacco

Type of neo-formation	Concentration (M) Sucrose	Growth substances*	Ratio of auxin: cytokinin	Other favourable conditions
Vegetative buds	8.8×10^{-2}	IAA 10^{-6} BA 10^{-5}	0.1	Light
Floral buds	8.8×10^{-2}	IAA 10^{-6} K 10^{-6}	1	Light
Callus	8.8×10^{-2}	2,4-D 5×10^{-6} K 10^{-7}	50	
Roots	$\mathbf{2.9 \times 10^{-2}}$	**IBA 10^{-5}** **K 10^{-7}**	**100**	**Darkness**

* K = Kinetin; IAA = Indole-3-acetic acid; IBA = Indole-3-butyric acid; 2,4-D = 2,4-Dichlorophenoxyacetic acid; BA = Benzyladenine. pH adjusted to 5.0–5.1

Explants are 1–10 mm^2 composed of 3–6 epidermal and sub-epidermal cell layers from the floral branches of *Nicotiana tabacum* cv. W38. (Data from Tran Thanh Van, 1977.)

earliest stages of neoformation induction. Specific combinations of cytoplasmic and nuclear activities must underlie the execution of particular morphogenetic programmes, and these are differentially activated by the particular chemical composition of the medium.

It is logical to believe that root development, whether during neoformation or embryogenesis, is determined by a unique and temporally ordered pattern of gene activity, or chreode (Waddington, 1957). Therefore, the rhizogenic chemical environment present in the in-vitro culture system induces a particular chreode for root formation and it may be tentatively suggested that this artificial chemical environment is in some way equivalent to the natural *in vivo* environment which determines the same chreode for root development in the embryo. The early embryo depends upon both the endosperm and the suspensor for its growth; these latter cells are a rich source of auxins, cytokinins and gibberellins whence they may be transported into the embryo. The availability of these growth regulators to the embryo, and their distribution therein, may play some part in the initial differentiation of the embryonic tissue-types. Thus, a high ratio of auxin to cytokinin could obtain at the point of attachment of the embryo to the suspensor and predispose the cells here to form root tissue; a different ratio, or different absolute concentrations, could obtain at the other pole of the embryo and there favour shoot and cotyledon development.

In conclusion, the fundamental determinant of a rhizogenic developmental pathway probably lies in a particular chemical milieu within and around cells that are capable of responding to it. As the primordial root grows, so epigenetic controls come into play to bring about differentiation of the cell-types specific to the root; this step is associated with the deployment of other specific tissue-determining chreodes that are subsidiary to the primary rhizogenic chreode. At the same time these resultant patterns of differentiation contribute to the epigenetic system and feed back to the meristem which is creating the organ so that the overall pathway of organ development becomes fixed, or canalized. Figure 8.1 illustrates this scheme as it may apply to root development.

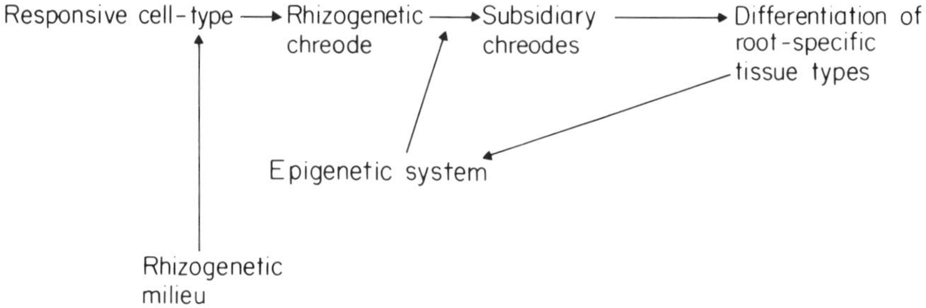

Fig. 8.1 Scheme to illustrate the development and canalization of root-forming tissue from a responsive cell-type (e.g. the proembryo or some other cell—such as a root stem-cell—capable of initiating root formation).

8.3 THE REACTIVATION OF ROOT GROWTH DURING GERMINATION

Within the seed of many plant species there is a dormant embryo containing a fully formed radicle and plumule. At this stage these two organs already possess most of the cell-types commonly found in the roots and shoots of the actively growing plant. In many instances germination is accompanied by a resumption of cell growth and division and the attendant cell differentiation within the apices of these organs.

Relatively little is known about the temporal sequence of reactivation of cell growth and division within the radicle. The early extension growth of the root consists of an exponential phase followed by a linear phase and somehow these are related to the development of the meristem and the elongating and mature zones in the young root. In their analysis of this problem, Kovalev & Obrucheva (1977) showed that in radicles of *Aesculus hippocastanum* the early exponential phase was due almost solely to the process of cell elongation while the change to a linear growth rate coincided with the entry of cells into division. Cell elongation at first occurred only in those cells proximal to where the meristem would appear, but later on, elongation occurred in this latter region also. Mature cells, i.e. cells which had completed their growth phase, were not recorded until the onset of the linear phase and thereafter the proportion of these cells in the radicle continued to increase. A rather different sequence of events was found to occur during the onset of root growth in *Quercus robur* (Kovalev & Obrucheva, 1976). Here, resumption of meristematic activity preceded activation of cell growth in the elongation zone. Later, at the onset of the linear phase of growth, cell elongation began to contribute more to root growth. Exactly how these two contrasting modes of growth reaction are controlled is not known, but the authors suggest that in *Quercus* the meristematic phase of growth is controlled by hormones whose source is in the seed, while the elongation phase is controlled by hormones made in the cells of the radicle.

A pattern of growth similar to *Quercus* was also found in adventitious roots of *Allium cepa* where, upon the resumption of root growth, cell division rate and mitotic index both achieved a constant value before a constant cell size was attained in the non-meristematic zone (Díez *et al.*, 1970). These authors (Díez *et al.*, 1971) suggest that in the early (exponential) phase of root growth there is a deficiency of auxin which limits the final size of the cells, since application of IAA (10^{-11}–10^{-9}M) to the roots during this phase caused the size of the mature cells to approach that found in roots in the linear growth phase. Pilet (1961) analysed roots of *Leus culinaris* of different age (length) for auxin content and found that young roots contain less auxin than older roots. He also suggests that the rate of root growth is regulated by their auxin content. Hormones probably do play a part in the cell growth and division associated with the onset of radicle emergence, but then so too would the development of an active hormone receptor and response system. At present it is not possible to say how the onset of root growth, particularly the apparently separate processes of cell division and enlargement, are controlled at the molecular level.

Cell growth and proliferation quickly achieve constant rates. It is the properties of the cells in steadily growing roots that will concern us from now on.

8.4 CELL DIFFERENTIATION IN THE ACTIVELY GROWING ROOT

All the cells of the actively growing root apex are generated by a meristem. The continual proliferation of cells within a meristem of constant size causes cells to be displaced into the zone of enlargement and from there into the zone of maturation. Actually, it is more correct to say that the apical meristem grows away from the cells it generates, but conceptually it is easier to consider the apex at a fixed point with cells being displaced, by divisions and growth, away from it.

The meristem may be considered to consist of two sets of cells. The first set is a stem-cell population, and it is from here that all cells in a root can trace their origin. In the roots of higher plants these stem-cells are located within the quiescent centre which lies at the pole of the root proper and whose members proliferate much more slowly than meristematic cells elsewhere. The second set of meristematic cells comprises the 'derivative' meristems, the cells of which are descendants of the stem-cells and make up the bulk of the meristematic cell populations of cap and root proper. It is from a derivative meristem that cells are displaced into the zone of elongation. When a root is in a steady-state of growth, the different properties of cells at points along the axis reflect the course of differentiation that cells undertake as they are displaced along the axis; the static image seen in cells of a root killed for observation is a reflection of their dynamic behaviour in the living tissue. Therefore, in order to follow the developmental history of any cell in the root it is sufficient to examine the properties of cells in the stem-cell zone, in the appropriate derivative meristem, and in the elongation and mature zones since this sequence reflects the course of differentiation taken by the cells. One may also view the steady-state root as a tissue along whose axis there are various sets of physiological conditions that support particular cellular activities, such as cell division and cell enlargement.

When seen in section, root apices of different species have different morphological appearances. Janczewski (1874), for instance, recognized five different morphological types of apex. Each type is due to specific patterns of cell division and growth in the meristem (Fig. 8.2). The types of cells that are recognizable are common from species to species. In all species the root apex is surmounted by a root cap. Cap cells for the most part differentiate in an acropetal sequence and they are generated by a cap meristem, or calyptrogen—a derivative meristem whose origin can be traced back to the quiescent centre/stem-cell zone. In the root proper, three major cell types are recognized: epidermis, cortex and stele; again, all the constituent cells have their origin in the stem-cell zone.

In discussing molecular aspects of differentiation it is only fair to say that there has been no study of the complete developmental sequence of any cell type in the root of any species that would allow the development of its structure and function to be fully interpreted in molecular terms. Such a study would require, at the least, a qualitative and quantitative description of the ultrastructure and molecular constituents of cells as they pass from a stem-cell state to maturity. What follows, therefore, is a collation of information from roots of a variety of species. In this connection it is convenient to describe separately the developmental changes that occur in (a) the cytoplasm and cell wall and (b) the nucleus.

8.4.1 CYTOPLASMIC DIFFERENTIATION

One of the most detailed studies of cytoplasmic changes during root cell differentiation pertain to *Zea mays* (Leech *et al.*, 1963) where observations were made on cells at various locations ranging from the quiescent centre up to about 10cm basal to it. Reference to this species, and this publication, serves as a good basis for a description of the process of cell differentiation in terms of the evolution of

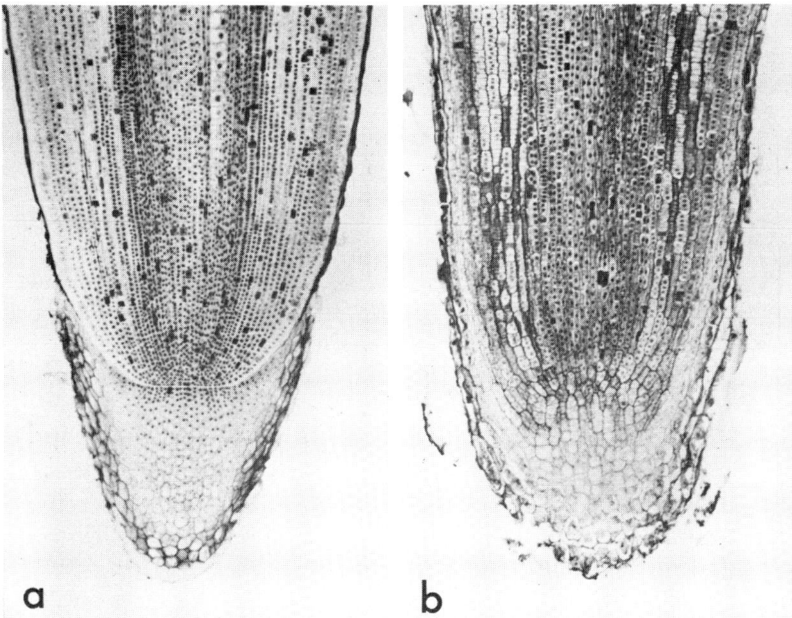

Fig. 8.2 Median longitudinal sections through root apices of (a) *Zea mays,* (b) *Tropaeolum majus* to illustrate contrasting organizations of the cells. In *Zea* an obvious boundary separates the cap from the root proper; in *Tropaeolum* no such boundary exists.

The apices have been impregnated with silver which causes the nucleoli to appear black. Some cells also are entirely black: these are cells in mitosis. Magnification: (a) $\times 61$; (b) $\times 98$.

cellular organelles, such as endoplasmic reticulum, dictyosomes and vacuoles. Each of the major tissues—epidermis, cortex and stele—will be described in turn. It is a basic assumption that each tissue develops as it does because of a particular pattern of activity of the organelles within its cells. This premise then allows certain, perhaps exaggerated, organellar changes also to be discussed, since a full understanding of organellar development and function, with all its variation, gives an insight into the molecular basis of cell differentiation and function.

8.4.1.1 *Epidermis*

The epidermis consists of cells whose dictyosomes are particularly active in producing materials from which are constructed both a thick wall (15μm thick in *Zea*) and a mucilaginous sheath at the root surface (Leech *et al.*, 1963; Clarke *et al.*, 1979). This surface material is part polysaccharide and part phenolic and serves as a protective skin around the root that is both strong and slippery enough to counter the abrasion that the root experiences as it grows through soil, and extensible enough to accommodate cell expansion. The polysaccharide component is secreted into the wall of meristematic cells only, while the phenolic component continues to be secreted into the zone of elongation (Clarke *et al.*, 1979). The dictyosomes in the epidermal cell lineage become active in secretion quite abruptly at the boundary of the quiescent centre. One cause of dictyosome activation is probably associated with the development of the enzyme complement within this organelle. Dauwalder *et al.* (1969) found that dictyosome-associated phosphatase activity increased during the maturation of root cap cells and correlated well with an intensification of polysaccharide synthesis. The same may be true of epidermal cells since the product of their dictyosomes is probably chemically similar to that made by the dictyosomes of the cap.

Coinciding with the secretion of polysaccaride material is the mitotic activity of the epidermal cells. Since one of the functions of the dictyosomes is also to contribute material to the new cell plate at cytokinesis there must be some differentiation of dictyosome behaviour during the meristematic phase of the epidermal cells. Mollenhauer & Mollenhauer (1978) investigated the behaviour of the dictyosomes in relation to cell wall and cell plate formation in the epidermis of *Zea.* The pattern of dictyosome activity changes as cells proceed through the mitotic cycle (Table 8.2); the size of the dictyosome vesicles becomes smaller when the cell enters mitosis from interphase, and at the same time they are diverted from their predominant site of fusion at the plasmalemma of the outer wall to the developing cell plate. The changing destination of dictyosome vesicles during the mitotic cycle is supported by the finding that less radioactively-labelled wall precursor material is incorporated into the outer epidermal wall during mitosis than during interphase, most of the material going into the cell plate (Dauwalder & Whaley, 1974). These results suggest that the contents of the vesicles may vary in composition with the phase of the cell cycle and also clearly indicate a cycle-associated change in the pattern of vesicle fusion with cellular membranes.

Table 8.2 Pattern of dictyosome activity in mitotically-active epidermis cells of *Zea mays*

Phase of mitotic cycle	Mean no. of vesicle fusions per 100μm length of wall:				Mean no. of vesicles per dictyosome cross-section	Mean diameter (μm) of dictyosome vesicles
	Inner wall	Transverse wall	Outer wall	Cell plate		
Interphase	1	1	45		3.2	0.15
Mitosis	3	5	13		1.3	0.11
Cytokinesis	4	3	18	254	1.7	0.11

In all cases the mean values recorded for interphase cells are significantly different from those for cells at mitosis and cytokinesis.
(Data from Mollenhauer & Mollenhauer, 1978.)

As epidermal cells mature some of the cells may form a root hair. At first there is a localized deposition of polysaccharides at the outer epidermal cell wall at the point where the hair will develop (Clarke *et al.*, 1979). In many species of plants this site is towards the apical end of the cell and indicates a polarity within the epidermal cell at the time when it is competent to form a hair. It has been suggested that protons flow into the apical end of the cell and here help to initiate growth of the hair (Weisenseel *et al.*, 1979). The earliest phase of hair growth involves the rupture of the thick surface layer of the outer wall of the cell. Growth of the hair is rapid (elongation to a length of a few millimetres may be complete in 12h) and occurs by the incorporation of dictyosome-derived wall material at the tip. The polarized deposition of this material may be related to the polarized flow of protons through the hair—protons entering at the growing tip and leaving at the base (Weisenseel *et al.*, 1979). When mature, the root hairs of some species (e.g. *Sorghum* spp.) are active in secreting mucilage which seems to be elaborated in the endoplasmic reticulum rather than in the dictyosomes (Werker & Kislev, 1978); the composition of this mucilage is unknown. Not all epidermal cells produce a hair; other factors involved in hair determination will be mentioned in Section 8.5.3. The ultrastructure of elongating root hairs of *Raphanus sativus*, and in particular the arrangement of microtubules just beneath the plasmalemma and their relationship to the cellulose microfibrils of the developing cell wall, has been described by Newcomb & Bonnett (1965).

8.4.1.2 *Cortex*

In the meristematic zone the cortex is a tissue recognizable by its cuboidal cells. Early in their life, these cells develop a conspicuous vacuole (Leech *et al.*, 1963). In the cortex of *Euphorbia*, and probably other plants as well, this cellular compartment originates from the dictyosomes. According to Marty (1978), the cisternae of the dictyosomes produce small vesicles (provacuoles) that coalesce to create a tubular labyrinth within the cytoplasm. At certain points many tubules anastomose to form a membranous cup (the autophagic vacuole) which engulfs and then digests a portion of the ground cytoplasm as well as the inner membrane of the provacuolar cup. The outer membrane remains and forms the tonoplast of the vacuole. The provacuole contains lytic enzymes (phosphatases, esterases, etc.) which cause the dissolution of the engulfed cytoplasm. The numerous vacuoles so formed, go on to fuse and produce a smaller number of larger vacuoles.

Another characteristic of cortical cells is the presence within them of starch-containing plastids called amyloplasts (Fig. 8.3a). It is not clear why some cells within the root contain amyloplasts whilst others do not, even though the proplastids within the latter cells possess the enzymatic system that is capable, under certain circumstances, of elaborating starch. For instance, in an investigation of starch synthesis in roots of *Pisum sativum*, Dyar (1950) noted that, after incubating sections of roots in glucose-1-phosphate, starch formed in all tissues

Fig. 8.3 Median longitudinal sections through root apices of (a) *Zizania aquatica*, (b) *Zea mays* to illustrate contrasting modes of origin of cell-files in the cortex. In *Zizania* new cell-files arise by longitudinal divisions of cells in the innermost layer of the cortex (arrowheads) and are recognized by a 'T' junction of the walls; in *Zea* new cell-lineages arise by longitudinal divisions of cells in lineages across the whole width of the cortex (arrowheads). Note that the cortical cells of *Zizania* are filled with starch grains. The material has been stained by the periodic acid—Schiff's reagent schedule. Magnification: (a) × 450; (b) × 380.

of the apex, the stele reacting more strongly than the cortex—the converse of the normal pattern of starch distribution. The presence or absence of starch within a tissue is probably related to the distribution of the enzyme invertase. There are two types of invertase, an acid and an alkaline invertase; in roots of *Pisum*, the former is nearly all confined to the cortex and is possibly located within the vacuole, the latter is most active in the stele (Lyne & ap Rees, 1971). Whilst no detailed correlation has been made at the cellular level between enzyme activity and starch content, it seems that acid invertase could be responsible for the equilibrium between free sucrose and starch (Lyne & ap Rees, 1971; Ricardo & ap Rees, 1970) since the level of its activity is inversely related to the free sucrose content. The presence or absence of acid and alkaline invertases also correlates with root form in a plant such as sugar beet (*Beta vulgaris*) where both fibrous roots and fleshy tap roots are developed. The differentiation of sucrose-storing parenchyma (of cambial origin) in the tap roots is accompanied by a loss of the acid invertase and the retention of the alkaline invertase (Silvius & Snyder, 1979).

The vacuole compartment within the cortex may play an important role in the rapid growth of these cells after they leave the meristem and in the elongation of the root as a whole. The manner by which the development and growth of the vacuole is regulated is still an open question, but in the early stages may depend upon the accumulation of sugars and inorganic salts, particularly potassium, within the cells (Scott *et al.* 1968; Mott & Steward, 1972). Subsequent vacuole growth may be the result of the high osmotic potential acquired early in its development.

Early during their development cortical cells increase in girth and the walls of some neighbouring cell-files separate from each other. This wall separation, or splitting, proceeds towards the growing tip like the opening of a zip and results in the development of intercellular spaces that are filled with air and run the length of the root. The wall at the edges of the cell is modified in advance of the split and can be seen as a layer of material resistant both to pectinase and cellulase, and which may also contain acid phosphatase (Roland, 1978). This layer is like a protective coat over the cell wall and constitutes a barrier to the leakage of solutes from the cell into the intercellular space. The separation of the neighbouring walls may be a consequence of some modification to the pectinaceous intercellular material, such as the removal of calcium from it. If so, this modification must be positionally controlled since it affects the lateral walls but not the transverse walls.

8.4.1.3 *Endodermis*

The endodermis is a layer of cells that surrounds the vascular cylinder and separates it from the cortex. A particular feature of the endodermis is the presence of the Casparian band that runs around the anticlinal cell wall. Endodermis development is usually divided into four phases which relate to successive modification of the cell wall, the cell itself during these phases going from a relatively undifferentiated meristematic cell to a vacuolated cell with thickened cellulosic walls traversed by pits (Robards *et al.*, 1973; Haas & Carothers, 1975; Scott & Peterson, 1979a,b).

The Casparian band develops in the first phase of development and appears electron-opaque in glutaraldehyde/osmium-fixed roots. It is believed to consist of polyphenolic and lipid substances intermingled with the cellulose of the primary wall. Later in development, suberin is deposited over the inner surface of all walls of the endodermal layer. In the final phase of development cellulose is deposited over the suberin and in some cases the two materials are deposited alternately to give a layered structure to the wall. Any special feature of the cytoplasm correlating with these modifications at the wall is difficult to see, though Scott & Peterson (1979a) recorded suberin-like material present in cisternae of the endoplasmic reticulum and also a clustering of cytoplasmic organelles close to the site of the forming Casparian band.

The endodermis of *Gossypium hirsutum* (cotton) has an exaggerated cytoplasmic differentiation relative to other species (Mueller & Beckman, 1976; 1978) and may give a clue to the physiological state of the cell that favours the characteristic modifications of the endodermal cell wall. In this species the endodermis is particularly rich in phenolic compounds. Phenols can be converted by polyphenol oxidase (PPO) to quinones which are constituents of the Casparian band. PPO is detectable in plastids only within the endodermal cells (the plastids of cortex, epidermis and stele do not react). Robards *et al.* (1973) noted that at the time of suberization 'vesicles with opaque contents' (phenols?) were in close proximity to the wall; also, in roots of *Hoja carnosa*, Olesen (1978) found an association between suberization of exodermal cell walls and the breakdown of plastids and a release of their electron-opaque contents. (Exodermis is a cell-type structurally and functionally rather similar to endodermis and which, in some species, has a wall structure rather like a Casparian band.) Thus, it is possible that the suberization of the Casparian band and, later, of the thickening cell wall, are both under the control of enzymes that are present in the plastids.

8.4.1.4 *Stele*

The stele is a relatively complex tissue and when mature contains at least three cell-types: xylem, phloem and vascular parenchyma. Certain of the parenchyma cells are capable of renewed division some time after leaving the apical meristem. In those species whose roots show secondary thickening, parenchyma is converted into a meristematic cambium the derivatives of which differentiate into secondary xylem and phloem, adding to the pre-existing complement of these cells. Furthermore, certain cells of the pericycle (which is the outer cell layer of the stele) are stimulated to divide and the new cells proliferate as a primordium from which the apical meristem of lateral roots is formed (see Section 8.6.2).

Little is known of the detailed chemical composition of the walls of the various cell-types in the stele of roots, and the way in which it changes during differentiation, because the relevant cells are difficult to isolate at different stages in their maturation for chemical analysis. However, histochemical tests have been used to demonstrate certain aspects of the changing chemistry of stelar cells as they differentiate (Obrucheva, 1969). Clearly, differences in wall composition must exist if only to account for the fact that the cells visibly differ in shape and wall thickness (cf. Jensen, 1961).

Cytoplasmic changes also occur during differentiation and these relate to the eventual function of the cells. The protoplast of the xylem is totally destroyed in the course of differentiation as a result of the activation of autolytic enzymes. Phloem degenerates to a more limited extent; the nucleus is autolysed and cytoplasmic strands, which assist in the long-distance transport of sugars, differentiate from an initially homogeneous cytoplasm.

8.4.1.5 *Chemistry of wall modification in relation to cell differentiation*

From what has been said above concerning the differentiation of the various cell-types it is clear that one aspect of differentiation, particularly as it affects the morphology of the cells, is expressed at the level of the cell wall in terms of its chemical composition (Jensen, 1961). In the stem-cells of the root we assume that the cell wall has a composition which is 'primary' and common to this class of cells in all roots. Pectin, hemicelluloses and cellulose are the basic constituents of this primary wall. The morphological characteristics of the various cell-types in the root that progressively differentiate from the stem-cell population are attributable to particular modifications of this primary cell wall. A scheme to illustrate these modifications is shown in Fig. 8.4. The relative proportions of wall constituents are controlled, in part, by the relative activities of epimerases, synthetases and other enzymes that interconvert nucleoside diphosphate sugars, since all these enzymes are involved in polysaccharide formation; moreover, their activities can be regulated by growth hormones. Further, there is evidence for an interrelationship between wall constituents and the plasma membrane which can feed-back information that regulates the type of polysaccharide synthesized in the cytoplasm (Northcote, 1979).

8.4.1.6 *Modifications of the basic cell-types within the root*

There are circumstances in which the developmental pathway of the cell-types of the root departs from the 'typical'. Only certain examples will be mentioned, but they seem to indicate that the number of sequential events connected with the differentiation of a mature, fully functional cell can be extended past the point at which the cell's development would normally be complete. Thus, it seems probable that certain stimuli can release additional information (genetic and epigenetic) within the cell which can be used for this additional differentiation. Such modifications, if they could be experimentally manipulated, might be useful probes of the molecular events associated with them.

The epidermis, by virtue of its proximity to the external environment, might be expected to show environmentally-induced modifications to its structure in order that its role as a buffer between the interior of the plant and the potentially harmful exterior milieu may be optimized. The epidermis of roots of *Atriplex hastata*, a halophyte, when exposed to 400mM NaCl solution, develops, along the outer tangential wall, ingrowths that are characteristic of transfer cells (Kramer *et al.*, 1978). An analysis of the ionic composition of the contents of such modified cells indicates that they concentrate K^+ and exclude Cl^-, and by so doing maintain a favourable $K^+ : Na^+$ ratio in the shoots. *Phaseolus vulgaris*, which is not a halophyte, also develops a transfer-cell morphology in both epidermal and vascular parenchymal cells in response to salt stress (Kramer *et al.*, 1977). Modification of certain cells of vascular parenchyma to a transfer-cell morphology has also been described in roots of *Hieracium florentinum* (Letvenuk & Peterson, 1976). In this case these cells were sited adjacent to xylem and phloem at points where the vascular cylinder of lateral roots links with that of the primary root. No transfer cells were found in vascular parenchyma in regions of root not associated with lateral roots.

The function of transfer cells is still an open question, but they develop in

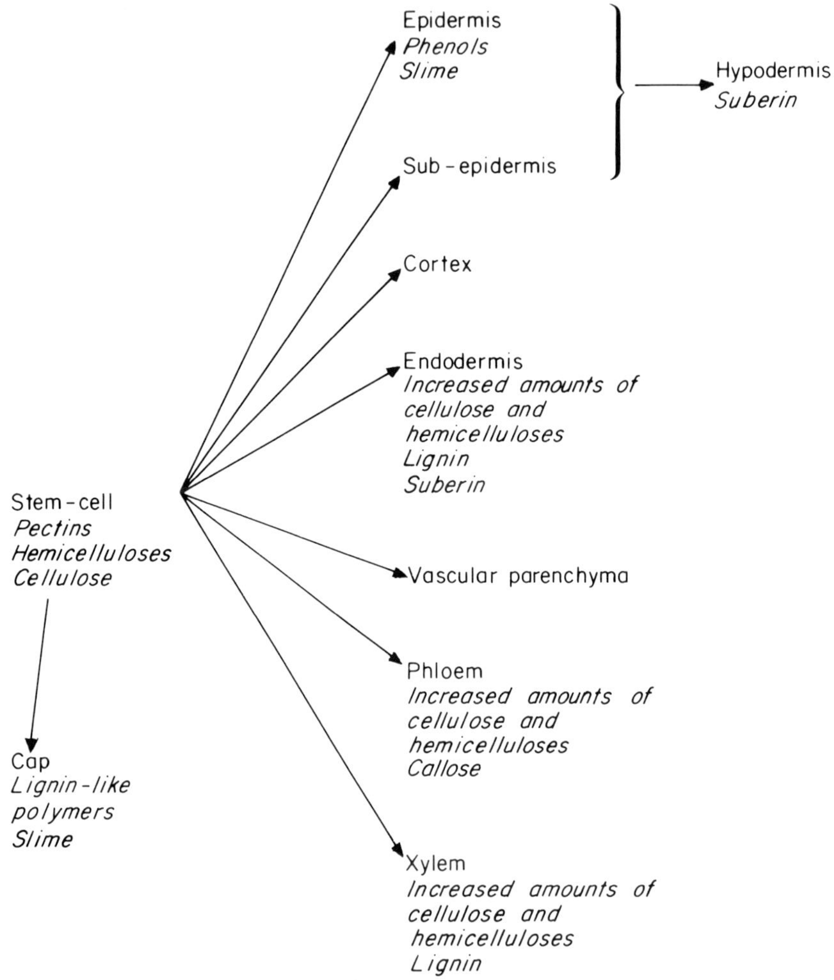

Fig. 8.4 Scheme to show how, in root apices, differentiation of cell types from a stem-cell is, in part, a consequence of modification of cell wall structure. The stem-cell population has a 'primary' wall structure of pectins, cellulose and hemicelluloses to which, as its descendents differentiate, are added various substances (these additions are indicated in italics).

It may be helpful to know more of the composition of the cell wall substances, though bearing in mind the fact that chemical analyses of these substances have been conducted on a wide variety of tissues that are usually not of root origin. *Pectins* are polymers consisting of D-galacturonic acid and methyl esters of D-galacturonic acid, L-arabinose, D-galactose and L-rhamnose. *Hemicelluloses* are polymers consisting of D-galactose, D-glucuronic acid, D-glucose, D-mannose, L-arabinose and D-xylose. *Cellulose* is a polymer consisting of β, 1 $\rightarrow$ 4 linked chains of D-glucose. *Suberin* consists of polymers of hydroxydecanoic acid, di- and tri-hydroxydecanoic acids, friedelin (a triterpenoid commonly found in cork) and other triterpenes. *Lignin* probably consists of polymers of p-hydroxycinnamyl alcohol, coniferyl alcohol and sinapyl alcohol. *Callose* consists

circumstances which lead to the belief that they are concerned with active transport of solutes over short distances. Similarly, the cause of their induction, either during normal development, or as a consequence of experimental procedures, is unknown. Gunning (1977) lists evidence indicating that their development is genetically determined. Internal and external environmental conditions which induce their development might act as a gene-switch (i.e., switch on 'transfer cell genes'). In the examples mentioned it may be significant that often it is ionic changes which induce the transfer cell since intracellular ionic balances have been implicated in regulating genetic activity in animal cells (Lezzi, 1970). On the other hand, an unusual ionic balance may upset the function and structure of the plasma membrane and the associated enzymes involved in wall synthesis in such a way that wall-building materials from the dictyosome vesicles can only be accepted at certain sites on the plasmalemma. Such a localized growth of the wall could lead to the building of the characteristic ingrowths.

The cortex of roots of many aquatic plants is modified to form large air-spaces (aerenchyma). In aquatic plants the air spaces can occupy from 30–60% of the total cross-sectional area of the root (Bristow, 1975). Aerenchyma is not confined to aquatic species since it can occur in the roots of plants growing in arid soil (Mulay & Saluja, 1975). The air-spaces may arise by the death of cells (when they do so, the tissue is known as lysigenous aerenchyma) and are continuous with those of the shoot and leaves. Their presence allows air to penetrate to the root tissue, which otherwise could become anoxic, and also may allow air and waste volatiles to diffuse out of the root into the rhizosphere. The development of aerenchyma in roots of *Zea mays* has been studied experimentally. McPherson (1939) showed that deprivation of oxygen encouraged the death and collapse of cortical cells resulting in the aerenchyma and was of the opinion that toxic products of anaerobic respiration in some way triggered these events. However, more recent work by Drew *et al.* (1979), also using *Zea*, does not confirm this. They showed that aerenchyma could be induced by bubbling air mixed with ethylene ($5\mu l \,.\, l^{-1}$) through the solution in which the roots were growing. Roots grown in this way are not deprived of oxygen but are exposed to an unusually (but not unnaturally) high concentration of ethylene. By analogy with other systems (e.g., petiole abscission zones) where ethylene causes cell separation,

---

of polymers of β, $1 \rightarrow 3$ linked chains of D-glucose. *Slime* consists of polymers that have some similarity to those found in the 'primary wall'; in *Zea mays* L-fucose is also present in the slime polymer.

During the differentiation of cell types from a stem-cell population and the subsequent cell maturation, the relative proportion of the sugars in the wall normally changes—a point which is illustrated by the results of analyses of hemicellulose sugars in the walls of xylem and phloem cells. Also, cell walls may contain polypeptides, but the composition and distribution of these is uncertain at present. Information for this scheme is derived, in part, from Pilet (1971).

ethylene within the root cortex may induce a cellulase that causes a dissolution of cell structure. Whether the mode of formation of the air spaces has any similarity to the formation of the intercellular spaces, mentioned on p. 195, is unknown. Since the air spaces form only in the cortex, and also show a spatial distribution within this tissue (intact groups of cells are interspersed with collapsed cells), their induction may be tissue-specific and at the same time subject to some sort of positional control within the tissue. In addition, the cell death associated with lysigenous aerenchyma formation may be an expression of a genetic control in much the same way as the death of the protoplast is an integral part of xylem cell differentiation.

8.4.1.7 *Root Cap*

This brief survey of differentiation of cell-types concludes with a description of the root cap. The root caps of terrestial plants show a common pattern of development. Their cells are generated by a meristem (calyptrogen) and are displaced both tipwards and laterally to the surface of the root cap during which process the cells differentiate. Differentiation is conspicuously evident from the changing appearance of the organelles as the cells pass from meristem to the surface of the cap (reviewed by Barlow, 1975; 1977a). There is good evidence that these changes are related to two important functions of the cap which are: (1) to perceive the gravitational field of the earth and provide a means whereby the root can orientate the direction of its growth with respect to the gravity vector (see Chapter 15), and (2) to synthesize a hydrophilic polysaccharide slime that lubricates the root in its passage through the soil.

The amyloplasts are apparently concerned with gravity perception. Plastids accumulate starch as the cells which house them are displaced from the calyptrogen. It is not known for sure why amyloplasts develop in the way they do in central cells of the cap (the statenchyme) and then degenerate in the outer cap cells. Huber *et al.* (1973) suggest that starch synthesis may be related to the removal, by mitochondria, of inorganic phosphate from the vicinity of the amyloplast; if so, the subsequent breakdown of starch may be related to further changes in the mitochondria later in cell development.

In contrast, information continues to accumulate on the synthesis of the slime which is a distinctive feature of the outer cap cells. The interconnected system of endoplasmic reticulum and dictyosomes is responsible for slime synthesis. The specific chemical composition of the slime, and its production by a particular group of cells, clearly indicate that this aspect of cap cell differentiation is a result of precisely regulated organelle differentiation. The macromolecular structure of slime is partially known and it consists of short $\beta,(1 \rightarrow 4)$-linked glucan chains surrounded by hydrophilic pectic material (Wright & Northcote, 1976). In caps of *Zea mays* one of the early precursors in the pathway of slime synthesis is a glycoprotein. Analyses of centrifuged cell homogenates showed that this glycoprotein is bound to a membrane fraction, a finding which led Green & Northcote (1978) to suggest that the slime polysaccharides are initially synthesized attached to proteins, possibly in the endoplasmic reticulum. The polysaccharide is then released from the protein before its eventual secretion from the root. The dictyosomes undoubtedly play a role in some aspect of slime synthesis and it is their vesicles which transport the slime to the plasmalemma (Fig. 8.5). There is evidence that before the dictyosome vesicles

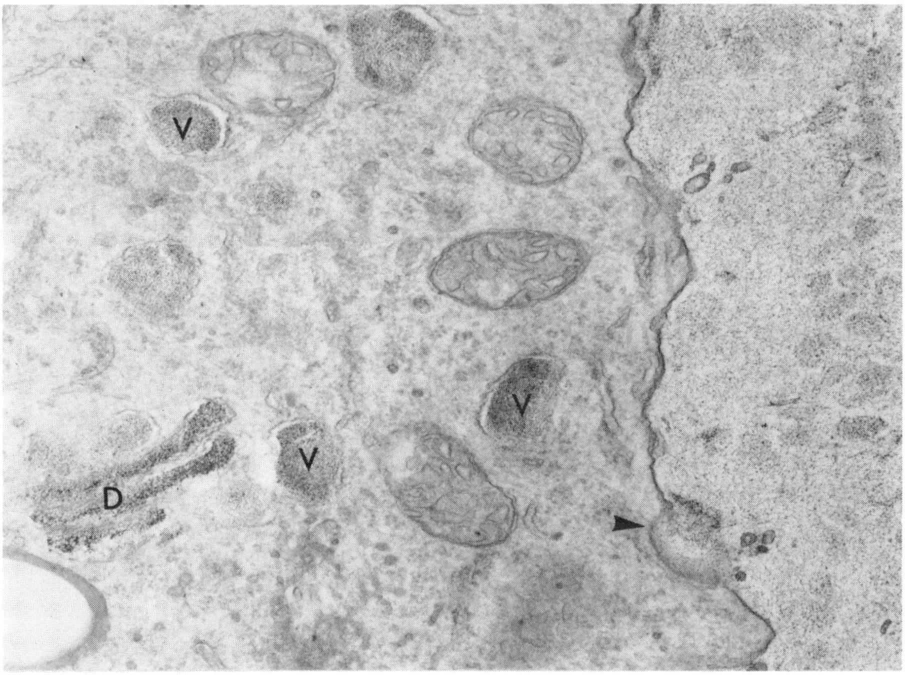

Fig. 8.5 Electron micrograph of a cell located near the periphery of the root cap of *Zea mays*. The section has been stained with silver proteinate (Thiery's reaction) which helps to identify polysaccharide material. The photograph shows dictyosomal cisternae (D) and vesicles (V) filled with polysaccharide slime material; one of the vesicles fusing with the plasmalemma (arrowhead); and the accumulation of the slime on the outside of the plasmalemma. Magnification × 47080. (Micrograph prepared by Dr J. A. Sargent.)

fuse with the plasmalemma their membrane structure is modified (Vian & Roland, 1972). This modification may be necessary to make the two membranes 'compatible' before fusion can occur.

There are experiments which indicate that the intracellular sugar content and composition may influence the pattern of dictyosome activity (Jones & Morré, 1973). Unfortunately, nothing is known of how sugar composition changes from cell to cell in the cap, but the synthesis and subsequent breakdown of starch during cap cell maturation could lead to local differences of free sugars within the cytoplasm of the cap cells. The breakdown of the starch in the amyloplasts may directly influence dictyosome activity and supply materials for slime synthesis.

The outer cells of the cap slough off from the surface and are lost. This property may be a result of the liberation, from the cap cells themselves, of enzymes that cause the cells to separate from one another (Hall *et al.*, 1965), though firm evidence for this is lacking. Paull & Jones (1976) have, however, detected a $\beta,(1 \rightarrow 3)$-glucanase in the apical 2mm segment of *Zea* roots and suggest that this enzyme participates in modifying the wall of the outer cap cell to cause the release of slime; it may also assist in the detachment of the cells from each other.

8.4.2 BIOCHEMICAL DIFFERENTIATION

As cells are displaced along the axis of the root their complement of enzymes and other proteins alters as they differentiate. A widely-used method of studying these changes is one where consecutive segments, of known length and cell number, are cut from zones along the length of the root and the activity of particular enzymes are then assayed in each segment (method reviewed by Heyes, 1977). It is assumed that the activity of the enzymes in the assay system reflects the activity within the tissue segments from which they were extracted. A picture can thus be built up of the manner in which enzyme activity changes with distance from the tip. Many of the enzymes that have been studied are ones whose biochemical function is known and so inferences can be drawn concerning their role in cell differentiation.

An example of biochemical differentiation along the root is apparent from studies on the enzymes of respiration. Respiratory activity (rate of O_2 uptake) of cells in roots of *Pisum sativum* increases in the cells as they are displaced through the meristem and elongation zone, and parallel studies have been made of the activities of the enzymes involved (Fowler & ap Rees, 1970; Sutcliffe & Sexton, 1974). Enzymes of both the glycolytic pathway and the Krebs cycle are most active in the meristem and young elongation zone, but in more basal regions of the root their activity declines somewhat. In contrast, enzymes of the pentose phosphate pathway have a relatively low activity in the meristem, but their activity increases in the elongation zone. Later, as the cells mature and stop growing 10–40mm behing the apex, the ratio of the activities of the two classes of respiratory enzymes tends to be constant. These results for enzyme activities are for whole root segments and take no account of any regional differences that may occur within the segment. However, attempts have been made to overcome this shortcoming. Sutcliffe & Sexton (1974) made histochemical analyses of longitudinal sections of pea root apices and showed that some of the enzymes studied in segments (e.g. malic dehydrogenase) had somewhat different levels of activity in stele and cortex. Similarly, in elongating and maturing regions of the root the relative activity of the pentose phosphate pathway was higher in the stele than in the cortex (Wong & ap Rees, 1971). The significance of the stronger activity of the pentose phosphate pathway in maturing stele may lie in the fact that this pathway generates $NADPH_2$ which is used as a reductant by many of the enzymes of lignin biosynthesis (Pryke & ap Rees, 1976).

Regional differences have also been found in sucrose metabolizing enzymes. According to Vieweg (1974), sucrose phosphate synthetase is found only in the stele of roots of *Zea mays* while sucrases predominate in the cortex. These enzymes are able to convert sucrose into glycosyl nucleotides, so it seems that the stele may be important for generating the precursors for the cell wall in all the different tissues of the growing apex from the sucrose coming into the root from the shoot. The cortex, on the other hand, may transform sucrose into material used in other metabolic pathways.

Quantitative differences have also been recorded in the isoenzyme patterns of malate and lactate dehydrogenases during the maturation of cells in roots of *Vicia faba* (Hadačová, 1974). Peroxidase is another enzyme that varies in its expression during cell maturation (Grison & Pilet, 1978); this enzyme can regulate numerous aspects of cell growth and differentiation which in some cases may be due to its power to destroy auxin. Other examples of variation in isoenzyme pattern are quite numerous; however, a given enzyme, such as galactosidase, may express variants during root differentiation of one species but not in another (Beneš & Hadačová, 1980). Where they are found the variations in isoenzyme patterns probably occur in response to changes in the local environment of the cells as they come to be located in different regions along the root. Perhaps these variations are an indication of a fine control of enzyme properties which thereby establishes the optimum pattern of metabolic activity in cells at given locations along the root.

During the development of any one of the various cell-types in the root one might expect the particular modifications, to the cell walls, for example, to show a correlate in the level, or activity, of a particular enzyme. Observations along these lines have been obtained with tissue cultures (see Northcote, 1979) and, by extrapolation, it would not be unexpected to find that the level of the enzyme phenylalanine ammonia-lyase increases in cells destined to differentiate as xylem, since this enzyme converts phenylalanine to a variety of compounds containing a phenylpropanoid nucleus that can be used in lignin synthesis (Khavkhin & Perelyaeva, 1970; Rubery & Northcote, 1968). Likewise, one might expect β,$(1 \rightarrow 3)$-glucan synthetase to act as a marker of the onset of phloem cell differentiation since this enzyme is involved in callose synthesis.

Another cause of biochemical differentiation must be differences in genetic activity: that is, differences must exist in the species of RNA synthesized and translated in different regions of the root. The base composition of RNA is known to change during root cell differentiation. For instance, in roots of *Pisum sativum* the purine : pyrimidine ratio of the RNA goes from near unity in the apex to 1.5 in the zone of maturity (Heyes, 1960); in roots of *Vicia faba* a similar change in ratio has been shown to be due to a change in the base composition of ribosomal RNA (Börner & Ramshorn, 1968). No-one has yet correlated the changes in RNA content with the types of protein made although it can be shown that the relative amounts of the species of proteins synthesized in young and old segments of roots are different (Ivanov *et al.*, 1974; Misharin *et al.*, 1974; Navarrete & Bernabeu, 1978).

A significant aspect of cell differentiation within the root relates to the rate at which cells grow. Cells in the elongation zone have a higher relative growth rate (that is, the increase in length per unit length per unit time) than cells in either the meristem or in the zone of maturation where cells cease their growth. One way in which cell growth is thought to be regulated is through changes in the extensibility of the cell wall with the actual lengthwise growth being effected by turgor pressure. Wall extensibility may, in turn, be regulated by protons extruded through the plasmalemma into the wall where they activate an enzymatic loosening of the wall material. Proton extrusion is effected by a plasmalemma-located ATPase and can be stimulated experimentally by IAA, and also by the diterpene glucoside, fusicoccin. Thus, differentiation entailing a change in the intrinsic growth rate of cells could be related to either the formation of a proton-extruding, plasmalemma-bound ATPase or the proton-activated wall-loosening enzymes. In connection with the former possibility, Gabella & Pilet (1979), analysing the response of consecutive segments of *Zea* roots to fusicoccin have shown that the degree of proton extrusion (judged by a pH change in the external medium) correlates with the amount of fusicoccin-induced growth exhibited by the segments. The fact that the greatest amount of growth (and proton extrusion) occurs in segments from the elongation zone, suggests that the increasing capacity of root cells to grow as they are displaced away from the apex may correlate with the development of the proton-extrusion system. The contribution of turgor pressure to changes in cell growth rate during differentiation may be related to the development of the vacuole and also the changing quantities of solutes, such as free sugars (see p. 194), which could assist in generating turgor within the cell.

An important functional aspect of roots is that they have a system for actively taking up ions from their surroundings. In *Zea mays* it is the maturing portion of the root apex that takes up Na and Rb (used in experimental work as a substitute for K) the fastest, while the meristematic zone has a slower rate (Eshel & Waisel, 1973). The patterns of the rates of uptake along the length of the root are markedly influenced by temperature and by the relative concentrations of the two ions in the medium. A rather similar pattern of K and Na uptake was found by Scott *et al.* (1968) for roots of *Vicia faba* though in this case it was the enlarging cells that had the highest rates of uptake. Travis *et al.* (1979) have attempted to analyse the development of another system in molecular terms and have followed changes in the activity of a plasmalemma-bound ATPase that imports K^+ into the cells of roots of *Glycine max*. Their results indicate that the activity of this ATPase is maximal in the meristem and declines in the elongating zone. It still has to be determined at what stage in cell maturation these plasmalemma-associated enzymes are first synthesised and how the enzyme activity is regulated. Clearly, there is much to learn about the development of the ion-uptake system.

8.4.3 NUCLEAR CHANGES DURING CELL DIFFERENTIATION

RNA molecules synthesized on the DNA of both nucleus and cytoplasmic organelles must play a primary role in determining the developmental pathways of cells. Thus, of fundamental importance to an understanding of cell differentiation is the elucidation of how progressive changes of the messages transcribed from the genetic material are brought about. The findings of Heyes (1960) and Börner & Ramshorn (1968) (see p. 203) show that the RNA species change in root cells during their maturation and, by inference, this reflects changes in the transcriptional activity of the DNA template. RNA polymerases can be detected in nearly all nuclei of the meristem and elongation zones (Olszewska & Kononowicz, 1979), apparently in rather constant amounts (Table 8.3). However, a

Table 8.3 Relationship between the capacity to synthesize DNA and RNA and the activity of DNA and RNA polymerases in nuclei of cells in different regions of the root cortex of *Zea mays*

Segment (mm from tip)	DNA synthesis*	DNA polymerase activity†	RNA synthesis‡	RNA polymerase activity§
0–1	39	7.3	23	17
2–3	24	9.1	17	18
4–5	12	8.2	6	20
6–7	7	6.8	4	16

* The percentage of nuclei labelled in autoradiographs following 2h of feeding with [^{3}H]thymidine.

† Mean number of silver grains derived from [^{3}H]thymidine and [^{3}H]cytidine triphosphates, seen in autoradiographs, following activation of the endogenous DNA polymerase.

‡ Mean number of silver grains, seen in autoradiographs, following 1h of feeding with [^{3}H]uridine.

§ Mean number of silver grains derived from [^{3}H]uridine monophosphate, seen in autoradiographs, following activation of the endogenous RNA polymerase.

(Data from Olszewska & Kononowicz, 1979.)

parallel study of RNA synthesis indicates that the capacity for this declines in the nuclei during cell maturation. Therefore, RNA polymerase activity must be progressively restricted in some way during cell maturation. The same situation also holds in regard to nuclear DNA polymerase: it is present in a roughly constant amount per nucleus during cell maturation, while the number of nuclei actually synthesizing DNA declines (Table 8.3).

Progressive structural alteration of the chromatin is one way in which the activities of DNA and RNA polymerases may be regulated in the course of cell maturation. Histones play a major role in giving structure to chromatin since they are the macromolecules which support the DNA double helix. Pairwise aggregations of the histone molecules H2A, H2B, H3 and H4 form an octomer around which is wound the DNA (1.75 turns per octomer). This unit, plus linker DNA, forms the nucleosomes (Kornberg, 1977; Sonnenbichler, 1979). H1 histone interacts with the linker DNA to stabilize the nucleosome chain. Phosphorylation, acetylation and certain other modifications of particular amino acids of the histone molecules alter the structure and transcriptional activity of the chromatin (Johnson & Allfrey, 1978; Sonnenbichler, 1979). The degree to which [^{3}H]actinomycin D binds to chromatin after selective removal of various histone fractions indicates the extent to which particular histones are responsible for modification of template activity. Results obtained by Kononowicz (1978) suggest that during the maturation of the cortex in *Zea* roots there is a progressive reduction in the number of binding sites for [^{3}H]actinomycin D on the chromatin (Table 8.4). The lysine-rich histones (mainly H1 histone) play a

Table 8.4 Binding of [^{3}H]actinomycin D and [^{3}H]sodium acetate to nuclei of cortical cells in different regions of the roots of *Zea mays*

Segment (mm from tip)	Control (mean no. grains)	Percentage increase in the number of silver grains due to [^{3}H]actinomycin D in autoradiographs after removal of the following histone fractions: Arginine-rich histones	Lysine-rich histones	All histones	Number of silver grains after binding of [^{3}H]-Na acetate
0–1	86	3	32	352	61
2–3	104	19	45	417	145
4–5	125	81	92	465	85
6–7	175	54	54	400	76

(Data of Kononowicz, 1978.)

greater role in this template repression than arginine-rich histones. Sodium [^{3}H]acetate incorporation has been used as a measure of histone acetylation (Kononowicz, 1978); this seems to be confined to the arginine-rich histones particularly in nuclei of the young elongating zone (Table 8.4). The degree of histone acetylation seems to be correlated with the intensity of RNA synthesis (Johnson & Allfrey, 1978; Sonnenbichler, 1979), so the high degree of acetylation in nuclei which are showing a decline in RNA synthesis is at first sight puzzling. However, the techniques are relatively crude and it could be that acetylation is occurring in only a small fraction of the total chromatin, whereas the assessment of the degree of RNA synthesis is for the nucleus as a whole.

Changes in nuclear morphology during differentiation have been documented, by electron microscopy, in a number of species. Nuclei in the cortex of roots of *Cucurbita pepo*, *Haemanthus katharinae* and *Pisum sativum* all show an increase in the proportion of condensed chromatin in the nucleus as cell maturation proceeds and this is coupled with a reduction in the size of the nucleoli and in the synthesis of RNA and protein (Chaly & Setterfield, 1975; Kononowicz *et al.*, 1976; Olszewska, 1976). Possibly it is the increasing compaction of the interphase chromatin during cell maturation that is responsible for the reduced activity of the RNA and DNA polymerases.

A comparison has not yet been made of nuclear ultrastructure in the stem-cells (quiescent centre) of the root and the derivative cell-types. RNA synthesis and nucleolar size are reduced in the former cells and, on the basis of the observations described above, it might be thought that the chromatin of their nuclei would be more condensed than in the nuclei of more metabolically active descendants. However, in *Zea*, the stelar portion of the quiescent centre actually has rather diffuse chromatin (Barlow, 1978a). This diffuse structure of the chromatin might indicate that a high proportion of chromatin is transcribed as RNA, although to be compatible with other lines of evidence the synthesis of this RNA must proceed at a slow rate.

The foregoing observations on nuclear structure could be construed as compatible with a hypothesis which proposes that as cellular development proceeds there is a progressive repression of genetic activity (Caplan & Ordahl, 1978). On the other hand, many people believe that development proceeds with the activation of previously repressed genes. If progressive repression occurs, then the stem-cell zones of roots and shoots would be sites where primary rhizogenic and caulogenic chreodes operate (cf. Fig. 8.1), but as cells move away from the stem-cell zone so positional information (epigenetic influences) differentially represses certain sets of genes resulting in the characteristic sequence of development of the various cell-types of the root. According to this hypothesis one might predict that the variety of RNA molecules would decline as development proceeds, i.e., stem-cells would have a greater variety of messenger RNA molecules than derivative and mature cells. Although genes concerned with basic cell metabolism presumably remain constantly active as long as the cell is alive, in some cells the complete repression of genetic activity could lead to cell death such as occurs in xylem development (Berjack & Lawton, 1973), and perhaps also during the formation of lysigenous aerenchyma. To demonstrate such a diminution in the diversity of RNA molecules during development requires the detection of single copies of RNA molecules. At present, the identification of single copies of tissue-specific RNA molecules is technically not possible.

Both the gene activation and gene repression models of development allow certain genes to be more active at certain stages of development than at others (Caplan & Ordahl, 1978). Gene amplification (i.e., selective replication of certain DNA sequences) is one way in which the need for a particular gene product can be modulated. There is little evidence that this process is a normal event in the differentiation of plant cells (see Buiatti, 1977), though there is some indication for the occurrence of gene amplification during metaxylem differentiation in roots of *Allium cepa* (Avanzi *et al.*, 1973; Durante *et al.*, 1977). In this case it is the ribosomal cistrons that are believed to be amplified. In general, there appears to be some coupling between cell growth, the production of ribosomes and the synthesis of rRNA, so it is conceivable that the fast growth of developing xylem cells relays information back to the genome thereby stimulating the differential replication of the DNA coding for the requisite ribosomal material.

8.5 THE MERISTEM

One of the major functions of the root system is to provide substances, such as water, inorganic elements and complex organic growth factors, that are essential for the growth of the shoot. As the shoot grows so the demand for these substances increases; the root system grows to cope with this demand. Roots grow by adding to themselves new cells which are formed in the meristem located at the root apex.

The control of the mitotic cycle that generates new cells in the meristem has two aspects that will concern us here. The first is at the level of the individual cells, the second is at the level of the meristematic cell population as a whole.

8.5.1 CONTROLS OVER DIVISION OF INDIVIDUAL MERISTEMATIC CELLS

8.5.1.1 *Energetic and genetic controls*

The root meristem receives sugars from the leaves which are used to construct new cells. Sugars also possess potential energy that can be liberated to drive growth itself; this energy is released by glycolysis and respiration, processes which, as already mentioned (Section 8.4.2) are highly active in meristematic cells.

The mitotic cell cycle has two steps—DNA synthesis (S phase) and mitosis (M)—that require more energy for their completion than do the pre- and post-DNA synthetic periods of interphase (respectively, G_1 and G_2 phases) (van't Hof, 1966). Evidence for these energy-requiring steps in the mitotic cycle originally came from the observation that when excised pea root tips were placed in a simple culture medium lacking sucrose (an energy source) cells were not arrested at just any point in the cycle, but all the cells in S and M at the time of sucrose deprivation completed these phases and accumulated in G_1 or G_2 phase. When sucrose was added to the medium, cells were rapidly stimulated to enter S from G_1, and M from G_2. Using the same experimental conditions, arrest in G_1 and/or G_2 phases was found to occur in the meristems of a number of species and this led van't Hof (1973) to propose that the boundaries between G_1 and S and between G_2 and M were principal control points in the mitotic cycle and that they might regulate proliferation in the meristem as a whole. Later, it was found that the energy source required to surmount these two principal control points also caused the synthesis of RNA and protein (Webster & van't Hof, 1970). Since S and M are quite distinct events in the cell cycle, one may postulate that different RNA and protein molecules are required to initiate S and M phases at these control points. By using appropriate experimental techniques it may be possible eventually to detect proteins which are unique to, and responsible for, the $G_1 \rightarrow S$ or $G_2 \rightarrow M$ transitions.

These remarks, however, beg the question as to whether genes are responsible for mitotic cycle progress and whether they are transcribed and translated in a temporal sequence that matches cycle events. Work with mitotic cycle mutants in unicellular organisms (particularly yeast) and cultured mammalian cells strongly suggests that there are genes regulating specific events in the cycle, and that they are expressed in a particular sequence (Hartwell, 1978; Simchen, 1978). Evidence indicating that something similar occurs in root meristems comes from work with inhibitors of protein and RNA synthesis (Giménez-Martín *et al.*, 1977; de la Torre & González-Fernández, 1979). These workers have shown: (1) that RNA synthesis during prophase is required for the nuclear membrane to break down at the end of prophase, (2) that protein synthesis is needed for cells to proceed from prophase to metaphase and also for S phase to be initiated, and (3) that nuclei in different stages of S phase have different sensitivities towards agents such as heat shocks, 5-aminouracil, 3′-deoxyadenosine and bromodeoxyuridine—agents which impair the function of the nucleus in directing its activities towards mitosis. Little more can be said about the role of the nucleus in controlling its own division cycle; the mechanisms that underlie the effects mentioned above are largely unknown, but at the same time indicate that the molecular controls over division might profitably be analysed further.

5.1.2 *Cell size and cell division*

Although the genes concerned with the mitotic cycle provide the molecular equipment with which the cell makes progress towards mitosis, there may also be an interaction between the growth of the cell and its commitment to division. In bacteria and yeast, cell mass and division appear to be intimately related since the coefficient of variation of the size of cells just about to divide is much lower than the coefficient of variation of newly-born cells. A similar situation obtains in root meristems (Ivanov, 1971) which suggests that in this system cell growth and division also have some close interrelationship. When cytokinesis generates unequally-sized daughter cells (Ivanov, 1971) the larger of the two daughters appears to have a shorter intermitotic period (Demchenko & Ivanov, 1978). The compensatory nature of the interrelationship between size and cycle duration of daughter cells tends to maintain the constancy of cell size at mitosis and, moreover, points to some step(s) in the sequence of events that lead to mitosis being regulated by cell size itself.

In unicellular organisms a critical cell mass is believed to cause the positive initiation of a deterministic phase of the nuclear cycle that culminates in cell division (Donachie, 1968). This principle may operate in cells of the root meristem where there must also be some coupling between cell growth rate and the temporal structure of the mitotic cycle. Within the meristem cells grow at different rates depending on their location: for instance, in *Zea* the calyptrogen has the fastest cell growth rate, the quiescent centre has the slowest. In all regions of the meristem the period from the start of S phase until mitosis is roughly constant (Barlow & Macdonald, 1973). The portion of the cell cycle that varies most extensively in duration is the G_1 phase. The G_1 phase may be variable only because the cell volume doubling time is longer than the sum of the S and G_2 phase durations. In the first cell-tier of the calyptrogen the cell volume doubling time has about the same duration as $S + G_2$ and therefore there may be no measurable G_1 phase; in the quiescent centre, cell volume doubling time is considerably longer than $S + G_2$ and consequently there is a long G_1 phase. Within the meristem, once S phase is initiated, genes for the accomplishment of mitosis are activated, and cell division automatically follows. A critical cell mass (or volume) may initiate S phase but, depending on cell volume doubling times, this will occur at different times following a mitosis. Thus, in the calyptrogen the critical mass must be present immediately after cytokinesis since the nuclei enter S phase immediately, but in the quiescent centre a critical mass is attained hours or days after cytokinesis because the cell growth rate is much lower.

Critical mass may be equivalent to a particular number of molecules (possibly protein) necessary for the initiation of S phase and the attainment of this number will be governed by the rates of synthesis and breakdown of the molecules. Cell growth is also coupled with rates of protein synthesis and so it is possible to see how cell growth rate and S phase initiation could be linked. The calyptrogen and quiescent centre have, respectively, the highest and the lowest rates of RNA and protein synthesis in the meristem (Barlow, 1978b; Clowes, 1958) as well as the highest and lowest rates of cell growth, yet, as mentioned, an approximately constant $S + G_2$ phase duration. Thus, the mitotic behaviour of these cells is probably regulated by their rates of protein synthesis and cell growth. It may be at these points that chemical factors (including hormones) present within the meristem exert their influence.

8.5.1.3 *Interrelationship of cell organelles and cell division*

The nucleus is not the only cellular component that synthesizes DNA and divides. Mitochondria and plastids also contain DNA, and it is reasonable to suppose that the division of these organelles is also related to the synthesis of their DNA and that these events in turn are related to the nuclear DNA replication and cell division cycles. It is possible to identify mitochondria in the act of dividing since during this process they become dumbell-shaped. In root meristems of *Allium cepa* the frequency of these division figures is greatest in G_2 and least in S phase (Sacristán-Gárate *et al.*, 1974). Thus, it is evident that nuclear and mitochondrial division cycles are coupled in some way.

There is evidence that the amount of mitochondrial material per cell increases during cell maturation in root apices (Table 8.5). The increasing level of

Table 8.5 Changes in the relative number of mitochondria during cell maturation in the root apex of *Hordeum vulgare* cv. Atlas

	Relative number, or mass, of mitochondria per cell in the following regions:		
Organelle	Stem-cell (Q.C.)	Derivative meristem (50–150μm)	Early elongation zone (300–450μm)
Mitochondria	100	204	217
Mitochondrial mass	100	194	328

Organelle numbers, or mass, are relative to that value found in the quiescent centre (Q.C.) which is arbitrarily set at 100. The results were obtained by counting the number of mitochondrial profiles seen in section with the electron microscope.
(Data of von Wangenheim, 1975.)

cytoplasmic protein synthesis during cell maturation may account for this increase in mitochondrial substance (Khavkin *et al.*, 1971) which, in turn, may account for the increase in respiratory capacity of cells as they mature (see p. 202). Von Wangenheim (1975; 1976) has argued that mitochondria can cause the repression of nuclear division activity and in this connexion he suggests that as cells are displaced away from the stem-cell zone their complement of mitochondria increases to such an extent that their effect upon division eventually causes the cells to cease further divisions. Thus, there is a possibility that the limit of meristematic activity is set by an imbalance between growth in the cytoplasm and nucleus, although obviously other mechanisms are also possible.

8.5.2 CONTROL OF DIVISION IN THE MERISTEM

As stated above, the meristem can be considered to consist of two types of cells: stem-cells and derivative cells. The stem-cell zone in the roots of higher plants can be equated with cells of the quiescent centre, and the derivative cells with the remainder of the meristematic cells (Barlow, 1978a). The stem-cells are a source of cells for the derivative meristems where cell numbers are amplified by a certain number of divisions to meet the demands of the tissue for functional differentiated cells. The number of amplification division cycles in the derivative meristem varies according to the location of the cells. For instance, in meristems of *Zea*, cells within the cortex complete about 8 amplification division cycles, while those in the metaxylem lineages complete only about three or four cycles; in the calyptrogen the number of cycles may be as few as only one or two. A consequence of these differences in the number of amplification divisions is that cells in the different locations ultimately achieve quite different sizes. Since the xylem cells cease division relatively close to the apex the period over which they can grow, by elongation alone, is correspondingly extended. Conversely, epidermal cells, which divide many more times, have a shorter elongation phase.

The number of amplification divisions that occur in the apex is under genetic control (Viell-Maek, 1978). As already mentioned (Section 8.5.1), the completion of the mitotic cycle depends upon the activity of certain genes and, since mitoses only occur in certain areas of the plant, the relevant genes are probably inducible. The question arises whether the cellular milieu is the inducing agent, and if so, what is its nature?

The growth regulators auxin and cytokinin are regarded as essential for cell division, although their precise roles in this process are not known. Other growth regulators, such as gibberellic acid and abscisic acid, may also influence the mitotic cycle, but it is not clear whether they are of primary importance in maintaining the cycle or whether they only modify cycles already in progress. Evidence for the involvement of auxins and cytokinins in cell division derives largely from work with callus and suspension cultures, and most workers in the field of growth regulators have little hesitation in translating the conclusions from these systems to organized meristems. This may be justifiable since the major differences between the two systems, with regard to cell proliferation, lies more in the spatial organization and metabolic cooperation of their cells than in any other property. Auxin (e.g. IAA) appears to have an involvement with the initiation of DNA synthesis and is also required for cell enlargement; in meristems the cell division cycle never proceeds in cells that are not also enlarging. As suggested earlier (Section 8.5.1.2) a critical size of cell may be important for the initiation of cycle events. Cytokinins (e.g., zeatin, 6-[4-hydroxy-3-methylbut-*trans*-2-enylamino]purine) seem to have some relationship to the actual process of mitosis and cell division.

Auxin (i.e., IAA or some hormonally-active derivative of IAA) has been detected in root apices (Rivier & Pilet, 1974) and therein regulates root growth (e.g., Elliott, 1977). The origin of the auxin in the root apex is thought to be either the shoot or the mature cells of the root whence it is transported to the root tip. There is less evidence that cytokinins are present in apices—what evidence there is is based on bioassays of fractions collected from ion exchange columns (Feldman, 1975; 1979a). Thus, although there may be considerable uncertainty concerning the mode of synthesis, the whereabouts, and the sites of action of chemicals that regulate cell proliferation, it is reasonable to suppose that they are present in the meristem and may be responsible at least for the establishment of a biochemical state that permits proliferation. It is interesting to note that cytokinins have an inhibitory effect on respiration (Miller, 1979) and have also been postulated to be synthesized in or around the quiescent centre (Barlow, 1976a; Feldman, 1975). The gradient of cytokinin suggested to be present in the apex and to be responsible for the control of meristematic activity (Barlow, 1976a) might also be responsible for the gradient of respiratory activity noted by Sutcliffe and Sexton (1974).

When root apices are exposed to conditions that do not favour meristematic activity the cessation of proliferative activity occurs first at the base of the meristem and moves towards the apex (Barlow, 1976b; Ivanov, 1968). This suggests that the most apical cells of the meristem have the greatest capacity for proliferation, perhaps because the chemical factors that favour division are concentrated here. There is evidence that the quiescent centre plays a role in governing root growth because, when it is excised from roots of *Zea mays*, the extension growth of such roots is much reduced compared with intact controls (Feldman, 1979a; b). The growth of roots whose quiescent centre has been removed can be enhanced by the addition of a cytokinin (zeatin). These results accord with, but do not prove, earlier suggestions that the quiescent centre and perhaps also the cells located on its basiscopic face, are a source of cytokinins that may regulate the number of amplification cell divisions in the remainder of the meristem (Barlow, 1976a; Feldman, 1975).

8.5.3 ASYMMETRIC DIVISIONS

It has already been mentioned (p. 209) that within the meristem cytokinesis does not always generate pairs of cells that are perfectly equal in size and that a consequence of this may be that the intermitotic periods of the two sister cells are different. However, there are cases where, although sister cells are equal in size, they still behave differently. For instance, Davidson and Pertens (1978) have shown that in the calyptrogen of maize roots the rate of nuclear growth in cells of tier 1 is greater than in the sister cells of tier 2, and the same pattern of growth operates within a cell of tier 1 experimentally rendered binucleate; the growth rates of the cells in these two tiers are also different (Barlow, 1977b). Similarly, inequalities of some factor inducing mitosis are found in roots of *Allium*, both within experimentally-induced polynucleate cells (González-Fernández *et al.*, 1971) and along the cells of a lineage (López-Sáez *et al.*, 1966; Ivanov, 1971). In the latter case, it is the more apically-located of the sibling cells which progresses most rapidly into mitosis. The nature of the gradient that operates over the short distances involved is not known, but, at least in the calyptrogen, and possibly in the other regions mentioned, the asymmetric growth correlates with an unequal partitioning of cytoplasmic RNA into the two daughter cells at cytokinesis (the faster-growing cell receiving more RNA).

But these inequalities of growth could equally reflect some gradient of a hormonal nature along the length of the cell lineage.

There are instances where an asymmetric division is associated with dissimilar paths of differentiation of the two sister cells. In roots of certain species unequal divisions occur in the epidermis and are associated with hair differentiation; the smaller sister cell (the trichoblast) does not undergo further divisions and develops as a hair, the larger sister continues to divide and remains as an epidermal cell. An example of this phenomenon is afforded by *Elodea canadensis* where the nucleus of the trichoblast becomes endopolyploid while the atrichoblast remains diploid (Dosier & Riopel, 1978); the trichoblast and atrichoblast also exhibit differences in their complement of enzymes (Dosier & Riopel, 1977). What actually occurs at the asymmetric mitosis that leads to this divergence of phenotype is not known; one must assume that the cytoplasm of each cell contains a different complement of 'factors', which favour divergent gene activities.

It might be inferred, from what has been said above, that the asymmetric mitosis in the epidermis in some way induces a new path of differentiation culminating in the trichoblast, whilst the atrichoblast continues in the developmental pathway held by the mother cell. However, there is some evidence of a tendency towards hair formation within the atrichoblast, but that this is usually suppressed after the asymmetric mitosis has occurred. Evidence for this comes from work of Ivanov & Filippenko (1976) who irradiated roots of three species of Festucoid grass in which normally an asymmetric mitosis terminates mitotic activity in the epidermis and results in trichoblast and atrichoblast. The dose of ionizing radiation was such that mitosis, but not cell enlargement, was suppressed. In the irradiated roots all the epidermal cells of the formerly meristematic apex formed hairs.

8.6 MOLECULAR BASES OF ROOT MORPHOGENESIS

The cells whose properties have been discussed in previous sections exist within an apex of definite structure. From the vast amount of work published concerning genetic activity, it is now possible to postulate the differentiation of cells in these terms. But the *pattern* of differentiation within a tissue, and the development of the *shape* of an organ (such as the root), are problems of a different order of complexity. While biochemical differentiation relies minimally on cell–cell interactions, this is the agency underlying morphogenetic processes. It is, at this point, appropriate to discuss two processes—the patterning of vascular tissue and the initiation and positioning of lateral roots—which are examples of coordinated cell differentiation, and then to conclude this section with a consideration of the generation of the shape of the root apex.

8.6.1 PATTERNS OF VASCULAR DIFFERENTIATION

Work with callus cultures has shown that exogenously applied sucrose and auxin interact with the parenchymatous callus mass to induce differentiated nodules of cells containing xylem, phloem and cambium (Wetmore & Rier, 1963; Jeffs & Northcote, 1967). These findings could be taken as evidence that the differentiation of these cell-types within the young stele tissue (procambial cylinder) of the root is regulated in a similar manner. The lineages of vascular cells show an unbroken continuity within the root apex as it grows, and it is unlikely that the inception of vascular cell differentiation within the apex occurs without there being some inductive influence from the already mature, or maturing, vascular cells.

Investigations with callus show that phloem formation is favoured by a high sucrose : auxin concentration ratio, while xylem formation is favoured by a lower ratio (see Table 8.6); cambium, which is indicative of meristematic ac-

Table 8.6 Concentrations of sucrose and auxin (IAA) which, when added to the top of a block of callus of *Syringa vulgaris*, induce vascular cell types within the callus tissue

Concentration (M)			
Sucrose	Auxin	Ratio of sucrose : auxin	Cell types observed within the callus
2.9×10^{-2}	2.9×10^{-6}	10^4	Cambium, few xylem elements
5.8×10^{-2}	2.9×10^{-6}	2×10^4	Cambium, xylem
8.7×10^{-2}	2.9×10^{-6}	3×10^4	Cambium, xylem, phloem
11.6×10^{-2}	2.9×10^{-6}	4×10^4	Cambium, phloem

All calluses were maintained on a medium that included 2.9×10^{-7}M IAA.
(Data derived from Wetmore & Rier, 1963.)

tivity, occurs at all ratios tested. Thus, in the root apex, it would be logical to infer that since phloem cells are transporting sucrose into the apex, those uncommitted cells within the procambial cylinder which have a direct continuity with the phloem would be greatly biased towards differentiating as phloem also. A similar argument could be presented for the differentiation of xylem if it were to be shown that xylem cells import auxin to the apex of the stele. Even if they do not, but given that auxin is present throughout the procambial cylinder, the differentiation and positioning of xylem could occur at some location related to the position of the phloem since this latter cell-type would be at a high-point of sucrose in a sucrose/auxin gradient.

The development of the pattern that blocks out xylem lineages within the root apex has been the subject of some experimental investigation, particularly in connection with the relationship between the number of xylem lineages and the structure of the meristem. Using roots of *Zea* cultured in a liquid medium, Feldman & Torrey (1975) found that the mean number of xylem lineages blocked out within the procambial cylinder could be influenced by the sucrose concentration in the medium. Moreover, the complexity of the vascular pattern was found to be positively related to the height of the quiescent centre, as judged from autoradiographs of roots fed with ^{3}H-thymidine. Later, Feldman (1977) found that roots of *Zea* showed a natural variability in the number of metaxylem lineages and again this could be correlated with the height of the quiescent centre (Table 8.7). It was proposed by these authors that the quiescent centre

Table 8.7 Metaxylem differentiation in roots of *Zea mays* cv. Kelvedon 33

	Metaxylem file no.					No. of roots sampled
	1–4	5	6	7	8	
% of roots with these numbers of metaxylem files	100	100	60	47	40	15
Mean distance (μm) from root cap junction at which the files originate	64	78	89	94	109	15
Mean height (μm) of QC in roots which develop these numbers of files	35*	48	60	71	84	not stated
Distance (μm) between adjacent files at the level of origin of a new file	40	52	56			1†

* By extrapolation (no roots had only 4 metaxylem files).
† A representative root in which 6 metaxylem files had differentiated.
(Data from Feldman, 1977.)

exerted some direct regulatory control on the pattern of xylem development, the efficacy of this control being related to the dimensions of this group of cells. However, it is misleading to suppose that the quiescent centre is the source of some substance that influences vascular development, since the height of the quiescent centre may, in its turn, bear some simple relationship to the diameter of the procambial cylinder within which xylem patterning is occurring. A relationship between the diameter of the procambial cylinder at the level at which vascular patterning is first established and the complexity of the pattern was demonstrated by Torrey (1955) using cultured excised root-tips of *Pisum*—the greater the procambial diameter, the more complex the pattern. Later, Torrey (1957) was able to induce up to three extra xylem arms (*Pisum* usually has a triarch vascular pattern) in the excised tips by the inclusion of IAA (5×10^{-6}M) in the medium. Torrey proposed that this hexarch condition was a consequence of an increased diameter of the procambial cylinder, and also suggested the endogenous auxin levels in the root apex may normally control the vascular patterning, albeit indirectly, through an effect on the planes of cell division and cell dimensions in the procambial cylinder.

There must also be some lateral influence on xylem positioning within the procambial cylinder, since in *Zea*, after a certain number of xylem elements are blocked out simultaneously in the procambial cylinder, additional xylem files are later inserted between these original files (which lie equidistant from each other in a circular arrangement in the procambial cylinders), but only when a certain distance has been generated between neighbouring files by further divisions and cell enlargement in the stele (Table 8.7). Furthermore, the symmetry of the experimentally-induced polyarch xylem patterns in Torrey's (1957) study, and the way in which asymmetric patterns evolve to symmetric patterns in roots regenerating after a disturbance to the apex (Barlow, 1977c), also suggests that vascular patterns, besides having some measure of self-determination by longitudinal inductive influences are also self-regulating in the lateral plane.

At the present time there is an interest in simulating pattern formation with the aid of the computer. Many of the models used in these studies derive from Turing's analysis of the way in which morphogens could interact within a homogeneous field to generate a stable pattern during the process of cell differentiation (see Bard & Lauder, 1974; Ernaux *et al.*, 1978). Although work of this nature relating to vascular differentiation is just starting (e.g. Ans, 1979), and although model building, if it is to be rewarding to the builder, tends to be so constructed that it will 'work', modelling should nevertheless lead to a clearer formulation of the questions posed concerning the molecular requirements of this complex area of morphogenesis.

8.6.2 PATTERNS OF LATERAL ROOT INITIATION

Most primary roots develop secondary (or lateral) roots, which in turn may produce tertiary roots. The initiation of the primordia from which the lateral roots develop occurs within the pericycle and/or endodermis (depending on the species) at points, often, but not always, opposite the protoxylem. Lateral root development is regulated in two ways: first, there is the determination of the position along the root axis at which primordia will develop and second, there is the induction of pericycle cells to enlarge and divide.

Much of the work on the chemical control of lateral root development has been performed with cultured root segments. However, the molecules that directly instigate the growth and mitotic activity of certain cells of the pericycle *in vivo* are not known, though a variety of substances such as adenine, biotin, thiamine and auxin all enhance the production of laterals in cultured root segments. The response of root segments to auxin can be quite striking: for instance, the auxin, naphthalene acetic acid (NAA), at a concentration of 5.4×10^{-7}M, can stimulate the formation of up to 60 lateral root primordia per centimetre length of segment of *Haplopappus ravenii* (Blakely *et al.*, 1972); only about two laterals per centimetre formed in the absence of NAA. In contrast, NAA could induce a maximum density of only five to six primordia per centimetre length in *Pisum sativum* (op. cit.). The experiments with *Haplopappus* also showed that the auxin had to be continuously presented to the root segment in order to sustain the growth of the laterals. In view of what is known of the effects of auxins and their distribution and transport along the root, it could be proposed that endogenous auxin moving acropetally in the root renders the pericycle cells competent for division, whilst cytokinins, moving in the xylem sap, and sugars moving in the phloem, work in concert with auxin to elicit the synthesis of DNA and the mitoses that establish the primordium. Such a scheme might also account for the positioning of primordia in relation to the xylem and phloem strands, but leaves unanswered the question of why there are discrete sites of cell proliferation (the primordia), instead of whole tracts of pericycle becoming active in proliferation. While it is possible that a young primordium produces an inhibitor that prevents the initiation of another primordium in its vicinity, it is also possible that, if growth factors or nutrients (e.g. amino acids) are limiting, the young primordium and the apical meristem of the parent root both act as sinks for these substances; consequently there is a local depletion of inducing substances so making it impossible for a new primordium to be initiated until their concentration at the pericycle returns to a favourable level.

Certain data have been interpreted to show that the apex of the parent root plays a negative role in the initiation of the primordium. For example, removal of the most distal 3 millimetres of the apex from three-day old pea seedlings can cause a rapid increase in the rate of initiation of lateral primordia (Böttger, 1978a). Böttger (1978b) assayed the de-tipped portion of the root for abscisic acid and xanthoxin and found that the amount of these substances fell during the period when laterals were initiated. The inference from these observations is that since these substances are believed to be transported basipetally away from the root apex it is possible that they could in some way check divisions in the pericycle until it is stimulated to divide by some other means. However, the uncertainty with which experiments on the chemical control of lateral root initiation can be interpreted emphasizes the need for further work in this area.

Another exciting aspect of the control of lateral root formation is revealed by studies of the root of the water fern *Ceratopteris* (Chiang, 1970; Chiang & Gifford, 1971). In the main root the pattern of cell division is very precisely controlled and it is possible to trace the ancestry of any particular cell-type through all the divisions following the derivation of the first cell of the lineage from the apical cell. The precursor of the lateral root is always a particular cell in the lineage forming the endodermis. However, the apical cell gives rise to three initials each of which creates the lineages from which the root is constructed, yet within only two of these lineages do lateral root initials form. Lateral roots are found opposite the protoxylem arms of a diarch stele. Thus it looks as though the stele may have some inductive influence over whether a particular endodermal cell becomes a lateral initial or not, even though, at the level at which the lateral initial is first recognizable, no protoxylem cell is apparently differentiated. The precise determination of differentiation within a cell lineage suggests that lateral primordium formation is just one aspect of the overall developmental sequence of the cells within the lineage. Primordium formation may, therefore, be programmed by the interplay of both genetic and positional information. It is not inconceivable that lateral primordium formation is part of a lineage-determined developmental sequence in angiosperms also.

In *Ceratopteris* the single cell that will give rise to each lateral primordium can be recognized by its larger size even before it has divided to contribute any cells to the primordium. This involves a change in the rate of cell growth as well as a change in the polarity of enlargement and division. How this is accomplished is unknown; it could reflect the first step in the genetic pathway leading the cell towards lateral root formation. A similar switch in cell polarity is seen in the group of cells that give rise to a lateral root primordium in angiosperms (Foard *et al.*, 1965).

8.6.3 CELL PATTERNS WITHIN THE ROOT MERISTEM

As mentioned (p. 190), the pattern of cells constructing the root apex can be categorized into a number of types that vary from species to species, although within any given species the constructional pattern is constant. An obvious distinction can be made between the patterns that constitute the so-called 'open' and 'closed' types of root construction (viz. Fig. 8.2). The constructional patterns are determined by the siting of cells with differing planes of division in the meristem, particularly in and around the quiescent centre. An example of how the sites and planes of division govern the patterning of just one cell-type—the cortex—is furnished by two species of grass, *Zea mays* and *Zizania aquatica* (Fig. 8.3). In *Zea* new layers of cortical cells arise by periclinal divisions in cells sited throughout the width of the most proximal portion of the cortical meristem (Fig. 8.3b); it is these divisions that cause most of the increase in width of the cortex (Seago & Wolniak, 1976). In the cortex of *Zizania*, and also of *Oryza*, where the situation has been described in detail by Kawata & Matsui (1977), the new layers arise by periclinal divisions only in the innermost layer of cells (Fig. 8.3a). One consequence of the latter type of division pattern is that the cortical cells are arranged in radial rows, each cell-file having four other cell-files in contact with it. In *Zea*, radial rows of cells are not so evident and each cell-file has five or six files in contact with it. The cortical patterning that generates radial cell rows seems to be a common feature of aquatic or semi-aquatic plants and it could facilitate the formation of intercellular air spaces and also allow them both to penetrate deeper into the apex and to increase the volume they occupy within the root.

One of the keys to understanding how the planes of cell division are regulated resides in the system of microtubules that lies within the cell cytoplasm in intimate association with the plasmalemma and cell wall. It is now firmly

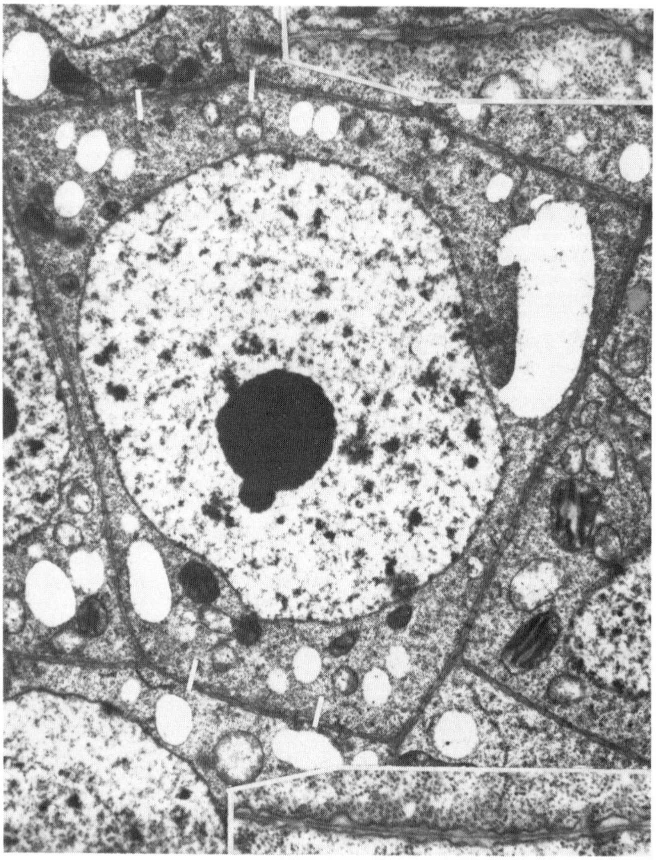

Fig. 8.6 Electron micrograph of a cell from the innermost cortical layer of the root of *Cyperus eragrostis*. The cell is at pre-prophase and a section of the pre-prophase band of microtubules (insets) is located at either end of the cell (between the bars). This band predicts where the new cell wall will be inserted at cytokinesis. The plane of division of this cell layer is similar to that of *Zizania aquatica* illustrated in Fig. 8.3a. Magnification × 8850; insets × 26500. (Micrograph by courtesy of Professor B. E. S. Gunning.)

established that a group of microtubules, which appears sometime prior to prophase and which girdles the interior of the cell, accurately predicts the site of insertion of the new cell wall at every forthcoming cell division in the meristem (Fig. 8.6), and does so even in those cases where the division is one which creates distinctly unequal cells (Gunning *et al.*, 1978). The new wall bisects the site occupied by this pre-prophase band of microtubules (PPB) even though the microtubules themselves disperse as the nucleus enters mitosis. During each cell division cycle the siting of a PPB seems not to be directly influenced by the plane of division that obtained in the mother cell. Thus the position of the PPB is presumably determined in response to influences of some sort operating upon or within the cell only during its progress to division. The actual assembly of the PPB may be part of a sequence of events evoked during the deterministic portion of the mitotic cell cycle since PPBs have never been described in cells that, even though they may be able to synthesize DNA as part of an endoreduplication cycle, are not going to divide. PPB formation may therefore usually be determined at some time after the initiation of S phase but before the onset of mitosis; this point merits further investigation.

The mechanisms whereby the position of the PPB is determined are unknown, but whatever they may be, they are primary determinants of morphogenesis. One external signal that may influence the plane of division is the pattern of strain within the cell wall. Consider the example of the modes of patterning of the cell-files in the root cortices mentioned earlier. The cells that exhibit periclinal divisions, whether they occur throughout the width of the cortex or are limited to the innermost cell layer, are those whose basal end is broader than their apical end (Fig. 8.6). The strain resulting from an asymmetric growth of the cell wall surface in the radial plane could be a condition that favours the positioning of the PPB so that a periclinal division results. A similar strain-inducing asymmetric surface growth, but in the circumferential plane, could favour radial division. The cortical cells that elongate longitudinally, but have relatively little lateral growth, tend to divide transversely.

The arrangement of cells in neighbouring files can influence the position at which a new cell plate is inserted. This is demonstrated by the fact that transverse walls in neighbouring cell files are very rarely in line but are usually slightly off-set from one another (Sinnott & Bloch, 1941). This off-setting of a new wall from previously established wall junctions is another important factor in morphogenesis and at the ultrastructural level is predicted by a similar off-setting of the PPB. Where a transverse wall joins with a longitudinal wall the latter, at the point where the walls meet, will tend to be mechanically more rigid and thus subject to less stretching than segments of longitudinal wall free of such transverse wall junctions. A stretched portion of wall and its underlying plasmalemma may also determine the site of assemblage of the PPB whose function could be one of somehow ensuring the attraction and interpolation of the new wall at these sites into the fabric of the mother cell wall. The contention is, then, that the planes of cell division may be determined by conditions in the wall of the mother cell which are sensed by whatever it is that causes the assemblage of the PPB (see Fig. 8.6).

The overall morphology of the apex might be determined by a few simple rules as indicated above. But how these rules could be put into effect at the molecular level is at present unknown. The cells of the apex physically interact with each other, and since the pattern of cell growth and morphogenesis is, within limits, constant and self-maintaining, apical morphology must also be an expression of a positive feedback between the already created three-dimensional cell patterns and the system that is creating new cells. It seems very much as though cell growth, and particularly its polarity, is of fundamental importance in morphogenesis, and that cell division is merely a follow-up process filling the space created by growth with new cells. One literal interpretation of Anton de Bary's well-known aphorism 'Die Pflanze bildet Zellen, nicht die Zelle bildet Pflanzen' could be taken as describing this latter situation quite well.

8.7 CONCLUDING REMARKS

The form of living organisms is determined by a Bauplan which resides largely in the hereditary material. All aspects of cellular behaviour—gene activity, cellular metabolism, cellular growth, morphogenesis—interlink to realize the Bauplan. The consequences of chemical and physical laws within the domain of living organisms—that is, what we loosely call 'molecular biology'—have an important role in this realization. Much of the current research concentrates on the minutiae of molecular phenomena, but this must also be associated with an active appreciation and comprehension of the form and function of the organism if molecular biology is to be an integral part of developmental biology. De Bary's aphorism seems to pinpoint a dilemma in the way in which research into developmental processes should be viewed; should we regard the organism or the cell as the fundamental determinant of growth? The apparent conflict between these two divergent lines of investigation dissolves when it is accepted that there are hierarchies of phenomena associated with the development of tissues, organs, and organisms. In this chapter the reader may have discerned these different strata of developmental phenomena; these phenomena, of course, apply not only to plants with their roots and shoots, but also to all organisms. Two strata predominate: the biochemical stratum, which is fundamental to development, because it is the ground-substance of living processes, and a 'higher' stratum of development within which the determinants of organic form operate. The Bauplan becomes more fully realized by activities at successively higher strata. Put another way, it seems as though the course of development is initiated by information residing within the genetic apparatus (the nucleus), but at some later stage a portion of this information is embodied in a system that is independent of this genetic apparatus yet works in harmony with it to realise the form of the organism. Roots are but one of a variety of developmental systems that are useful for discovering more about the way in which organic forms are constructed.

FURTHER READING

Green P. B. (1980) Organogenesis—a biophysical view. *Ann. Rev. Plant Physiol.*, **31**, 51–82.

Street H.E., Elliot M.C. & Fowler M.W. (1978) The physiology of roots. In *Interactions between Non-Pathogenic Soil Microorganisms and Plants*, (eds. Y.R. Dommergues & S.V. Krupa). pp. 69–130. Elsevier Publishing Co., Amsterdam.

Torrey J. G. & Clarkson D. T. (eds.) (1975) *The Development and Function of Roots*. Academic Press, London, New York & San Francisco.

CHAPTER 9

LEAF DEVELOPMENT

9.1 INTRODUCTION

Because they are the ultimate source of much of man's food and most of his past and present energy supplies, green photosynthetic leaves are one of the most important biological structures in nature.

Leaves have a relatively simple basic anatomy and consist of five to six layers of cells, serviced by a network of veins and surrounded by a cuticle which is penetrated, at intervals, by stomata. The photosynthetic cells of leaves are unique amongst the higher eukaryotes as they are autotrophic through much of their life and contain three separate sets of genetic machinery in three distinct compartments, the nucleus and cytoplasm, the plastids and the mitochondria. Each compartment is separated from the other by a double membrane and all cooperate in the functioning of the cell and during its division. Understanding the nature of this cooperation and how it evolved has challenged botanists for almost a century since Schimper (1885) proposed the concept of the autonomy and continuity of plastids.

This chapter is concerned with leaf development, which is a convenient term to cover all those activities associated with the generation and expansion of leaves. Since the prime function of leaves, which distinguishes them from other plant organs, is photosynthesis, this chapter relates leaf development to the formation and development of chloroplasts. The development of leaves is an expression of the cooperative functioning of the genome of cells and the plastome of plastids. It is therefore necessary to discuss in molecular terms the organization of, and interactions that take place between, the genome and the plastomes of leaf cells. Much of the information drawn upon comes from observations of spinach (*Spinacea oleracea*), a plant that has been used extensively by research workers for studies of both photosynthesis and leaf development.

9.2 THE DEVELOPMENTAL ANATOMY OF LEAVES AND OF PHOTOSYNTHETIC COTYLEDONS

Leaves arise as a lateral primordium on a stem when a small group of cells a short distance from the apical meristem begin to divide periclinally and at a much faster rate than the cells of the stem apex. The leaf primordium rapidly develops into a completely formed leaf which, in many plants, continues to

Table 9.1 Leaf growth in spinach, bean and wheat

Species	Age from sowing (days)	Leaf no.	Leaf length mm	Cell no. per leaf × 10^3	Chloroplast no. per cell	Chloroplast no. per leaf × 10^6
Spinach[1]	20*	5	20	7600	10–50	250
(*Spinacea*)	34	5	100	15 850	200	3170
Bean[2]	2*	1st pair	10	2000	8	16
(*Phaseolus*)	8	1st pair	100	50 000	50	2500
Wheat[3]	11	4	1	16		
(*Triticum*)	17*	4	45	990	46†	45
	25	4	275	4290	157†	673

* Approximate time of leaf emergence.
† Data for primary leaves.
[1] Possingham & Saurer, 1969, Saurer & Possingham, 1970.
[2] Bradbeer *et al.*, 1974; Dale, 1976; Dale & Murray, 1968; Whatley, 1974, 1978.
[3] Boffey *et al.*, 1979; Williams & Rijven, 1965.

expand during a programmed sequence in which cells are dividing until the leaf is one-half to two-thirds of its final area. The remaining growth is completed by cell expansion. The processes of cell division and cell expansion overlap one another so that at any stage of development the leaf consists of cells of varying volumes and in different phases of development.

The stem apex is commonly covered by enveloping leaves so that at least the initial phases of leaf formation occur in low light conditions. However, in many dicotyledonous plants the majority of leaf cells are formed from divisions which take place in emerged leaves, i.e. in the light, and only a small proportion of cells are elaborated in darkness. Details of relative numbers of leaf cells formed before and after emergence are given in Table 9.1 for two dicotyledonous plants, spinach and bean, and for the monocot wheat.

An important consequence of this pattern of leaf growth is that although the cells of the leaf initial and of the stem apex from which they are formed, contain proplastids (Fig. 9.1) both the dividing and the expanding cells of emerged

Fig. 9.1a E. M. Section of the stem apex of a spinach (*Spinacea oleracea*) plant showing outer cuticle, cu; proplastids, pp; mitochondria, m; nucleus, n. (× 51 300)

Figs 9.1b & c Proplastids in young unfolded leaves of a spinach plant showing invagination of the outer membrane, me; and osmophilic droplets, o (b × 19 000), and division (arrows) by binary fission. (c × 46 550).

Fig. 9.1d Etioplast, ep, with developing membrane system, mitochondria, m; in a young (8 days old) yellow cotyledon of a germinating spinach seed. (× 45 600)

Fig. 9.1e Proplastid of the stem apex region of an embryo of an imbibed spinach seed (× 46 550)

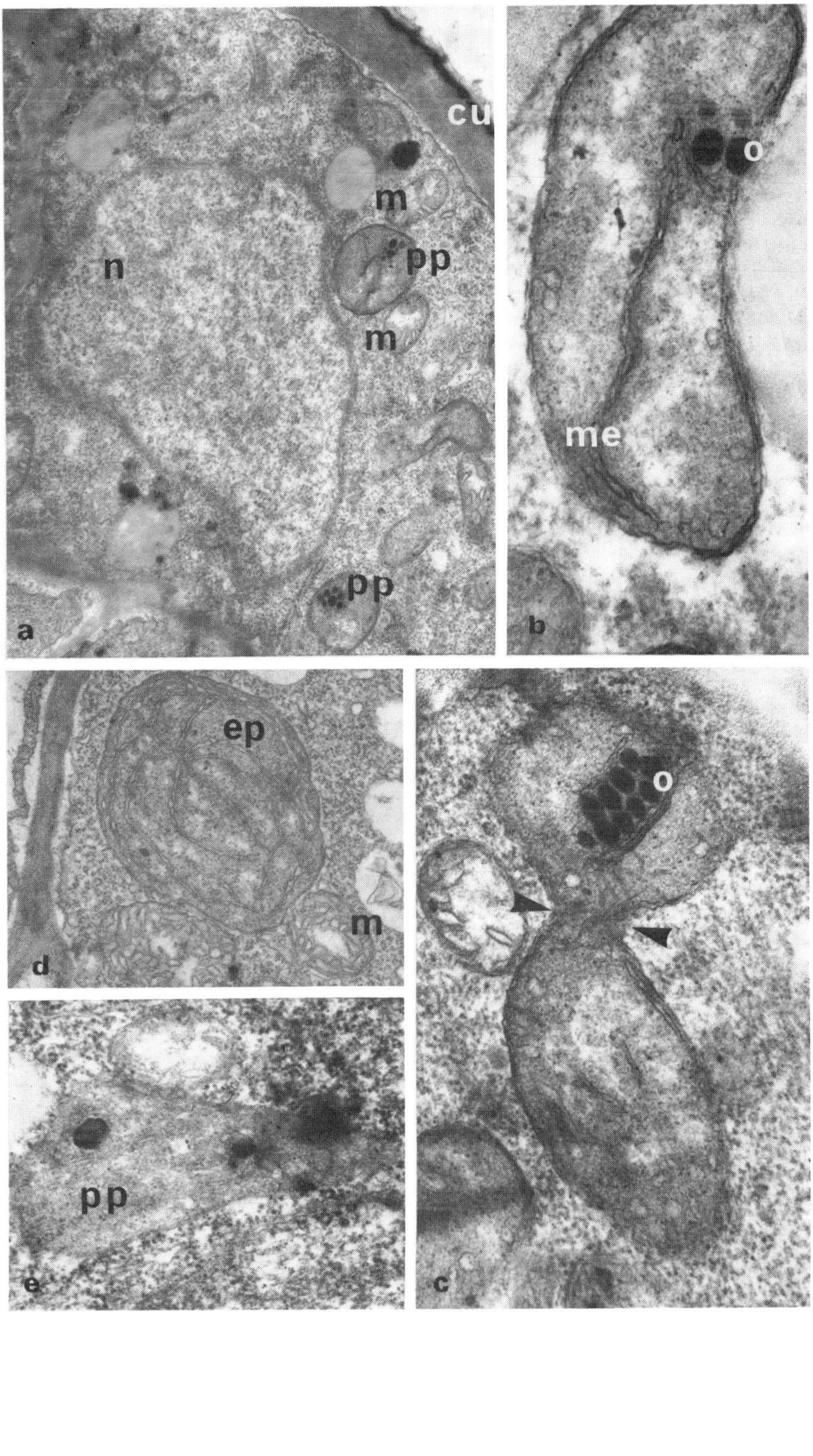
cu
m
pp
n
m
pp
a
o
me
b
ep
m
d
o
pp
e
c

leaves contain chloroplasts with grana thylakoids (Fig. 9.2). The majority of leaf chloroplasts in many dicots therefore arise directly from the division of chloroplasts and a relatively small proportion, those in the cells at the leaf tip, are generated from proplastids and/or etioplasts (Possingham & Rose, 1976).

On the other hand, the majority of the leaf cells of monocotyledons such as wheat, barley and maize, arise from meristematic cells located close to the leaf base and surrounded by enveloping leaves. These cells are poorly illuminated and contain proplastids and/or etioplasts. However, a relatively short distance from the basal meristem the expanding cells are green, despite the fact that they are surrounded by three to four layers of enveloping leaves. Boffey *et al.*, 1979, have established that chloroplast number per cell in the exposed primary leaves of wheat increases from 50 in young cells near to the basal meristem, to over 150 per cell in expanded cells. They consider that these chloroplasts arise from the division of young chloroplasts at various stages of development and well separated in distance and in time from the region of cell division. The evidence indicates that the majority of the plastids of the developing leaves of monocotyledons arise from the division of chloroplasts, as occurs in dicotyledons.

The cotyledons and embryo of seeds are formed as a result of cell divisions which take place in semi-darkness within the tissues of developing seeds. They contain proplastids similar in appearance to those of the stem apex and leaf initials (Fig. 9.1a & e). Cotyledon development stops when seed desiccation occurs, but recommences when water uptake and seed germination takes place. The growth of cotyledons during germination mainly involves cell expansion (Bain & Mercer, 1966a & b; Svrivastava & Paulson, 1968; Paulson & Svrivastava, 1968).

Many ultrastructural investigations have been made of the plastids of the cotyledons of germinating seeds (Kirk & Tilney-Bassett, 1978). These have established that during normal germination and while the cotyledons remain unemerged and in the dark, their proplastids become transformed to etioplasts with prolamellar bodies and a limited thylakoid system (Fig. 9.1d). When the cotyledons emerge into the light the etioplasts are rapidly transformed to chloroplasts and a thylakoid system containing chlorophyll develops. Biochemical studies have established that the capacity to carry out photochemical reactions and overall photosynthesis develops a short time (hours) after etioplasts are exposed to light.

9.2.1 DIVISION OF PLASTIDS

It is generally accepted that plastids increase in number by the division of pre-existing plastids and are passed on from cell to daughter cell during cell division (Schimper, 1885; Kirk & Tilney-Bassett, 1978). In lower plants such as the alga *Nitella* where the chloroplasts are embedded in the cytoplasm and do not stream, sequences in which one chloroplast passes through a dumbbell stage

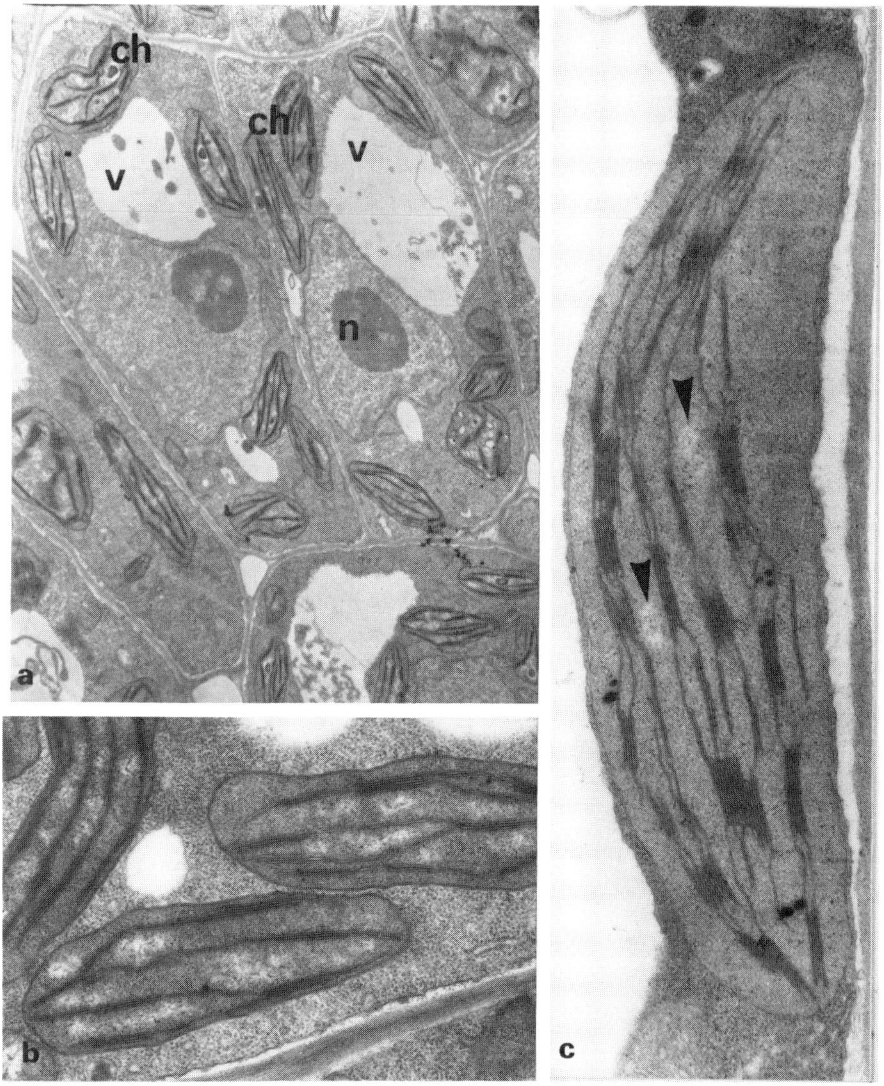

Fig. 9.2a E.M. of cells in youngest emerged leaves of spinach. The chloroplasts, ch, are immature and the central vacuole, v, is not fully formed (n, nucleus). (× 4800)

Fig. 9.2b E.M. section of immature chloroplasts in the basal cells of a 2cm spinach leaf showing numerous clear DNA containing areas within the stroma, and limited development of membranes. (× 20 000)

Fig. 9.2c E.M. section of a mature chloroplast of a 8cm spinach leaf showing two poorly defined DNA areas (arrows) and an extensive system of thylakoids. (× 20 000)

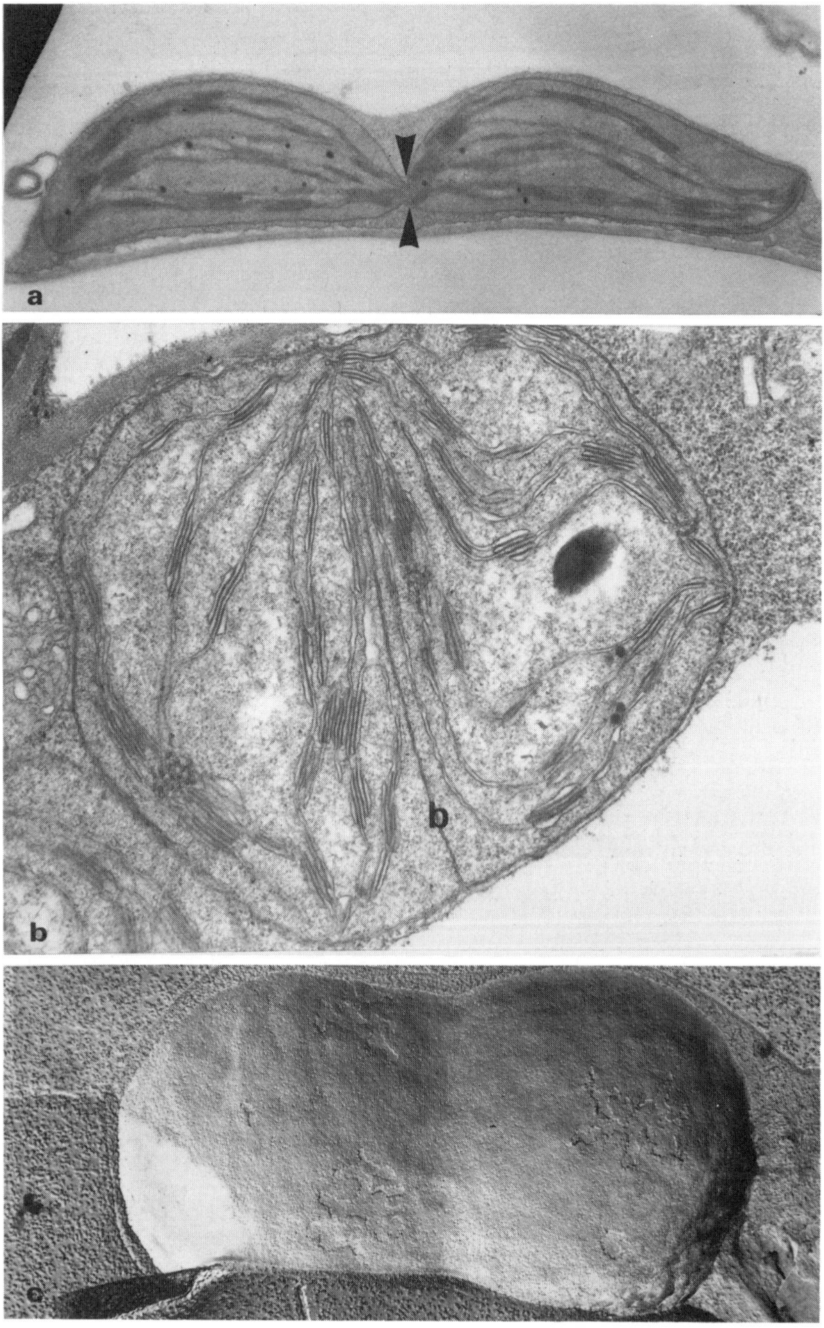
a
b
b
c

to give two daughter chloroplasts have been documented by cinematography (Green, 1964).

In higher plants, chloroplasts with central constrictions have been observed by light and electron microscopy in a number of species (Lance, 1958; Senser & Schotz, 1964; Cran & Possingham, 1972a; Leech *et al.*, 1973; Whatley, 1974) (Figs 9.3a & c). More rarely, plastids with profiles consistent with division by cross wall or baffle formation have been observed (Cran & Possingham, 1972b; Whatley *et al.*, 1979), (Fig. 9.3b).

Light microscope observations of developing leaf cells of spinach, tobacco and wheat have established that there is a general correlation between the chloroplast number increase and the presence of dumbbell-shaped chloroplasts (Boasson & Laetsch, 1969; Possingham & Saurer, 1969; Boffey *et al.*, 1979). More recently a precursor-product relationship has been established between the relative numbers of constricted dumbbell-shaped chloroplasts and new chloroplasts in spinach leaf discs cultured initially in low intensity green light and then transferred to high intensity white light (Chaly *et al.*, 1980).

Almost no quantitative data are available about the division of both proplastids and etioplasts because they are difficult to resolve and to count in the light microscope. Profiles which indicate that these organelles also divide by constriction division have been observed in many species. It is possible that these organelles as well as amyloplasts may divide by budding, which may be regarded as an asymmetric form of division (Buttrose, 1960).

9.3 DEVELOPMENT AND THE DNA CONTENT OF PLANT CELLS

The ultimate limit to the potential, development and variability of cells, tissues and organs, depends on both the quality and quantity of DNA characteristic of the particular species. As we currently understand these properties there are several important aspects when considering the amount of DNA per cell.

The first of these is the base sequence of the complete DNA of a species. Following recent technical advances, establishing this is now a feasible, though time-consuming and expensive, process. Accordingly, for some years it seems

---

Fig. 9.3a E.M. of a young spinach chloroplast during division by a central constriction (×8500)

Fig. 9.3b E.M. of a young spinach chlorplast during division by baffle (b) formation (×26 500)

Fig. 9.3c Freeze fracture replica of a spinach chloroplast during the early stages of constriction division (×21 000)

likely that DNA sequencing in higher plants will be confined to sequences of special interest such as control regions of the genome or to the DNA coding for proteins of particular interest.

The second of these is the concept of genomic complexity. When double-stranded native DNA is denatured to the single-stranded form, the rate at which the complementary strands recombine, or 'renature' is dependent on the length and number of DNA sequences present (Grierson, 1977). Analysis of the renaturation kinetics of denatured DNA gives an estimate of the minimum quantity of DNA which will contain at least one copy of every DNA sequence that occurs in the particular organism (Britten & Kohne, 1968).

The other main concept is that of ploidy—the number of copies of the base amount of DNA to be found in a cell. This concept has long been of particular interest in plant breeding, in studies of plant growth and development and more recently in certain situations related to chloroplast number per cell (Butterfass, 1979). In general terms, the concepts of genomic complexity and the 1C level of DNA are similar. However, differences involved in the measurement of these parameters will be discussed below. From existing data it is at least possible that the two measurements may not always be the same.

9.3.1 DNA CONTENT OF NUCLEI

Over the past 30 years two main approaches have been used to estimate the DNA content of plant cells. One has been to determine the DNA content of individual nuclei by direct micro-spectrophotometry of Feulgen-stained DNA. The other has been to extract and measure the DNA content of the tissue chemically by a colorimetric procedure and to relate this value to the number of cells in the tissue. Many of the figures derived in these ways for higher plants have been calibrated to a common standard and catalogued by Bennett & Smith (1976). Neither of these methods makes allowance for cytoplasmic DNA, the first because only the DNA content of the nucleus is measured and the second because the cell as a whole is considered.

The lowest amount of DNA found in the nucleus of a plant cell is the haploid or 1C amount present in the generative cells of ovules and pollen (Swift, in Bennett & Smith, 1976). In vegetative cells which are not polyploid, the diploid or 2C level is the amount of DNA in a cell at interphase.

More recently, estimates of genomic complexity and genome size based on DNA renaturation kinetics have begun to appear in the literature. Because the single copy component of the DNA (see Section 9.3.1.1) is the slowest to renature, and often the smallest fraction of the total DNA, the measurements made to date are not always accurate and most quote standard deviations of about 20%. Nevertheless, such measurements, which are independent of the amount of DNA per cell, are a more appropriate estimate of what we understand by the concept of a haploid level.

So far there are only a few reported cases in which the chemical and kinetic approach to measuring the DNA content of nuclei can be compared. In *Pisum sativum* a 1C amount of 5.1pg of DNA is recorded for root tip cells (Bennett & Smith, 1976) while kinetic measurements indicate that 4.6pg is the genome size (Murray *et al.*, 1978) suggesting that *Pisum* is a normal diploid. In *Nicotiana tabacum* a genome size of 1.7pg DNA (Zimmerman & Goldberg, 1977) is compatible with a 1C amount of 3.9pg and the observation that tobacco is a tetraploid plant (Bennett & Smith, 1976). In *Pennisetum americanum* the 1C amount of 0.22pg measured by Wimpee & Rawson (1979) is smaller than the genome size of 0.37pg that can be calculated from their renaturation results and, given the errors involved, the values may be the same. In *Glycine max* the genome size of 1.39pg DNA compares with 2n DNA amounts of 5–6.5pg (Gurley *et al.*, 1979) and thus match the cytological observation that soybean is a stable tetraploid (Sakai, 1951).

These few observations suggest that the kinetic estimates of genome size and the cytological estimates of 1C DNA amounts are in fairly good agreement for the species determined so far. It remains to be seen if this relationship holds for a majority of plants. Neither of these techniques take account of cytoplasmic DNA. The cytochemical measurements ignore mitochondrial and chloroplast DNA while the kinetic measurements include any mitochondrial or chloroplast DNA in the tissue or contaminating the crude nuclear preparations used in some studies. In root tissues these errors may be small, unless mitochondrial DNA is significant, but in leaf tissue with up to 20% chloroplast DNA (see Table 9.4) substantial errors are possible.

9.3.1.1 *Organization of the Nuclear DNA*

The kinetic measurements of DNA reassociation have also shown the presence of a continuum of DNA sequences of varying lengths and repetition frequencies in eukaryotes (Britten & Kohne, 1968). These studies have been discussed in more detail in an earlier volume in this series (Grierson, 1977) and only an outline will be presented here. The DNA sequences present in nuclei have been divided into three main classes based on their repetition frequency and these classes also appear to have some relationship to function.

The first class is the highly reiterated DNA which renatures most rapidly and comprises sequences which occur in several thousand copy numbers per genome, or even more, and in some cases appear as 'satellite' DNA bands in DNA preparations purified in CsCl gradients.

The middle repetitive sequences are those with an average copy number of around 100–2000 in the genome, and include the genes for cytoplasmic rRNA. The organization of the rRNA genes in the middle repetitive DNA of the plant genome has been described (Grierson, 1977) but in the absence of well defined functions and assays for particular sequences, other roles of the repetitive DNA classes are not clear. These two classes can also be grouped as one class of repetitive DNA.

The remainder of the genome is comprised of slowly renaturing, or 'single copy' DNA which is thought of as containing a single copy (or perhaps two copies) of the DNA sequence for each protein coded by the genome, and an indeterminate number of sequences, including spacer sequences, for which no function is known.

The complete sequencing of the nucleotides of an entire plant genome is some time away although short sequences of the genome with specific functions will be described in the near future. Pending the arrival of this level of analysis, the organization and distribution of reiterated and single copy DNA through the genome has been studied in several species. While no clear-cut differences have been defined, some interesting possibilities have arisen.

Flavell *et al.*, (1974) found that about 80% of the genome was reiterated DNA in plants with large 2C nuclear DNA contents (i.e. > 4pg per nucleus) while approximately 60% of the genome of plants with a 2C nuclear DNA content of 1.7–4pg was reiterated. They have suggested that differences in the level of reiteration accounts for some of the differences in 2C amounts. If further measurements show that these 2C DNA amounts are equivalent to the genome size of the plant, then potentially large differences exist in the amount of single copy DNA (and consequently in the variety of proteins and/or control sequences) in various plant species which would not have been predicted on the basis of the relatively small range of phenotypes through the plant kingdom. Bennett (1972) has suggested that the selective pressures which lead to a rapid mean cell generation time at the tissue and whole plant level, also select for low nuclear DNA content in order to achieve more rapid DNA replication.

There have been several studies on the interspersion of reiterated and single copy DNA in the genomes of various plants, and while there is still discussion of the quantitative interpretation of the methodology involved in the experiments (Hahn *et al.*, 1980, Kiper, 1980) several qualitatively different patterns of interspersion have been described in plant DNA.

In pea, single copy DNA sequences of about 300 nucleotides are common (Murray *et al.*, 1978) while, in tobacco, single copy sequences average 1400 nucleotides (Zimmerman & Goldberg, 1977) and in cotton 1800 (Walbot & Dure, 1976). In pearl millet, much of the single copy DNA is in sequences from 1400–8600 nucleotides long (Wimpee & Rawson, 1979). These single copy sequences are interspersed with repetitive DNA sequences also of varying lengths and in pearl millet 80% of the repetitive sequences are at least 5000 nucleotides long (Wimpee & Rawson, 1979). In tobacco (Zimmerman & Goldberg, 1977) and pea (Murray *et al.*, 1978) their average length is 300 nucleotides. In soybean 65–70% of the single copy DNA is in sequences of 1100–1400 nucleotides alternating with repetitive DNA sequences 300–400 nucleotides long (Gurley *et al.*, 1979).

It is stressed that in the experiments summarized above the classification of sequence classes by reiteration frequency does not allow direct correlation of specific functions with specific sequences, which will only be clarified by direct base sequencing of the DNA. Nonetheless, it is clear that single copy DNA sequences of 300 nucleotides, such as are found in pea, are too short to code for most proteins. It has already been shown that several mammalian proteins are coded for by DNA sequences whose continuity is broken by insertions (the so-called 'introns') of non-coding DNA. It seems possible that the occurrence of introns may be related to the interspersed DNA fragments and Sun *et al.*, (1981) have reported three introns in a gene for the French bean protein, phaseolin G_1 globulin.

9.3.1.2 *Expression of the Nuclear DNA*

The alteration and regulation of the expression of the genome can be affected at many stages in cellular metabolism and development, but the most fundamental changes and controls are probably effected at the level of transcription and translation of mRNA (but see Chapter 2). For a restricted number of abundant plant proteins the level of transcription or translation of a specific mRNA has been followed. For example, in barley aleurone layer the auxin induced accumulation of α-amylase mRNA was demonstrated (Higgins *et al.*, 1976). In *Lemna* the mRNA levels for RuBP carboxylase are affected by light intensity (Tobin, 1978).

In studies of the complexity and composition of the mRNA population of tobacco leaves, Goldberg *et al.*, 1978, and Kamalay & Goldberg (1980) separated polyribosomal mRNA and nuclear RNA. They concluded that the leaf polyribosomal mRNA contained transcripts of 27 000 different average size (1250 nucleotides) structural genes, which represent about 5% of the single copy DNA of the genome. These transcripts could be classified into classes equating to 9, 52 and 39% of the transcribed DNA with averages of 4500, 300 and 17 copies of each mRNA transcript in the respective classes. About 8000 of these transcripts were common to all organs in the plant, and leaves and petals had a very similar range of transcripts. While the leaf had some transcripts in common with other organs (stem, root, ovary, anther) many were unique to the leaf (and petal). The total number of transcripts found in the various mRNA populations in this study was about 60 000, and analysis of the heterogeneous nuclear RNA (HnRNA) population of the nuclei suggested that all these transcripts were found in the nuclear HnRNA. This work implies an important level of control exists between transcription of the nuclear DNA and the translation of a finished transcript by the polyribosome.

The polyribosomal mRNA in parsley leaves appears to be transcribed from about 6% of the single copy DNA of the genome and the transcripts distributed in about the same proportions (Kiper *et al.*, 1979) although only 13 000 structural genes were reported. The difference in structural gene numbers between tobacco and parsley leaves may not be significant as disagreement exists about how to calculate this number (Hahn *et al.*, 1980; Kiper, 1980; Goldberg & Timberlake, 1980).

In either case the existence of such large amounts of different mRNA transcripts in polyribosomal populations indicates that there may be thousands of as yet unidentified and separated proteins in leaves and in other plant tissues. The relationship between mRNA in polyribosomes and the number and amount of proteins found in plant cells is not known, and may be affected by such factors as translation rates, protein turnover rates and whether or not all polyribosomal mRNAs are translated.

9.3.2 THE PLASTOME

In contrast to the nuclear genome, the plastome is well characterized physically and chemically, and in all higher plants so far investigated the general properties of chloroplast DNA listed below are similar. It is a circular molecule with a contour length of about 40μm corresponding to a mol. wt. of 10^8 daltons, or 150 000 base pairs with a base composition of about 38% GC. Kinetic measurements have not shown reiterated DNA in the plastome although restriction enzyme analysis has shown that in spinach, and chloroplasts of some other higher plants, one sequence of 24 500 base pairs containing the rRNA gene is repeated twice. The evidence leading to these conclusions has been extensively discussed by Kirk & Tilney-Bassett (1978) and Bedbrook & Kolodner (1979). There are some quite wide variations in the properties of algal plastomes which are also discussed in the above reviews.

As more precise analytical techniques are applied, differences in detail between the chloroplast DNAs of various higher plant species are emerging, accounting for minor differences found in the properties described above. Restriction enzyme digests of chloroplast DNAs show markedly different fragment patterns between species (Vedel *et al.*, 1976; Atchison *et al.*, 1976; Timothy *et al.*, 1979) and even between varieties of the same species (Frankel *et al.*, 1979) indicating both sequence changes, where restriction sites are missing or have appeared, and differences in lengths of sequences between restriction sites.

Lamppa & Bendich (1979a) studied the relationship of chloroplast DNAs of various species to pea chloroplast DNA by following the renaturation kinetics of the various chloroplast DNAs in the presence of a pea (^{3}H) chloroplast DNA probe. The sequence similarities, as assessed by hydroxyapatite chromatography, were greatest in the case of broad bean chloroplast DNA at 78% and varied through a range of dicotyledonous and monocotyledonous plants to 37% in barley. When the similarities were assessed by more rigorous S_1 nuclease digestion all the species tested showed a similarity of about 30% except for the more closely related broad bean. These results all imply a strong selective pressure to retain at least 30% of very similar nucleotide sequences and an even higher amount of less closely related sequences in chloroplast DNA of various species. These differences between chloroplast DNA from different species could represent significant divergence in the proteins coded. On the other hand, they may reflect nothing more than the rearrangement of information in the plastome and/or the substitution of alternative codons for amino-acids.

These properties of chloroplast DNA indicate that while there has been a selective pressure to alter the base sequence of the plastome, the functions coded for by the plastome are so specific that major changes are not viable. The converse of this interpretation is that all of the plastome has some essential function.

There is no evidence of extensive intramolecular heterogeneity of chloroplast DNA within a species or in any one plant. It is of interest here that in the C_4 plant *Panicum maximum* Walbot (1977) has reported that the chloroplast DNA of mesophyll and bundle sheath chloroplasts cannot be distinguished by kinetic methods. However, it seems likely that at least two species of chloroplast DNA exist in the same leaf in the case of some plastome coded chimeras (see Table 9.4) although the 'sorting out' of the plastids means that the chloroplast DNA within the cells of chimeric areas is homogeneous. With current methods of scoring plastome mutations, chimeras of most of the other plastome coded functions (see Table 9.3) would not be observed.

9.3.2.1 *DNA in other plastids*

All classes of plastids are thought of as having common origins and bearing a relationship to proplastids (Kirk & Tilney-Bassett, 1978). The low levels of chloroplast DNA reported in pea (Lamppa & Bendich, 1979b) and spinach (Scott & Possingham, 1980) roots are presumably due to the level of chloroplast DNA in proplastids being low, as these organelles are common in electron micrographs of root-tip cells. On the other hand Bennett & Radcliffe (1975) found the equivalent of 600 copies of the plastome per etioplast in deoxyribonuclease-treated etioplasts from peas.

DNA-containing areas have been seen in electron micrographs of chromoplasts from *Narcissus* petals and show similar chemical properties to chloroplast DNA (Kowallik & Herrmann, 1972; Herrmann, 1972). Isolated DNA occurred as chloroplast DNA-like circles in *Narcissus* (Falk *et al.*, 1974) and *Tulipa* (Wuttke, 1976) chromoplasts.

9.3.2.2 *Amount of DNA per chloroplast*

For an understanding of chloroplast growth and development it may be important to know the number of plastome copies per chloroplast. The situation may be analogous to the influence of the number of genome copies per nucleus on cellular development (Price *et al.*, 1974). It is clear that chloroplasts are polyploid (the term polyenergidic is often used to describe the chloroplast equivalent of polyploidy) but the extent and variability of the phenomenon within and between species, during ontogeny and through the various plastid transformations, is not defined. Unfortunately there is no quantitative technique similar to Feulgen staining of nuclei that enables measurements of the amount of chloroplast DNA in a tissue or plastid preparation. Estimates can be made by using a combination of light and electron microscopy to observe the DNA areas within the plastid and chemical measurements of the amount of chloroplast DNA in a preparation. The observations made in these various ways are still not completely reconciled and will be discussed in some detail below.

9.3.2.2.1 *Light microscopy of chloroplast DNA:* The DNA fluorochrome 4,6-diamidino-2-phenylindole (DAPI) can be used to stain chloroplast DNA *in situ* for observation by epifluorescence microscopy (Coleman, 1978; 1979; James & Jope, 1978; Scott & Possingham, 1980). Hoechst 33258, a benzimidazol derivative, is also an effective fluorochrome for cytoplasmic DNA (Coleman, 1978).

Plastid nucleoid areas stained with DAPI appear as blue fluorescent dots on the field of the chloroplast (Fig. 9.4c) and epifluorescence observations of the same sections for red chlorophyll fluorescence show the plastid nucleoids in the stroma of the chloroplasts. In any one spinach cell, all chloroplasts appear to have roughly equal numbers of plastid nucleoids. Observations of the distribution of fluorescent stain in mature spinach chloroplasts suggest that the nucleoids are not linked, whereas the immature plastids of two centimetre leaves may have only one semi-continuous DNA containing area (Fig. 9.4d). In tobacco chloroplasts James & Jope (1978) observed that some plastids had more nucleoids than others and that the nucleoid areas appeared to be linked by filaments. The continuous linking of the plastid genome has been described in the algae *Sphacelaria* (Bisalputra & Bisalputra, 1969) and *Ochromonas* (Gibbs *et al.*, 1974).

9.3.2.2.2 *Electron microscopy of chloroplast DNA:* Electron micrographs of chloroplasts show translucent areas, usually criss-crossed with DNA fibrils (Figs 9.2b and 9.2c) first described by Ris & Plaut (1962). These areas are in the stroma. An association of these areas with lamellae was suggested from autoradiography of chloroplast DNA after uptake of ^{3}H-thymidine into *Spinacia* leaf discs (Rose & Possingham, 1976). In preparations from lysed chloroplasts of *Spinacia* (Woodcock & Fernandez-Moran, 1968; Yoshida *et al.*, 1978) and *Beta* (Herrmann *et al.*, 1974) long strands of DNA equivalent to between one and ten copies of the plastome can be seen associated with membrane-like material.

In *Spinacia* chloroplasts it is difficult to decide where the plastid nucleoid finishes and stroma begins, particularly in older leaves (cf. Figs 9.2b and 9.2c). The density of packing of chloroplast DNA in nucleoid areas is not known, but it could be that in any one species, or plastid type, the volume of the plastid nucleoid is proportional to its DNA content, similar to the relationship of

---

Fig. 9.4a A 70μm fresh section from an 8cm spinach leaf showing a mature cell with approximately 150 chloroplasts, viewed by interference contrast microscopy. ($\times$950)

Fig. 9.4b A 60μm fresh section from a 2cm spinach leaf showing cells containing approximately 20 chloroplasts per cell viewed by interference contrast microscopy. ($\times$950)

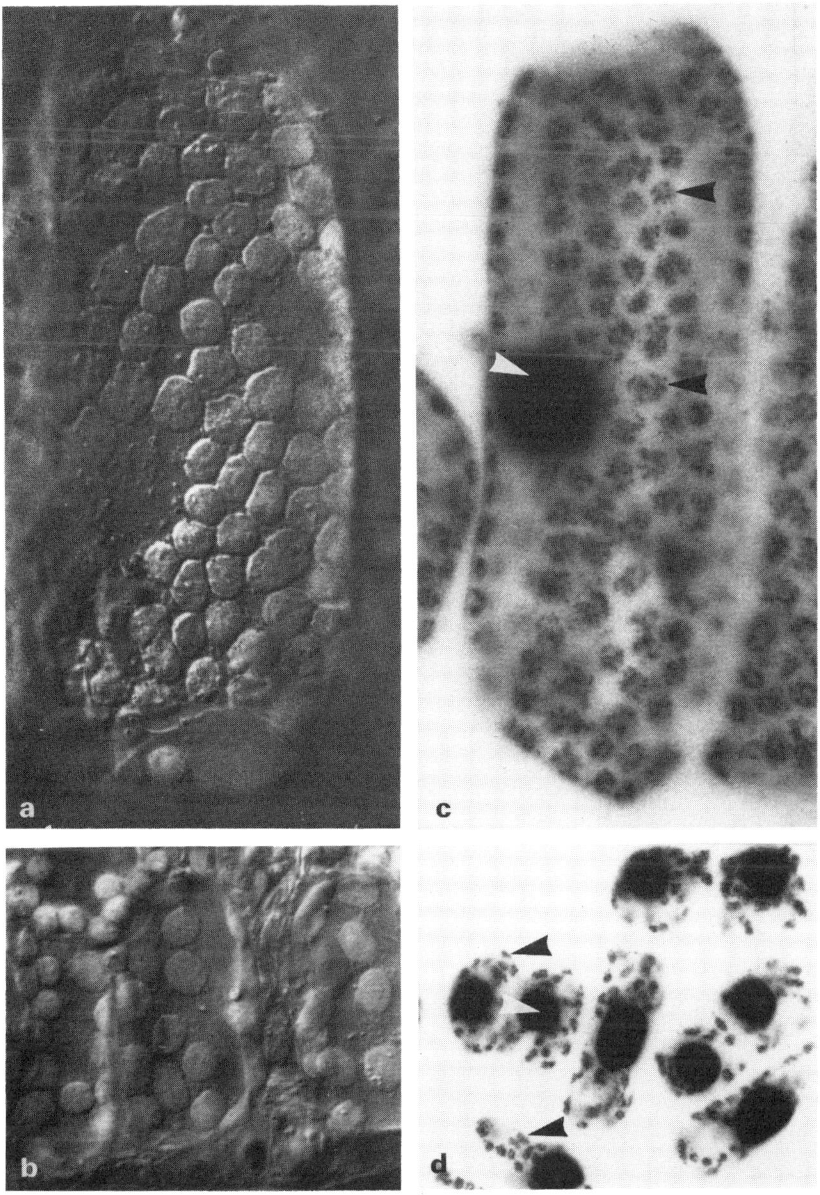

Fig. 9.4c A mature cell from an 8cm spinach leaf stained with the DNA fluorochrome DAPI and viewed in a fluorescence microscope. DNA staining occurs over the nucleus and in discrete areas within the chloroplasts. (× 960)

Fig. 9.4d Cells from a 2cm spinach leaf stained with DAPI and viewed in a fluorescence microscope. DNA staining of both chloroplasts and nuclei. (× 960)

nuclear DNA content and nuclear volume (Baetcke *et al.*, 1967). Herrmann *et al.*, (1974) have used this argument to link observations of the size and number of plastid nucleoids in *Beta* chloroplasts to conclude that larger chloroplasts contain more DNA areas and hence more chloroplast DNA than smaller chloroplasts. As discussed below (9.3.2.2.3) certain chemical measurements do not agree with this conclusion, and we may suspect that while the plastid nucleoid areas are possibly larger in older spinach chloroplasts, they may be less densely packed with DNA (Fig. 9.2c).

9.3.2.2.3 *Chemical measurements of chloroplast DNA*: The chemical methods of measuring chloroplast DNA have all been based on the misleadingly simple principle of estimating the amount of chloroplast DNA in a tissue and equating this amount with the number of chloroplasts.

Owing to the similarities in physicochemical properties between chloroplast DNA and nuclear DNA in most species, it has been difficult to separate the chloroplast DNA from the nuclear DNA. However, in two species, *Cucumis melo* (Kadouri *et al.*, 1975) and *Antirrhinum majus* (Herrmann *et al.*, 1975) a satellite DNA which comprised at least 20% of the total DNA was identified as chloroplast DNA.

Another approach has been to purify chloroplasts and digest away the nuclear DNA with deoxyribonuclease before lysing the chloroplasts, and estimating their DNA content, by a variety of means. In these experiments it is difficult to provide reliable criteria that all the nuclear DNA and none of the chloroplast DNA has been removed during either the chloroplast purification or the deoxyribonuclease digestion. The results of some of these experiments are summarized in Table 9.2.

A third approach is to prepare a radioactive chloroplast DNA probe and measure its renaturation kinetics in a total DNA extract from a tissue in order to obtain an absolute estimate of the amount of chloroplast DNA in the tissue. Using this method, the amount of chloroplast DNA in *Pisum sativum* varied from 0.4% in roots to 12% in leaves (Lamppa & Bendich, 1979b) and in *Spinacia* from 1.3% in roots to 21% in leaves (Scott & Possingham, 1980). In the *Spinacia* study cell areas and numbers per leaf and chloroplast areas and numbers per cell were measured and the amount of chloroplast DNA per chloroplast was estimated on this basis. In small leaves with small cells containing about 20 small, partly developed, chloroplasts Figs 9.4b & d) there were, on average, about 200 copies (32×10^{-15}g DNA) of the plastome per chloroplast while in older leaves with large cells containing about 170 fully developed chloroplasts (Figs 9.4a & c) there were about 30 copies (5×10^{-15}g DNA) of the plastome per chloroplast. These results are summarized in Fig. 9.5. When the chloroplast DNA was allocated to chloroplasts solely on the basis of chlorophyll measurements, then the same trend emerged although the absolute plastome copy numbers were 50% higher (Scott & Possingham, 1980) than those shown in Fig. 9.5. A similar trend was also observed when the DNA

Table 9.2 Chloroplast content of various photosynthetic tissues. Results abstracted from the literature have been grouped according to the general principle of the way in which they were obtained. All determinations were on leaf material unless otherwise indicated

Species	Chloroplast DNA (% of total)	Plastomes per chloroplast	Reference
(1) Methods based on separation of Chloroplast DNA from total DNA extracts			
Antirrhinium majus	> 20		Herrmann *et al.*, 1975
Cucumis sativus	21		Kadouri *et al.*, 1975
Nicotiana tabacum	9		Tewari & Wildman, 1966
(2) Methods based on purification of chloroplasts before isolation of DNA			
Spinacia	4–6	11–53	Mares & Possingham, unpublished
Phaseolus coccineus	0.9	5–20	Pascoe & Ingle, 1978
Antirrhinum majus		5–20	Pascoe & Ingle, 1978
Cucumis sativus	2.3	5–20	Pascoe & Ingle, 1978
Swiss chard			
Vicia faba			
Pisum sativum		220	Bennett & Radcliffe, 1975
Pisum sativum etioplasts		400–800	Bennett & Radcliffe, 1975
(3) Methods based on renaturation kinetics of chloroplast DNA in total DNA extracts			
Pisum sativum	8–12	100–270	Lamppa *et al.*, 1980
Pisum sativum roots	0.4		Lamppa & Bendich, 1979b
Spinacia	21	30–200	Scott & Possingham, 1980
Spinacia roots	1.3		Scott & Possingham, 1980

content of isolated deoxyribonuclease-treated spinach chloroplasts was measured, but in this case the absolute plastome copy numbers were about a quarter of those shown in Fig. 9.5 (Mares & Possingham, unpublished observations).

In a study of pea leaf growth, Lamppa *et al.*, (1980) have found an exactly similar trend. The number of chloroplasts in expanding pea leaf cells rises from 24 to 64 while the number of plastomes per chloroplast falls from 270 to 100. They also report no increase in leaf cell ploidy during this period.

Leaves of *Beta vulgaris* have a similar growth pattern to those of spinach, and in preliminary experiments (Scott & Possingham, unpublished observations) using a spinach chloroplast ^{3}H-DNA probe to measure levels of *Beta* chloroplast DNA, a similar trend was found to that reported for spinach.

It should be noted that despite the decrease in chloroplast DNA content per chloroplast which occurs with age, as chloroplasts expand and divide there is a net synthesis of chloroplast DNA on a per leaf or per cell basis (Scott & Possingham, 1980).

Numerous studies have shown the tight coupling between cell division and net DNA synthesis in all organisms. Only after DNA replication does cell

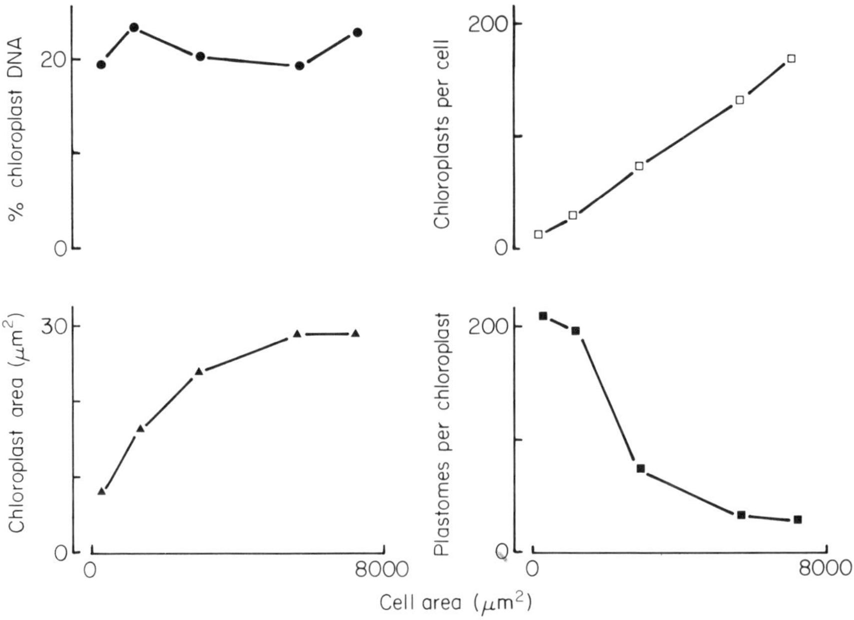

Fig. 9.5 The development of chloroplasts during the growth of spinach leaves. The growth of spinach leaves is depicted graphically as an increase in cell area on the x axis where the base of small (2cm) leaves have cells with an average area of $300\mu m^2$ and the top halves of larger leaves (10 cm) have cells with an average area of $7200\mu m^2$. As the cells expand there is no change in their percentage content of chloroplast DNA (●), while the chloroplast area (▲) and chloroplast number per cell (□) increases, and the number of copies of the plastome per chloroplast decreases (■). (Adapted from Scott & Possingham, 1980)

division take place. In expanding leaves of spinach and pea (see above) the synthesis of chloroplast DNA is included in this tight coupling of nuclear DNA synthesis and cell division, but the bulk of the chloroplast division parallels the independently synchronized processes of cell and leaf development (cf. Fig. 9.5). Thus in the chloroplast situation the nexus between net DNA synthesis and division can be broken. Nonetheless, because of continuing cell division there is net DNA (including chloroplast DNA) synthesis per leaf. In spinach leaf discs, where cell and chloroplast development parallels the leaf situation closely, the incorporation of ^{3}H-thymidine into chloroplast DNA is consistent with semi-conservative replication of chloroplast DNA in all chloroplasts and the even distribution of the chloroplast DNA to daughter chloroplasts (Rose *et al.*, 1974). Thus while the link between chloroplast division and chloroplast DNA amount may be broken, a link with chloroplast DNA synthesis may remain. It should be noted that these arguments are consistent with the relationships between DNA amounts and chloroplast numbers described by Butterfass (1979) (see also Section 9.4.2).

All the estimates of chloroplast DNA discussed so far are the results of average measurements of populations. It is possible that if we could measure the chloroplast DNA content of individual chloroplasts a new relationship might be revealed. In *Acetabularia* some chloroplasts have no detectable DNA (Woodcock & Bogorad, 1970; Coleman, 1979). However, observations with DAPI and the studies of ^{3}H-thymidine uptake into chloroplasts of spinach leaf discs (Rose & Possingham, 1976) suggest that chloroplast DNA is evenly distributed between all the chloroplasts of a cell in higher plants.

9.3.2.3 *Expression of the Plastome*

Kirk & Tilney-Bassett (1978) have analysed the genetics and biochemistry of the plastid extensively. One of the most striking conclusions to arise from reading their book is that nearly all the steps of the biochemical pathways of the plastid that have been analysed are controlled by nuclear genes (Table 9.3). In some pathways one or two steps are controlled by the plastome and they are listed in

Table 9.3 Chloroplast functions coded by nuclear DNA. Most of the characteristics of the plastid for which a coding site has been identified are of nuclear determination. These assignments have been arrived at by genetic analyses. This table lists these characteristics by groups and a † indicates where biochemical evidence of the nuclear determination is available as well. In addition, a * marks where the group includes a plastome coded characteristic(s) which will be found in Table 9.4. Where the number of separate characters analysed in each group is indicated they probably correspond to individual gene products. The information in this table is adapted from Kirk and Tilney-Bassett, 1978.

Characteristic	Number of genetic characters analysed in the group
Steps in pigment biosynthesis	
chlorophyll	14
carotenoids in chloroplasts	5
carotenoids in chromoplasts	6
Photosynthetic mutants	
†*Electron transport chain	15
†*CO_2 fixation	5
*Timing of plastid formation	
Starch formation in amyloplasts	9
*(?) Ribosomal protein formation	
Variegations	
*Chimeras	
†*Chlorophyll-protein complexes	
Temperature sensitivity of plastids	
†Aminoacyl tRNA synthetases	

Table 9.4. Similarly, nearly all the genetic variation related to plastid morphology and performance is accounted for in the nuclear genome and only a relatively few characteristics can be assigned to the plastome (Table 9.4).

In compiling Table 9.4, plastome controlled, or coded, characteristics have been grouped according to the method of analysis that led to the assignment of plastome function. With one exception, the characteristics analysed do not overlap from group to group.

The exception is the RuBP carboxylase protein which has been identified as a plastome-coded protein by conventional genetic analysis (Kawashima & Wildman, 1972), by analysis of the products of plastid protein synthesis (Blair & Ellis, 1973) and by translation of a mRNA which maps to a specific sequence on the plastome (Coen *et al.*, 1977).

The different phenotypes observed in Table 9.4 range from morphological to biochemical characteristics and may in some cases be the same plastome function observed at different levels of expression. For example, it is quite possible that the same plastome sequence may code for the phenotypes observed as a chimera (Table 9.4, 1.1.1), chlorophyll apoprotein synthesis (Table 9.4, 1.2.1), a photosystem deficiency (Table 9.4, 1.2.2) and one of the unidentified membrane polypeptides (Table 9.4, 2.2.3).

The circular chloroplast DNA molecule of about 1×10^8 daltons, after allowing for the inverted repeat sequence of about 1.6×10^7 daltons (Bedbrook & Kolodner, 1979) has enough information to code for two sets of plastid rRNA, a complete set of tRNAs and about 150 proteins of average mol. wt. 30 000, assuming that the DNA sequence is read once only. After including the 80 or 90 unidentified polypeptides referred to by Ellis & Barraclough (1978, Table 9.3, 2.2.3) the functions listed in Table 9.3 account for a good proportion of the potential information in the plastome.

While the capacity of the plastome seems miniscule, compared to that of the genome, its qualitative importance to leaf function must be emphasized. In young wheat leaves for instance, plastid ribosomes (rRNA, Table 9.4, 2.1.1) account for more than half the protein synthetic capacity of the leaf and most of their product (Brady & Scott, 1977) is the large subunit of RuBP carboxylase (Table 9.4, 1.2.3).

9.3.3 OTHER CYTOPLASMIC DNA

The other cytoplasmic DNA of which the existence is established in higher plants is mitochondrial DNA. Relatively little is known about plant mitochondrial DNA although mitochondrial DNA of other organisms is well characterized. Plant mitochondrial DNAs tend to have a density of 1.707g ml^{-1} in CsCl and are found as circles of average length 30μm, equivalent to a mol. wt. of about 30×10^6 (Vedal & Quétier, 1978). For any particular plant species restriction enzyme digests showed characteristic patterns for mitochondrial DNA, indicating that the mitochondrial genome of a species is stable. However

Table 9.4 Expression of the Plastome. Properties and characteristics of the plant which are a function of expression of the plastome have been listed according to the method by which their plastome origin was recognized

	Characteristic	Species	References
	1 CHARACTERS ASSIGNED BY GENETIC ANALYSIS		
	1.1 *Morphology and development—at the anatomical level*		
1.1.1	Assorted variegations and chimeras in leaves and flowers	Widespread	Kirk & Tilney-Bassett, 1978
1.1.2	Leaf size, leaf shape and development	*Oenothera* *Hypericum* *Acacia* *Pelargonium* *Rhododendron* *Silene* *Geranium* *Medicago*	Kirk & Tilney-Bassett, 1978 Sager, 1972
1.1.3	Male sterility	*Nicotiana*	Chen *et al.*, 1977 Frankel *et al.*, 1979
	1.2 *Biochemical pathways—at the subcellular level*		
1.2.1	Chlorophyll synthesis:		
	Chlorophyll apo-protein synthesis	*Antirrhinum* *Pelargonium*	Kirk & Tilney-Bassett, 1978
1.2.2	Photosynthetic electron transport:		
	Chlorophyll/photo-system II protein complex	*Antirrhinum* *Pelargonium*	Kirk & Tilney-Bassett, 1978
	Chlorophyll/photo-system I protein complex	*Antirrhinum*	Kirk & Tilney-Bassett, 1978
1.2.3	CO_2 fixation:		
	Large subunit of RuBP carboxylase	*Nicotiana*	Kawashima & Wildman, 1972
	2 CHARACTERS ASSIGNED BY BIOCHEMICAL ANALYSIS (These functions have often been found in only one or two species but are almost certainly universal)		
	2.1 *By direct analysis of the plastome*		
2.1.1	Plastid rRNA	Many species	Scott & Smillie, 1967 Kirk & Tilney-Bassett, 1978 Herrmann & Possingham, 1980
2.1.2	tRNA (14 out of 20 amino acids)	*Spinacia*	Steinmetz *et al.*, 1978 Driesel *et al.*, 1979
2.1.3	Large subunit of RuBP carboxylase	*Zea* *Chlamydamonas*	Coen *et al.*, 1977 Malnoe *et al.*, 1979
2.1.4	'Photogene 32'	*Zea*	Grebanier *et al.*, 1978
2.1.5	Membrane proteins	*Chlamydomonas*	Rochaix *et al.*, 1979
	2.2 *By analysis of the products of plastid protein synthesis*		
2.2.1	Large subunit of RuBP carboxylase	*Pisum*	Blair & Ellis, 1973
2.2.2	3 coupling factor subunits (out of 5)	*Pisum*	Mendiola-Morgenthaler *et al.*, 1976
2.3.3	About 90 unidentified peptides	*Pisum*	Ellis & Barraclough, 1978
2.2.4	Cytochrome f	*Pisum*	Doherty & Gray, 1979

the range of DNA restriction fragment sizes suggests there is intramolecular heterogeneity within the DNA population and more recently Quetier & Vedel (1979) reported a heterogeneous length distribution of circular DNA in higher plant mitochondria. They concluded that the mitochondrial genome may be composed of several distinct molecules.

Timothy *et al.*, (1979) found similar difficulties in explaining the DNA fragments obtained in restriction enzyme digests of mitochondrial DNA from *Zea* Teosinte races, and could only make comparisons on the basis of the higher molecular weight of DNA fragments. They concluded there were four different groups of mitochondrial genomes in these races.

It is possible that these confusing DNA fragment patterns generated by restriction enzyme digestion of mitochondrial DNA (Quetier & Vedel, 1979; Timothy *et al.*, 1979) reflect an unstable mitochondrial genome. Alternatively, the genome may be much larger than originally supposed and thus incorporate all the restriction fragments observed in the one molecule.

In a range of male sterile *Zea mays* species, Kemble & Bedbrook (1980) found small supercoiled circular DNA molecules of around 2000 base pairs (1.3×10^6 daltons) and suggest that this may be a plasmid type molecule associated with mitochondria in all the varieties examined. They also found three other small DNA species of mol. wt. 1.0 to 1.6×10^6 in the mitochondria variously distributed in the varieties examined.

The interaction of the genome and the mitochondrial genome in plants is presumably similar in principle to that already well characterized in organisms such as yeast (Tzagoloff *et al.*, 1979) and to that of the genome and the plastome (see Section 9.4).

There have been many observations of plant mitochondria in close association with chloroplasts. These associations are usually related to metabolic integration of a particular pathway. In tobacco, mitochondria and chloroplasts were observed in close adpression at times when chloroplasts were accumulating starch (Wildman *et al.*, 1974). In the bundle sheath cells of C_4 plants, where the decarboxylation of aspartate is of the NAD-dependent malic enzyme type, the association of large numbers of mitochondria with chloroplasts is observed (Downton, 1971; Hatch *et al.*, 1975). In the barley mutant 'albostrians' apparent fusion of mitochondria and plastids has been observed (Wellburn & Wellburn, 1979) in the etiolated parts of the leaves where the plastids lack protein-synthesizing capacity.

As well as the close coordination between the two cytoplasmic and the nuclear genomes there may be an association between the plastome and the mitochondrial genome.

9.4 THE INTERACTION OF PLASTOME AND GENOME

The existence of plastomes in organs other than leaves was discussed in section 9.3.2.1. However, virtually all observations of the nature and function of the plastome in plants relate to leaves. Accordingly, discussions of the interactions of the plastome and genome are mainly confined to leaf functions.

There are several aspects of plastome–genome interaction to be considered, although our knowledge of most of them is slight. There is the role of the genome in synthesizing the plastome in situations in which plastid continuity is maintained in the apparent absence of plastid protein-synthetic capacity (see 9.4.3). In uni- or bi-parental inheritance of plastome-coded functions the plastome(s) may be maintained, eliminated (if of paternal origin) or subsequently sorted out. All of these phenomena, while not necessarily involving expression of the plastome in the form of an RNA transcript or a protein product, involve communication of some sort between the plastome and the genome. For example, the absence of a plastome product such as the large sub-unit of RuBP carboxylase may provide a negative communication (Hallier *et al.*, 1978).

The genome-plastome interaction is almost certainly tightly coordinated and can be considered at several levels—all of which will eventually have to be reconciled. We shall consider biochemical and developmental approaches to this problem.

9.4.1 INDEPENDENCE OF PLASTIDS

The idea that some plastid functions are not dependent on the genome dates to the turn of the century when Correns and Baur observed cytoplasmic inheritance of some plastid characters in plants. The expansion of these ideas to include the concept of autonomy or semi-autonomy of plastids peaked with the discovery of chloroplast DNA and tended to ignore the already large body of evidence (Kirk & Tilney-Bassett, 1978) relating many chloroplast functions to nuclear genes. The discussion of the coding functions of the genome and plastome in the preceding section indicates that the plastid is dependent on the genome and shows that the autonomy of higher plant plastids is limited.

As a result of mutations, proplastid-like plastids without plastid ribosomes have been found in the achlorophyllous parts of variegated leaves of *Pelargonium zonale* (Borner *et al.*, 1972 Knoth *et al.*, 1974), *Hordeum vulgare* (Knoth & Hageman, 1977), and the 'iojap' mutant of *Zea mays* (Walbot & Coe, 1979). In these systems the loss of plastid ribosomes was assessed by electron microscopy and gel electrophoresis of rRNA, and while it is difficult to provide conclusive evidence of a total and absolute loss, it is certain that insignificant levels of plastid ribosomes remain. The *Pelargonium* mutants had DNA (presumably chloroplast DNA) in their plastids (Knoth *et al.*, 1974).

The 'iojap' mutation of maize is expressed as a vertical leaf variegation and plastids from the achlorophyllous tissue contained DNA (Walbot & Coe, 1979). Walbot and Coe showed, by restriction enzyme analysis that the chloroplast DNA of the affected mutant was almost identical to that of the wild type chloroplast DNA, with less than 2.2% of the bases altered. They were not convinced that the differences they observed were more than a microheterogeneity of chloroplast DNA relating to the different nuclear backgrounds of the parent stock.

In these studies plastid numbers per cell and cell numbers per leaf during development of the variegations were not closely monitored. However, it is clear from studies of the growth of similar leaves in green plants (see Possingham, 1980) that nearly all ribosome-free plastids found in these achlorophyllous tissues would have arisen during cell and plastid division that occurred after loss of plastid ribosomes. It therefore follows that the synthesis of the new plastid material, including chloroplast DNA itself, is independent of plastid (ribosome) protein synthesis.

In rye leaves, plastid ribosomes could not be found in the plastids of temperature-stressed plants. The loss of plastid proteins known to be synthesized on plastid ribosomes was associated with the loss of plastid ribosomes, while the synthesis of those plastid proteins of cytoplasmic origin and replication of the plastids continued (Feierabend & Schrader-Reichhardt, 1976).

Genetic analysis of many characters that affect plastid function and formation have been reviewed by Kirk & Tilney-Bassett (1978). In all cases where the structure of the chloroplast, and in most cases where the synthesis of plastid components, was concerned, nuclear genes were involved. A comparison of Tables 9.3 and 9.4 for instance, shows the vast difference between plastome and nuclear genome-coded biochemical functions of the plastid.

These observations make it clear that the concept of plastid autonomy is limited. The plastid genome is only three to ten times the size of the mitochondrial genome of most organisms and plastid autonomy may be almost as limited as mitochondrial autonomy. In yeast, for example, the replication of both the mitochondrion and its DNA are under the control of the nuclear genome (Tzagoloff *et al.*, 1979).

The plastid genes whose function has been analysed at the biochemical level are mainly known to affect the biochemistry of photosynthesis (Table 9.4, 1.2.1–3). Thus, despite the evidence of effects on leaf shape in several species and male sterility in tobacco (see Table 9.4, 1.1.2), it is tempting to speculate that the function of the plastid genes is primarily related to the photosynthetic function of chloroplasts. It is possible that chloroplast DNA has virtually no coding role in plastid precursors such as proplastids, although plastid DNA may be required in some unknown role, not necessarily a coding function, to act as an initiator or as part of the structural template, to ensure the synthesis of plastids.

Feierabend and Schrader-Reichhardt (1976) have already proposed that the plastid envelope can be synthesized by the cytoplasm. While it is well known that subunits of quantitatively important plastid proteins such as RuBP carboxylase and coupling factor protein are nuclear coded, it has been suggested that, as well, the nuclear-cytoplasmic system controls the synthesis and replication of chloroplast DNA.

In higher plants there is no evidence that plastids can be formed *de novo* or in fact that aplastidic cells exist. Thus, although the genome may code for all the main constituent molecules of a proplastid, it seems that these organelles cannot be formed *de novo* but must arise only from the division of pre-existing organelles. If this is the case, the basic structural template may be the limiting membrane, a ground substance protein and plastid DNA.

These considerations lead us to the following conclusions about plastid synthesis:

(1) A basic plastid which contains chloroplast DNA and is closely related to the proplastid can be maintained and is capable of growth and replication in the absence of plastid protein synthesis. The synthesis of the constituent molecules of this skeletal organelle are controlled by nuclear DNA. This suggestion does not conflict with the concept that the chloroplast DNA sequence is independent of the nucleus and that the function of chloroplast DNA can be expressed independently of nuclear DNA as well as in cooperation with it, nor does it exclude the possibility that under these conditions an RNA polymerase may still transcribe the chloroplast DNA.

(2) Extending this line of reasoning, chloroplast DNA may have no direct role either in the cells containing proplastids or in the synthesis of the proplastids. There would probably be no necessity, for instance, for the synthesis or export of either plastid protein or RNA molecules to act as (de)repressors, messengers etc., in the cytoplasm or nucleus.

(3) It is only in etioplasts, chloroplasts etc., that once the differentiation of proplastids has begun, the transcription and translation of chloroplast DNA is involved.

It is possible that the conclusions drawn above may be invalidated by the discovery of further plastome functions. It is established, for example, that many components of the nitrate reduction and pentose phosphate pathways are located in plastids (Kirk & Tilney-Bassett, 1978) but there is no evidence that any of these components are synthesized on plastid ribosomes or coded by chloroplast DNA.

9.4.2 THE EFFECT OF NUCLEAR AND CHLOROPLAST DNA AMOUNTS

In the previous section the relative amounts and the organization of plastid and nuclear DNA in cells were discussed. There is a considerable body of information about the influence of nuclear DNA amount on plastid number. This covers work on the absolute quantity of DNA per nucleus and the degree of polyploidy of nuclei. These effects have been extensively reviewed by Butterfass (1979) who came to the conclusion that the number of chloroplasts per cell is controlled and limited by the amount of nuclear DNA. This conclusion is based on comparisons made between species of related taxa and between cells or tissues that are of equivalent function and stage of development. It is different from the situation described in Fig. 9.5 which can be thought of as a secondary developmental pattern superimposed on the basic controls described by Butterfass. Some of the more important examples of the relationships examined by Butterfass are summarized below.

In guard cells a doubling of chromosome number is associated with a 70% increase in chloroplast number in a range of plant species, and guard cell plastid numbers have been used as a crude index of ploidy by plant breeders. A similar relationship has been found in intra-specific studies of natural polyploids and allopolyploids, when the increase in chloroplast number with doubling of ploidy is 41% and 67% respectively (Butterfass, 1979). In studies with trisomic beet plants Butterfass has identified the contribution of individual chromosomes to increased chloroplast numbers. A similar relationship between increased chloroplast number per cell and nuclear DNA amount can be demonstrated (Butterfass, 1979) but the correlation is less significant.

In mature spongy parenchyma cells that had been induced to divide in sugar-beet leaves, cells with a range of ploidy values were obtained (Butterfass, 1963). In these experiments chloroplast numbers per cell were closely related to the ploidy levels found, and ranged from 20–40 chloroplasts in a diploid cell, up to 180 in a hexadecaploid. Measurements of chloroplast DNA are not available for this study.

The relationships summarized above, between nuclear ploidy (and DNA amount) and chloroplast number are well established but as yet there is not enough experimental evidence to include chloroplast DNA in these correlations. The levels of chloroplast DNA in leaf cells are higher than in roots and the percentage of chloroplast DNA in leaf cells of both spinach and pea changes little during leaf development (Scott & Possingham, 1980; Lamppa *et al.*, 1980). Root and leaf chloroplast DNA levels as a percentage of total cellular DNA can differ by as much as ten-fold (Scott & Possingham, 1980) but the transition point between them has not been defined or localized in the plant, nor have the factors which control this change been identified.

There are many more copies of the plastome per leaf cell than of the genome and even in root cells the difference is substantial (Table 9.5). This difference is also reflected in the number of genes for complex proteins such as RuBP carboxylase where each gene of the small subunit of the enzyme in the nucleus is matched by a very large number of genes for the large subunit of the enzyme distributed amongst the chloroplasts of the cell. In pea the ratio is 1 : 4000 and in spinach 1 : 800 (Table 9.5). Even on a per chloroplast basis there are 5–20 times as many genes of the large subunit per chloroplast as there are of the small subunit per nucleus. An explanation for this difference based on mRNA requirements would have to involve such factors as differential mRNA stabilities or transcription rates. The difference in the number of rRNA cistrons between cytoplasm and nucleus is much smaller, at most two-fold, in spinach and pea leaves (Table 9.5). Chloroplast 70S ribosomes generally account for 30–50% of total leaf cell ribosomes and the large number of plastome copies may reflect a need for a great number of plastid rRNA genes and ribosomes for the synthesis of RuBP carboxylase and other plastid proteins.

It is usual to find that the larger the protein the larger the polyribosome involved in its synthesis since more ribosomes can be packed on the message

Table 9.5 Relativities of genome and plastome amounts in pea and spinach

	Pea	Spinach
DNA per nucleus (pg)	9.7[1]	2.8[2]
Genome size (pg)	5.0[3]	0.29[4]
Estimated ploidy level in leaves	diploid (?tetraploid)[5]	octaploid (?)
Genome copies per leaf cell	2	8
Plastome copies per leaf cell	up to 8000[5]	up to 6000[2]
Plastome copies per root cell	250[5,*]	250[2,*]
Cytoplasmic rRNA genes per cell	7800[6]	6000[4]
Chloroplast rRNA genes per leaf cell	up to 16 000[5,†]	up to 12 000[2,†]

[1] Bennett & Smith (1976) for root tips
[2] Scott & Possingham (1980) for leaf cells
[3] Murray *et al.*, (1978)
[4] Ohja (1978)
[5] Lamppa *et al.*, (1980)
[6] Ingle & Sinclair, (1972)
† Two rRNA genes per plastome
* Assuming root and leaf cells have similar ploidy levels

(Tissieres, 1974). It might be expected that a protein the size of the large subunit of RuBP carboxylase would be synthesized on a polyribosome containing 20–30 ribosomes, but it has been suggested only two or three ribosomes are involved (Howell *et al.*, 1977). Such a difference between plastid and cytoplasmic protein synthesis may generate a requirement for more plastome copies in order to maintain a relationship between plastid and cytoplasmic protein synthetic rates.

It has been argued that the large plastome number in leaves is related to the great stability and remarkable similarity of the plastome from all species. It is easy to see how somatic mutations in a single copy of chloroplast DNA would be lost even in roots. However, the number of copies of the plastome transmitted in the germ line is not known and in the event of this being low, changes in chloroplast DNA might be expressed at this stage of development.

9.4.3 SYNTHESIS OF PLASTID COMPONENTS

The interaction of the plastome and genome has been studied by a range of different approaches in relation to the synthesis of various plastid components. It has been suggested above that the genome alone may control the synthesis of a basic plastid skeleton. It is in the construction of a functional chloroplast that the plastome is involved as well. In a review of chloroplast protein synthesis, Ellis (1977) suggested two main principles which can be applied to these studies. The first is that the function of organellar ribosomes is to synthesize some subunits of multicellular proteins where the other subunits are products of cytoplasmic ribosomes. This principle is certainly still applicable to those few plastid proteins whose synthesis is well understood. There are, however, a large number of proteins synthesized on organellar ribosomes (Table 9.3) whose assembly into, and function in, chloroplasts is not understood. The other principle is that cytoplasmic products control organelle synthesis but the converse does not occur.

These principles provide a simple rationale that fits most of the evidence currently available. However, their interpretation should not be allowed to hinder the search for examples of plastome expression, albeit at a secondary level, in which the genome is not directly involved (for further discussion see 9.4.4).

9.4.3.1 *Antibiotic studies*

The existence of antibiotics which would inhibit protein synthesis *in vitro* on either 70S or 80S ribosomes provided a ready tool to determine if particular plastid and cytoplasmic proteins were synthesized *in vivo* on chloroplast (70S) or cytoplasmic (80S) ribosomes. D-threo-chloramphenicol, lincomycin, streptomycin and spectinomycin have been used as inhibitors of 70S ribosomal protein synthesis and cycloheximide for 80S ribosomes. Unfortunately, in most species studied, particularly higher plants, these antibiotics often have direct effects on metabolic processes other than protein synthesis and this can give rise to non-specific indirect effects on protein synthesis. Some of these effects are summarized by Ellis (1977).

Nonetheless, these studies provided the first evidence that chloroplast protein synthesis produced functional proteins. In *Chlamydomonas*, Chua and Gilham (1977) used inhibition of protein synthesis by chloramphenicol and cycloheximide to demonstrate that 9 out of 33 polypeptides from thylakoid membranes were synthesized on chloroplast ribosomes. Kirk and Tilney-Bassett (1978) list 40 separate proteins from plants and algae whose synthesis has been studied using antibiotics and conclude that 10 of them are synthesized wholly or in part in the chloroplast.

No evidence has yet appeared to suggest that nuclear-coded mRNA can cross the plastid envelope and be translated on the plastid ribosomes *in vivo*, although there is some evidence that this barrier can be broken *in vitro* (Geetha *et al.*, 1980). Consequently the identification of chloroplast ribosomes as a site of synthesis of a chloroplast protein is strong circumstantial evidence that the protein is coded for by plastid DNA.

9.4.3.2 *Plastid proteins*

The mechanisms of synthesis of RuBP carboxylase, or Fraction I protein, are the best understood of all plant proteins and have been reviewed by Ellis (1977) in an earlier volume in this series. The protein is a hexadecamer of mol. wt. 550 000 built up of eight small (mol. wt. 14 000) and eight large (mol. wt. 55 000)

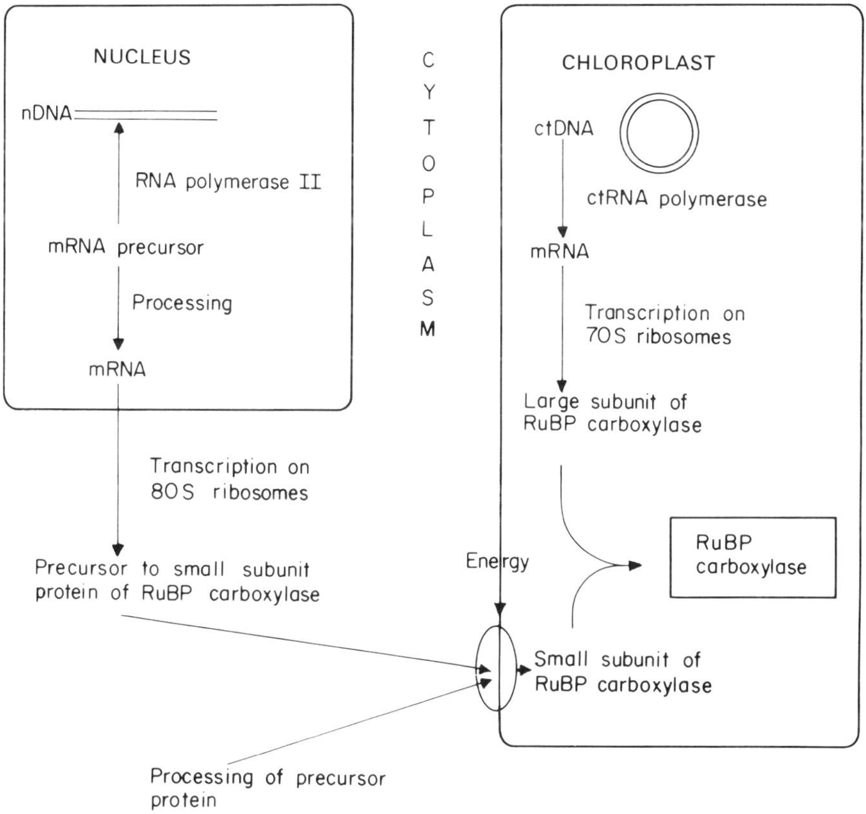

Fig. 9.6 A model for the synthesis of RuBP carboxylase showing the need for coordinated control of plastome and genome expression.

subunits. The synthesis of the subunits and the assembly of RuBP carboxylase are summarized in the model in Fig. 9.6. The mRNA for the large subunit of RuBP carboxylase is transcribed, presumably by a chloroplast RNA polymerase, from the chloroplast DNA, and the protein is translated on 70S chloroplast ribosomes. The gene for this protein in *Zea mays* has been sequenced (McIntosh *et al.*, 1980) and contains information for the direct transcription of a mRNA without introns and translation of the large subunit of RuBP carboxylase without a precursor form.

The small subunit of RuBP carboxylase is synthesized as a higher mol. wt. precursor protein (mol. wt. 20 000) in the cytoplasm on 80S ribosomes and is processed to its final size during passage through the chloroplast envelope (Highfield & Ellis, 1978; Cashmore *et al.*, 1978; Chua, 1980). Studies of the cloned cDNA of the mRNA for the precursor protein show that the additional amino acids of the precursor protein are on the NH_2 terminus of the small subunit protein (Bedbrook *et al.*, 1980). The specific pathway of synthesis of the mRNA for the precursor protein is not known but it is assumed that it arises from the processing of a primary mRNA sequence transcribed from the nuclear DNA by RNA polymerase II, as is the general case for plant mRNAs (Hall, 1979; Becker, 1979).

The precursor to the small subunit of RuBP carboxylase is processed during uptake into maize chloroplasts, whereas the cytoplasmically synthesized chloroplast protein ferrodoxin does not appear to have a precursor form and is taken up directly into the chloroplast (Chua, 1980). As most chloroplast proteins are nuclear coded, it follows that they are synthesized in the cytoplasm and have to cross the chloroplast envelope, or be inserted into it. Studies of several chloroplast proteins of known and unknown function suggest that transport into the chloroplast is energy dependent, irrespective of any post-translation processing (Grossman *et al.*, 1980a; b; Chua, 1980).

The separation of the synthesis of the two subunits of RuBP carboxylase into the cytoplasmic and chloroplastic compartments offers many possibilities for control and synchronization of the assembly of the protein. Cycloheximide inhibits the incorporation of amino acids into the small subunit of RuBP carboxylase before incorporation into the large subunit is inhibited in soybean leaf cells (Ellis & Barraclough, 1978). In plants lacking plastid ribosomes, synthesis of the small subunit of RuBP carboxylase continued (Feierabend & Wildner, 1978). Both these observations are compatible with the cytoplasmic control principle of chloroplast protein synthesis enunciated by Ellis (1977) (Section 9.4.3).

On the other hand, a plastome mutant of *Oenothera* was shown not to synthesize the nuclear-coded small subunit of RuBP carboxylase as well as the plastid-coded large subunit (Hallier *et al.*, 1978), implying that plastome-coded factors were interfering with the synthesis of nuclear coded proteins.

The interaction of the nuclear genome and the plastome is dynamic and changes during leaf development. In the expanding second leaf of wheat, rapid synthesis of RuBP carboxylase was accompanied by a high rate of incorporation of radioactive amino acids into the protein. After expansion ceased the pattern of incorporation of radioactive amino acids into RuBP carboxylase changed relative to other soluble proteins and decreased faster than the rate of decrease in chloroplast polyribosomes (Brady & Scott, 1977). During expansion of the primary leaves of spinach the incorporation of ^{35}S-methionine into the large subunit of RuBP carboxylase changes relative to the incorporation into a thylakoid polypeptide, also known to be plastome-coded, and these changes are reflected in the translatable mRNA population of the chloroplast (Silverthorne & Ellis, 1980).

9.4.3.3 *Chloroplast DNA*

We have already discussed the probability that chloroplast DNA is synthesized in a replicating plastid system in the absence of plastid protein synthesis (9.4.1).

Not enough is known of the DNA polymerase enzymes found in plastids so far to determine if they are replication or repair enzymes (Bedbrook & Kolodner, 1979) but the thesis described above would require the chloroplast DNA replicative enzymes to be nuclear coded and cytoplasmically synthesized.

The synthetic mechanisms for replication of chloroplast DNA may be of most importance in the fertilized embryo. Maternal inheritance of chloroplast DNA, predominant in most higher plants, involves the loss of paternal chloroplast DNA either at some late stage during pollen development, germination or ovule penetration or an early stage of fertilization. The alga *Chlamydomonas* also shows maternal inheritance and a possible mechanism for the elimination of the paternal chloroplast DNA following fertilization involves its degradation at the gamete stage by restriction enzymes, while the maternal plastome is protected from restriction by methylation of the DNA (Burton *et al.*, 1979; Royer & Sager, 1979).

9.4.4 SYNTHESIS OF CYTOPLASMIC COMPONENTS

While many aspects of the synthesis of chloroplasts show the interaction of plastome and genome, possible roles for the plastome in the synthesis of cytoplasmic components has received little attention. The cytoplasmic synthesis of proteins destined for incorporation in the plastid either independent of (Section 9.4.1) or in cooperation with (Section 9.4.3) the plastome, is well recognized. But what of other cytoplasmic proteins and metabolic pathways? Is the plastome involved, maybe in a trigger role, in the synthesis of enzymes intimately associated with chloroplast function but localized in the cytoplasm, or other organelles? The large number of different gene products found in leaves when compared to roots, for instance, (Section 9.3.1.2) argue for the existence of organ- or function-specific controls, and a plastome product(s) would be well suited to fill this role in leaves.

It is clear that the interaction of the plastome and genome is capable of fine tuning. RuBP carboxylase, for instance, is absent from mesophyll chloroplasts of C_4 plants and present in the bundle sheath chloroplasts of the same plants (Huber *et al.*, 1976) when both sets of chloroplasts presumably have the same genome as shown in maize (Walbot, 1977). Indeed the existence of at least three different combinations of enzymes for the transport of CO_2 between the mesophyll and bundle sheath chloroplasts in the C_4/C_3 pathways of C_4 plants (Hatch *et al.*, 1975) is further evidence of the range of different possibilities.

The possibility of plastome gene products being transported to the cytoplasm has been raised in the alga *Euglena gracilis* by McCrea & Hershberger (1978). They found a small fraction of tRNA associated with cytoplasmic polyribosomes which hybridized with an amount of chloroplast DNA equivalent to 8–9 tRNA cistrons. In similar experiments Schwartzbach *et al.*, (1979) could not confirm this result.

9.4.5 THE ORIGIN OF CHLOROPLASTS

Because plastids are self-replicating semi-autonomous organelles and because there is close interaction between the genome and the plastome in recent years, the possibility that chloroplasts may have an endosymbiotic origin has been seriously considered. However, any hypothesis that explains the origin of chloroplasts by invoking the incorporation of a foreign genome from such organisms as blue-green algae or photosynthetic bacteria, must consider the origin of the large nuclear input into the synthesis of chloroplasts. It seems that all the theories of chloroplast origin have given insufficient emphasis as to how the information and machinery for the synthesis of so many plastid molecules came to be located in the nucleus. It is possible that the algal genome may have been cleaved and partitioned between the two sites of plastid and nucleus. However it is not clear how such a large amount of what may originally have been algal genome became transferred from the plastid compartment to the nucleus. Alternatively, it could be that two separate but complementary genomes have been evolved for the development and functioning of chloroplasts, part coming from the alga and part from the host plant.

It is not the purpose of this chapter to consider the various hypotheses of chloroplast origin, but such questions remain to be answered. For a range of views on the origin of plastids see the reviews by Stanier (1970), Cavalier-Smith (1975) and Taylor (1979).

9.4.6 FORMATION OF CYTOPLASMIC HYBRIDS

Following fusion of protoplasts from two different *Nicotiana* species, *N. glauca* and *N. langsdorfii*, parasexually derived hybrid plants were found which had chromosome numbers greater than the n = 48 characteristic of the parents. This suggested that elements of the nuclei of both parents were present in the individual hybrids. Analysis of the small subunit of RuBP carboxylase in the hybrid also showed proteins characteristic of both parents. On the other hand, in most of the hybrid plants the large subunit of RuBP carboxylase characteristic of only one of the parents was found suggesting only one of the parent plastomes had survived in any individual hybrid (Kung *et al.*, 1975; Chen *et al.*, 1977).

In parasexual hybrids of *N. tabacum* × *N. debneyi* chromosome numbers were normal (n = 48) and phenotypically the plants were of one or other parental type. The chloroplast DNA in individual hybrid plants was identified by either restriction enzyme analysis or phenotypic markers and also found to be of one or other parental type (Belliard *et al.*, 1978; 1979). Presumably the chloroplast DNA was selected by a sorting out mechanism from the original mixed population in the fused cells, but the random assortment of genome and plastome types recorded suggested there was no incompatibility between plastome and genome in these intraspecific crosses.

Given the close relationship which has been shown to exist between the expression of the genome and plastome in chloroplast development, and doubtless exists in such mechanisms as maternal inheritance, it is surprising that the plastome and genome are so readily interchangeable. Perhaps this is evidence that the close relationship between the chloroplast DNAs of different species revealed by physicochemical studies extends to the operational details of the plastome–genome interaction.

FURTHER READING

KIRK J. T. O. & TILNEY-BASSETT R. A. E. (1978) *The Plastids. Their chemistry, structure, growth and inheritance.* Elsevier/North Holland Biomedical Press, Amsterdam, New York, Oxford.

HERRMANN R. & POSSINGHAM J. V. (1980) Plastid DNA—The Plastome. In *Results and Problems in Cell Differentiation*, Vol 10: *Chloroplasts*. pp. 45–96 (ed. J. Reinert.) Springer-Verlag, Vienna

POSSINGHAM J. V. (1980) Plastid replication and development in the life cycle of higher plants. *Ann. Rev. Plant Physiol.*, **31**, 113–129.

BEDBROOK J. R. & KOLODNER R. (1979) The structure of chloroplast DNA. *Ann. Rev. Plant Physiol.*, **30**, 593–620.

CHAPTER 10

LEAF SENESCENCE

10.1 INTRODUCTION

For purposes of analysis and discussion it is convenient to divide the lifespan of a leaf into a sequence of events (Fig. 10.1). After initiation of the leaf at an apex, the first phase is one of development involving cell division, expansion and differentiation, the end of which is marked by the completion of expansion of the lamina. There follows a period of maturation in which the leaf undergoes gradual changes in composition and functional capacity leading ultimately to a phase of rapid change in the appearance and functioning of the lamina and concluding in most cases with the abscission of the leaf. Upon this simply-stated developmental theme there have arisen a great array of variations in response to the forces of natural selection. These variations affect the length of time occupied by development, maturation and senescence and hence the lifespan of the leaf and are probably for the most part genetically determined.

The longevity of leaves is immensely variable from a few days in some desert ephemerals to over thirty years in the bristle cone pine (*P. aristata*). The morphology and method of growth of leaves also varies greatly. In plants such as *Welwitschia* and the Gramineae and Cyperaceae growth continues from an intercalary meristem at the base of the leaf with the consequence that the distal portions of the leaf may develop, mature and senesce while new elements of the same leaf are still being initiated. The diversity of leaves is also manifest in their anatomy, particularly with respect to the number of dermal layers, the relative proportions of supporting, vascular and chlorophyllous tissues and the extent of hairs, glands, and other specialized cells. These points are made at the outset because we shall be concerned subsequently with events at the cellular and metabolic levels, the evidence for which has been gathered largely from studies on the leaves of a few commonly cultivated plants selected primarily on grounds of abundance, convenience and economic importance. This concentration of effort has doubtless helped to advance knowledge of the subject but we do well to remember that the species in question, cereals, tobacco, radish, *Perilla*, *Tropaeolum*, *Phaseolus vulgaris* and a few others, all have relatively short-lived leaves of a mesophytic character with a high proportion of chlorenchyma to vascular and supporting tissues. We should also remember that many of the studies referred to involve at some point the sacrificing of the leaves for purposes of extracting and analyzing specific constituents; when this is done the results obtained are averages derived from many different types of cell and from populations of cells which are probably not behaving synchronously with respect to their maturation and senescence.

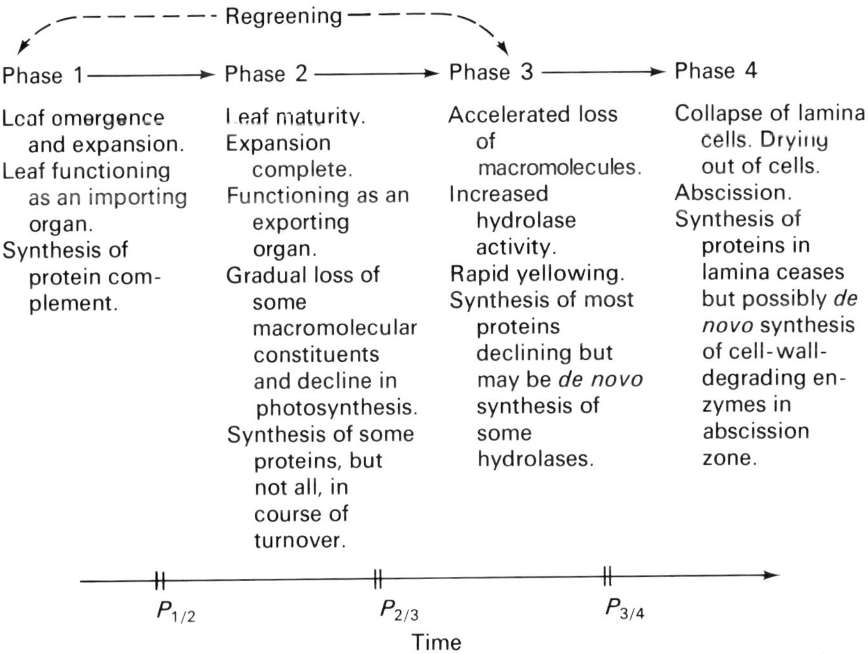

Fig. 10.1 A scheme in which the lifespan of a leaf is divided into phases for purposes of reference and description. Although somewhat arbitrary because the cells of the leaf do not behave synchronously and there is a continuum of changes, the four phases do correspond to visually recognizable events in the life of the leaf. $P_{1/2}$, $P_{2/3}$, and $P_{3/4}$ represent periods of transition between the phases P_1, P_2, P_3, and P_4.

10.2 PHYSIOLOGICAL CHANGES ACCOMPANYING LEAF SENESCENCE

10.2.1 PHOTOSYNTHESIS AND PHOTORESPIRATION

Measurements of photosynthesis have been made on leaves of different ages from a large number of species; some measurements are available for plants grown in controlled environments, many for plants grown in greenhouses, some for cultivated plants in the field and a few for plants in the wild. Four general points can be distilled from this array of studies. (i) The lifespan and photosynthetic activity of leaves may be profoundly modified by the conditions under which they develop (Schwabe, 1970; Woledge, 1972). (ii) The pattern of photosynthesis throughout the life of a leaf is subject to ontogenetic factors so that the temporal relationships of changes may differ between successive leaves. (iii) The combination of environmental and developmental variables which operate under field conditions severely constrain the confidence limits which can be attached to measurements made under these conditions. (iv) The observed pattern of change in photosynthesis varies according to the type of senescence which the leaf is undergoing. For this purpose four main types of senescence may be distinguished (Simon, 1967): sequential senescence associated with relative position on the stem, monocarpic senescence associated with reproductive development of the plant, senescence associated with seasonal rhythms such as winter or dry seasons and senescence following leaf excision.

Analyses of the decline in net photosynthetic rate during leaf senescence suggest three main causes, a decreased stomatal conductance, a decrease in the residual or mesophyll conductance of the leaves and an increase in the relative rate of photorespiration, but it is important to recognize that the relative contribution of these changes may differ between species. Thus in a study of tropical grasses and legumes (Ludlow & Wilson, 1971) decreased stomatal conductance was the main source of declining photosynthesis in the grasses whereas decreasing mesophyll conductance was more important in the legumes. Both stomatal and mesophyll conductance declined in the course of leaf senescence in *Lolium perenne* (Woledge, 1972) and soybean (Woodward & Rawson, 1976) but declining mesophyll conductance accounted for most of the decrease in photosynthesis in *Perilla frutescens* (Woolhouse, 1968) and *Medicago sativa* (Hodgkinson, 1974). Changes in the relative rate of photorespiration during leaf senescence are reported by Fraser and Bidwell (1974) and Čatský *et al.*, (1976).

There are both practical and theoretical reasons why photorespiration as a function of leaf age should be studied in greater detail. On the practical level measurements of photorespiration are notoriously difficult to make and it would be valuable to have corroboration of the changes in rate which have been reported to occur as leaves senesce. From the theoretical standpoint changes in the rate of photorespiration relative to that of photosynthesis (Thomas *et al.*, 1978), are of interest because they do not rest easily alongside the growing body of opinion that photorespiratory efflux of CO_2 arises primarily from a cyclic process of glycollate oxidation which is generated in the oxygenase reaction of RuBP carboxylase/oxygenase and coupled to the reductive pentose phosphate pathway of CO_2 assimilation via the chloroplast pool of phosphoglycerate (PGA) (Fig. 10.2). Opinions differ concerning the range of variation in the relative proportions of oxygenase to carboxylase activity under carefully defined conditions of assay, but it is the experience of many workers that the relative activities are constant with leaf age (Thomas *et al.*, 1978) and over a wide range of genotypes: if this is indeed so then an explanation of the changing relative rates of photosynthesis and photorespiration must presumably be sought in other aspects of these linked carboxylation and oxidation cycles.

10.2.2 ACTIVITY OF PHOTOSYNTHETIC ENZYMES

10.2.2.1 *Carbon dioxide fixation*

Woolhouse (1968) correlated the increasing mesophyll resistance of senescing leaves of *Perilla* with a decline in Fraction 1 protein content (equivalent to RuBP carboxylase protein) and suggested that the level of RuBP carboxylase

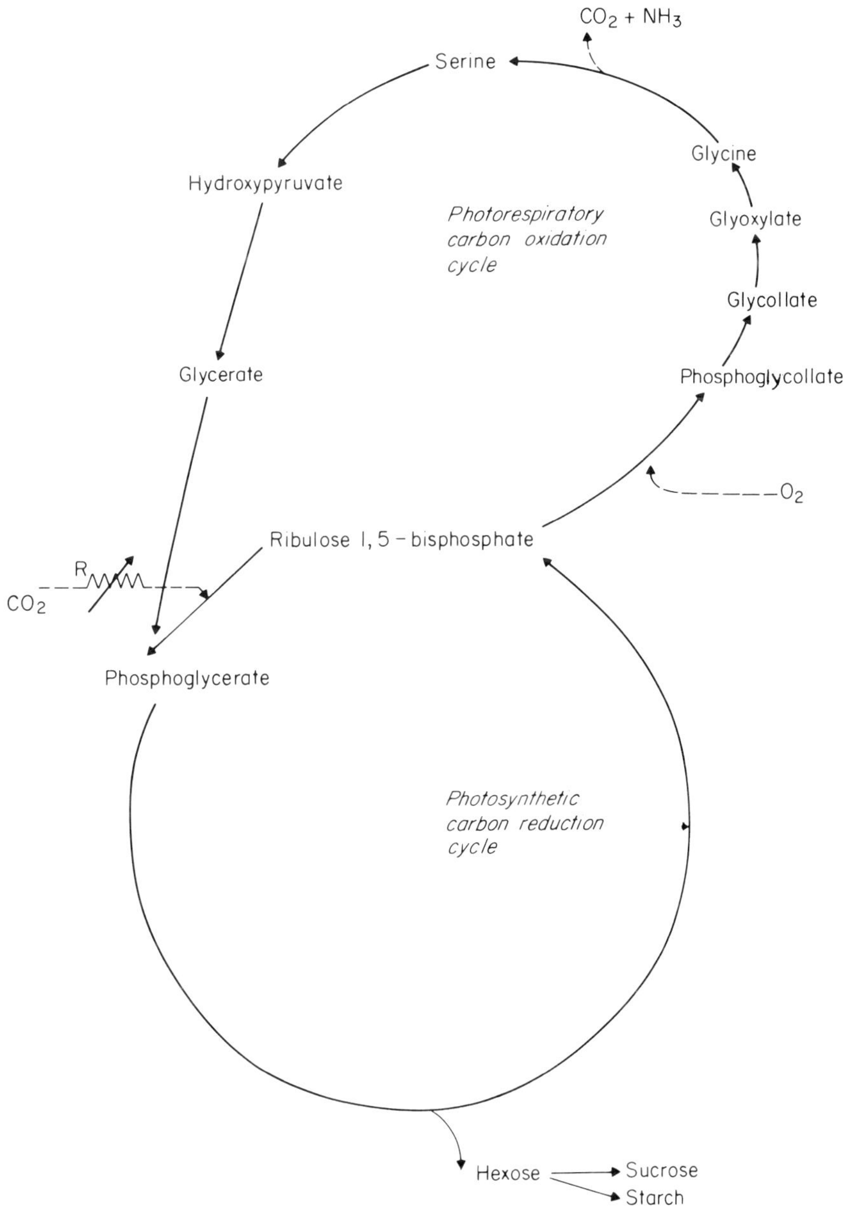

Fig. 10.2 A scheme showing the coupling of the photosynthetic carbon reduction cycle (RPC) to the photosynthetic carbon oxidation cycle (PCO). Phosphoglycollate generated in the oxygenase reaction of RuBP carboxylase is further metabolized through the sequence of intermediates shown to yield CO_2, NH_3 and glyceric acid. The reactions of the RPC and the formation of glycollate take place in the chloroplasts; glycollate oxidation and transamination of glyoxylate to glycine take place in the peroxisomes; the oxidation of glycine to serine, CO_2 and NH_3 is carried out by the mitochondria, reduction of hydroxypyruvate to glycerate occurs in the cytosol from where the glycerate is returned to the chloroplast via a dicarboxylic acid transporter. Reassimilation of NH_3 probably occurs in the chloroplasts. The stomatal aperture is represented by the variable resistance R.

activity, referred to as the carboxylation resistance, may be a major component of the mesophyll resistance. RuBP carboxylase expressed per unit of leaf area, has been shown to decrease in the course of senescence in leaves of tobacco (Kawashimia & Mitake, 1969), *Phaseolus vulgaris* (Heyes & Dale, 1971) cucumber (Callow, 1974), *Perilla* (Batt & Woolhouse, 1975) and wheat (Peoples & Dalling, 1978; Hall *et al.*, 1978). The cause of this decreased RuBP carboxylase activity is not entirely clear; most of the studies referred to report a concomitant decrease in fraction 1 protein content but in tobacco, cucumber and wheat, inactivation as well as loss of the enzyme appears to be involved since the specific activity of the enzyme is also reported to decrease. Other enzymes of the reductive pentose phosphate cycle (RPC) have been studied in senescing leaves of *Perilla* (Batt & Woolhouse, 1975) and it is of interest to note that the changes in activity are not similar for all enzymes of the cycle.

It is tempting to present these studies as further evidence that the decline in RuBP carboxylase is a major factor limiting the photosynthetic rate of senescing leaves. It must be emphasized however that correlations of photosynthetic rate with measurements of *in vitro* activity of the enzyme do not constitute proof of a causal relationship. Accurate assay of RuBP carboxylase activity is notoriously difficult and is dependent on such factors as pH, CO_2 concentration and Mg^{2+} concentration (Walker, 1976; Jensen & Bahr, 1977), which may all be altered by other changes which are taking place in the senescing leaf; the supply of RuBP substrate for the enzyme may also be a limitation to its activity *in vivo* (Jensen & Bassham, 1968). Also important here is evidence that other enzymes of the RPC, notably D-fructose-1,6-bisphosphate 1-phosphohydrolase and sedoheptulose-1,7-bisphosphate 1-phosphohydrolase may be even more important than RuBP carboxylase in the regulation of CO_2 fixation (Heldt *et al.*, 1978). These enzymes require activation by a process of photoreduction which is coupled to the electron transport chain of the chloroplasts via reduced ferredoxin (Schürmann & Buchanan, 1975); it may be therefore that the declining activity of the RPC in senescing leaves is at least partially determined by concomitant changes taking place in the electron transport system of the chloroplasts.

10.2.2.2 *Electron transport*

Electron transport could constitute a major limitation to photosynthesis in senescing leaves because it is needed for the production of NADPH and ATP used in the RPC and for the movement of ions from thylakoid space to stroma, where they activate enzymes of the RPC such as RuBP carboxylase (Jensen & Bahr, 1977). In addition electron transport may also regulate the RPC through the coupled reductive activation of the regulatory phosphatase enzymes of the cycle referred to in the preceding section (Heldt *et al.*, 1978). Several thylakoid proteins are reversibly phosphorylated in the light (Bennett, 1977), amongst which a component of the chlorophyll a/b binding complex which regulates transfer of excitation energy is particularly conspicuous (Bennett, 1979a). The protein kinase which phosphorylates thylakoid polypeptides is bound to the thylakoid membrane and is activated by light-dependent electron transport (Bennett, 1979b) thus providing a further point at which changes in the capacity of the electron transport system could profoundly modify the functioning of the photosynthetic apparatus. There have been many attempts to study electron transport during foliar senescence (Šesták, 1977) but these have mostly been confined to the Hill reaction and have often involved chloroplast preparations of a dubious quality. Few studies have dealt with the overall rate of non-cyclic electron transport and they have not included concomitant measurements of the rate of photosphorylation (Haraguchi & Schimizu, 1970; Anderson *et al.*, 1972).

Jenkins and Woolhouse (1980) carried out a detailed study of non-cyclic electron transport and photophosphorylation during senescence of the primary leaves of *Phaseolus vulgaris*. Rates of electron transport declined by 80% from completion of leaf expansion to late senescence (Fig. 10.3) but photophosphorylation remained coupled to electron transport with a mean P : 2e

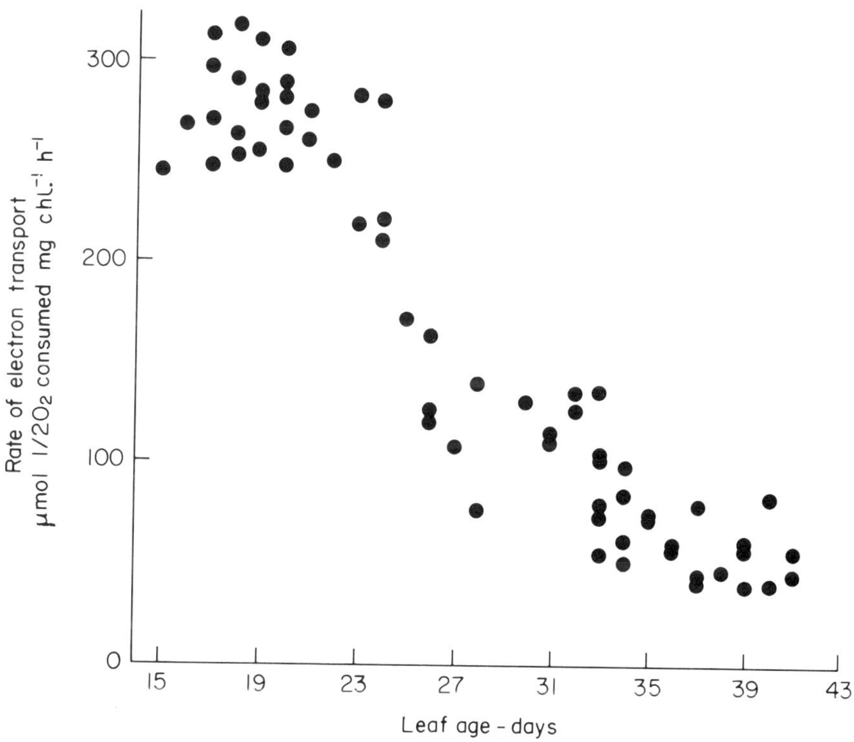

Fig. 10.3 Rates of coupled electron transport from water to methyl viologen in isolated chloroplasts from leaves of *Phaseolus vulgaris*. Each point refers to one measurement on a chloroplast preparation from a single primary leaf. (From Jenkins & Woolhouse, 1981).

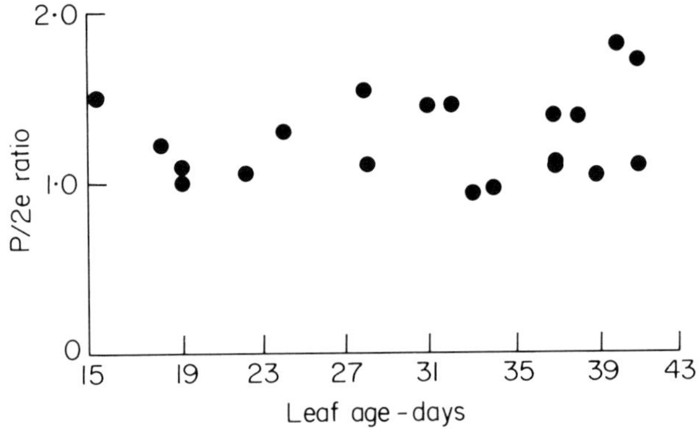

Fig. 10.4 Rates of phosphate esterified per 2 electrons transported from water to methyl viologen in chloroplasts from primary leaves of *Phaseolus vulgaris*. (From Jenkins & Woolhouse, 1981).

ratio of 1.3 (Fig. 10.4). Other attributes of the isolated chloroplasts which remained little altered in senescing leaves were the ADP : O ratio of 1.7; the photosynthetic control ratio of 5.0 and the 45% stimulation of electron transport by 10mM NH_4Cl; these observations suggest that the quality of the chloroplast suspensions was good even in the case of older leaves. Partitioning of the electron transport chain with a series of electron donors and acceptors (Fig. 10.5) showed that the capacity for electron flow through photosystem I (PSI) did not change much in the course of leaf senescence (Fig. 10.6). Electron flow through photosystem II (PSII) declined by only 30% during leaf senescence so that taking these observations together it is suggested that the main change limiting electron flow in the senescing leaves is between the two photosystems rather than within them—an inference which gains further support from studies of fluorescence in this system (Jenkins *et al.*, 1981). From this work it would appear that there are concomitant changes in both the electron transport and the carbon assimilating enzymes of senescing leaves; there may be elements of a common causality to these changes (Section 10.5) but we do not know yet the extent to which each contributes to the overall decline in the rate of photosynthesis in senescing leaves.

10.2.3 RESPIRATION

There have been many studies of respiration in leaves as they age; a typical example is that of Hover and Gustafson (1926) showing that in maize, sorghum, oats and wheat the rate of respiration decreased gradually from full expansion to midway through the maturation phase but then increased again in older leaves. In *Perilla frutescens* undergoing sequential senescence the rate of respiration was constant through the maturation phase and then rose sharply for a

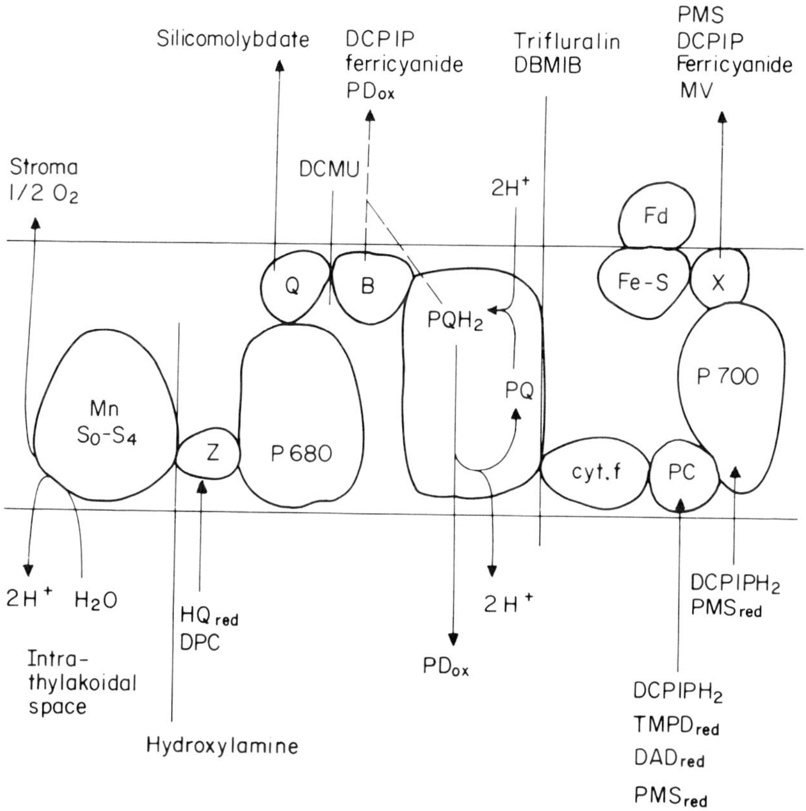

Fig. 10.5 A proposed arrangement of components of the electron transport chain in the thylakoid membrane (after Trebst, 1974) together with indication of sites of action of co-factors and inhibitors used to analyse the electron transport chain. Inhibitors: hydroxylamine, DCMU (3,3-4-dichlorophenyl-1,1-dimethyl urea), trifluralin and DBMIB (2,5-dibromo-3-methyl-6-isopropyl-p-benzoquinone). Electron donors: HQ_{red} (reduced o-hydroquinone), DPC (1,5-diphenylcarbazide), DCPIP H_2 (reduced 2,6-dichlorophenol indophenol), DAD_{red} (reduced diaminodurine), PMS_{red} (reduced phenazine methosulphate). Electron acceptors: silicomolybdate, PD_{ox} (oxidized phenylenediamine), MV (methyl viologen), TMPD (N,N,N′,N′-tetramethyl-p-phenylenediamine).

brief period in a manner reminiscent of the climacteric rise observed in some fruits (Hardwick *et al.*, 1968); similar patterns of climacteric-like respiration in leaves are observed in *Tropaeolum* (Eberhardt, 1955) and barley (Yemm, 1956) but are not recorded in the work of Richardson (1957) on *Acer pseudoplatanus*, Hoffman (1962) on sunflower or Smillie (1962) on *Pisum sativum*. It is apparent from many of the studies in which photosynthesis and respiration have been measured on the same leaves, that the two processes do not change in a similar manner in senescing leaves; most commonly the decline in rate of photosynthesis commences before the decrease in respiration and shows a more rapid

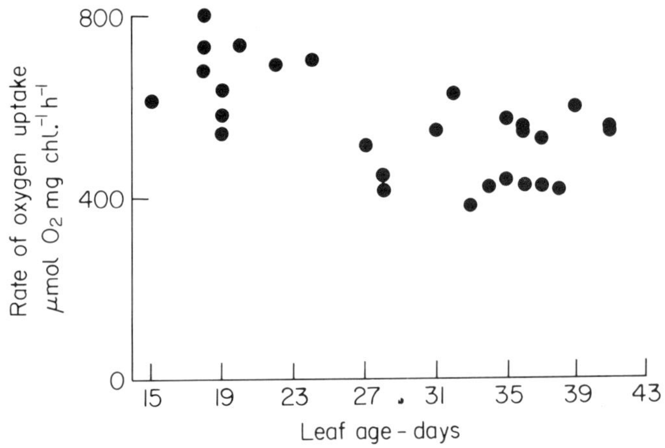

Fig. 10.6 Rate of electron flow through photosystem I (reduced TMPD → MV, Fig. 10.5)—measured in terms of rate of autoxidation of reduced methyl viologen by isolated chloroplasts from primary leaves of *Phaseolus vulgaris*. (From Jenkins & Woolhouse, 1981).

rate of decrease. This difference in photosynthesis and respiration rates is correlated with differences in the pattern of ultrastructural changes during leaf senescence (Barton, 1966; Butler & Simon, 1970). Chloroplast changes begin first; the predominant features are loss of parallel alignment of the grana, loss of compactness in the appearance of the grana associated with swelling of the thylakoids, loss of ribosomes and other particles from the stroma and the appearance of electron-dense plastoglobuli. Changes in the ultrastructure of leaf mitochondria begin much later than in the chloroplasts; they are characterized by an increased separation of the inner and outer membranes and some swelling of the cristae but even at a late stage of leaf senescence a majority of the mitochondria may still appear relatively intact. Detailed studies of the biochemistry of mitochondria in senescing leaves have so far failed because of lack of a technique for obtaining intact mitochondria freed from chloroplast fragments and other cellular contaminants.

For these reasons there is a lack of definitive evidence on whether respiration becomes uncoupled in senescing leaves or whether there is a shift from cyanide-sensitive to cyanide-insensitive respiration as in some ripening fruits.

A crucial question which is still not adequately answered is whether respiration is necessary for leaf senescence; James (1953) found that anaerobic conditions prevented leaf senescence and concluded that respiratory energy was needed. Tetley and Thiman (1974) took up this problem using excised leaves of *Avena sativa* maintained in darkness; respiration rose by 2.5 fold over three days and cytokinins, which depressed the respiratory increase, also delayed senescence. It was concluded that the increased respiration was uncoupled on the flimsy grounds that dinitrophenol increased the respiration rate to the level observed during prolonged darkness and cyanide did not inhibit the respiration. Further indirect support for the idea that respiratory energy is needed for senescence is drawn from the finding that senescence is also delayed by inhibitors of protein synthesis, which also requires respiratory energy. This general conclusion that respiration is necessary for senescence may ultimately prove correct but requires much more rigorous experimental evidence than exists at present. It should also be borne in mind that we now know that ethylene is necessary for leaf senescence (Lieberman, 1979), involving oxygen-dependent mechanisms which are not concerned with respiration (see Section 10.5.1).

10.3 THE GENETIC BASIS AND BIOCHEMISTRY OF LEAF SENESCENCE

10.3.1 GENETIC CONTROL OF SENESCENCE

It is a matter of common observation that there is intraspecific variation in the timing of leaf senescence in deciduous plants. This variation is genetically determined, so that in some species cultivars can be obtained which differ in the time of onset and duration of their autumnal senescence. We have at present little idea of the physiological mechanisms underlying this variation in seasonal senescence, so that although we may readily agree that the process is a natural part of the developmental programme of the plant this does not lead us directly to the nature of the controlling genes.

One of the most promising approaches to the more detailed analysis of the genetic basis of senescence is the use of mutants, such as the form of *Festuca pratensis* which is unable to break down chlorophyll as the leaves senesce (Thomas & Stoddart, 1975). The non-yellowing characteristic is due to a single nuclear gene (NY); it appears to be confined in its effects to aspects of metabolism which are closely concerned with chlorophyll catabolism since other general features of senescence such as loss of RNA and protein, changing enzyme activities and patterns of isozymes remain unaltered in the mutant (Pearson *et al.*, 1978; Thomas, 1975; Thomas, 1977; Thomas & Stoddart, 1975; 1977). Electron micrographs of chloroplasts of the NY mutant show the thylakoid membranes becoming released from the granal arrangement but not undergoing the usual pattern of degradation (Thomas personal communication), suggesting that the mutation may involve some aspect of lipid degradation so that the chlorophyll is not released to sites at which it is normally degraded.

Thomas and Stoddart (1980) give references to ten enzymes which undergo changes in isozyme complement in the course of leaf development and senescence. For some of these enzymes the mechanism of inheritance is known but in no case is it yet clear at what level the isozyme changes are controlled or what their significance is in the senescence process.

Another feature of the genetic control of senescence is that interactions must presumably occur between the nucleus, chloroplasts and mitochondria in order to coordinate events. Yoshida (1961) provides an intriguing example which suggests involvement of the nucleus in regulating chloroplast senescence in *Elodea densa*. When leaves were plasmolysed in calcium chloride solution, protoplasts of some of the mesophyll cells separated into two halves, one with and one without a nucleus. When the leaves were subsequently cultured the chloroplasts of enucleate portions of the protoplast remained green and accumulated starch for at least five days whereas chloroplasts in the protoplasts containing nuclei disintegrated and were broken down over the same period. The question thus arises of how these controls operate; it will be convenient to return to this question when we have looked a little further into the biochemical events of the senescence process.

10.3.2 BIOCHEMICAL CHANGES IN SENESCENCE

10.3.2.1 *Nucleic acid and protein turnover*

Borodin (1876) and Schulze (1878) suggested that there was a continuous breakdown and re-synthesis of protein in plant tissues; the idea gradually gained precision in the theory of the 'protein cycle' (Mothes, 1933; Gregory & Sen, 1937), in which the proteins were held to be in a dynamic state, being continuously synthesized and broken down, even in circumstances where the total protein content of the tissue may not be changing. With the advent of radioactive and heavy isotope techniques renewed attempts have been made to measure protein turnover in leaves, but the problems of methodology and interpretation remain formidable. There are several published estimates for the turnover of total leaf protein but for the purposes of defining crucial steps in onset of synthesis or enhancement of degradation during senescence, measurements of turnover of particular proteins are needed. At the present time such turnover studies have only been reported for RuBP carboxylase, from which it appears that in *Perilla* (Woolhouse, 1967) and barley (Peterson *et al.*, 1973; Peterson & Huffaker, 1975), this protein is not turned over once expansion of the leaf is completed whilst in *Zea mays* turnover of the RuBP carboxylase continues in the mature leaf, albeit with degradation exceeding synthesis in the later stages (Simpson & Varner, 1980). The study by Simpson and Varner makes use of an extremely promising technique involving the use of ^{3}[H]-acetic anhydride applied *in vivo*. With the simultaneous development of immunological and affinity column techniques to assist the purification of leaf proteins it may be hoped that good measurements of turnover of specific proteins will be rapidly extended.

In respect of nucleic acid turnover in leaves the position is if anything worse than with proteins; DNA is not turned over in mature leaves (although some repair of DNA seems likely), whilst turnover of all RNA fractions is normal (Callow *et al.*, 1972; Callow & Woolhouse, 1973) but precise measurements of the rates of turnover of the RNA fractions from leaves have not been reported.

Against this limited background knowledge we must now turn to consider the supplementary evidence for regulation of the loss of leaf constituents during senescence in terms of arrested synthesis (Fig. 10.7a) and enhanced degradation (Fig. 10.7b).

10.3.2.2 *The shut-down of nucleic acid and protein synthesis in senescing leaves*

The main evidence of a shutting-down of synthetic activities in the course of leaf senescence comes from studies of the photosynthetic apparatus. We have seen (Section 10.2) that there is a preferential loss of activity of some enzymes of the RPC, at least one component of which, the large subunit of RuBP carboxylase, is coded for and synthesized within the chloroplast. We have also noted the declining rate of electron transport in senescing leaves of *Phaseolus*, associated with a lesion in the PQ-cyt f region of the electron transport chain (Fig. 10.5); cytochrome f is another component synthesized within the chloroplast (Chua & Gillham, 1977). There is also evidence of differential rates of loss amongst the thylakoid proteins of *Glycine max* (Bricker & Newman, 1980).

In leaves of *Perilla* there is a loss of capacity for chloroplast protein synthesis in the course of leaf senescence, although cytoplasmic protein synthesis is maintained (Callow *et al.*, 1972). Similar preferential loss of chloroplast activity has been noted in leaves of cucumber and *Phaseolus vulgaris*. The decline of protein synthesis in the chloroplasts is associated with a loss of polysomes and the cessation of chloroplast RNA synthesis (Callow & Woolhouse, 1973). In *P. vulgaris* the arrest of chloroplast rRNA synthesis at the time of completion of leaf expansion has been traced to the loss of chloroplast RNA polymerase, an enzyme which appears to be coded for in the nucleus and synthesized on cytoplasmic ribosomes (Bunger & Feierabend, 1980), suggesting that the regulation of this enzyme may be a key to the initiation of shut-down in the plastid metabolism (Ness & Woolhouse, 1980 a and b); in this context it is of interest that when leaves are caused to re-green by appropriate manipulation of the plant, there is an early resumption of chloroplast RNA polymerase activity (Ness & Woolhouse, 1980b).

The chloroplast genome is made up of 45μM double circles of DNA which code for two proteins present in large amounts, the large subunit of RuBP carboxylase and a 32 000 dalton thylakoid protein (Siddell & Ellis, 1975), and at least 100 other proteins which are made in much smaller quantities. Silverthorne and Ellis (1980) have shown that the relative proportions of the two major proteins synthesized by the chloroplast changes in the course of leaf development in spinach, due to changes in the amounts of translatable RNA for each. Bogorad *et al.*, (1980) have isolated a protein of mol. wt. 26–29 000 which specifically stimulates transcription of maize chloroplast DNA by the maize chloroplast RNA polymerase and have proceeded to show that transcription of different parts of the chloroplast genome, isolated by restriction endonucleases, is differentially influenced by this ‘stimulator’ protein. The chloroplast genome

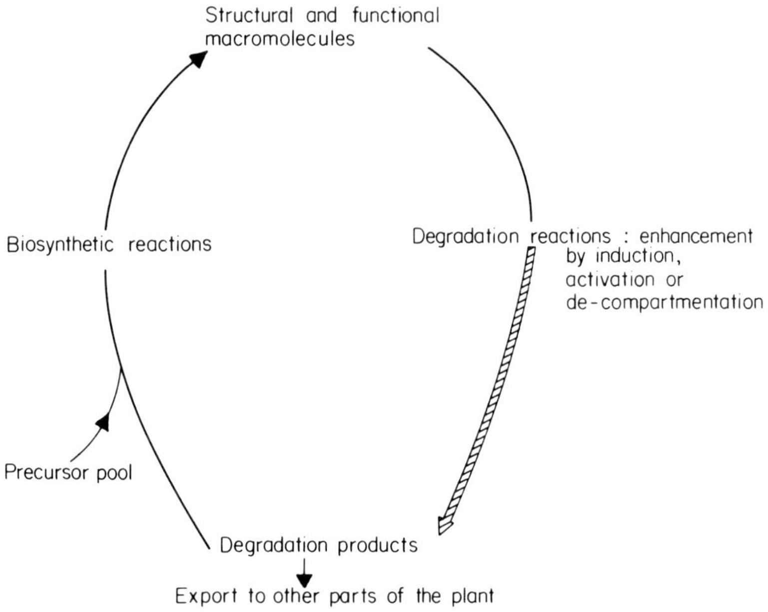

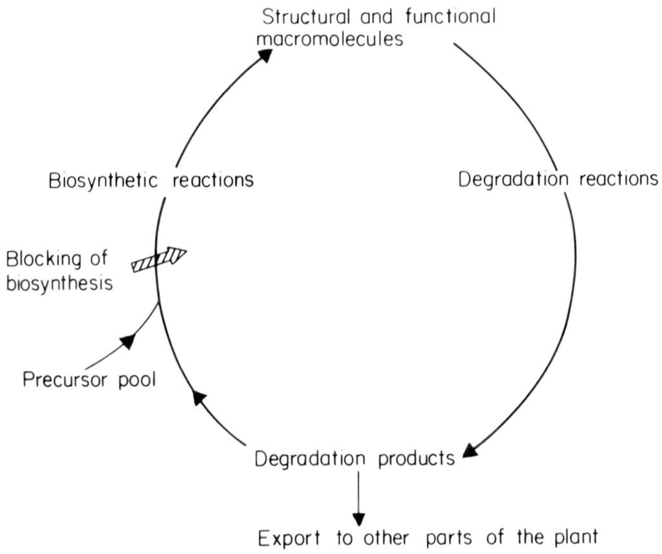

Fig. 10.7 (a) Scheme for the turnover of macromolecular constituents of the leaf involving some recycling of degradation products, in which net loss of constituents arises from loss of function in elements of the biosynthetic system.

(b) Scheme for the turnover of macromolecular constituents of the leaf in which net loss of constituents arises from enhanced activity in enzymes of the degradation system. The schemes in (a) and (b) are not of course mutually exclusive.

also codes for two sets of ribosomal genes per DNA molecule (Bedbrook *et al.*, 1977; Jenni & Stutz, 1978; Chu & Tewari, 1979) and for about 30 tRNA species (Burkard *et al.*, 1980; Keller *et al.*, 1980). Figure 10.8 provides a working hypothesis for a mechanism by which modulators like the stimulator protein described above, may be produced by nucleus and cytoplasm and passed into the plastid. The model is based on the envelope-carrier mechanism which appears to be involved in the transport of the small subunit of RuBP carboxylase into chloroplasts (see Chapter 9) and may well serve as a general mechanism. Through a mechanism of this type we may envisage a ready means whereby the chloroplast RNA polymerase activity could be modified (Silverthorne & Ellis, 1980; Bogorad *et al.*, 1980) or totally arrested (Ness & Woolhouse, 1980b) in a manner commensurate with the involvement of the nucleus as inferred by Yoshida (1961).

10.3.2.3 *The acceleration of hydrolysis of constituents in senescing leaves*

Hydrolase activity. Table 10.1 is a selection of reports of hydrolytic enzymes which degrade macromolecules which have been reported to increase in activity in extracts of senescing leaves; included in the same table for purposes of comparison is a list of hydrolases released by the aleurone layer of barley, which may also be regarded as a senescing system. The list of enzymes is notably similar in the two organs although in the case of leaves the list had to be compiled from widely scattered reports on a range of different species. It is a matter of regret that there does not appear to have been a systematic study of the spectrum of hydrolases in the leaves of a single species, a matter of more than academic interest for several reasons. Firstly it is important to recognize that the extent of the hydrolytic degradation of the constituents of senescing leaves varies greatly with the species. A good example of this may be seen in an oakwood (*Quercus robur*) carpeted with bluebells (*Endymion non-scriptus*). In the case of the oak, hydrolysis affects much of the cytoplasmic constituents but the skeletal remains of the dead leaves still account for a substantial proportion of the dry matter whereas in the case of the herbaceous bluebell there is an almost total dissolution of the leaf material in the later stages of senescence. Another reason for interest in the whole pattern of increased hydrolase activity for one species would be to know whether these enzymes represent a sort of senescence operon (using the term operon very loosely) which may be turned on in the germinating seed, the senescing leaves or any other tissue where a regulated senescence process is taking place. Alternatively it may be that the similar hydrolases in seeds and leaves represent tissue-specific isozymes, as demonstrated for the five different forms of alcohol dehydrogenase in the various tissues of maize (Scandalios, 1977). Much more detailed work on the hydrolytic enzymes of senescing leaves is needed before these questions can be answered.

As a prelude to examination of the regulation of hydrolytic activity in senescing leaves it is well to ask whether the frequently-observed increases in

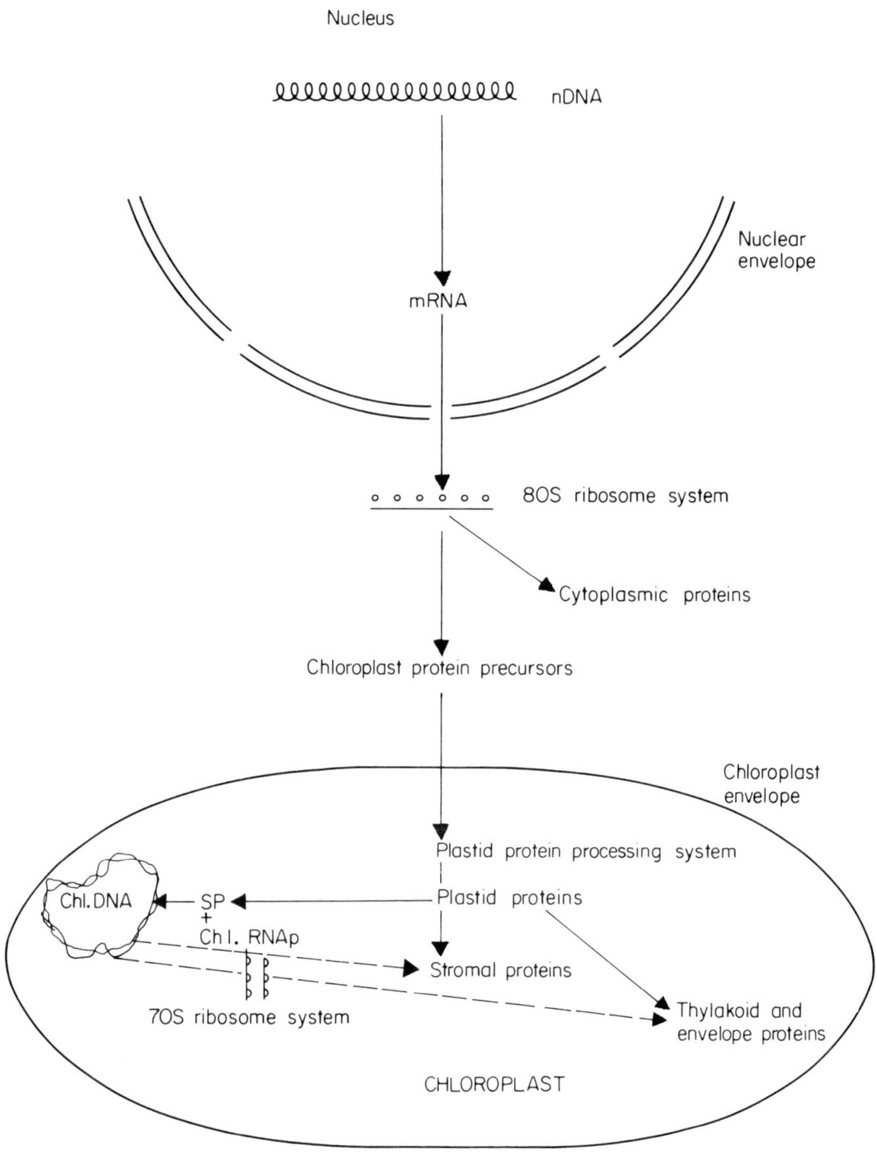

Fig. 10.8 Postulated mechanism for regulation of interactions between nucleus and chloroplast. Proteins synthesized within the cytoplasm which are destined for the plastid and identified by a precursor segment which is removed after entry into the plastid stroma. SP indicates a stimulator protein modifying the extent of transcription of particular genes in the Chl DNA. Chl RNAp indicates chloroplast RNA polymerase, an enzyme encoded in the nucleus.

Table 10.1 Hydrolytic enzymes reported to show increased activity in senescing leaves. Included for purposes of comparison (see text) is a list of hydrolytic enzymes released from the aleurone layer of barley

Leaves	
Protease	
Tobacco	Anderson & Rowan, 1965
Mung bean (cotyledons)	Baumgartner & Chrispeels, 1977
Pea	Storey & Beevers, 1977
Wheat	Dalling *et al.*, 1976
	Peoples & Dalling, 1978
	Peoples *et al.*, 1979
	Waters *et al.*, 1980
	Feller, 1978, 1979
	Nair *et al.*, 1978
	Wittenbach, 1978
Maize	Feller *et al.*, 1977
Barley	Peterson & Huffaker, 1975
	Miller & Huffaker, 1979
Oats	Martin & Thiman, 1972
	Drivdahl & Thiman, 1977, 1978
Perilla	Kannangara & Woolhouse, 1968
Lolium temulentum	Thomas, 1978
Ribonuclease	
Rhoeo	De Leo & Sacher, 1970
Rhoeo	Sacher & Davies, 1974
Phaseolus vulgaris	Phillips & Fletcher, 1969
Oat	Udvardy *et al.*, 1967
Acid Invertase	
Lolium temulentum	Pollock & Lloyd, 1978
Chlorophyllase	
Barley and oats	Sabater & Rodriguez, 1978
Esterase	
Festuca pratensis	Thomas & Bingham, 1977
Acid Phosphatase	
Perilla	Kannangara & Woolhouse, 1968
Aleurone layers of barley	
α-amylase	*De novo* synthesis, Varner, 1975
protease	*De novo* synthesis
ribonuclease	*De novo* synthesis (in part)
β-glucosidase	*De novo* synthesis (in part)
phosphatase	Release + *de novo* synthesis
pentosanase	Release + *de novo* synthesis, Dashet & Chrispeels, 1977
peroxidase	Release
esterase	Release

activity are in fact necessary at all for senescence to occur. In many cases the hydrolase activity present before the onset of senescence is more than adequate for the observed rate of breakdown and the activity during senescence is increased only by a factor of two or three (Waters *et al.*, 1980), which may be of much less significance than a change in compartmentation bringing enzymes and substrates into contact.

10.3.2.4 *Hydrolase regulation*

The possible levels of control of the hydrolytic activity of senescing leaves, according to the present state of knowledge, are summarized in Figure 10.9. The diagram offers a vast array of possibilities but they may for convenience be broadly divided under three headings: (i) *de novo* synthesis of the enzymes; reactions and factors a to k (ii) activation and in-situ regulation; reactions m and n, and (iii) compartmentation; reactions o and p. It is important to consider briefly the evidence for these various levels of control.

10.3.2.4.1 *De novo synthesis. Reaction (a).* Several workers have treated excised leaves with actinomycin without affecting the rate of senescence (von Abrams, 1974; Thomas, 1975; Thomas & Stoddart, 1977), from which the conclusion has been drawn that a transcription control is not involved in senescence. It is not certain whether the inhibitor is able to gain access to the nuclei under the conditions of these experiments. Attempts to demonstrate qualitative differences in the RNA produced in young and old barley leaves by means of DNA : RNA hydridization have also proved negative (Srivastava, 1972), though here again uncertainties surround the efficiency of the technique employed.

Reactions (b) and (c). Virtually nothing is known concerning mRNAs in leaves, whether they contain interrupted sequences and require processing or whether they become polyadenylated. Maize seedlings have been shown to be an active source of RNA polyadenylating enzyme and the polyadenylation inhibitor cordycepin has been shown to delay senescence in leaf discs of *Nicotiana* (Takegami & Yoshida, 1975), providing an interesting lead worthy of further investigation.

Reactions and factors of the ribosome cycle (d to k). The ribosome cycle in eukaryotes is still imperfectly understood but it seems reasonable to suppose that we are here concerned with a vast array of possible controls on protein synthesis which will include: availability of membrane surfaces upon which polysome assembly can occur, availability of intact ribosomal subunits, appropriate supplies of RNA and the corresponding aminoacyl tRNA synthetases and availability of active initiation, translocation and termination factors.

The cytoplasmic ribosome cycle in mature leaves has not been analysed at this level of detail and so we have no direct evidence concerning rate-limiting reactions involving any of these factors. Callow *et al.*, (1972) obtained active 80S-based polysome fractions from leaves of *Perilla* throughout maturation and

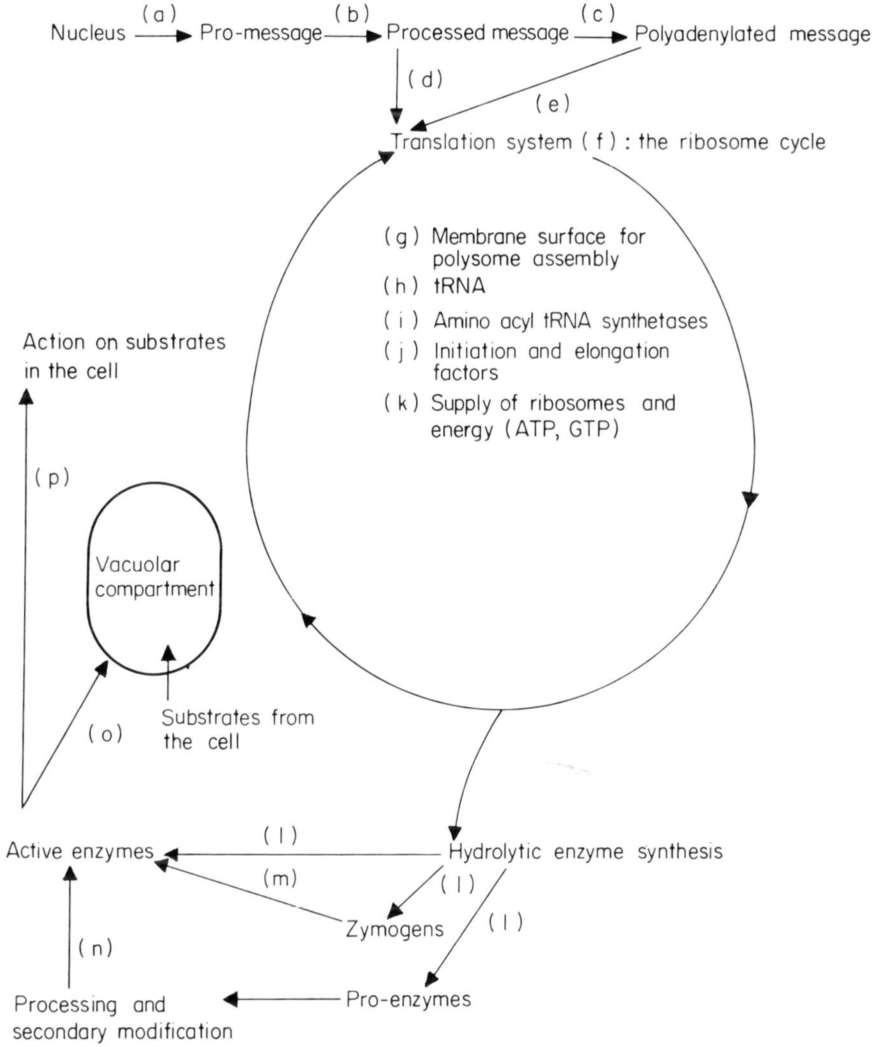

Fig. 10.9 Diagram indicating some of the multiple controls which may be involved in the regulation of the increased hydrolytic enzyme activity associated with the later stages of leaf senescence. Reactions and factors a to 1 are concerned with aspects of the *de novo* synthesis of the enzymes, reactions m and n denote in-situ regulation of the enzymes by processing or secondary modification or release from a zymogen; reaction o indicates enzyme regulation by compartmentation; reaction p denotes enzyme regulation at the substrate level.

senescence up to the time of abscission, long after the 70S-based plastid system had become inactive. Most of the evidence on the involvement of protein synthesis in senescence is however indirect and depends upon the credence which one is prepared to give to the results of applying inhibitors. It has been the general experience that inhibitors of protein synthesis such as MDMP (Thomas, 1976) and cycloheximide (Knyple & Mazurczyk, 1972; Makovetzki & Goldschmidt, 1976; Martin & Thiman, 1972; Peterson & Huffaker, 1975) arrest the progress of senescence in excised leaf tissue. It is tempting to conclude that the key to this effect is the blocking of synthesis of degradative enzymes which mediate the autolysis of cell constituents; it must be borne in mind however that these inhibitors will block the synthesis of *all* enzymes being made on 80S ribosomes and this may be sufficient to arrest other aspects of metabolism, whether involved in senescence or not, rather than just the synthesis of degradative enzymes.

The vicarious reader may cull from the literature numerous papers having a possible bearing upon the ribosome cycle in leaves, such as changing patterns of leucyl-tRNA and leucyl-tRNA synthetases, changes in lability of rRNA under electrophoresis and lowered protein : rRNA ratios in the ribosomes (Thomas & Stoddart, 1980), but there is insufficient evidence to accord to any of these observations a specific significance in the process of senescence.

The only conclusive evidence for regulation by synthesis *de novo*, of a hydrolase which increases during leaf senescence concerns a ribonuclease. When leaf discs of *Rhoeo discolor* are incubated in darkness there is a loss of RNA and protein accompanied by an increase in ribonuclease and acid phosphatase activities (De Leo & Sacher, 1970). The increase in activity of these enzymes is suppressed by IAA, kinetin and inhibitors of RNA and protein synthesis and is promoted by abscisic acid. Sacher and Davies (1974) used a density labelling technique to show that the increased ribonuclease activity is accompanied by *de novo* synthesis of the enzyme. The precise level at which this enzyme synthesis is being controlled is not known and there is clearly a great need to extend this type of work to the other enzymes which show increased activity in the course of leaf senescence.

10.3.2.4.2 *Activation and in situ regulation.* Many forms of activation of pre-existing enzymes have been reported; they include release from an inhibitor complex, processing of a precursor molecule and conformational change to an active state following such modifications as reduction of sulphydryl groups, phosphorylation, glycosylation, deamidation or other modifications of specific groups. There is no evidence at present for the involvement of modifications of this kind in the regulation of hydrolase activity in the course of leaf senescence but it should be borne in mind that the majority of the enzymes concerned are extremely poorly characterized from a biochemical standpoint.

10.3.2.4.3 *Hydrolase control by compartmentation.* As may be seen from Table 10.1, the reports of hydrolases associated with senescence in leaves are increasing; likewise we have seen that the substrates for these enzymes, the nucleic acids and proteins in particular, are being lost at different rates within a given organelle and the time course of losses is different for different organelles. Difficult questions therefore arise as to whereabouts these hydrolytic enzymes are located in the cell, whether the enzymes move to the components which are to be degraded or vice versa and, if the former, then why it is that the hydrolase, having arrived in a particular organelle, degrades one protein relatively quickly but not its neighbour. It is tempting to comment that controversy reigns supreme; more instructive however is to examine the sources of confusion. The problem really boils down to the difficulties which are being encountered in making the transition from studies of total activity of enzymes, in which the central objective is to pulverize everything as thoroughly as possible, to the gentlest possible entry to the cells, usually by means of protoplasts, followed by delicate fractionation techniques in which the most rigorous criteria for detection of cross-contamination of fractions are applied.

Matile introduced the concept of the vacuole in plant cells as a lysosome-like organelle which served as a prime repository for the hydrolytic enzymes. Reviewing the subject (Matile, 1978), he turned to the question of the vacuole as an autophagic compartment into which cellular constituents destined for breakdown were transferred and commented:

> 'The autophagic nature of vacuoles is suggested by numerous ultrastructural observations concerning the presence within vacuoles of structural cytoplasmic material.The bad preservation of such vacuolar inclusions is likely to be due to the action of digestive enzymes. Membrane processes seem to be responsible for the sequestration of cytoplasmic material in vacuoles. On the other hand the engulfment by the invaginating tonoplast results in the formation of intravacuolar vesicles containing a portion of cytoplasm. On the other hand the transfer of cytoplasm into the digestive compartment coincides with the formation of vacuoles: portions of cytoplasm are encircled by components of the endomembrane system which ultimately fuse together and form an autophagic vacuole. These observations suggest that the compartmentation of vacuole hydrolases is never abolished. Yet autophagy cannot be understood unless the morphological observations are complemented with corresponding knowledge about the biochemical events. At the moment we are dealing with a mosaic of somewhat isolated ultra-structural observations of autophagic vacuoles; assessments of turnover reactions, localizations of enzymes and correlations between catabolic processes and hydrolase activities. What is needed is an integrated study of one organism which covers all the relevant aspects.'

This comment remains true. In the intervening period Boller and Kende (1979) have provided impressive evidence that most of the hydrolytic enzymes generally regarded as involved in senescence are located in the vacuoles in leaves of tobacco. Peoples and Dalling (1979) on the other hand report an acid hydrolase co-sedimenting with the chloroplasts from leaves of wheat; though L. Wittenbach (personal communication) has suggested that this may be an artefact of cross-contamination. Huffaker and Miller (1978) propose the cytoplasm as a source of endoproteolytic activity in leaves of barley with the chloroplast envelope serving as a barrier between the hydrolase and the chloroplast contents. Hampp and de Fillipis (1980) have provided evidence for two chloroplastic endoproteases which they claim are involved in the transformation of prolamellar bodies during greening; endoprotease activity in the chloroplast stroma is also associated with the removal of a peptide of 65 amino acid residues from cytoplasmically-synthesized chloroplast proteins following their entry into the organelle (Smith & Ellis, 1979) and there is evidence that up to 65% of the aminopeptidase activity (exoprotease) of barley leaves may be present in the chloroplasts (Matile personal communication). It remains to be seen whether any of these proteases are involved in the observed degradation of thylakoid and stomal proteins referred to earlier, but clearly blue-green bacteria, which lack vacuoles, are capable of differential protein turnover and so some measure of internal autonomy with respect to chloroplast protein degradation cannot be ruled out on the evidence presently available. For the alternative view mention should also be made of the observed stability of the photosynthesis of isolated chloroplasts of oat (Choe & Thiman 1974; 1975; 1977) and spinach (Knapf & Jacobi, 1975) which was interpreted as evidence for a requirement for cytoplasmically-located proteases to bring about chloroplast breakdown.

Clearly there is such a ferment of activity over this question of the location and regulation of the enzymes of macromolecular turnover and senescence that much new information may be expected in the near future. Meanwhile it is perhaps sobering to reflect that amidst all this enthusiasm for proteases in particular, it is the lipid fraction of the cell which is in the most rapid state of turnover, not least within the chloroplast. Moreover the degradation of chlorophyll, the most characteristic attribute of a senescing leaf, remains a complete mystery.

10.4 CONTROL MECHANISMS UNDERLYING FOLIAR SENESCENCE

10.4.1 GENERAL CONSIDERATIONS

It is beyond the scope of this review to discuss the environmental, developmental, and correlative aspects of leaf senescence (Woolhouse, 1974; Nooden & Leopold, 1978; Thomas & Stoddart, 1980) but some consideration is required of the hormonal mechanisms through which the metabolic changes are brought about. Chibnall (1939) made the important observation that formation of roots on excised leaves of *Phaseolus vulgaris* led to the arrest of their senescence and to increased protein synthesis, suggesting the possibility that substances formed in the roots may be necessary to sustain the viability of the leaves. Much more is now known about compounds which can control foliar senescence but any clear understanding of the mechanisms involved seems far distant.

No detailed survey of the accumulated evidence concerning substances which may modify foliar senescence will be attempted here (for reviews see Osborne, 1967; Woolhouse, 1974; Nooden & Leopold, 1978; Thomas & Stoddart, 1980). Rather, some specific observations have been selected, which suggest new avenues of approach to this problem.

10.4.2 HORMONE EFFECTS

All five of the well characterized plant hormones or groups of hormones (see Fig. 3.1) have been implicated in some way in foliar senescence.

The first observations were of the effects of cytokinins in delaying senescence in excised leaves of *Xanthium* (Richmond & Lang, 1957) and the effects of this class of hormones in delaying senescence have now been described for many species, both for excised leaves and for leaves attached to the plant. In leaves of *Prunus* species IAA delays senescence (Osborne, 1967); gibberellins will delay the senescence of excised leaves of *Taraxacum officinale* but the cytokinins are without effect (Whyte & Luckwill, 1966).

In *Rumex* species and *Tropaeolum majus* cytokinins or gibberellins are equally effective in delaying senescence over similar ranges of concentration (Beevers, 1968). The case of *Tropaeolum* is particularly interesting because it is found that the GA content of the leaves decreases in the course of their senescence, providing circumstancial evidence that the effects of particular hormones may be related to their rates of turnover, which will affect the rate at which they fall to the threshold necessary to sustain metabolism (Woolhouse, 1967; 1968; 1974; 1978).

Abscisic acid has been found to accelerate senescence in leaves of many species (Paranjothy & Wareing, 1971). Colqhoun and Hillman (1972) failed to find any correlation between ABA content and the stage of senescence in leaves of *Phaseolus vulgaris* though it is not of course to be expected that any such correlation between total content of the hormone and degree of senescence should of necessity exist; it could for example be serving a trigger function in the early stages of senescence and thereafter changing in amount in a manner unrelated to the progress of the senescence syndrome.

Excised leaves of oat incubated in darkness show a five-fold increase in abscisic acid content which may be involved in promoting the rapid senescence which is observed (Gepstein & Thiman, 1980). We have noted earlier that the aleurone layer shares many attributes of a senescing tissue and it is of interest that abscisic acid exerts an action in this system which raises interesting possibilities for further work with leaves. Mozer (1980b) has clearly shown that both GA and ABA added to barley aleurones induces the appearance of unique mRNAs which are translatable *in vitro*. GA treatment alters *in vivo* protein synthesis by stimulating one set of proteins, the hydrolases, exclusively. ABA induces the synthesis of several new proteins, whose functions are as yet unknown, and prevents synthesis of α-amylase. Addition of both GA and ABA together to aleurones reduces the tissue to a pattern of protein synthesis resembling that of aleurones to which no hormones were added, but assays of the mRNAs of these aleurones by means of *in vitro* translation showed that the GA-induced mRNAs were still produced. Thus it must be concluded that ABA is acting to block the translation of the GA-induced messages *in vivo*. It is not yet clear whether this effect of ABA on translation is mediated via the ABA-induced proteins; it would be of particular interest to know whether ABA is operating in a similar manner in senescing leaves.

The relationship of ethylene to the process of foliar senescence is of particular interest and suggests some intriguing avenues for further research. It has been suggested that ethylene may be without effect on the rate of senescence in some species, (Thiman, 1978); other evidence however is more positive. It has been shown for example that ethylene will induce senescence in leaves of tobacco and that ethylene is produced by senescing leaves of this species. Excised leaves of evergreens such as *Ilex aquifolium* (holly) and *Hedera helix* (ivy), may be maintained in darkness for many months without undergoing senescence but can be induced to senesce rapidly by exposure to ethylene. It has also been found that treatment of leaves of these species with ethylene will promote senescence but if the gas is then withdrawn while the senescence is yet incomplete the senescence is arrested (Woolhouse unpublished).

It has recently been shown that the pathway of ethylene biosynthesis from methionine involves S-adenosyl methionine and 1-aminocyclopropane-1-carboxylic acid (ACC) as intermediates with a recycling of the methylthiol group as a unit into methionine (Fig. 10.10) (Adams & Yang, 1979). Using excised leaves of tobacco it was shown that there is a rapid rise in ethylene production prior to the final rapid phase of senescence (Aharoni *et al*., 1979a). Treatment of attached leaves of *P. vulgaris* with 1mM ACC led to ethylene production and rapid yellowing and senescence.

Auxin treatment of many plant tissues leads to an enhanced production of ethylene (Abeles, 1973) which has been shown to result from increased activity of ACC synthase, the enzyme which converts SAM to ACC (Fig. 10.10), (Adams & Yang, 1979). AVG, which inhibits ACC synthase, and Co^{2+}, which inhibits the conversion of ACC to ethylene, delay the senescence of excised leaves and Ag^{+} ions and high CO_2 concentrations, which inhibit ethylene action, give rise to similar effects (Aharoni & Lieberman, 1979; Aharoni *et al*., 1979b). It is of interest to note from Fig. 10.10 that molecular oxygen is also needed for the conversion of ACC to ethylene, as it is also for ethylene action (Beyer, 1976). These results suggest alternative explanations for the delaying of leaf senescence under anaerobic conditions, or at least an additional mechanism to respiratory energy requirements, as a source of this effect.

These recent findigns concerning the role of ethylene in regulation of leaf

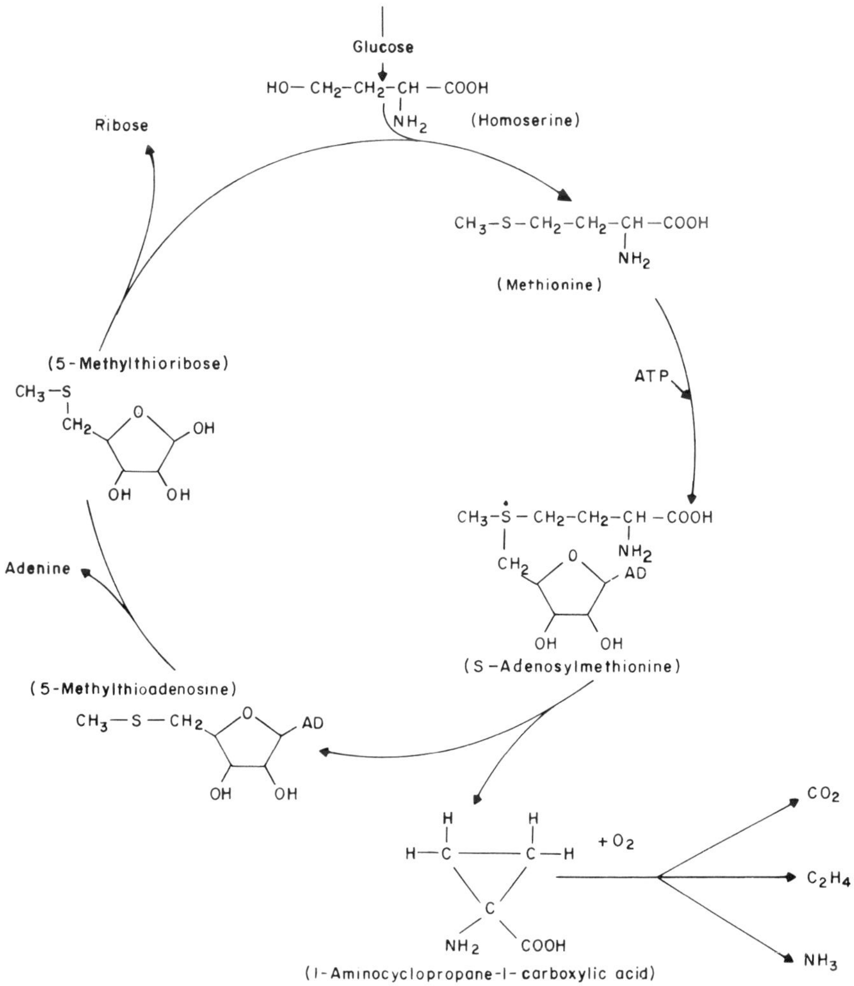

Fig. 10.10 The proposed methionine cycle for the biosynthesis of ethylene (Adams & Yang, 1979). Ad = adenosine. Note the oxygen requirement for conversion of ACC to ethylene.

senescence suggest the urgent need for a re-examination of many of the well known interactions between ethylene and the other plant hormones (for review see Lieberman, 1979) in the specific context of leaf senescence.

10.4.3 EFFECTS OF CALCIUM AND POLYAMINES ON FOLIAR SENESCENCE

Deficiency of calcium is the cause of a physiological disease syndrome which is found particularly in horticultural crops with bulky storage organs. Apples, brussels sprouts, carrots, cauliflowers and celery frequently develop these calcium disorders, probably because the low surface to volume ratios of the storage organs reduces the flow of xylem sap to a level at which the transport of calcium into the tissues is limiting. The symptoms of Ca deficiency in these tissues include breakdown of cell walls, increased permeability and leakiness of the cell membranes (Simon, 1978) leading to a waterlogged appearance of the tissues with necrotic browning in the later stages, a syndrome which resembles cellular senescence in many of its aspects.

Calcium treatment may delay senescence in excised leaves of *P. vulgaris* (Pooviah & Leopold, 1973) and delay abscisson in petiolar explants of the same species; it is also used routinely in the pre-treatment of excised roots and discs of storage tissue prior to studies of ion uptake, on the grounds that it diminishes leakiness of the tissues. In these various treatments Ca^{2+} has been held to be functioning primarily as a stabilizer of the plasma membranes of the cells. Calcium has recently been found to stimulate production of ACC and ethylene from SAM leading to the suggestion that plasmamembrane stabilization may be essential to the ACC synthase enzyme system (Lieberman, 1979).

Calmodulin, the calcium-binding protein involved as a 'second messenger' in the regulation of phosphodiesterase enzymes in animals (Wang & Waisman, 1979) has also been identified in plant tissues (Anderson & Cormier, 1978) where it has been implicated in the activation of NAD kinase. These findings should stimulate further study of the role of Ca and Ca-binding proteins in the regulation of metabolic events associated with plant senescence and in those physiological disorders of calcium nutrition which resemble the senescence syndrome.

Excised leaves of *Avena sativa* maintained in darkness showed increased ribonuclease activity after 1 hour, protease activity after 6 hours and chlorophyll degradation after 16 hours. These changes were delayed by treatment with 1–10mM polyamines, cadaverine, putrescine, spermine and spermidine, (Kaur-Sawhney & Galston, 1979). In the light the effect was reversed as the polyamines accelerated senescence. Calcium ions (1–10mM) supplied together with polyamines diminished the polyamine effect suggesting the possibility of a common ionic attachment mechanism as a basis for Ca^{2+} and polyamine action. These observations raise the intriguing possibility that polyamines may be normal elements of metabolic control in plant tissues as they are known to be in animals and micro-organisms (Bachrach, 1973).

10.5 CONCLUDING REMARKS

Attempting to review the biochemistry and molecular biology of leaf senescence is almost as harrowing an experience as trying to work on it. The plain fact is that for the study of regulatory mechanisms at this level, leaves are heterogeneous and inconvenient organs of appalling complexity. Faced with these difficulties many investigators will prefer the calmer waters of more clearly definable systems. Yet it is undeniable that if one wishes to know about the senescence of leaves one will ultimately have to work on leaves, and for an appraisal of the great importance of leaf duration one need consult no further than the nearest available crop physiologist (Woolhouse, 1978b; 1980).

FURTHER READING

BEHNKE J. A., FINCH C. E. & MOMENT G. B. (1978) (Eds) *The Biology of Ageing*, Plenum Press, London.

NOODEN L. E. & LEOPOLD A. C. (1978) Phytohormones and the endogenous regulation of senescence and abscission. In *Phytohormones and Related Compounds*, A Comprehensive Treatise, Vol. 2. (Eds Letham B. S., Goodwin T. C. & Higgins A.) Elsevier/North-Holland Biomedical Press, Amsterdam.

THOMAS H. & STODDART J. L. (1980) Leaf senescence. *Ann. Rev. Plant Physiol.*, **31**.

MOLECULAR ASPECTS OF REPRODUCTIVE DEVELOPMENT

CHAPTER 11

GAMETOGENESIS, FERTILIZATION AND EMBRYO DEVELOPMENT

11.1 INTRODUCTION

The developmental stages from meiosis to seed formation play an important role in the life of most higher plants. For the many species propagated only by seed, effective functioning of these pathways is essential for survival. Even when asexual propagation is possible, the sexual process creates new gene combinations which can result in novel and sometimes superior plant types. Moreover, the events of the sexual cycle frequently influence the growth and development of vegetative plant parts.

The reproductive stages of the life cycle also offer striking examples of many basic developmental processes—e.g. polarity, pattern formation, determination, interactions of tissues, cell–cell recognition, nuclear–cytoplasmic interactions. Coverage of such a large area of plant development must inevitably be selective and incomplete. This chapter offers a brief account of events from meiosis through embryogenesis in angiosperms, points out questions that remain unresolved, and discusses a few areas of active research in more detail.

11.2 OBSTACLES TO PROGRESS

In spite of years of effort by plant physiologists, cytologists, and geneticists, our understanding of gametogenesis and fertilization is still quite limited. Use of modern biochemical, ultrastructural, immunological and cell culture techniques is increasing but a variety of difficulties still slows progress.

Obtaining suitable experimental material can pose serious problems. Flowering plants are usually needed; the seedlings so convenient for many physiological and biochemical studies are of little use. Mature plants are often large, occupying substantial amounts of space. In many locations, flowering material can be obtained all year-round only if extensive greenhouse or growth chamber facilities are available. The alternative is to collect samples or do research only during a limited period of natural flowering.

Even when suitable plants are available, access to the cells and tissues of interest may prove difficult. Study of the male gametes is fairly straightforward since large, relatively homogeneous populations of developing pollen grains can be obtained from a single anther. Even so, many anthers are so small that analysis of tissue interactions is awkward. Formation and fertilization of the female gametes involves only a few cells, usually located within other tissues and separated from similar cells. Collection of sufficient material for biochemical work may be laborious, inconvenient, or even impossible. Some stages of the sexual cycle occur so rapidly that they are hard to capture for study.

An additional problem is our limited physiological or genetic control over many stages of the sexual cycle. Flowering can often be regulated by photoperiod or other environmental factors; morphology of floral parts may be altered by application of growth hormones. It is more difficult to influence specific later stages of sexual development. Mutants can provide essential tools for analysis of many developmental processes, but they are either not available or not extensively used in studies of most stages of gametogenesis and fertilization. One reason is that mutations reducing viability of gametes tend to be eliminated unless they have agricultural value or have deliberately been selected and maintained. The experimental systems which have received the most attention in recent years (male-sterility and self-incompatibility) are ones in which genetic controls are available and understood.

The agricultural importance of angiosperms has led to use of a wide range of species and varieties as experimental material. Considerable diversity in reproductive pathways has been revealed as a result, but no one system or species has received the intensive study that usually insures the most rapid progress.

11.3 MICROSPOROGENESIS: AN OVERVIEW

The life of the male gametophyte (pollen) encompasses three distinct phases: an initial one of development within male sporophytic tissue (microsporogenesis); a brief independent stage during which it travels to a compatible stigma or a less hospitable surface; a final phase of rapid growth through female sphorophytic tissue. The general description of microsporogenesis given below is largely based on information and references in Vasil (1967; 1973), Stanley & Linskens (1974), Mascarenhas (1975), Heslop-Harrison (1978a) and Kapil & Tiwari (1978).

11.3.1 MEIOSIS

Microsporogenesis occurs within anthers, organs differentiated into a central mass of sporogenous cells and peripheral cell layers, the innermost of which is a nutritive tissue known as the tapetum. Synchronous meiotic divisions of the sporogenous cells (pollen mother cells or microsporocytes) yield tetrads of haploid cells, each of which can develop into a male gametophyte.

Many types of plant cells undergo *mitosis* in response to phytohormones or other treatments, but induction of *meiosis* in cells outside the anther (or ovary) has not been achieved. Initiation of meiosis seems to require the anatomical setting of the flower. Sporogenous cells in premeiotic anthers removed from buds and cultured do not begin meiosis; those in anthers excised before the leptotene stage of meiotic prophase revert to an abnormal mitosis (Stern & Hotta, 1973). Once well underway, however, meiosis can continue and reach completion even in isolated microspore mother cells. Such observations have led to the suggestion that some stimulus from vegetative plant parts is involved in the initiation of meiosis. The amount of vegetative tissue required apparently can vary considerably. Premeiotic anthers on excised flowering branches of tomato developed normally only after a graft to a healthy flowering stock was performed (Gregory, 1940); however, anthers in tiny flowers formed *in vitro* on thin cell layers from tobacco stems contain meiotic cells (Tran Thanh Van & Trinh, 1978).

Many interesting mutations specifically affecting meiosis have been identified (see review in Baker *et al.*, 1976). In the *ameiotic* mutant of maize, meiotic divisions do not occur. Instead, sporocytes undergo asynchronous abnormal mitotic division and then degenerate. Other mutations have less drastic effects; they usually involve alterations in chromosome behaviour during the first meiotic prophase. A later defect is seen in the *divergent spindle* mutant of maize. Meiotic spindles fail to converge normally at the poles, so that chromosomes also fail to converge. The microspores formed are often multinucleate. Few meiotic mutants have been analyzed at the molecular level.

11.3.2 CYTOPLASMIC CHANGES AFTER MEIOSIS

The tetrad of cells formed as a result of meiosis soon separates into four haploid microspores. These increase in volume and become vacuolate, with the nucleus and cytoplasm typically at one end. Then, asymmetric mitosis in the microspores produces two quite dissimilar cells, the vegetative and the generative cells. The vegetative cell is the precursor of the pollen tube that emerges as pollen grains germinate. It is larger than the generative cell, stains less densely, and contains the majority of the microspore organelles. Starch and lipids accumulate in the vegetative cell cytoplasm as pollen matures. The generative cell is typically smaller and has fewer organelles. Plastids are especially sparse. The generative cell wall eventually separates from the original microspore wall, so that the generative cell is located within the cytoplasm of the vegetative wall. Division of the generative cell, either before or, more commonly, after pollen is shed, produces the two sperm cells which participate in fertilization. The male gametophyte thus consists of only three cells, two of them within the third.

11.3.3 WALL DEVELOPMENT

A complex sequence of wall changes parallels the cytoplasmic developments described above. The end result is a thick, multilayered and often elaborately sculptured wall around the cytoplasm of the pollen grain.

11.3.3.1 *Callose*

A distinctive feature of several stages of male (and female) gametogenesis is the rapid synthesis and breakdown of callose walls. Callose, a β-1, 3 polyglucan, is usually identified by fluorescence microscopy of material stained with aniline blue (see review by Kapil & Tiwari, 1978). The microspore mother cells, initially joined into a syncytial structure by plasma channels, become separated by massive callose walls by the end of meiosis. These walls have been characterized as 'molecular filters' isolating the tetrads from external substances (Heslop-Harrison, 1964).

The walls between and around the tetrads also consist of callose. Separation of the tetrads into individual microspores depends on the release of the enzyme callase from the tapetum (Mepham & Lane, 1969). Incubation of tetrads with callase rapidly releases spherical wall-free protoplasts, which might be useful material for studies of pollen wall synthesis (Bhjowani & Cocking, 1972). Callose makes a final brief appearance along the wall between newly formed vegetative and generative cells.

11.3.3.2 *Exine*

The outer portion of the pollen wall is the elaborate multilayered exine, (Fig. 11.1) composed largely of sporopollenin, a substance so resistant to chemical degradation that structural analysis is difficult. It is generally considered to be a

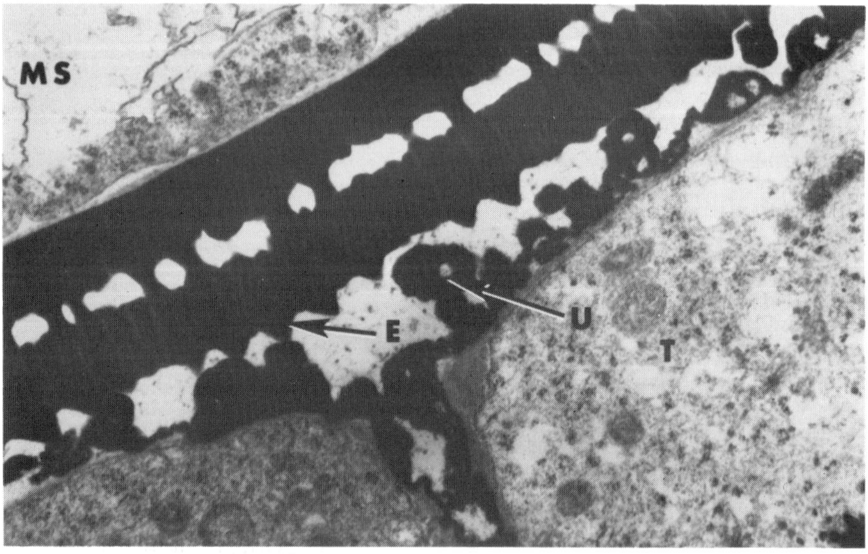

Fig. 11.1 Electron micrograph of walls on microspore (MS) and tapetal cells (T) of maize. The exine (E) consists of two darkly staining layers. Note the channels running through the exine and the spines on the surface. Ubisch bodies (U) are present on the walls of the tapetal cells. × 20 000. (From Lee *et al.*, 1980).

complex polymer of carotenoids and carotenoid esters. A complicated and sometimes disputed terminology has been used to describe the different components of the exine.

Formation of the exine begins while tetrads are still encased in callose. At that time, a fibrillar cellulose layer (the primexine) is laid down between the plasmalemma and the callose wall. Radially directed rods (the probacula) extend outward through the primexine. After release of the tetrads, sporopollenin is deposited on both sides of the primexine and on the sides, heads and feet of the probacula. The interior of the exine communicates with the outside via large pores over areas between bacula or via micropores through the roof (tectum) formed by coalescence of the heads of the bacula.

Precursors for the formation of sporopollenin are probably derived from the tapetum. In one major type of tapetum (the secretory type), some areas of the tapetal walls form structures coated with sporopollenin (Ubisch bodies or orbicules) but the relation of these structures to pollen wall development is not well understood. The tapetum definitely contributes lipid, carotenoid and hydrophilic materials to the outside of the exine.

The surface features of the exine (elevations, spines, and pores) are so distinctive that they can be used in the identification of modern and fossil pollen species (Fig. 11.2). The functional significance of this elaborate patterning is unknown. The origin and control of the wall patterning also raises major questions. Patterning begins to appear soon after meiosis in the primexine stage. Primexine is not laid down over some areas of the plasmalemma, usually those to which endoplasmic reticulum is closely appressed. These primexine-free areas are the future pores, areas of decreased sporopollenin deposition (Vasil, 1973). The callose wall around young tetrads may also play a role in the patterning of sporopollenin deposition by providing a template for the deposition of the primexine matrix (Waterkeyn & Bienfait, 1970).

11.3.3.3 *Intine*

After the microspores are released from the tetrad, an additional wall layer known as the intine forms between the plasmalemma and the developing exine. The intine consists of pecto-cellulosic materials derived from the cytoplasm of the microspore. The mature pollen thus is surrounded by two major wall layers (the intine and the exine) which differ in composition, structure, origin and time of formation.

11.3.3.4 *Wall Proteins*

Many loosely bound proteins are located in cavities of the exine and in the intine (Heslop-Harrison *et al.*, 1973). The exine proteins, which are derived from the tapetum, leach out within seconds after hydration of pollen. The intine proteins, transferred from the microspore cytoplasm, are released more slowly. At least some of these proteins are enzymes.

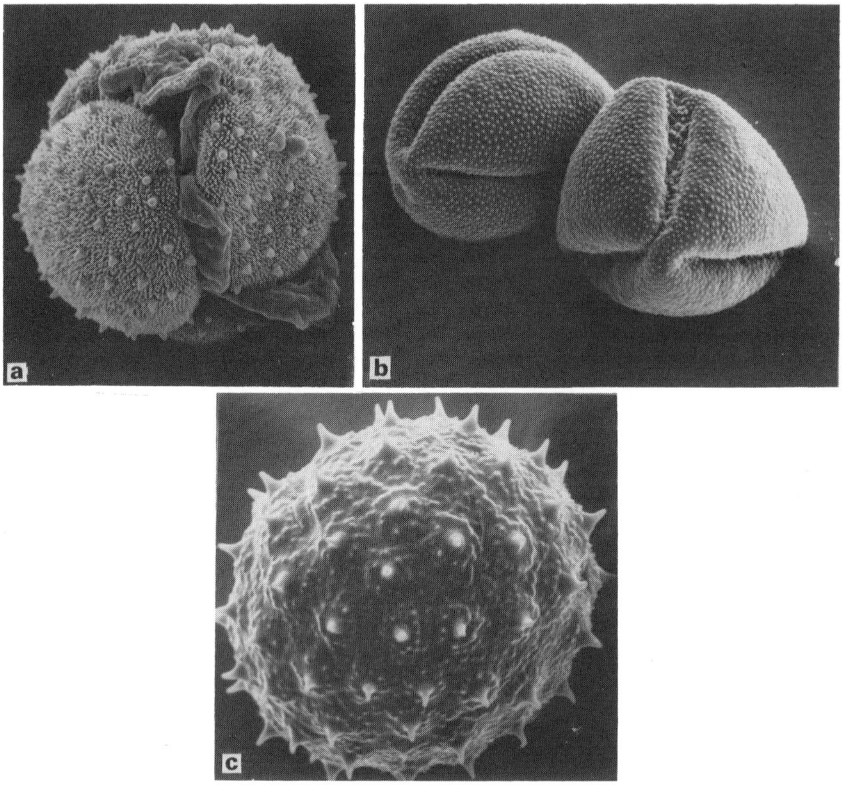

Fig. 11.2 Scanning electron micrographs showing surface features of pollen grains. a. *Drosera*. b. *Papaver*. c. *Malva*. All ×2100. (By courtesy of M. K. Hausmann, Section of Plant Biology, Cornell University.)

Pollen wall proteins may play a role in germination. Germination of *Petunia* pollen *in vitro* is substantially reduced by a brief elution in distilled water; a protein component of the diffusate restores germination (Kirby & Vasil, 1979). Wall proteins may also be involved in self-incompatibility reactions (see Section 11.7.2). In addition, they can trigger allergic reactions in sensitive humans.

11.4 HAPLOID TISSUES AND PLANTS FROM MICROSPORES

Microsporogenesis normally yields haploid pollen grains capable of pollination and fertilization, but microsporocytes sometimes follow other developmental pathways. Microspores cultured *in vitro* may become haploid embryoids or masses of haploid callus, instead of pollen grains. Plants can often be recovered from such embryoids or callus. Production of plants from the male gametes (androgenesis) was first described in *Datura* (Guha & Maheshwari, 1964). It has now been achieved in many other species (Keller & Stringam, 1978). The microspores are usually cultured within intact anthers but culture of isolated microspores has also been successful (Nitsch, 1974).

Factors which influence androgenesis include genotype of the plant material, nutrient and cultural conditions, high or low temperature treatments of the anthers, and developmental stage of the microspores. The most favourable stage varies with the material, but usually is about the time of the first microspore mitosis. If division into the vegetative and generative nucleus has already occurred, the vegetative cell is typically the one which gives rise to the haploid callus or embryoids. Normally this cell would never divide again. In *Hyoscyamus niger*, the generative cell (which usually forms the sperm) is the one that proliferates *in vitro* (Raghavan, 1979).

Androgenesis is often considered a shift of normal microspores to a new developmental pathway and a dramatic demonstration of the flexibility of plant cells. It has recently been claimed that all haploid tobacco plants produced *in vitro* are derived from atypical microspores capable of dividing repeatedly even while still on the plant (Horner & Mott, 1980). In this view, continued division and androgenesis depend on providing permissive conditions to a special subpopulation of microspores rather than inducing a developmental shift in the normal ones.

11.5 MALE-STERILITY

A second variant pathway in microsporogenesis leads not to formation of haploid callus or plantlets but to abortion of the pollen. While many factors can impede pollen development, genetically controlled inability to form functional pollen is perhaps of most interest. Male-sterility of this type occurs in many plant groups.

Cytological studies of various species have demonstrated that pollen abortion can occur at many different stages, from meiosis to well after the microspore mitosis. The blocks need not always be lethal; microsporocytes in tomato mutants which normally fail to progress through meiosis can divide mitotically *in vitro* (Zamir *et al.*, 1980). Rather little is known about the reasons for the breakdown in male-sterile material but a disturbed relation of the tapetum and the sporogenous tissue is frequently implicated. In *Petunia* for example, faulty timing of callase release from the tapetum prevents normal separation of the microspores from the tetrads (Izhar & Frankel, 1971).

Male-sterility may be inherited either via nuclear genes (genic male sterility) or via the maternal cytoplasm (cytoplasmic male-sterility). Cytoplasmic male-sterility is a convenient way to prevent self-pollination in the production of hybrid crop plants. Male-sterile cytoplasms are receiving attention from molecular biologists as well because they provide material for comparative study of organelle genomes in higher plants.

11.5.1 THE MALE-STERILE CYTOPLASMS OF MAIZE

The most extensive studies of cytoplasmic male-sterility have been done with maize. Four general groups of cytoplasms, known as N, T, C, and S, have been recognized (Duvick, 1965). Plants with N (normal maintainer) cytoplasm produce functional pollen; plants with T, C, or S cytoplasms do not unless nuclear genes for restoration of fertility (Rf genes) are present. The T (Texas), C, or S male-sterile cytoplasms were initially distinguished by their response to different Rf genes and by the genetic control of the restoration process. All pollen produced by T or C cytoplasm plants either homozygous or heterozygous for the appropriate Rf genes is functional (sporophytic restoration). Only 50% of the pollen grains produced by S cytoplasm plants heterozygous for Rf genes are fertile (gametophytic restoration). In plants with gametophytic restoration, the genotype of the haploid microspore determines its fate; in the sporophytic type, the genotype of the diploid tissue of the anther is the controlling factor. T, C, and S cytoplasms also differ in their stability in different nuclear backgrounds or environments and in the timing and cytology of pollen abortion.

Light microscope studies of the three male-sterile cytoplasms have documented differences in the timing of pollen abortion and in the development of the tapetum. Meiosis is normal in all three. In the T and C anthers, abortion occurs early, at the late tetrad stage in some C anthers and at the uninucleate microspore stage in T and other C anthers. (Warmke & Lee, 1977; Lee *et al.*, 1979). Various tapetal abnormalities are also apparent. In S anthers, pollen continues to the binucleate stage and then aborts abruptly (Lee *et al.*, 1980). The tapetum behaves as it does in N anthers. This absence of tapetal abnormalities fits in with the observed gametophytic control of fertility in S cytoplasm.

11.5.1.1 *Comparisons of Mitochondrial DNA in Maize Cytoplasms*

A basic question about maize (and other) male-sterile cytoplasms is the cellular location of the cytoplasmic genes involved. Current evidence strongly favours a mitochondrial site. Analysis of chloroplast DNA by agarose gel electrophoresis after digestion with restriction endonucleases has revealed only minimal differences between the four cytoplasms (Pring & Levings, 1978). In contrast, mtDNAs (mitochondrial DNAs) from the cytoplasms can readily be distinguished; the banding patterns seen after digestion with a variety of endonucleases show multiple and characteristic differences (Pring & Levings, 1978). Some differences between mtDNA of different sources of N and C cytoplasms can also be seen (Levings & Pring, 1977; Pring *et al.*, 1980) but the inter-cytoplasmic differences are much more striking. We now need to learn more about the differences in sequence and organization of mtDNA that the digestion results reflect.

Separation of undigested maize mtDNA on agarose gels yield several bands identified as low molecular weight DNA (Table 11.1) (Kemble & Bedbrook, 1980). One (1.94 kilobasepairs) is common to all four cytoplasms. Another

Table 11.1 Low molecular weight DNA species in mitochondria of maize with normal or male-sterile cytoplasms

DNA length (kilobasepairs)	Configuration	Cytoplasm N	T	C	S
6.2	linear	–	–	–	+ (S-1)
5.2	linear	–	–	–	+ (S-2)
2.35	?	+	–	+	+
1.94	supercoiled circular	+	+	+	+
1.57	circular	–	–	+	–
1.42	circular	–	–	+	–

Data from Kemble & Bedbrook (1980) and Kemble *et al.*, (1980).

(2.35kbases) is found in N, C, and S but not in T. Two circular molecules (1.57 and 1.42kbases) are present only in C cytoplasm while two linear molecules (6.2 and 5.2kbases) are seen only in S cytoplasm. Each cytoplasm thus has a unique complement of low molecular weight mtDNAs. Low molecular weight mtDNA patterns can be used to identify the cytoplasm present in a small amount of seedling maize material, thus circumventing the field studies of mature maize otherwise required to distinguish N, C and S cytoplasm (Kemble, 1980).

11.5.1.2 *Relation of S Cytoplasm Plasmids to Male-Sterility*

Differences in mtDNA of the maize cytoplasms suggest but do not prove that these DNAs are involved in the control of male-sterility. Studies of the behavior of the low molecular weight mtDNAs in the S system offer additional evidence for such an involvement.

Plants with the Vg strain of S cytoplasm can spontaneously revert to fertility. Restoration is sometimes inherited via the cytoplasm and sometimes via nuclear genes which can be mapped to many different chromosomes. These genetic results were originally interpreted in terms of an episomal factor whose cellular location determined whether material was sterile or had a cytoplasmic or nuclear pattern of restoration (Laughnan & Gabay, 1975).

Recent studies have suggested that the low molecular weight mtDNA found only in S cytoplasm (Pring *et al.*, 1977) may be the postulated episome. The two plasmid-like DNAs (known as S-1 and S-2) are no longer apparent in cytoplasmically restored S maize (Levings *et al.*, 1980). Restriction endonuclease analysis of the mtDNA in these restored S lines reveals bands not seen in the sterile lines. Labelled probes prepared from S-2 hybridized to some of these new bands. The DNA reorganization that gave the new bands may thus have resulted from insertion of S-2 sequences. S-1 and S-2 may prove to be transposable elements, correlated with male sterility when present in free form and with cytoplasmically inherited fertility when present only in high molecular weight mtDNA. Whether nuclear-controlled reversion to fertility involves an interaction of S-1 and S-2 with chromosomal DNA is not yet known. A plant plasmid which could integrate into organelle or nuclear DNA would have great potential as a vehicle for genetic engineering experiments.

11.5.1.3 *Mitochondrial Proteins in Maize Cytoplasms*

The observed differences in mtDNA of the maize cytoplasms have led to comparisons of proteins synthesized by isolated maize mitochondria (Forde *et al.*, 1978). Most of the approximately 20 polypeptides synthesized are similar, but each male-sterile cytoplasm has some distinctive feature. T, for example, lacks a 21 000 molecular weight polypeptide common to the other three cytoplasms and contains a unique 13 000 molecular weight product. Synthesis of this unique polypeptide is suppressed when a nuclear Rf gene is present, but there is no simple correlation between the degree of suppression and restoration (Forde & Leaver, 1980).

11.5.1.4 *T Male-Sterility and Sensitivity to Helminthosporium maydis race T.*

The maize cytoplasms are proving to be good material for studies of mitochondrial genomes. It is apparent, however, that work on DNA and protein differences between fertile and sterile cytoplasms has as yet provided few clues about how these differences lead to the observed deviations from normal gametogenesis. Some ideas about the possible cellular and biochemical basis of sterility in the T system come from studies of the interaction of T cytoplasm maize with the leaf blighting fungus *Helminthosporium* (*Bipolaris*) *maydis*, race T (HmT).

This pathogenic fungus is especially virulent on maize with T cytoplasm. Both male-sterile T lines and ones whose fertility has been restored by nuclear genes are sensitive. The sensitivity led to the 1970 epiphytotic of southern corn leaf blight in the U.S. and to heightened concern about genetic uniformity in crop plants. A toxin produced by the fungus (HmT toxin) is responsible for the increased susceptibility of T cytoplasm; material with N, C, or S cytoplasm is unaffected by the toxin (see review in Gregory *et al.*, 1977).

Further evidence for the close connection between T cytoplasm and sensitivity to HmT toxin comes from the work of Gengenbach *et al.*, (1977) on regeneration and analysis of maize plants from T cytoplasm callus selected *in vitro* for resistance to HmT toxin. These plants were resistant to the toxin and to *H. maydis* race T and were also no longer T male-sterile. The toxin resistance and male-fertility were cytoplasmically inherited. Mitochondria from these plants still had the low mol. wt. DNA characteristic of T cytoplasm, indicating that the simultaneous shift to resistance and fertility did not simply involve selection of pre-existing N cells or mitochondria (Kemble & Bedbrook, 1980).

Since T male-sterility and sensitivity to HmT toxin appear to be controlled by closely linked or identical cytoplasmic genes, information on the mechanism of one process may help explain the other as well. Studies with HmT toxin are more convenient since they can be done using easily obtained somatic tissue from young plants instead of sporogenous cells from mature plants. The availability of a chemical probe is a further experimental advantage.

Mitochondria are the most likely sites of the initial interaction of HmT toxin with sensitive T cells. The cytoplasmic inheritance of sensitivity, the distinctive features of T mtDNA and the observed biochemical and ultrastructural effects of HmT toxin on T mitochondria all support this view. HmT toxin rapidly uncouples mitochondria isolated from T (but not N, C or S) cells (Miller & Koeppe, 1971) and causes distinctive ultrastructural changes, including swelling and disruption of the cristae (York *et al.*, 1980). Other subcellular fractions show no clearcut response to toxin treatment. The first ultrastructural change seen in toxin-treated T cells and protoplasts is mitochondrial damage (York *et al.*, 1980). Even earlier (within 1–2min.) comes a sharp drop in the ATP content of T leaf protoplasts (Walton *et al.*, 1979). Inability to generate ATP after mitochondrial damage could account for most subsequent changes seen in toxin-treated T material.

Ultrastructural studies of anther development in the various maize cytoplasms implicate mitochondria in T male-sterility as well as in interaction with HmT toxin (Warmke & Lee, 1977). The first signs of abnormality in T cytoplasm anthers are alterations in the mitochondria, first seen in the tapetum shortly after meiosis and then in the microspores. The changes are very similar to those seen in T somatic cells treated with HmT toxin. No mitochondrial alterations precede the breakdown of developing pollen in C and S cytoplasm anthers (Lee *et al.*, 1979; 1980). It therefore seems plausible that mitochondria in T cytoplasm cells have distinctive cytoplasmically inherited defects or alterations rendering them sensitive both to HmT toxin and to some possible similar substance produced during normal anther development (Flavell, 1974). Isolation of such an 'anther toxin' would bring together the work on T male-sterility and sensitivity to HmT toxin in a very satisfying way.

Nevertheless, other mechanisms of sterility involving the mitochondria can be envisioned. Warmke and Lee (1978) reported a very rapid 20 to 40-fold increase in mitochondrial numbers in both N and T anthers just before the T microspores begin to degenerate. They speculate that the T mitochondria are unable to survive the stress of this period of rapid replication.

11.5.2 PROTOPLAST FUSION AS A TOOL FOR ANALYSIS OF MALE-STERILITY

Because paternal cytoplasmic genes are usually not expressed after sexual crosses, conventional breeding methods do not permit analysis of the interaction, recombination, or dominance relations of cytoplasmic genes. The new technique of asexual protoplast fusion now makes it feasible to combine two different cytoplasms and to study the phenotypes and organelle DNA of plants regenerated from the fused protoplasts.

Belliard *et al.* (1978) recovered plants after fusion of leaf protoplasts from fertile and cytoplasmic male-sterile varieties of *Nicotiana tabacum* carrying different nuclear and cytoplasmic markers. On the basis of leaf shape (nuclear marker) and flower morphology (cytoplasmic marker), some of these plants were identified as hybrid only for cytoplasmic traits (cybrids). The chloroplast-DNA (ctDNA) and fraction I protein from different cybrid plants corresponded to that from one or the other parent; both parental types were never seen in a single plant. A sorting-out of chloroplast genomes apparently occurs sometime after fusion. The degree of male-sterility in the cybrids was not related to the type of ctDNA found, indicating that ctDNA alone does not control male-sterility in this material.

Analysis of mtDNA from the cybrids showed that they contained both parental types of mtDNA as well as some new DNA bands (Belliard *et al.*, 1979). The most sterile plants were those whose mtDNA most closely resembled that of the male-sterile parent. In this tobacco system as in maize, cytoplasmic male-sterility appears more closely linked to mtDNA than to ctDNA. Analogous work with maize is not yet possible since regeneration of plants from maize protoplasts has not yet been achieved.

11.6 POLLEN GERMINATION AND GROWTH

11.6.1 POLLEN *in vivo*

Mature pollen released from anthers can be transported to the female stigma by air currents, insects, humans, or other vectors. Germination begins with hydration of the desiccated grains with fluids from the stigma or the surrounding atmosphere (Heslop-Harrison, 1979). The cytoplasmic membranes are reorganized and the pollen tube emerges through the germinal pore on the pollen wall. The shift from inactivity to active metabolism and tube growth occurs rapidly; rye pollen, trinucleate when shed, germinates within a few minutes and grows 2–3μm s^{-1} (Heslop-Harrison, 1980).

Growth of pollen tubes *in vivo* can be monitored by fluorescence microscopy using the dye aniline blue (Fig. 11.3). This technique reveals the callose present in tube walls or in plugs within the lumen of the tube (Kapil & Tiwari, 1978). Remnants of the pollen grain remain on the surface of the stigma as the tube, growing only at the last few microns of the tip, penetrates this surface. *Tropaeolum* pollen excretes a cutinase which probably facilitates passage through the cutinized surface layer of the stigma (Shayk *et al.*, 1977). This cutinase, which is different from typical fungal cutinases, is not synthesized during germination, but is already present in ungerminated pollen. Similar enzymes are likely to be

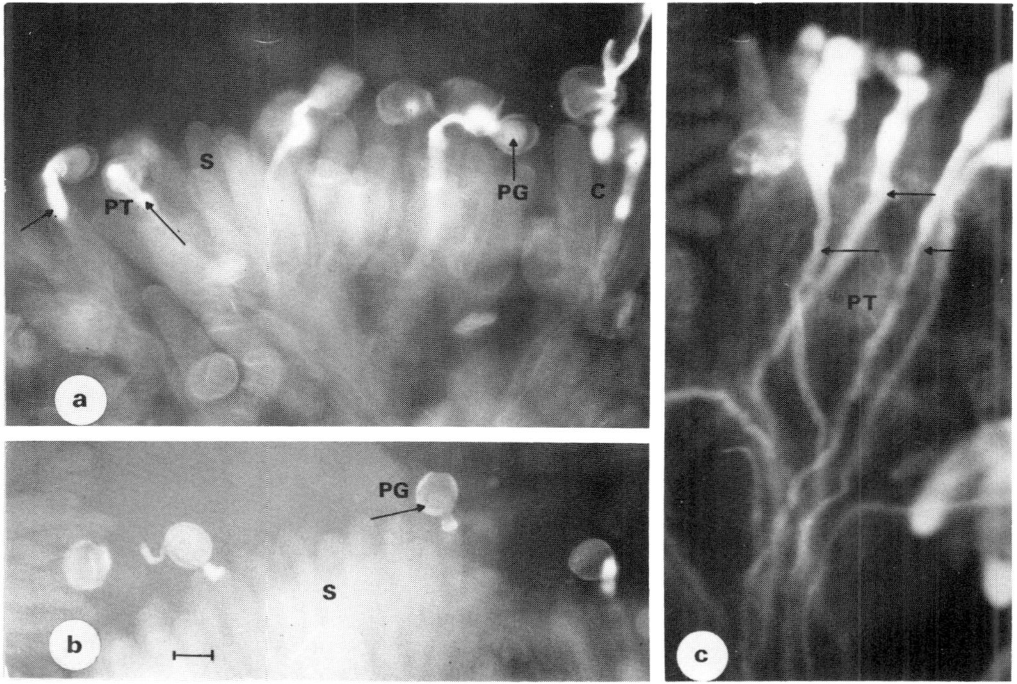

Fig. 11.3 *Brassica* pollen on *Brassica* stigma. Squash preparations stained with aniline blue and viewed by fluorescence microscopy. Calibration bar equals 40μm. a. Compatible pollination showing germination of pollen grains (PG) and growth of pollen tubes (PT) on stigma papillae (S). A brightly fluorescent callose plug (C) is seen in one pollen tube. b. Incompatible pollination. Pollen grains have germinated but tubes have not elongated or penetrated the stigma. c. Compatible pollination showing pollen tubes penetrating into and through stigma. (By courtesy of Dr. T.E. Ferrari, Department of Plant Breeding, Cornell University.)

involved in movement of pollen through other cutinized stigmas. Once into the stigma, the pollen tube grows over or between the walls of the stylar cells toward the female gametes in the ovule. Whether chemotropic or physical factors guide the tube towards its destination is uncertain.

11.6.2 POLLEN *in vitro*

Binucleate pollen of many species can germinate and grow in nutrient media containing carbohydrates, calcium, boron and an osmoticum. Only in a few species (e.g. *Tradescantia*) does the pollen tube elongate as much *in vitro* as in the style (Mascarenhas, 1975). Secretions from the stigma or style can contribute to vigorous and continued pollen tube growth. Pollen which is trinucleate when shed (e.g. cereal pollen) tends to develop less well *in vitro* than binucleate pollen.

Pollen germinated *in vitro* is much more convenient material for biochemical studies than pollen *in situ*. One striking observation from such studies is the insensitivity of germination and early tube growth to Actinomycin D (Mascarenhas, 1975). Apparently a supply of stable mRNAs and other RNAs is pre-formed during microsporogenesis. The effect of inhibitors of protein synthesis depends on the developmental and metabolic state of the pollen at the time of hydration (Hoekstra & Bruinsma, 1979). Vegetative and generative nuclei from pollen grown *in vitro* can be isolated for biochemical comparisons (Mascarenhas, 1975). Pollen *in vitro* has also been used in tests of pollen viability, nutrition, and chemotropism, and in the analysis of incompatible pollen–stigma interactions.

11.7 SELF-INCOMPATIBILITY

Pollen often fails to develop normally after a self-pollination. Such self-incompatibility (SI) occurs in some 250 genera in over 70 families of angiosperms (Heslop-Harrison, 1978a). Like male-sterility, SI is a barrier to inbreeding. The impediment in SI is, however, not absence of viable pollen but rather the specific interaction of functional pollen with a self or sexually related stigma or style.

11.7.1 GENETIC CONTROL OF SELF-INCOMPATIBILITY

Genetic studies indicate that SI is controlled by multiple alleles (S-alleles) of nuclear genes present at one or more loci. Numerous S-alleles with characteristic dominance, co-dominance, and mutual weakening interactions have been identified (Wallace, 1979).

Two general systems of genetic control exist. In the sporophytic type (SSI), first analyzed in *Nicotiana* by East and Mangelsdorf (1925), the S-allele genotype of a plant determines the response of the pollen produced. All pollen grains from a heterozygous plant have the same incompatibility phenotype regardless of the S-allele received during meiosis. The behavior of SSI pollen on a particular stigma depends on the relative activities of the S-alleles. S-allele interactions are often not identical in male and female tissues (Wallace, 1979). In gametophytic systems (GSI), the genotype of the pollen itself determines its incompatibility response. Pollen grains from heterozygous plants have two different phenotypes. The female tissues in heterozygous GSI plants exhibit simultaneous co-dominant activity of both S-alleles, so incompatibility results after pollinations with pollen bearing either S-allele.

Some general correlations between the genetics and the anatomy and physiology of incompatibility have been noted (Table 11.2). While exceptions are not uncommon, these correlations suggest where and when to look for important initial interactions. In most sporophytic systems, incompatibility is already apparent on the stigma at or soon after the time of pollen germination. In most, but not all, gametophytic systems, initial germination is normal but growth of the pollen tube is inhibited somewhere in the style.

Self-incompatibility barriers can sometimes be overcome by altering the physiological or developmental stage of the female tissue. Heat treatment (50°C) of *Lilium* styles (GSI) permits self-fertility (Hopper *et al.*, 1967). So does pollination of immature stigmas (bud pollination) in many SSI species. A more drastic approach is to excise the stigma and style and to apply pollen directly to ovules *in vitro* or *in situ* (Rangaswamy & Shivanna, 1967). *Petunia* (GSI) and *Brassica* (SSI) plants have been produced from self-incompatible parents in this way.

11.7.2 MOLECULAR ASPECTS OF SELF-INCOMPATIBILITY

In spite of many detailed physiological and cytological studies of self-incompatibility in various species (reviewed in de Nettancourt, 1977 and Heslop-Harrison, 1975), important questions about the molecular control of the process still remain largely unanswered. These questions can be expressed as follows:

(1) What are the products of the S-alleles in pollen and the female tissues?

(2) When and where are these products synthesized? The origin of the S-allele products in the pollen of SSI species is particularly puzzling. The sporophyte may determine the incompatibility response of the haploid pollen via proteins transferred from the diploid tapetum to the pollen exine. Alternatively, factors influencing incompatibility may be transferred from the cytoplasm of the diploid microsporocytes to all four haploid microspores.

(3) How do the male and female gene products interact to allow recognition of self?

(4) How does this interaction lead to inhibition of pollen germination and/or growth?

Considerable progress is being made on the problem of S-allele products in the stigmas of *Brassica*. Glycoproteins leached from stigmas with different S-

Table 11.2 Comparison of cytology and physiology of self-incompatibility in gametophytic and sporophytic systems

	Gametophytic Systems	Sporophytic Systems
Pollen	binucleate when shed germinates readily *in vitro* relatively long-lived	trinucleate when shed germinates poorly *in vitro* short-lived
Stigma	usually smooth or with low papillae copious exudate ('wet')	papillate no copious exudate ('dry')
Site of inhibition	style	stigma
Time of inhibition	pollen tube growth down style	pollen germination or early pollen tube growth
Representative families and genera	Solanaceae (*Nicotiana, Petunia, Lycopersicon*) Liliaceae (*Lilium*) Rosaceae Leguminoseae (*Trifolium*) Onagraceae (*Oenothera*)	Compositae Cruciferae (*Brassica*)

Exceptions: Graminae have trinucleate pollen and dry stigmas but gametophytic control; *Oenothera organensis* is a gametophytic species with inhibition on the stigma rather than in the style.

Adapted from Heslop-Harrison (1978a).

allele genotypes can be distinguished electrophoretically and serologically (Nasrallah *et al.*, 1972). Several lines of evidence summarized in Ferrari *et al.*, (1981) indicate that these substances are closely involved in the self-incompatibility response. A specific stigma glycoprotein is strictly correlated with each S-allele. Plants heterozygotic for S-alleles have the stigma glycoproteins corresponding to both alleles. Young stigmas, in which incompatibility is not expressed (hence the success of bud pollination) contain little stigma glycoprotein, while maturing stigmas have increasing amounts. A mutant *Brassica* which is self-compatible has low glycoprotein levels.

The stigma glycoproteins from *Brassica* with the S_7S_7 genotype (Nishio & Hinata, 1979) and from *B. oleracea* with the S_2S_2 genotype (Ferrari *et al.*, 1981) have recently been purified and characterized. Both are glycoproteins with similar molecular weights (about 54 000), carbohydrate : protein ratios (about 1.2), and amino acid composition. Pretreatment of S_2 *B. oleracea* pollen with the S_2-glycoprotein prevents growth of this pollen on normally compatible stigmas. Additional information about structure, synthesis, localization, and biological activity of stigma glycoproteins will probably be available soon.

Studies of S-allele products in pollen are advancing more slowly. Pollen wall diffusates from *Oenothera organensis* (GSI) contain S-allele specific antigens (Lewis *et al.*, 1952), presumably proteins of gametophytic origin. In *Oenothera*, unlike most GSI species, incompatibility occurs on the stigma so pollen wall proteins could well be involved. Such involvement has not been directly demonstrated, however.

The nature of the S-allele products in SSI pollen has also not been resolved. Many proteins of sporophytic (tapetal) origin are rapidly released from the exine during germination. Application of wall diffusates or pollen-free tapetal extracts to incompatible (but not compatible) stigmas results in formation of callose in the stigmas after several hours (Heslop-Harrison, 1978a). Appearance of the callose rejection response is far removed from the initial incompatible interaction, but characteristic of incompatible pollinations. Attempts to correlate specific antigens in the wall diffusate with specific S-alleles in *B. oleracea* have been unsuccessful.

In many cell–cell recognition systems (e.g. host-parasite, host-symbiont), binding of surface carbohydrates of one partner by lectins of the other is an essential step (Sequiera, 1979; Dazzo, 1980). Whether the recognition between S-allele products of pollen and female tissues will have a comparable mechanism remains to be seen. Several models have been proposed to explain how recognition could lead to diminished pollen growth (Ferrari & Wallace, 1977; Heslop-Harrison, 1978b) but the application of existing data to the models is controversial. The fact that many pollens germinate and grow *in vitro* does suggest that incompatible interactions involve an active inhibition and not just failure to stimulate pollen development. Elucidation of the mechanisms of incompatibility will require much further work but the results should be relevant to many issues in developmental biology.

11.8 FEMALE GAMETOGENESIS AND FERTILIZATION

The female gametes are formed within ovules, organs consisting of central sporogenous tissue surrounded by nutritive layers and integuments. Unlike anthers, which are directly exposed to the atmosphere, ovules are enclosed within the ovary at the base of the style.

Light microscope studies have provided a wealth of information about the cytology of female gametogenesis in different species (Maheshwari, 1963 and references therein). The four haploid cells formed after meiosis in an ovule behave quite differently from the tetrads formed after meiosis in an anther. Three of them degenerate; the one survivor undergoes three mitoses to produce the female gametophyte or embryo sac. This structure consists of seven cells of four different types in a characteristic arrangement. The egg cell, which after fertilization becomes the zygote, is flanked by two synergids. Towards the center of the embryo sac is the large central cell with two, often partially fused, nuclei (the polar nuclei); this cell probably provides nutrition for the less metabolically active egg cell. After fertilization, the central cell gives rise to the triploid endosperm. At the opposite end of the embryo sac are three antipodal cells which usually, but not always, degenerate. The embryo sac is surrounded by a multilayered diploid tissue, the nucellus.

Recent ultrastructural studies have provided details about the cytology of these various cells and have answered some questions about the events immediately before and after fertilization which could not be resolved by light microscopy. The work of Jensen and his collaborators on cotton has been particularly instructive (Jensen, 1972).

After cotton pollen germinates on the stigma, the pollen tube grows between the walls of the stylar cells and degenerating nucellar cells toward the embryo sac. The tube takes about 12–18 hours to cover this distance, but one of the two synergids begins to degenerate some six hours before the pollen tube reaches it. Apparently some stimulus acts at a distance to trigger breakdown of a synergid. The pollen tube enters the egg sac via the degenerating synergid. Entry is through the filiform apparatus, an elaborate extension of wall material into the cytoplasm of the synergid. Once in the cytoplasm of the synergid, the tube stops growing and discharges its cytoplasm and the two sperm through a subterminal pore. The cytoplasmic discharge contains many small polysaccharide particles which distinguish it from the cytoplasm of the synergid.

The sperm are reduced but complete cells, each bounded by a plasma membrane and containing a nucleus and mitochondria but no plastids. They are probably passively transported to the egg cell and central cell. One sperm nucleus enters the egg cell and fuses with its nucleus, forming the zygote; the other fuses with the polar nuclei in the central cell. This unique process is termed double fertilization.

There is no evidence that any cotton sperm cytoplasm enters with the nuclei. After most sexual crosses, only maternal cytoplasmically inherited traits are expressed; usually only maternal organelle DNAs are seen in plants derived from crosses in which maternal and paternal organelle DNAs can be distinguished (Conde *et al.*, 1979). Jensen's cytological observations indicate that in cotton paternal organelle DNA does not even enter the zygote. In *Plumbago zeylanica*, an angiosperm which lacks synergids, male cytoplasmic organelles can be seen in the zygote and the central cell after entry of the sperm nuclei (Russell, 1980). Plastids from the male parent are found both in the zygote and in mature plants of *Oenothera* (Meyer & Stubbe, 1974). Further comparisons of the transmission and replication of the male cytoplasmic organelles in plants with strictly maternal or biparental inheritance of cytoplasmic traits would be valuable.

Many questions about the mechanism of double fertilization remain unanswered. Unfortunately, electron microscopy of this part of the plant life cycle is particularly difficult and time-consuming. Many tissue samples must be fixed and sectioned to obtain one good illustration of ephemeral processes like discharge of sperm, entry of sperm into the egg and central cell, and nuclear fusion. Physiological and biochemical studies of the cytological phenomena revealed by light and electron microscopy are equally desirable but perhaps even more difficult. Techniques now available for *in vitro* culture of unfertilized ovules of cotton (Beasley & Ting, 1974) and other species and for fertilization *in vitro* may facilitate such work. Some synergids and polar nuclei in unfertilized cotton ovules cultured *in vitro* undergo ultrastructural changes reminiscent of those seen after fertilization (Jensen *et al.*, 1977). Isolation of mutants blocked at different stages of pollen tube penetration and fertilization would also be helpful.

11.9 EMBRYOGENESIS

After fertilization and fusion of the sperm and egg nuclei, the diploid zygote divides to form the embryo and auxiliary structures. Different taxonomic groups have characteristic patterns of initial divisions (Wardlaw, 1955), but these patterns may not be essential since embryos from cell cultures may achieve a normal form without following them (Street, 1977).

The general features of embryo development in dicots can be illustrated with *Capsella bursa-pastoris*, a species used in many anatomical and physiological studies (Raghavan, 1976). The first division of the *Capsella* zygote produces two dissimilar daughter cells, a small densely protoplasmic one from which the embryo develops and a larger, more vacuolated basal cell anchored to the end of the embryo sac through which the pollen tube entered. The basal cell undergoes several transverse divisions to produce the suspensor, a short filament of cells between the developing embryo and the basal cell. The basis of the polarity established at the first division of the zygote is not clear. In some species (e.g. cotton) the organelles move to the end of the zygote that will become the embryonic cell but in others, including *Capsella*, there is no obvious cytological polarity in the undivided zygote (Raghavan, 1976).

The first signs of tissue differentiation in the *Capsella* embryo appear at the 16-cell stage. At this time, an inner group of cells which will form meristematic tissue can be distinguished from outer cells that will form epidermal layers. Further divisions of the embryonic cells result in morphological stages named for the shape of the embryo (globular, heart-shaped, torpedo, cotyledonary, etc.). A globular embryo consists of some 50 cells and is less than 80μm long; a mature embryo is about 2mm long. After the globular stage, the sites of the cotyledons, the root axis and other morphological and histological features can be identified.

Ultrastructural, physiological and biochemical events accompanying the morphological changes in embryos have been reviewed by Raghavan (1976). Unfortunately, the very earliest embryonic stages, which seem of particular developmental interest, are the hardest to obtain for experimental manipulation and biochemical analysis.

Very small embryos are also difficult to culture *in vitro*. Pre-globular stages have not been grown successfully. Globular embryos have more stringent nutritional and osmotic requirements for plantlet formation than larger ones. Excised globular embryos of *Capsella* need addition of either 12–18% sucrose or high levels of salts or, preferably, a carefully balanced mixture of plant growth hormones to a basal nutrient medium (Raghavan & Torrey, 1963). Larger embryos can grow on a medium containing only salts, 2% sucrose and vitamins. Hormone levels that promote division in globular embryos inhibit some later morphogenetic events. These changing nutritional requirements probably reflect increasing synthetic capacities in the embryos. Young *Cattleya* orchid seedlings (which correspond developmentally to globular embryos) require nitrogen in the form of ammonium. Ability of the seedlings to grow after transfer to medium with nitrate as sole nitrogen source is paralleled by the appearance of nitrate reductase in the seedlings (Raghavan & Torrey, 1964).

The possible contribution of the suspensor to successful culture of young embryos should not be ignored. Ultrastructural studies of suspensor cells in pre-globular and globular *Capsella* embryos indicate that these cells may function in absorption and translocation of nutrients to the embryo (Schulz & Jensen, 1969). Monnier (1978) has suggested that damage to the suspensor during excision accounts for at least some of the difficulties in growing very small embryos. Removal of the suspensor greatly impedes development of heart-shaped and early cotyledonary stage *Phaseolus coccineus* embryos but has little effect on larger ones (Cionini *et al.*, 1976). Low concentrations of gibberellic acid (10^{-8}–10^{-6}M) can substitute for the presence of the suspensor on the young embryos. Rates of RNA synthesis are much higher in bean suspensors than in the early embryonic cells (Clutter *et al.*, 1974). Suspensors may be supplying a variety of hormones and nutrients to young embryos.

Embryogenic cell cultures, particularly ones which can be synchronized to produce large uniform populations of healthy globular, heart-shaped, or torpedo embryos (Fujimura & Komamine, 1979) may make biochemical studies of embryogenesis more convenient. Extrapolations of results from somatic to zygotic embryos must be made with caution, however. Crouch (1979) used the sensitive technique of rocket immunoelectrophoresis to compare production of storage proteins in embryos from zygotes, microspores, and hypocotyls of *Brassica napus*. Zygotic embryos contain little storage protein until some 25 days after fertilization when they already are at a cotyledonary stage. In microspore-derived embryos, the same protein is apparent at a much earlier morphological stage.

11.10 CONCLUSIONS

Reproduction in angiosperms involves a remarkable variety of cell types and developmental phenomena. Some of the latter (e.g. asymmetric division to give dissimilar progeny) are analogous to ones familiar from other developing systems. Others, such as pollen wall formation and double fertilization, are unique.

Sophisticated modern techniques are starting to be applied to this part of the plant life cycle. The information presented about a few areas, notably cytoplasmic male-sterility, is likely to be superseded soon, but links between the molecular analyses and events at the cellular and tissue levels are still not well-established. Progress in other areas seems even slower. All-in-all, questions about the formation and behaviour of the male and female gametes outnumber the answers available. Molecular and developmental biologists seeking new and challenging problems will find reproduction in angiosperms a fruitful and uncrowded field for study.

FURTHER READING

HESLOP-HARRISON J. (1975) Incompatibility and the pollen–stigma interaction. *Ann. Rev. Plant Physiol.*, **26**, 403–25.

KAPIL R. N. & TIWARI S. C. (1978) Plant embryological investigations and fluorescence microscopy: an assessment of integration. *Int. Rev. Cytol.*, **53**, 291–331.

LEVINGS C. S. III & PRING D. R. (1979) Molecular bases of cytoplasmic male-sterility in maize. In *Physiological Genetics*, pp. 171–93, (Ed. Scandalios J. G.). Academic Press, New York.

MASCARENHAS J. P. (1975) The biochemistry of angiosperm pollen development. *Bot. Rev.*, **41**, 260–314.

NETTANCOURT D. DE (1977) *Incompatibility in Angiosperms.* Springer-Verlag, Berlin and New York.

RAGHAVAN V. (1976) *Experimental Embryogenesis in Vascular Plants.* Academic Press, New York.

STANLEY R. G. & LINSKENS H. F. (1974) *Pollen.* Springer-Verlag, Berlin and New York.

VASIL I. K. (1967) Physiology and cytology of anther development. *Biol. Rev.* **42**, 327–73.

VASIL I. K. (1973) The new biology of pollen. *Naturwissenschaften* **60**, 247–53.

CHAPTER 12

SEED MATURATION AND DEPOSITION OF STORAGE PROTEINS

12.1 INTRODUCTION

Mature seeds are rich in protein and other reserve materials and, for this reason, play a vital role in human and animal diets. An increasing amount of research is now being directed at gaining a better understanding of the biological factors which determine the amount and composition of those proteins which are synthesized and accumulated during seed maturation. The total protein fraction of the seed is made up of many individual protein species which can be broadly classified into two groups. These are the very heterogeneous metabolic proteins, such as the enzymes and the ribosomal, chromosomal and membrane proteins, and the more homogeneous fraction known as storage proteins. The primary biological function of this latter group is to provide fixed carbon and nitrogen to support the growth of the seedling during germination and in this chapter we are mainly concerned with cellular events involved in their accumulation in the maturing seed. It is true that during germination the metabolic proteins may also be hydrolysed and re-utilized by the germinating seedling but this role is secondary to their multitude of specific functions performed in the maturing seed.

Unfortunately there is no straight-forward and entirely satisfactory definition of seed storage proteins. In this chapter the term will be used to cover those proteins and protein fractions which make a significant contribution to the seed's total protein, which are not found elsewhere in the plant and whose *primary* function is to support seedling growth during germination.

Metabolic proteins are thus excluded from our definition but another class of proteins occupies a grey area. These are the lectins which, although not unique to the seed, are frequently concentrated there and can make a significant contribution to the total seed protein, as in *Canavalia* species where concanavalin A accounts for 25% of the seed protein (Hague, 1975). One seed lectin, from *Phaseolus vulgaris* clearly merits consideration as a storage protein. It accounts for 10% of the seed protein. It is localized in protein bodies and is rapidly broken down during germination (Bollini & Chrispeels, 1978).

Because of their importance in human and animal nutrition, the storage proteins of the cereal and legume crop plants have been the most extensively studied and in this chapter we will largely concentrate on representatives from these two groups.

12.2 SUBCELLULAR STRUCTURE IN RELATION TO STORAGE PROTEIN DEPOSITION

It is important to consider molecular events involved in storage protein synthesis and deposition against a background knowledge of the related changes in subcellular morphology. In the course of seed development the storage tissue (endosperm or cotyledon) usually goes through three successive phases. The first phase is one of cell division, followed by a phase of cell expansion and then seed desiccation and maturation. Storage protein synthesis and deposition are essentially confined to the phase of cell enlargement. It is during this phase that discrete storage protein deposits are first seen which then build up progressively to the levels found in mature seeds. These deposits are usually localized in discrete organelles known as protein bodies. Depending on the seed type, other reserve material, mainly carbohydrate (mostly as starch grains) or lipid (as lipid bodies), also accumulates during the cell expansion phase and, along with protein bodies, makes a substantial contribution to the cell profile of mature seeds (Figs. 12.1 & 12.2).

The protein bodies of mature seeds are largely confined to the cotyledon or endosperm cells. Some protein bodies are also found in the aleurone layers of cereal grains. These are distinctly different in protein composition and morphology from the protein bodies of the starchy endosperm and contribute relatively little to total seed protein (Pernollet, 1978). They will not be considered further here. The protein bodies of cotyledon and endosperm cells in mature seeds are usually 1–5μm in diameter and surrounded by a single unit membrane. In the mature seed they may occupy a large proportion of the cross-sectional area of the cell (Fig. 12.2). In some seeds, such as wheat, the surrounding membrane is no longer apparent in the mature grain. This arises from the fact that the endosperm at this stage is essentially dead tissue in which a great deal of disorganization of cellular structure has taken place. It is important to note that not all storage proteins are necessarily deposited inside protein bodies. In maize and barley, the prolamins (zein and hordein, respectively) are in protein bodies but the glutelins, which contribute quite significantly to the total storage protein fraction, form a protein matrix outside the protein bodies.

12.2.1 PROTEIN BODY FORMATION

In spite of numerous developmental studies, the origin of protein bodies is still not fully understood. This uncertainty arises because protein bodies, as recognized by light and electron microscopy in the mature seed, make their appearance rather late in the phase of storage protein accumulation by which time more than half of the total storage protein has been synthesized. In addition storage proteins are difficult to identify, by microscopy, in immature seeds. There has been a tendency in many studies in this field to assume that any intracellular deposit which reacts positively with protein stains, or electron

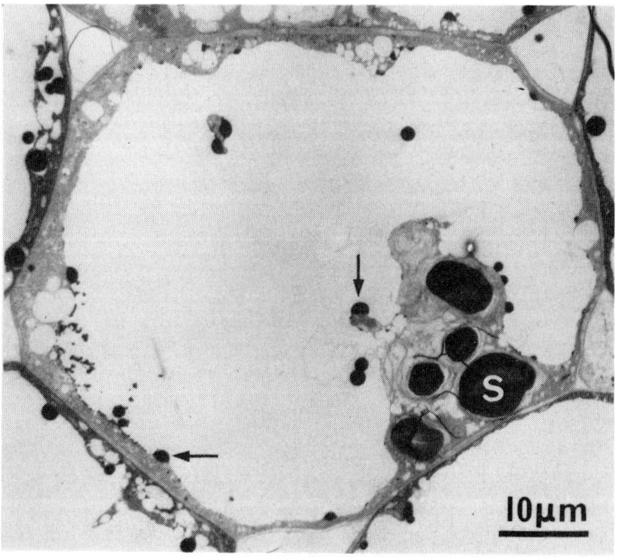

Fig. 12.1 & Fig. 12.2 Changes in vacuole and protein body structure and identification of storage protein deposits during maturation of pea seeds.

Fig. 12.1 Shows an electron micrograph of a cotyledon cell at 10 days after flowering, approximately one day after the storage protein accumulation phase has begun. A single large vacuole containing small peripheral protein deposits (arrows) occupies a major part of the cell profile (Courtesy of S. Craig).

Fig. 12.2 opposite Shows pea cotyledon cells at 12, 15 and 22 days after flowering. Protein deposition is complete at day 26. In sections a, c and e, storage protein (legumin) was specifically localized by an indirect immunofluorescence technique. The same sections were then treated with the general protein stain, aniline blue-black (b, d and f). Protein deposits in vacuoles and protein bodies react selectively with fluorescent antibody. Storage protein deposits accumulate in the vacuoles and protein bodies throughout development. The vacuoles fragment progressively and give rise to protein bodies in the near-mature cells (compare days 12 and 15 with day 22) (after Craig *et al.*, 1980b; reproduced with permission). S = starch grains, N = nucleus. In Fig. 12.2 the bar represents 20μm.

dense material which is digestible with proteases, is indeed storage protein. Protein bodies can be readily isolated from many mature and near mature seeds and shown to contain storage proteins, but no equivalent isolation of intact structures containing storage proteins from very immature seeds has been reported. Examples of positive identification of storage proteins at this early stage are rare. Graham and Gunning (1970) and Craig *et al.*, (1979b), using fluorescent antibodies as a probe, have identified storage proteins within protein bodies of *Vicia faba* and *Pisum sativum*, respectively, in cotyledon cells near the end of the protein accumulation phase and Craig *et al.*, (1980b) have used the same approach to identify the discrete, dense deposits on the inner side of the vacuolar membrane as storage protein at a much earlier developmental stage (Fig. 12.2).

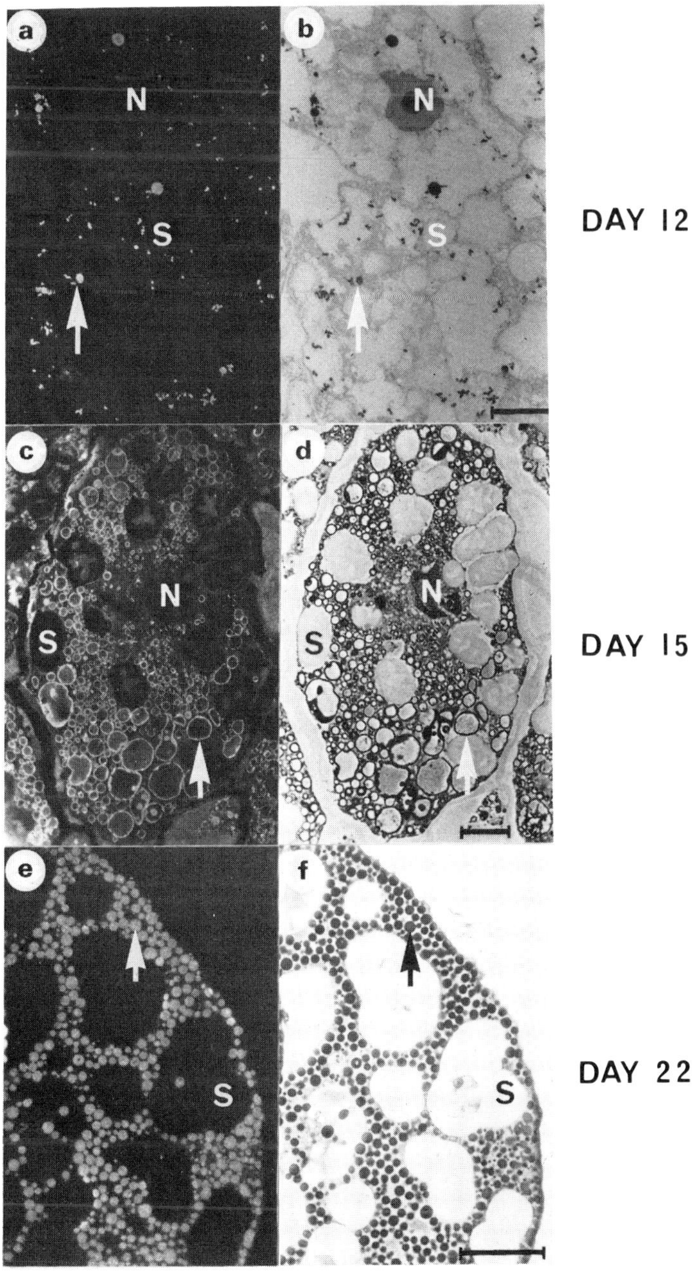

Fig. 12.2

12.2.1.1 *Protein body formation in legumes*

The usual observation in dicotyledonous species is that during seed development, the one or two large vacuoles with their protein deposits are replaced by progressively smaller and more numerous vacuoles which also contain localized protein deposits (Bain & Mercer, 1966a; Briarty *et al.*, 1969; Craig *et al.*, 1979a). By the end of the protein accumulation phase, these small vacuoles are completely filled with protein and approximate the size and appearance of the protein bodies of the mature seeds (Fig. 12.2). This sequence has been variously interpreted as a progressive subdivision of the original vacuoles to give rise to the smaller protein bodies (Öpik, 1968; Briarty *et al.*, 1969; Rest & Vaughan, 1972) or as the 'disappearance' of the central vacuole and its replacement by a new vacuolar apparatus which gives rise to the final form of protein body as seen in mature seeds (Harris & Boulter, 1976; Neumann & Weber, 1978). The new vacuolar system has been thought to originate either from evaginations of the rough endoplasmic reticulum (RER) (Neumann & Weber, 1978) or in the dictyosomes (Golgi bodies) and their associated vesicles (Harris & Boulter, 1976). It is a common, but not universal, observation that the onset of storage protein accumulation is accompanied by a proliferation of rough endoplasmic reticulum (RER), dictyosomes and small vesicles (0.3–0.5μm in diameter). These vesicles were thought to give rise, by fusion, to about 75% of the mature protein bodies in cowpea (Harris & Boulter, 1976), although dictyosome vesicles have also been thought of as vehicles for transporting storage proteins to their site of deposition in the large vacuole of peanut and broadbean (Dieckert & Dieckert, 1976; Harris, 1979). The tentative nature of these proposals is illustrated by the recent observation by Harris (1979) that the large population of small cytoplasmic vacuoles seen in *V. faba* is an artefact of thin sectioning. High voltage electron microscopy of thick sections from developing cotyledon cells has shown that the apparently discrete vesicles are in fact sections through a network of interconnected, tubular, smooth endoplasmic reticulum (ER) which is continuous with the RER. These new data reopen the question of the existence of an alternative vacuolar system.

The recent work of Craig *et al.*, (1979a, 1980a,b) has greatly clarified the relationship between the one or two large vacuoles seen early in the cell expansion phase and the protein bodies of mature seeds of *P. sativum*. They found progressive change from one or two vacuoles per cell with an average diameter of 39μm at eight days after flowering to approximately 170 000 protein bodies with 1μm average diameter per cell at 20 days with a very sharp reduction in vacuolar diameter between days 10 and 15 (from 39–3μm). Protein accumulation ceased 24 days after flowering. Serial sections of single cotyledon cells at 12 days after flowering revealed that the many discrete vacuoles apparent in any one section were in fact all a part of one or two large vacuolar structures. It appears that the apparent early breakdown of the large vacuoles is due to the fact that they undergo progressive and extensive modification to form a highly convoluted, multi-lobed, irregular structure which generates the increasing number of progressively smaller profiles seen in cross section. Serial sections showed that the actual fragmentation of this complex structure to yield a population of discrete protein bodies has begun by day 15 and presumably continues until the end of the protein accumulation phase.

This model is appealing in that is accounts for the ontogeny of protein bodies by a single mechanism. A similar model has been proposed by Yoo and Chrispeels (1980). However the possibility that protein bodies may originate by different routes in different species or even within the same cell must be borne in mind.

12.2.1.2 *Protein body formation in cereals*

No comparably detailed studies have been made of developing cereal cells. It appears that two mechanisms of protein body formation may be involved in some cereals. In the early stages of grain development in wheat, small, membrane-bound protein bodies are found within a plastid membrane which also encloses ribosomes (Morton *et al.*, 1964; Barlow *et al.*, 1974). Later in development protein bodies appear to arise directly from the ER by swelling and finally segmentation (Barlow *et al.*, 1974). Khoo and Wolf (1970) and Larkins and Hurkman (1978) also propose a direct formation of protein bodies in maize by a localized swelling of the endoplasmic reticulum as a result of storage protein accumulation within the lumen. Protein bodies may increase in size by fusion of these protein granules. Larkins and Hurkman (1978) observed continuity of membrane between endoplasmic reticulum and protein bodies in maize.

12.3 PHYSICOCHEMICAL CHARACTERISTICS OF STORAGE PROTEINS

12.3.1 INTRODUCTION

From his pioneering work on the characterization of seed proteins, Osborne (1924) recognized that seed proteins from all sources fell into one of four solubility classes, namely albumins, globulins, prolamins and glutelins. His findings are still used as the basis of many current procedures for preparing these fractions. If finely ground seed meal is extracted with salt solution (0.5–1.0M NaCl) and the extracted protein dialysed against water, the globulins (salt-soluble, water-insoluble) are precipitated and the albumins (salt- and water-soluble) remain in solution. Successive extraction of the residual meal with aqueous alcohol (50–70% ethanol) yields the prolamins and finally with acid or alkali yields the glutelin fraction. After surveying the seeds of many species he observed that in general the major storage proteins of legumes are globulins while those of many cereals are prolamins and glutelins (see Table 12.1). There are exceptions to this generalization. For example, in oats globulins are the major

Table 12.1 Characteristics of Some Seed Storage Proteins

Protein class	Some examples (source)	Approximate average size	Solubility characteristics	Characteristics of amino acid composition
Albumins	Lectins (Canavalia sp., *P. vulgaris*) 'Storage' albumins (*Ricinus communis, Vigna radiata*)	12 000 to 100 000	Water-soluble	
Globulins	Legumin (*P. sativum, V. faba*) Glycinin (*G. max*) Oat globulin (*Avena sativum*)	360 000 (12–13S)	Salt soluble (0.5–1.0M NaCl)	High in glutamine, asparagine, arginine; low in methionine, cysteine
	Vicilin (*P. sativum, V. faba*) Conglycinin (*G. max*) Phaseolin (*P. vulgaris*)	180 000 to 200 000 (7–9S)		
Prolamins	Gliadin (*Triticum aestivum*) Zein (*Z. mays*) Hordein (*H. vulgare*)	16 000 to 60 000	Soluble in 70% aqueous alcohol (with reducing agent)	High in glutamine, proline; low in arginine, histidine, lysine
Glutelins	Glutenin (*T. aestivum*)	Subunits (~ 30 000 to 50 000) aggregating up to several million	Acid- or alkali-soluble	

storage protein class with only small amounts of prolamins and glutelins (Peterson & Smith, 1976).

Osborne (1924) and later Danielsson (1949) were struck by the homologies of solubility class, molecular size and isoelectric point which existed among the seed proteins of a diverse range of legumes on the one hand and cereals on the other. For instance, in most legumes there appeared to be only two holoproteins within the globulin fraction. These differed in a number of properties including molecular weight, electrophoretic mobility and salt solubility. The more sophisticated analytical techniques available today have revealed heterogeneities of both mass and charge at both the holoprotein and subunit level within these broad protein classes and these will be discussed in more detail below. However it is important to remember that there is nevertheless, a great deal of overall similarity in physicochemical properties between the major storage proteins within the cereals and within the legumes. This may reflect the physicochemical criteria which storage proteins must meet in order to fulfill their biological role as reserve proteins for the developing embryo during germination.

Albumins are not generally regarded as important contributors to the storage protein fraction of seeds. Although the albumin fraction may make a significant contribution to total seed protein it is generally assumed to contain a wide range of metabolic proteins none of which are unique to the seed. However, in castor bean, 40% of the protein in the protein bodies is albumin (Youle & Huang, 1978) and the embryonic axes of mung bean contain a similar 'storage' albumin (Manikam & Carlier, 1980). This fraction has a distinctive and restricted subunit composition on sodium dodecyl sulphate-polyacrylamide gels (SDS-PAG), is rich in glutamine and is rapidly broken down during germination. It could be that further work on other species may show the need to reassess the role of albumins as storage proteins. However in this chapter we will confine our attention to the remaining seed protein classes.

Although the separation of storage proteins by solubility class is not entirely satisfactory in the operational sense, due to appreciable overlap in solubility behaviour between components of the different classes, the major categories recognized by Osborne serve as a useful basis to discuss the physicochemical characteristics of the storage proteins of some of the more important cereal and legume crop plants.

12.3.2. GLOBULINS

Legume seeds are characteristically high in protein, usually being in the range between 20–40% protein on a dry weight basis as compared to 7–16% for cereal grains. The major components of this seed protein in legumes are the globulins. Osborne (1924) and later Danielsson (1949) recognized two major holoprotein species within the globulins. These could be distinguished on the basis of differential precipitation with ammonium sulphate, differential heat coagualability, isoelectric precipitation and mobility during moving boundary electrophoresis and ultracentrifugation. One globulin fraction (named legumin) had a sedimentation constant of 11–13 S over a range of species, molecular weight of 331×10^3 daltons (from *P. sativum*) and was insoluble at pH 4.7 in 0.2M NaCl. The other fraction (vicilin) had an S value of 7–9, a molecular weight of 186×10^3 daltons and remained in solution at pH 4.7 (Danielsson, 1949).

Subsequent work has extended the range of species studied and largely confirmed the existence of two major globulin fractions in legumes and some other dicotyledonous seeds and their conformity to the above physicochemical characteristics. The proportion of 'legumin-like' and 'vicilin-like' protein varies greatly between genera, between species and within species, but in general two major globulins are seen with these characteristics. An exception appears to be *P. vulgaris* and *P. aureus* in which the 12S globulin component is a very minor contributor (Derbyshire & Boulter, 1976).

The 12S, legumin-like components have been found to consist of two classes of subunit. One class of about 35–40 $\times 10^3$ daltons is acidic and the other class of approximately 20×10^3 daltons is basic. It has been proposed that acid and basic subunits are covalently linked through disulphide bonds and six such pairs are combined in each molecule of holoprotein (Derbyshire *et al.*, 1976). The data from a wide range of seeds including *P. sativum*, *V. faba*, *Glycine max*, *P. vulgaris*, *P. aureus*, *Arachis hypogea*, *Cucurbita* sp. and *Avena sativa* are consistent with this general model. The 7–9S, vicilin-like globulin fraction is usually more complex than legumin and no general model of holoprotein structure has been formulated.

In spite of the overall similarity in physicochemical characteristics within the legumin-like and vicilin-like globulins in a range of legume seeds there is increasing evidence that these homologies are far from absolute. When different criteria are applied and more sophisticated fractionation methods are used many differences are seen at all levels of comparison. For example, using antiserum to legumin or vicilin from *V. faba*, a survey of 12 tribes in the Leguminosae subfamily Faboideae showed that, while some degree of immunological relatedness was seen extensively in the genera of the Fabeae and Trifolieae, it was fairly rare in all other tribes examined, including Phaseoleae (*Glycine*, *Phaseolus*, and *Vigna* species) (Dudman & Millerd, 1975). However, a degree of chemical homology can exist even when this is not detected by immunological means. Gilroy *et al.*, (1979) have found that there is sequence homology between about 25 of the first 30 N-terminal amino acids of the basic subunits of the 12S globulin from *V. faba* and *G. max*, and Hara *et al.*, (1978) have found a similar sequence in the first 6 amino acids from the basic subunit of the 12S globulin from pumpkin.

Considerable heterogeneity at the holoprotein level is seen within the same species. In fact there is no certainty as to the precise number of holoproteins which make up the 12S and 7S globulin fractions in any one species. For example Thomson *et al.*, (1978; 1979; 1980) using a wide variety of fractionation techniques, have shown that the legumin and vicilin fractions of *P. sativum* are each composed of multiple holoproteins. The number and type of holoprotein varies within the species and genetic evidence makes it clear that these variants are genotype-specific and are inherited in a co-dominant manner (Thomson & Schroeder, 1978).

The heterogeneity at the holoprotein level is matched by further heterogeneity at the subunit level. For example, the basic subunits of legumin from *P. sativum* can be resolved into at least five components and the acidic subunits into a further three or four major components (Thomson *et al.*, 1978), and up to seven or eight minor ones (Casey, 1979b). A similar degree of complexity has been reported for glycinin, the 12S globulin from *G. max* (Moreira *et al.*, 1979). In a detailed study of the relationship between the different members of the acidic and basic families of subunits of glycinin, it was shown (Moreira *et al.*, 1979) that the N-terminal amino acid for all four basic subunits was glycine, whereas the six acidic subunits had phenylalanine, arginine, leucine or serine. Very close sequence homology was found in the first twenty N-terminal amino acids of all four basic subunits with a highly significant but less close homology between the acidic subunits. This work gives substance to the suggestion that individual members within a particular group of storage protein subunits are products of a set of homologous genes which have evolved from a common ancestral gene.

The vicilin-like globulins are even more complex and heterogeneous than the 12S species when examined by solubility, electrophoretic behaviour and immunological specificity (Derbyshire *et al.*, 1976). The best characterized 7S globulin is β-conglycinin from *G. max* (Thanh & Shibasaki, 1978). The six distinct forms of β-conglycinin were found to consist of specific combinations of each of three subunits. The vicilin fraction from *P. sativum* was resolved into four distinct antigenic species by two-dimensional immunoelectrophoresis (Millerd *et al.*, 1978) and into seven holoprotein components on acid-PAG (Thomson *et al.*, 1980). One explanation for this heterogeneity of the 7S holoproteins is in their complex subunit composition. No less than twelve polypeptides were found in varying proportions in the vicilin species that were fractionated on acid-PAG. Each vicilin species had a distinctive quantitative and in some cases qualitative subunit composition and those differences were genetically determined (Thomson *et al.*, 1980). However a still greater degree of heterogeneity is indicated by the fact that in any one vicilin species the sum of the molecular weights of the component polypeptides exceeds the molecular weights of that holoprotein. Thus, the vicilins in *P. sativum* appear to be a heterogeneous collection of many closely related holoproteins which are made up of varying proportions of an extensive range of subunits. Recently, Croy *et al.*, (1980b) and Casey and Sanger (1980) have isolated a 7S protein from *P. sativum* and *P. fulvum*, respectively, which contains only a single species of subunit of approximately 70×10^3 daltons.

An additional contribution to the heterogeneity of the 7S globulins may stem from the fact that they are glycoproteins. Although not highly glycosylated in comparison with many animal glycoproteins, values of around 1% carbohydrate have been commonly reported for vicilin-like proteins (Basha & Beevers, 1976). Values up to about 6% have been reported for β-conglycinin from *G. max* (Thanh & Shibasaki, 1978), glycoprotein 11 from *P. aureus* (Racusen & Foote, 1971) and β-conglutin from *Lupinus angustifolius* (Eaton-Mordas & Moore, 1978). The composition of the carbohydrate fraction has been studied in a range of holoproteins and Davey and Dudman (1979) have also examined the distribution of carbohydrate between the component subunits of vicilin from *P. sativum*. Three out of four distinct vicilin fractions were heterogeneous with respect to reaction with concanavalin A, with about 30% of the molecules not binding. Among the polypeptides of the four vicilins, the 14×10^3 subunit consistently contained a large proportion of the carbohydrate with variable amounts on the 50×10^3 and 30×10^3 dalton subunits.

Some controversy exists regarding the association of carbohydrate with 'legumin-like' proteins. While some authors found low, but significant levels of carbohydrates (Basha & Beevers, 1976; Davey and Dudman, 1979), more recent reports have failed to confirm this (Casey, 1979b; Gatehouse *et al.*, 1980).

As presently understood, the globulins are quite heterogeneous within the 7S and 12S holoprotein classes. There appears to be a fairly large number of distinct genetic entities. Within each class there may be several related genetic entities as discussed for the basic and acidic subunits. Some of the microheterogeneity may be generated by varying levels of post-translation modification such as glycosylation.

12.3.3 PROLAMINS

Although very minor contributors to seed proteins of legumes, prolamins are a major, probably the major constituent of most cereal grains, exceptions being oats, as noted above, and rice, in which glutelins predominate (Juliana & Boulter, 1976). Prolamins are characterized by their solubility in aqueous alcohol, their high content of amides (especially glutamine) and proline and their low content of the basic amino acids arginine, histidine and lysine. Their molecules have very low net charge. The distinction between prolamins and glutelins is not always clear depending as it does on the precise conditions under which the aqueous ethanol extraction is done. Miflin and Shewry (1979) argue that exhaustive extraction with warm (60°C) aqueous isopropanol in the presence of a reducing agent is essential to extract all the prolamins. Furthermore, Bietz and Wall (1973) found that, after reduction and alkylation, about 62% of wheat glutelin (glutenin) was soluble in 70% ethanol and the extracted protein had a characteristic prolamin subunit pattern.

All prolamin fractions so far examined in detail show extensive heterogeneity especially when fractionated in a two-dimensional system involving electrophoresis in one direction and isoelectric focussing in the other. The prolamins of wheat (gliadins) consist mainly of single polypeptide chains with mol-

Fig. 12.3 Two dimensional separation of gliadins from wheat cv. Chinese Spring showing 43 individual components. The identity of chromosomes which control the synthesis of specific components is also shown. Separation was by starch gel electrophoresis in aluminium lactate, pH 3, in the vertical dimension and isoelectric focussing between pH 3 (left) and pH 10 (right) in the horizontal direction (from Wrigley & Shepherd, 1973; reproduced with permission).

ecular weights around 36 $\times$ 10^3 daltons with other components ranging from 11.4 to 78 $\times$ 10^3 present in far lower amounts (Bietz & Wall, 1972) when run under reducing and denaturing conditions. The 16 or so components detected in one dimension are increased to 46 when this fractionation is followed by isoelectric focussing (Wrigley & Shepherd, 1973 and Fig. 12.3). Given that there is also extensive intervarietal variation in gliadin components it seems likely there are many hundreds of different gliadin components in wheats in general. N-terminal sequencing and peptide mapping suggest that there are close similarities between many of these components (Bietz *et al.*, 1977).

Comparable heterogeneity occurs in the prolamins of maize (zeins) and barley (hordeins). By one dimensional electrophoresis of the zeins in SDS-PAG major components are found with molecular weights of 23 $\times$ 10^3, 21 $\times$ 10^3 and

13.6×10^3 daltons, with two less prominent components of 22×10^3 and 9.6×10^3 daltons. Fractionation in a second dimension by isoelectric focussing revealed a total of 28 components, mostly derived from the high molecular weight species (di Fonzo *et al.*, 1977). Isoelectric focussing of the hordeins reveals up to 20 components (Miflin & Shewry, 1979). In spite of this microheterogeneity, it is important to note that extensive conservation of amino acid sequences has occurred among the gliadins even between different genera. Thus, a purified C-hordein component from barley and an ω-gliadin component from wheat were found to be homologous at 23 out of 27 residues at the N-terminus (Shewry *et al.*, 1980a).

The prolamins are made up of a large number of entities which when analyzed in detail by biochemical and genetical techniques are reduced to a small number of related families. The relationships of members within families are closer than those between families. The genes coding for the prolamins seem to occur as clusters of closely linked genes. The clusters may occur on different chromosomes or some distance apart on the same chromosome. Despite their spatial separation, their synthesis is closely coordinated and, in the case of maize, all of the zeins are under the quantitative regulation of the opaque-2 gene (Gianazza *et al.*, 1977; Soave *et al.*, 1978; Mecham *et al.*, 1978; Oram *et al.*, 1975; Shewry *et al.*, 1980b).

12.3.4 GLUTELINS

The precise definition and identity of the glutelin fraction is uncertain because of the wide range of procedures that have been used in its preparation. The glutelins are essentially the alkali- or acid-soluble proteins remaining after extraction of albumins, globulins and prolamins. It seems clear that with less than complete extraction, some components of the prolamin fraction have been included in the glutelins. When warm isopropanol and mercaptoethanol are used exhaustively to remove prolamins, the somewhat diminished residual fraction, largely glutelins, can be characterized as having a higher lysine and lower glutamate plus glutamine and lower proline content than the prolamins (Miflin & Shewry, 1979). These authors suggest that at least in wheat, maize and barley, but not in rice and oats, the components of this exhaustively-extracted glutelin fraction are not true storage proteins, but are derived from membranes from the endoplasmic reticulum (ER) and other subcellular structures. However these proteins still make a significant contribution to the total protein of cereal grains, and until proved otherwise, should continue to be regarded as storage proteins.

Wheat glutenin is the most widely studied glutelin. Preparative methods are so varied that the final glutenin fraction has been found to account for between 10 and 50% of the total seed protein (Kasarda *et al.*, 1976). Glutenin is a highly complex association of about 15 subunits ranging in molecular weight from 11.6×10^3–133×10^3 daltons. These are associated into complex aggregates, probably via intermolecular disulphide bridges, whose molecular weights vary from 40×10^3 to many millions. The subunits in the 33×10^3–50×10^3 dalton range make up the major part of these complexes and are therefore felt to be important in determining the general properties of the complex

The glutelin fraction of other cereals has not been as well studied as wheat glutenin. In maize, di Fonzo *et al.* (1977) subdivided the glutelins into two fractions, the first consisting of the components soluble in alkali (pH 10) plus mercaptoethanol and the second those remaining proteins soluble in the same solvent plus SDS. The first fraction (G1) resolved into five bands on SDS-PAG ranging from 10×10^3–66×10^3 daltons. The second fraction (G2) was far more heterogeneous and quite distinctive. It consisted of approximately 20 components on SDS-PAG, ranging in apparent molecular weight from $> 100 \times 10^3$–10×10^3 daltons. Taken together G1 and G2 accounted for 24% of the total protein of the maize endosperm.

Despite the uncertain composition of the glutelin fraction it seems likely that it may contain important storage proteins in addition to contaminating prolamins and possibly membrane proteins. Further analysis is required to reveal the precise nature of this fraction.

12.4 BIOCHEMISTRY OF STORAGE PROTEIN SYNTHESIS

12.4.1 CARBON AND NITROGEN METABOLISM

The developing seed is dependent upon the rest of the plant for the carbon and nitrogen with which to make the needed amino acids for storage proteins. In general sucrose is the major source of carbon and asparagine and glutamine are the major sources of nitrogen for the developing seed (Pate, 1980). The cereals and legumes both derive the fixed carbon for amino acid biosynthesis from the upper leaves close to the developing fruit (Wardlaw, 1968; Pate & Flinn, 1973; Kipps & Boulter, 1974). The glumes of wheat are a significant source of carbon skeletons for amino acids and may also be the site of amino acid synthesis (Lee, 1978). The nitrogen destined for seeds is more generally derived from the whole plant (Neales *et al.*, 1963; Pate & Flinn, 1973; Kipps & Boulter, 1974). In the case of peas, the major transport form occurs as asparagine (Lewis & Pate, 1973) and it has been shown that the seeds can synthesize their own amino acids when provided with a nitrogen source (Kipps & Boulter, 1974; Macnicol, 1977). Wheat heads, at least in culture conditions, can also synthesize a range of amino acids when supplied with glutamine, asparagine or ammonium nitrate (Donovan & Lee, 1978).

12.4.2 ROLE OF AMINO ACID LEVELS

It is generally assumed that the amino acid pool of the developing seed may be an important regulatory element in the synthesis of storage proteins. This is difficult to demonstrate clearly and although many free amino acid pool measurements have been made and correlated with the composition of the major proteins being synthesized (e.g., Jennings & Morton, 1963a,c; Kapoor & Gupta, 1977; Brandt, 1976) these measurements cannot be used to show that one or more amino acids are limiting protein synthesis. For example, in the lysine-rich mutant of barley (Risø 1508) the free lysine pool is either the same or lower than that of the normal depending upon the stage of seed maturation (Brandt, 1976). Flux through a particular amino acid pool and compartmentation of amino acids are important factors which need to be considered in any model proposing to establish a regulatory role for amino acids.

One way to study the regulatory role of amino acids in protein synthesis would be to limit the synthesis of individual amino acids. This may be possible in the cases of methionine and cysteine by regulating the supply of fertilizer sulphur to the plant. It has been shown that under low levels of S nutrition the levels of non-protein cysteine and methionine are reduced considerably in lupin (Blagrove *et al.*, 1976), cowpea (Evans *et al.*, 1977), and wheat seeds (Byers & Bolton, 1979). Levels of sulphur-rich storage proteins of lupin, pea, and wheat are greatly diminished under low S plant nutrition (see Section 12.6.2). This suggests that limiting amino acids, in these cases, methionine and cysteine, may participate in the regulation of the level of storage proteins in some seeds.

12.4.3 ROLE OF tRNA LEVELS AND AMINOACYL tRNA SYNTHETASES

The potential regulatory role of tRNA and aminoacyl-tRNA synthetases in storage protein synthesis has been examined in only a few cases. Norris *et al.*, (1973) found no specific changes in total amino acylating activity of wheat endosperm during maturation although there was a general increase in activity during reserve protein accumulation. Furthermore, they found no changes in lysyl- or prolyl-tRNA or their isoaccepting species during the ripening of wheat (Norris *et al.*, 1975) although there are dramatic changes in the proline and lysine content of the proteins being synthesized. In developing cotton seed no change was observed in the levels of tRNAs for fifteen amino acids (Merrick & Dure, 1972). However, in the maturing maize endosperm which is actively synthesizing zein, which in turn is rich in glutamine, leucine, alanine, and proline and deficient in lysine, Viotti *et al.*, (1978a) found increased levels of tRNA for glutamine, leucine and alanine and changes in the relative levels of each of their isoacceptors but no change in those for lysine and proline. It is therefore possible that the tRNAs for some amino acids are modulated and hence are involved in regulating the synthesis of specific proteins whereas others such as lysyl- and prolyl-tRNA in wheat and maize are much less subject to fluctuation.

12.4.4 TOTAL RNA AND DNA

Changes in the level of RNA and DNA have been reported for many developing seeds. As a general rule, RNA and DNA increase in parallel with protein accumulation in the legumes (Millerd & Whitfeld, 1973; Scharpé & van Parijs, 1973; Smith, 1973; Madison *et al.*, 1976; Millerd & Spencer, 1974; see Fig. 12.4) while in the cereals, the nucleic acid levels reach their maximum value at about the beginning of storage protein accumulation (Jennings & Morton, 1963b; Ingle *et al.*, 1965; Viotti *et al.*, 1975). The DNA synthesized during cell expansion and reserve accumulation in legumes is probably a result of endoreduplication of the whole genome (Millerd & Whitfeld, 1973). Specific amplification of particular regions of the DNA, e.g., those involved in storage protein synthesis, would probably not have been detected by the methods used.

The function of the extra DNA is unknown but, as judged by the cereals, it does not seem to be obligatory for the synthesis of reserve protein, nor is the final protein level in legumes (e.g., soybeans) closely linked with DNA content (Madison *et al.*, 1976). In the legumes, there is a positive association between cell volume increase and DNA level although in the cereals, where there is also a marked increase in cell volume during the accumulation phase of seed maturation (Jennings *et al.*, 1963), the DNA content remains essentially constant during the period of storage protein synthesis (Jennings & Morton, 1963c).

Although the template activity of the total DNA does not change during seed maturation, the endogenous RNA synthesizing activity per unit of DNA declines with development (Millerd & Spencer, 1974) suggesting that RNA polymerase activity is limiting and that not all of the extra DNA is used as template. In summary, there is to date no clear evidence to suggest that the extra DNA in legume seeds is directly involved in storage protein accumulation.

Similarly, much of the RNA in legumes may not be directly involved in protein synthesis. In cereals, the maximal RNA levels are attained prior to the increase in the rate of storage protein synthesis and in fact the latter occurs when the RNA level has dropped to about 30% of maximum (Ingle *et al.*, 1965; Viotti *et al.*, 1975).

12.4.4.1 *Role of messenger RNA*

The role of mRNA level in the regulation of storage protein synthesis is receiving attention in both the cereals and legumes. The earliest approaches employed cell-free protein-synthesizing systems programmed with polysomes isolated from developing seeds. These 'chain completion' assays are largely non-initiating and they provide a good measure of the proteins being synthesized at the time of isolation. In contrast, the heterologous, cell-free translation systems, dependent upon the addition of mRNA, are used to provide a measure of the mRNA level in a tissue without the complications of the homologous controls such as specific initiation factors, aminoacyl tRNAs and their synthetases and other factors involved in regulating translation. For example, the 'chain completion' assay showed that the polysomes in the endosperm of the opaque-2 mutant of maize synthesize only 37% as much zein as the polysomes from

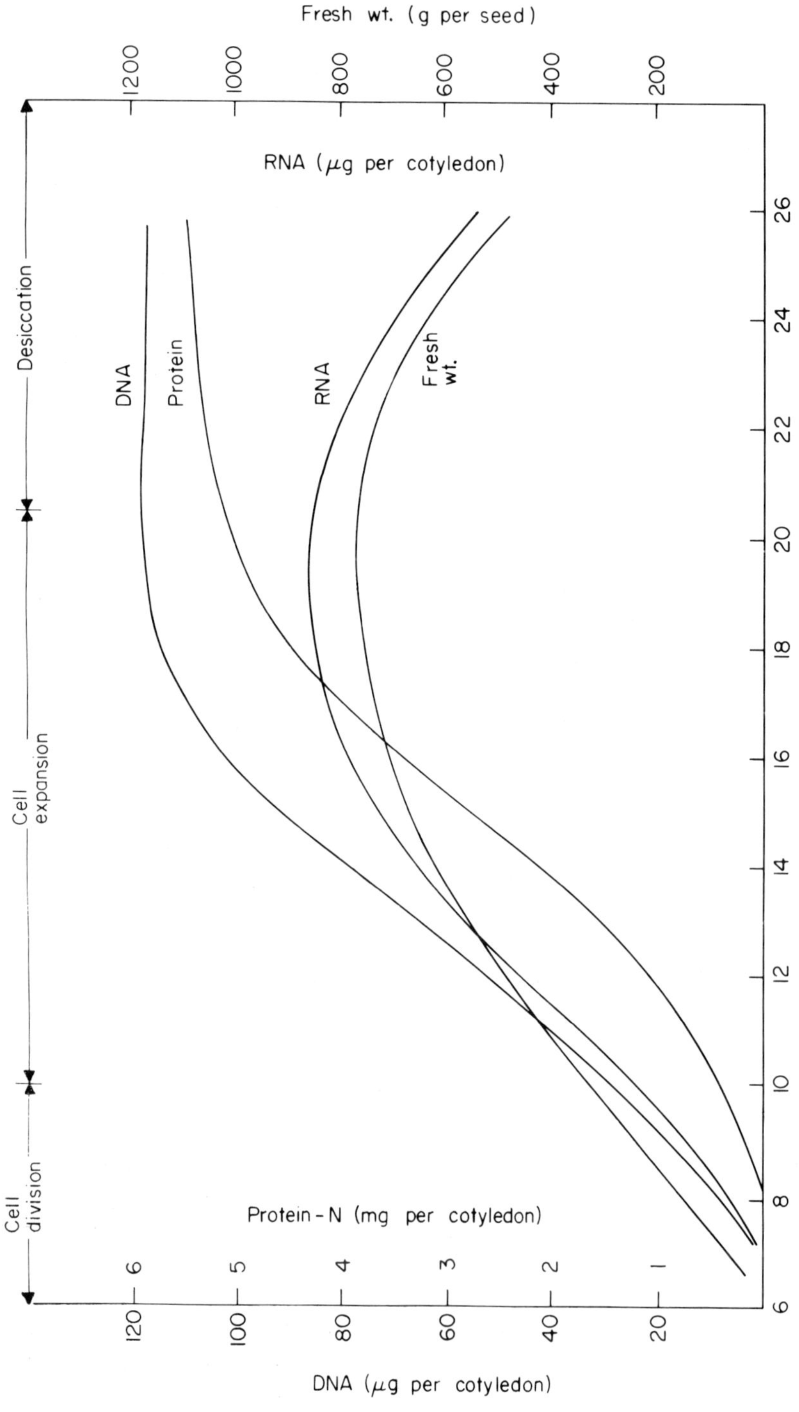

Fig. 12.4 Changes in DNA, RNA, protein and fresh weight in developing pea seeds (after Millerd & Spencer, 1974; reproduced with permission).

normal endosperms (Jones *et al.*, 1976). The lower level of zein in the mutant is therefore due to reduced protein synthesis and not increased protein turnover.

Synthesis of specific proteins has been accomplished in heterologous systems by translation of total, polysomal or poly(A)RNA from developing seeds of maize (Burr *et al.*, 1978; Larkins *et al.*, 1976; Wienand & Feix, 1978; Melcher, 1979; Park *et al.*, 1980), barley (Brandt & Ingversen, 1978; Matthews & Miflin 1980), soybean (Beachy *et al.*, 1980), French bean (Hall *et al.*, 1978), field bean (Puchel *et al.*, 1979) and pea (Higgins & Spencer, 1977; Evans *et al.*, 1979). These initiating systems can be used to quantify mRNA levels for specific proteins by assuming that the proportion of the specific protein synthesized *in vitro* is a direct reflection of the amount of its mRNA in the sample. Although this is not always true, it is often a good approximation. Zein constitutes 70–90% of the *in vitro* translation product of maize polysomal RNA in a heterologous cell-free system (Larkins *et al.*, 1976). Thus, it is possible that the mRNA for zein is over 70% of the total mRNA population and that mRNA level is likely to be a major factor in the regulation of zein synthesis. A similar role for hordein mRNA in barley is suggested by work with the high-lysine mutant, Risø 1508 (Brandt & Ingversen, 1978). In this instance, there is evidence that hordein synthesis is also controlled at a number of other points including processing of the newly synthesized molecules (Cameron-Mills & Ingversen, 1978).

Similarly, in legumes it seems likely that the level of translatable mRNA for storage protein plays a major role in the regulation of reserve protein synthesis. Higgins and Spencer (1977) found that 40–60% of the mRNA population of developing pea seeds was involved in storage protein synthesis and this is in reasonable agreement with the fact that storage protein synthesis is about 30–40% of the total protein synthesis of developing cotyledons as determined by *in vivo* pulse-labelling studies combined with immunoaffinity chromatography (D. Spencer & T. J. V. Higgins, unpublished observations). The major storage protein in French beans is also thought to be largely under the control of translatable mRNA level (Hall *et al.*, 1978). Factors which determine the level of translatable mRNA are discussed in Section 12.6.4.

12.4.5 TRANSLATIONAL FACTORS OTHER THAN mRNA

The protein-synthesizing components and their relative efficiencies during seed development have rarely been studied. There is some evidence to suggest that the activity of some factors in the post-polysomal fraction of pea cotyledons changes the efficiency of protein synthesis during maturation (Beevers & Poulson, 1972). Ribosomal activity and the soluble factors which are involved in protein synthesis have yet to be assigned their regulatory role in reserve protein synthesis but Payne *et al.*, (1971) provided preliminary evidence to suggest that ribosomal activity changed during seed maturation in *V. faba.*

12.4.6 DEVELOPMENTAL ASPECTS OF PROTEIN BIOSYNTHESIS AND DEPOSITION

12.4.6.1 *Patterns of holoprotein and polypeptide accumulation*

It is probably because relatively little is known about the native structure of the major storage holoproteins of cereals that there are few studies of their developmental biology. The solubility classes, albumins, globulins prolamins and glutelins have been studied during development in wheat (Jennings & Morton, 1963a), barley (Brandt, 1976), oats (Peterson & Smith, 1976) and maize (di Fonzo *et al.*, 1977). It is a general feature that the proportion of albumins declines with seed formation. In oats, the globulin and prolamin fraction become an increasingly large proportion of the total seed protein whilst in barley, these and the glutelin fraction increase with maturation. In most cases there are differential rates of accumulation, the prolamins usually accumulating earlier than the other fractions. It is generally assumed that storage proteins, once synthesized, are not turned over but there are no data on this point. If differential turnover of various storage proteins does occur it would clearly affect the rate of accumulation of specific protein fractions.

The accumulation of the prolamin fraction in barley and maize has been studied at the polypeptide level. All of the constituent polypeptides maintain a constant ratio to one another during seed formation (Brandt, 1976; di Fonzo *et al.*, 1977), suggesting that the synthesis of all the prolamin polypeptides is regulated in a coordinate fashion. This may also occur in the barley prolamins. Whether this regulation occurs at the transcriptional or the post-transcriptional level is not known.

The rate of accumulation of the storage globulins in legumes has been studied in detail. The 7S protein (vicilin) of *P. sativum* and *V. faba* begins to accumulate earlier than the 11S protein (legumin) (Danielsson, 1952; Wright & Boulter, 1972; Millerd & Spencer, 1974; Guldager, 1978). The known molecular heterogeneity of the 7S proteins (see Section 12.3) is often reflected in the pattern of their accumulation during development. The 7S protein of soybean has been divided into three major electrophoretic entities each of which has a distinctive rate curve of accumulation (Hill & Breidenbach, 1974). These proteins undergo some form of modification during seed maturation which alters their reaction to antibodies (Ochiai-Yamagi *et al.*, 1978). The nature of the changes is presently unknown. Individual serological entities among the 7S proteins of pea accumulate at characteristic stages of seed development (Millerd *et al.*, 1978) and minor changes in the electrophoretic mobility of some of the vicilins (Millerd *et al.*, 1978) as well as the legumins (Guldager, 1978) have led to the suggestion that modifications to these holoproteins may also occur during seed maturation (Fig. 12.5).

In the case of pea storage proteins, there are data to suggest that the earlier appearance of vicilin relative to legumin is a reflection of the earlier synthesis of

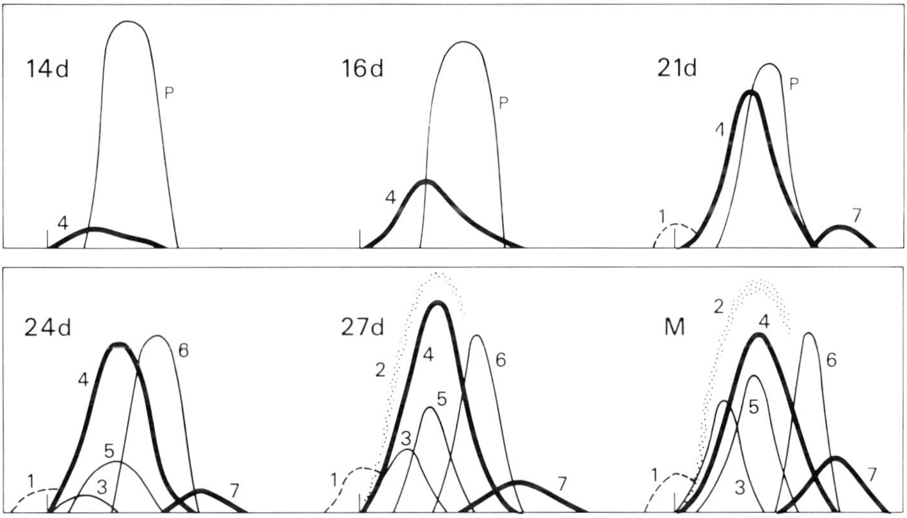

Fig. 12.5 Developmental patterns of pea cotyledonary storage proteins revealed by crossed immunoelectrophoresis against anti-serum to protein body extract from mature seeds. Cotyledons were extracted at 14, 16, 21, 24 and 27 days after flowering and at seed maturity. Precipitin lines 2, 3, 4 and 5 are due to vicilin-related proteins and line 7 is due to legumin. Line 1 is due to pea lectin and line P or 6 is of unknown origin. Under the growing conditions used, storage protein accumulation is maximal at 30 days after flowering (Reproduced with permission; after Millerd *et al.*, 1978).

their component polypeptides (Spencer *et al.*, 1980) and is not due to legumin being more unstable than vicilin. It therefore seems likely that in this case the storage protein mRNAs are not transcribed and translated as a block but rather that expression of genes for individual proteins is activated by controls that differ, at least in time.

Polypeptide patterns of legume storage proteins have been charted for many species during seed development. General features are that the major polypeptides of the 12S globulins only vary their proportions very slightly during maturation (Ochiai-Yamagi *et al.*, 1978; Thomson *et al.*, 1979) whereas the component polypeptides of the 7S proteins change their relative proportions fairly dramatically (Wright & Boulter, 1972; Hill & Breidenback, 1974; Sun *et al.*, 1978; Ochiai-Yamagi, 1978; Thomson *et al.*, 1979; Spencer *et al.*, 1980). This is consistent with the data mentioned earlier and in Section 12.3 which show that there are a number of 7S proteins each with a characteristic rate of accumulation *in vivo*.

12.4.6.2 *Evidence of precursor forms from pulse labelling studies*

In most studies the accumulated polypeptides have been monitored at progressive developmental stages. A refinement of this approach is to pulse-label seeds for short periods at various times during seed maturation. This approach permits one to assess the current protein synthesizing activity of the seeds at defined developmental stages and to determine whether storage proteins are synthesized via precursors with short lifetimes. Such studies, which have been initiated for pea (Spencer & Higgins, 1980; Spencer *et al.*, 1980) and French bean (Bollini & Chrispeels, 1979), show that storage protein synthesis is the major protein-synthesizing activity of these seeds during the accumulation phase of seed maturation and at least in peas there is evidence of extensive post-translational modification of the proteins. For example, a one to two hour pulse-labelling of developing pea cotyledons has shown the existence of at least three large precursors (of about 60–65 × 10^3 daltons) from which are derived both the acidic and basic subunits of legumin. During a subsequent two hour chase period there was a decrease in labelling in the high molecular weight precursors accompanied by an increase in both acidic and basic subunits of legumin (Spencer & Higgins, 1980 and Fig. 12.6). A single high molecular weight legumin precursor has been reported by Croy *et al.*, (1980a).

Further modification occurs to the major basic subunit of legumin days and even weeks after its synthesis. In the immature seed, the major basic subunit detected by either protein staining or pulse-labelling has an apparent molecular weight of 19 × 10^3 daltons on SDS-PAG. Much later, during the desiccation phase, the 19 × 10^3 dalton component is progressively modified in such a way that its apparent molecular weight increases to the 20 × 10^3 daltons characteristic of mature seed (Spencer *et al.*, 1980).

Many storage proteins are glycosylated (see Davey & Dudman 1979) and the major glycopolypeptide (14 × 10^3 daltons) of pea vicilin appears only after an extended chase when immature pea cotyledons are labelled with ^{14}C-glucosamine (J. Badenoch-Jones, personal communication). The polypeptides of soybean 7S protein also appear to undergo extensive slow post-translational modification including glycosylation (Sengupta *et al.*, 1981). The function of these modifications remains to be elucidated but they may play a role in maintaining the integrity of storage proteins in protein bodies during seed maturation.

12.4.6.3 *Evidence of precursor forms from cell-free systems*

Cell-free systems have revealed two other classes of storage protein precursor distinct from those discussed in Section 12.4.6.2. One of these is represented by transient forms which are processed during translation. They have been detected in the prolamins of barley (Brandt & Ingversen, 1978; Matthews & Miflin, 1980) and maize (Burr *et al.*, 1978; Larkins & Hurkman, 1978; Wienand & Feix, 1978; Melcher, 1979; Park *et al.*, 1980) and in the storage globulins of pea (Higgins & Spencer, 1981) and soybean (Sengupta *et al.*, 1981). The precursors migrate more slowly than the mature polypeptides on SDS-PAG and it is currently thought that they contain an extra sequence (Burr *et al.*, 1978; Larkins

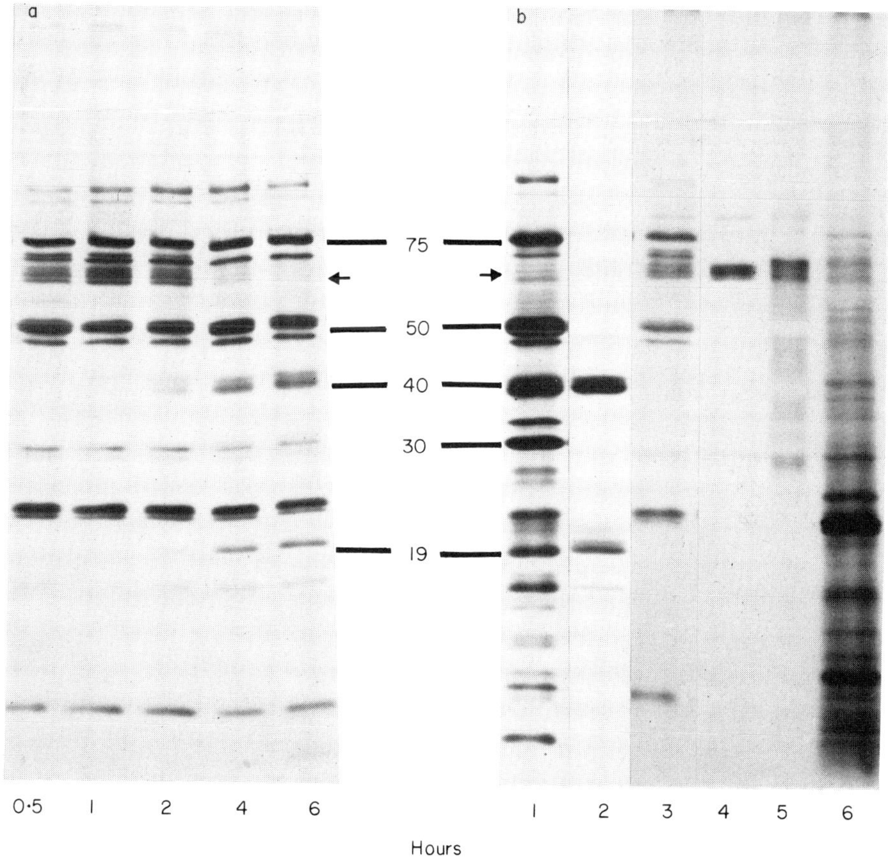

Fig. 12.6 Evidence of a large precursor of the small and large subunits of pea legumin. a. Cotyledons dissected from pea seeds at 15 days after flowering were labelled with ^{14}C-amino acids for the periods indicated. After 3h incubation, the 4 and 6h treatments were transferred to non-radioactive medium. Proteins were extracted, fractionated on SDS-PAG and fluorographed. Note transient appearance of a set of bands at approximately 60 000 daltons (arrow) which precede the appearance of radioactive legumin subunits at 40 000 and 19 000 daltons. b. Identification of legumin-related proteins by immunoaffinity chromatography using legumin-specific immunoglobulin G coupled to Sepharose, followed by fractionation on SDS-PAG. Track 1, non-radioactive proteins extracted from 15 day cotyledons; track 2, those proteins from the extract used in track 1 which bound to the anti-legumin column; track 3, radioactive proteins from cotyledons labelled for 1h with ^{14}C-amino acids; track 4, those proteins from the extract used in track 3 which bound to the anti-legumin column; tracks 5 and 6, radioactive products from *in vitro* translation of total polysomal RNA from 15 day cotyledons; track 6, total reaction mixture; track 5, those polypeptides which bound to the anti-legumin column. Note selective binding of polypeptides of approximately 60 000 daltons in both *in vivo* and *in vitro* products (arrow head). The number on vertical axis indicate approximate $M_r \times 10^{-3}$. (After Spencer & Higgins, 1980; reproduced with permission).

et al., 1979) equivalent to the signal peptide on bacterial and animal proteins destined for secretion or membrane insertion. Evidence to support this proposal has been obtained by adding microsomal membranes to a mRNA-dependent cell-free system in the presence of pea mRNA. Some of the storage protein precursor polypeptides are modified to the correct size and the modified polypeptides are sequestered within vesicles (Higgins & Spencer, 1981 and Fig. 12.7). Similarly, by leaving the polysomes attached to the endoplasmic reticulum of maize, Larkins and Hurkman (1978) showed that zein polypeptides of the correct size were synthesized. Translation of zein mRNA by the *Xenopus* oocyte system, which is known to possess the ability to modify proteins during and after translation, resulted in correct-sized prolamins being synthesized (Larkins *et al.*, 1979). The formation of the large (60–65 × 10^3 dalton) precursors, which contain both acidic and basic subunits of legumin (described in 12.4.6.2), has

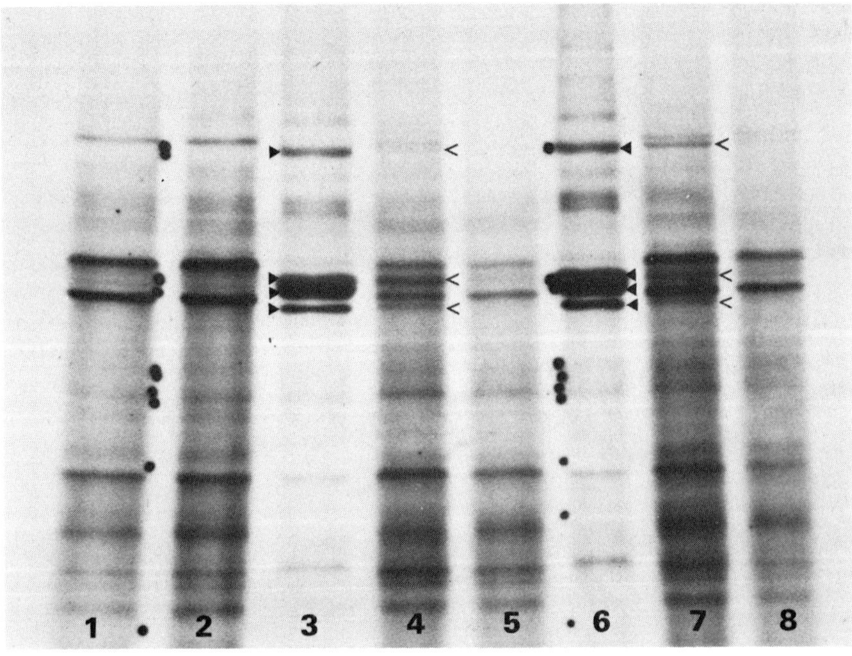

Fig. 12.7 Evidence of modification of pea storage proteins in cell-free systems in the presence of microsomal membranes. Total polysomal RNA from cotyledons 11 days after flowering was translated in a cell-free system in the presence and absence of microsomal membranes and the products fractionated on SDS-PAG. For comparison, proteins from 11 day old cotyledons, pulse-labelled for 1h with ^{14}C-amino acids were also fractionated (tracks 3 and 6). Tracks 1, 2, 4, 5, 7 and 8 show *in vitro* products. Dog pancreas microsomal membranes were included in the reaction mixtures in tracks 4 and 5, pea cotyledon microsomal membranes in those in tracks 7 and 8 and Triton X-100 in those in tracks 2, 5 and 8. New polypeptides arising from membrane treatment are indicated by open arrow heads and polypeptides known by immunoaffinity chromatography to be vicilin-related by closed arrow heads. (After Higgins & Spencer, 1981; reproduced with permission).

also been demonstrated in an mRNA-dependent, *in vitro* translation system (Spencer & Higgins, 1980; Croy *et al.*, 1980a and Fig. 12.6).

The other forms of precursor that have been detected in an *in vitro* system are the polypeptides of the 7S globulin of *P. vulgaris* (Hall *et al.*, 1978). In this case, the cell-free translation products migrate more rapidly than the authentic polypeptides. It is not known whether microsomal membranes modify these products but it has been suggested that they represent unglycosylated forms of the authentic polypeptides. The full story of the biosynthetic intermediates in storage protein synthesis is only beginning to be unravelled. There are differences between the translation products programmed with mRNA and those made by polysomes detached from membranes (Higgins & Spencer, 1977; Hall *et al.*, 1978; Sun *et al.*, 1975; Brandt & Ingversen, 1978; Matthews & Miflin, 1980). Furthermore these are often different from the authentic polypeptides synthesized *in vivo* (Higgins & Spencer, 1977; Brandt & Ingversen, 1978; Matthews & Miflin, 1980; Sengupta *et al.*, 1981) and from those found in mature seeds. Clearly, storage protein polypeptides undergo a number of modifications either during or immediately after translation. These include the removal of a putative signal sequence and glycosylation both of which may occur during or soon after translation while the polypeptides are in close contact with the ER. The latter is known to possess glycosylating activity (Nagahashi & Beevers, 1978).

In summary, the data from cell-free translation systems, *in vivo* pulse labelling studies and polypeptide and holoprotein accumulation patterns during development all indicate that the biosynthesis and deposition of storage protein is a highly complex process involving modifications to the protein at a number of stages. The function of these changes is largely unknown, but the presence of a sequence equivalent to a signal fragment is likely to be involved in the recognition of the ER and in the early steps in intracellular transport and sequestration.

12.4.7 FEATURES OF STORAGE PROTEIN mRNAS

Among mRNAs for storage protein polypeptides which have been partially purified so far are those coding for the prolamins of maize. The two high molecular weight chains of zein (23×10^3 and 21×10^3 daltons) are coded for by two separate mRNAs of similar size and not from a dicistronic mRNA (Burr *et al.*, 1978; Larkins & Hurkman, 1978; Wienand & Feix., 1978). The mRNAs are 1000–1200 nucleotides long whereas only about 600–700 are required to code for the polypeptides. Excluding the poly(A) segment known to be present on these and most other storage protein mRNAs, this means that about 300–500 nucleotides are untranslated.

As mentioned above, mRNA for storage proteins contain poly(A) and may be 'capped' at the 5′ end (Beachy *et al.*, 1978; Burr *et al.*, 1978; Hall *et al.*, 1978). So far, nothing is known about the regulation of mRNA synthesis (e.g. existence of precursors), turnover or lifetime. Some of the storage protein mRNA may survive into the dry seed (Mori *et al.*, 1978) but the physiological significance of this is obscure. It may reflect incomplete degradation of the mRNA during seed drying.

The mRNA for storage protein is usually found in large polysomes and attached to the ER or occasionally to protein bodies (Brandt, 1976; Jones *et al.*, 1976; Burr & Burr, 1976; Larkins & Hurkman, 1978; Viotti *et al.*, 1978b; Bollini & Chrispeels, 1979; Higgins & Spencer, 1981; Sengupta *et al.*, 1981).

Future studies on the regulation of storage protein synthesis are likely to be greatly aided by the availability of pure mRNA for defined protein species. Currently, the mRNAs for a number of storage proteins are being used to make DNA copies (Viotti *et al.*, 1979; Pedersen *et al.*, 1980; Evans *et al.*, 1980) which are in turn being used for making large quantities of the nucleic acids by recombinant DNA methods (Wienand *et al.*, 1979; Park *et al.*, 1980; Sun *et al.*, 1981). These will constitute a powerful tool which can be used to answer questions relating to quantification of mRNA levels, the switch-on of mRNA synthesis, the nature of the primary transcription product and possible precursors, the number of copies of specific storage protein genes in the genome and their arrangement relative to the rest of the DNA.

12.5. INTRACELLULAR TRANSPORT OF STORAGE PROTEINS

In a previous section (12.2) we discussed the ontogeny of protein bodies and the uncertainties that exist in this area. There are even greater gaps in our knowledge of a closely related area, namely, the molecular and structural events involved in the intracellular transport of storage proteins from their site of translation to their ultimate site of accumulation, which is usually the protein body. By what route does this occur and at what points along this route do post-translational modifications, including holoprotein assembly, occur? Although there are no clear answers to most of these questions, it is worth canvassing the possibilities in terms of what is currently known about storage proteins and their biosynthesis.

12.5.1 SITES OF POST-TRANSLATIONAL MODIFICATIONS

A brief survey of the known translational and post-translational modifications which occur in the biosynthesis of at least some storage proteins would include removal of a presumed signal sequence, glycosylation, cleavage of high molecular weight polyprotein precursors, holoprotein assembly and modification of some of the polypeptides even after holoprotein assembly and deposition in protein bodies. As an example of these extensive modification events we can point to the synthesis of vicilin and legumin in peas (see 12.4.6.2) which together include cases of each of the above. On the other hand the prolamins, such as zein, which consist of single polypeptide chains, may only undergo removal of a signal sequence. Excision of the signal-peptide sequence and the initial steps of glycosylation, if any, take place on the RER. By analogy with animal systems (Case, 1978) it is possible that extensions to the carbohydrate side-chains would occur in association with the dictyosomes.

Using autoradiographic techniques, Bailey *et al*., (1970) found that in *V. faba* unidentified, radioactive proteins synthesized on the ER were transported to the protein body within 25 minutes. Pulse-chase experiments with intact pea cotyledons, followed by subcellular fractionation, have shown that the newly-synthesized storage protein polypeptides, first detected in the ER, are transported to the protein bodies within 30 minutes. These include the high molecular weight precursor polypeptides of legumin, but not the individual acidic and basic subunits (M. J. Chrispeels, T. J. V. Higgins & D. Spencer, unpublished observations). It seems clear that both the cleavage of the precursors to yield the subunits and the subsequent, much-delayed changes to the basic subunits (see 12.4.6.2) must occur after intracellular transport and deposition in the protein body.

We have no definitive information on the site and timing of the assembly of the more complex storage holoproteins, such as the globulins of legumes and the glutelins of cereals. The simplest model would envisage the coordinated synthesis of the component polypeptides at a site such as the ER and their rapid assembly into holoproteins in the lumen of the ER but, at the moment, all other possibilities must be borne in mind including the delay of holoprotein assembly until after transport of the individual polypeptides to the protein body.

12.5.2 THE ROUTE OF INTRACELLULAR TRANSPORT

The route by which the newly-synthesized protein arrives at its final site of deposition will clearly depend on a number of factors and will probably be different for different species and for different storage proteins within the one species if only because sites of synthesis and deposition may vary. For instance, in maize and barley the globulins and glutelins, which make up approximately 50% of the total storage protein, are found in an unenclosed matrix around the protein bodies within which the prolamins are deposited. It is possible that the globulins and glutelins are not involved in any transmembrane movement.

For most storage proteins, including the prolamins of maize and barley, the weight of evidence suggests the ER is the site of synthesis and the protein body the site of deposition. The least complex situation in this category is that which applies for wheat, maize and barley where protein bodies are believed to arise directly from the ER (see 12.2.1.2). The only intracellular movement required here would be through the ER membrane and thence via the cisternal space to the sites within this space where ER swelling and protein accumulation is occurring.

The situation must be more complex for other species studied where it seems probable that most protein bodies are vacuolar in origin. Since there is no evidence of a unique association of ribosomes with the tonoplast, synthesis on RER is the most likely site and this raises the question of the intracellular route from this point to the site of deposition in the vacuolar protein bodies. There is no definitive evidence on this question and the little information that is available can only serve as a basis for conjecture. We will list here some of the possibilities. It seems likely that those storage proteins which are synthesized on the RER are first transported through the ER membrane into the cisternal space. This suggestion is supported by the fact that storage proteins synthesized by *in vitro* systems containing microsomal membranes are protected from the action of proteases (Cameron-Mills & Ingversen, 1978; Higgins & Spencer, 1981). Several possibilities have been suggested for the subsequent route. One possibility is that storage proteins are transported out of the ER within small vesicles derived from ER membrane. These vesicles could either coalesce to form protein bodies or fuse with the vacuolar membrane and thereby deposit storage protein within the vacuole. However, more recent electron microscopy of thick sections (see Section 12.2.1.1) leaves this model in some doubt. Another possibility that has been suggested is a transport route through the ER cisternal space into the dictyosomes and their associated vesicles. This is analogous to the situation in the acinar cells of the pancreas which secrete large amounts of hydrolytic enzymes made on the RER (Case, 1978). In the case of storage proteins, it is suggested that the dictyosome vesicles then travel to and fuse with the vacuole or protein body membrane releasing the storage protein within the organelle (Dieckert & Dieckert, 1976; Harris, 1979). The evidence in favour of this route is a moderate increase in the number of dictyosomes in developing cotyledon cells (Dieckert & Dieckert, 1976; Harris, 1979), although other reports specifically note that no such increases were seen (Bain & Mercer, 1977; Savelbergh & van Parijs, 1971; Craig *et al.*, 1979a). Other support comes from the observation that some dictyosome vesicles, and similar vesicles apparently fused to the tonoplast, have the same degree of electron density in stained electron microscope sections as the deposits within the vacuolar protein body. These deposits are removed by pre-treatment with proteolytic enzymes (Dieckert & Dieckert, 1976).

Clearly the resolution of these alternatives and a solution to the related problem of the origin of protein bodies depend on the application of more sensitive techniques for localizing small amounts of storage proteins within the cell. Until recently, all ultrastructural studies have lacked any positive identification of particular electron dense deposits as storage protein, and, as mentioned earlier, such evidence is quite rare even in light microscope studies. The best prospect for a more precise insight into all these questions at the present time seems to be the development of immunocytochemical methods that can be applied at the electron microscope level. Two such methods, using ferritin-labelled antibody (Baumgartner *et al.*, 1981) and gold-labelled antibody (Craig & Millerd, 1981) have recently been reported.

12.6 REGULATION OF STORAGE PROTEIN SYNTHESIS

Many aspects of the regulation of storage protein synthesis have been mentioned in previous sections of this chapter and will therefore be mentioned only briefly here. It is nonetheless worthwhile to consider the broad categories of known and potential factors which may regulate the synthesis of storage proteins.

12.6.1 GENETIC FACTORS

Genetic factors have a major effect on the amount and the composition of the storage protein fraction. Mutations, such as opaque-2 in maize and 1508 and Hiproly in barley, affect a regulatory locus which controls the level of the entire prolamin fraction. Although some differential effects may be seen on the component polypeptides of this fraction, the overriding effect is a suppression of prolamin synthesis. The individual globulins are clearly under similar genetic control in legumes. Within *P. sativum*, the relative amounts of legumin and vicilin can vary widely between lines in a genetically-determined fashion (Thomson & Doll, 1979).

At the level of defined holoproteins and subunits a significant amount of intraspecific variation for both globulins and prolamins has been described (see Section 12.3). These variations, both qualitative and quantitative, have been shown for a number of the globulins to be under genetic control and to be inherited in a co-dominant fashion. This implies the involvement of structural rather than regulatory loci. The molecular basis for these variants is not well understood, and neither is the basis of the microheterogeneity which generates the bewildering complexity of 46 related gliadins, 28 zeins, and 6 acidic and 6 basic subunits of the 12S globulins. In the case of the gliadins and the basic subunits of 12S globulins, sequence studies indicate a degree of heterogeneity which provides a reasonable basis for this microheterogeneity and supports the notion of a common ancestral structural gene. However, it should be borne in mind that variable degrees of post-translational modification (see Sections 12.4.6.2 and 12.4.6.3) may also contribute to this microheterogeneity.

Analysis of the number of genes for storage proteins in the haploid genome is still at an early stage. Estimates for the number of zein genes range from 1–5 (Pedersen *et al.*, 1980) up to 120 (Viotti *et al.*, 1979). It is still too early to assess the role of gene number in storage protein synthesis and accumulation. Analysis of a gene for bean phaseolin showed the presence of three intervening DNA sequences (Sun *et al.*, 1981),

12.6.2 ENVIRONMENTAL FACTORS

The effects of the plant's environment on the size and composition of the storage protein fraction are superimposed upon the genetically determined characteristics. The most striking examples of environmentally-induced modifications are those caused by variations in the level of the plant's mineral nutrient supply. Not surprisingly, the level of nitrogen fertilization has a marked effect on the total protein content of the seed up to a genetically pre-determined level in non-legumes while in legumes, the nitrogen available from symbiotic nitrogen fixation (see Chapter 16) seems sufficient to support seed protein synthesis to its full genetic potential (Johnston *et al.*, 1977).

Mineral nutrition of the plant also affects the composition as well as the size of the seed protein fraction. For instance, in barley the high molecular weight, low-lysine hordein components were preferentially increased (Koie *et al.*, 1976) at high fertilizer nitrogen levels, and at high levels of applied N, P and K in dwarf wheats there was a relatively greater increase in the storage proteins (prolamins and glutelins) than in albumins (Abrol *et al.*, 1971).

The effects of nutrient deficiencies on storage proteins can be strikingly selective. A deficiency of sulphur in peas causes a big reduction in legumin and a minor vicilin component accompanied by a relative increase in the major vicilin fraction. Both legumin and the minor vicilin are known to be relatively rich in sulphur-amino acids, unlike the major vicilin component (Randall *et al.*, 1979, Millerd *et al.*, 1979). Similar striking effects of sulphur deficiency have been observed in lupins (Blagrove *et al.*, 1976) and wheat (Wrigley *et al.*, 1980). This immediately suggests some form of translational control, possibly through the supply of sulphur-amino acids. Other effects of plant nutrition are not so readily interpreted. Deficiencies of phosphorus and potassium cause a significant increase in the relative amount of legumin and the same minor vicilin component in peas (Randall *et al.*, 1979).

Temperature of plant growth has also been reported to affect differentially the relative proportions of protein fractions in both wheat (Kolderup, 1975) and peas. In the latter there was a preferential increase in legumin in some genotypes at the optimum growth temperature (A. Millerd, personal communication).

12.6.3 HORMONE EFFECTS

No direct role of plant growth regulators in the synthesis of seed storage proteins has yet been reported. On the other hand, considering the general modulating effects that these compounds, as a group, have on various stages of plant development, it seems quite probable that they are involved in the regulation of storage protein synthesis. There are a number of reports of the presence of plant growth regulators, particularly gibberellins, cytokinins and abscisic acid, in developing seeds and changes in their levels during development. Cytokinins are present in developing seeds but may not be synthesized there (Hahn *et al.*, 1974; van Staden & Button, 1978). Skene (1970) found a correlation between the increase in one gibberellin and the cell expansion (protein accumulation) phase of cotyledon development in peas. Abscisic acid has been implicated in the development of seeds of cotton (Ihle & Dure, 1972) and *P. vulgaris* (Hsu, 1979) and a direct correlation has been found between abscisic acid levels and storage protein synthesis in *P. vulgaris* (I. M. Sussex and R. K. M. Dale, personal communication). Excised embryos continued to synthesize the vicilin-like storage protein for at least 14 days in culture. Addition of abscisic acid caused a three- to five-fold increase in the accumulation of this protein. Transfer to medium without abscisic acid resulted in a decreased rate of synthesis indicating that the hormone was not acting as a trigger for synthesis but was needed to be continuously present for the effect to be seen.

12.6.4 TRANSCRIPTIONAL AND POST-TRANSCRIPTIONAL CONTROL

Storage proteins are unique to the seed and the relevant genes must therefore only be expressed as cytoplasmic mRNAs during seed development, but how this specificity of expression is controlled remains a mystery, as indeed it is in most other eukaryotic systems. Davidson and Britten (1979) have identified three prevalence classes of mRNAs, namely those with 1–15, 15–300 and up to 10^5 copies per cell, respectively. They propose that structural genes for the first two classes are transcribed constitutively in all tissues of an organism and that events at the post-transcriptional rather than transcriptional level determine which nuclear RNAs will be processed and transported from the nucleus into the cytoplasm to code for proteins appropriate to a particular cell. They go on to propose a model for the regulation of these post-transcriptional events which involves the RNA transcripts of repetitive DNA sequences. The third, 'superprevalent' category of mRNAs implies large increases in transcription initiation rates during differentiation and they speculate that special, direct triggers may control this step for each structural gene. Just which prevalence class the storage protein mRNAs belong to is not known for certain, however, in view of the evidence summarized in Section 12.4.4.1, it seems likely that mRNAs for zein, hordein and storage proteins from *P. sativum* and *P. vulgaris* may be in the 'superprevalent' category.

As discussed earlier (Section 12.4.6.1), the timing of the synthesis of individual storage protein components during seed maturation can be quite complex and would presumably require an equally complex set of controls to operate in the nucleus at either the transcriptional or post-transcriptional level. The actual rate of synthesis of any one storage protein polypeptide must ultimately depend, to a large extent, on the level of mRNA available in the cytoplasm. This level would be, in turn, a function of the above nuclear events together with the rate of degradation of the mRNA in the cytoplasm. With more sensitive and more specific analytical probes becoming available the developing seed is a potentially promising system in which to study these regulatory mechanisms in higher plants.

12.7 CONCLUDING REMARKS

This chapter attempts to highlight what is not known about storage protein synthesis in developing seeds as well as summarizing what is understood. This field of research, which had been rather neglected by biochemists and others for many years, is at an exciting stage and one can confidently expect a lot of new information to come forth in the near future. For instance, the advent of recombinant DNA technology opens the way to answering many questions concerning the storage protein genes and their messenger RNAs. The role of plant growth regulators in storage protein synthesis is beginning to receive more attention and may provide another means for manipulating the storage protein content of seeds. The basis for the present apparent microheterogeneity of many storage protein fractions will be better understood when we have more information from sequence studies and other analyses of the extent and nature of the relatedness of the many similar components that can be found in a single protein fraction and of the post-translational events, chemical and structural, that are involved in the accumulation of storage proteins.

FURTHER READING

Briarty L. G. (1978) The mechanisms of protein body deposition in legumes and cereals. In *Plant Proteins* pp. 81–106. (ed. G. Norton). Butterworths, London.

Derbyshire E., Wright D. J. & Boulter D. (1976) Legumin and vicilin, storage proteins of legume seeds. *Phytochemistry*, **15**, 3–24.

Dure, L. S. (1975) Seed formation. *Ann. Rev. Plant Physiol.*, **26**, 259–78.

Kasarda D. D., Bernardin J. E. & Nimmo C. E. (1976) Wheat proteins. In *Advances in Cereal Science and Technology* pp. 158–236. (ed. Y. Pomeranz) American Association of Cereal Chemists Inc., St. Paul.

Millerd A. (1975) Biochemistry of legume seed proteins. *Ann. Rev. Plant Physiol.*, **26**, 53–72.

Spencer D. & Higgins T. J. V. (1981) Molecular aspects of seed protein biosynthesis. In *Commentaries in Plant Science*, Vol. 2 pp. 175–189 (ed. H. Smith) Pergamon Press, Oxford.

CHAPTER 13

NUCLEIC ACID AND PROTEIN SYNTHESIS DURING GERMINATION

13.1 INTRODUCTION

A familiar quotation states 'Tall oaks from little acorns grow', yet a more correct phrasing would recognize that the lofty oak actually develops from the even tinier seed embryonic axis. The ability to grow resides entirely in this miniaturized structure. At the same time, mobilization of reserve materials that constitute the rest of the seed, is required for growth to continue beyond the initial stage.

Germination of the embryonic axis proceeds in three phases (Fig. 13.1). In the first phase, water is rapidly absorbed and a number of metabolic activities are greatly increased. Subsequently, there is a second phase which lasts for several hours, and seems to be preparatory to growth. Rates of certain complex processes are steadily enhanced in this period, yet the axis gains only minimally in fresh weight. Growth is manifest during the third period, first by a sharp increase in cell expansion, and is later augmented by DNA synthesis and cell division.

The first part of this chapter deals with changes in nucleic acid and protein synthesis occurring in the embryonic axis during the phase that is preparatory to growth, on into the period when growth has commenced. A further major consideration is a determination of the extent to which particular changes in the metabolism of macromolecules are critical to the onset of growth. The second half of the chapter is concerned with the role of protein synthesis and activation in the mobilization of the food reserves from the supporting tissues of the seed. Not surprisingly, biochemical systems responsible for providing the flow of nutrients to the embryonic axes exhibit their greatest burst of development at precisely the time that the axes are growing most rapidly (frequently 2–4 days).

13.2 THE EMBRYONIC AXIS

13.2.1 RATES OF PROTEIN SYNTHESIS DURING EARLY GERMINATION

The seed axis contains all of the components required for protein synthesis including an extremely high level of ribosomes (Marcus & Feeley, 1964). The lack of polyribosomes in extracts of dry axes indicates, however, that protein synthesis does not occur appreciably until the axes are imbibed. Two processes

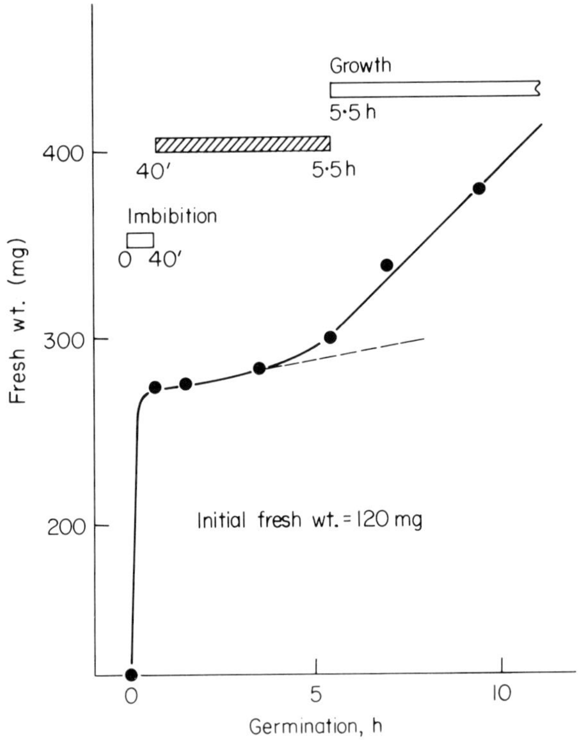

Fig. 13.1 Fresh weight changes in isolated wheat embryos during early germination. Preweighed embryo samples were imbibed for the times shown, blotted and weighed.

seem to be important for the rapid increase in protein synthesis: the attachment of ribosomes to mRNA (Spiegel & Marcus, 1975) and the formation of large amounts of ATP (Brooker *et al.*, 1977). Experiments with the inhibitors α-amanitin (which inhibits mRNA synthesis by RNA polymerase II) and cordycepin (which inhibits addition of poly(A)sequences to mRNA) demonstrated that the rapid initial increase in polyribosomes could occur in the absence of both mRNA transcription and polyadenylation. Thus the seed embryonic axis must contain a substantial level of functional preformed mRNAs (Spiegel & Marcus, 1975). Analyses of the rates of synthesis of new poly(A)RNA (Huang *et al.*, 1980) indicated further, that even in the absence of the inhibitors, protein synthesis must occur initially with the preformed mRNAs.

Are there changes in the rates of protein synthesis in the period subsequent to the initial intake? A number of studies with axes of lima bean (Klein *et al.*, 1971, Walton & Soofi, 1969), wheat (Marcus *et al.*, 1966), and rye (Sen *et al.*, 1975), have indicated that there is a steady rise in the rate of incorporation of exogenously supplied amino acid into protein throughout germination. A common feature of these studies is the use of carrier-free radioactive amino acids with occasional correction for the differences in rates of uptake of radioactivity. Not taken into account, however, is the possibility that different endogenous levels of amino acids might generate considerable variation in the specific activity of the amino acid precursor. A reasonable approach to circumvent this problem is to incubate embryos in a high level of exogenous radioactive amino acid such that the endogenous pool is rendered negligible (Hod & Hershko, 1976). Results from an experiment of this type are documented in Fig. 13.2 where a comparison between 5h and 1.5h embryos is made with respect to their incorporation of 8mM ^{14}C-leucine into protein. The endogenous level of leucine, about 20nmol per 75mg of embryos, varied little between 1 and 5h of germination. In the experiment shown, within 10min the 1.5h embryos contain 120nmol of leucine and the 5h embryos contain 220nmol. The radioactive amino acid thus exceeds the endogenous level by 6-fold and 11-fold respectively. Under these conditions, the incorporation of radioactivity into protein is linear with time, and is essentially independent of the level of radioactive leucine taken up by the embryos. The radioactive amino acid taken up by the embryos is, for the most part, in a pool accessible to protein synthesis, since transfer out of the labelling solution into water (after 10min) results in continued incorporation at a rate similar to that of embryos that remain in the radioactive precursor. Overall, it seems reasonable to conclude that the high exogenous concentration is providing a precursor pool of known specific activity, thus allowing an apparently reasonable analysis of protein synthesis rates throughout early germination. Data obtained with such an assay (Fig. 13.3) show a 5-fold linear increase between 1 and 5h, in essential agreement with the earlier studies.

Recent experiments, however, suggest a reconsideration of the interpretation of the *in vivo* incorporation data. Briefly summarized, the findings are as follows: (a) comparison of polyribosome profiles of wheat embryos germinated for 1.5 and 4.5h indicate that there is only a 1.4-fold greater content of polyribosomes in the 4.5h sample; (b) highly active cell-free extracts prepared from 4.5h embryos incorporate ^{14}C-leucine *in vitro* at a rate only 1.5–1.8fold greater than similar preparations from 1h embryos. In the presence of excess tobacco mosaic virus RNA (saturating the system for mRNA), both extracts have similar rates of incorporation; (c) relative transit time measurements (Palmiter, 1972) indicate a difference of only 1.3-fold between 1.5 and 5.5h embryos suggesting that there is only a minor difference in the rate of ribosome movement across the mRNA chain. These results are clearly at variance with the 5-fold rate increase indicated by the *in vivo* labelling studies.

In explanation, it appears that notwithstanding the large excess of endogenous radioactive leucine, the differences in the *in vivo* rates of incorporation between the embryo samples may be primarily due to differences in the specific activity of the *immediate* precursor to protein synthesis. Such a conclusion requires that there be at least two separate amino acid pools in the wheat embryos, and that the pool providing the precursors for protein synthesis be filled preferentially by endogenously formed, non-radioactive leucine (Fig. 13.4).

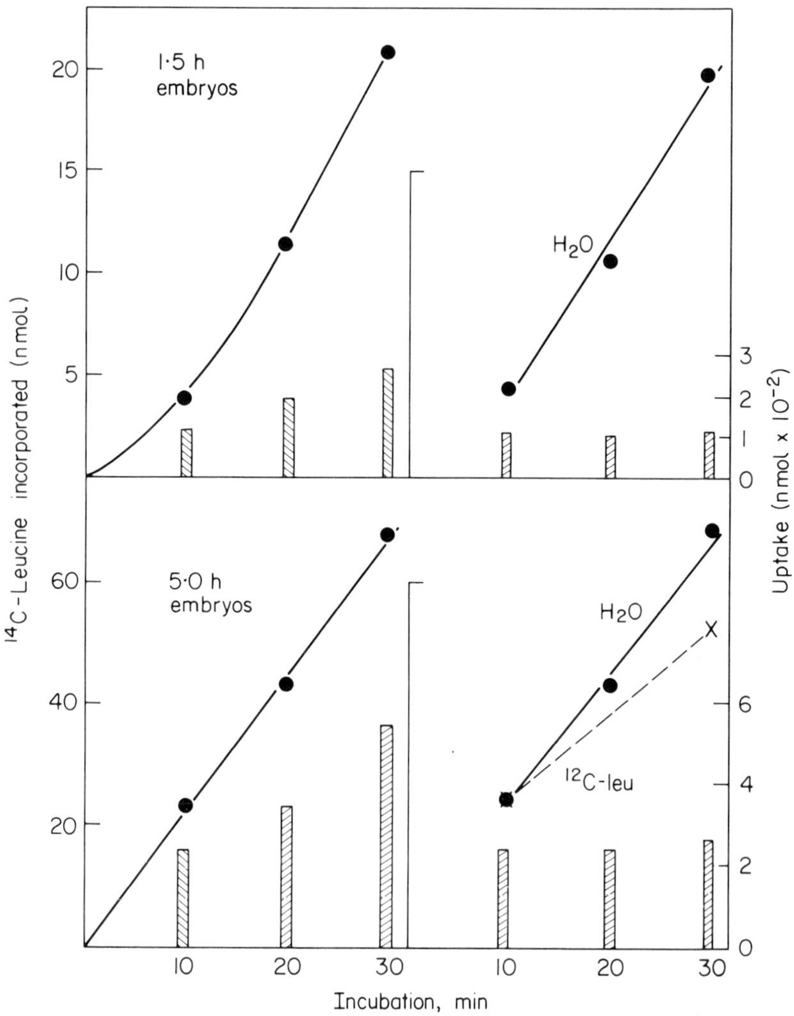

Fig. 13.2 Incorporation of ^{14}C-leucine is independent of uptake. Wheat embryos (75mg) were imbibed either for 1.5 or 5h and then incubated in 8mM ^{14}C-leucine (left hand panels), or were transferred to water or 8mM ^{12}C-leucine after a 10min incubation in ^{14}C-leucine (right hand panels). The uptake data (bars) are TCA-soluble radioactivity and the incorporation data (solid circles), TCA-insoluble radioactivity. In another experiment with 1.5h embryos, the ^{14}C-leucine in the incubation medium was raised to 12mM at the same specific radioactivity. The uptake into the embryos increased by 50% while the rate of incorporation into protein was similar to that of an 8mM incubation.

The 5h embryos either have a greater rate of flux of exogenous amino acid into the protein precursor pool or early embryos feed a greater quantity of non-radioactive amino acid into the precursor pool. Alternatively, the 5h embryos may have a protein precursor pool which can be more readily 'expanded' by exogenous amino acids. In any event, the net result is a higher specific activity of

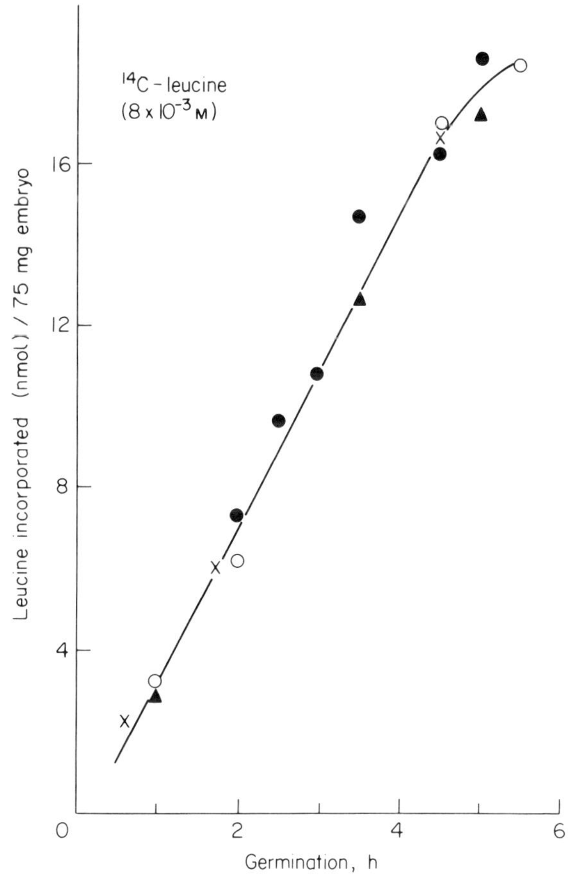

Fig. 13.3 Rate of incorporation of ^{14}C-leucine into wheat embryos at different times during early germination. Embryos were imbibed for different periods of time, incubated in ^{14}C-leucine for 12min, and TCA-insoluble radioactivity was determined as in Fig. 13.2.

the amino acid precursor in the 5h embryos. The explanation is clearly speculative. Yet clearly the conclusion that there is a substantial increase in the rate of protein synthesis during early germination of seed embryos, is far from certain. Furthermore, it seems not unreasonable that a change in the rate of flux into a specific precursor pool could be closely related to the mechanism(s) signalling the onset of growth in the embryo. This possibility makes it desirable to seek a thorough investigation of the 'expansion' phenomenon.

13.2.2 SYNTHESIS OF SPECIFIC PROTEINS

When protein synthesis is inhibited, growth of the embryonic axis is arrested (Walton & Soofi, 1969; Klein *et al.*, 1971; Brooker *et al.*, 1977). This suggests

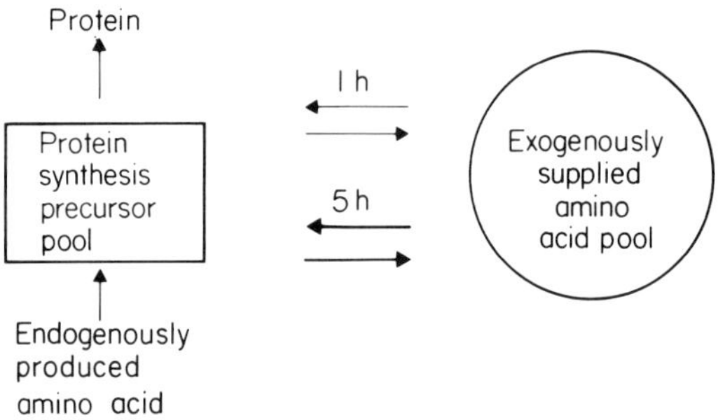

Fig. 13.4 Diagramatic representation of the protein precursor pool. The steady state level of free amino acid is maintained by a balance between endogenous production of the amino acid and its utilization for synthesis of new protein. Exogenously supplied amino acid forms a separate pool that exchanges (in a reversible manner) with the protein precursor pool. It is postulated that the rate of this exchange increases steadily during the quiescent period of germination.

either that the germinating axis must begin the synthesis of proteins specifically linked to growth, or that growth requires a general, constant renewal of proteins. Two approaches have been used to ascertain the extent to which germination is accompanied by changing patterns of protein synthesis. In the first approach, copy-DNA (cDNA) was prepared using reverse transcriptase and mRNA from polyribosomes of 45min embryos as template, and its ability to hybridize to mRNA from 45min and 5h embryos was analyzed. The results, showing a similar extent of hybridization, indicated that there is little change in nucleotide diversity between the mRNAs during the period prior to the start of the growth phase (Brooker *et al.*, 1978). Direct analyses of the pattern of proteins synthesized *in vivo* and *in vitro* (in response to mRNA isolated from embryos at different times of germination) have confirmed this conclusion. The array of protcins synthesized throughout the initial 16h of germination are generally similar in migration in SDS-PAG (Cuming & Lane, 1979; Caers *et al.* 1979). Notwithstanding this general conclusion, several recent studies have demonstrated convincingly that there are specific changes in the patterns of protein synthesis. The mRNA of ungerminated wheat embryo is enriched for sequences coding for ribosomal proteins (Cuming & Lane, 1979). In mung beans, the unimbibed axis contains an mRNA coding for a 12 000 dalton protein. The level of this mRNA drops rapidly during early germination (Carlier *et al*, 1980). In wheat embryos, there is a high-turnover protein whose rate of synthesis increases greatly after 6h, concomitant with growth of the axis (Fig. 13.5). Finally, in a series of detailed 2-dimensional analyses of *in vivo* and *in vitro* protein synthesis in wheat embryo, a number of proteins have been found that are synthesized only after 6h of germination (Thompson & Lane, 1980).

13.2.3 RATES OF RNA SYNTHESIS DURING EARLY GERMINATION

Cheung and Suhadolnik (1978) analyzed RNA synthesis in germinating wheat embryo and concluded that there was a significant increase in overall RNA

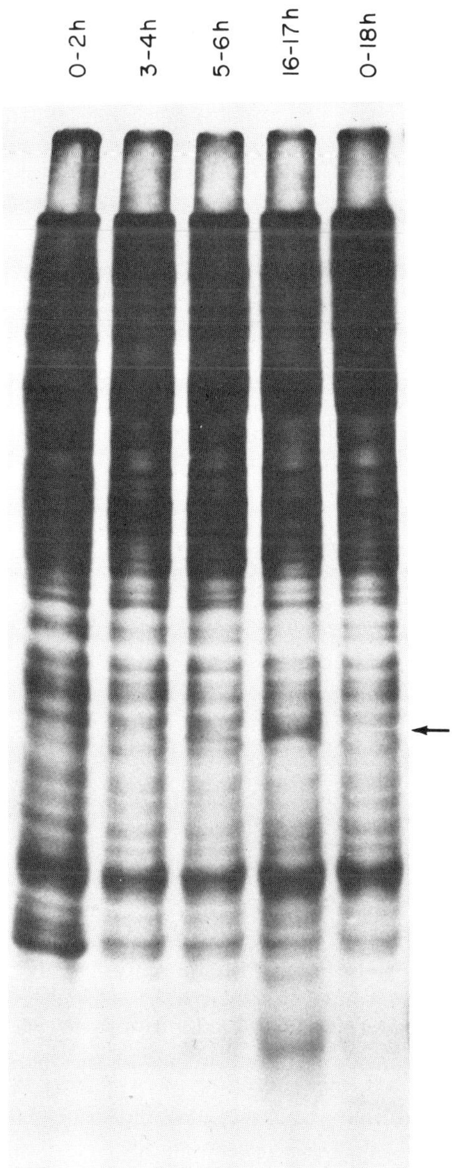

Fig. 13.5 Synthesis of a growth-associated protein by germinating wheat embryonic axes. Isolated embryos were incubated in H_2O then in 10μCi per μl of ^{35}S-methionine for the interval indicated above each lane. Tracer levels of methionine were used except in the last lane where 8mM was used continuously. This latter experiment shows that the growth-associated protein (mol.wt. 27 500) does not accumulate, relative to other labelled proteins. Proteins were separated on a gel of 15% polyacrylamide in the presence of SDS (Rodaway & Marcus, unpublished observations).

synthesis between 1 and 5h. Huang *et al.* (1980) and Aspart *et al.* (1980) studied this change in greater detail in axes of soybean and radish, in addition to wheat embryos. An important observation in these studies was the finding that the processing of the procursor to rRNA was accelerated substantially in the later stages of the second phase of germination. Figure 13.6 demonstrates this result in an experiment in which the radioactivity present in rRNA precursors and products was determined throughout an 8h incubation. During early germination, a large fraction of the radioactivity was retained in the 2.3, 1.4, and 0.9×10^6 dalton rRNA precursors. At a later stage in germination, the rRNA precursors turned over much more rapidly, and the radioactivity in the 1.3 and 0.7×10^6 dalton rRNAs reached 80% of the total rRNA (including the pre-rRNA) radioactivity within 2h. The mechanisms used by the embryos to regulate the rate of transcription and the rate of processing of RNA are unknown. The idea that there is amplification of the rRNA genes during germination (Chen & Osborne, 1970) has been shown to be incorrect, since the fraction of DNA capable of hybridizing with either 26S or 18S rRNA remains the same throughout early germination (Ingle & Sinclair, 1972). Two other ideas, that RNA synthesis rates are regulated by the level of UTP or by the capacity to methylate pre-rRNA, have also been rendered unlikely (Huang *et al.*, 1980). In soybean axes the increase in the cellular level of UTP substantially precedes the increase in the rate of RNA synthesis. The evidence with regard to methylation is indirect. Radioactive rRNA found in cytoplasmic ribosomes synthesized early in germination was shown to be methylated to the same extent as that found in ribosomes synthesized late in germination. In contrast, considerable accumulation of undermethylated ribosomal-rRNA can be demonstrated in model systems when methylation is inhibited (Caboche & Bacchelerie, 1977).

One possible mechanism for regulating transcription and/or processing of rRNA is the existence of a high turnover protein that either functions directly in rRNA synthesis or regulates it. This suggestion is based on the demonstration in radish axes that inhibition of protein synthesis leads to the rapid inhibition of rRNA synthesis, but has a much delayed effect on mRNA synthesis (Aspart *et al.*, 1980). An attractive candidate would be the enzyme(s) that process the rRNA precursors. Though characterized in prokaryotes, such activities have yet to be described in higher organisms.

13.2.4 PREFORMED mRNA VERSUS NEWLY SYNTHESIZED mRNA

The presence of conserved mRNA in the axes of dry seeds, initially observed in wheat embryos (Spiegel & Marcus, 1975), has now been established for radish (Delseny *et al.*, 1977), cotton (Hammett & Katterman, 1975) and rye (Peumans & Carlier, 1977). At the same time, it seems certain that substantial synthesis of both rRNA and mRNA begins immediately upon hydration (Sen *et al.*, 1975; Spiegel *et al.*, 1975; Payne, 1977). Preliminary analyses with wheat embryos (S. Rodaway, unpublished results) indicate that the turnover of the preformed

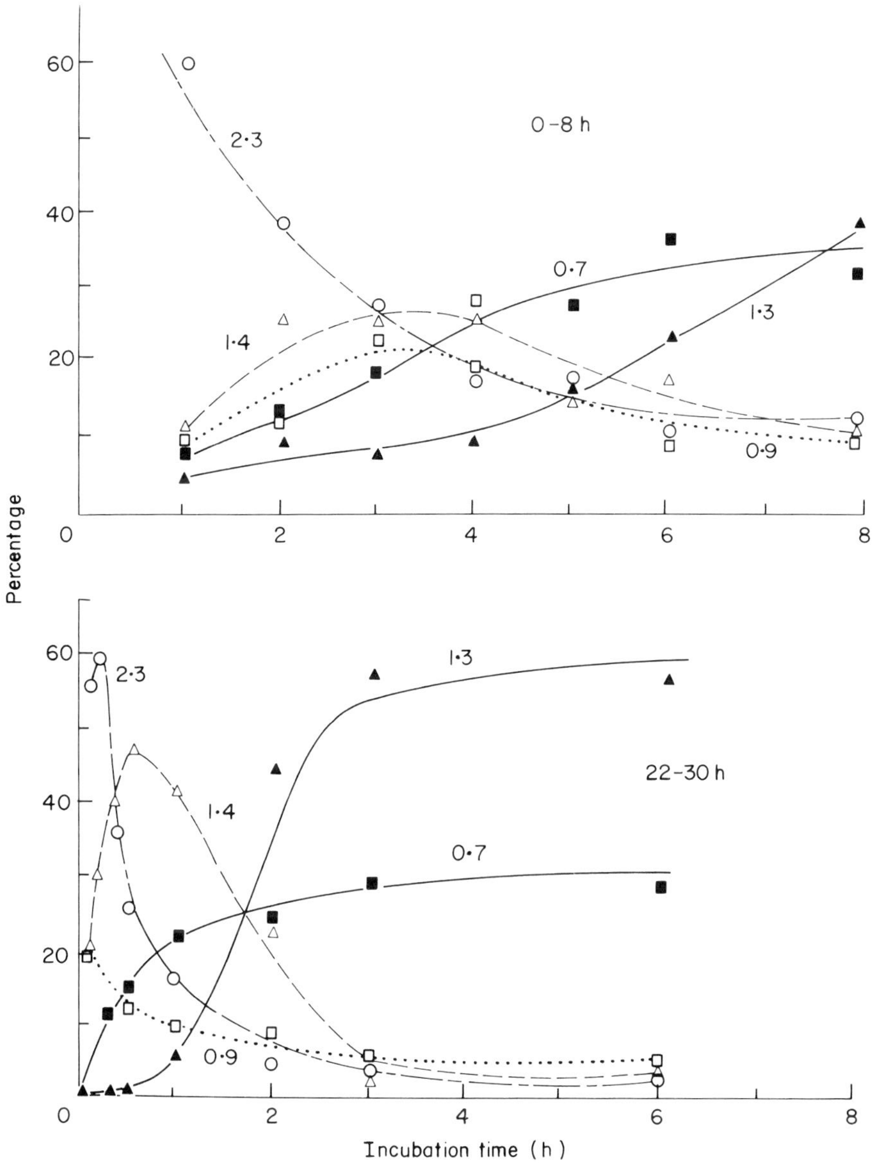

Fig. 13.6 Kinetics of accumulation of radioactivity into different RNA species. RNA synthesized by radish embryonic axes during the 0–8h period (upper panel) and by seedlings during the 22–28h period, were analyzed by electrophoresis and their percent distribution computed. Open circles are the 2.3×10^6 dalton pre-rRNA, the open squares and triangles are the 0.9 and 1.4×10^6 pre-rRNAs, respectively, while the closed squares and triangles are the 0.7 and 1.3×10^6 rRNAs (from Aspart *et al.*, 1980).

mRNAs and their replacement by new mRNAs is exceedingly rapid, about 90% before two hours. Since the bulk of the proteins encoded by the preformed and the newly synthesized mRNAs are similar, it is not clear why such a turnover should occur. Another intriguing phenomenon, described in greater detail below, is that the turnover of the preformed mRNAs seems to be closely linked to the synthesis of new mRNAs. The basis of this observation is that there is little change in the amount of poly(A)RNA when synthesis of new mRNA is substantially curtailed. Mechanisms to accomplish such a coupling of synthesis and turnover are completely unknown.

13.2.5 IS RNA SYNTHESIS NECESSARY FOR EARLY EMBRYO GROWTH?

The ability to resolve high molecular weight RNA into fractions containing predominantly mRNA and rRNA allows one to ascertain the ability of inhibitors to affect the specific synthesis of these classes of RNA and to correlate these results with the effects of the inhibitors on growth. Table 13.1 presents data with three such inhibitors, 5-fluorouridine (FUrd), cordycepin (3′deoxy-adenosine), and 6-azauridine (Datta & Rodaway, unpublished observations). 5-Fluorouridine strongly inhibits synthesis of cytoplasmic rRNA with little effect on the synthesis of poly(A)RNA. At the same time, both growth and protein synthesis are essentially unaffected by the FUrd treatment. Thus the

Table 13.1 Effect of 5-fluorouridine, cordycepin and 6-azauridine on germinating wheat embryos

Inhibitor	Conc	Growth	Protein synthesis	rRNA synthesis	mRNA synthesis
		% of control			
5-Fluorouridine	2mM	100	90	30	95
1.0h–7.5h	4mM	95	85		
Cordycepin	75μM	97	92	42	55
40 min–7.5h	150μM	78	81	31	45
	500μM	32	42	17	19
6-Azauridine					
0 Time–1.5h	5mM	26	33	35	71
40 min–2.5h	7.5mM	32	40	65	91

Embryos were incubated with the analogues and analyzed at the time periods indicated below. In the case of 6-azauridine, the embryos were transferred to water either at 1.5 or 2.5h as indicated. The analyses for protein synthesis were as in Fig. 13.3 and were carried out at 5.5h in the azauridine and fluorouridine series, and at 6h in the cordycepin experiments. The RNA synthesis experiments analyzed radioactivity in mRNA [i.e. poly(A)RNA] and RNA [i.e. RNA lacking poly(A) sequences] (Huang *et al.*, 1980) after a 40min labelling period. The midpoint of the RNA labelling period was 4.5h in the azauridine and fluorouridine series, and 5.0h in the cordycepin experiments. Growth was determined as in Fig. 13.1 after 7.5h.

FUrd results establish that embryo growth does not require rRNA synthesis. The results with cordycepin are less clear-cut. With cordycepin, low levels cause considerable inhibition of mRNA synthesis with little effect on embryo growth, suggesting that new mRNA may not be a requirement for early growth. High levels of cordycepin (500μM) substantially inhibit growth, but whether this is due to the further inhibition of mRNA synthesis, or to another effect of the analogue (Kredich, 1980) is uncertain. In any event, the absolute level of poly(A)RNA is only slightly affected by cordycepin, being the same in the 150μM cordycepin sample as in the control and decreasing by only 22% in the 500μM cordycepin-treated embryos. It seems therefore that if the growth inhibition at 500μM cordycepin is due to its stronger action on mRNA synthesis, the effect must be more subtle than a simple deprivation of mRNA. The third set of experiments, those with azauridine, represent an inhibitor functioning in a novel way at the level of RNA. This compound, generally used to reduce the level of cellular UTP (Chen & Jones, 1979), is metabolized in wheat embryo to the 5′-triphosphate and is incorporated into RNA (Rodaway & Marcus, 1980). Under appropriate conditions, it strongly inhibits growth, while having a negligible effect on mRNA synthesis. While its mechanism of inhibition has not been elucidated, the experiments of Table 13.2 suggest strongly that the action of azauridine is a

Table 13.2 Reversal of azauridine inhibition by cordycepin

	Conditions	Growth	Protein synthesis
		% of control	
(1)	10mM azauridine, 40min–2.5h	38	48
	10mM + 100μM cordycepin		70
	10mM + 200μM cordycepin		81
(2)	7.5mM azauridine, 40min–2.5h	41	38
	7.5mM + 200μM cordycepin	58	94

Embryos were imbibed in water for 40min at 25° and transferred either to water (control), to azauridine, or to a mixture of azauridine and cordycepin. At 2.5h all samples were transferred to water. Protein synthesis was analyzed at 5h as in Fig. 13.3 and growth was determined at 7.5h as in Fig. 13.1.

consequence of the synthesis of an azauridyl-RNA. When cordycepin is added simultaneously with the azauridine, it prevents to a considerable extent the azauridine inhibition of growth and protein synthesis. This result suggests that azauridine is being incorporated into a nonfunctional RNA that either replaces or competes with an RNA that functions in protein synthesis. The first of these ideas is reminiscent of the regulation of mRNA turnover. That is, when cordycepin represses RNA synthesis (thereby preventing the azauridine inhibition), it may stabilize an RNA that would otherwise be degraded and replaced by the azauridyl RNA.

Taken together, these data indicate that early embryo growth can occur despite substantial inhibition of the synthesis of new mRNA and almost complete prevention of rRNA synthesis. Nevertheless, a key species of RNA does seem to be required for the embryonic axis to progress normally into the growth phase.

13.2.6 DNA SYNTHESIS

Synthesis of DNA, as noted earlier, only occurs some considerable time after the axis commences growth. Once DNA synthesis starts, its rate increases rapidly over a short time period (Mory *et al.*, 1972). In the ungerminated wheat embryo there is a high level of DNA polymerase activity (Castroviejo *et al.*, 1975), suggesting that the formation of this enzyme may not be the rate-limiting step (see, however, Robinson & Bryant, 1975). An interesting possibility is that the level of the deoxynucleotide triphosphates may regulate DNA synthesis. Seeds of some species already have a 4C level of DNA, and in these seeds mitosis occurs before there is a large increase in DNA synthesis. It seems that in the formation of such seeds, the axes have arrested their development at the G2 phase. This phenomenon is, however, an infrequent event (see Bewley & Black, 1978).

13.3 MOBILIZATION OF RESERVES

The increased metabolic demand of the germinating embryonic axis necessitates fluid access to substrates important in maintaining the growing state. Since the seed is a compact reproductive unit, it has evolved to store food reserves in virtually every tissue of which it is composed. Frequently these reserves are sequestered within special organelles such as protein bodies, lipid storage organelles, starch grains, and the cell wall (see Chapter 12).

13.3.1 CONTROL MECHANISMS OF ENZYME SYNTHESIS AND ACTIVATION

Before it begins to elongate, the embryonic axis appears to rely predominantly on reserves stored within the axis itself. The growing axis, on the other hand, utilizes primarily the food reserves stored in adjacent tissues. These must be converted by a series of hydrolytic cleavages and metabolic events into compounds which can be transported through the seed and taken up by the axis as it grows. In most instances, seed tissue external to the axis must develop new enzyme activities in order to accomplish the mobilization of these reserves. Investigations on the biochemistry of germination have therefore provided opportunities for studying control mechanisms leading to synthesis of new proteins *de novo* and to the activation of existing proteins, as well as for analyzing the enzymology of utilization and mobilization of seed storage products.

The post-imbibitional development of tissues of the endosperm and cotyledons occurs coincidentally with the initiation of seedling growth. Although considerable biochemical and cytological changes in adjacent tissues take place even in the absence of the embryonic axis, the presence of the axis may be required for a 'normal' developmental sequence (Bewley & Black, 1978). This is probably because the embryonic axis acts both as a sink for transported products and sometimes as a source of diffusible factors, such as hormones, which influence the biochemistry of nearby seed tissues.

The initiation of reserve breakdown requires a transformation of cells which are merely hydrated, into cells capable of synthesizing the enzymes that are rate-limiting in reserve hydrolysis. There are thus two sets of proteins involved, the hydrolases themselves, and proteins required for cellular changes that lead to synthesis of the hydrolases. Within this latter group are proteins required for membrane proliferation and the development of new organelles for activating transcription or translation of RNA, or for regulating the activities of constitutive metabolic pathways. Usually one observes an elevation in the level of a particular protein, this being brought about by one of several mechanisms, including the activation of a zymogen, the removal of a negative control factor, or *de novo* synthesis.

A number of enzymes arise *de novo*, during germination or early seedling growth. That an enzyme does not pre-exist (except perhaps at a very low level) can be proven by meeting several criteria. One test resides in the ability to shift the average density of an enzyme population to a new value by the incorporation of ^{18}O or deuterium into these proteins following imbibition. For example, when storage tissue is incubated in H_2O^{18}, amino acids generated from storage protein hydrolysis incorporate ^{18}O at the carboxyl position. When these 'heavy' amino acids are subsequently used for *de novo* protein synthesis, some carbonyl groups in the protein remain tagged by ^{18}O, and a population of proteins of greater than usual density is generated. Similarly, deuterated water (^2H_2O) can be used *in vivo* to deuterate the α-carbons of free amino acids. Once the amino acid is incorporated into protein, the deuterium is metabolically stable in this position, and the density of the protein is increased. Some instances of *de novo* enzyme synthesis are listed in Table 13.3.

Another method for showing that a protein has been synthesized *de novo* involves the incorporation of radioactive amino acids into all portions of the protein in question. Thus, Varner and Chandra (1964) provided indirect evidence that most of the α-amylase formed during treatment with gibberellic acid had been synthesized *de novo* since ^{14}C-threonine was incorporated into 25 out of 31 tryptic peptides.

Density labelling can also be used to test whether accumulation of an inactive precursor precedes an eventual increase in enzyme activity. In the case of

Table 13.3 Enzymes synthesized *de novo* as shown by density labelling

Enzyme	Tissue	Source	Method	Reference
Isocitrate lyase	Cotyledons	*Citrullis vulgaris*	^2H_2O	Hock, 1970
Endopeptidase	Cotyledons	Mung bean	^2H_2O	Chrispeels *et al.*, 1976
Phosphatase	Cotyledons	Pea	^2H_2O	Johnson *et al.*, 1973
Sucrose-6-phosphate synthetase	Scutellum	Rice	^2H_2O	Nomura & Akazawa, 1973
Isocitrate lyase, Malate synthetase	Cotyledons	Peanut	^2H_2O	Longo, 1968
α-Amylase	Aleurone + GA	Barley	O^{18}	Filner & Varner, 1967
			^2H_2O	Hardie, 1975
Protease	Aleurone + GA	Barley	O^{18}	Jacobsen & Varner, 1967
β-1,3-Glucanase, RNase	Aleurone, no GA	Barley	^2H_2O	Bennett & Chrispeels, 1972
Limit dextranase	Aleurone + GA	Barley	^2H_2O	Hardie, 1975
α-Glucosidase	Aleurone + GA	Barley	^2H_2O	Hardie, 1975
Peroxidases	Embryo	Barley	^2H_2O	Anstine *et al.*, 1970

induction of α-amylase synthesis in barley aleurone tissue, most of the enzyme activity appeared between 12 and 24h after hormone addition. To ascertain whether a pro-enzyme was synthesized before 12h and activated subsequently, a different density labelling experiment was devised. As expected, when normal ^{12}C-amino acids were given from 0–12h and heavy ^{13}C-amino acids from 12–24h, the α-amylase was of greater density than if ^{12}C was used continuously (control). If, instead, ^{13}C-amino acids were given from 0–12h and ^{12}C-amino acids from 12–24h, the density was like that of the ^{12}C-control. This proved that no pro-amylase had been formed during the 12h lag period, then later activated (Ho and Varner, 1978).

Given that certain key enzymes are indeed formed *de novo*, how are these syntheses regulated? One possibility is that certain genes are more actively transcribed directly following imbibition or after hormone treatment. Some support for this suggestion comes from the finding that the synthesis of a major non-histone chromosomal protein (mol.wt. 37 000) begins between 24 and 48h of germination of winter wheat embryos (Sugita *et al.*, 1979), and that phosphorylation of six nucleoproteins is specific to a short period following imbibition (2–5h) of isolated barley embryos (Trewavas, 1979d). On the other hand, the initial transcript could pre-exist, but require either additional processing or the development of regulatory protein factors to enhance the translation of a stored mRNA.

Since dry seeds (Hammett & Katterman, 1975; Ferrer *et al.*, 1979), and embryonic axes (Brooker *et al.*, 1978) contain large amounts of pre-formed mRNA that are translatable *in vitro*, it has been speculated that there are specific classes of 'germination' mRNA that are synthesized during seed maturation, held over as stored mRNA, then activated shortly after re-hydration of the dry seed. This 'stored' mRNA is sometimes distinguished from 'residual' mRNA in that the latter, though carried over into the dry seed, is thought to have no special function in germination. Dure and co-workers (see Dure, 1979) have suggested that the carboxypeptidase C of cotton cotyledons is such a stored mRNA, and that its activation involves polyadenylation during the first day of imbibition. This conclusion has been reached by using inhibitors such as actinomycin D, 3′-deoxyadenosine, and 3′-deoxycytidine. In addition, labelling studies with ^{32}P and ^{3}H-adenosine showed significantly more label in the poly(A) end of the mRNAs, relative to that in the coding portions after one day of germination (Harris & Dure, 1978). Direct probing of the mRNA population present in the developing seed with a cDNA prepared against the carboxypeptidase mRNA would help to substantiate this conclusion and allay various criticisms (Payne, 1976; Bewley & Black, 1978).

When isolated barley aleurone layers are treated with gibberellic acid after being exposed to H_2O for two or three days, they begin to synthesize α-amylase, with a lag time as short as 2h. If poly(A)RNA is extracted from these tissues and translated *in vitro*, α-amylase (detected by an antibody) begins to appear among the translation products in close relation to its rate of synthesis *in vivo* (Higgins *et al.*, 1976). Thus the GA_3-induced synthesis of α-amylase appears to provide a case of gene activation during seed germination. Alternative explanations present themselves, however. The α-amylase mRNA may be synthesized during seed maturation or during the period of H_2O-incubation, but become more competitive in translation after GA_3 is added. Another possibility is that GA_3 may slow the turn-over of α-amylase message, thus increasing its steady state level. With the availability of purified α-amylase mRNA (Mozer, 1980a) it should be possible to prepare α-amylase cDNA and to determine whether synthesis of the initial α-amylase transcript preceeds or follows the addition of hormone.

13.3.2 SITES OF RESERVE HYDROLYSIS

Protein reserves are frequently, though not always, stored in discrete organelles called protein bodies (see Chapter 12). The dense proteinaceous matrix within the organelle often surrounds inclusions such as the phytin globoid, carbohydrate and protein-carbohydrate globoids, or crystalloids composed of a single protein component. Zein, a major storage protein found in maize protein bodies, seems to be synthesized on ribosomes of the rough endoplasmic reticulum (RER). These areas of the ER, which seem to expand as protein products accumulate, probably give rise to the protein body (Larkins *et al.*, 1976a; Burr & Burr, 1976 and Chapter 12). Also deposited in association with the protein bodies are enzymes capable of hydrolyzing the storage products within. Protein bodies of unimbibed cotton contain phosphatase (Dure & Waters, 1965) and those of 36h germinated mung bean cotyledons contain proteases, α-mannosidase and carboxypeptidase (Table 13.4). Usually the hydrolytic activity of these enzymes is not manifest until two or three days after imbibition, despite the close association of enzyme and substrate. The enzymes may be spatially separated from their substrates or, more likely, the synthesis of a key enzyme may be necessary to initiate the hydrolysis, and render more sites available to the hydrolases stored in the dry seed. In the case of mung bean, the limiting enzyme for proteolysis may be vicilin peptidohydrolase. After 36h of incubation, protein bodies of the cotyledons already contain several acid hydrolases (van der Wilden *et al.*, 1980), yet the digestion of the protein body contents appears only to follow the synthesis of this particular enzyme. As has been shown by immunofluorescence, this synthesis begins about three days after imbibition (Baumgartner *et al.*, 1978). Once protein body reserves are fully mobilized, the protein body may take on a lysosomal role and degrade other organelles, thus hastening the senescence of the cotyledon cells. Protein bodies of barley aleurone appear to fuse following digestion of their contents, leading to formation of a large central vacuole (Jones & Price, 1970).

Wheat glutens are examples of proteins that are stored as a vast matrix rather than within a protein body. Gluten degradation is again controlled by the synthesis of of endoproteases; for although some proteolytic activity pre-

Table 13.4 Enzyme activities in protein bodies from 36h mung bean cotyledons

Enzyme	% of Total protoplast activity
α-mannosidase	78.4
Carboxypeptidase	82.1
N-Acetyl-β-glucosaminidase	75.3
Acid phosphatase	78.7
Phosphodiesterase	77.1
Ribonuclease	75.5
Phospholipase D	78.5
Leucine aminopeptidase	4.7
β-Amylase	< 0.1
NADH-cytochrome C reductase	< 0.1

Protein bodies were isolated from protoplasts by fractionation on a 5/25% discontinuous Ficoll gradient (from van der Wilden *et al.*, 1980).

exists in the endosperm, the activity of the endoproteinases increases 8-fold during the period of gluten hydrolysis (Preston & Kruger, 1979). If cycloheximide is added during steeping, the level of pre-existing endoprotease falls, suggesting either that these proteases are constantly turning over during germination, or that the stored proteases are especially unstable. In pea cotyledons, a secondary rise in protease activity at day five appears to result from enzyme activation rather than *de novo* synthesis (Yomo & Srinivasan, 1973). Certain proteases in the pea radicle during the first day of germination differ from those present during the fourth day (Murray *et al.*, 1979), and it has been suggested that this change may bear some relation to the disappearance of two peptides (mol.wt. 13 000 and 84 000) following day one (Murray, 1979). Thus what is often taken to be a general increase in total proteolytic activity may in reality be due to specific changes in individual, perhaps less abundant, proteases.

What are the end products of protein mobilization? In theory, either oligopeptides or amino acids might be the final degradative products, but the actual species that are transported out of the storage cells have yet to be identified. In the case of barley aleurone, both large peptides and proteases are released into the starchy endosperm where hydrolysis is presumably completed (Melchers & Varner, 1971).

Uptake by the scutellum, through which endosperm products enter the monocot embryonic axis, is somewhat better described. After imbibition, the scutellum of the barley grain develops a dipeptide transport system, the capacity of which peaks at three days and then declines. This system, which develops exclusively in the scutellum, is not influenced by the presence of other seed tissues (Sopanen, 1979). By way of contrast, leucine uptake in isolated scutella, which can develop an increase of up to 20-fold in carrying capacity, is moderated by the presence of the endosperm (Sopanen *et al.*, 1980). At least two systems may operate in proline uptake. Proline can be taken up either as a dipeptide (pro-X or X-pro) or as free proline produced by the combined activities of two carboxypeptidases plus acid protease in the starchy endosperm (Mikola & Mikola, 1980).

Carbohydrates are stored in starch grains, in cell walls, and sometimes as minor components of protein bodies. In cereals and many economically important legumes, starch is the major food reserve of the seed. Starch grains may occlude the ruptured cells of cereal 'starchy' endosperm or may appear as smaller grains in cotyledons when endosperm is reduced or lacking. Since starch grains grow by apposition, enzymes utilizing ADP-glucose or starch phosphorylase are found tightly bound to and often embedded (by entrapment) within the starch grains (Gunning & Steer, 1975), but such an association may have no role in germination.

In barley, the β-amylases of the starchy (non-living) endosperm are stored as zymogens. They can be activated artificially by treating the gluten fraction of dry seeds with sulfhydryl reagents or protease (Rowsell & Goad, 1962), although these enzymes are normally activated by proteases released from the aleurone cells. In the rice grain, β-amylases first arise *de novo* in the scutellum (as shown by immunoprecipitation). Later on, a much greater increase in β-amylase activity develops in the endosperm by zymogen activation (Okamoto & Akazawa, 1980). Another apparent zymogen is the amylopectin-1,6-glucosidase of pea cotyledons (Mayer & Shain, 1968). Sometimes carbohydrases are associated with the protein body. Two examples are β-amylase and α-glucosidase of sorghum (Adams *et al.*, 1975).

As indicated in Table 13.3, several carbohydrases are formed *de novo*. Synthesis of α-amylase by barley aleurone is of particular interest, first because it is probably the key enzyme in starch mobilization and secondly because at certain times it constitutes about 50% of the protein being synthesized by aleurone layers. As discussed earlier, this synthesis is under the control of gibberellic acid (Higgins *et al.*, 1976). α-Amylase may also be synthesized by the embryonic axis of pea, where the presence of the cotyledon exerts an influence on starch metabolism by the axis (Davis, 1979). When the cotyledons are present, starch accumulates in the axis and axis α-amylase is low. But when cotyledons are removed, the starch disappears and the α-amylase content of the axis increases 10-fold by day two (Fig. 13.7).

Cell walls also contribute carbohydrate to the young seedling. This is especially the case when cell walls are rich in hemicelluloses such as the galactomannans (Reid, 1971) and mannans (Keusch, 1968). In *Trigonella foenum-graecum* where 30% of the seed reserves are the galactomannans of the endosperm, this hemicellulose appears to have a dual role: it acts both as a reserve polysaccharide and it traps sufficient water to prevent desiccation during temporary drought stress (Reid & Bewley, 1979). Aleurone cell walls,

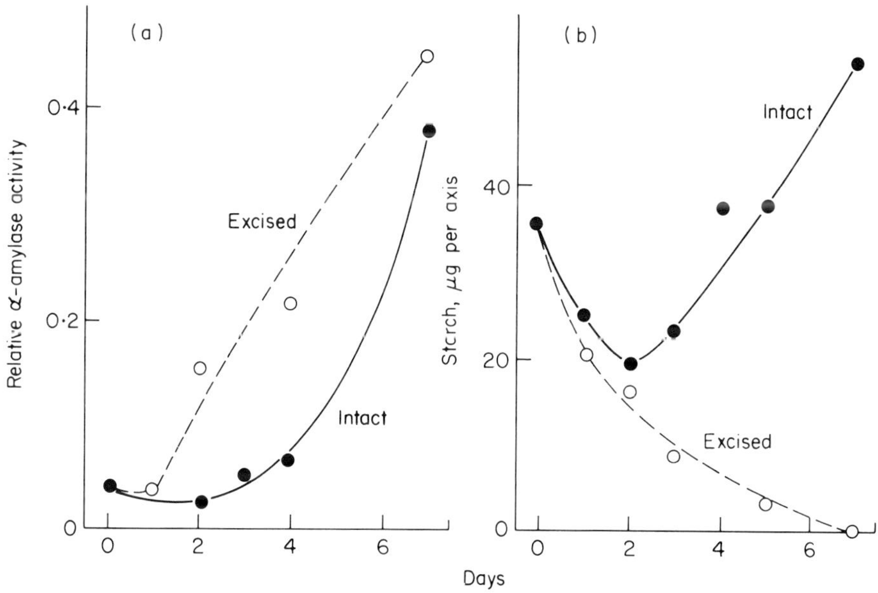

Fig. 13.7 Influence of cotyledons on α-amylase activity and starch levels in pea embryonic axes. Embryonic axes were analyzed from intact seeds (solid line) or as isolated axes cultured in the absence of an organic carbon source (broken line) (data redrawn from Davis, 1979).

which in barley are 85% arabinoxylan (McNeil *et al.*, 1975), are degraded by pentosanases (Taiz & Honigman, 1976), on the one hand to provide channels through which secreted enzymes can be released, but perhaps on the other to provide arabinose and xylose for axis growth.

Various oligosaccharides can act as a more immediately accessible store of monosaccharide units. In grains of fenugreek, 80% of the verbascose and stachyose are metabolized within 48h (Sioufi *et al.*, 1970). Similarly, whole soybean seeds, which initially contain 3% stachyose, 0.4% raffinose and 5% sucrose, have almost completely degraded these compounds with 4 days (East *et al.*, 1972). The embryonic axis of soybean is depleted even more rapidly since all of the reserve oligosaccharide is catabolized by 15h (Fig. 13.8). A difference in the maltose content of mature corn endosperm of one variety may explain its longer period of edibility after harvest (Ferguson *et al.*, 1979), probably by the function of maltose as an osmoticum.

Lipid mobilization is unique in the sense that it frequently requires the *de novo* biogenesis of a new organelle, the glyoxysome, to manage the heavy demand for gluconeogenesis in oil-storing seeds. Triglyceride breakdown in castor bean endosperm originates within the spherosome (lipid storage body) at a time well past the onset of growth of the germinated embryonic axis, and is thus, like starch and protein degradation, perhaps more important for growth of

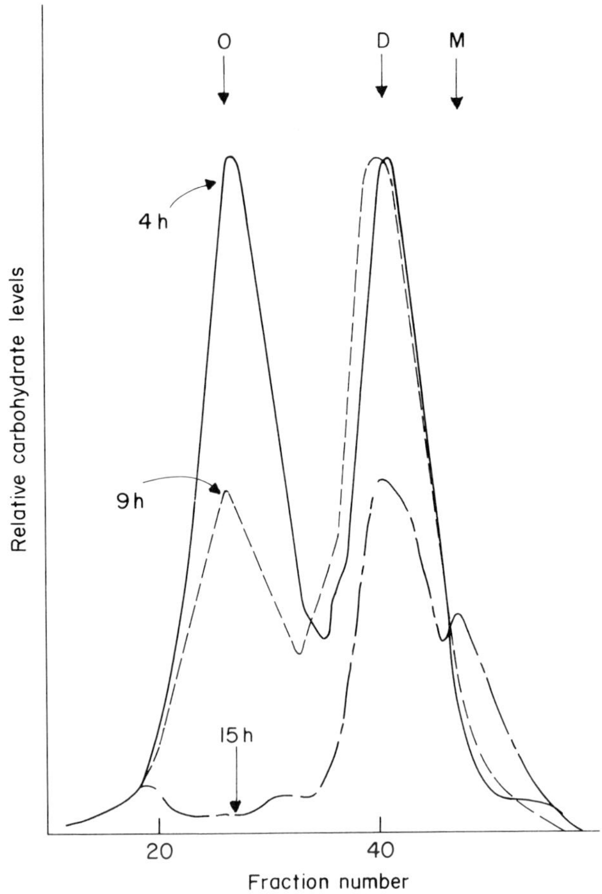

Fig. 13.8 Mobilization of oligosaccharides in germinating soybean embryonic axes. Trichloroacetic acid-soluble oligosaccharides (O, primarily stachyose), disaccharide (D, primarily sucrose), and monosaccharides (M) were separated by size on BioGel P2, and were quantitated by the phenol–sulfuric acid reaction. Isolated embryonic axes were germinated for 4, 9 or 15h before analysis (Rodaway S. & Sobel N., unpublished results).

the seedling than for germination *per se*. An acid lipase activity appears early on, and later, alkaline lipases are in evidence (Muto & Beevers, 1974). The alkaline lipases are most responsible for generating monoglycerides and their increased activity parallels the onset of lipolysis. The association of alkaline triglyceride lipase with 'spherosomal appendices' in rapeseed cotyledons (Rosnitschek & Theimer, 1980) suggests that vesicles carrying the lipase might fuse with spherosomes as a prelude to degradation. Fatty acids are oxidized by the glyoxysome to acetyl CoA and succinate, then the succinate is metabolized to malate and oxaloacetate by the mitochondrion. These products are substrates for cytoplasmic synthesis of sucrose, which can be transported to other tissues. In castor bean endosperm new lipids are also synthesized, and contribute to developing new membranes. ATP citrate lyase, found predominantly in proplastids, is instrumental in this process (Fritsch & Beevers, 1979), utilizing glyoxysomal citrate as a source of acetyl CoA for *de novo* fatty acid synthesis in the plastid (Rapp & Randall, 1980). This enzyme, like several others in fatty acid metabolism, has initially low activity which rises considerably after the third day of imbibition. A recently proposed pathway suggests how both sucrose and fatty acids are generated (Fig. 13.9).

The key enzyme in the glyoxysomal pathway is isocitrate lyase, which is synthesized *de novo* in a number of species (Beevers, 1979). The proportion of malate synthetase that is formed *de novo*, however, may vary considerably. For example, malate synthetase from cotton is stored in the dry seed, while in castor bean both malate synthetase and glyoxysomal malate dehydrogenase are formed *de novo* following imbibition (Miernyk *et al.*, 1979). There is a strong increase in the biosynthesis of poly(A)RNA in castor bean endosperm after one day of imbibition and, when glyoxysome production is maximal (three to five days) 15% of the imunoprecipitated products of *in vitro* translation are glyoxysomal in character (Table 13.5). Thus considerable new mRNA synthesis may be directed toward glyoxysome biogenesis. At least four glyoxysomal enzymes require some sort of post-translational modification involving removal of a peptide and/or glycosylation (Riezman *et al.*, 1980). The GDP-man mannosyl transferase (a glycosylation enzyme) of both pea (Nagahashi & Beevers, 1968) and castor bean (Marriott & Tanner, 1979) exhibits an increase of up to a 50-fold (castor bean) in activity, which peaks at the same time as maximum glyoxysome formation. Glyoxysome biogenesis may begin with formation of a pre-glyoxysomal region of the endoplasmic reticulum. Bowden and Lord (1977) have shown that antibody to glyoxysomal proteins will also precipitate peptides that initially are found in a microsome fraction and later appear in mature glyoxysomes.

How important is the glyoxysomal activity to germination *per se*? An analysis of the effect of itaconate, a specific inhibitor of isocitrate lyase, on the germination of numerous species, has shown that seeds which synthesize isocitrate lyase fail to germinate when itaconate is added (Kahn & McFadden, 1979). Wheat seed, which normally has little if any isocitrate lyase activity, is able to germinate normally in its presence. This suggests that seeds which normally exhibit increased isocitrate lyase activity require it for germination, and that metabolism of triglycerides within the axes of many species may be required for the initiation of growth.

The endosplasmic reticulum also undergoes dynamic changes following imbibition. In castor bean endosperm, net synthesis of phospholipid accompanies early seedling growth (Lord, 1978). A different situation obtains in mung bean cotyledons where there is a net decline in endoplasmic reticulum even though synthesis is taking place (Gilkes & Chrispeels, 1980). The greatest change is in the disappearance of the tubular form of ER, leaving behind a cisternal form

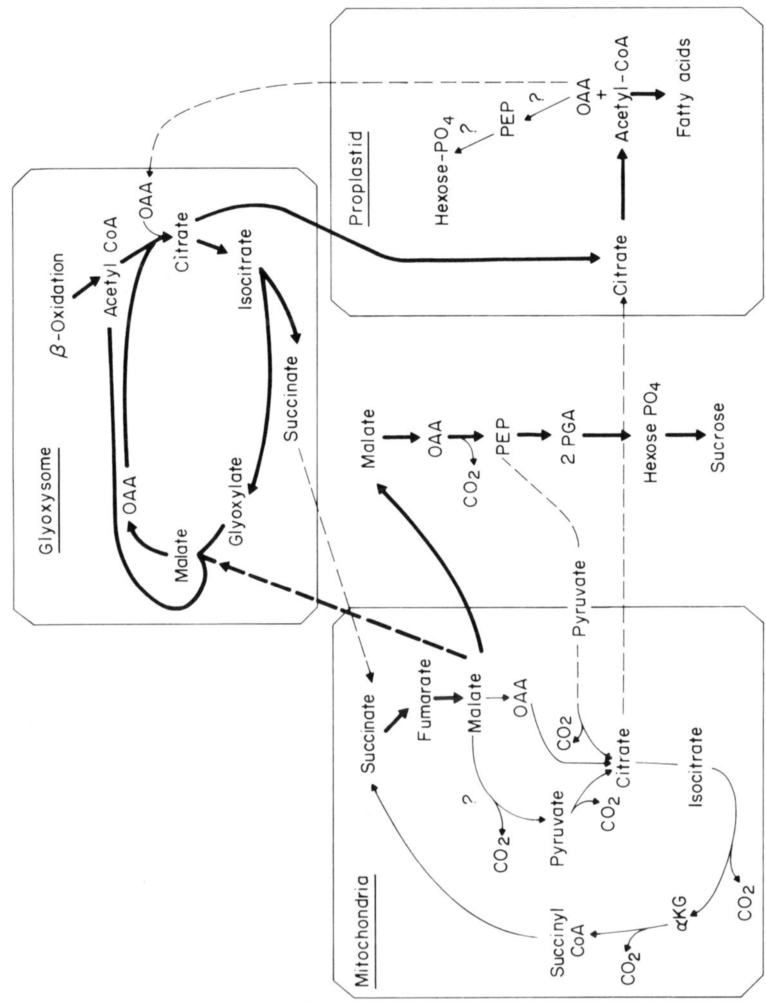

Fig. 13.9 Subcellular and metabolic relations of fatty acid breakdown and the synthesis of sucrose and new fatty acids in castor bean endosperm (Rapp & Randall, 1980).

Table 13.5 Immunoprecipitation of glyoxysomal proteins from total *in vitro* translation products

Tissue age	Total products	Immunoprecipitated products	%
Days	Radioactivity (cpm × 10^{-3})		
Seed, dry	95	1.2	1.7
Seed, imbibed	148	10	6.7
1	400	31	7.8
2	539	48	9.0
3	760	107	14
4	876	133	15
5	624	106	17
6	268	24	8.9

Poly(A)RNA extracted from castor bean endosperm was translated in the wheat germ system. The products were precipitated by antibody raised against total matrix proteins separated from isolated glyoxysomes (Roberts & Lord, 1979).

(Harris & Chrispeels, 1980). The net decrease in ER-phospholipid is apparently the result of postponed dismantling of an ER system which had proliferated during the final stages of seed maturation (Gilkes & Chrispeels, 1980).

Cotyledons of germinating seeds exhibit an increase in the proteins of their inner mitochondrial membrane (Sato & Asahi, 1975), including the assembly of pre-formed succinate dehydrogenase into the inner membrane (Nakayama *et al.*, 1980). At the end of imbibition, mitochondria isolated from pea cotyledons had little succinate dehydrogenase or cytochrome oxidase activity, but within 24h these enzyme activities were elevated to a level which was half the maximum eventually attained (James & Spencer, 1979). Cyanide resistant respiration developed after two days. Morohashi and Bewley (1980) have shown in pea cotyledons that renewed development of mitochondrial enzyme activity occurs in two phases. The first phase requires only the hydration of the cotyledon while the second requires the presence of the embryonic axis.

13.3.3 GIBBERELLINS AND SEED GERMINATION

Exposure of a number of dormant seeds to exogenous gibberellins stimulates their germination (Villiers, 1972), but the relation of gibberellins found in some dry seeds to the germinative process is unclear. The gibberellins present in dry seed tissue are frequently the less active forms, i.e. irreversibly hydroxylated in the 2-β position (e.g., GA_8, GA_{29}, GA_{51}), or inactive conjugates (Hedden *et al.*, 1978), and are found in generally much reduced levels, compared to those reached during seed development (Frydman *et al.*, 1974). This suggests that in some seed tissues gibberellins are 'de-activated' prior to desiccation of the developing seed. Germinating bean seeds retain one inactivating system, rapidly converting radioactive GA_1 fed during seed maturation (Yamane *et al.*, 1975), or after 30h of germination (Nadeau & Rappaport, 1972), to its inactive form. The major GA_{20} derivative found in ungerminated dwarf pea, a novel α,β-unsaturated ketone, disappears altogether during germination (Durley *et al.*, 1979). In contrast, in germinating embryos of barley grains, gibberellins are formed *de novo*, probably in the scutellum (Radley, 1967) as early as two hours after imbibition (Stoddart *et al.*, 1973). Thus, biosynthetic gibberellins, produced concurrently with germination and early growth of the embryonic axis, are most likely the species which are active in the germinating seed.

At present it would appear that the greatest value of gibberellin production lies in its effects upon mobilization of food reserves from storage organs of the seed, which in turn supports growth of the young seedling. Whether gibberellins have a direct role in the onset of growth of the embryonic axis has not been established; added gibberellins may have no overt effects since endogenous gibberellin synthesis may have already saturated the response.

The storage tissue showing the greatest response to added gibberellic acid is the aleurone layer of cereal endosperm. The response of this tissue is basically two-fold: to establish a system for secretion of new and pre-existing hydrolases, ions, and peptides, and to bring about the *de novo* synthesis of enzymes required for hydrolysis of reserves both within and external to the aleurone cells. A distinction has been made, using wheat aleurone, between the effects of hydration, factors from the embryo, and factors from the starchy endosperm on aleurone development during germination (Laidman *et al.*, 1974). Hydration alone allows activation of phytase as well as the development of general cellular activities. Cytokinins replace a factor from starchy endosperm in triglyceride mobilization and IAA plus glutamine have an embryo-like role in both triglyceride metabolism and induction of neutral lipase activity. GA_3 replaces the embryo in development of glyoxysomal enzyme activities and in the induction of hydrolases responsible for protein and carbohydrate mobilization (see also Jones & Stoddart, 1977).

In barley aleurone, pre-incubation in water alone allows the synthesis of β-1,3-glucanase, ribonuclease, and acid phosphatase. About 8h after GA_3 is added, *de novo* synthesis of α-amylase and protease achieve a constant rate. The release of all of these enzymes from the aleurone layer is under the control of GA_3, although different mechanisms may be involved. Secretion of the accumulated β-1,3-glucanase can occur after 4h of GA_3-treatment (Taiz & Jones, 1970), while α-amylase and protease are secreted soon after their synthesis (Jacobsen & Varner, 1967). Ribonuclease activity, however, which accumulates even further after GA_3-treatment, is not released until the aleurone cells have become highly vacuolated (Chrispeels & Varner, 1967; Jones & Price, 1970), and, perhaps, senescent. Acid phosphatase is synthesized and secreted into the periplasmic space even in the absence of the hormone (Ashford & Jacobsen, 1974). Release of the acid phosphatase, from its entrapment behind the cell wall, is effected by GA_3 through its stimulation of pentosanase biosynthesis; the pentosanases create cleavage furrows in the cell wall and allow the acid phosphatase to escape (Taiz & Honigman, 1976).

A membrane fraction having properties of the ER may have a role in the secretion of α-amylase. In pulse-labelling studies about half of the α-amylase present in barley aleurone after 20h of GA_3-treatment is in an ER-like fraction. Kinetic analyses have shown that only the α-amylase from the ER-like fraction chases into the external medium, suggesting that particulate α-amylase is the true precursor of secreted α-amylase (Locy & Kende, 1978). Immunofluorescence microscopy has also been used to localize α-amylase in the vicinity of the ER (Jones & Chen, 1976). How does gibberellic acid influence the membrane system of the cell and at the same time lead to *de novo* synthesis of new enzyme activities? The simplest suggestion would be that in addition to the hydrolases, some proteins synthesized in response to GA_3 are responsible for organizing the secretory processes of the aleurone cells, probably via eventual modification of protein or glycoprotein components of the ER and/or plasmalemma. Alternatively, the hormone effect on hydrolase synthesis may involve a separate mode of action and an entirely different receptor than is used for the enhancement of secretion.

FURTHER READING

BEWLEY J. D. & BLACK M. (1978) *Physiology and Biochemistry of Seeds.* Vol. 1. Development, Germination and Growth. Springer-Verlag, New York.

KHAN A. A. (ed.). (1977) *The Physiology and Biochemistry of Seed Dormancy and Germination.* Elsevier-North Holland Biomedical Press, Amsterdam.

KOZLOWSKI T. T. (ed.) (1972) *Seed Biology*, vol. II. Academic Press, New York.

LAIDMAN D. L., COLBORNE A. J., DOIG R. I., & VARTY K. (1974) The multiplicity of induction systems in wheat aleurone tissue. In *Mechanisms of Regulation of Plant Growth.* pp. 581–90 (eds. Bieleski R. L., Ferguson A. R. & Cresswell M. M.). The Royal Society of New Zealand, Wellington.

MAYER A., & POLJAKOFF-MAYBER A. (1975) *The Germination of Seeds*, 2e Pergamon Press, Oxford.

RUBENSTEIN I., PHILLIPS R. L., GREEN C. E. & GENGENBACH B. G. (1979) *The Plant Seed: Development, Preservation, and Germination.* Academic Press, New York.

MOLECULAR ASPECTS OF PERCEPTION AND RESPONSE TO ENVIRONMENTAL SIGNALS

CHAPTER 14

PHOTOMORPHOGENESIS

14.1 INTRODUCTION

The definition of photomorphogenesis is not a straightforward matter. Most photobiologists tend to use the term to cover all studies of the control of growth and differentiation by light. However, the plant's responses to light, as mediated by specific photoreceptors, include both developmental and non-developmental phenomena, probably sharing, at least in part, common mechanisms. Thus photomorphogenesis, in practice, transcends the normal limits of morphogenesis.

Some authorities have attempted to define photomorphogenesis in molecular terms; that is, as the photocontrol of gene expression, e.g. Mohr (1978) stated 'photomorphogenesis can be considered a model process in our efforts to understand the control of gene expression during development'. Such a definition represents a useful rule of thumb but by no means an axiom. Inasmuch as differently differentiated cells from the same organism harbour the same DNA but synthesize different mRNA molecules and different proteins (whether that differentiation is promoted by light or by any other means is immaterial), it is inevitable that differential gene expression, and indeed differential gene transcription, will be, quantitatively, a major contributor to photomorphogenesis and it is given appropriate emphasis in this chapter. However, just as the occurrence of normal morphogenesis in enucleated *Acetabularia* (Chapter 6) has demonstrated convincingly that the control of development need not be equated with the control of gene transcription, so too do numerous examples of the effect of light on developmental processes testify that photomorphogenesis cannot be reduced to a study of the photocontrol of gene transcription (see Section 14.3.4).

Acetabularia is, of course, an extreme example, an exceptional case. Many of the examples of photocontrol to be discussed here represent extreme or unusual situations too and, provided caution is exercised, these examples can, as with *Acetabularia*, reveal far more than numerous 'typical' cases. In other words, whilst we expect typically, light-promoted regulation of gene expression (and indeed gene transcription) to be a major contributor to photomorphogenesis we may well obtain more valuable information concerning the nature of the primary events from the unusual examples in our attempts to elucidate the chain of processes linking photoreception to photoresponse. A word of caution is necessary. Whilst it is inevitable that the most precise information comes from situations in which a response is displayed in an exaggerated or atypical manner the corollary to this is that some of these extreme examples may be red herrings in our pursuit of the mechanisms of photomorphogenesis. Alas, only when we have a much clearer understanding of those mechanisms will we be in a position to recognize the false clues.

14.2 PHOTORECEPTORS FOR PHOTOMORPHOGENESIS

14.2.1 PHYTOCHROME

The photoreceptors involved in photomorphogenesis absorb light primarily in the blue and red (extending to far-red) parts of the spectrum. It is clear now that the photoreceptor in both the red and far-red parts of the spectrum is phytochrome, although controversy raged for some considerable time as to the extent of phytochrome involvement in the responses of plants to prolonged irradiation with far-red light. The principal criterion for recognizing phytochrome involvement in a response remains the classical red light promotion, far-red reversibility test, as illustrated in Table 14.1. Numerous examples of developmental

Table 14.1 Control of lettuce seed germination by red and far-red light. (From Borthwick *et al.*, 1952)

Irradiation	Percentage germination
Red	70
Red/far-red	6
Red/far-red/red	74
Red/far-red/red/far-red	6
Red/far-red/red/far-red/red	76
Red/far-red/red/far-red/red/far-red	7
Red/far-red/red/far-red/red/far-red/red	81
Red/far-red/red/far-red/red/far-red/red/far-red	7

responses to low doses of red light, reversible by subsequent far-red light, are to be found in the literature and a summary is given in Table 14.2.

Phytochrome exists in two stable forms, Pr and Pfr, inter-convertible by light, thus

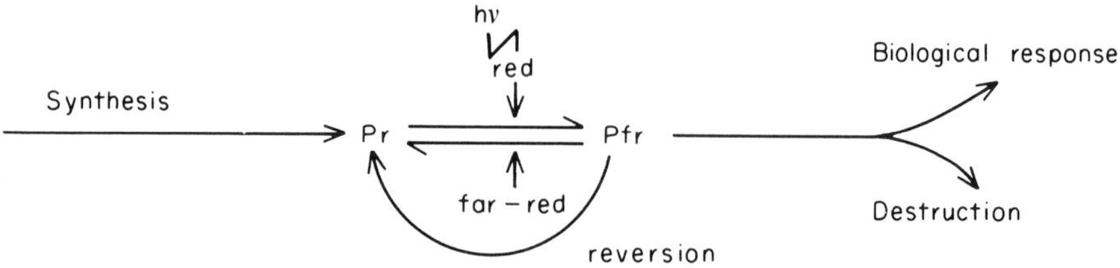

Table 14.2 Some red-promoted/far-red-reversible responses

Non-vascular lower plants
Oospore germination
Cell elongation
Filament branching
Spore germination
Cell number per protonema
Growth rate of thalli
Chloroplast size
Chloroplast division rate
Chloroplast phototactic movement
Chlorophyll content
Pteridophyta
Spore germination
Growth of rhizoid and filament
Gymnospermae
Hypocotyl hook formation
Longitudinal stem growth
Angiospermae
Seed germination
Growth rate (almost every aspect)
Hook opening
Hyponastic leaflet movement
Closing leaflet movement
Geotropic reactivity
Phototropic reactivity
Photoperiodism (light break effects)
Stabilization of vernalized condition
Auxin-induced lateral root initiation
Leaf primordia formation
Cytoplasmic viscosity
Dry weight
Protein content
Transport of sucrose
Starch/sugar degradation
Anthocyanin biosynthesis
Flavonoid biosynthesis
Carotenoid biosynthesis
Chlorophyll content
Ascorbic acid content
ATP concentration
Activities of enzymes
Change in electric potentials

The Pfr form is generally believed to be biologically active, Pr inactive, and (in young, etiolated seedlings at least) the Pfr form is much the less stable of the two. The loss of Pfr is usually referred to as destruction. Pfr is also lost in a dark reaction in which it reverts to Pr. It is thought that synthesis of phytochrome occurs entirely in the Pr form. The promotion of a red light response is explained by the formation of the (presumed) active form Pfr by the irradiation, far-red reversibility by the removal of the Pfr before sufficient time has elapsed for the coupling of the next stage in the chain of events leading from phytochrome to the ultimate response. This explanation leads to the expectation that light doses greater than those required to saturate conversion of Pr to Pfr should be supra-optimal for the responses also. This is not the case. The effects of prolonged irradiations at high fluence rates can be considerably greater than would be expected in terms of the accumulated presence of Pfr during the irradiation. During such prolonged irradiations, considerable loss of phytochrome occurs as a result of Pfr destruction. This makes it much harder, often even impossible, to measure phytochrome spectrophotometrically after plant tissues have been in the light for prolonged periods. However it has been deduced mathematically that the steady-state (equilibrium) level of Pfr settles down to a value that depends neither on the wavelength nor on the irradiance (fluence-rate). Just the opposite happens to the photoresponses, which are strongly dependent on the wavelength (usually with far-red light becoming more effective than red) and become progressively irradiance dependent, i.e. the extent of the response is a function of the fluence rate of the light which the plant is receiving. It is no wonder that there have been repeated suggestions that photoreceptors other than phytochrome must be involved in these 'high-irradiance responses'.

The most convincing evidence against these views comes from work published by Hartmann in 1966 (Hartmann, 1966, 1967). Hartmann was investigating a well-known phenomenon of widespread occurrence, the inhibition of stem extension by light; in this case the lettuce hypocotyl was the test material. He found that, over a wide range of wavelengths, the response was proportional to the logarithm of the fluence rate and that the wavelengths having the greatest quantum effectiveness (i.e. those at which a given effect is produced at the lowest fluence rate) were in the far-red region with a maximum of 716nm (Fig. 14.1a). Neither light of 660nm or longer wavelength far-red light (which would be maximally effective in converting Pr to Pfr and vice versa) was found to have a significant effect. In short the action spectrum does not look much like the absorption spectrum of either Pr or Pfr (Vol. 14, Fig. 14.1). Hartmann devised an elegant experiment which proved that phytochrome is the photoreceptor involved. He irradiated seedlings simultaneously with light of two wavelengths 660nm and 760nm, each of which was individually ineffective in inhibiting hypocotyl extension. Addition of an increasing fluence-rate of red light (which alone yields an equilibrium of 75% Pfr, 25% Pr) to a fixed high fluence rate of 760nm light (which yields more than 99% Pr) caused first an increase in inhibition to a maximum, followed by a decrease (Fig. 14.1b). The ratio of red to far-red light in the bichromatic mixture causing maximum inhibition was calculated to produce precisely the same proportion of Pfr to total phytochrome, (i.e. about 3%) as the optimum wavelength with monochromatic irradiation

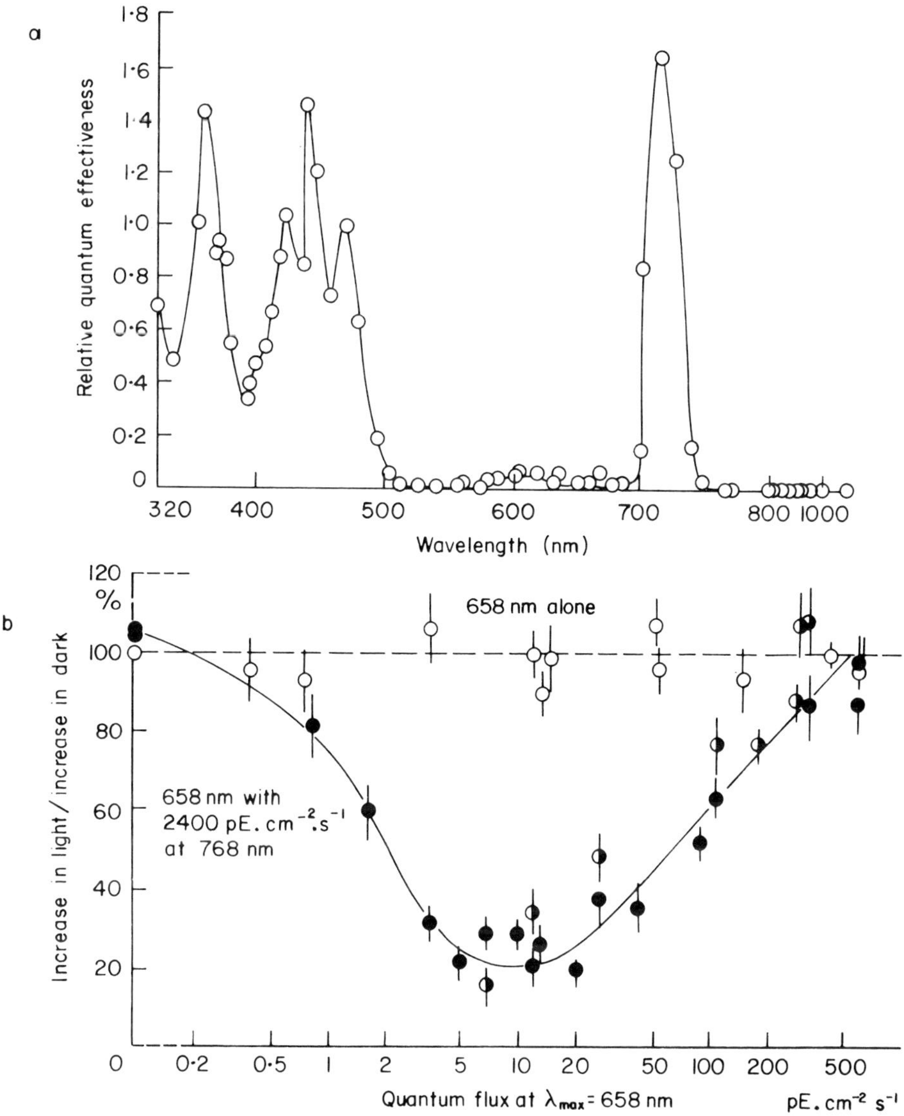

Fig. 14.1 (a) Action spectrum for the photo-inhibition of lettuce hypocotyl extension growth (from Hartmann, 1966). (b) Effect of simultaneous bichromatic irradiation with a fixed fluence rate of 768nm and various rates of 658nm light, on lettuce hypocotyl extension (from Hartmann, 1967)

(716nm). Thus, phytochrome was proved to be the photoreceptor for the 'high-irradiance-reaction', at least in this part of the spectrum.

This elegant solution leaves open the crucial question of *how* phytochrome is involved. Clearly, there is no simple correlation between the level of the presumed active form, Pfr and the response. Some of the suggested resolutions of this problem are described in Chapter 14. Vol. 14, but it is worth noting in passing that recent work has indicated the necessity for two interacting components to explain both red—far red reversible responses and high irradiance responses of phytochrome (Mohr *et al.*, 1979; Johnson, 1980).

14.2.2 THE BLUE LIGHT PHOTORECEPTORS

Hartmann went further and attempted to prove that the action in the blue region of the spectrum (Fig. 14.1b) could also be accounted for in terms of phytochrome alone. This argument has to be viewed in the context of clear evidence for one or more separate blue-light-absorbing pigments involved in phototropism (see chapter 15) as well as the sole responsibility of such pigments for photomorphogenesis in fungi and protista. Recently Thomas and Dickinson (1979) have provided clear evidence that the inhibition of hypocotyl extension promoted by blue light involves another receptor pigment in addition to phytochrome. These workers adopted the ingenious approach of using high-pressure sodium lamps to provide a high fluence rate of light (λmax 590nm) which can be absorbed by phytochrome but which is without any blue-light component. To this main source they added varying, but comparatively low, fluence rates of blue light, insufficient to lead to significant absorption by phytochrome (which would have been detected as a depression of the phytochrome photoequilibrium). Nevertheless, the effect of the added blue light was a strong inhibition of hypocotyl extension (Fig. 14.2). It would be difficult indeed to explain these results in terms of phytochrome. The extent of involvement of blue light receptors *vis à vis* phytochrome seems to vary from species to species. At one extreme for example, inhibition of lettuce hypocotyl extension shows a very strong blue light response with comparatively little phytochrome component. At the other, mustard seedlings (*Sinapis alba*) have been shown to have little, if any, blue light response which cannot be attributed to phytochrome, although they do exhibit blue-light-mediated phototropic reactions. What is certain is that, in higher plants, phytochrome has taken over many of the morphogenetic roles of the blue light receptor in fungi.

14.2.2.1 *The nature of the blue light photoreceptor*

Whereas phytochrome was purified as long ago as 1964 and an extensive literature describing its properties has built up, the blue light receptor has proved far more elusive, even though apparently ubiquitous (Presti & Delbrück, 1978). Its precise nature remains a matter of controversy so needless to say research has hardly begun to tackle the question of its mechanism of action. The action spectra for a wide range of physiological responses are remarkably uniform (Fig. 14.3), suggesting that the same receptor might be involved. However, two pigments, flavins and carotenoids, both have absorption spectra which display many, but not all, of the requisite features of the action spectra.

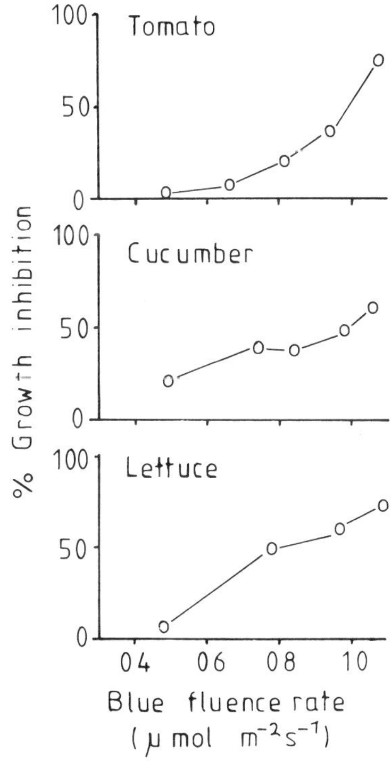

Fig. 14.2 Fluence response curves for blue light inhibition of hypocotyl extension in the presence of a high fluence rate background from low pressure sodium lamps (fluence rate 210μmol m^{-2}s^{-1}) (from Thomas & Dickinson, 1979)

Carotenoids were first in the field. As long ago as 1936 they were proposed as the receptor pigments for *Avena* phototropism (Wald & Dubuy, 1936). The evidence was, first, that carotenoids were present in the phototropically sensitive region and, secondly, had absorption spectra similar to phototropic action spectra. The structure and absorption spectrum of β-carotene *in vitro* is shown in Fig. 14.4. It has absorption peaks at 450 and 480nm and a shoulder at 420nm in good agreement with the various action spectra (Fig. 14.3). There are three main objections to the candidacy of carotenoids. First, there is no absorption peak in the near ultra-violet, corresponding to the 370nm peak in the action spectra. Secondly, the excited singlet states of carotenoids have very short lifetimes and the intersystem crossing to the metastable triplet state is inefficient (Song & Moore, 1974) suggesting that carotenoids would be inefficient as photoreceptors. Thirdly, there have been a number of recent reports of situations where blue light responses persist in mutant plants deficient in carotenoids.

Many of the criticisms can be met. For example, a ready explanation for differences between absorption spectra of chromophores *in vitro* and *in vivo* is that these are the consequences of differences in the molecular environment, or

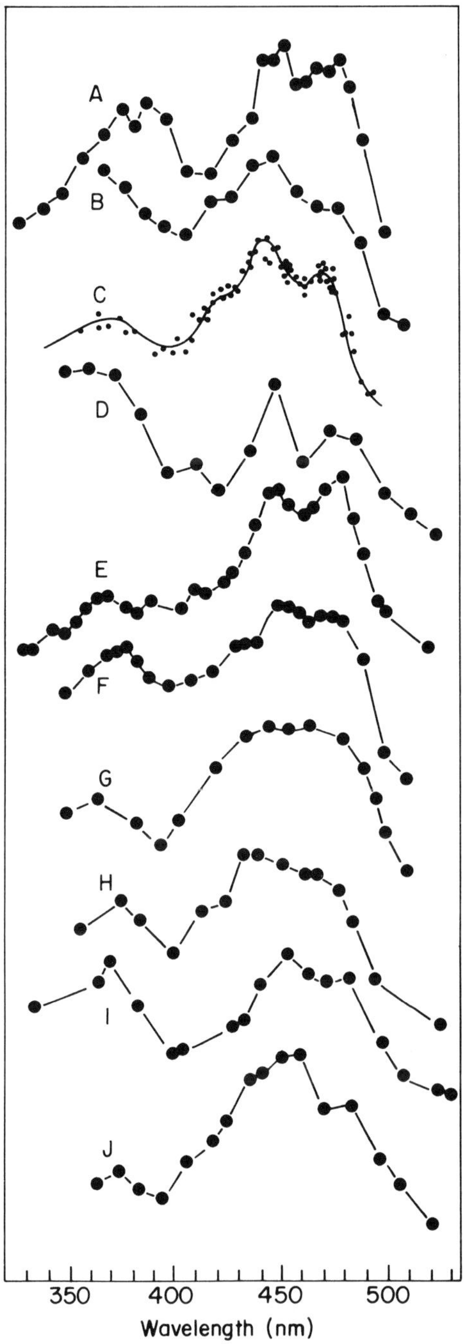
A
B
C
D
E
F
G
H
I
J
350
400
450
500
Wavelength (nm)

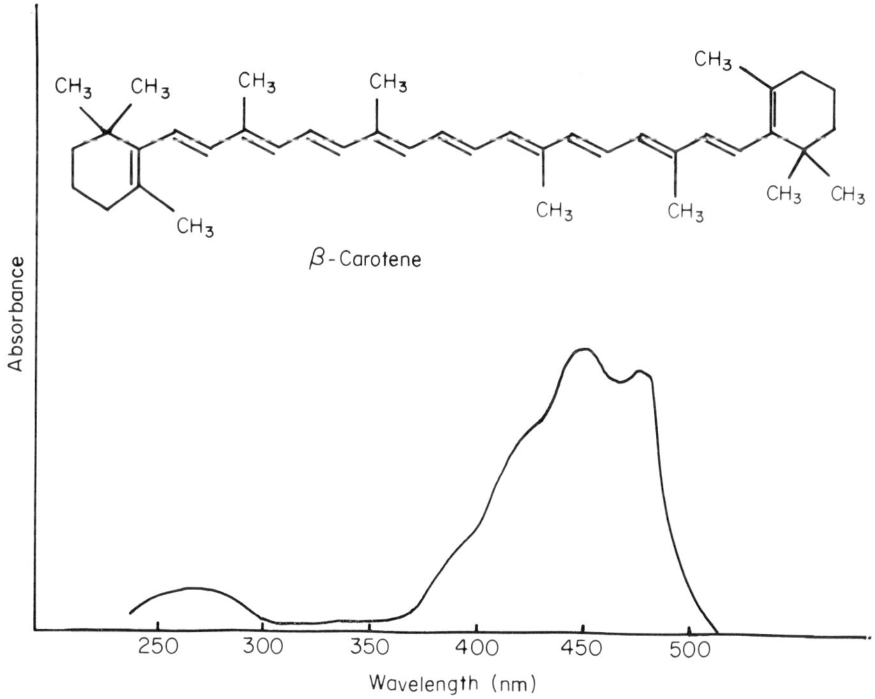

Fig. 14.4 Structure and absorption spectrum of β-carotene

that the chromophore is aggregated or associated with a protein *in vivo* (cf. phytochrome where the absorption spectrum of the chromophore alone is very different from the complete chromo-protein). It has been shown that aggregates of some carotenoids could be found in solution and that these absorb in the UV (Song & Moore, 1974). There is also a *cis* isomer of β-carotene which has an absorption peak *in vitro* at 340nm. There is however no evidence for either this isomer, or for stacked carotene aggregates, *in vivo*. The second argument against carotenoids, poor intersystem crossing (i.e. inefficient transfer of excitation energy), is certainly overcome in the so-called 'photosynthetic units' where energy is transferred with up to 80% efficiency from carotenoid receptors to

---

Fig. 14.3 Examples of action spectra for some physiological responses of several organisms to blue light: (A) phototropism in the fungus *Phycomyces*, (B) phototropism in the fungus *Pilobolus*, (C) phototropism in seedlings of the plant *Avena*, (D) phototaxis in the protist *Euglena*, (E) stimulation of carotenoid synthesis in the fungus *Neurospora*, (F) stimulation of carotenoid synthesis in the fungus *Fusarium*, (G) stimulation of carotenoid synthesis in the bacterium *Mycobacterium*, (H) enhancement of respiration in the alga *Chlorella*, (I) chloroplast rearrangement in the moss *Funaria* and (J) entrainment of the circadian rhythm of pupae emergence in the fruit fly *Drosophila*. Compiled by Presti & Delbrück (1979)

chlorophyll. Again, however, there is no evidence for the existence of such units outside the photosynthetic apparatus.

The absorption spectrum and structure of riboflavin is shown in Fig. 14.5. It has characteristics complementary to the β-carotene spectrum; a peak in the near-UV part of the spectrum (as observed in the action spectra), but a rather poor likeness to the blue 'triple-peak' typical of the responses. Again, however, it is possible that the absorption spectrum would acquire the necessary 'fine structure' *in vivo*. Sun *et al*., (1972) have shown that, in solid solutions of ethanol at 77°K, an absorption spectrum very like the action spectra is obtained (Fig. 14.6) and these workers argue that their experimental conditions may mimic the situation *in vivo*, where flavin chromophores are likely to be rigid components of flavo-proteins. There is evidence from a wide range of species from fungi to grasses that flavin can act as the photoreceptor for the *in vivo* reduction of a b-type cytochrome (Poff & Butler, 1974; Briggs, 1976) and it has been suggested (Jesaitis *et al*., 1977; Briggs, 1976) that a flavo-protein-cytochrome b complex, associated with the plasma membrane, may be the receptor for phototropism (and by implication, for other blue light responses). There is as yet little evidence of any relationship between flavin-mediated cytochrome b reduction and physiological responses, but it is of interest that nitrate reductase, itself a flavo-protein containing a b-type cytochrome, can be activated *in vitro* by blue light in

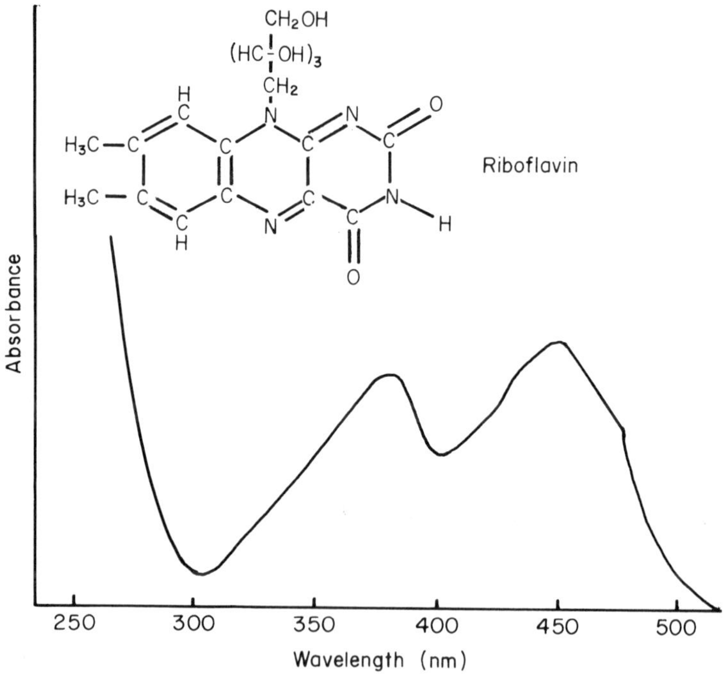

Fig. 14.5 Structure and absorption spectrum of riboflavin in aqueous solution

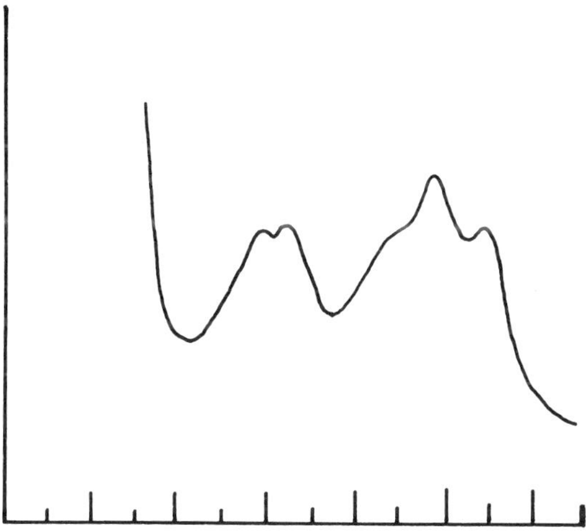

Fig. 14.6 Absorption spectrum of riboflavin in ethanol at 77K (from Sun *et al.*, 1972)

a reaction stimulated by the addition of exogenous flavin but not by β-carotene (Fig. 14.7). Recently, Roldán and Butler (1980) have shown that the action spectrum for photo-activation of nitrate reductase *in vitro* strongly resembles the typical blue light action spectrum *in vivo*. Furthermore it has been shown (Klemm & Ninnemann, 1979) that the nitrate reductase protein is involved in blue-light-mediated conidiation in *Neurospora*.

Thus, the evidence points to the likelihood of a flavo-protein as the receptor both for phototropism and for blue light mediated morphogenesis, although the possibility that carotenoids are involved in some responses cannot at present be ruled out. Clearly, at this stage in our understanding, we can go no further in elucidating the chain of events from the blue light photoreceptors to the ultimate displays, and in discussing these chains we return primarily to a consideration of phytochrome mediated photomorphogenesis.

14.3 FROM PHOTORECEPTOR TO PHOTORESPONSE

A useful rule of thumb (no more than that) is that the shorter the time between light absorption and an observed photoresponse, the more likelihood of a relatively direct relationship between photoreceptor and response, but almost invariably less certainty that the response will be part of a pathway leading ultimately to events recognizable as photomorphogenesis. (As will be seen, many rapid phytochrome effects fall into this category.) It is clear too that the longer the time scale, the further we move along the chain of biochemical events

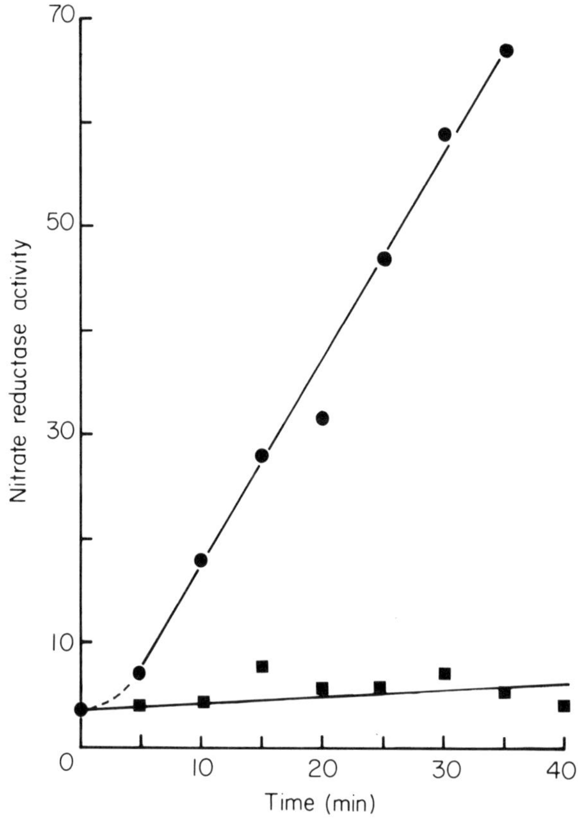

Fig. 14.7 Activation *in vitro* of nitrate reductase from *Spinaceae oleracea.* The partially-purified enzyme was irradiated with blue light in buffer containing either flavin adenine nucleotide (circles) or β-carotene (squares) (unpublished data)

following photoreception and the more clearly we are able to recognize that photomorphogenesis is occurring, then the greater is the difficulty in tracing the responses back to the photoreceptor. Indeed it becomes progressively harder to be sure even that the morphogenetic effect is a part of a direct chain from the photoreceptor pigment rather than a side effect of another phytochrome-mediated response. The problem seems superficial at the philosophical level but presents a real dilemma for the research worker attempting to elucidate the mechanisms of photomorphogenesis. On the one hand he wants to be sure that the phenomenon he is dealing with is (or is coupled to) a real 'physiological' response, and on the other, that the response is causally related to the initial photo-event in question. With respect to phytochrome, a lot is known about rapid phytochrome-mediated photoresponses and also something about light-mediated photomorphogenesis, but because most of the experimental evidence falls discretely into one or other of these classes rather little is yet known about the mechanism of photomorphogenesis. This dichotomy is to be found also in the two principal concepts regarding phytochrome action, namely

(1) that phytochrome has a multiplicity of primary effects of which the most crucial with respect to photomorphogenesis is a direct interaction with the genome (Fig. 14.8) and

(2) that phytochrome interacts with cellular membranes to modify their properties and thereby brings about a variety of photoresponses (Fig. 14.9).

These two concepts almost paraphrase the dichotomy outlined above and hence the two hypotheses have by and large been allowed to go their separate ways, mutually unimpeded, to the great disadvantage of our understanding of photomorphogenesis.

14.3.1 RAPID RESPONSES

A wide range of rapid phytochrome responses has now been described in the

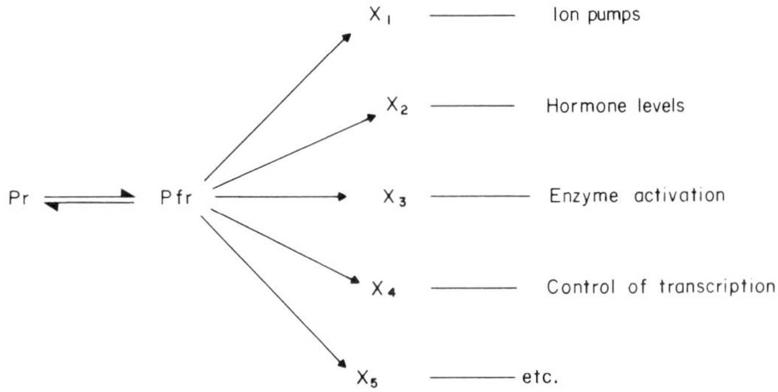

Fig. 14.8

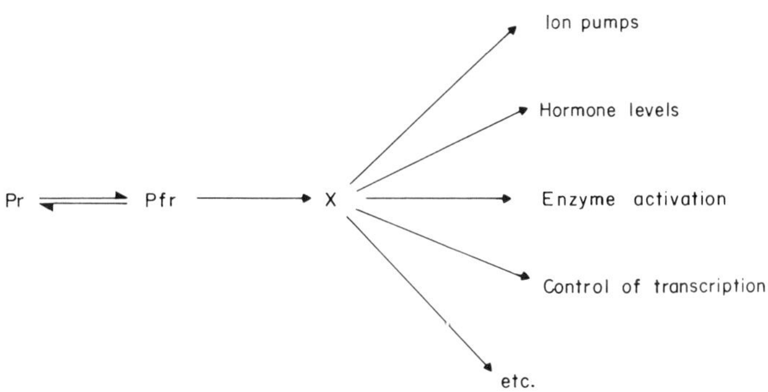

Fig. 14.8 and 14.9 Two models of phytochrome action

literature. Most of these rapid effects involve, or could imply, changes in membrane function promoted by light. However, it is often difficult to establish the relevance of these responses to physiologically important events. Indeed some of them could be trivial manifestations of differences between Pr and Pfr with no physiological consequences whatsoever.

14.3.1.1 *Phytochrome pelletability: primary event in phytochrome action?*

There is no more striking example of this dilemma than the phenomenon of light-mediated phytochrome pelletability (see Pratt, 1978). It is a measure of our ignorance concerning phytochrome action that, despite the considerable efforts of many research groups, it is still possible, on the one hand, that this phenomenon represents the long sought-after primary action of phytochrome and, on the other, that it is a trivial artifact of no physiological moment. The basic observation is this: brief irradiation of intact etiolated seedlings with red light results in a substantial increase in the proportion of phytochrome which becomes associated with the pellet following homogenization of the tissue in a buffer containing divalent cations and low-speed centrifugation. For example, in *Avena*, 90% of the phytochrome is soluble in homogenates from dark grown seedlings whereas 60% is found in the pellet following *in vivo* irradiation with red light. Divalent cations (preferably Mg^{2+}) are required in the extraction buffer otherwise pelletability is not enhanced by the light treatment. The response is very fast. By using very short irradiation times at high fluence rates and with rapid tissue homogenization techniques it has been possible to show (Pratt & Marmé, 1976) that the half-life for the association of phytochrome with the pelleting cellular constituents is about 40s at 0°C but less than 5s at 22°C. This is one of the fastest known phytochrome responses and hence there is a strong possibility that pelletability could represent the first step leading to phytochrome responses. Hendricks and Borthwick (1967) had already postulated the existence of a biologically active receptor for phytochrome and it is not surprising that discovery of the pelletability phenomenon was soon followed by claims that it might represent an association of phytochrome with just such a receptor, integrally associated with a membrane (Marmé, 1974). The essentials of the initial experimental observations have since been confirmed in a wide range of tissues, a process which was complicated by the presence of many artifactual data (see Pratt, 1978, for a review). However, the physiological significance of the data is far less clear.

There are two principal questions which have yet to be adequately answered, even before the physiological importance of light-induced phytochrome pelletability can be evaluated. First, does the phenomenon displayed as pelletability *in vitro* have a counterpart in the same or a similar phenomenon *in vivo*? Secondly, does the observed association represent a real, specific association between phytochrome and a specific membrane fraction(s) or does it merely represent a non-specific aggregation of phytochrome with any material pelletable *in vitro*? The best answers to these questions to have emerged so far seem to be the immunocytochemical studies of Pratt and his co-workers. In these studies an antiserum raised against purified phytochrome was used to detect phytochrome in sections of dark-grown and red-irradiated oat seedlings. It was claimed that red light caused a subsequent redistribution of phytochrome. However, the impression from the figures was that, although an aggregation of phytochrome molecules occurred, it was not necessarily with cell membranes. Furthermore, such sequestering *in vivo* was not found in all species tested.

Studies of the biochemical characteristics of the association *in vitro* confirm that phytochrome does not become an integral membrane protein: the association is broken down by high ionic strength, addition of chelating agents or simply by withdrawal of divalent cations (Quail, 1978). Whilst this does not exclude a possible 'loose' association of Pfr with a membrane, the more cautious interpretation of the available data would be that the pelletability phenomenon probably does not represent an association with the major cellular membranes, but nevertheless seems to have a real counterpart *in vivo*, albeit of unknown physiological significance.

There is an additional problem which appears to block the simple, attractive hypothesis that the association of phytochrome with a 'receptor' (observed as pelletability) represents the initial event of phytochrome action. If the pelletability response does indeed represent the association of phytochrome with a receptor then this association cannot alone be the trigger for phytochrome action. The reason is that although pelletability is far-red reversible the dissociation of phytochrome *in vivo* (as measured by subsequent pelletability *in vitro*) has a half life of 25min at 22°C, whilst many responses have much shorter times of escape from far-red reversibility. Thus it seems that the difference between Pr and Pfr would not only be a prerequisite for the association with the presumed receptor but the presence of Pfr would remain a requirement for phytochrome action after such an association was formed.

14.3.1.2 *Photocontrol of bio-electric potentials*

Prior to the discovery of irradiation-induced pelletability, the most rapid known phytochrome effects were those involving changes in membrane potential. One of the first observations of such changes was not made by direct measurement of electrical potentials at all. Tanada (1968) placed root tips of etiolated barley or mung bean seedlings in a complex solution in a negatively charged glass dish. Under these conditions the sections were observed to float or sink in the medium but not to adhere to the glass. However, within 30s of red light treatment they were found to adhere by their tips, the adhesion being reversible by subsequent far-red irradiation. Subsequently Racusen and Miller (1972) were able to measure the potentials required for adhesion and release by placing electrodes in the bathing solution and adjusting the potential accordingly.

Using this method changes in electrical potential were confirmed. The same year (1972) Newman and Briggs showed directly that phytochrome could mediate electrical changes in oat coleoptiles. Using a 'flowing-drop' technique in which electrical contact with the plant was made by two 'liquid bridges' on the ad- and abaxial sides of the coleoptile they showed that a potential change of between 5mV and 10mV could be observed within 15s of the onset of irradiation. The response was reversible by far-red light, which alone had no effect.

The most rapid changes yet observed in membrane potential are those reported by Racusen (1976) who used micro-electrodes inserted into oat coleoptile parenchyma cells to measure (intracellularly) differences in membrane potential. Differences of the same magnitude as those found by Newman and Briggs were obtained but with considerably shorter lag phases, between 1s and 5s for red irradiation, and between 5s and 30s for far-red reversal at 25°C (Fig. 14.10). Thus the red light mediated depolarization measured by Racusen is at least as fast a response as that involved in the pelletability phenomenon, especially when one remembers that the half-life is calculated for pelletability

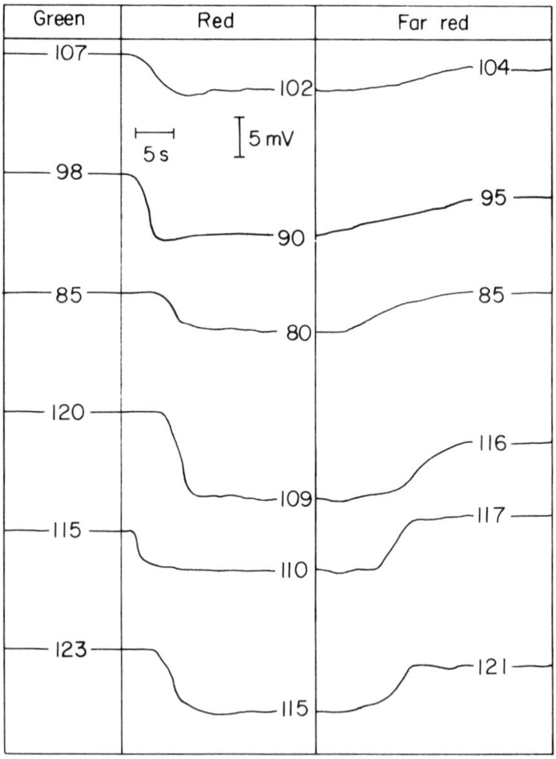

Fig. 14.10 Membrane potential changes in oat coleoptile parenchyma cells during successive exposures to continuous green, red and far red light (from Racusen, 1976)

starting from the *end* of a 5s red pulse whereas the electrical change is observed between 1 and 5s after the *onset* of red irradiation. Indeed the half-life for the slowest dark relaxation reaction in the formation of Pfr from Pr is itself a few seconds (Linschitz *et al.*, 1966) so the electrical change following Pfr formation is almost instantaneous. Furthermore the far-red mediated repolarization is nearly two orders of magnitude faster than the reversal of Mg^{2+}-induced pelletability. Thus this response is not a result of the association of phytochrome with the membrane or whatever other phenomenon the pelletability response represents *in vivo*. Indeed Racusen showed that red/far red reversibility of electrical potential could be repeated up to at least three times with no significant differences in amplitude. Since little or no further increase in pelletability would occur with the second or subsequent red pulses it seems that neither the depolarization nor repolarization are consequences of phytochrome 'pelleting' but rather these data indicate that the electrical changes represent a feasible alternative to the pelletability phenomenon as a candidate for the primary action of phytochrome. (There could, of course, be an even earlier effect, as yet undiscovered!)

Although the mechanisms responsible for the electrical changes are unknown, the changes are likely either to involve or to result in electrolytic movement, hence providing a plausible basis for subsequent physiological action, a basis, at present lacking, for the pelletability response. It is worth noting too that Racusen was able to show that, when a pair of electrodes were impaled each into two different cells in a column of cells, the depolarization and subsequent repolarization could be transmitted through at least six cellular partitions in the longitudinal direction (but not transversely), providing a possible explanation for plants' apparent ability to respond to the 'average' signal from many cells.

The two most rapid responses yet discovered and outlined in this chapter are experimental embodiments of the two theories of the nature of the primary action of phytochrome; (1) that it involves interaction of Pfr with a receptor (not necessarily the same receptor for each response) and (2) that it involves interaction of phytochrome with a membrane leading to modification of membrane properties. As with the concepts, so with the experimental observations: we are as yet unable to say with any certainty whether either phenomenon is relevant to photomorphogenesis as we observe it. It is in this rather unsatisfactory state that we move now to events which, though perhaps not so close to the initial photoconversion of Pr to Pfr, have clearly recognizable physiological consequences.

14.3.2 PHYTOCHROME ACTION IN *Mougeotia*

One of the most intriguing effects of phytochrome involves the photocontrol of chloroplast rotation in *Mougeotia*, a phenomenon studied extensively by Haupt and his co-workers in Erlangen. *Mougeotia* cells each contain a single flat

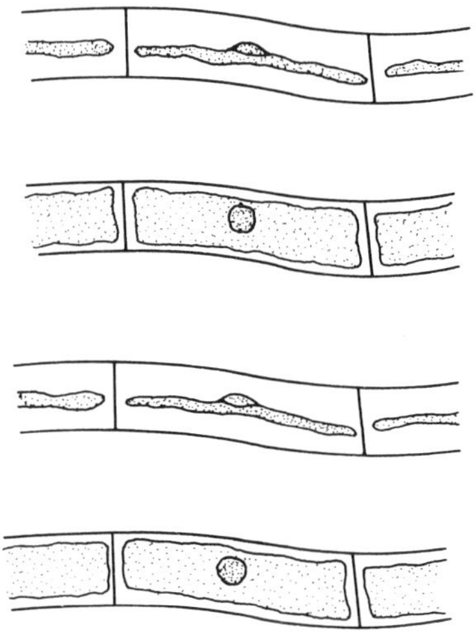

Fig. 14.11 The *Mougeotia* chloroplast and the effect of various light treatments in its orientation. The direction of irradiation is normal to the plane of the paper (from Haupt, 1972)

chloroplast whose orientation is controlled by phytochrome in such a way that the flat face of the chloroplast is presented to the light following a pulse of red light whereas far-red light leads to an edge-wise orientation (Fig. 14.11). That is to say, it is the physiological display of this phytochrome response which is reversible by far-red light. By using very short irradiations with microbeams of polarized light Haupt has shown that the phytochrome involved in controlling this response is almost certainly located in the plasma membranes and is orientated differently as Pr and as Pfr (Fig. 14.12). It is slightly ironic that the strongest evidence for a membrane location of physiologically important phytochrome has been obtained from an organism in which phytochrome has never been directly detected.

The phytochrome-controlled movement of the *Mougeotia* chloroplast is a highly specialized (and indeed non-morphogenetic) response but the Erlangen group has suggested that the mechanism mediating this response may be of more widespread occurrence. Dreyer and Weisenseel (1979) have shown that incorporation of radio-labelled calcium ions into *Mougeotia* cells occurs rapidly following brief (30s) red light irradiation; the uptake was far-red reversible. It is suggested that the calcium ion might be a common messenger in some (or all) phytochrome-mediated responses i.e. that phytochrome might act by mediating a change in intracellular Ca^{2+} concentration, either by affecting Ca^{2+} uptake or Ca^{+} extrusion at the plasmalemma. The reasons cited by the authors for proposing Ca^{2+} as a messenger are based on a number of observations that Ca^{2+}, which plays a central role in cellular activities, causes a number of effects similar to phytochrome responses, e.g. promotion of localized growth, activation of

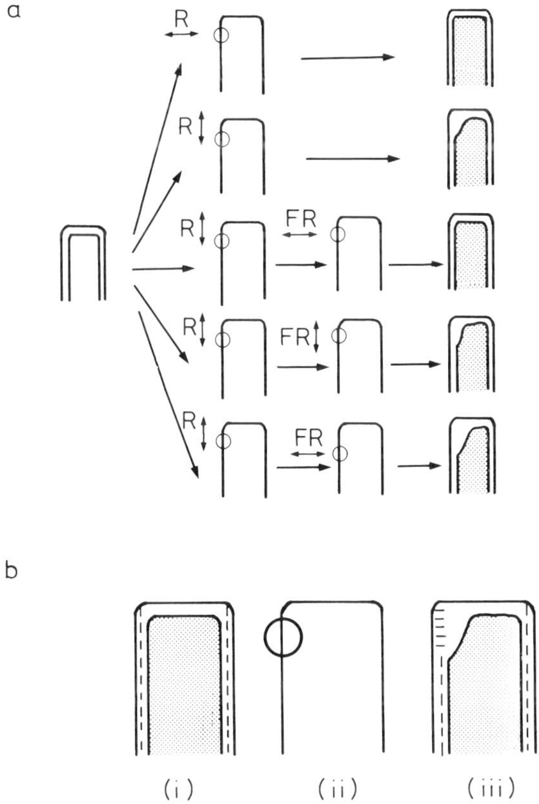

Fig. 14.12 (a) Movement of the *Mougeotia* chloroplast following irradiation, normal to the plane of the paper, with polarized light. The electrical vector is indicated by the double arrows (from Haupt, 1972). (b) Interpretation of the effect of red light with respect to the orientation of phytochrome molecules.

enzymes and (intriguingly) regulation of membrane permeability. It is worth noting that the red light-mediated influx of Ca^{2+} into the *Mougeotia* cell has not been shown to be the *cause* of the observed light-mediated chloroplast rotation. However it has been shown (Wagner & Bellini, 1976) that this phenomenon is sensitive to changes in Ca^{2+} concentration.

Effects of phytochrome on calcium movement are not limited to *Mougeotia*. For example, Weisenseel and Ruppert (1977) have shown that phytochrome-mediated membrane depolarization in *Nitella* is accompanied by changes in Ca^{2+} concentration. Recently Hale and Roux (1980) have demonstrated photoreversible calcium fluxes (i.e. the direction of calcium flow is reversed with far-red light) in protoplasts obtained from oat coleoptile tips. Following incubation in Ca^{2+}-containing medium in the dark, the calcium concentration in the medium was found to be enhanced following one minute red irradiation (i.e. the intracellular Ca^{2+} concentration decreased, the opposite of the response observed in *Mougeotia*), and the effect was reversible after one minute far-red irradiation. These workers suggest that changes in calcium ion concentration in the cell wall are part of the mechanism of phytochrome control of cell extension. Certainly the coupling between phytochrome and cell extension can be very rapid. For example, in maize coleoptiles, the effect of 5s red light on subsequent extension in the dark is far-red reversible only during the first 20s following the red irradiation (Warner *et al.*, 1980). An even more rapid escape from photoreversibility is observed in the same tissue when cell wall extensibility is measured using an extensometer and is accompanied by a decrease in the water potential of cells (Warner & Ross, 1981). It is intriguing but not necessarily prudent to recall the role of calcium in cell extension (Cleland & Rayle, 1977) and to note another similarity to the *Mougeotia* situation, namely the phytochrome control of 'water permeability' (Weisenseel & Smeibidl, 1973). The Erlangen group has also suggested that the rapid effects on membrane potential referred to in the previous section may be the result of increased Ca^{2+} binding at the cell membrane. The attentive reader will note the absence of evidence for chains of cause and effect or even a precise time sequences in these events. (For example, do phytochrome mediated changes in Ca^{2+} concentration precede or follow membrane permeability changes? See Section 14.3.1.2.) Thus the idea of Ca^{2+} as a second messenger in phytochrome action, like the pelletability hypothesis, is no more than an attractive idea at present.

Calcium is not the only ion whose intracellular concentration is modified by phytochrome action. Leaflet movement in *Albizzia julibrissin* and *Samanea saman* is controlled by phytochrome: movement follows after about 10min red or far-red light treatment and is accompanied by large fluxes of K^+ and Cl^- ions out of the ventral motor cells of the pulvinus (Fig. 14.13). These fluxes are preceded by substantial changes in trans-membrane potential (Racusen & Satter, 1975) which are observed within 5s. Clearly, in this case the potassium (or chloride) movement represents the mechanism of leaflet movement (cf. stomatal opening). It is not the first, nor even necessarily an early event in phytochrome action. There are several other reports of movement of potassium and other ions in the literature, but in no case is there strong evidence that such movements represent a primary action of phytochrome.

14.3.3 PHYTOCHROME AND HORMONES

The idea that phytohormones might act as 'second messengers' in phytochrome action is not a new one. The central role of plant hormones in the control of plant growth and development makes them attractive candidates for such a function and there are numerous instances where effects of phytochrome can be mimicked by the exogenous application of hormones. Examples include the regulation of activity of many enzymes, anthocyanin synthesis, germination,

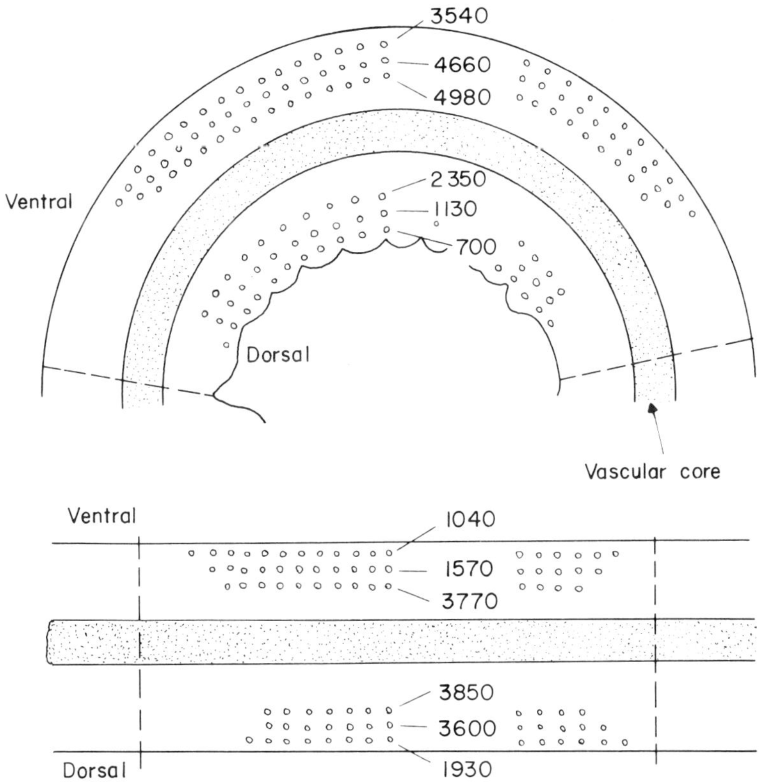

Fig. 14.13 Leaflet movement and K^+ flux in pulvinal motor cells of *Samanea* leaves. The figure shows the K^+ distribution in longitudinal sections when the pulivnus was closed in the dark (top) or open in the light (bottom) (from Galston & Satter, 1976)

stem extension rate and of cereal leaf unrolling. Furthermore there are many reports of phytochrome modulation of plant hormone levels (usually as measured by bio-assay). The many similarities between the effects of phytochrome and hormones make it almost too easy to assume that hormones must be intermediates in the photocontrol of these phenomena. However it is not always easy to distinguish experimentally between this assumption (1) and other possibilities (2 and 3) outlined below.

1. PHYTOCHROME ⟶ HORMONE ⟶ A → B → C → D ⟶ ACTION

2. PHYTOCHROME ⟶ A → B → C ↘ G ⟶ ACTION; HORMONE ⟶ D — E — F ↗ G

3. A ⟶ B ⟶ C ⟶ D ⟶ E ⟶ ACTION (PHYTOCHROME acting at A ⟶ B; HORMONE acting at C ⟶ D)

The problem has been approached experimentally in two ways. The first involves attempts to separate the biochemical events involved in the response. One of the best examples of this approach concerns the photocontrol of cereal leaf unrolling. When cereal seedlings are grown in darkness their leaves are tightly rolled. Upon transfer to the light the leaves unroll as the cells of the upper mesophyll expand. The response can be promoted by a short pulse of red light whose effect is far-red reversible (Virgin, 1962), indicating phytochrome control. A similar response can be obtained by the addition of gibberellin or cytokinin in the dark (Beevers *et al.*, 1970). Furthermore there is a rapid increase in gibberellin level following irradiation with red light, followed by a sharp decline. It was subsequently shown that equally rapid increases in gibberellin level followed irradiation of homogenates of barley leaves with red light (Reid *et al.*, 1972). Taking up the suggestion of Railton and Reid (1974) that the chloroplast might be an important site of gibberellin synthesis, Saunders and Cooke (1975) and Evans and Smith (1976) demonstrated that irradiation *in vitro* of partially purified preparations of intact etioplasts, from wheat and barley leaves respectively, led to substantial increases in bio-assayable gibberellin in the medium. Recent work of Hilton and Smith (1980) has shown that highly purified intact etioplasts both contain phytochrome and are able to release gibberellin-like substances upon red irradiation (Fig. 14.14). Thus it is possible that, *in vivo*,

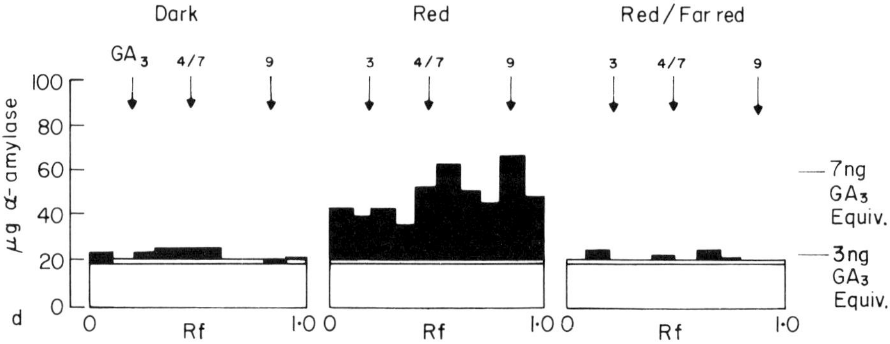

Fig. 14.14 Thin layer chromatograms of gibberellin-like substances in a purified etioplast fraction from barley leaves. The gibberellin levels are detected using a bioassay in which the level of α-amylase is a function of the GA-like activity (from Hilton & Smith, 1980)

the efflux of gibberellins mediated by phytochrome is the cause of the cereal leaf unrolling. As in the previous section however it is the time scale which poses one of the greatest problems for this interpretation, for the red-light-induced unrolling response *in vivo* can still be reversed with far-red light long after the release of the gibberellins from the etioplasts and indeed after the peak in giberrellin level *in vivo* has passed.

The other sort of approach to the question is the use of quantitative studies of the interaction between the effects of red light and exogenous phytohormones. For example, both red light and gibberellin stimulate the germination rate of some varieties of lettuce. Black (1969) demonstrated a strong multiplicative effect of sub-optimal amounts of red light and GA_3 in promoting germination. This is the result expected if phytochrome and gibberellin act independently of each other but on a common pathway (as in [3] above). A similar approach was used by Bühler *et al.*, (1978) to test the hypothesis (Burg, 1973) that ethylene acts as a mediator of phytochrome action with regard to anthocyanin synthesis. Again, it was found that the effects of light and ethylene were multiplicative, similarly indicating independent effects of phytochrome and ethylene on the same pathway. There are also many instances of additive interactive effects between gibberellins and cytokinins and phytochrome (e.g. in extension growth of internodes), suggesting that, often, phytochrome and these hormones act independently *and* on different pathways (as in [2] above) but leading to similar physiological responses (see Wareing & Thompson, 1976). It should be noted however that these conclusions are based almost entirely on experiments involving exogenous application of hormones, and which may therefore not reflect changes occurring *in vivo* at the relevant sites of action. Even so it is obvious that considerable caution is necessary in the interpretation of experiments designed to test the hypothesis of hormones as second messengers. Nevertheless, even if hormones do not always form part of a direct route between photoreceptors and responses, it seems hardly possible that the often substantial changes in hormone levels mediated by phytochrome should be without developmental consequences.

14.3.4 PHYTOCHROME AND GENE EXPRESSION

The molecular biologist interprets morphogenesis as the spatial and temporal regulation of gene expression, thereby reducing the central problem of developmental biology to the study of the mechanisms of regulating gene expression. Even though not everyone would accept this approach it is not surprising that there is an enormous body of work concerning the effects of phytochrome and other photoreceptors on nucleic acid and protein metabolism. All too often in the past, the control of gene expression has been more or less equated with control of gene transcription. The doubtful wisdom of this will be obvious to the reader of Chapters 2, 5 and 6 in this volume. The potential sites for control of gene expression are, of course, far more varied, viz.

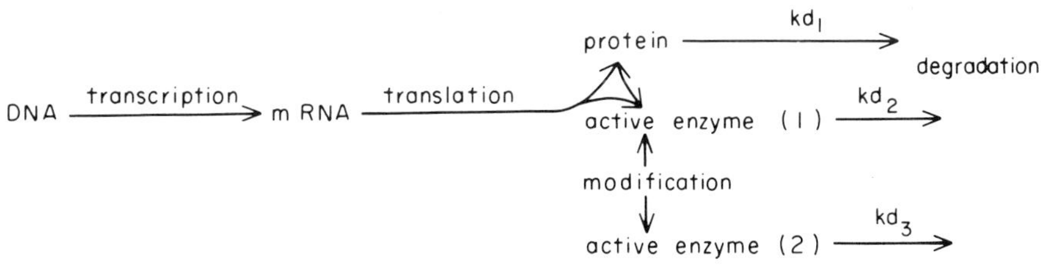

Even this large number of opportunities for control represents a gross underestimate of the number of possible mechanisms; the 'translation' step could for example be regulated at any point from amino-acylation of tRNA to the release of completed polypeptides from the ribosome (see Vol. I, Chapter 10).

The tendency to regard gene expression as tautologous with gene transcription is derived, no doubt, from the work with bacterial control systems, culminating in Jacob and Monod's (1961) elegant work on the lactose operon of *Escherichia coli*, where specific interactions between regulatory proteins and the genetic material have been demonstrated. It was surely the impact of this elegant study which inspired the model of phytochrome action of Mohr (1966). His proposal was, essentially, that there were genes with which phytochrome could interact directly (presumably *via* specific regulatory proteins) to bring about photomorphogenesis. The subsequent discovery of numerous rapid phytochrome responses eliminates this mechanism as the sole mode of phytochrome action, though not as the single pathway to photomorphogenesis. Mohr has more recently argued (e.g. Mohr, 1978, p. 113) that we should consider there to be a multiplicity of primary actions of phytochrome, one or more of which involves an interaction of phytochrome with the genome. The question of possible multiple primary actions *per se* is discussed in the final section of this chapter, but the extent to which the change in emphasis modifies the original theory is obvious and is well illustrated by a consideration of the experiments required to test it. Thus, in its original form the theory would be disproved by the demonstration of a single instance of phytochrome control of gene expression at a stage later than transcription. In its present form, taken to its extreme limit of interpretation, to discount it the critic would have to prove that there was no single instance of direct interaction between phytochrome and the genome; of course an impossible task.

It is now clear that light modifies gene expression at a number of different levels and over a wide range of time scales. However, even the fastest phytochrome-mediated enzyme response is considerably slower than the rapid effects described in Section 14.3.1 and there does not seem to be any strong evidence available at present which precludes the participation of one or more of them in the photocontrol of gene expression.

A change in gene expression is usually detected operationally by changes in enzyme activity. A very large number of enzymes have been shown to be under phytochrome control (see Schopfer, 1977, for a near exhaustive list). Indeed so many enzymes in the mustard seedling were found to be controlled by phytochrome that the Freiburg group found it necessary to verify that there are indeed some enzymes whose activity is unaltered by phytochrome action! The fastest phytochrome-mediated modulation of enzyme activity to have been observed so far *in vivo* is that of nitrate reductase in the cotyledons of *Sinapis alba*. In response to a short pulse of red light an increase is detectable within five minutes (Fig. 14.15). As might be expected the response is insensitive to cycloheximide, whereas a nitrate-induced increase in activity, which can be ob-

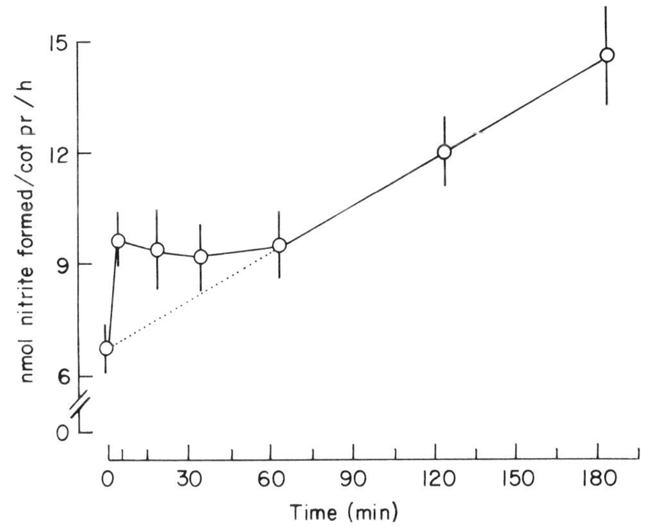

Fig. 14.15 Effect of 5min red light on nitrate reductase activity in 40h old etiolated mustard cotyledons (from Johnson, 1976)

in identical tissue of the same age, is almost completely blocked by the drug. This and the time scale clearly indicate a post-translational phenomenon.

The enzyme phenylalanine ammonia-lyase (PAL) has been extensively studied as a model example of the photocontrol of gene expression. Again activation by light has been demonstrated, but not ubiquitously. For example in gherkin hypocotyls light causes an increase in activity after a lag phase of about an hour, which reaches a maximum within four hours of the onset of irradiation and declines. Blue light (which may be acting through phytochrome and a blue-light receptor) is the most effective but far-red light (which acts only through phytochrome) will also elicit the response. Using the technique of density labelling, with deuterium oxide as the heavy isotope being supplied at the same time as the onset of irradiation, Attridge and Smith (1974) were able to show that the light-mediated increase in activity occurs in total absence of incorporation of density-label. In comparison, the dark control, treated with ^2H_2O at the same time incorporated substantial amounts of the label during the same period. This has been interpreted as light-mediated release from an inactive form of enzyme and is supported by evidence for inactive PAL in this tissue in the dark (Attridge & Smith, 1973) and the isolation of a proteinaceous inactivator of the enzyme (Billett *et al.*, 1978). In other systems the situation is more complex. For example, in *Sinapis alba*, where phytochrome is certainly the photoreceptor for a somewhat slower response there is considerable controversy concerning the effect of far-red light on the rate of PAL synthesis (see Smith *et al.*, 1977, and Schopfer, 1977, for two different views of this). Finally, it has been shown that, in cell suspension cultures of *Petroselinum hortense*, white light causes an increase in PAL activity commensurate with an increase in the rate of translation of the mRNA for PAL (see Hahlbrock and Grisebach, 1979). Part of the discrepancy between the various results is probably caused by the fact that the rate of PAL synthesis is known to be regulated also by the reaction products (Johnson *et al.*, 1975). Hence, a plausible explanation for the apparently conflicting data: light regulates the activity of the total enzyme pool (as demonstrated in gherkin), the reaction products regulate the rate of synthesis. This hypothesis predicts that the actual rate of PAL synthesis depends on the product level—itself a major variable depending both on the activity of PAL and of the subsequent enzymes in the pathway. If this hypothesis is correct then, in the long term, phytochrome-mediated changes in PAL activity (and indeed in that of other enzymes) would be expected to involve *indirectly* changes in their rates of synthesis: essential if plant cells are not to be packed full of inactive enzymes!

The model studies on PAL in mustard were accompanied by investigations into another enzyme, ascorbate oxidase. Here there seemed to be no problem. The same density-labelling technique used for PAL showed clearly a marked stimulation of incorporation of density-label in response to far-red light (Attridge, 1974; Acton *et al.*, 1974), apparently indicating increased synthesis promoted by phytochrome. However even here there is now room for doubt. A recent study, using immunological methods, has shown that the level of immunoreactive ascorbate oxidase protein is quite unaffected by the far-red light treatments (Newbury & Smith, 1981).

Despite the difficulties illustrated by these examples it is quite clear that the rate of protein synthesis can be stimulated within quite a short time (< 2h) by light acting through phytochrome. Smith (1976) has shown that in *Phaseolus vulgaris*, the overall proportion of ribosomes present as polysomes (and hence assumed to be active in protein synthesis) is increased by a short pulse of red light (reversible by far-red) or by continuous far-red light (Fig. 14.16). Even when synthesis of poly-adenosine-containing RNA (known to be predominantly mRNA) was about 90% inhibited with cordycepin, the effect of light on polysome levels was altered hardly at all. It seems that even those changes in enzyme activities which may be the result of this increased protein synthesis are controlled predominantly at the level of translation rather than transcription.

What of transcriptional control? In 1977 it was possible for Smith to argue convincingly that 'transcriptional control would appear to have only a minor role, if any, in the photo-regulation of plant development'. This statement was based in part on the strong evidence for control at stages later than transcription but also on the sheer experimental difficulty of estimating messenger RNA quantitatively. The methodology for such estimation is now becoming available and is based on isolation of mRNA and its translation in a cell free system containing the factors necessary for protein synthesis. The isolation of mRNA has been made possible following the discovery that many mRNAs contain sequences of poly(adenylic) acid which enables them to be purified using affinity-chromatography methods, e.g. with oligo-deoxythymidine-cellulose.

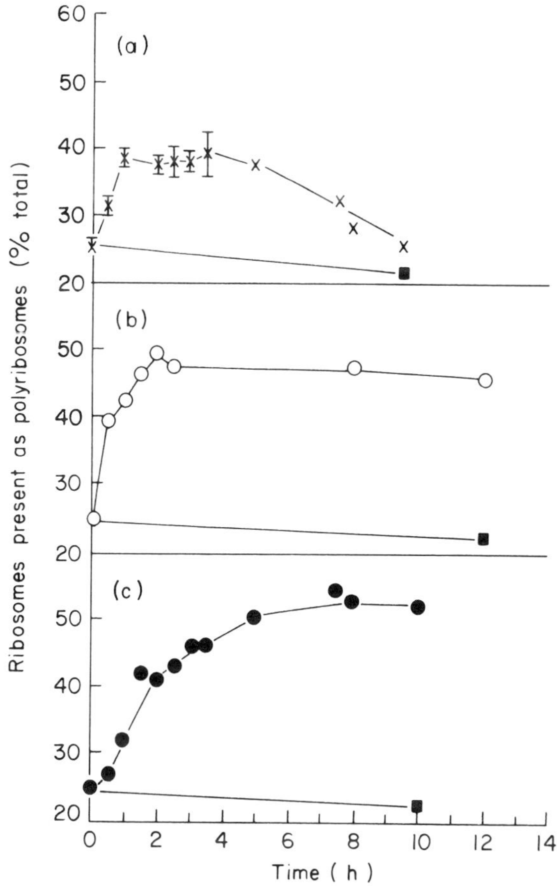

Fig. 14.16 The effect of light on polyribosome proportions in bean leaves. Leaves were treated either (a) with 10min red light and returned to darkness, (b) with continuous far-red light or (c) continuous white light (from Smith, 1976)

The first evidence consistent with phytochrome control of transcription has recently been obtained using these methods (Apel, 1979). It has been shown that the increase in the rate of formation of the light-harvesting chlorophyll a/b protein, following a pulse of red light, is accompanied by the appearance in the poly(adenylic) acid-containing fraction of a messenger RNA which could be translated *in vitro* to produce a product, recognizable immunologically as the light harvesting protein (Fig. 14.17). There is a lag phase of about half an hour prior to the increase in mRNA activity, of course leaving ample time for the interpolation of other events more closely linked with the photoconversion of Pr to Pfr.

It is too early to say how many other phytochrome mediated phenomena will involve transcriptional control. Indeed it is even slightly premature to

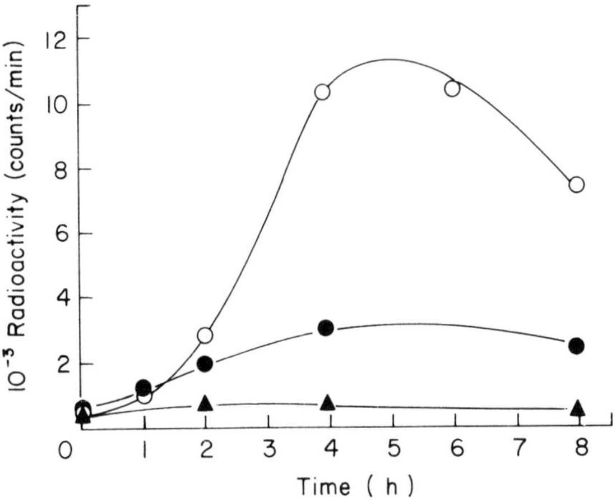

Fig. 14.17 Time course of changes in the relative concentrations of the mRNA coding for the approprotein of the light-harvesting chlorophyll a/b protein, estimated from immuno-precipitable protein synthesized *in vitro*. Open circles: following 15s red light; closed circles: after 15s red followed by 5s far-red light; triangles: after 5min far-red alone (from Apel, 1979)

interpret Apel's data in this way: there is the possibility that the increase in messenger activity he observed is the result of a reduced rate of turnover of the message rather than increased synthesis. Even so it seems likely that proof of transcriptional control will be increasingly forthcoming, now that the methodology is available.

The present situation concerning the photocontrol of gene expression leaves many questions open. Nevertheless we may draw two crucial conclusions. First, the regulation of gene expression by light (and indeed by phytochrome) cannot be equated with the control of transcription nor even with control of enzyme synthesis. Regulation at every level is involved. Secondly, there is no compelling evidence for a direct interaction of phytochrome with the genome.

14.4 PHOTOMORPHOGENESIS IN NATURAL RADIATION ENVIRONMENTS

For a number of reasons it is much easier to investigate problems of photomorphogenesis by using etiolated rather than light-grown plant material. To start with, it is much simpler to detect phytochrome in seedlings which have not been exposed to the light for any length of time. Indeed once plant tissues contain even small amounts of chlorophyll it becomes very difficult to measure phytochrome spectrophotometrically because of the overlapping absorption spectra of phytochrome and chlorophyll. Also, the phytochrome content of plants exposed to the light declines considerably because of Pfr destruction. If these were not problems enough, the variations in daylength, light intensity, temperature, etc., which characterize natural environments, and which all exert major effects on basic cellular metabolism, especially photosynthesis, create formidable difficulties for the study of photomorphogenesis under realistic growth conditions. In recent years however, photobiologists have begun to take seriously the fact that plants do not normally grow in the dark or under continuous far-red light and have attempted to understand how photomorphogenesis may be affected by natural variations in the radiation environment.

There are, basically, four aspects of the natural radiation environment which plants might detect through their photoreceptors and exploit. These are:

(1) The spectral quality of the radiation during and at the end of the day;

(2) The fluence rate of the light—the number of quanta incident per unit time;

(3) The duration of light in a day or, perhaps no less important, the duration of darkness.

(4) The direction of the actinic beam (see also Chapter 15).

14.4.1 LIGHT QUALITY AND PLANT DEVELOPMENT

It is only during the last few years that the real significance for photomorphogenesis of the variation in the spectral quality of natural radiation has been perceived. In 1972 Smith began what has become a thorough and fascinating study which clearly demonstrates the significance of this natural variation for phytochrome action. The basic observations exploited by Smith and his co-workers are so well known that, with hindsight, it seems remarkable that they previously received so little attention from photobiologists. From Fig. 14.18 it can be seen that the spectral distribution of daylight (upper curve) is modified strikingly by transmission of the light through one or more leaves. The leaf acts primarily as a long-pass filter, transmitting much of the light above 700nm (far-red and beyond) but absorbing most of the visible radiation. The importance, with respect to phytochrome action, of these differences in spectral quality is that the proportion of red to far-red light (absorbed prediminantly by Pfr and Pr respectively) is altered significantly by them. The proportion of phytochrome which would be present as Pfr (known as the phytochrome photoequilibrium and often abbreviated as ϕ) under different natural conditions is shown in Table 14.3. It can be seen that leaf-shade lowers ϕ considerably. There is also an end-of-day change in ϕ due to an increase in the proportions of both blue and far-red content of the light, for purely physical reasons (Holmes & Smith, 1977a). The end-of-day effect is relatively small but might nevertheless be important in the control of circadian rhythms and photoperiodic timing (see Section 14.4.3).

By utilizing artificial light sources in which a variable amount of far-red light is added to a constant background of photosynthetically-active white light it has been shown that a wide range of plants respond in a variety of ways to such

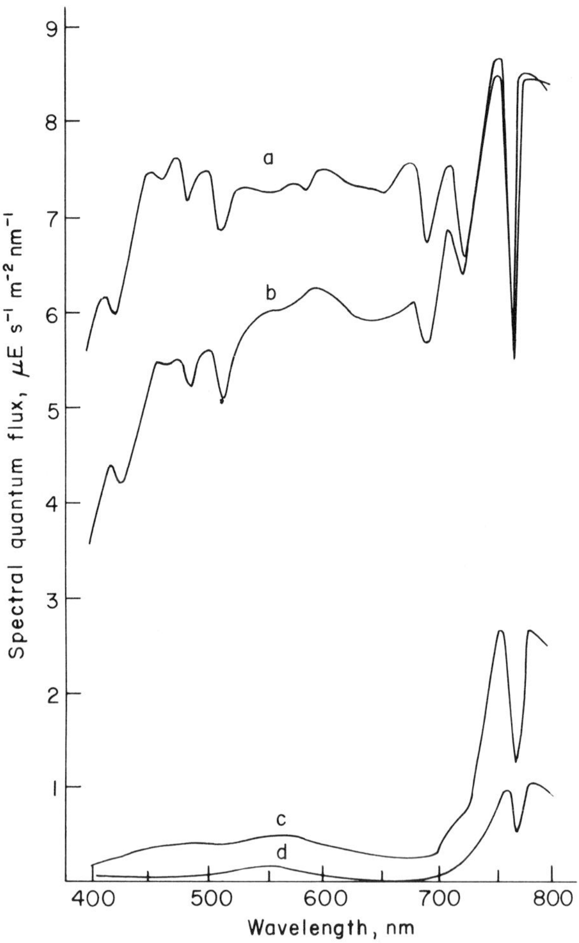

Fig. 14.18 Spectral energy distribution of daylight above and within a wheat canopy: (a) above the canopy; (b) sunfleck; (c) shade 60cm above ground; (d) shade at ground level. The canopy height was 85cm (from Holmes & McCartney, 1976)

changes in light quality. For example, enzyme levels, chlorophyll content, leaf area, the dry weights of leaves and stems and the germination rate of light-sensitive seeds have all been shown to be functions of the phytochrome photo-equilibrium as affected by these light treatments.

One of the main differences due to naturally-occurring variation in light quality is in the rate of stem extension, illustrated here in *Chenopodium album* and *Impatiens parviflora* (Fig. 14.19). Characteristically, stem extension growth is logarithmic and Morgan and Smith (1978) demonstrated that an inverse linear relationship between ϕ and the logarithmic extension rate can be observed (Fig. 14.20) in a range of species. Such a response can be readily explained

Table 14.3 Examples of red/far-red ratios and phytochrome photo-equilibria typical of some natural and artificial irradiation conditions. (From Holmes & Smith, 1975 with additions)

Irradiation conditions	Red/far-red ratio	Phytochrome photo-equilibrium
Sunlight (midday)	1.1	0.65
Sunlight (dusk)	0.75	0.5
Under sugar-beet canopy	<0.05	<0.1
Under wheat canopy	0.2	0.3
Beneath beech wood (June)	0.25	0.3
Beneath oak wood (June)	0.4	0.35
Beneath birch wood (September)	0.6	0.45
Fluorescent white light	13.5	>0.7
Incandescent white light	0.7	0.5

ecologically as an adaptation for avoiding shade by growing out of it. Indeed Smith has proposed that the function of phytochrome under natural conditions is to detect mutual shading and to modify the plant's development accordingly.

The stem extension response is, at least in part, a fairly fast one. Morgan *et al.*, (1980) used a sensitive transducer to detect very small changes in growth rate and irradiated the internodes or leaves of single plants with supplementary far-red light, via a fibre-optic probe. They showed that addition of far-red light to the internode led within 15min to an increase in extension rate, which could be partially reversed when the far-red light was switched off (Fig. 14.21). Only after 2–3h irradiation with the supplementary far-red light could an increase in extension rate be observed if only the leaf was irradiated. Thus, from the kinetics it seems that at least the rapid constituent under natural conditions may be interpretable in terms of the early phytochrome responses discussed in Sections 14.3.2 and 14.3.3. Here, as there, it seems unlikely that regulation of gene transcription or translation is involved in the initial phase of the response.

14.4.2 RESPONSES TO FLUENCE RATE

It is clear that not only the spectral quality of light but also the fluence rate is of importance in shaping the growth and development of plants. Indeed, prior to the work of Smith and his colleagues most emphasis had been placed on studies utilizing neutral-density screens to alter irradiance without affecting spectral quality. Such changes may be expected to affect not only phytochrome but also the photosynthetic pigments and blue-light-absorbing photoreceptors and, needless to say, many of the differences observed using such methods are, directly or indirectly, the result of changes in the rate of photosynthesis. The question arises; are the plant's developmental photoreceptors, especially phytochrome, capable of detecting fluence rate as well as spectral quality, in plants grown for prolonged periods in the light?

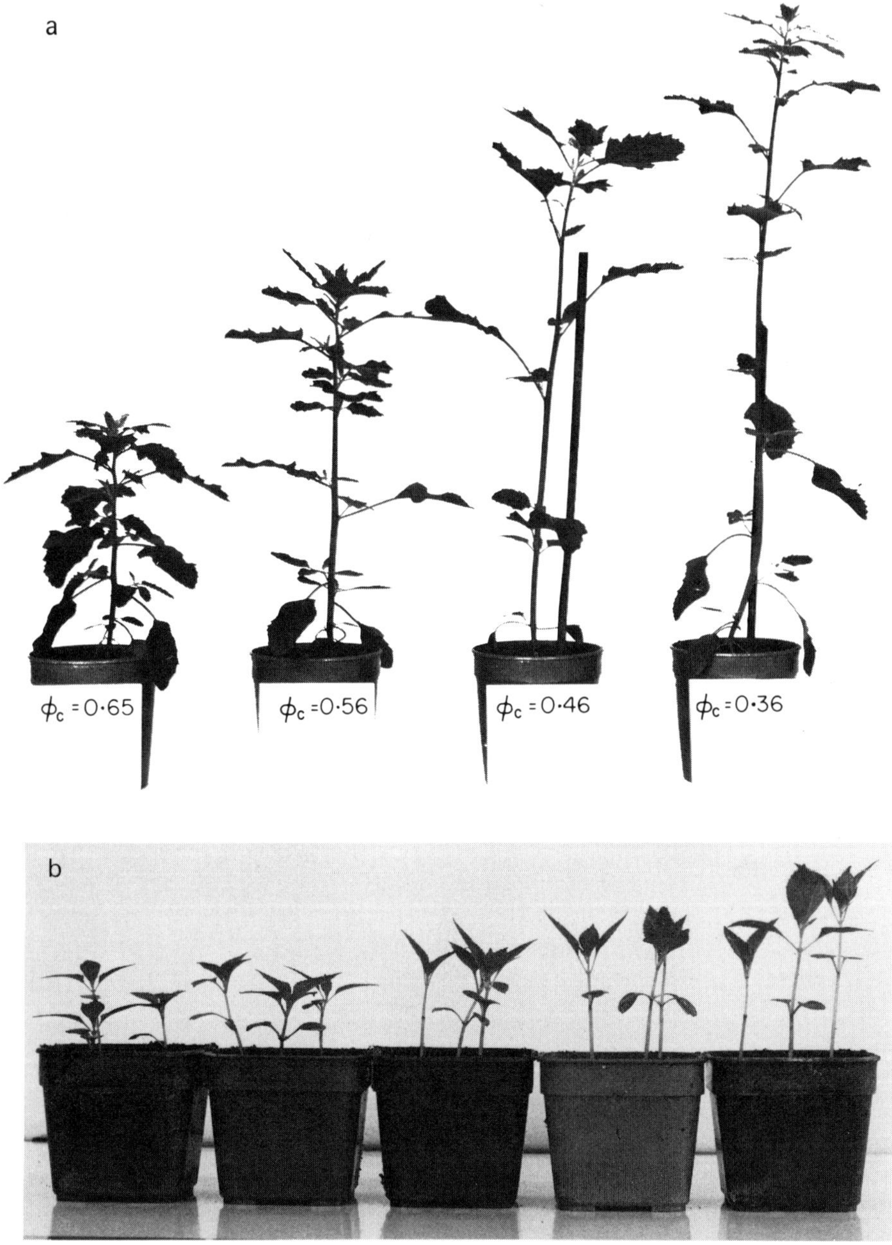

Fig. 14.19 Effect of phytochrome photoequilibrium on plant growth under simulated natural conditions (a) *Chenopodium album* seedlings grown for four weeks under standard white light sources then transferred for ten days to white light of equal photosynthetically active radiation (PAR) but with added far-red to establish the stated photoequilibria (ϕ) values. (b) *Impatiens parviflora* plants from a woodland habitat, grown for only four days under artificial light, again of equal PAR, with, from right to left in this case, progressively increasing fluence rates of far-red light added. The photoequilibrium in etiolated test material ranged from $\phi = 0.67$ on the left $\phi = 0.10$ on the right.

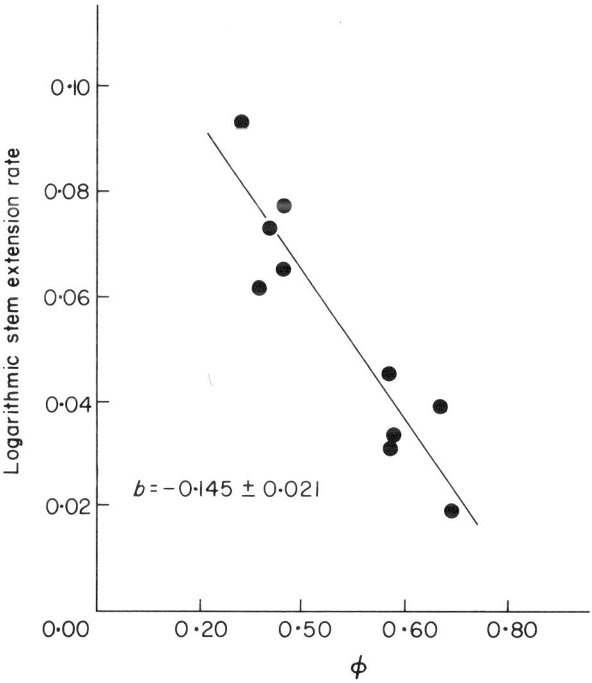

Fig. 14.20 The relationship between phytochrome photoequilibrium (ϕ) and stem extension rate in *Chenopodium album* (from Morgan & Smith, 1978)

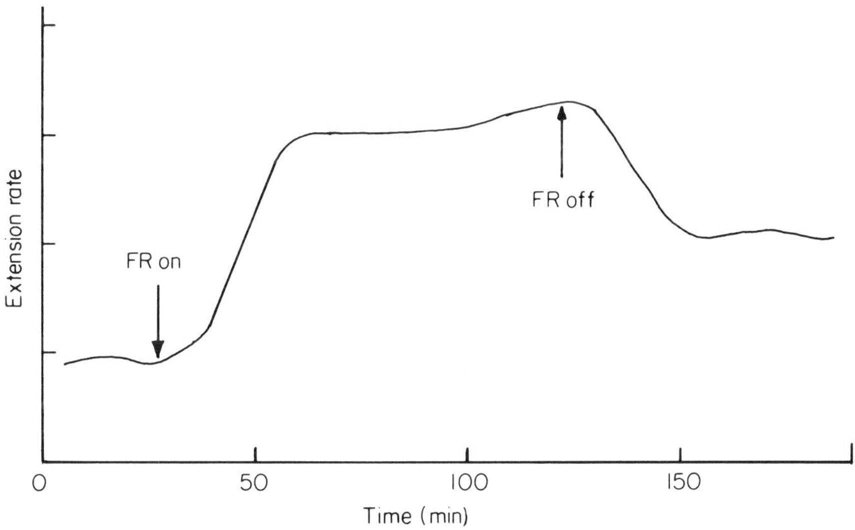

Fig. 14.21 Extension rate of *Sinapis alba* seedling during irradiation with white light (fluence rate $23 \mu mol\, m^{-2} s^{-1}$) to which far-red light (fluence rate $27 \mu mol\, m^{-2} s^{-1}$) was added where indicated. The phytochrome photoequilibrium was altered from 0.65–0.46 by the addition of the far-red light (from Morgan *et al.*, 1980)

The experiments described in the previous section strongly suggest that the plants are responding there predominantly to changes in light quality. However, it was mentioned in Section 15.2.1 that the responses of etiolated seedlings to continuous light were strongly irradiance dependent. Furthermore, the procedure of adding increasing fluence rates of far-red light to lower ϕ means that, even though the photosynthetically active radiation is constant, the lower the photo-equilibrium yielded by the light source the greater is the fluence rate of light which can be absorbed by phytochrome. Hence, it is pertinent to ask: to what extent is phytochrome in light grown plants measuring light quality directly and to what extent is the irradiance detected?

Physiologically, it seems obvious that it is the spectral quality that is being detected by phytochrome. Consider Fig. 14.19 again. The plants in the simulated shade are extending faster and we assume this to be an adaptation which may enable them to out-top the imagined surrounding vegetation indicated by the shade light quality. In fact however, the simulated light shade is of identical photosynthetically active content. Even more obvious, at least in the case of *Impatiens*, is the effect of the different light treatments on the leaf angle. In the plants in 'artificial sunlight' the leaf angle is lower than the simulated shade. When fully expanded the leaves would be horizontal in the shade and drooping in the sunlight and it may be that this response is an adaptation to reduce the incident irradiation in sunlight and protect the plant against water loss. However, in the growth chamber the total radiant energy (including the added far-red) is three or four times greater in the simulated shade light than in the simulated 'sunlight'. Since similar responses occur in nature when shade light is also much less bright, it is obvious that it is the spectral quality and not the fluence rate to which the plant is responding.

Convincing though this may seem as an explanation of the physiological relevance of light quality vs quantity, it is not conclusive if our concern is about the *mechanisms* of phytochrome action in light grown plants. It does not answer the important question: can phytochrome still detect changes in fluence rate in light grown plants or must these be perceived by other photoreceptors?

Whitelam *et al.*, (1979) attempted to answer this question using the photocontrol of nitrate reductase in cauliflower curd as a model system. The plant material was selected because it is a rare example of a light-grown plant tissue which does not contain chlorophyll, so one photoreceptor is eliminated from the experiment. Using the same light sources as described in the previous section it was shown that there is a typical, linear relationship between ϕ and the response (Fig. 14.22). However, it was found that the response obtained was identical if 48h irradiation was given or a short, non-saturating pulse of red light, just sufficient to produce the same photoequilibrium (Table 14.4). Thus small differences in ϕ yield different responses but for the same value of ϕ the response is unaffected by a difference of four orders of magnitude in the light dose. Thus, whilst it would be premature, based on the limited data as yet available, to assume that phytochrome is not able to respond to the fluence rate

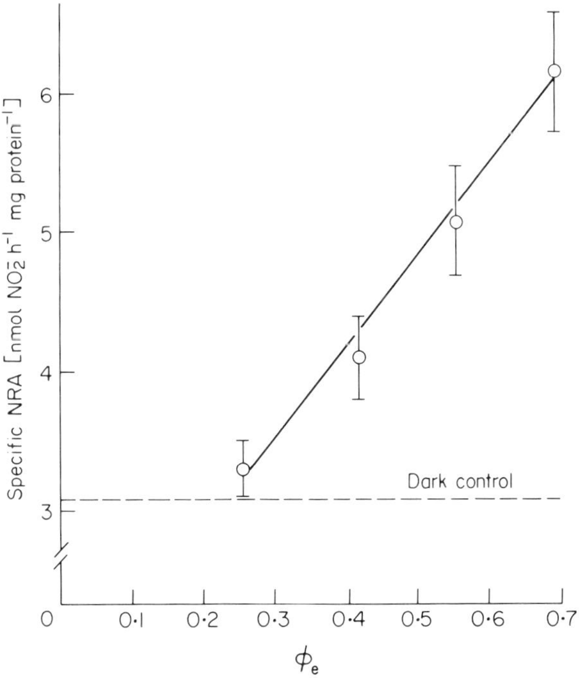

Fig. 14.22 The relationship between phytochrome photoequilibrium (ϕ) and nitrate reductase activity in excised florets of cauliflower curd (from Whitelam *et al.*, 1979).

in plants growing under natural conditions, it does seem likely at present that phytochrome detects predominantly light quality leaving the perception of irradiance to the photosynthetic and blue-light-absorbing pigments. Unfortunately next to nothing is known at present about the role of the blue light receptor in plants grown under natural radiation conditions.

Table 14.4 Summary of the effects of varying the fluence of actinic light on nitrate reductase activity relative to a dark control, from cauliflower curd. (From Whitelam *et al.*, 1979)

Light source	Photon fluence rate μmol m^{-2}s^{-1}	Duration	ϕ	Relative NRA
Broad band	130	48h	0.26	1.10 ± 0.04
Red	6	1min	0.29	1.01 ± 0.02
Broad band	130	48h	0.42	1.31 ± 0.09
Red	6	5min	0.42	1.27 ± 0.03
Broad band	130	48h	0.69	1.86 ± 0.15
Red	6	8h	0.66	1.87 ± 0.22
Broad band	300	2min	0.70	1.90 ± 0.30
Broad band	300	8h	0.70	2.01 ± 0.20

14.4.3 RESPONSES TO THE DURATION OF LIGHT

The seasonal changes in the length of day (and night) are potentially important sources of information to both plants and animals. The ability to detect and respond to such changes confers the opportunity of avoiding or adapting to impending unfavourable conditions, for example extreme heat or cold, and to exploit an approaching favourable season. Plants use the information obtainable from measurement of daylength to time many of their developmental stages, e.g. flowering, seeding, dormancy and the acquisition of low-temperature hardiness. The basis for such a photoperiodic response is the ability to distinguish between light and darkness. It is known that the photoreceptor for this response in plants is phytochrome. Indeed, one of the earliest phytochrome action spectra to be carried out was for the photoperiodic induction of flowering in *Xanthium* (Parker *et al.*, 1946).

It is thought that two factors are involved in the time-measuring mechanism. The first is some sort of mechanism actually to detect the light/dark transition, and the second an interaction with circadian oscillations to insert the information into the plant's basic timing mechanism. Many plants change from short-day to long-day behaviour and vice versa after a critical day length has been attained. This day length can often be very precisely determined. For example, in *Xanthium* a single day shorter than 15.5h will provide the signal for flowering which then persists for up to 12 months, even in continuous light.

The nature of the 'darkness-detecting' mechanism is not understood but two experimental approaches indicate the existence of such a mechanism and indicate that it is the night length which is actually measured. First, where there is a sharp transition betweeen long-day behaviour and short, as with *Xanthium*, it has been shown that if long days are coupled with long nights (instead of short) the plant behaves as if it were in short-days. Short days coupled with short nights yield a long day response. Secondly, a short pulse of red light given in the middle of the long night negates the effect of the long night. The effect of such a red pulse is far-red reversible (Table 14.5). Both these experiments indicate that the plant can recognize the duration of the dark period. There seem to be three main suggestions for the mechanism of detection:

(1) Loss of Pfr during the darkness until a critical level is reached acts as a timer. If the level falls below the critical level, the signal for change of behaviour is given;

(2) Cessation of phytochrome cycling or reduction to a critical level at dusk and restarting at dawn provides markers for both end and start of day;

(3) Changes in the red/far-red ratio at dusk and dawn lead to changes in Pfr level providing two synchronization events.

The first hypothesis, which is also the oldest, differs from the others in that it does not necessarily involve an interaction with circadian rhythms. The dusk signal is proposed to be followed by the loss of Pfr and the critical night length could simply be the time taken for Pfr to fall to a critical level. The basic

Table 14.5 Effect of daily interruptions of the dark period with several consecutive irradiations with red and far-red in sequence on flower initiation of *Xanthium* and Soybean. (From Wareing & Phillips, 1978)

Treatment	Mean state of floral development in *Xanthium* (Arbitrary units)	Mean no. of flowering nodes in *Biloxi* soybean
Dark control	6.0	4.0
R	0.0	0.0
R,Fr	5.6	1.6
R,FR,R	0.0	0.0
R,FR,R,FR	4.2	1.0
R,FR,R,FR,R	0.0	—
R,FR,R,FR,R,FR	2.4	0.6

assumption of this hypothesis; that there is consistently a loss of Pfr during the night, is in some doubt, however. Whilst in controlled environment rooms, there is an immediate clear-cut transition between light and darkness the situation in nature is otherwise. Not only is there a gradual change from light to darkness but also the darkness is often less than complete. Recently, Holmes and Wagner (1980) compiled a number of measurements from the literature indicating that, for example, the fluence rate in moonlight would be sufficient to prevent significant loss of Pfr during a moonlit night. This conclusion is supported by earlier experimental observations. Takimoto and Ikeda (1961) showed, for example, that *Pharbitis* plants regard light as 'darkness (with regard to photoperiodic phenomena at least) when the fluence rate falls below about $1 Wm^{-2}$. Plants transferred to complete darkness from daylight when the fluence rate had fallen to this level behaved identically, with regard to the critical daylength, to those left in the light until the fluence rate had fallen naturally to $0.01 Wm^{-2}$. At a fluence rate of $1 Wm^{-2}$ photoconversion of Pr to Pfr is saturated within minutes so it seems that the plant can register 'darkness' even when there is no loss of Pfr due to failure to photoconvert.

Similar difficulties also hinder the phytochrome cycling hypothesis. Again, at $1 Wm^{-2}$, the threshold in *Pharbitis*, there would be considerable cycling occurring. Furthermore, the attainment of a critical cycling rate, which potentially offers a way out of this difficulty, would occur at very different times from one day to the next, since, depending on cloud cover, etc., the morning and evening fluence rates vary greatly from day to day, even more, one imagines, than the ultimate level of 'darkness'.

The most consistent change which invariably occurs at dawn and dusk is an alteration of the red/far-red ratio (Fig. 14.23). Whilst the maximum change of

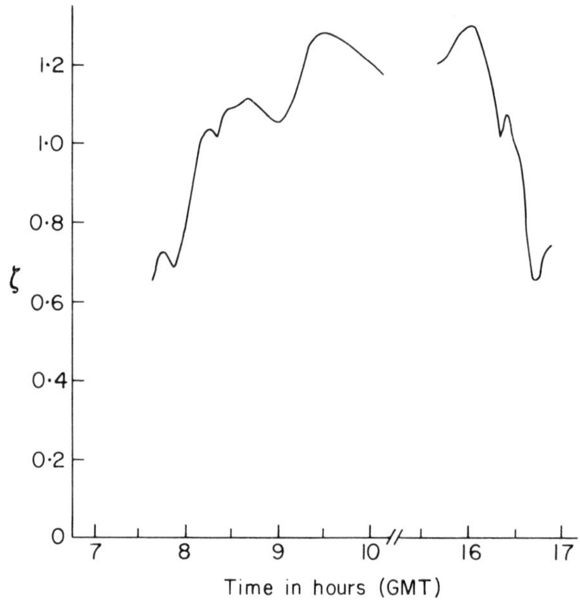

Fig. 14.23 An example of the changes which normally occur in red/far-red ratio at dawn (07.53 GMT) and dusk (16.32 GMT) (from Holmes & Smith, 1977b).

phytochrome photo-equilibirum (and hence, of Pfr) which would occur as a result of such changes is only from about $\phi = 0.60$ to $\phi = 0.45$, the regular occurrence of such small changes might represent more valuable information than more erratic changes of greater magnitude.

Because we have been unable to measure phytochrome in green plants and because photoperiodic phenomena are not observed in etiolated tissues, our understanding of the role of phytochrome in the perception of time is inevitably less advanced than its applied importance merits. The imminent availability of quantitative immunological assays of phytochrome in green plants (see Hunt & Pratt, 1980) may help, but we shall not really begin to make progress in this field until it is possible to measure both phytochrome and phytochrome photo-equilibria in green tissues, a prospect which is not immediately in view.

14.5 OUTLOOK

In the previous sections discussion has been concentrated mainly on attempts to establish plausible chains of events between light perception and the ultimate biological displays. The present state of our knowledge of phytochrome action may be summarized baldly thus: there is not a single instance in which we can implicate with reasonable certainty any specific intermediate effector between phytochrome and the ultimate biochemical mechanism of a biological response.

A number of early responses have suggested themselves as possible candidates for primary events in phytochrome action but evidence that these events are the essential mediators of known responses is lacking, despite intensive effort. At the other end of the photomorphogenetic response chain, the mechanism of control of gene expression has itself been shown to involve several different levels of control. Again, there is little evidence as to the extent, if any, to which these different mechanisms share a common pathway from phytochrome (or indeed from any other photoreceptor). It is impossible at present to rule out the possibility that there is nothing necessarily in common between any two phytochrome-mediated responses other than the initial photo-event. This view has been cogently argued by Mohr and his colleagues at Freiburg (Frosch *et al.*, 1977; Mohr, 1979). The argument is based on observations that different phytochrome responses in the same organ show profoundly different characteristics in two important respects. First the times after germination at which different responses in the mustard cotyledon become accessible to phytochrome control vary from response to response. Secondly, the dose response curves linking phytochrome to the display are completely different in character. Prior to this coupling-point any effect of light on a subsequent response can be reversed by far-red light; i.e. the chain from Pfr to the response has not until then reached a point from which it cannot be recalled following reconversion of Pfr back to Pr. Thus, the effect of any light treatment given any time between sowing and 27h after on the enzyme PAL can be fully reversed by far-red light but on ribulose bisphosphate carboxylase, only up to 15h. Thus between 15h and 27h the next step of the chain between phytochrome and the response can occur for carboxylase but not for PAL. Hence, it is argued, the step immediately following Pfr formation must be different for the two responses: why else should it occur for one but not for the other? However, there is an escape route away from this conclusion. The argument so far rests on the assumption that none of the events in the action sequence leading from Pfr to the response, not even the 'primary events' would themselves be *reversed* if Pfr is removed. It will be noted however that both of the candidates for primary events discussed in Section 14.3.1, the pelletability phenomenon and the changes in electrical potential, are actually reversible following conversion of Pfr back to Pr. Hence, either of these events might still be the single long-sought after initial phytochrome mediated action.

The data do illustrate a more fundamental point. The 'starting points' at which these and other responses become coupled to the phytochrome system seem to be fixed and are not moveable by light treatments. Hence, whilst phytochrome regulates these phenomena and somehow brings about photomorphogenesis once 'competence' has been attained the mechanisms whereby the expression of particular genes becomes accessible to phytochrome control are not controlled by phytochrome. It is salutary to conclude by noting that little as we know about the mechanism of realization of photomorphogenesis, we know even less about the underlying mechanisms determining the pattern of photomorphogenesis.

FURTHER READING

MOHR H. (1972). *Lectures on Photomorphogenesis.* Springer-Verlag, Berlin.

MOHR H. (1978). Pattern specification and realisation in photomorphogenesis. *Bot. Mag. Tokyo Special Issue*, **1**, 199–217.

PRATT L. H. & MARMÉ D. (1976). Red light-enhanced pelletability. *Plant Physiol.*, **58**, 686–92.

PRESTI D. & DELBRÜCK M. (1978). Photoreceptors for biosynthesis, energy storage and vision. *Plant, Cell and Environment* **1**, 81–100.

QUAIL P. H. Phytochrome Action in '*The Molecular Biology of Plant Cells*' (ed. H. Smith) Blackwell Scientific Publications, Oxford.

SMITH H. (1975). *Phytochrome and Photomorphogenesis.* McGraw-Hill, U.K. Ltd., Maidenhead.

SMITH H. (1981). (Ed.) *Plants and the Daylight Spectrum.* Academic Press, New York, London.

CHAPTER 15

GRAVITROPISM AND PHOTOTROPISM

15.1 INTRODUCTION

The adaptive advantage for an organism to be capable of perceiving its position in space is evident. The widest choice of such behaviour in response to the environment is found of course in animals. In animals, unlike most plants, information is processed by the nervous system in milliseconds, leading to immediate reorientation without need for structural body changes. Sessile plants, nevertheless, can reorient various organs in response to stimuli. The 'sensitive plant', *Mimosa pudica*, for example, folds its leaves when touched or exposed to heat. Caused by cell turgor changes, this response is visible in less than a second, and might be compared to sensory responses in animals. This class of response is usually termed 'nastic' rather than 'tropic', and there are photonastic effects resulting from light stimulation whereas those resulting from physical contact are termed thigmonastic. Another important class of plant response, known as tropic responses, requires growth for expression although the perception of the stimulus may be rapid. Phototropism and gravitropism[1] are reorientations resulting from light and gravitational stimuli that usually require at least one of the many types of growth for them to be expressed. Some photoinduced reorientations do not require growth for expression, but these are functionally similar to growth-mediated phototropism and also involve perceiving the direction of light falling on the plant.

The range of organisms used in experimentation has been limited. Darwin (1881) performed beautiful experiments on phototropism with grass coleoptiles, and that fashion has continued ever since. More recently, the *Phycomyces* sporangiophore has been extensively employed in studies of the responses to light stimuli (Bergmann *et al.*, 1969). Coleoptile gravitropism has apparently attracted more investigation than root gravitropism. Green plants beyond the seedling stage have, until recently, been virtually ignored. Organisms may have evolved parallel yet different mechanisms, and different organs on the same plant may have separate means to perceive unilateral stimuli, amplify the signal, and then reorient. For this reason one must be wary of over-generalizations. Conclusions reached from observing coleoptiles must obviously be applied with care to other material.

The subject of this chapter is covered exhaustively in a strongly recommended and encyclopaedic volume (Haupt & Feinleib, 1979).

[1] *Gravitropism*, as opposed to *geotropism* is the term preferred for etymological reasons (see Volkmann & Sievers, 1979).

15.1.1 CHOLODNY–WENT AUXIN REDISTRIBUTION THEORY

Darwin's classic experiments represented the first experimental evidence for transmission of a tropic stimulus. He showed that, in grass coleoptiles, perception of the phototropic stimulus takes place in the tip, while the curving zone is several millimetres further down. During the following few decades it was shown that asymmetrically replacing an excised coleoptile tip could induce curvature without light. Finally, 57 years ago, Went (1925; 1929) in a well-known series of experiments, demonstrated that auxin, collected from excised coleoptile tips in agar blocks, could also induce curvature. About the same time Cholodny (1924; 1927) postulated that this is the mechanism of phototropic and gravitropic curvature with auxin being redistributed by the stimulus. This postulate, known as the Cholodny–Went theory, has become almost universally accepted by plant physiologists.

Despite the substantial evidence for this theory, it nevertheless has its limitations. In particular, problems arise when attempting to understand systems other than the coleoptiles for which the theory was first developed. It is even argued that the observed auxin differential cannot explain the observed growth differential between the two sides of a bending coleoptile.

The Cholodny–Went concepts have so coloured scientific thought that in later sections data must be considered as to how they relate to them. The recent controversy (Firn & Digby, 1980) over the applicability of the Cholodny–Went model is discussed in more detail in later sections.

15.2 THE PERCEPTION OF THE STIMULI

The same organ can respond to gravity, light or to both simultaneously. Structural observations of a grass coleoptile or a *Phycomyces* sporangiophore after reorientation cannot ascertain whether light or gravity was responsible. Equilibrium responses between the two tropisms can be induced experimentally. This suggests that the actual mechanisms, controlling the growth leading to the reorientation after gravity and light perturbations, are similar. It is obvious, however, that the perception mechanisms of the different physical stimuli must differ. Even for the same stimulus, a common perception mechanism may not apply to all organisms and organs studied.

15.2.1 GRAVITROPIC PERCEPTION

In total darkness (or uniform light), uniform humidity, and in the absence of obstacles, primary roots grow vertically downwards (positive gravitropism). Secondary roots may orient themselves at a fixed angle from the vertical. Studies on stimulus perception and orientation in secondary roots are sparse.

Each organ has a 'preferred' direction of growth, which is, of course, consistent with adaptation to the surroundings. Gravity acts on the sensory mechanism all the time, but as long as the sensitive tissue is oriented in the 'preferred' direction no macroscopic bending occurs.

It is not yet possible to say what is the first change to be initiated by exposure to the gravitational field in a new direction. Any discussion of gravitropic perception at the cellular level will therefore be partly speculative. Nevertheless, some data have accumulated and we can state under what conditions the receptor apparatus must operate.

As gravitational force is a physical stimulus, it is useful to consider how a physicist might view what happens when the angle between the direction of elongation (growth) and the gravitational field is changed. The first event is a change in the forces acting on components of the sensitive tissue. The prevailing steady state is perturbed, and this change can be expressed in several ways: the cell constituents which are sufficiently more dense than their surroundings will move downwards, the protoplast will press on a different cell wall, and tensions in the organ as a whole will change. Through these changes gravity does work on the system, and the energy thus transferred could be transduced to a physiological signal which initiates bending.

Physiological data are certainly needed to conclude which mechanism occurs. If a root or shoot is stimulated for a short time and then returned to its former orientation, a curvature will be produced, provided that the period of stimulation exceeds a threshold. This minimum period to induce a response is termed the *presentation time*. Presentation times in the order of minutes or less have been reported, for example three minutes for *Asparagus* epicotyls, (Hawker, 1932); two minutes for *Lepidium* roots (Audus, 1962), and, also for *Lepidium* roots, twelve seconds (Iversen & Larson, 1973). The presentation time measured depends on the criterion chosen for a visible response.

Recent estimates based on extrapolation to zero curvature gave values of 18 seconds for *Artemesia* roots and 20 seconds for *Avena* coleoptiles (Johnsson & Pickard, 1979). Although the accuracy of the extrapolation can always be questioned, it is clear that less than a minute is necessary for perception to begin, initiating an irreversible signal. What is this rapidly-induced signal?

15.2.1.1 *The statolith hypothesis : higher plants*

The possibility that dense cell constituents enable higher plant tissues to perceive gravity was suggested at the turn of the century by Haberlandt (1900) and by Němec (1900). Some internal, movable, sensor body would be needed which is analogous to the statoliths of invertebrate statocysts which are multicellular gravity-sensitive organs. The redistribution of dense amyloplasts in sensitive cells of nodes of the grass *Echinochloa* after reorientation is shown in Fig. 15.1 (Parker, 1979). The bundle sheath cells are clearly visible in the electron micrographs, showing the two orientations of the samples. In almost all gravisensitive higher plant organs, reorienting amyloplasts are found. Statolith redistribution precedes that of auxin (Fig. 15.2).

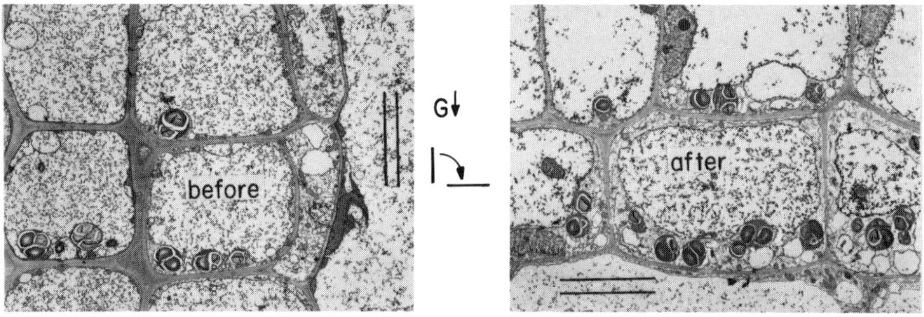

Fig. 15.1 Statoliths sedimenting in *Echinochloa* leaf sheath bases. Parallel lines indicate orientation of bundle sheaths. (Reproduced with permission from Parker (1979). Original print supplied by Dr. M. L. Parker.)

A simultaneous observation of two processes does not, however, imply *ipso facto* that one causes the other. Indeed, under certain conditions, amyloplasts sediment without causing any gravitropic curvature. This was shown in maize root caps by McNitt and Shen-Miller (1978). In the strain of maize studied, a red-light pretreatment is necessary for gravitropic curvature. Electron microscope studies of root cap cells showed sedimented amyloplasts in both red-light treated and dark-control material; on the other hand dictyosomes (Golgi

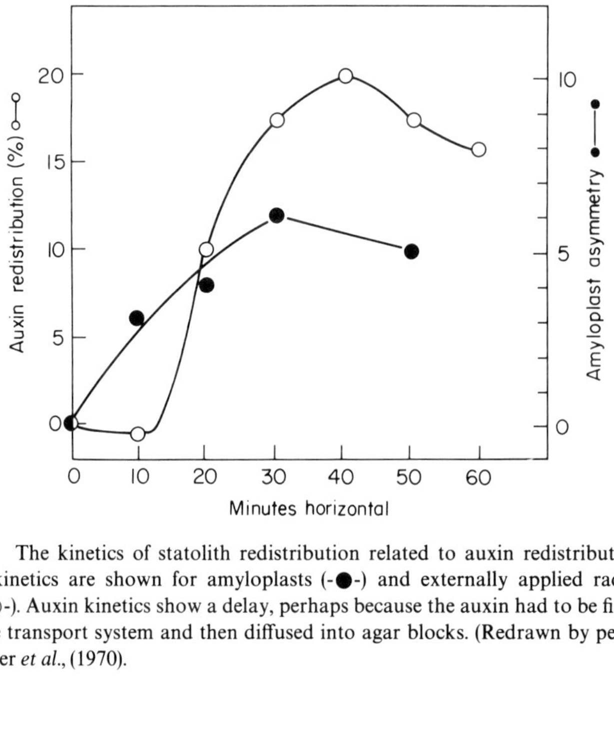

Fig. 15.2 The kinetics of statolith redistribution related to auxin redistribution. The parallel kinetics are shown for amyloplasts (-●-) and externally applied radioactive auxin (-○-). Auxin kinetics show a delay, perhaps because the auxin had to be first taken up by the transport system and then diffused into agar blocks. (Redrawn by permission from Filner *et al.*, (1970).

apparatus) were apparently found preferentially at the top of the red-light pretreated cells but not in the dark controls. Dictyosome rearrangement may be only a secondary consequence of gravitropic stimulation in certain cases, so this observation does not mean that the statolith hypothesis must be replaced by a dictyosome model. Another conclusion can be inferred: although amyloplast displacement alone is not always sufficient to initiate gravitropic curvature, it seems to occur whenever there is to be curvature.

If amyloplasts are important for graviperception, their absence should cause disorientation. Two lines of experimentation have been used to examine this question. One is the use of mutants. Maize mutant strains low in starch exhibit reduced responses to gravity (Filner *et al.*, 1970). The degree of curvature was related to amyloplast size. Conversely, phototropism is not affected by starch depletion meaning that the coleoptile, while not perceiving gravity, can still be induced to bend. Another related experimental approach is 'destarching' with gibberellin and kinetin. When coleoptiles were completely depleted in starch there was no gravity response; under conditions where some starch was left, a reduced response was obtained (Pickard & Thimann 1966).

15.2.1.2 *Localization of perception in the responsive organ*

It is difficult to isolate a part of an organ and state with certainty that this part is essential for perceiving gravity. Each plant and each organ may have its special sensitive regions. In an effort to localize perception, Juniper *et al.*, (1966) removed the caps of primary roots of maize, obliterating graviperception. After some time graviperception returned, but the recovery preceded the start of cap regeneration and could be associated with the stimulation of amyloplast development in the meristem by decapping (Barlow, 1974; Barlow & Grundwag, 1974). In negatively gravitropic organs (shoots and hypocotyls) the situation is less clear. In sunflower and pea hypocotyls the sensitive region is localized in the peripheral cell layers (Iwami & Masuda, 1976; Firn & Digby, 1977). In the grass *Echinochloa*, perception and response both occur in the nodes (Wright & Osborne, 1977). Often, cells are considered to be perceptive if they contain sedimenting amyloplasts. Such cells, when found in sensitive tissues, are termed statocytes. Other criteria are obviously still needed to provide an independent test of the statolith hypothesis.

15.2.1.3 *Alternatives to the statolith hypothesis*

Pickard & Thimann (1966) suggested that the protoplast presses in the direction of gravity. The pressure of the protoplast against the wall is, however, probably influenced more by hydrostatic relations than by gravity. Any change in water relations would interfere with gravitropic sensitivity. Also, such a mechanism would not explain inhibition of gravitropism by starch depletion (see Section 15.2.1.1).

On a still larger scale, gravity could alter tensions in an entire organ. The force on a region of tissue is proportional to its mass, and changing the orientation of the organ could modify tensions. Still, it is difficult to see how such changes could be coupled to metabolism and growth. Darwin & Darwin (1881) found that a negatively gravitropic shoot curves upwards even when supported along its entire length, which should negate the tension concept. Tension also seems to be incompatible with the observation that in roots, perception occurs in the cap and the cap is composed of loosely bound cells which in experimental situations are not under tension (see 15.2.1.2). An internal sensor seems necessary.

It must be emphasized again that there is little reason to insist that all organisms perceive gravity in the same way. The *Phycomyces* sporangiophore is sensitive to gravity, yet contains no microscopically recognizable statolith particles (Dennison, 1961). It is possible here that modification of tensions and subsequent deformation of the stalk, trigger the gravitropic response.

15.2.1.4 *Presentation time and the statolith model: theoretical considerations*

With the amyloplast statolith model as a useful working hypothesis, it is instructive to consider whether or not it is able to account for the short presentation times.

To obtain some idea of the problems involved, consider a simple physical model. This model was first used by Audus (1962) who showed that of all cell organelles, amyloplasts (because of their large size and high density) are the most likely to sense gravity. Two physical processes determine how far the statolith moves in a given time interval: settling due to gravity, and random Brownian motion of the statolith particle. We will assume that the amyloplast is a sphere of radius $\mathrm{r} = 2.5\mu\mathrm{m}$ and density $\rho = 1.5 \times 10^3 \mathrm{kg\,m^{-3}}$ moving in a uniform medium of viscosity $\eta = 2 \times 10^{-2} \mathrm{kg\,m^{-1}\,s^{-1}}$ (or 20 centipoise in cgs units). This last assumption is the most restrictive; cytoplasm is not uniform, and if the statolith should collide with a membrane or other intracellular structures, the viscosity will be much higher locally. According to Stokes Law, a particle sedimenting in a viscous medium quickly reaches a constant velocity,

$$v = \frac{F}{6\pi\eta \mathrm{r}} \qquad \text{Eqn 15.1}$$

where v is the velocity in $\mathrm{m\,s^{-1}}$, F is the gravitational force acting on the particle, and $6\pi\eta\mathrm{r}$ is the coefficient of friction given by the values of η and r chosen above for the medium and the amyloplast. The gravitational force is

$$F = \tfrac{4}{3}\pi \mathrm{r}^3(\rho_a - \rho_w)\mathrm{g} \qquad \text{Eqn 15.2}$$

where $\mathrm{g} = 9.8\,\mathrm{m\,s^{-2}}$ is the acceleration of gravity. The rest of the expression is just the effective particle mass: $\frac{4}{3}\pi\,\mathrm{r}^3$ is the amyloplast volume, $\rho_a = 1.5 \times 10^3$

kg m^{-3} (its density) and $\rho_w = 10^3 kg\,m^{-3}$, the density of cytoplasm taken to be that of water. Substituting the data from Eqn 15.2 into Eqn 15.1 we get $v = 0.34\mu m\ s^{-1}$. At this rate, the amyloplast would move $6\mu m$, just over its diameter, in an 18s presentation time. Experimentally observed amyloplast sedimentation rates can be as low as $0.017\mu m\,s^{-1}$ ($1\mu m\,min^{-1}$), (Iversen *et al.*, 1968). This would be the non-random motion caused by gravity, neglecting thermal motion and cytoplasmic streaming.

We can now calculate the random thermal (Brownian) motion and see how much time the statolith must travel to overcome it. Einstein showed that if Brownian displacements Δy in direction y are sampled at time intervals of Δt,

$$\frac{(\Delta y)^2_{av}}{2\Delta t} = \frac{kT}{6\pi\eta r} \qquad \text{Eqn 15.3}$$

where k is Boltzmann's constant, and T the absolute temperature, so that $kT = 4 \times 10^{-21}$J at 25°C. $(\Delta y)^2_{av}$ is the square of the observed displacement averaged over all time intervals. If the statolith sinks a distance h metres in Δt s, we can set $h^2 = (\Delta y)^2_{av}$ as the limiting case when the stimulus vanishes into the average noise level. Since the (viscosity-limited) velocity is constant,

$$v = \frac{h}{\Delta t} \qquad \text{Eqn 15.4}$$

Using $v = 0.34\mu m\,s^{-1}$, we can calculate from Eqn 15.3 and Eqn 15.4 that $h = 0.025\mu m$ and $\Delta t = 0.07s$. If we take the observed velocity $v = 0.017\mu m\,s^{-1}$, we get $h = 0.5\mu m$ and $\Delta t = 30s$. The observed presentation time $\Delta t = 18s$ falls within this range: for the estimated Stokes law velocity, the statolith would overcome noise in only 0.1s, while 30s would be needed at $1\mu m\,min^{-1}$.

What does this show? In the calculations it was assumed that the threshold presentation time includes only the time needed to overcome random Brownian motion. Mathematically, this means identifying Δt in Eqn 15.3 with presentation time.

If sedimentation requires most of the presentation time, then the cellular amplifier (cf. Section 15.3) must act rapidly, if not instantaneously. This does not mean that the above calculation and related ones (Audus, 1962; Johnsson & Pickard, 1979) contradict the amyloplast statolith hypothesis. The presentation time does, however, correspond to a stimulation of the order of the noise level. This limitation suggests that some cell component (e.g. plasmalemma, endoplasmic reticulum or cytoskeleton) perceives the proximity and displacement of amyloplasts in a specific way, setting the transduction mechanism in motion.

15.2.1.5 *Graviperception in rhizoids of the alga Chara*

The *Chara* rhizoid is a cylindrical cell $30\mu m$ in diameter and it can reach a length of several centimetres. Rhizoids anchor the fresh water alga to the bottoms of lakes and streams; it is therefore easy to see why positive gravitropism is an advantage in *Chara*.

The graviperceptive apparatus in *Chara* is fairly well understood (see Sievers & Volkmann, 1979). Between 10–20μm from the rhizoid tip there is a group of 50–60 membrane bound compartments 2μm in diameter. Electron microprobe analysis and other physical methods have shown that the compartments contain solid $BaSO_4$ (Schröter *et al.*, 1975). The rhizoid grows by incorporation of substances into the cell wall which seem to be transported by Golgi vesicles. When the rhizoid is turned away from its favoured vertical growth direction, the dense $BaSO_4$-containing statoliths sediment to the lower side. The presence of the statoliths is correlated with a local inhibition of growth, possibly because Golgi vesicles cannot reach this region of the plasmalemma. The upper side grows faster than the lower side and the rhizoid bends. The statoliths reach the lower plasmalemma in about three minutes, and the response is first apparent as an asymmetry in the tip after ten minutes (Sievers & Schröter, 1971). The kinetics of sedimentation in *Chara* are thus easily correlated with response, unlike the case for higher plant statocytes (see 15.2.1.4).

15.2.1.6 *Possible mechanisms of statolith action in higher plants*

Let us assume, as in 15.2.1.4 that the amyloplast statolith model is correct, and look at some of the speculative mechanisms that have been postulated recently. The first transduction step in gravitropism is the conversion of statolith movement to a biochemical signal. Roots are sensitive to orientation changes of less than 10°, and during the presentation time, the amyloplast may only move a fraction of its own diameter. The question is thereby raised as to how such small movements are transduced into molecular signals strong enough to initiate the growth changes.

The higher plant root is more complicated that a *Chara* rhizoid; in higher plants it is not enough to create asymmetry in individual cells. Proposed mechanisms that produce asymmetry at the tissue level are thus attractive. Juniper (1976) suggested that amyloplasts open and close plasmodesmata, thereby influencing cell-to-cell transport. If adjacent cells are chemically or electrically polarized by the stimulus, symplast continuity would allow a group of cells, as a unit, to be polarized. An entire sensitive region of tissue would eventually reflect the response of the individual perceptive cells. This hypothesis poses a question available to experiment: is symplast continuity necessary for tropic response? One way to test this is to disrupt plasmodesmatal connections. Drake and Carr (1978) disrupted plasmodesmata by severely plasmolyzing *Avena* coleoptile sections using 80% w/v sucrose. The protoplasts separated from the walls except for a few 'Hechtian strands' of protoplasm. Following a stepwise rehydration which required 3h to regain full turgor, the sections were stimulated gravitropically. Negative gravitropic curvature then commenced at a normal rate. There is no certainty, of course, that the symplast was still disrupted after turgor recovery in Drake and Carr's (1978) experiments. Further measurements of electrical coupling (Drake *et al.*, 1978) suggest that symplastic transport was seriously interfered with.

Another possibility is that the amyloplasts transmit information through pressure on the endoplasmic reticulum. In cress roots in the vertical position, amyloplasts rest on a distal endoplasmic reticulum complex, and displacing the amyloplasts induces stimulation (Iversen & Rommelhoff, 1978). If pressure of an amyloplast initiates local changes in membrane structure, a slight, rapid release of pressure could have immediate biochemical significance. Induction of asymmetry at the tissue level is explained for cress roots (Volkmann & Sievers, 1979) according to the geometry of sensitive tissue: the endoplasmic reticulum is located such that when the root is turned horizontally, only statocytes on one side (where amyloplasts now press in the outward radial direction) are maximally stimulated. Such an arrangement remains to be experimentally demonstrated in other plants, as Perbal (1978) did not find this system in lentil roots.

Perbal (1978) found that in lentil roots, amyloplasts reach a region a fraction of a micrometre from the plasmalemma, where conceivably, they are in contact with sensitive structures. These could be the endoplasmic reticulum or some fibrillar network as suggested by Filner *et al.*, (1970). Perhaps there is a specific, enzymatic interaction between amyloplast membrane proteins and cytoskeletal filament proteins. This interaction could be transmitted to the plasmalemma, explaining the bio-electric changes known to accompany gravitational stimulation (Woodcock & Wilkins, 1969).

15.2.2 THE PERCEPTION OF PHOTOTROPIC STIMULI

The preception of the light stimulus cannot be a simple matter. A small number of quanta must be 'recognized' by a specific pigment system often in the presence of other pigments. In many responses, phototropism occurs in bright light with only a small asymmetric difference in fluence rate. In some responses such as suntracking, not only is the unilateral nature of the light recognized, but its angle of incidence as well (Schwartz & Koller, 1978). Thus, it is necessary to know what the perceiving pigment(s) are, where they are located, how they are oriented and how the measurement of the light differential is subsequently translated into a growth response.

15.2.2.1 *Cryptochrome, the elusive phototropic photoreceptor*

A first approach in identifying a photoreceptor is to ascertain the action spectrum of the biological response. If only one pigment acts as a photoreceptor, and if it is not masked by large amounts of other pigments present in the tissue, it may be easy to isolate and identify. Action spectra measured for phototropism (Thimann & Curry, 1961; Delbrück & Shropshire, 1960) fit those of a widespread group of blue light-initiated physiological changes, most prevalent in lower plants (cf. Chapter 14); the initiation of bidimensional growth in fern gametophytes, stomatal opening, and control of some secondary product syntheses, seem to be under control of the same pigment as phototropism. No one has been able conclusively to isolate the active pigments that fit the action spectra, but theories abound as to possibilities (Gressel, 1979; Senger, 1980).*

The absorption spectra of flavins and some *cis*-carotenes resemble the action spectra. For lack of a more definite term, the 'blue/near-UV-light-absorbing-photoreceptor' has been termed *cryptochrome*, a simple pun on its cryptic nature and prevalence in cryptogamic plants. Cryptochrome is very primitive from an evolutionary point of view. It occurs in bacteria which move towards light (phototaxis). In many fungi, cryptochrome controls sporangiophore phototropism as well as inducing the formation of the sporangiophores themselves. The adaptive advantage of this dual control is clear: if mycelia of a soil fungus were exposed to light they would give rise to propagules that resist the adverse conditions of light and low humidity. Once a sporangiophore has formed, positive phototropism and negative gravitropism keep it above ground, allowing the spores to be dispersed by wind and water. Positive phototropism in green plants, involving orientation of growth direction (by cell elongation and division plane), maximizes photosynthetic yield in weak light. Green plants also reorient leaves towards light by turgor changes (photonasty) without the irreversible modifications associated with growth. This is also cryptochrome-modulated.

Cryptochrome also modulates production of protective reactions to high light intensities in a variety of organisms. The activity of many flavoenzymes is modified by high light fluence rates (Montagnoli, 1977) and needs protection when an organism emerges from the soil. Motile organisms are negatively phototactic at high fluence rates. Non-mobile organisms produce carotene (Rau, 1979) and melanin, which absorb light in the blue region, minimizing photodegradation of flavo-enzymes.

It may not be simple to isolate and characterize the particular cryptochrome controlling phototropism. There are many carotenes and flavins with similar absorption spectra, all 'competing' for the few photons required for perception of the stimulus. A lower limit concentration for cryptochrome has been calculated in Bergman *et al.*, (1969). The pigment was assumed to have the same molar absorbance (extinction coefficient) as visual pigments, and it was speculated that 500 photochemically effective excitations would be required for a *Phycomyces* sporangiophore to differentiate statistically between sides during unilateral illumination. On this basis, they concluded that 3×10^9 receptor molecules per growing zone would be required. In concentration this would be 3×10^{-7}M. Earlier, using different assumptions Briggs (1964) calculated a minimal required concentration of photoreceptor of 10^{-9}M in coleoptiles. As low as these figures may seem, some photobiologists consider them to be quite high, as

* However to prove the lack of universal generalizations in biology, it has been shown that the phytochrome system also controls phototropism in mosses and ferns.

they could account for a large proportion of the total flavins, assuming cryptochrome is a flavin. If the estimates are correct, they would support the possibility that cryptochrome is a conglomerate of more than one species of flavoenzyme (Gressel, 1979). Because of the earliness in evolution and the multiplicity of flavo-enzymes which react to light, it is highly possible that cryptochrome is a (or a group of) sensitive enzyme(s) which has taken on the secondary function of being a photoreceptor (this is reviewed in greater detail in Gressel, 1979).

One way to approach the problem is to accumulate evidence for cryptochrome being a carotenoid or a flavin using specific inhibitors of each pigment class. A major drawback is that the compounds used to block flavin reactions also can block other reactions in which flavins are important, i.e. in respiration. It has been ignored by some users of inhibitors that non-respiring plants may have problems in bending to light. Schmidt *et al.*, (1977) circumvented this by using an interesting control in maize seedlings. Both the gravitropic and the phototropic responses give a maximum curvature of 35° in 5h. They assumed that the processes following gravity and photoperception are mutual. If an inhibitor can specifically block phototropism but not gravitropism, it should be effective at some point just after the sensory transducer rather than blocking all curvature. NaN_3 at 10^{-4}M, a potent flavin inhibitor, had no effect on the extent of the gravitropic curvature but reduced phototropism by 60%. Other inhibitors of flavins such as KI and phenylacetate had differential (albeit less clear-cut) effects. They thus concluded that cryptochrome is probably a flavin.

The general consensus is that cryptochrome is a flavin and many flavins and flavoenzymes have been proposed, each based on indirect evidence (see Senger, 1980; Gressel, 1979). The alternative view that cryptochrome may be a carotene has recently been reviewed lucidly by Shropshire (1980). His arguments supporting the possibility are based on ubiquity, absorption spectra, known carotenoid photochromicity, localized increased concentration in photosensitive regions, and the known function of the carotenoid retinal, in animal vision, as well as some genetic evidence. In addition, it was recently reported that the depletion of carotenes by a specific inhibitor of their synthesis (norflurazon) also depletes at least one blue light response. The residual response was directly correlated with the amount of carotene remaining (Hartmann & Schmid, 1980). While none of this is conclusive evidence for catotenoids, neither is the flavin evidence conclusive. In fact, both may be correct. There may be both flavin and carotene cryptochromes.

15.2.2.2 *Dose–response relationships in phototropism*

There have been a considerable number of experiments performed on the dose–response relationships in phototropism. Photobiologists generally accept the Bunsen–Roscoe Law of photochemical equivalence to hold in all cases. This reciprocity is certainly expected for strictly photochemical reactions, and results in the absolute amount of the photoproduct being proportional to the product of the duration of illumination and the fluence rate. However, it cannot be expected to be valid in all cases within an organism. For example, if a threshold concentration of a photochemical product must be reached within a certain time, or if the steady state degradation rate of the photochemical product is high, subsequent biological responses will not display reciprocity. Even more problematic are phenomena in which the biological sequence of events occurs rapidly and the duration of illumination extends beyond the beginning of observable effects. Such cases are especially suspect if the measurements of the effect are made at a certain invariant time after the start of exposure to light. An example of these problems, from some of the first such experiments, is presented in Fig. 15.3. Here we see an almost linear increase in *Avena* curvature with increasing fluence rate, when a low fluence rate is used. When a high fluence rate is used, negative phototropism occurs even though with lower fluence rates, the same total fluence causes positive phototropism. This anomaly of obtaining 'first positive', 'first negative' and then 'second positive' curvatures (Fig. 15.3a) led to the construction of various models with different photochemical rate constants affecting a two-step reaction (Zimmerman & Briggs, 1963a). The constants in the equations were constructed to include reciprocity and to generate a theoretical curve matching those measured (Zimmerman & Briggs, 1963b). Almost implicit in the two positive curvature situations is the possibility of two separate pigment systems absorbing the blue light. The evidence for and against such possibilities in the context of phototropism are discussed at length by Dennison (1979).

As if merely to complicate matters further, a third positive curvature was found. Blaauw and Blaauw-Jensen (1970) tested reciprocity to its limits by illuminating from 0.01s to 40min over 5 logarithmic scales of fluence rate. Their 3-dimensional graph (Fig. 15.3b) demonstrates how non-reciprocal responses may be.

15.2.2.3 *Cryptochrome and phytochrome*

In higher plants both the extent and rate of the phototropic response are often modulated by other photomorphogenic effects of light, including growth responses modulated by cryptochrome, and possibly more important, those governed by phytochrome (see review by Dennison, 1979). A pre-illumination of *Avena* coleoptiles with red light increases the fluence requirement tenfold for the first positive cryptochromal phototropic curvature but decreases it by about the same order of magnitude for the second positive curvature (Zimmerman & Briggs, 1963a).

What is the relevance of studying the phototropic response of *Phycomyces* or of grass coleoptiles to unilateral blue light (often with a depressive red pretreatment) to the real world of the green plant growing in mixed light which may be slightly unilaterally imbalanced? This is hard to answer at present. It takes a few hours with 1–2 Wm^{-2} to obtain a maximum curvature of green

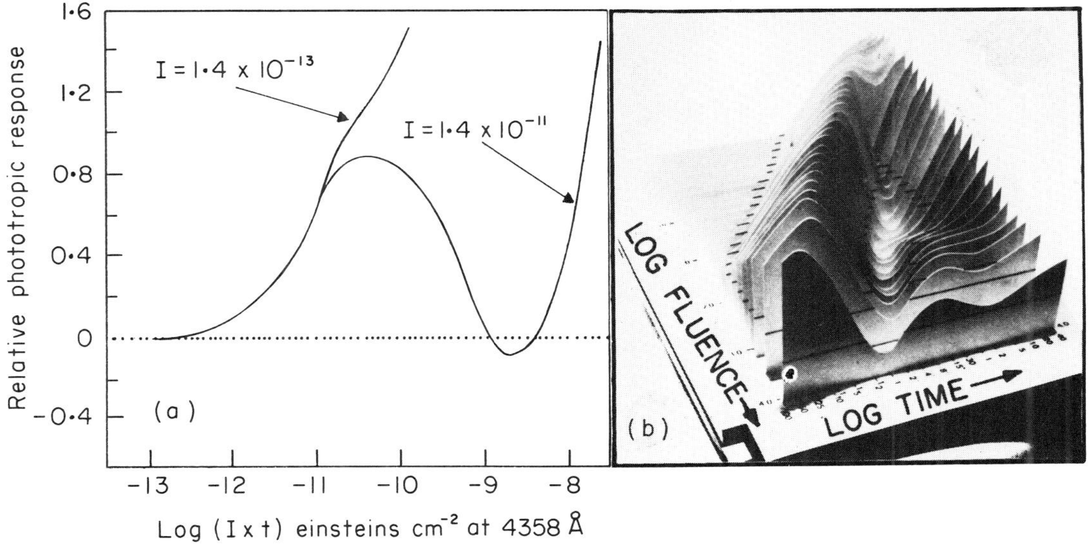

Fig. 15.3 The dose response anomalies in phototropism. (a) A simple anomaly (redrawn from Zimmerman & Briggs, 1963b with permission) (b) A complicated anomaly (with permission from Blaauw & Blaauw-Jensen, 1970). An original print was provided by Dr. Blaauw).

sunflower or cucumber plants (Shuttleworth & Black, 1977). This is many times more light than is required to saturate the bending of an etiolated *Avena* coleoptile. This does not mean that it is felt that phototropism is irrelevant to the green plant. It undoubtedly plays a role in canopy spreading and filling by plants in their natural environment.

15.2.2.4 *The localization of cryptochrome*

A series of processes must commence following the light sensing by cryptochrome. In Section 15.3 the need for amplification of the signal prior to response is discussed. Need the signal be transmitted any distance or is cryptochrome localized where it will also act?

Cryptochrome photoreceptors may be localized differently when modulating different functions. Furuya (1978) has elegantly described this in two, related but separate, reactions in the blue-light-controlled change from threadlike to bidirectional growth of fern sporelings. Using very narrow beams of light, it was found that the photoreceptor for the signal to change the plane of division was localized in the region of the nucleus and was not dichroically oriented. The photoreceptor perceiving the signal to expand the growth of the tip was located at the tip, and was affected differentially by light polarized in different planes. This does not necessarily mean that the photoreceptor pigments are different; they could be the same molecules which, because of their specific localization, transduce the environmental signals *via* different specific secondary acceptors or amplifiers.

One hundred years ago, Darwin & Darwin (1881) reported that the coleoptile tip recognized the unilateral illumination and that tissue subtending the tip elongated differentially. More recently it was shown that the apical 350μm of coleoptiles is about twice as sensitive as the subapical 350μm which is, itself, more sensitive than the more basal regions where bending occurs (Meyer, 1969a). In *Phycomyces*, the reactive area begins 0.5mm below the sporangium and the zone of bending is just beneath that. The length of the lag between a pulse of blue light and the first measurable bending increases as a function of the distance from the sensitive region where the stimulus is given.

15.2.2.5 *The subcellular localization and orientation of cryptochrome*

Knowledge of the subcellular localization of cryptochrome could contribute considerably to our understanding of its mechanism of action. Difficulties have been encountered when attempting to localize cryptochrome within the *Phycomyces* sporangiophore. Meistrich *et al.*, (1970) used a blue laser beam to graze the organ. The assumption was that by grazing the side at different points, varying fluences could be applied to the different subcellular fractions. The experiments were performed in iso-refractive oil to prevent light refraction. The results clearly showed that neither the vacuole nor the cell wall bore the photoreceptor. A higher sensitivity was found in the cytoplasm than in the plasmalemma, a result which conflicts somewhat with current consensus and bias of opinion. A differential response to light polarized in different planes has been found in *Phycomyces* (Jesaitas, 1974), indicating that the photoreceptor molecules are oriented with respect to each other. It has been assumed that the receptor must be on a fixed membrane fraction. Considering that there is cytoplasmic streaming, the only membranes which do not move are thought to be the plasmalemma and the tonoplast. This argument may not be completely valid, as for example streaming mitochondria remain oriented along one axis and their cristae would have a statistical differential dichroic orientation. The fact that blue light can cause immediate hyperpolarization of certain areas on bean stems, which are not phototropic photoreceptor regions (Hartmann & Schmid 1980), has been used as support of the plasmalemma hypothesis. Still, the hyperpolarization might originate in a photochemical reaction in the cytoplasm or organelles adjacent to the plasmalemma, and then be propagated to the plasmalemma. Up to now, dichroic orientation has not been demonstrated in an adult green plant.

At temperatures near the membrane phase transition, the normal semi-ordered structure of the lipid phase breaks down. If the photoreceptor is indeed oriented in the membrane, a strong inhibition of the observed effect of polarized light should occur simultaneously with the phase transition. Surprisingly, no one has reported changes in dichroism at the non-toxic temperatures of the membrane phase transitions which are different for each membrane type.

There is a simple, general argument that can be raised against a dichroic orientation of cryptochrome, especially as the maximum difference in effect of light polarized in cross planes was only 24% in *Phycomyces* (Jesaitas, 1974). There is always the strong much-ignored possibility, that the cell wall or some inert pigments act as a polarizing filter in one plane, leaving fewer photons available in that plane for stimulating a randomly-oriented cryptochrome.

15.2.2.6 *The unilateral perception of light*

From the extremely low fluence rate requirements it is quite obvious that very few molecules of cryptochrome need be triggered. This is often done in the presence of other pigments which 'compete' for the photons. As etiolated stems or *Phycomyces* sporangiophores are not always yellow, it is obvious that there is little cryptochrome present. With *Phycomyces* it has been found that more than 99% of the incident quanta are not absorbed by the stalk. Consequently, how is the plant able to detect which side has received more light than the other with only a small number of quanta, few photoreceptor molecules, and most of the photons passing completely through the sporangiophore? Also, the plant must be able to recognize this few-photon differential, whether the unilateral illumination occurs in the dark or in bright sunlight.

The simplest hypothesis would be that the plant somehow perceives that the

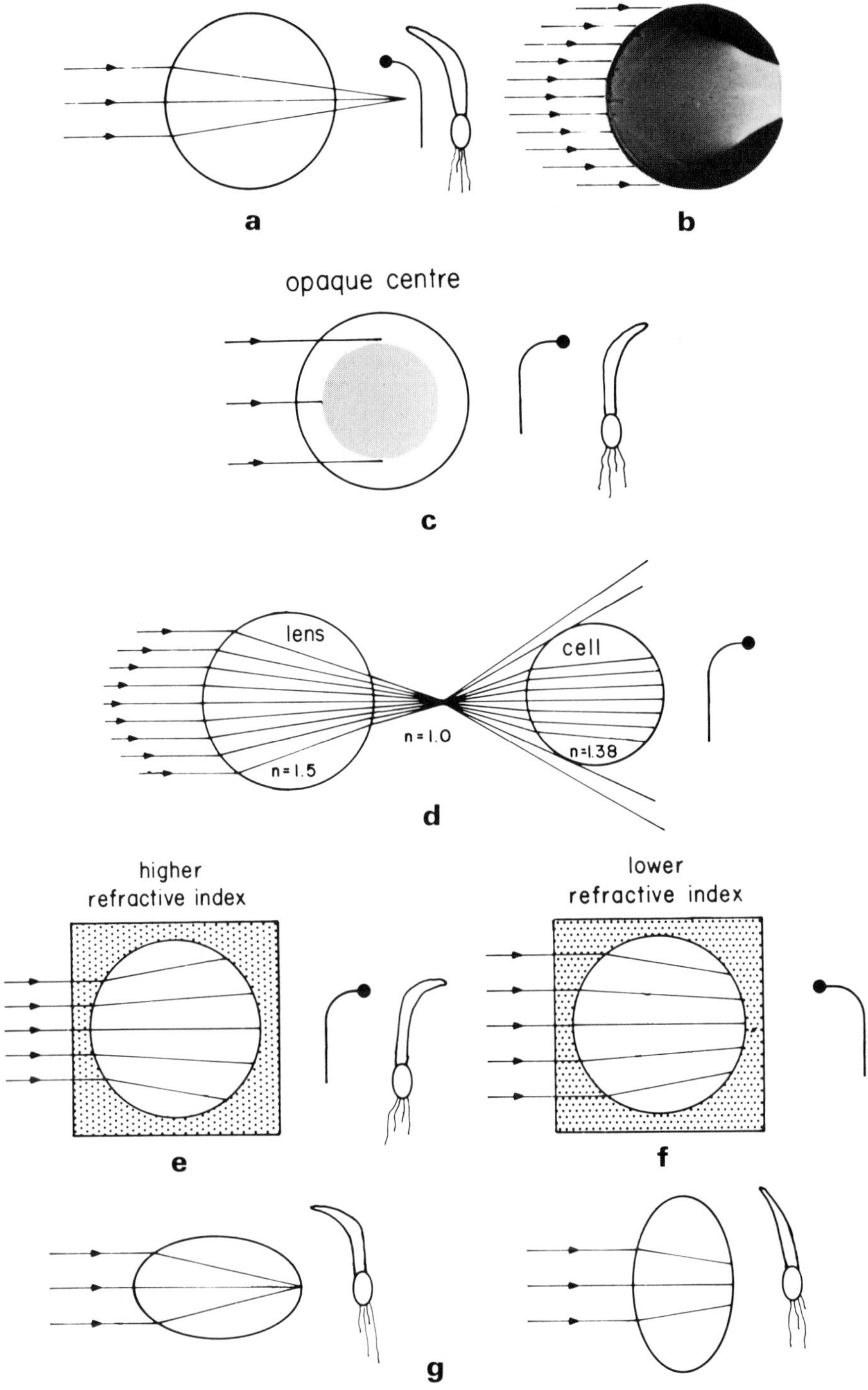
a
b
opaque centre
c
lens
cell
n=1.0
n=1.5
n=1.38
d
higher
refractive index
lower
refractive index
e
f
g

side nearest the light received more light. This could be achieved by having cryptochrome oriented outwards in a double membrane and having a second inert pigment embedded directly behind it in the membrane. The inert pigment would screen the cryptochrome from light coming from within the organ. This arrangement would allow the cryptochrome to absorb only light coming from outside the organ. A similar type of screening has been found in *Euglena* and *Volvox* where the stigma pigments shade the phototactic receptive cryptochrome from one side (Schletz, 1979; Lenci & Colombetti, 1978).

This simplest of hypotheses may not always apply. For example, it is known that, in a few cases, the impinging light is *concentrated* on the far side of the illuminated organ by a cylindrical lens, formed by the tissue, which intensifies the light gradient and leaves flanking areas unilluminated (Fig. 15.4a,b). Two principal experimental approaches have been used to test the lens hypothesis: first, by making the centre of the biological lens opaque and second by dispersing light with lenses or media of different refractive indices inserted between the light source and the organism.

The first line of evidence supports the lens hypothesis and strongly suggests that the sporangiophore of the *Phycomyces* acts as a lens. The evidence is as follows; many cryptochrome effects have an action spectrum peak in the UV at 280nm in addition to peaks in the blue. In *Phycomyces*, 280nm light causes negative phototropism and not the expected positive curvature. The cylindrical central vacuole of *Phycomyces* contains gallic acid which absorbs UV light; i.e. it is not only a lens but also a filter. The UV blocking results in a higher fluence

---

Fig. 15.4 Intensification of unilateral light with a lens. Each diagram explains bending of coleoptiles or sporangiophores, as indicated by the pictures.

(a) A schematic diagram of a lens that is transparent for light of wavelength above 330nm.

(b) A simulation of the light pattern. The picture was taken through opal diffusing glass, looking into the end of a glass cylinder filled with water and illuminated from the side. (By permission, from Dennison, 1979—Original print provided by Dr. Dennison).

(c) Vacuoles that are optically opaque (endogenous gallic acid) with illumination at 280nm or vacuoles filled with opaque dyes (Shropshire, 1962; Meyer, 1969b). The near side is stimulated giving negative phototropic curvature.

(d) Light is dispersed by a cylindrical lens before entering sporangiophore. (Redrawn from Shropshire, 1962, with permission). Index of refraction of each region indicated.

(e) Light dispersed by immersion in high refractive index medium; a negative curvature obtained (data of Castle, 1933; Shropshire, 1962; Meyer, 1969b).

(f) Light affected by immersing sporangiophore in low refractive index medium. A stronger positive curvature was obtained (data from Shropshire, 1962).

(g) Light from different sides is converged differently through the elliptical *Avena* coleoptile giving different curvatures with equal light fluxes (data of Meyer 1969a but explanation of Shropshire (1962) and Dennison. (1979).

Note: Meyer (1969a) explained the data without invoking a lens; light from the narrow end must cross a longer path in the tissue, giving a greater attenuation.

on the near side of the sporangiophore giving rise to the negative phototropism (Fig. 15.4c). Similarly enhanced negative curvatures in blue light have been found in *Phycomyces* which had been allowed to take up tetrazolium dye (Shropshire, 1962), or were super-abundant in carotenes (Jesaitas, 1974), and in *Avena* coleoptiles which had taken up an opaque Sudan dye and were immersed in paraffin oil (Meyer, 1969b). These results are interpreted as being due to making the lens optically opaque resulting in cryptochrome on the near side absorbing more light than that on the far side.

Two lines of optical evidence have been used to demonstrate a lens which concentrates light on the far side of an illuminated organ. The first was to illuminate through compounds of refractive index greater than that of the organism. Castle (1933) placed *Phycomyces* sporangiophores in paraffin oil and obtained a negative phototropic curvature (Fig. 15.4e). Similar experiments in which *Avena* coleoptiles were immersed in oil gave a negative curvature on illumination (Meyer, 1969b). A more sophisticated demonstration of the lens effect was achieved by preparing media covering a refractive index range by mixing fluoro-hydrocarbons and perfluorotributylamines (Shropshire, 1962). In this way, it was possible both to enhance the positive curvature with a low refractive index medium (Fig. 15.4f) and obtain a negative curvature with a high refractive index medium (Fig. 15.4e). The point of no curvature where the effectiveness on the near and far surfaces must be equal is at a refractive index of 1.295. It is therefore possible to calculate how the focussing advantage balances the attenuation due to tissue absorption and scatter. Such experiments with liquids of varying refractive index need to be done both with coleoptiles and with green organs showing phototropic sensitivity, in which the chlorophyll should act as an opaque filter, to assess the likelihood that a lens also acts there. The solubility of oxygen in fluids with varied refractive index used by Shropshire (1962) varies inversely with the refractive index. This means (formally) that there were two variables to the experiment, oxygen tension and refractive index. Still, anaerobic conditions alone or inhibitors of aerobic respiration are not known to stimulate negative phototropic responses (see Section 15.2.2.1).

Another technique to test for the lens effect was to place a fine cylindrical glass lens in front of the organism so that the light, instead of being collimated, is dispersed from a point between the lens and the organism (Fig. 15.4d). Despite the internal lens, more photons hit the near side than the far side, and the result at least in *Phycomyces* is a negative curvature. Similarly, a study of the elliptical structure of the coleoptile confirmed that illumination from the flatter side gave less curvature than from the sharper side as would be expected (Fig. 15.4g), despite the mechanical considerations indicating that the resistance to bending is greater on the sharper side.

There is some evidence that the lens is not universal in nature. Paraffin oil did not reverse the direction of bending of the green alga *Vaucheria* (Kataoka, 1975) nor could a lens be shown during certain stages of growth of the fungus, *Pilobolus longipes* (Jacob, 1964).

15.3 TRANSDUCTION AND AMPLIFICATION OF THE PERCEIVED SIGNAL

In the tropic responses, as in many other photobiological and photomorphogenic responses, the photoreceptor or gravireceptor and final effect are separated by a set of unknown reactions, which may be considered as a 'black box'. It is useful in this case to use the classical analogies with electronics. When an apparatus picks up a signal in one form and displays it in a different form, it is easy to find out what happens: we look at the plans and see how the components performed the task. When sound vibrations reach a microphone, they are converted to oscillations of electric potential (transduction); the small oscillations are then made larger by introducing energy from a power supply (amplification) and recorded (actually another transduction step), or transduced again into louder sound through a speaker.

The circuit diagrams for the mechanism of gravitropic or phototropic curvature are not available, but we can deduce the need for transducers and amplifiers. When a photoreceptor molecule absorbs a quantum of light, it enters an excited state. The unstable product is chemically active: i.e. light signals are transduced to chemical signals. In graviperception, reorientation of sensitive tissue effects a rapid change in structure. Unlike phototropism, where the primary transducer is a photoreceptor which should operate according to the principles of photochemistry, it is not yet possible to say how the signal is transduced in graviperception. Perhaps one of the statolith hypotheses discussed in Section 15.2 will provide the directions towards an answer.

When tropic curvature is produced by a growth rate differential across the organ, metabolic energy is expended in a way defined by the signal, i.e. the signal is amplified. Auxin for example, is an amplifier: gradients of the growth effector determine how much energy will be expended as growth in each cell of the affected tissue. There may well be, in addition, a preamplifier stage: Galston (1974) calculated that if auxin is indeed the growth effector in coleoptile phototropism, the overall quantum yield for setting up the hormone gradient must be more than 4000. But the quantum yield of a primary photochemical reaction cannot exceed unity: the signal must be amplified. There are many ways that light-sensitive enzymatic systems can simultaneously perceive, transduce and biochemically amplify weak light signals. This is done by changing catalytic efficiency following a photo-induced conformational change. (Martinek & Berezin, 1979).

It is obvious that the processes occurring during the amplification may vary. A process as 'simple' as the bending of a unicellular *Phycomyces* sporangiophore has a time lag of 4–6 minutes. Many processes could occur during that short time including transcription of new mRNAs and/or translation of additional enzymes. Enzyme activation could be lag free. There is no known case yet where the signal is perceived, transduced, amplified, transmitted and effected at a rate which precludes macromolecular syntheses or hormone transmission.

Stomatal opening, which is also a blue-light-modulated process and in which there are turgor changes, is without an apparent lag (Skaar & Johnsson, 1978). In such lag-free cases bioelectric transmission and amplification cannot be precluded. Not enough cases have yet been studied to be sure, but from the few that have it may be postulated that there is no lag when the tropic response is achieved solely by water extrusion. Where ion uptake/osmosis and where irreversible growth must occur, a measurable lag has always been found.

It is odd that no research has been reported to test whether nucleic acid transcription or the translation of proteins is implicated in the latent period before growth ensues. With phototropism, not even the less aesthetic experiments using inhibitors of transcription and translation have been reported. In a closely related system, the cryptochrome-modulated synthesis of carotenoids in *Phycomyces*, there are two levels of effect which can be measured as two separable dose–response curves. The low fluence response ($\sim 30\ \mathrm{Jm}^{-2}$ for half maximum saturation) is not affected by actinomycin D and cycloheximide which are supposed to inhibit transcription and translation respectively. The added carotenoid synthesis modulated by the high fluence response ($\sim 1000\ \mathrm{Jm}^{-2}$) is suppressed by these inhibitors (Jayaram *et al.*, 1979). Unfortunately some recent genetic data (Jayaram *et al.*, 1980) indicate that the cryptochrome for phototropism may not necessarily be the one(s) controlling carotenogenesis.

15.4 THE LOCAL CONTROL OF TROPIC EFFECT

Once the stimulus is perceived in the specialized area of the plant and then transduced, amplified and transmitted (translocated) to the responsive area, the organ or organism can change its position and orientation in space. The induced movements are different for different organisms; cilia or flagella can be activated in unicells, the motor pulvinus in leaves or the movement can be true growth, encompassing irreversible changes in the morphology of roots, stems, coleoptiles or sporangiophores. It is difficult to separate conceptually the transmission of the stimulus and its overt effects.

15.4.1 NON-GROWING SYSTEMS: SUNTRACKING, PHOTONASTY AND PHOTOTAXIS

In more orthodox definitions of phototropism (cf. Dennison, 1979) the movement of unicells and leaves towards light are excluded, as no growth occurs. The direction of movement of these organisms and organs is affected by light and at least semantically could be called phototropism although photonasty and phototaxis are the terms usually used. Physiologically they are among the more important responses. In plants with suntracking leaves, this photoeffected movement acts during a highly significant proportion of the life cycle and adds considerably to the photosynthetic productivity of the organism. In the motile micro-organisms, the ability to respond unilaterally to light can be a matter of life and death. The motility of micro-organisms is beyond the scope of this text, but is reviewed by Lenci and Colombetti (1978), and is discussed in a few chapters in Haupt and Feinleib (1979). Suntracking of leaves is not discussed therein.

The pulvinus at the base of the leaf blade or petiole (in different species) is responsible for the light-tracking movement. Turgor changes in the pulvinus control the direction of bending, and the motor cells seem to be connected to different parts of the leaf. If the two halves of a *Lavatera* blade are illuminated obliquely at different angles, the leaves can be made to curl in or out. It is interesting that the kinetics of bending up are much slower than the kinetics of bending down (Schwartz & Koller, 1978). By analogy with many other leaf movement systems, one may assume the operation of a K^+-pump in the motor cells of the pulvinus. One method of ascertaining the operation of a K^+-pump as in the suntracking movements is to use lanthanum. La^{3+} inhibits K^+ transport in animal as well as plant systems, and inhibits nastic movements in plants involving K^+ influx. Wainwright (1977) has reported that La^{3+} also inhibits the suntracking movements in *Lupinus* but there are reasons to suspect the specificity. Only K^+ influx should be affected by La^{3+} but efflux is considered passive and should be unaffected. La^{3+} affected pulvinar movement in all directions, both those which should be due to efflux and due to influx.

The case for auxin involvement in the suntracking movements is tenuous. Auxin applications have been shown to modulate the rate of nyctinastic (circadian) leaf movements (Satter *et al.*, 1972). The case for auxin modulation of suntracking has only been demonstrated indirectly using triiodobenzoate which suppresses *all* tracking movements, up and down. This compound is thought to prevent auxin transport but its specificity in doing so can be strongly questioned. The rate of auxin movement from the blade lamina to the pulvinus would have to be exceedingly rapid to account for the seemingly lag-free downward movement.

15.4.2 TROPIC BENDING IN GROWING SYSTEMS

Just as the sites of perception differ in gravitropism and phototropism (Section 15.2), it is also apparent that the sites of tropic growth may vary.

Roots in soil bend only near the apex. This facilitates redirection around obstacles in the soil. Herbaceous shoots are more subject to change than roots in gravitation and light status brought about by snow, wind, rain and shading by other plants. Thus it is advantageous for shoots to bend tropically in the middle as well as near the tip.

It is quite obvious that many tropic curvatures are due to a differential rate of growth on opposite sides; but which is it, stimulation or inhibition of growth? One side may grow faster than the normal unstimulated plant organ, or the other side may grow more slowly. Both effects may even occur simultaneously;

in *Phycomyces* phototropism, a 7% decrease in growth rate on the illuminated side and a 7% increase on the far side were found (Shropshire, 1974). This occurred irrespective of the rate of bending itself. The rate of bending was a function of the diameter of the spore stalks. By growing *Phycomyces* at 2°C until a few hours before experiments, Shropshire (1971) induced thick sporangiophore formation; by dense planting he also induced thinner than normal organs. The thinner sporangiophores bent faster than the thicker ones, probably because of their mechanical, rather than optical, properties.

In higher plants, examples can be found of acceleration, deceleration and cessation of growth following a tropic stimulation (Fig. 15.5). These differences

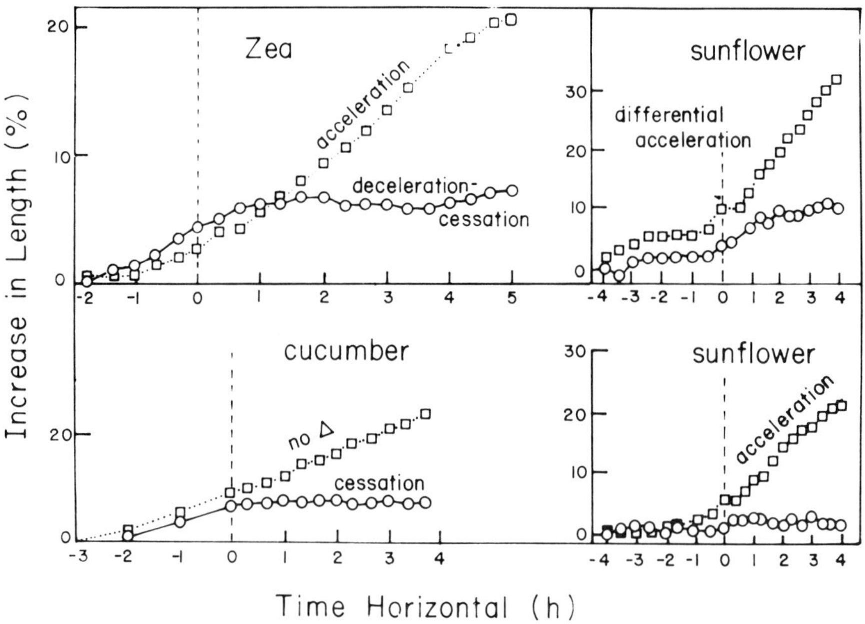

Fig. 15.5 Examples of acceleration, deceleration and cessation of growth following gravitropic stimuli (from Digby & Firn, 1979, by permission). ○, growth of upper side; □ growth of lower side, expressed as percent difference from starting length of a marked region of the growing coleoptile or hypocotyl.

in growth pattern may provide information on the distribution and concentration of the immediate effector of the response, which will be discussed in 15.4.2.2.

15.4.2.1 *Tropic growth in plants lacking auxins*

Phototropic fungi clearly lack auxins and their differential tropic growth must be effected by some other compound. There are suggestions that this function is filled by the compound with properties of the 'second messenger' in animal systems: cyclic-AMP. Both the existence of, and functions for, cyclic-AMP in higher plants are controversial (Ownby *et al.*, 1975). There are, however studies linking cyclic-AMP and its phosphodiesterase to the photoresponses of *Phycomyces*. A transient decrease of about 50% in cyclic-AMP content was found within 60s of light stimulation (Cohen, 1974), well before the sporangiophore bending can be measured. Cyclic AMP phosphodiesterase can be photoactivated *in vitro* (Cohen & Atkinson, 1978) which places it as one of many candidates for cryptochrome. Circumstantial support for this possibility stems from a *mad D Phycomyces* mutant which is non-phototropic. It has a cyclic-AMP phosphodiesterase with an abnormal pH dependency and the transient cyclic-AMP decrease is much smaller than that of the wild type (Cohen, 1978).

15.4.2 *Tropic growth in plants containing auxins*

The Cholodny–Went theory of auxin redistribution in tropic growth has until recently long been accepted without major criticism. The role of auxin in cell extension growth is well documented. Auxin redistributions have been measured following tropic stimulations (Table 15.1) and IAA has the appropriate properties for it to act as the transmitted stimulus, amplifier and effector of tropic curvature. It is appealing to assume that the effector of bending is a small molecule, i.e. an auxin, some other hormone or an inhibitor. A nucleic acid or a protein would be far too bulky to be a translocated effector. Great theories, however, attract challenges, and the Cholodny-Went theory is now seriously in question, notably by two groups (Bruinsma, 1977; Digby & Firn, 1976; Firn & Digby, 1980) who have asked pertinent and cogent questions about the assumptions regarding the involvement of auxins in the growth processes of positive and negative gravitropism and phototropism. The involvement (or lack thereof) of auxin may be different in all three phenomena. The debate revolves around the following questions. Is a translocated compound necessary? Can auxin be redistributed quickly enough? Can enough auxin be redistributed? Do inhibitors play a part? Are other hormones that induce elongation involved?

Firn and Digby (1980) review the evidence for the need for transport and decide that the cells perceiving both positive and negative gravitropic stimuli are also the cells that elongate to cause bending, and that a hormone is not required. These conclusions are difficult to accept unreservedly, as no one (for obvious reasons) has microbeamed a gravitational field to treat just part of a stem or root *in situ*, leaving the remainder of the root in a different gravitational field. Peeling the peripheral cell layers from sunflower hypocotyl sections (Firn & Digby, 1977) abolishes gravitropic response without stopping growth. Also, hypocotyl sections (bisected longitudinally) with intact peripheral cell layers, grew faster with the epidermis down. This response corresponds to negative

Table 15.1 Distribution of auxin after gravitational or photostimulation

Plant Material	Stimulus	Applied (app) Extracted (ext) IAA	% of total auxin in bending side	Time after stimulation	Parameters detmn.*	Reference
Shoots						
maize coleoptiles	gravity	app	56	35–50min	cpm	Filner *et al.*, (1970)
(dark grown)	light 1st +	ext	63	3h	activity	Briggs, (1963)
	light 2nd +	ext	70	3h	activity	Briggs, (1963)
	light	app	75	3h	cpm	Pickard & Thimann, (1964)
Echinochloa (green node)	gravity	ext	61	15min	IAA-sf	Wright *et al.*, (1978)
		ext	78	30min	IAA-sf	Wright *et al.*, (1978)
Helianthus (light grown)	gravity	app	48	6h	cpm	Phillips & Hartung, (1976)
Internode segments *Helianthus* Hypocotyls	light	ext	52	30min	IAA-sf	Franssen & Bruinsma, (personal communication)
Cucumber (hypocotyls)	gravity	app	56	3h	cpm-chr	Iwami & Masuda, (1976)
Pea (light grown) *intact*	gravity	app	54	3h	cpm	Iwami & Masuda, (1976)
(internodes) hollow	gravity	app	59	3h	cpm	Iwami & Masuda, (1976)
epidermis	gravity	app	81	1h	cpm-chr	Iwami & Masuda, (1976)
epidermis	gravity	app	78	2h	cpm-chr	Iwami & Masuda, (1976)
epidermis	gravity	app	55	3h	cpm-chr	Iwami & Masuda, (1976)
Roots						
Pea (dark grown) intact roots	gravity	app	40	1.5h	cpm	Konings, 1967
Maize (dark grown,	gravity	ext	26	1h	activity	Suzuki *et al.*,
red pretreated)		ext	23			
blackcurrant (light grown)	gravity	ext	38	4h	GLC	El-Antably, (1975)

* Parameters determined: cpm: distribution of unidentified radioactivity; cpm-chr: distribution of cpm co-chromatographing with IAA; activity—auxin by bioassay; IAA-sf: IAA determined spectrofluorometrically.; GLC:IAA determined by gas-liquid chromatography.

shoot gravitropism, and suggests that transport of a hormone from one side to the other is not necessary for shoot gravitropism. One could always argue, of course, that this experiment may not be valid for intact plants.

In roots, graviperception apparently occurs in the root cap, as discussed in Section 15.2.1.4, while curvature is limited to a zone a few millimetres behind the apex, implying transmission of the stimulus.

There is ample evidence (15.2.2.2) that the phototropically perceptive cells in coleoptiles are not those that elongate; translocation is therefore implied and auxin is a good candidate for a translocated compound. There is always a distinct lag or latent period between stimulus and the onset of growth. The known rates of movement of small solutes are sufficient to allow auxin translocation while leaving ample time for other processes of secondary amplification in preparation for growth.

Some of the major interconnected questions raised are as follows. Is enough auxin redistributed? Is it redistributed before or after the effect? Is there always evidence for the redistribution? In which tissues does redistribution occur? From the accumulated data (Table 15.1) it is obvious that only two groups have tried to ascertain the extent of auxin redistribution during the lag period or just after it. Some groups did not observe auxin redistribution at all. In the partial kinetic analysis of Iwami and Masuda (1976), the earlier the measurement was made, the greater the observed differential redistribution. Comparing the work where no redistribution was found with experiments performed during the lag period, one is not surprised; Iwami and Masuda (1976) also hardly found a difference three hours after stimulation. The greatest differences were early in the latent period (Table 15.1).

Most groups supporting Cholodny–Went have measured the redistribution of exogenous applied auxins which may have little to do with the endogenous redistribution proposed by Cholodny and Went. Wright *et al.*, (1978) have measured a redistribution of endogenous auxins using a sensitive spectrofluorometric technique (Table 15.1).

Firn and Digby (1980) doubt that the usual 60 : 40 redistribution found could account for the differential growth because of the linear growth response to the logarithm of the concentration. They believe that the small measured differentials should be insufficient to cause curvature. In the one test that matters, the *Avena* curvature test, bending is a direct linear function of auxin dose, at the concentrations expected in an endogenous system. With the small differentials in concentration generated by a redistribution, the differences between linear dose/linear growth and logarithmic dose/linear growth are small enough to become immaterial. The major question remains; can a 60 : 40 or 80 : 20 auxin redistribution induce sufficient differential growth to cause bending? Theoretically, the answer is affirmative, as shown in Fig. 15.6. Taking recent *Avena* coleoptile bioassay data, we calculated the curvatures that should be caused by two different auxin asymmetries. A 60 : 40 redistribution can produce curvature at certain auxin levels. The major problem in interpreting such data is that endogenous concentrations reaching sensitive cells are difficult to measure. Quantitative correspondence between bioassay (response to applied auxin in agar blocks) and tropic curvature might not be found even if the Cholodny–Went model provided a complete explanation of tropic growth differentials. Still, the Cholodny–Went theory cannot be faulted on the basis of the differences in concentration being too low to cause bending.

Whether higher plants perceive unilateral light with a lens, or whether the near side of green tissue perceives, the perception would have to be in the outer

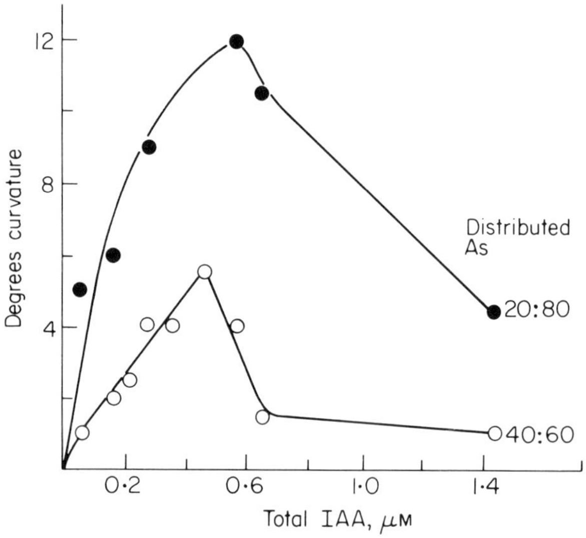

Fig. 15.6 *Avena* coleoptile curvatures predicted for asymmetrically distributed IAA at ratios of 20 : 80 (●-●) or 40 : 60 (○-○). Calculated from bioassay data of Jacobs (1979) by assuming that the total curvature is the algebraic sum of the curvatures of each side alone.

layers of cells. It is interesting to note that the tissues most affected and active in the ultimate growth response are in the outer epidermal layer (Table 15.1, Iwami & Masuda, 1976; Firn & Digby, 1977). Firn and Digby (1977; 1980) more or less concur with Iwami and Masuda (1976) that the outer cells effect the curvature in shoots, except that they prefer to leave out any mention of hormones as being involved in the curvatures, whereas Iwami and Masuda believe that the auxin redistribution is not a mere coincidence. There is no reason to believe that auxin is the *sole* effector of curvature in higher plants; indeed there is some evidence to the contrary. Inhibitors may play a role, either alone or in tandem with auxin effects. It would be exceedingly difficult to invoke redistribution of auxin to explain the tropic responses where growth on one side of a seedling ceases completely (Fig. 15.5).

Alternative theories include mechanisms in which unknown growth stimulators or growth inhibitors act unilaterally. A glance at Table 15.2 suggests that, if hormones are indeed necessary to initiate curvature, different combinations of growth promotors and inhibitors will be found in each plant studied. Recent data from the Bruinsma group (Table 15.2) showing xanthoxin redistribution supports this idea.

Endogenous gibberellin levels in the upper and lower halves of geostimulated roots and photostimulated shoots are distributed according to the growth differential. Of course, this does not imply, as pointed out by Firn *et al.*

Table 15.2 Alternatives to auxin as translocated inducers of tropic curvatures: redistribution of various hormones after gravitational or photostimulation.

Plant Material	Stimulator of tropism	Hormone applied (app) or extracted (ext)	% in faster growing side	Time after stimulus	Reference
Helianthus hypocotyls	light	xanthoxin (ext)	35	30min	Franssen & Bruinsma (personal communication)
Maize roots	gravity, after induction with red light*	abscisic acid (ext)	62	1h	Suzuki *et al.*, (1979)
		unidentified inhibitor, ext, bioassay	32	1h	Suzuki *et al.*, (1979)
Blackcurrant, light grown plants	gravity	abscisic acid (ext)	17	4h	El-Antably (1975)
	gibberellins		84	4h	
Vicia faba	gravity	abscisic acid (app)	25	30min	El-Antably & Larsen (1974)
		gibberellins (ext)	66	30min	
Helianthus light grown shoot tips	light	gibberellins (diffused)	88†	20h	Phillips (1972)
Helianthus light grown internode segments	gravity	radioactive GA_1 (app)	48	6h	Phillips & Hartung (1976)

* light pretreatment is necessary for gravitropic response in some maize cultivars, See 15.2.1.1.

† Firn *et al* (1977) consider this redistribution to be a consequence of differential growth not a cause.

(1977), that the redistribution caused asymmetric growth. The same objection applies to all hormone redistributions, and was the basis for questioning the original Cholodny–Went model.

There is considerable evidence for the involvement of inhibitors in effecting positive gravitropism in roots. The original Cholodny–Went hypothesis stated that, following gravistimulation, auxin accumulates in supraoptimal concentrations on the lower side thus acting as an inhibitor. Some reports support this (Konings, 1967 and Table 15.2). Kundu and Audus (1974) found two inhibitors in root tips, one in the cap (which co-chromatographed with abscisic acid) and one in the meristem. Neither inhibitor co-chromatographed with IAA. This evidence, while not excluding a regulatory role for auxin, suggests that other tip-produced inhibitors bring about the growth differential responsible for gravitropic curvature in roots. Recent cap and meristem removal and replacement experiments with maize primary roots (Gibbons & Wilkins, 1970; Pilet, 1973; Shaw & Wilkins, 1973) suggest that an inhibitor is asymmetrically exported to the growing zone by the tip. Generating an asymmetry by removing part of the cap, or asymmetrically replacing an intact cap, causes curvature even in vertical roots. If the caps are removed from vertical roots and replaced (symmetrically) with caps from six hour gravistimulated roots, they bend. Six hours is a long time in such experiments. As growth commenced much earlier, kinetic studies are needed to show that redistribution of inhibitor actually precedes bending.

Pilet's group found that a mica barrier reduced curvature when inserted horizontally into the entire tip (cap and meristem) but did not do so when inserted into the cap alone (Pilet, 1977). Also, larger effects of asymmetrical tip replacement were found if cap and meristem regions were used, and the quantitative bending data can be explained by assuming that inhibitors are exported from the cap, and redistributed in the apex. Shaw and Wilkins (1973), using the same techniques, also found evidence for redistribution of inhibitor in the apex. This idea remains to be reconciled with the root cap statocyte model: if perception occurs in the cap and not in the apical meristem, how does the apex initiate lateral transport of a growth inhibitor reaching it from the cap? Curvature (though reduced) still occurs when lateral transport in the meristem is blocked, presumably because the cap itself exports more inhibitor on its lower side.

Measurements of hormone concentrations by sensitive chemical methods in parts of the quickly-frozen intact root, could confirm the inhibitor hypothesis in a system free from possible artefacts caused by cutting and replacing root caps or tips. It will also be of interest to follow hormone levels over minutes, rather than hours.

15.5 CONCLUDING REMARKS

After many years of stagnation, there has been a renewed interest in the molecular basis of plant tropisms.

The perception mechanism of gravitropism has been clarified by excellent electron microscopic analyses indicating that amyloplasts in higher organisms, and $BaSO_4$ grains in the alga *Chara*, act as statoliths. In gravitropic fungi the mode of graviperception has yet to be elucidated.

The molecule(s) perceiving unilateral light has been 'named' but very little indeed is known about it. Majority consensus, more based on hunch and personal preferences than on hard data has it that 'cryptochrome' is probably a flavin. The possibility of its being more than one compound, flavin, carotenoid or other, cannot be ruled out. It is likely to be easier to gain chemical insight into cryptochrome action in mould sporulation or fern gametophyte growth, than in fungal or plant phototropism, since the former are more amenable than is phototropism to genetic and biochemical analyses.

Much of the information relating to amplification of the perceived signal remains in the proverbial 'black box' that the physicists use to denote an unknown mechanism. This is particularly true for the primary amplification steps. Unchecked possibilities abound for the amplification mechanism: they involve changes in transcription, translation, enzyme activation, as well as electrochemical gradients and even lesser understood or unknown steps. There are some temporal correlations with changes in known 'secondary messengers' such as cyclic AMP and related esterases, and plant growth promoting hormones such as auxins, as well as plant growth inhibitors. That the Cholodny-Went theory implicates changing auxin distributions cannot be general and is the subject of much scientific debate. There is not yet sufficient evidence to exclude auxin redistribution; the Cholodny–Went theory is still a viable hypothesis, yet it is surely not the only effector mechanism. Perhaps it should be modified to state that 'in higher plants there is a changing distribution of growth regulators following a change in plant orientation'.

The biophysical approach to phototropism, mainly by the Delbrück group using *Phycomyces*, has been the major force keeping research in phototropism going. It has contributed much to our knowledge but told little of what is happening in the higher plant or about the molecular mechanisms involved.

The recent resurgence of interest in all aspects of tropisms, especially in higher plants should soon make this discussion of the subject outdated. Considering that tropism research using modern scientific methods was started by Darwin, progress has not been as rapid as one might have hoped.

FURTHER READING

Firn R. D. & Digby J. (1980) The establishment of tropic curvatures in plants. *Ann. Rev. Plant Physiol.* **31**, 131–48.

Gressel J. (1979) Blue light photoreception. *Photochem. Photobuol.*, **30**, 749–754.

Haupt W. & Feinleib M. E., (eds) (1979) *Physiology of Movements. Encyclopaedia of Plant Physiology* N.S. 7, 725 pp. Springer Verlag, Berlin.

Presti D. & Delbrück M. (1978) Photoreceptors for biosynthesis, energy storage, and vision. *Plant, Cell Envir.*, **1**, 81–100.

Senger H. (ed.) (1980) *The Effect of Blue Light on Plants and Microorganisms*, Springer Verlag, Berlin (in press).

INTERACTIONS WITH OTHER ORGANISMS

CHAPTER 16

PLANT-RHIZOBIUM INTERACTIONS IN SYMBIOTIC NITROGEN FIXATION

16.1 INTRODUCTION

The capacity for biological reduction of atmospheric nitrogen to ammonia is restricted to some bacteria, blue-green algae and fungi. Some higher plants have acquired the ability to associate with the nitrogen-fixing organisms. Among various forms of nitrogen-fixing associations that exist in nature, legumes and *Rhizobium* spp. have developed a highly specialized intracellular association where these organisms can live symbiotically. A functional root nodule, an organ *sui generis* and a site of symbiotic nitrogen fixation, is developed by cooperative interactions between the legume host and *Rhizobium*. Fidelity of these interactions culminates in an effective state where bacteria develop the capacity for reduction of nitrogen to ammonia which is then assimilated by the plant for its growth. Thus an effectively nodulated legume is fully autotrophic for nitrogen, a limiting factor in plant nutrition.

The symbiotic association between *Rhizobium* and legumes, in which two organisms interact in such a way as to influence and coordinate the expression of both procaryotic and eucaryotic genes, has evolved over millions of years. It is apparent that this relationship is influenced not only by the genetics of both organisms but also by the physiological factors prevailing during the development of the nodule. While the mechanism and biochemistry of nitrogen fixation is relatively well understood, little is known about the various interactions that take place between the two partners particularly during the early stages of development of the infection and establishment of symbiosis. This chapter will describe some of these interactions at the cellular and molecular level, with an emphasis on the contribution and involvement of the host in this unique biological process in nature. Most of the examples are derived from the extensively studied soybean (*Glycine max*) nodule system. However, an attempt has been made to provide a general overview of the legume-*Rhizobium* symbiosis.

16.2 DEVELOPMENT OF THE ROOT NODULE

16.2.1 INFECTION OF LEGUME BY *Rhizobium*

It has been established that specific groups of legumes are receptive to a particular species of *Rhizobium*. This has formed a basis for the taxonomic classification of Rhizobia and the development of cross-inoculation groups in legumes, although this specificity has recently been questioned (Section 16.8). Initiation of root nodule development requires the successful entry of Rhizobia into the host root cell, which generally takes place through the root hair (Bond, 1948). The first visible plant response to *Rhizobium* is the curling or contorted growth of the root hair at the point of infection (Sahlman & Fahraeus, 1963; Libbenga & Harkes, 1973) followed by the development of a tube-like structure known as the infection thread. The subsequent growth of this infection thread and the pattern of infection of the root cortical cells is determined by the host. In the club-shaped nodules with apical meristem e.g., pea (*Pisum sativum*), there is continuous infection of cells below the meristem; however, in round nodules e.g., soybean (*Glycine max.*) where meristematic activity is uniformly distributed, infection is confined to the early stage of the nodule development. In those legumes such as *Arachis* and *Lupinus* spp., where an infection thread is not observed, it appears that infection proceeds through intercellular spaces, and nodules generally arise in the junctions of lateral roots (see Dart, 1974; 1977, for a detailed description of morphological events in the process of infection).

16.2.1.1 *Host-Rhizobium specificity*

Recognition of a specific host by *Rhizobium* is a prelude to infection. In recent years, a significant body of data has accumulated which suggests that specific plant proteins, phytohaemoglutinins (lectins), may be involved in this recognition process. It was first suggested by Hamblin and Kent (1973) that lectin from *Phaseolus* can bind *R. phaseoli* to its roots at sites suitable for infection. Lectin prepared from soybean seeds binds specifically to soybean-nodulating bacteria, *R. japonicum* (Bohlool & Schmidt, 1974) and that from clover binds to *R. trifolii* (Dazzo & Hubbell, 1975), suggesting that these molecules provide a host-*Rhizobium* specificity. Since lectins are generally multivalent, they can serve as a 'linker' between two receptor molecules, one on the host root and another on the *Rhizobium* surface providing a correct interface between the two organisms. The biochemical receptor for lectin binding appears to be an O-antigen moiety of a lipopolysaccharide of *Rhizobium* (Wolpert & Albersheim, 1976). The role of the bacterial O-antigen in the process of host recognition was demonstrated by a *R. japonicum* mutant strain (SM2) which lacks this antigen, rendering the mutant non-nodulating (Maier & Brill, 1978). Similarly, a *R. leguminosarum* mutant strain that produced a low level of polysaccharide failed to nodulate pea (*Pisum sativum*), its respective host (Sanders *et al.*, 1978). A spontaneous revertant of this mutant regained the ability to produce extracellular polysaccharide along with its ability to nodulate. It has been shown that ineffective strains [i.e. strains which infect but do not yield functional nodules (see Section 16.3)] and non-infective strains of *R. trifolii* differ in their capsular polysaccharide composition (Dazzo & Hubbell, 1975). Moreover, a mutant strain of *Azotobacter vinelandii* that is unable to fix nitrogen has been transformed into Nif^+ with DNA from *R. japonicum* and some of the transformants have acquired the ability to synthesize O-antigen-related polysaccharide (Maier *et al.*, 1978a). A purified capsular polysaccharide from *Rhizobium* has been shown to bind to the root hair of soybean (Hughes *et al.*, 1979). So it appears that the presence of the surface polysaccharides is obligatory for the participation of *Rhizobium* in the infection of the plant and the lectins may provide specificity in their recognition.

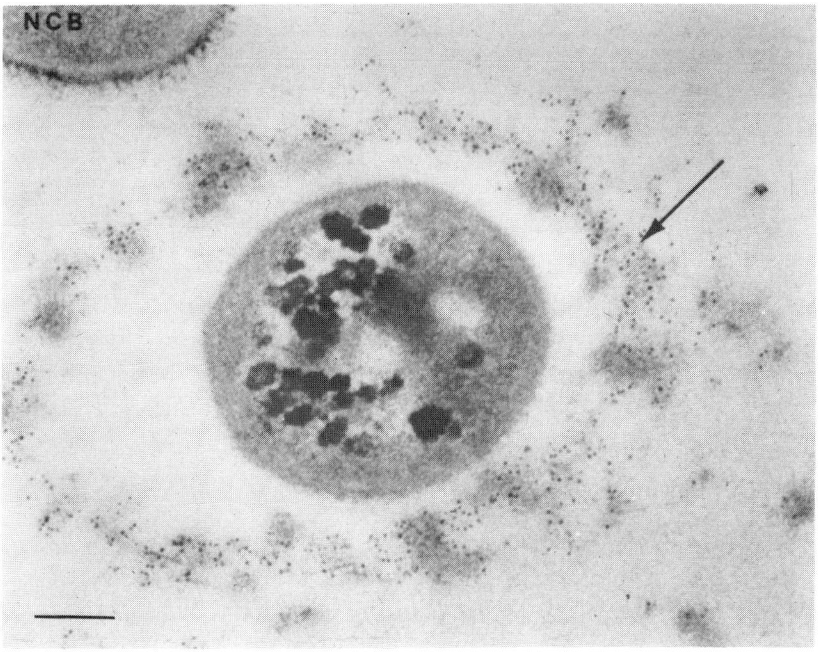

Fig. 16.1 Binding of the ferritin-conjugated soybean lectin to wild type *Rhizobium japonicum* (strain 61A76). Note the binding of ferritin to capsulated bacteria (arrow) only. NCB, non-capsulated bacteria, Bar = 0.5μm. (From Bal *et al.*, 1978).

Ultrastructural observations indicate (Fig. 16.1) that lectins bind specifically to the capsular form of *Rhizobium* and the binding is almost uniform on the capsule polysaccharide (Bal *et al.*, 1978, Calvert *et al.*, 1978) although polar binding has also been observed (Bohlool & Schmidt, 1976). These capsular forms of bacteria were also found to attach to the root hair (Bal *et al.*, 1978), and the polar attachment of *Rhizobium* seems to correlate with the polar binding of lectin to the capsular material (Bohlool & Schmidt, 1976; Dazzo *et al.*, 1976). Direct evidence for the role of lectins in binding bacteria to the root hair has been obtained in clover (Dazzo & Brill, 1977; Dazzo *et al.*, 1978), where a hapten of clover lectin, 2-deoxy-D-glucose, has been shown to prevent specific binding of *R. trifolii* to clover root hairs. A lectin-like protein, trifolin, involved in the recognition of *R. trifolii* has been isolated and purified from clover (Dazzo

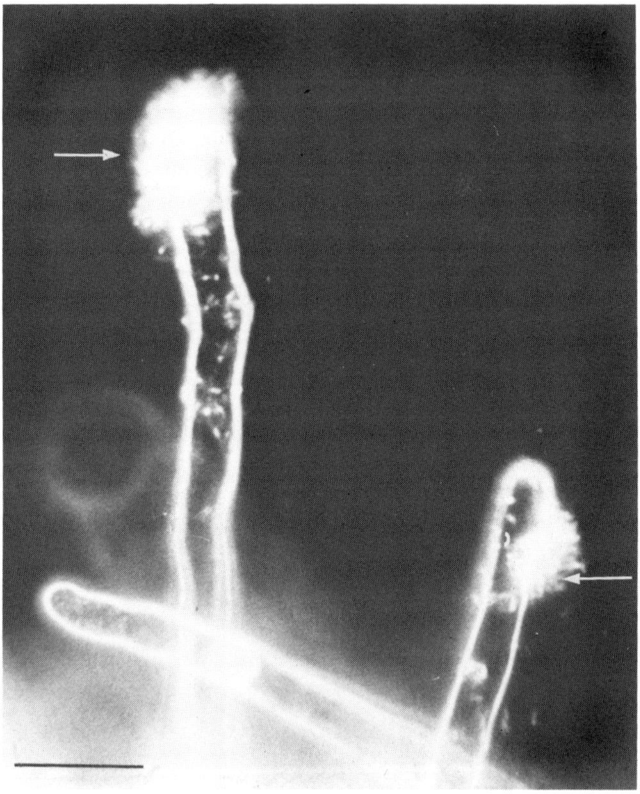

Fig. 16.2 Binding of agglutinated *Rhizobium trifolii* to the tips of clover root hairs. Examined by dark field illumination. Bar = 50μm, (Dazzo *et al.*, 1978).

et al., 1978). It was shown (Fig. 16.2) that trifolin can agglutinate *R. trifolii* which can then bind specifically to the root hair of its host, clover. This protein can be eluted from the clover root with 2-deoxy-D-glucose suggesting that it is anchored to the surface carbohydrate. The observation (Pueppke *et al.*, 1978) that the concentration of soybean lectin decreases in older root tissue at a time when nodulation can occur efficiently is inconsistent with the role of lectins in the infection process; however, this low level of lectins may be specifically localized in fewer cells. The strains of *Rhizobium* that do not bind lectins are incapable of nodulation but there is no quantitative relationship between the infectivity and lectin-binding properties of the bacteria. Furthermore, some isolines of soybeans, that do not contain detectable amounts of lectin, nodulate well with *Rhizobium* (Pull *et al.*, 1978), suggesting that the lectin requirement may not be absolute and other alternative mechanism(s) for recognition between host and *Rhizobium* may exist.

In a population of *R. japonicum*, less than 1% of the bacteria appear to differentiate into a capsular form (Bal *et al.*, 1978); thus only a very small percentage of Rhizobia may actually participate in the infection process. A similar situation exists in infection by virulant forms of pathogenic bacteria. Particular nodule-associated bacteria often accumulate at or near the root of their respective host plant. Currier and Strobel (1977) showed that a phenomenon of chemotaxis is involved in attracting specific Rhizobia in trefoil roots and is mediated by a high molecular weight (60 000) glycoprotein, chemotactin. It has been suggested that this protein may provide some specificity for attraction for *Rhizobium* by a host.

16.2.1.2 *Involvement of cellulolytic and pectic enzymes in infection*

In contrast to the pathogenic micro-organisms which generally enter the host cells by dissolving their cell wall with hydrolytic enzymes, *Rhizobium* does not appear to destroy the wall of the root hair completely at the point of attachment. Instead, the cell wall of the host is invaginated at the site of infection and, due to the deposition of new cell wall material, an infection thread is formed

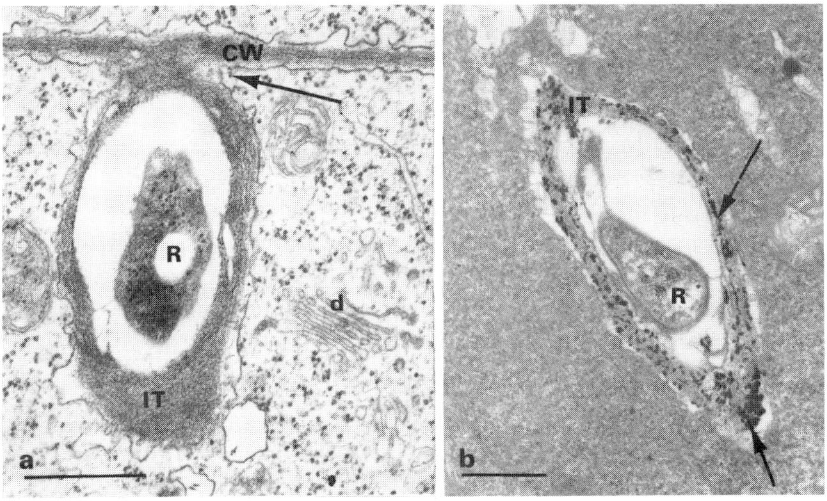

Fig. 16.3 Ultrastructure of young nodule of soybean. a, tangential section of the infection thread entering the cell; note the continuity of the host cell wall (cw) and the plasma membrane with the infection thread (IT) wall. b, Electron-microscopic localization of cellulolytic activity in the infection thread of soybean root nodules. Arrow indicates the reaction product of cellulase. Bar = 0.5μm; R, *Rhizobium*; d, dictyosomes, (From Verma *et al.*, 1978b).

through which *Rhizobium* enters the host cell (Nutman, 1956). A continuity of the host cell wall with the infection thread (Fig. 16.3a) has been observed in several legumes. Obviously an alteration in the cell wall structure is a prerequisite for such an invagination; however, no cellulase activity has been detected in free-living *Rhizobium* (Mosse, 1964). It was proposed that a polygalacturonase enzyme might be involved in this process since activity of this enzyme was found to increase when plant roots were inoculated with the nodulating strain of bacteria as compared to the non-nodulating strain (Ljunggren, 1969). However, involvement of cell wall hydrolysing enzymes in the process of rhizobial infection remained controversial (see Dart, 1974; 1977). This is primarily due to the fact that infection is a localized event and such a localized increase in enzyme activity may not be detectable when the total enzyme activity of the root is measured (cf. Hunter & Elkan, 1975).

Verma *et al.*, (1978a; b) investigated the possibility of localized involvement of cell wall hydrolysing enzymes and the roles of the two partners in the process of infection. They demonstrated that activities of both cellulase and pectinase enzymes increase during early development of the nodule in soybean, and that these activities are differentially compartmentalized, i.e., pectinase is primarily present in bacteria while most of the cellulase is in the host cell. Cellulolytic activity has been specifically localized in the cell wall matrix of the infection thread (Fig. 16.3b). A low level of pectolytic enzyme activity has also been demonstrated in free-living *Rhizobium* (Hubbell *et al.*, 1978; Verma *et al.*, 1978a). The composition of the primary cell wall of plants is such that pectic polysaccharides are easily accessible to hydrolytic enzymes known to be secreted by pathogens (Bauer *et al.*, 1973). Thus the pectolytic enzyme activity present in free-living *Rhizobium* may be sufficient to cause a pliability in the cell wall at the point of infection. A further increase in this activity may help in the penetration of the infection thread through cortical cells of the root. Since the pectolytic activity present in the host does not change and most of the activity is associated with the bacteroids, Verma *et al.*, (1978a) believe that it is the bacterial pectinase enzyme which is primarily responsible for initiation of infection. Eventual dissolution of the infection thread wall however, takes place with the help of cellulase that appears to be produced by the host. This enzyme has been shown to be secreted through the plasmalemma in hormone-treated plant tissues (Bal *et al.*, 1976). Thus, a cooperative action of both host and *Rhizobium* is involved in the initiation and development of the infection during root nodule formation. Since cellulase induction is under phytohormone control (Fan & Maclachlan, 1967; Verma *et al.*, 1975), and *Rhizobium* can produce these hormones (see below), the production of cellulase in nodules may be indirectly controlled by *Rhizobium*. Such a cooperative action of both partners is desirable for the initiation of localized infection and the organized and coordinated development of root nodules. Alternatively, if both enzymes were produced by one organism, nodulation might occur randomly over the entire root surface. The process of infection may be aborted in the event of the uncontrolled production of these enzymes.

16.2.1.3 *'Release' of Rhizobia into the host cell*

The manner in which the wall of the infection thread is dissolved and bacteria 'released' into the host cell appears to be host-specific. Hubbell *et al.* (1978) have suggested a general role of the cell-wall-hydrolysing enzymes in the infection process. In soybean, most of the infection thread wall is dissolved by cellulolytic enzyme(s) (Verma *et al.*, 1978a; b), and any residual cellulose matrix that remains is sequestered into membrane vesicles through endocytosis (Bassett *et al.*, 1977a). These cellulose microfibrils eventually disintegrate but vacuoles containing fibrous material have often been observed (Basset *et al.*, 1977a; Bal *et al.*, 1980) in mature nodules. In pea and clover, where the infection thread contains a large amount of mucopolysaccharide material secreted by *Rhizobium*, the 'release' of bacteria from the infection thread first occurs into a droplet of an amorphous material (Fig. 16.4) and the Rhizobia slowly move out of this matrix into the host cell cytoplasm (Newcomb, 1976). The amount of this

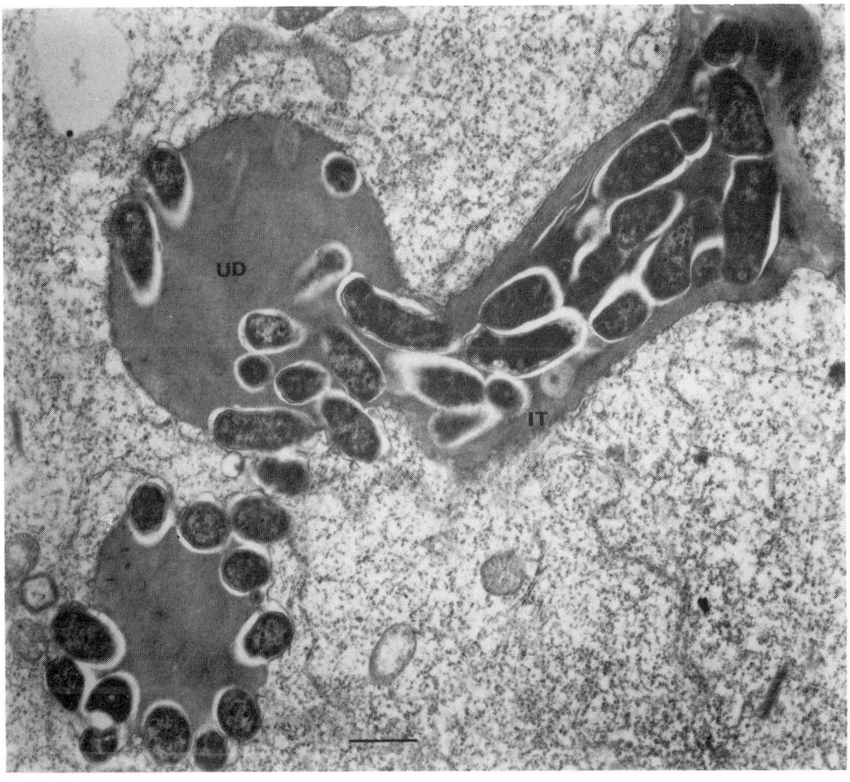

Fig. 16.4 Electron micrograph of an infected cell of pea in early phase of growth. Note the 'release' of Rhizobia from infection thread (IT) into an unwalled-droplet (UD) of mucopolysaccharide which is bounded by the plasmamembrane of the host. Rhizobia are 'released' from this mucopolysaccharide material into the host cell and are enclosed singly in a membrane envelope. Bar = 1 μm. (From Newcomb, 1976).

mucopolysaccharide is dependent upon the *Rhizobium* spp.; however, its increased production during the infection process may be stimulated by the host. A successful infection results in the 'release' of bacteria from the infection thread into a membrane vesicle derived from the host plasma membrane (see below). Lack of cellulose deposition on the membrane surface facing the bacteria is due either to the inactivation of the enzyme cellulose synthetase, or the cellulose microfibrils are continuously deposited but they are digested by cellulase activity present in the space between *Rhizobium* and the membrane envelope (Verma *et al.*, 1978b). It is likely that the cellulose synthetase enzyme is not synthesized since the membrane envelope undergoes substantial changes following endocytosis. In an ineffective nodule of soybean developed by *R. japonicum* strain 61A24, loose cellulose microfibrils appear to be continuously synthesized in the membrane envelopes (Fig. 16.5) resulting in a significant increase in the size of these envelopes which eventually disentegrate liberating Rhizobia into the host cell cytoplasm (see also, Werner *et al.*, 1980).

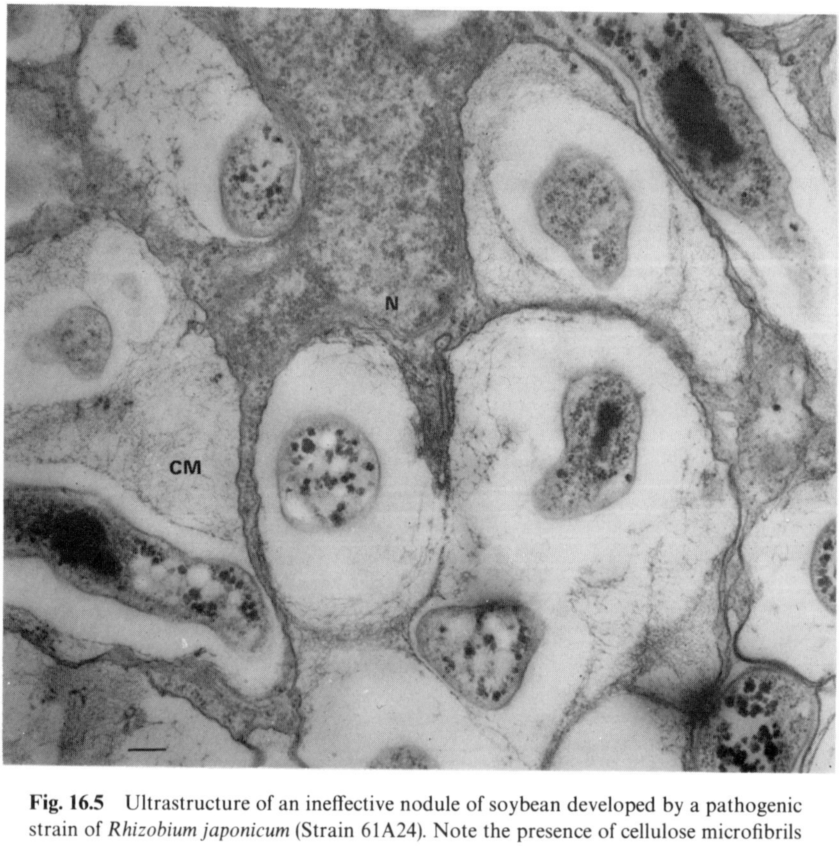

Fig. 16.5 Ultrastructure of an ineffective nodule of soybean developed by a pathogenic strain of *Rhizobium japonicum* (Strain 61A24). Note the presence of cellulose microfibrils (cm) inside the membrane envelopes enclosing the bacteria. Bar = 0.5μm; N, nucleus. (Verma & Bal, unpublished data).

It is interesting to note that the outer cell wall membrane of the *Rhizobium* also undergoes significant changes during development of the root-nodule symbiosis (Bal *et al.*, 1980). The outer cell wall membrane of free-living *Rhizobium* is sloughed off during early development of the nodules and appears to be replaced by a new membrane which may be more compatible with the host intracellular environment. The membrane of the *R. leguminosarum* bacteroids has been shown (de Veris, 1980) to contain an inducible C_4-dicarboxylic acid transport system which allows them to utilize these organic acids produced by the host.

16.2.2 ORIGIN AND ROLE OF PHYTOHORMONES IN NODULATION

Plant growth is regulated by phytohormones. The presence of cytokinins and auxins in root nodules and their role in stimulation of mitotic activity has been known for some time (Thimann, 1936; Wipf & Cooper, 1940). Indoleacetic acid (IAA) and kinetin have been demonstrated in nodule tissue of various legumes (Dullart, 1967; 1970; Pate, 1977; Syono & Torrey, 1976). High levels of cytokinin activity have also been found in root nodules of non-leguminous plants (Henson & Wheeler, 1977). Several species of *Rhizobium* e.g., *R. leguminosarum*, *japonicum* and *trifolii* are able to convert tryptophan into IAA. It has been suggested that tryptophan produced by roots of legumes is converted to IAA by bacteria and thus auxins may be synthesized symbiotically (Kefford *et al.*, 1960). Pea nodules contain, in addition to zeatin, a naturally occurring cytokinin, isopentenyladenine and its riboside (Syono & Torrey, 1976). Free-living *R. leguminosarum* has been shown to produce several cytokinins (Phillips & Torrey, 1972). It has been suggested that isopentenyladenine and isopentenyladenosine produced by *Rhizobium* may serve as a precursor for zeatin (Miura & Miller, 1969) in the host cell.

The involvement of phytohormones in the development of infection and formation of root nodules is apparent from several studies. It has been observed that as the infection thread progresses through the root hair into the cortical tissue, cell division and proliferation activity starts ahead of the infection thread (Libbenga & Harkes, 1973). That this stimulus is due to auxin and cytokinin may be deduced from an observation on auxin-treated pea tissue *in vitro* (Libbenga *et al.*, 1973). Such treatment in the presence of cytokinins also causes endoreduplication (Torrey, 1961), a phenomenon commonly observed in nodule meristematic tissue. Nodule initiation has been shown to be inhibited by abscisic acid, a substance that inhibits polyploid mitosis in pea roots cultured in the presence of auxins and cytokinins (Phillips, 1971). Furthermore, cytokinin contents of the nodules have been shown to change in pea with the age of the nodule, the highest concentration being in two-week-old nodules when mitotic activity is at its peak (Syono *et al.*, 1976). If the host utilizes IAA produced by *Rhizobium*, its intracellular concentration may be controlled by the plant IAA-oxidase. Auxins are known to induce hydrolytic enzyme activity such as cellulase that appears to be essential for cell proliferation (Fan & Maclachlan, 1967). It has been observed that pea epicotyls treated with IAA followed by the application of Rhizobia at the cut surface, develop an asymbiotic association, although no infection threads were observed in this study (Verma *et al.*, 1978c).

Some mutant strains of *Rhizobium* fail to be released from the infection thread (Bergersen, 1957); however, it is not known if there is any defect in auxin and/or cytokinin biosynthesis capacity of these strains. Lack of IAA production may prevent the host from generating sufficient hydrolytic enzyme(s) for dissolving the infection thread wall. Indeed, a mutant of *R. japonicum* which produced a low level of IAA has been shown to be ineffective (Hubbell & Elkan, 1967).

16.2.3 INTRACELLULAR COMPARTMENTALIZATION OF *Rhizobium*

Following dissolution of the infection-thread wall Rhizobia may be considered 'released' into the host cell cytoplasm though they remain bound in membrane envelopes. Thus from a physiological point of view, bacteria remain 'extracellular' (Dixon, 1969). In an effective (nitrogen-fixing) root nodule, no bacteria have been observed to be free inside the host cell. The view that the membrane enclosing the bacteria is derived from the plasma membrane of the host cell (Bergersen & Briggs, 1958; Dixon, 1967) has now been widely accepted (see Verma *et al.*, 1978b). The proliferation of bacteria in the host cell is accompanied by the division of the membrane envelope enclosing them, although the two processes are not strictly coordinated. The number of bacteria per envelope is determined by the host; for example, pea generally has one bacterium per envelope (Newcomb, 1976), while in soybean up to eight bacteria have been observed (Bergersen & Briggs, 1958; Bal *et al.*, 1980). Furthermore, the same strain of *Rhizobium* may be compartmentalized into different numbers per envelope depending upon the host it has invaded; e.g., *R. lupini* (Strain D-25) produces infection on *Lupinus luteus* yielding only one bacterium per envelope while the same strain infects *Ornithopus sativus* resulting in several bacteria per envelope (Kidby & Goodchild, 1966). The larger number of bacteria per membrane envelope may be due to the cessation of membrane proliferation before the bacteria stop dividing. In a *R. leguminosarum* mutant strain, the membrane envelope does not grow fast enough and as a result bacteria become 'naked' in the host cell (Newcomb *et al.*, 1977). Whether a cell is capable of dividing following the endocytosis of *Rhizobium* from the infection thread is also determined by the host. No cell division has been observed in pea (Newcomb, 1976) while soybean and other legume cells appear to divide during early stages of infection. In nodules where infection threads are rare, spread of bacteria through host cells takes place through cell division.

That the membrane enclosing the bacteroids retains some of the characteristics of the plasma membrane has been shown by recent ultrastructural and biochemical studies (Tu, 1975a; Robertson *et al.*, 1978a; b; Verma *et al.*, 1978b).

Both membranes have similar thicknesses (8.4nm), are stained with phosphotungstic acid-chromic acid (PTA)—a characteristic stain for plant plasma membrane (Roland *et al.*, 1972), are sensitive to nonionic detergents (e.g. Nonidet P-40), have similar patterns of localization of ATPase activity (Verma *et al.*, 1978b) and show similar freeze-fracture face particle decoration (Tu, 1975a; Robertson *et al.*, 1978b); also, there are similarities in their peptide composition (Verma *et al.*, 1978b). Development of this membrane envelope is essential but not sufficient for effective symbiosis. Disintegration of this membrane, as usually occurs in the older nodules, marks a reduction in the nitrogen fixation capacity and the beginning of senescence (Tu, 1975b). Although the host determines the synthesis of the membrane envelope, a mutation in *Rhizobium* that leads to the development of ineffective (non N_2-fixing) nodules, changes the pattern of development of this membrane (MacKenzie & Jordan, 1974; Bassett *et al.*, 1977b; Verma & Bal, unpublished). In an ineffective association of alfalfa (*Medicago sativa*) developed by mutant strains of *R. meliloti*, disintegration of bacteria is accompanied by the proliferation of the endoplasmic reticulum (MacKenzie & Jordan, 1974). In ineffective nodules of soybean developed by a *R. japonicum* strain (61A24) disintegration of the membrane is evident in the infected cells of mature nodules (Fig. 16.5). However, in another ineffective association in soybean, Basset *et al.*, (1977b) found destruction of the bacteroids prior to that of the membrane envelope. A similar situation has been observed in pea nodules developed by a mutant strain of *R. leguminosarum* (de Veris, 1980). Thus development of this intracellular compartment enclosing the bacteroids appears to be influenced both by the host and *Rhizobium* strains.

Biosynthesis of the membrane envelope is an early event in nodule development as it keeps pace with the proliferation of bacteria. Once the bacterial division ceases, there appears to be little synthesis or turnover of this membrane. This is demonstrated by the very low levels of incorporation of radioactive amino acids into the mature membrane envelopes (Verma & Zogbi, unpublished data) and also by reduced activity of the endoplasmic reticulum and Golgi apparatus generally observed in mature nodules (Newcomb, 1976; Robertson *et al.*, 1978a). Based upon the area of the membrane envelope in soybean root nodules, there is about 20–40 times more membrane as compared to the plasma membrane of the uninfected cell (Verma *et al.*, 1978b). It is not known how the bacteria induce proliferation of this membrane—it may be a normal response of the host against an invading organism. The development of such an intracellular compartment may play a role in segregating eukaryotic and prokaryotic metabolism and the attendent incompatibilities between these two organisms during development of the symbiosis. This membrane must harbour specific transport system(s) for the passage of metabolites to and from bacteroids. The recent isolation of the bacteroids enclosed in the membrane envelope—a functional compartment of the root nodules—and subsequent purification of this membrane (Robertson *et al.*, 1978b; Verma *et al.*, 1978b) should allow studies on the function of this membrane in root nodule symbiosis. Failure to develop this membrane envelope in the host cell may result in a 'parasitic or saprophytic' stage (Tu, 1975b; Verma *et al.*, 1978c).

16.3 EFFECTIVENESS OF NODULES IN NITROGEN FIXATION

Development of a nodule structure does not ensure its effectiveness in fixing nitrogen; ineffective (unable to fix nitrogen) or inefficient nodules may be formed due to the genetic incompatibility of the host or *Rhizobium* as well as to other physiological factors. The same strains of *Rhizobium* may form effective nodules on one plant and ineffective ones on another (Nutman, 1956). Similarly, the host plants can be classified into nodulating and non-nodulating types with respect to a specific *Rhizobium* sp. Among the nodulating forms, a plant may develop effective nodules with one strain and ineffective ones with another. Furthermore, the effective nodules vary in efficiency of nitrogen fixation (Nutman, 1969; Caldwell & Vest, 1977, see also Section 16.4). Thus there is a continuous interaction in the compatibility of these two organisms during the development of effective symbiosis. Various properties of the host and *Rhizobium* that account for the development of ineffective nodules in legumes have been summarized by Nutman (1956; 1969) and Brill (1974) respectively. Ineffective nodules produced either due to the incompatibility of host or *Rhizobium* result in the eventual senescence and breakdown of the nodule tissue which appears to be manifested by nitrogen starvation (MacKenzie & Jordan, 1974). Ineffective nodules of soybean are found to contain low levels of the plant glutamate dehydrogenase and glutamine synthetase and high levels of aspartate and alanine aminotransferases. The glutamine synthetase activity of the bacteroids on the other hand increases several-fold in these nodules (Werner *et al.*, 1980).

16.3.1 GENETIC FACTORS

The plant genome plays an important role in the development and the effectiveness of the nodules in nitrogen fixation. The plant determines the size, shape, distribution and morphology of the nodule including intracellular organization of the infected host cell. The same strain of *Rhizobium* (e.g. *R. lupini*, strain D-25) forms collar-shaped nodules on *Lupinus luteus* containing rod shaped bacteroids enclosed singly in the membrane envelope while it forms cylindrical or sometimes branched nodules on *Ornithopus sativus* with several bacteroids per membrane envelope (Kidby & Goodchild, 1966). Similarly, the ultrastructure of nodules formed by the same strain of *Rhizobium* on *Vigna spp.* differs greatly from that on *Arachis hypogaea* (see also, Dart, 1977).

The lack of nodulation in a plant exposed to a compatible *Rhizobium* can be due to a complete lack of infection or to the abortion of the infection thread. The mechanism that determines the non-nodulating response is not understood but it may involve surface recognition by *Rhizobium*, inability of the host cell wall to invaginate, lack of growth of the infection thread and premature, or lack of, release of *Rhizobium* from the infection thread. This character has been linked in red clover (*Trifolium pratense*) to a single recessive gene *rr* along with a cytoplasmic factor (Nutman, 1969). In soybean, Williams and Lynch (1954) have identified a recessive gene condition, 'no no', (now known as rj_1 rj_1) which does not allow infection to take place (Caldwell, 1966). Similarly, two genes in pea, sym_1 and sym_2, confer resistance to *Rhizobium* infection (Holl, 1973). Several plant genes have been linked to ineffectiveness of the nodules in fixing nitrogen. In soybean, three such genes (Rj_2, Tj_3, Tj_4) have been identified (Caldwell & Vest, 1968; Vest, 1970; Vest & Caldwell, 1972). All these genes that render the nodules ineffective in soybean are dominant. While Rj_2 appears to be independent of Rj_1, it appears to be linked to RJ_3 (Caldwell & Vest, 1977). A biological relationship exists between *Rhizobium* strains that induces the ineffective response by Rj_2. This appears to be related to the particular serogroups of *Rhizobium*, C_1 and 122, indicating the involvement of some surface interaction. Nutman (1957) has described two recessive genes, 'ie' and 'i_1' in red clover that confer ineffective symbiosis when inoculated with certain strains of *R. trifolii*. These two genes appear to segregate independently. The i_1 ineffectiveness is related to the inability of bacteria to differentiate into bacteriods, while the ineffectiveness due to 'ie' is primarily a result of abnormal cell growth in nodules. Some selective strains of *R. trifolii* can form moderately effective nodules on these mutant plants. A gene, sym_3, has been identified in pea that allows the production of normal nodules including the presence of nitrogenase, but the enzyme does not function (Holl, 1973). A diploid species of *Lotus*, *L. uliginosus* which forms effective nodules with a specific bacterial strain, when crossed with a tetraploid species *L. carniculatus* that is effective with another strain of *Rhizobium*, was found to be dominant and also showed a dose effect of genes (Gershon, 1961). This led to the suggestion that host specificity for effectiveness involves more than one plant gene. According to a conservative estimate, ten host genes are involved in symbiotic nitrogen fixation (Holl & LaRue, 1976); however, this estimate is likely to increase when considering various interactions of host/*Rhizobium* at the molecular level (see also Sections 16.5 and 16.6).

16.3.2 PHYSIOLOGICAL FACTORS

Temperature, intensity and duration of light and concentration of soluble nitrogen affect the development, structure and function of nodules. The optimum temperature for nodulation varies with host and *Rhizobium* spp. The initial infection process has most exacting temperature requirements while the actual nitrogen-fixation activity exhibits a wide optimum. The role of light is related to the supply of photosynthates—an integral component of root nodule symbiosis (see below). Few nodules develop in the dark and the ones that do, degenerate. Soluble nitrogen decreases or even prevents nodulation and when added to fully developed effective nodules, inhibits nitrogen fixation with a subsequent degeneration of leghaemoglobin and the bacteroidal tissue (Bisseling *et al.*, 1978). The duration of darkness that has deleterious effect on a nodulated plant varies with host, e.g., in *Pisum sativum*, 24 h darkness destroys leghaemoglobin and nitrogen fixation activity (Virtanen *et al.*, 1947), while in *Glycine max*, several days of darkness are required before leghaemoglobin is affected (see also, Bergersen, 1977).

Although there is no specific requirement for mineral elements for nodulated plants as compared to non-nodulated ones, molybdenum, manganese, copper, cobalt, boron, calcium, and phosphate all affect nodulation and nitrogen fixation. Molybdenum is a constituent of nitrogenase and nitrate reductase of bacteroids (Klucas *et al.*, 1968; Chenial & Evans, 1960) and cobalt is a constituent of vitamin B^{12}, although the role of vitamin B^{12} is not clear in root nodules. Calcium and boron appear to be involved in the initial infection and the development of the nodules (see Pate, 1977). An auxotrophic mutant strain of *Rhizobium trifolii* that requires riboflavin for growth has been shown to produce ineffective nodules on red clover (*Trifolium pratense*) but effectiveness can be restored by addition of flavin mononucleotides to the growth medium (Schwinghamer, 1970). The same auxotroph displayed different levels of ineffectiveness on other hosts suggesting that this requirement can be partly fulfilled by the host.

16.4 SYMBIOTIC NATURE OF NITROGEN FIXATION IN ROOT NODULES

There are two main requirements for biological reduction of atmospheric nitrogen: (a) continuous availability of carbohydrate for the metabolic energy-ATP; and (b) protection of the *Rhizobium* nitrogenase from oxygen since it is an oxygen-sensitive enzyme. Although these requirements can be met experimentally and free-living *Rhizobium* can be induced to fix nitrogen in culture (Kurz & LaRue, 1975; McComb *et al.*, 1975; Pagan *et al.*, 1975; Reporter & Hermina, 1975), the legume–*Rhizobium* association provides an environment in the nodules where these requirements are met by the host. The role of photosynthesis in nitrogen fixation in nodules has been realized for a long time. After various periods of darkness nodules show a progressive loss in their capacity for nitrogen fixation. That this loss is due to the limitation of photosynthates is evident from its reversal with light treatment. It has been shown (Bach *et al.*, 1958) that ^{14}C-labelled photosynthates are rapidly translocated to the nodule and radioactivity is found to be distributed among organic acids, amino acids and carbohydrates. *Rhizobium* metabolizes these compounds to generate reductants and ATP for the nitrogenase reaction. Hardy *et al.*, (1968) observed a diurnal rhythmn in nitrogen fixation by soybean nodules. These nodules continue to fix nitrogen at a reduced rate in the dark as well as when they are detached from the root for a long time, indicating that they have higher reserves of carbohydrates. The soybean bacteroids contain up to 50% of their dry weight as polyhydroxybutyrate (Klucas & Evans, 1968) that is synthesized during the early stages of nodule development and bacteroidal differentiation. This carbon source may be available for metabolism during darkness or under restricted supply of photosynthates (Bergersen, 1977). Bacteroids have also been shown to utilize organic acids, malate and succinate, produced by the host during symbiotic nitrogen fixation (de Veris, 1980).

There is a high demand for O_2 in nitrogen-fixing tissue (Tjepkema, 1971) and there appears to be a limitation in the availability of O_2 in the micro-environment of the infected cells. Although these conditions may be well suited for the induction and preservation of bacteroid nitrogenase, in order to maintain a continuous flux of O_2, host cells produce an oxygen-binding protein, leghaemoglobin, in the infected cells. The role of leghaemoglobin in transport of O_2 has been well documented (Yocum, 1964; Bergersen, 1971; Tjepkema, 1971; Appleby, 1974; Wittenberg *et al.*, 1974). Although there has been a controversy regarding the intracellular site of leghaemoglobin in nodules, it is now clear from immunological and biochemical data (Verma & Bal, 1976; Verma *et al.*, 1978b; 1979) that this protein is localized in the host cell cytoplasm (Section 16.5.1.1). In this location, it maintains a sufficiently high pO_2 in the host cell cytoplasm for oxidative phosphorylation and provides a low but sustained rate of O_2 supply to the bacteroids. Appleby *et al.*, (1975a) have suggested the involvement of cytochrome P-450 in conjunction with oxyleghaemoglobin in an efficient oxidative phosphorylation in *Rhizobium* bacteroids. Furthermore, leghaemoglobin provides a mechanism for adaptation of nitrogenase activity during altered rhizosphere pO_2 (Criswell *et al.*, 1976). In non-legume nodules where leghaemoglobin is not detected, another mechanism for the control of O_2 supply may be present.

A general pathway for symbiotic nitrogen fixation in legume root nodules is shown in Fig. 16.6. Although there are some variations in the organization of the nodule structure and specific requirements for symbiosis, the basic mechanism is the same. A key point in the maintenance of symbiosis following successful development of the nodule structure, appears to be the carbon–nitrogen balance of the nodule. As discussed above, most of the carbon is provided by the host as photosynthates (sugars and organic acids) which are converted to reducing power and ATP in bacteroids, while nitrogen fixed by the bacteroids is assimilated and taken up by the host. The membrane envelope which encloses the bacteroids forms an intracellular compartment in which bacteria are transformed into bacteroids that actively fix nitrogen. Ammonia is the primary product of symbiotic nitrogen fixation (Bergersen, 1965). Although the bacteroids contain enzymes for the assimilation of ammonia into amino acids (see Bergersen, 1977), most of the ammonia appears to be secreted directly into the medium under nitrogen fixation conditions (O'Gara & Shanmugam, 1976). It thus appears that ammonia is assimilated primarily by the enzymes of the plant system (Brown & Dilworth, 1975) according to the following pathway:

Bacteroids: $N_2 + ATP + \text{Reductant} \xrightarrow{\text{Nitrogenase}} NH_4^+ + ADP + P_i$

Host Cell: $NH_4^+ + ATP + \text{Glutamate} \xrightarrow[\text{Synthetase}]{\text{Glutamine}} \text{Glutamine} + ADP + P_i$

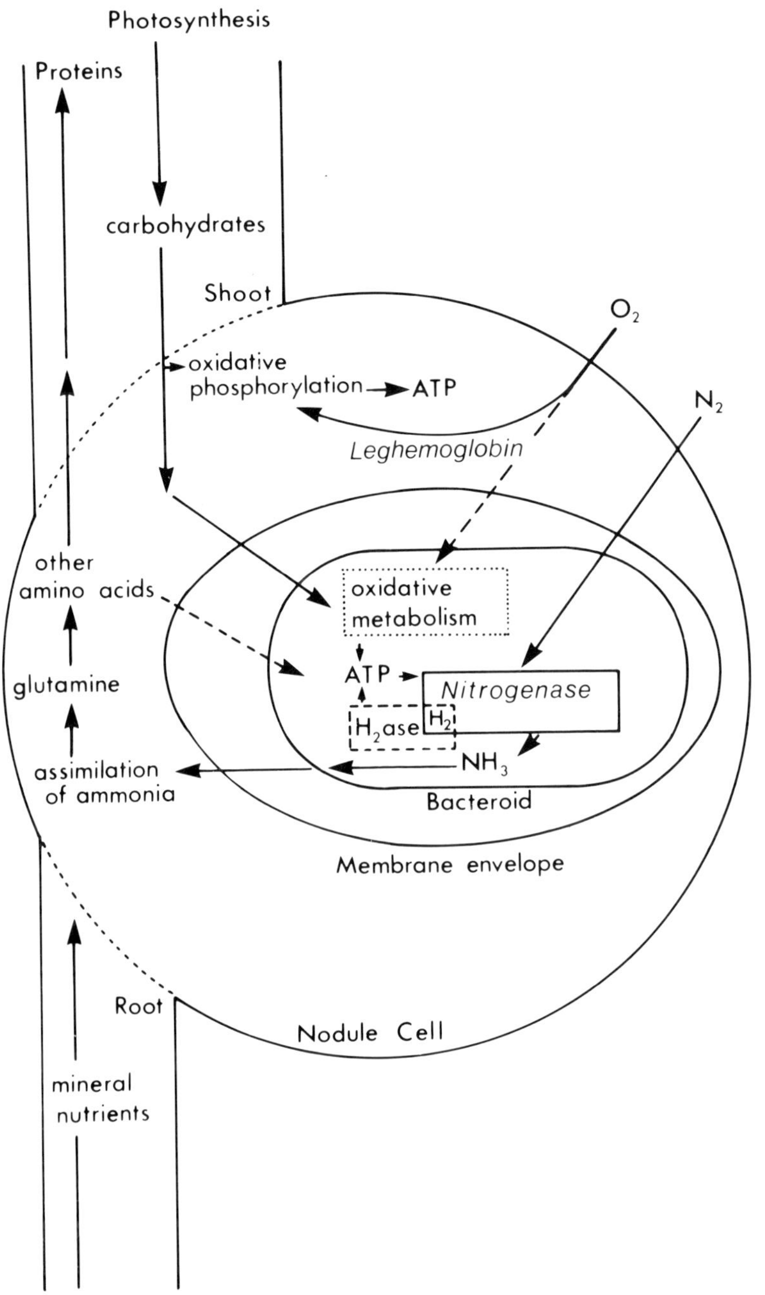

Fig. 16.6 A general pathway of symbiotic nitrogen fixation in legume root nodules.

Estimates of the consumption of carbohydrates in nodules vary from 0.8–19 mg/mg of N_2. Minchen and Pate (1973) have shown that in pea (*Pisum sativum*), nodules command 32% of the net photosynthetic gain of carbon to the shoot, out of which 16% is allocated for growth, 37% for respiration and 47% for translocation of amino compounds. Thus factors affecting photosynthesis directly influence the rate of nitrogen fixation and a prolonged retardation in the supply of carbohydrates may irreversibly damage the nodule tissue. In ineffective nodules, nitrogen starvation leads to the senescence of nodular cells.

A high concentration of hydrogen is known competitively to inhibit reduction of nitrogen (Wilson, 1940). During the nitrogen fixation reaction about 30% of the electron flux through the nitrogenase enzyme is consumed in the reduction of protons to hydrogen (Evans *et al.*, 1978). Some *Rhizobium* strains exhibit a capacity to oxidize hydrogen with hydrogenase and utilize the energy derived from this reaction (Dixon, 1972; Maier *et al.*, 1978b). It is possible that the evolution of hydrogen by nitrogenase provides a reductant for oxygen thereby reducing the concentration of oxygen in the vicinity of nitrogenase. In those associations where *Rhizobium* exhibits hydrogenase activity, there is a higher efficiency of nitrogen fixation (Schubert & Evans, 1976). Expression of this enzyme activity in *Rhizobium* appears to be determined by the host (Dixon, 1972). This is shown by a strain of *Rhizobium* that upon infection of *Pisum sativum* and *Vicia bengalensis* contains hydrogenase activity while on *V. faba*, it lacks this activity.

Attempts to develop a forced association between *Rhizobium* spp. and plant cells in tissue culture have not been very successful. Holsten *et al.*, (1971) have claimed to have developed a symbiosis between root callus tissue of *Glycine max* and *R. japonicum*, that was based upon the observations that *Rhizobium* in this association was able to reduce C_2H_2 (a measure of nitrogenase activity). Similarly, other *Rhizobium* spp. show acetylene reduction activity in association with callus cells of legumes and non-legumes (Phillips, 1974; Carlson & Chaleff, 1974; Child, 1975). The ultrastructural observations of such infected cells do not support the symbiotic nature of these associations and there is no evidence that nitrogen fixed in these systems is directly utilized by the host in a manner similar to that in the legume nodules (Verma *et al.*, 1978c). It appears that in these associations, the host provides the necessary nutrients for growth of bacteria, and the proliferation of bacteria inside the host cell (which is killed due to the infection) creates a low pO_2 where *Rhizobium* can derepress its nitrogenase genes. The activity of nitrogenase is very low and no leghaemoglobin and the membrane envelopes enclosing the bacteria have been detected in these associations. A similar situation appears to occur in associations between nitrogen-fixing bacteria (*Spirillum* spp.) and some non-leguminous plants, e.g. corn and grasses (cf. Bulow & Dobereiner, 1975; Smith *et al.*, 1976). Most of the nitrogen fixed under these conditions is utilized by the bacteria for their own growth and only through death and decay of the bacteria may it become available to the plant. Obviously, such asymbiotic associations do not favour the host for carbon nitrogen balance and thus, their potential usefulness remains in doubt.

16.5 EXPRESSION OF SPECIFIC HOST GENES

16.5.1 LEGHAEMOGLOBIN

The presence of a myoglobin-like protein, leghaemoglobin, in effective (nitrogen-fixing) root nodules of legumes was first reported by Kubo (1939). This protein represents about 25–30% of the total soluble protein in nodules. There is a direct correlation between the presence of leghaemoglobin and the nitrogen-fixation capacity of the nodules (Virtanen *et al.*, 1947, Smith, 1949). Induction of leghaemoglobin appears to be obligatory for the development of effective symbiosis in legumes. Although the role of this protein as an oxygen carrier has now been well established (see above) its mode of induction and the regulation of its biosynthesis is not fully understood.

16.5.1.1 *Intracellular site of leghaemoglobin*

Since the early observation of Smith (1949) that leghaemoglobin is confined to the central nodule cells which contain bacteroids, there has been a controversy over its intracellular location. By using electron microscopic autoradiography of *Serradella* nodules grown in the presence of ^{59}Fe, Dilworth and Kidby (1968) concluded that this protein is located within the membrane sac enclosing the bacteroids. Similar observations were made by Bergersen and Goodchild (1973) using a diaminobenzidine reagent. Since these results were based upon the localization of haem (prosthetic group of leghaemoglobulin known to be secreted by the bacteroids, Cutting & Schulman, 1969), these results led to the conclusion that leghaemoglobin is located in the space between the membrane envelope and bacteroids. In contrast, Verma and Bal (1976) used ferritin-conjugated antibodies to leghaemoglobin and showed the intracellular location of this protein to be restricted to the cytoplasm of the infected cells of the nodules. In those preparations where membrane envelopes enclosing the bacteroids were broken, no ferritin-conjugated antibody was found inside these vesicles (Fig. 16.7). This was further confirmed by isolating the membrane envelopes enclosing the bacteroids and demonstrating that no antibody-reactive material is present in these vesicles (Verma *et al.*, 1978b). The proposed function of leghaemoglobin in root nodules does not require that this protein be in a direct contact with bacteroids as has been suggested by Bergersen (1977).

16.5.1.2 *Leghaemoglobin genes*

Indirect evidence has suggested that leghaemoglobin may be encoded by the host genome (Dilworth, 1969; Verma *et al.*, 1974). A direct demonstration of the origin of leghaemoglobin genes has been obtained by purification of a complementary DNA to leghaemoglobin-mRNA (Lb-cDNA) and its hybridization to soybean DNA from uninfected tissue (Baulcombe & Verma, 1978). In soybean,

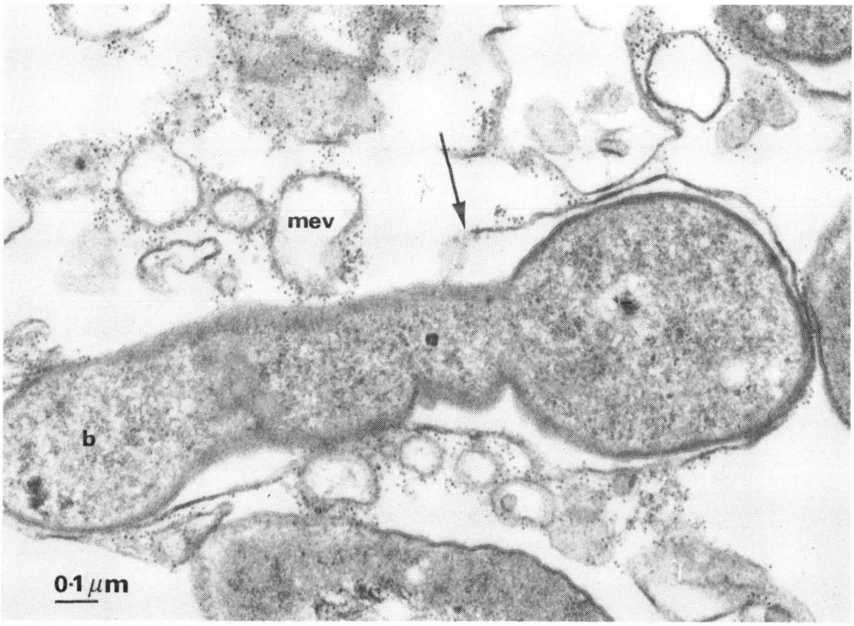

Fig. 16.7 Localization of leghaemoglobin in soybean root nodules by ferritin-conjugated antibodies to leghaemoglobin. Note the presence of ferritin in the host cell cytoplasm and its complete absence inside the membrane envelopes enclosing the bacteroids even in those sections where the membrane envelope is broken (arrow). MEV, membrane envelope vesicles. (From Verma *et al.*, 1978b).

there are three major leghaemoglobins (Appleby *et al.*, 1975b). Although all of these are very similar in their structure (Ellfolk & Sievers, 1971; Hurrell & Leach, 1977), the differences in their primary sequence show that they are different gene products. Analysis of the hybridization kinetics of Lb-cDNA to soybean DNA also indicated repetition of leghaemoglobin genes in the soybean genome (Fig. 16.8). The Cot$\frac{1}{2}$ value (330) for Lb-cDNA hybridization was found to be somewhat lower than the unique sequence DNA (cf. hybridization of uninfected root-cDNA) indicating that leghaemoglobin genes may be repeated in the soybean genome. A more accurate estimate of the repetition of leghaemoglobin genes was obtained by titrating increasing amounts of Lb-cDNA with a fixed amount of soybean DNA from both uninfected and infected root tissues (Baulcombe & Verma, 1978). This showed that about 12 pg of Lb-cDNA were hybridized with 1μg soybean DNA representing about 40 copies of leghaemoglobin genes per haploid genome (1.4 pg). This value does not change during nodulation. It is not known whether all the genes for major leghaemoglobins are repeated equally. Several-fold reiteration in leghaemoglobin genes coupled with a high level of ploidy in the infected cells may be responsible, in part, for the abundant synthesis of this protein in infected cells of the nodules. Fuchsman

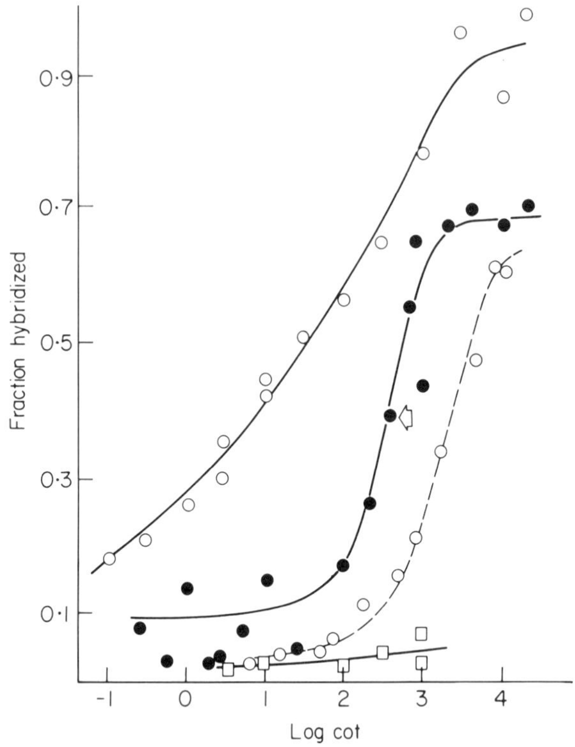

Fig. 16.8 Hybridization kinetics of leghaemoglobin-cDNA with soybean embryo DNA (○—○) and *Rhizobium* DNA (□—□). The hybridization of control root-cDNA (○--○) and reassociation kinetics of soybean DNA (○—○) have also been shown for comparison (data redrawn from Baulcombe & Verma, 1978).

and Appleby (1979a) have resolved four major and four minor leghaemoglobins in soybean. Four major primary translation products of polysomes can be observed in soybean (Verma *et al.*, 1979); the minor leghaemoglobins may arise due to some post-translational modifications and/or ligand bindings. Since the synthesis of LbS (electrophoretically slow-moving component of leghaemoglobin) and LbF (fast moving, which includes several subcomponents, Appleby *et al.*, 1975b; Fuchsman & Appleby, 1979a) changes during root nodule development (Verma *et al.*, 1979), leghaemoglobin genes appear to be differentially regulated. There is a closely-related family of leghaemoglobin genes in soybean and two of these genes have been cloned by Verma and his associates (Sullivan *et al.*, 1981). This approach should allow a detailed understanding of the organization and regulation of leghaemoglobin genes. Since the uninfected plant does not synthesize detectable amounts of leghaemoglobin, derepression of its genes appear to be under the control of *Rhizobium*.

16.5.1.3 *Induction of leghaemoglobin in relation to nitrogenase*

Although a direct correlation between the presence of leghaemoglobin and nitrogenase activity has been established in legume root nodules, little is known about the coordination of induction of these two molecules in the host and *Rhizobium* respectively. This correlation has been based primarily upon the assay of leghaemoglobin as hemochromogen which parallels nitrogenase activity in nodules. By using antibodies against leghaemoglobin and the analysis of the *in vitro* translation product of the nodule polysomes, Verma *et al.*, (1979) have detected synthesis of leghaemoglobin apoprotein three to four days before the onset of nitrogenase activity which occurs in soybean on the tenth day after infection with *Rhizobium* (Fig. 16.9). These data also show the difference in the sensitivity of two assays (see also Bergersen & Goodchild, 1973). Similar results have been obtained in pea root nodules (Bisseling *et al.*, 1979). These observations suggest that the two molecules may be regulated through independent mechanisms, although functional nitrogenase has not been detected in nodules without the presence of leghaemoglobin. The induction of the two nitrogenase components does not appear to be strictly coordinated (Bisseling *et al.*, 1979).

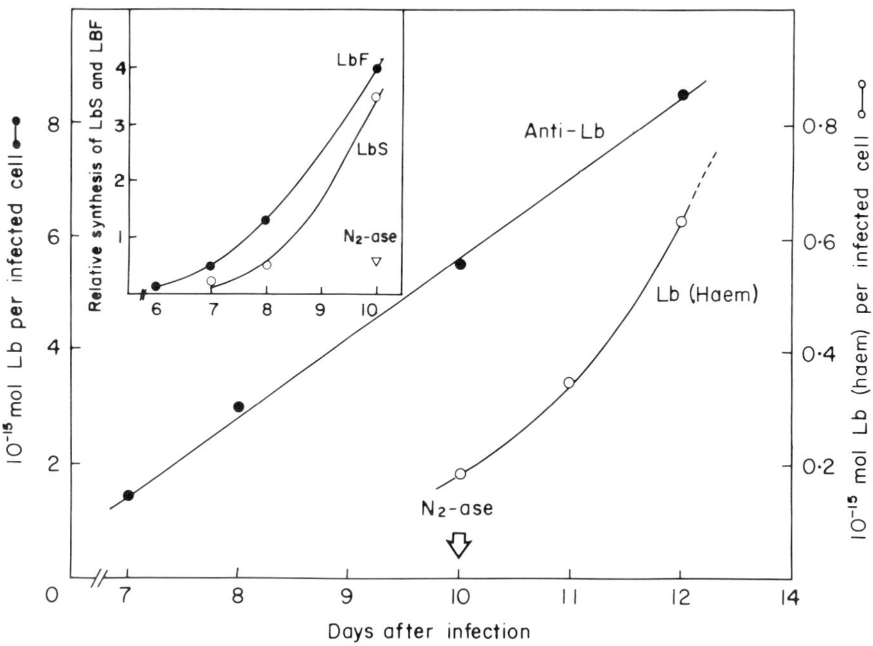

Fig. 16.9 Biosynthesis of leghaemoglobin in relation to nitrogenase during root nodule development. Leghaemoglobin was determined by haemochromogen (Lb-haem), by rocket immunoelectrophoresis (anti-Lb) and by *in vitro* translation of nodule polysomes followed by analysis of LbS and LbF on disc-gel electrophoresis. (From Verma *et al.*, 1979).

To determine further the interdependence of leghaemoglobin and nitrogenase induction, Verma *et al.*, (1981) used several strains of *Rhizobium* which form nodules that are unable to fix nitrogen (ineffective). One of these strains, SM5 (Maier & Brill, 1976), has a mutation in the nitrogenase protein and is thus unable to synthesize functional nitrogenase. Nodules developed with this mutant as well as with other ineffective strains, such as SM4 and 61A24, were assayed both for the presence of apo-leghaemoglobin (measured by antibody against leghaemoglobin) and the appearance of the leghaemoglobin mRNA as assayed by hybridization with Lb-cDNA and a cloned sequence of leghaemoglobin. Such analyses suggested that all ineffective nodules contain leghaemoglobin mRNA sequences in their polysomal poly(A)RNA, although at much reduced levels. Among the ineffective nodules (produced by different *Rhizobium* strains) the concentration of leghaemoglobin sequences varies over several orders of magnitude. This, along with the observation that leghaemoglobin appears before nitrogenase during nodule development, further suggests that the induction of leghaemoglobin genes is independent of nitrogenase; however, the magnitude of induction is controlled by the *Rhizobium* strain. In some ineffective nodules of soybean (e.g. developed by strain 61A24) only one of the major electrophoretically distinguishable leghaemoglobins, LbF, is synthesized in mature nodules but it does not accumulate indicating that the stability of leghaemoglobin may be dependent upon the effectiveness of the nodules in fixing nitrogen (Verma *et al.*, 1981). In effective nodules of soybean, LbF accumulates more than LbS, even though the synthesis of the latter predominates in mature nodules (Verma *et al.*, 1974; 1979). This suggests that there are different half-lives for various leghaemoglobins. It has been shown in lupin root nodules that the half-life of leghaemoglobin is about 16 days (Coventry & Dilworth, 1976). Since there is a temporal difference in the induction of leghaemoglobins in soybean, and their synthesis changes during root-nodule development, they may perform different functions in root-nodule symbiosis. The O_2 binding of various leghaemoglobins has been shown to be different (Fuchsman & Appleby, 1979b). A change in the oxygen requirement during nodule development may provide an explanation for the multiplicity of leghaemoglobins or they may simply represent variants encoded by a related family of genes. The mechanism of derepression of leghaemoglobin genes may involve the secretion of some 'inducer' from bacteria or may result due to the reduced oxygen concentration in the infected cells.

16.5.2 NODULINS

Considering the complexity of the host-*Rhizobium* interactions, it is apparent that besides leghaemoglobins other 'nodule-specific' plant proteins must be involved in root nodule symbiosis. Analysis of the soluble cytoplasmic proteins from uninfected roots and nodules developed by various mutant strains of *R. japonicum* by sodium dodecyl sulfate—polyacrylamide gel electrophoresis has shown the presence of several polypeptides specific to nodules (Legocki & Verma, 1979; 1980). Among these is a polypeptide of about 35 000 mol. wt. which is present in all nodules developed by effective as well as ineffective strains of *Rhizobium*. This peptide has been referred to as 'Nodulin-35' (Legocki & Verma, 1979) and appears to be the second most abundant protein in the host cell. Purification of this peptide was accomplished by preparative SDS-gel electrophoresis and antiserum was raised against this protein. Reaction of the anti-Nodulin-35 serum with soluble proteins from various ineffective nodules as well as from control root tissue showed that this protein is 'nodule-specific' and is present in all nodules irrespective of their effectiveness in fixing nitrogen. Lack of this protein in free-living *Rhizobium* and bacteroids suggests that it may be of plant origin. This conclusion is supported by immunoprecipitation of the *in vitro* translation product of nodule polysomes with antibodies against Nodulin-35 as well as inhibition of its synthesis by cycloheximide, which inhibits protein synthesis by 80S plant ribosomes. It is hoped that further characterization of Nodulin-35 will demonstrate the role of this protein in the development of root-nodule symbiosis.

In an attempt to identify other 'nodule-specific' host proteins, antiserum raised to total soluble proteins from nodules (anti-nodule serum), was reacted with soluble proteins from control root and nodules in tandem-crossed immunoelectrophoresis. These results showed (Fig. 16.10) that there are several

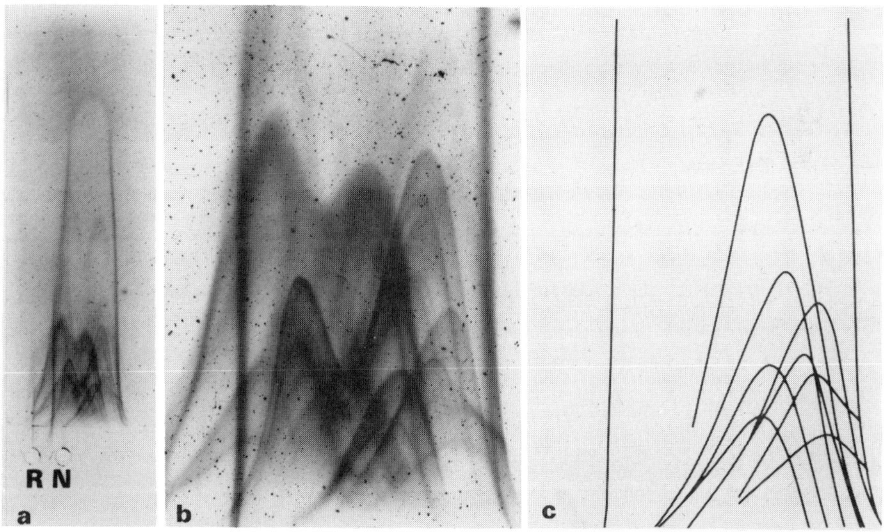

Fig. 16.10 Identification of 'nodule-specific' proteins by tandem-cross immunoelectrophoresis. Soluble cytoplasmic proteins from root (R) and nodules (N) were electrophoresed in the gel containing antibodies against nodule proteins. b is an enlargement of a and c outlines the precipitation arcs obtained from nodule proteins which do not have counterparts in root proteins (Data from Legocki & Verma, 1980).

proteins which are present in nodules but are not detectable in root tissue. Most of these proteins, including Nodulin-35 (see above), are synthesized on 80S type host polysomes. In order to obtain an antiserum specific to these proteins, the anti-nodule serum was absorbed by root proteins and the remaining fraction of antibodies, which did not react with proteins from uninfected roots, was considered 'nodule-specific' (Legocki & Verma 1980). *In vitro* translation products of root nodule polysomes were reacted with this antiserum and the resulting immunoprecipitate when analysed on two dimensional SDS-gel electrophoresis showed several 'nodule-specific' peptides, Nodulins (Legocki & Verma, 1980). These proteins are not detectable in the translation product of the control root polysomes. It is interesting to note that most of these proteins are of relatively low molecular weight.

The mechanism of induction of Nodulins in root nodules remains unclear. Since these proteins are not synthesized in uninfected roots, it follows that induction of 'nodule-specific' host genes are under the control of procaryotic function(s). This may involve the transfer of specific products from bacteroids into the host cytoplasm or alternatively may result from changes in the microenvironment of the host cell. Free-living *Rhizobium* can be stimulated by restricted aeration to produce several 'bacteroid-specific' products, including nitrogenase (Avissar & Nadler, 1978; Daniel & Appleby, 1972; Pagan *et al.*, 1975). Since Nodulins are produced after infections with *Rhizobium*, the host origin and their relationship to the effectiveness of the nodules in fixing nitrogen should remain the focus in their further characterization. An understanding of the coordination of induction of these proteins with that of leghaemoglobin and nitrogenase may help in elucidation of the interaction of *Rhizobium* with host genes.

16.6 ESTIMATION OF THE NUMBER OF HOST GENES INVOLVED IN SYMBIOTIC NITROGEN FIXATION

Classical genetic experiments have suggested the involvement of several host genes in nodulation. Some of these genes appear to be involved in the recognition of the bacteria while others may be responsible for the development and function of the nodules (Verma, 1980). In order to estimate the total number of structural genes active in nodules, Auger *et al.*, (1979) determined the total sequence complexity of mRNA from uninfected root and the mature nodule tissues by RNA-cDNA hybridizations. Using homologous and heterologous hybridizations with RNA from these two tissues the extent of specific expression of host genes was determined.

Such experiments are shown in Fig. 16.11a and b. The poly(A)RNA sequences were resolved into three abundance classes in both uninfected root and nodule tissue. The most abundant component in nodules has been characterized (Baulcombe & Verma, 1978; Auger *et al.*, 1979) and contains

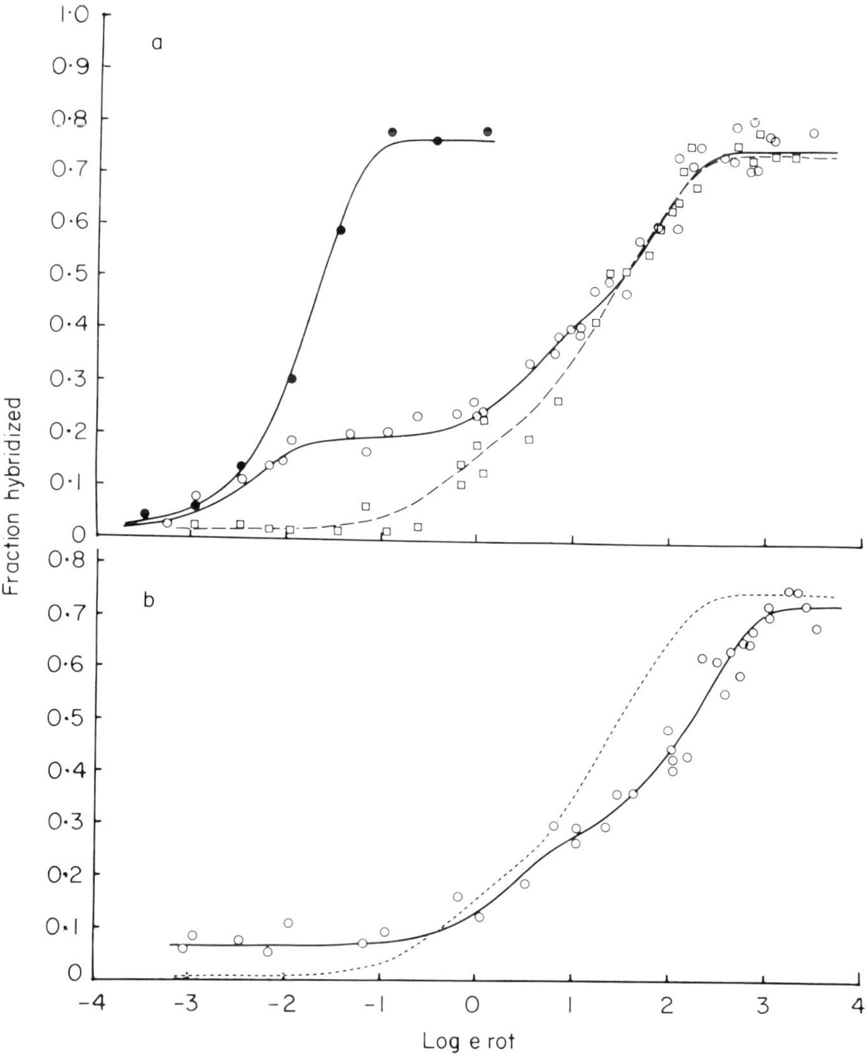

Fig. 16.11 (a) Homologous hybridization of nodule-DNA (○) and Lb-cDNA (●) with nodule polysomal poly(A)RNA, and root-cDNA with root-polysomal poly(A)RNA (□). (b) Cross hybridization of nodule-cDNA with root mRNA (○) and root cDNA with root mRNA shown by dotted line (redrawn from Auger *et al.*, 1979).

primarily leghaemoglobin sequences. Similarities between the mRNAs of the major leghaemoglobins permit extensive cross hybridizations and they thus appear as a single kinetic component (Section 16.5.1). This super-abundant component is not detectable in control root tissue and increases during root-nodule development. The middle abundant component, which comprises about 500 sequences in nodules, is much more abundant in nodules than that in roots while the final kinetic transitions of RNA from both tissues, indicating the least abundant mRNA fraction, show almost the same number of sequences and constitute similar fractions of the cDNA. Thus, it appears that, within the experimental limitations of this technique, both uninfected root and nodule tissues contain similar number of structural genes (18 000–22 000) in their polysomal poly(A)RNA populations and the number of sequences that may be specifically induced in nodules is limited.

A 'nodule-specific' cDNA probe for abundant sequences other than leghaemoglobin has recently been prepared by hybridization of nodule cDNA with root mRNA followed by hydroxyapatite fractionation (Auger & Verma, 1981). This probe consists of about 40 sequences which are most abundant in nodules after leghaemoglobin, and constitutes less than 7.5% of the total nodule mRNA. These 'nodule-specific' sequences are encoded by the soybean genome and appear on root-nodule polysomes only after infection. In addition, a set of host gene transcription products, which appear to be under phytohormone control, are increased several fold in root-nodules (Auger & Verma, unpublished). It is not known if these 'nodule-specific' sequences encode for Nodulins. Molecular cloning of these sequences would allow the characterization of their products as well as facilitating the studies on coordination of their induction in various effective and ineffective nodules.

16.7 POSSIBLE INVOLVEMENT OF *RHIZOBIUM* PLASMID IN SYMBIOSIS

It is now well established that various species of *Rhizobium* contain large (100–200 × 10^6 mol. wt.) plasmids (Prakash *et al.*, 1980). The number, size and the homology among *Rhizobium* plasmids does not appear to be constant. Several lines of evidence suggest that the *Rhizobium* plasmid harbours some of the genes required for effective nodulation (see Johnston *et al.*, 1979). Beringer (1976) has listed some *Rhizobium* genes that are extrachromosomal. By hybridization of a cloned 'nif-gene' from *Klebsiella* to restriction enzyme-digested *R. leguminosarum* plasmid it has been shown that nitrogen-fixation genes are localized on the plasmid (Nuti *et al.*, 1979); however, it may not be true for all species of *Rhizobium*, particularly slow growing species, e.g., *R. japonicum* (Haughland & Verma, unpublished data). Some of the *Rhizobium* properties such as effectiveness and infectivity are relatively unstable. Treatment with a plasmid curing agent resulted in a loss of ability of *R. trifolii* to initiate infection (Higashi, 1967). Strains of *R. trifolii* and *R. phaseoli* grown together in liquid medium, followed by selective elimination of *R. trifolii* by specific phages, resulted in some phage-resistant bacteria that could infect clover. Based on these experiments along with the acridine orange curing data, Higashi has suggested that host specificity involves a plasmid. Host specificity has been transferred from *R.*

trifolii to *R. phaseoli* (Higashi, 1967) and from *R. leguminosarum* to other strains of *Rhizobium* (Johnston *et al.*, 1978). By using transposons to mark these plasmids, it would be possible to locate insertions for physical and genetic analysis.

Rhizobium and *Agrobacterium* are reciprocally transformable and it is now well established that the latter harbours a large plasmid that is involved in oncogenicity in crown gall tumorigenesis (Chilton *et al.*, 1977; see also Chapter 18). Attempts to find homology between DNA from soybean and its effective strain of *Rhizobium* have been negative (Rake, 1972) indicating that there is no stable integration or that it is below the detection level. However, the region of crown-gall Ti-plasmid that is actually integrated into the plant host involves only a very small segment of DNA (see Chapter 18). Although the two systems are physiologically very different, the possibility of a similar mechanism for transformation cannot be ruled out at present. Data from preliminary experiments indicate that there is no extensive homology between plasmids from *R. japonicum* and *Agrobacterium* (Haugland & Verma, unpublished).

16.8 SUMMARY

In spite of the lack of detailed information at the molecular level, a temporal map of various interactions leading to symbiotic nitrogen fixation can be drawn as shown in Fig. 16.12. It should be pointed out that although *Rhizobium* can fix nitrogen in culture under low pO_2 and a suitable supply of nutrients, development of a conducive environment by the host is an obligatory requirement for effective nitrogen fixation by *Rhizobium* and its utilization by the plant. A coordinated induction of both host and *Rhizobium* genes is required for successful development of this process. The initial recognition process that appears to be mediated by plant lectins and surface lipopolysaccharide of *Rhizobium* allows the initiation of infection on the root hair. The development of infection leading to the intracellular compartmentalization of bacteria is mediated by both host and *Rhizobium* as shown by their contributions towards cell wall hydrolysing enzymes required for this process. The phytohormones appear to be major inducers of plant genes in the development of the root nodule symbiosis. A localized cellulase activity is apparently produced in response to phytohormones generated during the infection process. 'Release' of bacteria from the infection thread is followed by an increase in plasma membrane biosynthesis which keeps pace with the proliferation of bacteria. In an effective nitrogen-fixing nodule no bacteria have been found to be free in the host cell cytoplasm. Proliferation of this membrane is determined by the host, and results in compartmentalization of a single bacterium or more than one bacterium per envelope.

As a result of an increased population of bacteria and a high demand for oxygen, the intracellular environment of the host cell becomes hypoxic. This has a major influence on the differentiation of bacteria into bacteroids in which

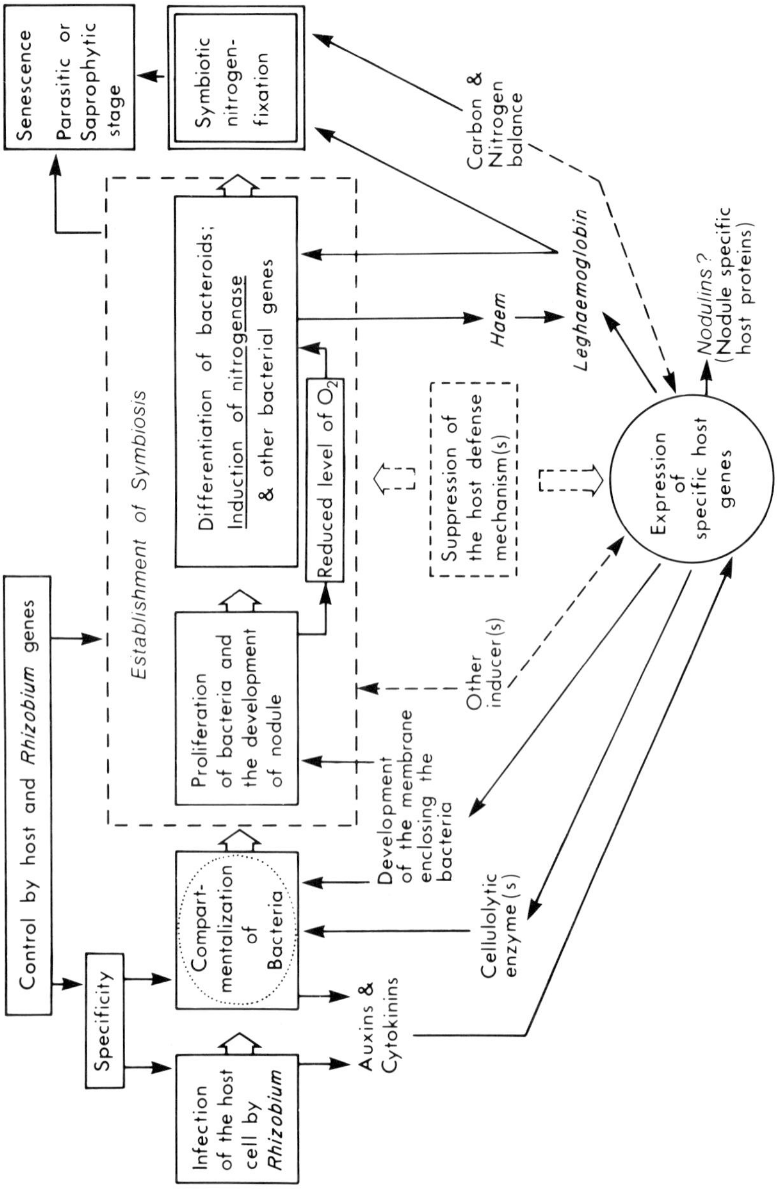

Fig. 16.12 Schematic diagram of host-*Rhizobium* interactions during development of root nodule symbiosis.

form they synthesize nitrogenase, several specific cytochromes, and produce a large quantity of haem. Reduced pO_2 also appears to influence the expression of host genes. Concomitant induction of leghaemoglobin and other enzymes is consistent with this notion. It is apparent from the data on the biosynthesis of leghaemoglobin and nitrogenase that induction of these molecules is independent of each other although the function of nitrogenase is dependent upon the availability of leghaemoglobin in nodules. The requirement for molecules like leghaemoglobin in symbiotic nitrogen fixation has been discussed earlier (Section 16.5.1) whereas that for Nodulins is unknown.

Once the nodule structure is formed and both nitrogenase and leghaemoglobin are induced the sustained rate of symbiotic nitrogen fixation depends upon the carbon–nitrogen balance of the nodule and thus photosynthesis and respiration in bacteria play an important role in this process. Interruption of any of these coordinated events leads to the disintegration of the symbiotic state resulting in the senescence and death of the infected cell.

16.9 AREAS OF FUTURE RESEARCH

The understanding of molecular mechanisms underlying the establishment of plant-*Rhizobium* symbiosis leading to the fixation and utilization of atmospheric nitrogen for growth is a challenge to the plant molecular biologist. Our knowledge of various interactions in the development of root-nodule symbiosis would be facilitated by the use of known mutant strains of *Rhizobium* which form ineffective nodules that are aborted at different stages of development. While some progress is being made on the biochemistry and genetics of the microsymbiont, very little is known about the host component, an integral partner in the process of symbiotic nitrogen fixation. Progress is also impeded by the lack of sufficient host genetics. It is not known how many structural genes of the plant or *Rhizobium* are obligatory for symbiosis and how many of them are related to the development of the root-nodule structure and their function in symbiotic nitrogen fixation. Furthermore, the mode of expression of the host genes and the influence of the *Rhizobium* genome on their expression is not at all understood. The role of the *Rhizobium* plasmids needs to be established and an analysis of various plasmid-borne genes including their physical organization should be made. The possibility of transfer of DNA between *Rhizobium* and the host plant, similar to that occurring in *Agrobacterium*-crown gall system, remains to be explored.

The total metabolic contribution of the host for maintenance and growth of bacteria in nodules is poorly understood specifically with respect to the supply of energy for nitrogen fixation. It is not known how various metabolic process are compartmentalized in the infected cells. Information about this could be obtained by studying the role of the membrane envelope enclosing the bacteroids. Plants generally exert a 'defence' against the invading pathogen; however, such a mechanism if it existed in nodules, would have to be suppressed. Nothing is known about the involvement of the 'host defence' mechanism(s) in the development of this symbiosis. Identification and characterization of various host and *Rhizobium* genes and their products is essential for any significant progress to be made towards an understanding and improvement of this novel association in nature. This would also form a foundation for future genetic engineering work towards manipulation of non-leguminous plants for their association with nitrogen-fixing organisms in a manner similar to legumes.

FURTHER READING

BERGERSEN F. J. (1971) Biochemistry of symbiotic nitrogen fixation in legumes. *Ann. Rev. Plant Physiol.* **22**: 121–40.

DART P. J. (1977). Infection and development of leguminous nodules. In *A Treatise on Dinitrogen Fixation.* Section III, biology, Hardy R. W. F. & Silver W. S. pp. 367–472, John Wiley and Sons, N.Y.

AUGER S., BAULCOMBE D. & VERMA D. P. S. (1979) Sequence complexities of the poly(A)-containing mRNA in uninfected soybean root and the nodule tissue developed due to the infection by *Rhizobium. Biochem. Biophys. Acta*, **503**, 496–507.

NUTMAN P. S. (1965) The influence of the legume in root nodule symbiosis. A comparative study of host determinants and functions. *Biol. Rev. Camb. Philos. Soc.*, **31**, 109–51.

VERMA D. P. S. (1980) Expression of host genes during symbiotic nitrogen fixation. In *Genome Organization and Expression in Plants*, pp. 439–452. Ed. Leaver C. J. Plenum Pub. Corp.

BRILL W. J. (1974) Genetics of N_2-fixing organisms. In *The Biology of Nitrogen-Fixation.* Ed. Quispel A. pp. 639–60, Elsevier N.Y.

VERMA D. P. S., KAZAZIAN V., ZOGBI V., & BAL A. K. (1978b) Isolation and characterization of the membrane envelope enclosing the bacteroids in soybean root nodules. *J. Cell Biol.*, **78**, 919–36.

JOHNSTON A. W. B., BERINGER J. E., BEYNON J. L., BREWIN N., BUCHANAN-WOLLASTON A. V. & HIRSCH P. R. (1979) Plasmid and the *Rhizobium*-legume symbiosis. In *Proceedings First International Symposium Nitrogen Fixation*, Madison.

CHAPTER 17

MOLECULAR ASPECTS OF FUNGAL INFECTION

17.1 INTRODUCTION

Plants suffer from hundreds of important diseases caused by fungi, bacteria, viruses, viroids, mycoplasmas, nematodes and even certain higher plants. It is difficult to obtain precise estimates of the effects of these pathogens on crop yields, some authorities estimate that collectively such pathogens consume enough food to feed 5×10^8 people, i.e. approximately 1/8 of the world population. More specifically, fungal pathogens have been responsible for some of the major epidemics of the last decade, for example Southern Corn Leaf Blight in the U.S.A., Dutch Elm Disease in N. America and Europe, coffee rust in S. America and rice blast in Asia.

The major goal of applied plant pathology is the control of such diseases. Traditionally this has been accomplished by the often empirical use of pesticides and the development of resistant cultivars. Increasingly, however, the traditional approaches have been found wanting. Plant breeding has been carried out in a background in which the function of not one gene for disease resistance is understood. More restrictions are being placed on the use of pesticides, more intensive cultural practices exacerbate disease spread. Therefore, it would seem logical that more effort should be expended in the fields of biochemical and physiological plant pathology to improve our understanding of the molecular basis of disease in general, but specifically the mechanisms of pathogenesis and host resistance. Indeed plant pathology is a fertile field for those who are interested in multidisciplinary approaches to research and there is a particular need for biochemists and geneticists. The fundamental information obtained should be directed towards developing more rational and integrated approaches to disease control using chemical agents; breeding programmes may be conducted more easily and effectively if the breeder can make use of biochemical characteristics as indices of resistance. Ultimately use may be made of the exciting prospects of genetic engineering.

The aim of this chapter is to examine the interaction between plant hosts and fungal parasites at the molecular level, paying particular attention to mechanisms of pathogenesis, specificity and resistance. Inevitably much valuable material must be excluded, for example control of host primary metabolism and nutrient interrelationships will not be discussed. For these, and other important aspects not dealt with here, the reader is advised to consult recent texts such as Horsfall and Cowling (1978), Heitefuss and Williams (1976), Wheeler (1975).

17.2 SYMBIOTIC NUTRITION

All fungi are heterotrophic in their nutrition, i.e. since they do not possess chlorophyll they are unable to convert CO_2 into organic compounds, other than as a minor fraction of their carbon requirements through non-photosynthetic carboxylating mechanisms. Rather, previously-elaborated sources must be exploited either *saprotrophically* (i.e. from non-living materials) or through some form of *symbiosis* (i.e. carbon compounds derived directly from another living organism by virtue of intimate contact). A major distinction in the latter group is that between parasitic fungi that kill their hosts quickly, living saprotrophically on the dead remains ('*necrotrophs*') and those that derive nutriment from living host cells ('*biotrophs*'). Within each group the terms 'obligate' and 'facultative' may be further used in an ecological context to distinguish between those fungi that require the host plant to complete their life-cycle and those that do not (Lewis, 1973). Previously the same terms were also used to describe the ability of parasitic fungi to grow in axenic culture, an increasingly meaningless distinction as more and more previously 'un-culturable' fungi are being successfully cultured.

Table 17.1 Major characteristics of necrotrophic and biotrophic pathogens (after Dickinson & Lucas, 1977)

Necrotrophs	Biotrophs
Host cells killed rapidly	Host cells not killed in compatible associations until late in the disease
Toxins and cytolytic enzymes produced	Few or no toxins produced, cytolytic enzymes restricted to penetration stages
No specialized parasitic structures formed	Specialized parasitic structures (haustoria) typically formed
Wide host range	Narrow host range
Able to grow saprotrophically away from host, cultured with ease	Require host to complete life-cycle, unable to grow away from host in nature, difficult to culture

Table 17.1 lists the major general characteristics of biotrophs and necrotrophs, although few individual fungi possess all characteristics of that group, especially within the biotrophs. For example, the biotrophic smut fungus *Ustilago maydis* is readily cultured and penetrates living host cells without specialized haustoria. *Phytophthora infestans* (potato leaf blight) is generally considered to be semi-biotrophic. This parasite is readily cultured, produces small, almost rudimentary haustoria, can infect hosts other than potato such as the closely related tomato, and secretes cell wall degrading enzymes.

17.3 DISEASE STRATEGIES: MOLECULAR CONFRONTATION

Parasitism is conventionally discussed as a form of warfare in which the invading parasite has at its disposal a range of weapons, employed to varying degrees against the complex defences or even counter-offensives of the host. The outcome of an interaction between host and parasite is determined at many different levels. The various methods of defence employed by plants are the subjects of later sections; here it is pertinent to examine the molecular basis of parasite offensive strategies. Following contact between the appropriate organ of the host and the parasite, mediated by spore germination inhibitors, stimulators and chemotactic agents, entry into the plant may be achieved by direct penetration, or indirectly through lenticels, hydathodes, stomata or wounds. In the case of the former, penetration is achieved by secretion of cell wall degrading enzymes (CWDE), with or without the assistance of mechanical forces. The invading parasite may kill the host cell or, as in the case of biotrophs, the host plasma membrane remains intact and new membrane may even be synthesized to accommodate invaginating haustoria.

17.3.1 PRIMARY PENETRATION

17.3.1.1 *Cuticular penetration*

The structural component of plant cuticles is the polyester cutin derived from hydroxy and epoxy fatty acids. An exemplary series of studies by Kolattukudy's group has provided elegant and convincing evidence for the involvement of cutinase enzymes in cuticle penetration. Cutinase enzymes have been isolated from six different necrotrophs growing on cutin as sole carbon source, purified to homogeneity and characterized (Purdy & Kolattukudy, 1975; Soliday & Kolattukudy, 1976; Lin & Kolattukudy, 1980). The cutinases characterized are glycoproteins of approximate molecular weight 25 000. Those from *Fusarium solani pisi*, *F. roseum culmorum*, *F. roseum sambucum* and *Ulocladium consortiale* contain mannose, glucosamine and glucuronic acid 0-glycosidically linked to protein via serine, β-hydroxyphenylalanine (*F. solani pisi* and *F. roseum culmorum*), threonine (*F. solani pisi* and *U. consortiale*), and β-hydroxytyrosine (*F. solani pisi*). The two β-hydroxy amino acids are peculiar to these enzymes alone. Strong evidence for the involvement of cutinase in penetration of pea leaves by *F. solani pisi* was obtained in two ways. Cutinase secretion by germinating spores and infection pegs was demonstrated using immunocytochemical procedures involving ferritin-labelled antibodies raised to the cutinase I isozyme, (Shaykh *et al.*, 1977). Secondly, both specific antisera raised against cutinase I, and diisopropylfluorophosphate, a potent inhibitor of cutinase, prevented infection of host leaves (Maiti & Kolattukudy, 1979). Since cutinase does not appear to be involved in host metabolism the latter demonstration immediately suggests possibilities for the control of disease through the use of 'enzyme-targeted fungicides'.

17.3.1.2 *Cell wall degrading enzymes*

A pathogen confronted by a plant cell wall faces a complex barrier of various carbohydrate polymers requiring the induction and secretion of specific enzymes for their degradation. Thus, a primary wall contains microfibrillar cellulose, hemicelluloses (xylans, xyloglucans, mannans and heteropolymers of glucose, galactose and mannose) and the pectic substances (primarily α1,4-linked rhamnogalacturonans, β1,4-linked galactans and 1,3- and 1,4-linked arabanans). In addition, the hydroxyproline-rich protein ('extensin') is present. Studies by Albersheim and associates have led to a cell wall model in which cellulose microfibrils are H-bonded to xyloglucans, the reducing ends of which are covalently linked to pectic polymers. The latter, in turn, are bound to the cell wall protein through arabinogalactans, the whole thus forming a complex network of cross-linked cellulose fibres. More detailed consideration of this model may be obtained by consulting Albersheim (1978) and Bauer (1977).

A very substantial body of evidence shows that virtually all pathogenic fungi are capable of recognizing the cell wall polymers and secreting polysaccharidases either locally, to assist them in wall penetration, or more extensively as primary offensive agents in diseases such as those caused by soft-rot and vascular wilt fungi (reviews by Bateman, 1976; Bateman & Basham, 1976; Byrde & Archer, 1977). Thus, many fungi cultured on cell walls as substrates *in vitro* will synthesize and secrete arabinanases, galactanases, xylanases, glucanases etc. However, it appears that in the majority of cases little degradation of native cell wall substrates *in situ* can be achieved by such enzymes, without the prior removal of at least some portion of the pectic substances exposing other substrates to attack (although for exceptions see Cooper *et al.*, 1978). Thus when cultured on plant cell walls *in vitro* many fungi secrete CWDE in a defined temporal sequence, pectic-degrading enzymes being produced first, followed by hemicellulases and finally cellulase, a sequence also detected in some infected tissues (Heale & Gupta, 1972). Progressive degradation presumably increases the spectrum of substrates available for enzyme induction. A further consideration is the apparent inverse relationship between the molecular weights of CWDE of some fungi and their ability to attack wall polysaccharides (Knee, 1977). Carpita *et al.*, (1979) have shown that the average diameters of pores or capillaries in higher plant cell walls is of the order 4–5nm. Since a globular protein of only 17 000 molecular weight has a diameter approximately twice this size, many CWDE known to be of much higher molecular weight clearly have limited access to cell wall materials.

17.3.1.3 *Control of CWDE synthesis*

The synthesis of CWDE is subject to control by catabolite repression, i.e. synthesis is reduced in the presence of readily metabolized carbon sources. In the presence of the appropriate polysaccharide substrates the enzymes are rapidly synthesized and secreted but since the polysaccharides of the cell wall are insoluble they are unlikely to be the inducers *per se*. Rather they probably provide low molecular weight inducers after slight degradation by low levels of parasite basal enzymes. After synthesis has been initiated progressive degradation by the new enzyme molecules ensures that the process is autocatalytic. Cooper and Wood (1975) showed that the most effective inducers for polysaccharidases are low (non-repressive) levels of the specific monomeric or dimeric sugars from which the polysaccharide is composed. Thus polygalacturonase is induced by D-galacturonic acid, xylanase by xylose, cellulase by cellobiose. D-galacturonic acid actually induces pectin trans-eliminases, hydrolases and methyl esterases (Section 17.3.1.4) suggesting coordinate control by a single regulatory gene. When the level of monomeric inducer increases to a level exceeding the requirement of the fungus for carbon substrates then catabolite repression operates.

The significance of catabolite repression in the diseased plant is far from clear. Potentially high levels of sugars *in vivo* might repress the synthesis of CWDE and thus modify disease progression. Could it be, for example, that the restriction of CWDE of rusts to the primary penetration phase is the result of catabolite repression due to the high levels of sugars available in the free space of leaf tissues? In a classical study Horton and Keen (1966) showed that severity of onion pink root (*Pyrenochaeta terrestris*) and activities of polygalacturonase and cellulase in infected tissues were inversely related to the amount of sugar in host tissues, as affected by spraying with glucose or maleic hydrazide or by removing cotyledons.

Differential induction of CWDE by wall polymers is unlikely to account for varietal- and race-specific forms of resistance; cell walls of resistant and susceptible cultivars are generally very similar if not identical in polysaccharide composition and are equally effective as inducers *in vitro* (Cooper & Wood, 1975). However, cell wall composition may determine differences in general resistance between tissues of the same plant. The greater degree of resistance of roots of *Phaseolus vulgaris* to *Colletotrichum lindemuthianum* has been correlated with the higher amounts of arabinose polymers in root cell walls compared to the more susceptible leaves, internodes and hypocotyls. Root cell walls were less effective as inducers of CWDE *in vitro*. Friend (1976) has suggested that esterification of galacturonan or galactan components of cell walls by hydroxycinnamic acids (see also 17.4.3.1) might limit cell wall degradation.

17.3.1.4 *Pectic enzymes*

Most attention has been paid to CWDE that degrade the pectic substances, a wall fraction consisting of a 'backbone' of α1,4 linked galacturonic acid (polygalacturonic or pectic acid) interspersed with 1,4 linked rhamnose. The carboxyls may be methyl-esterified to form pectins, (Fig. 17.1). Six major enzyme activities are recognized according to the nature of bond cleavage (hydrolytic or β-eliminative), preferred substrate (pectin or pectic acid) and whether bond

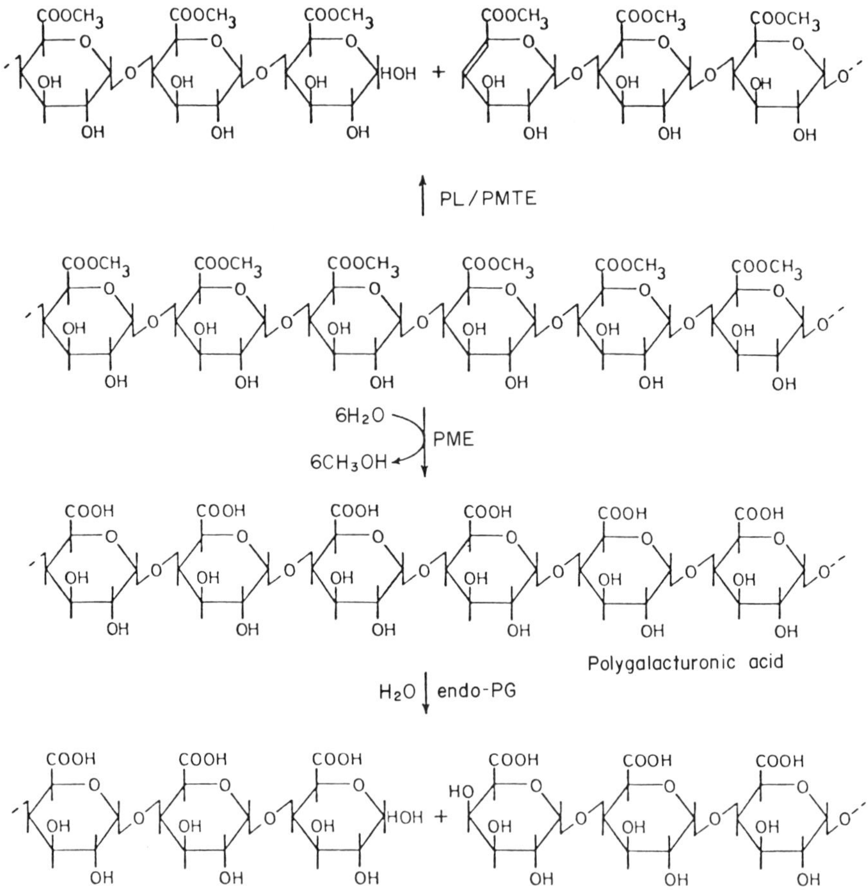
PL / PMTE
6H₂O
PME
6CH₃OH
Polygalacturonic acid
H₂O
endo-PG

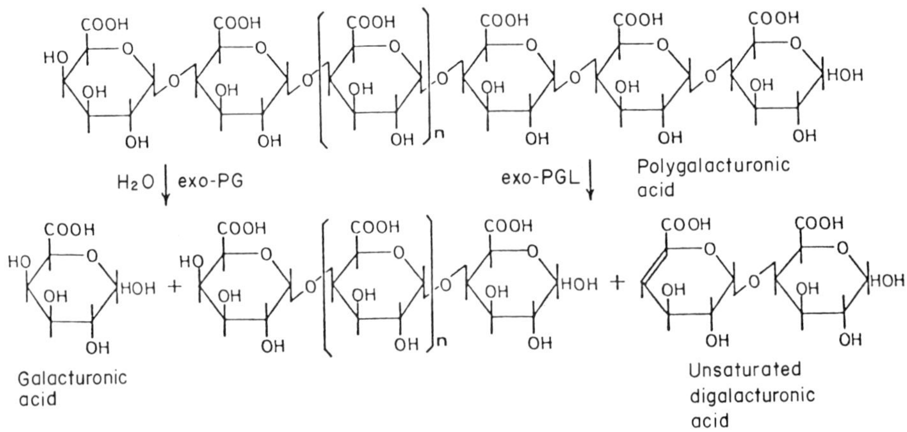
H₂O
exo-PG
exo-PGL
Polygalacturonic acid
Galacturonic acid
Unsaturated digalacturonic acid

cleavage is terminal (exo-acting) or random (endo-acting), (Fig. 17.1). The action of pectin methyl esterase (PME) appears to facilitate degradation of methylated polymers by enzymes whose preferred substrate is pectic acid. Polygalacturonases (PGs) hydrolyze pectic acid to the monomeric galacturonic acid (exo-PG, acting from the non-reducing end) or to oligogalacturonides (endo-PGs, generally unable to hydrolyze substrates smaller than the dimer or trimer). Most PGs have acidic pH optima and are unable to hydrolyze substrates rendered insoluble by Ca^{2+}. It is now generally accepted that enzymes previously characterized as polymethylgalacturonases are of doubtful existence. The lyase enzymes (PGL/PATE, PMGL/PMTE) all release unsaturated products through β-climinativc bond cleavage. Endo-PGLs release unsaturated digalacturonic acid or trigalacturonic acid from pectate, exhibit high pH optima (8.5–9.5) and are stimulated by Ca^{2+}. Exo-PGL, unlike exo-PG degrades pectate from the reducing end of pectate chains, releasing unsaturated digalacturonic acid. Highly-esterified pectins are degraded only in a random manner by endo-PLs (endo-PMTE), no exo-PLs have been discovered. Endo-PLs exhibit pH optima in the range 5.0–8.5 and are stimulated by Ca^{2+}.

Pectolytic enzymes have been implicated in many diseases, especially those that involve tissue maceration, as in various rotting diseases, (Bateman, 1976). The principal enzymes involved appear to be endo-PGs and endo-PLs, on grounds of *in vitro* production by the pathogen, isolation from infected tissue and effectiveness of highly-purified enzymes on intact tissue. Cases are known of multiple enzyme production, e.g. *Sclerotinia fructigena* (apple brown rot) secretes both endo-PG and two isozymes of endo-PL (endo-PMTE). The endo-PG in this case is somewhat anomalous since it will not macerate apple tissue (Byrde & Fielding, 1968). Maceration by both pathogens and pectolytic enzymes is also accompanied by increased host cell permeability and eventually necrosis. A large number of hypotheses have been advanced to explain this (Bateman, 1976; Wood, 1976) but the experimental evidence appears to indicate that the killing of host cells results from a loosening of primary wall structure so that the cell is unable to retain the protoplast which is under osmotic stress. This is consistent with the fact that perfectly viable protoplasts can be isolated from tissue following treatment with crude fungal CWDE providing the tissue is first plasmolyzed.

Pectic enzymes are also frequently implicated in wilting diseases caused by vascular pathogens (Mussell & Strand, 1976). Pectic enzymes characteristic of the pathogen can be detected in diseased plants, the amounts often bearing a direct relationship to disease severity. Pure enzymes such as endo-PG and

---

Fig. 17.1 Enzymic degradation of pectin and polygalacturonic acid. Enzymes shown are; pectin methyl esterase (PME), pectin lyase (PL) or pectin methyl-trans-eliminase, (PMTE), endo-polygalacturonase (endo-PG), exo-polygalacturonase (exo-PG) and exo-PGL (exo-polygalacturonate lyase). Not shown, endo-polygalacturonate lyase (endo-PGL or endo-polygalacturonate-trans-eliminase, PATE).

endo-PL generate typical wilting symptoms when applied to some plants. However, such evidence linking pectic enzymes to wilt pathogenesis is circumstantial, direct proof of a causal relationship is largely lacking. An important approach to this problem is through the use of enzyme-deficient mutants. Induced mutants of *Verticillium dahliae* deficient for endo-PG, PTE + PATE and PME activities, both *in vitro* and in the plant, were able to produce normal wilt symptoms in cotton plants following stem inoculation (Howell, 1976). This suggests that these enzymes may not be necessary for wilt development, although it does not rule out the possibility that they may be required for root penetration prior to vascular colonization.

17.3.2 TOXINS

In plant pathology the term 'toxin' is used to describe injurious molecules produced by micro-organisms and active at low physiological concentration: enzymes and growth regulators are conventionally excluded. Toxins are chemically diverse, generally of low molecular weight (but see 17.3.2.2) and belong to, or are derived from; phenolics, terpenoids, amino acids and glycosides. Knowledge of the chemistry of toxins is fundamental to understanding their mode of action and therefore discovering the molecular basis of disease. Some toxins are used in screening for resistant plants in breeding programmes.

The terminology of toxin classification is unfortunately confusing, here discussion will be confined largely to those toxins that have been clearly implicated in a causal role in pathogenesis, i.e. as 'pathotoxins' *sensu* Wheeler (1975). Pathotoxins may be separated into 'specific' or 'selective' forms where sensitivity of host plants to the toxin is correlated with susceptibility to the pathogen, and 'non-specific' or 'non-selective' forms. The clear implication of the host-specific pathotoxin concept is that the sum total of the symptoms of a given disease may be explained in terms of sensitivity to the toxin in question. To date some seven diseases are known where, arguably, this criterion has been met, and certain of these find considerable favour as 'model systems' in which the sequence of events in pathogenesis may be studied. However, as yet, our understanding of the mechanisms of toxin action at the molecular level is poor. Here only a few selected examples will be discussed, wider reviews may be consulted (Rudolph, 1976; Scheffer, 1976; Strobel, 1976; Wheeler, 1978).

17.3.2.1 *Host-specific pathotoxins*

The most potent and specific pathotoxin known is victorin, a complex peptide/terpenoid toxin of *Helminthosporium victoriae* still to be completely characterized. Victorin affects only oat varieties containing the single 'Victoria' gene and the ratio of its effectiveness on susceptible versus resistant cultivars is of the order 4×10^5, less than 0.1ng cm^{-3} will cause symptoms on a susceptible host. The major physiological effect of victorin, indeed many other toxins, is an increase in tissue permeability to ions and water. However, despite considerable study (see Wheeler, 1975; 1978) there is, as yet, no precise understanding of the action of victorin at the cellular or molecular levels. Current interpretations favour a disruptive effect on energy metabolism crucial to the maintenance of plasma membrane function. Attempts to isolate membrane receptors have proved fruitless although a role for cell wall components has been suggested.

In contrast, spectacular success has been claimed in the elucidation of the molecular basis of the action of a toxin from *Helminthosporium sacchari* the causal agent of eye-spot disease of sugar cane (reviewed Strobel, 1976). A fraction of toxin activity was identified as a derivative of the simple sugar α-D-galactose (the C1 cyclopropyl moiety is tentative). The toxin induces most of the characteristic symptoms of the disease when applied to leaves of susceptible (S) clones but not those of resistant (R) clones and can be recovered from diseased leaves. Crude plasma-membrane-containing preparations from S leaves were able to bind ^{14}C-labelled helminthosporoside (the trivial name of the toxin) but similar preparations from R leaves had no binding activity. Toxin binding was suppressed by α-galactose and α-galactosides such as raffinose, acting as competitive inhibitors, (Strobel, 1973a). A membrane-bound toxin-binding protein (TBP) was isolated from membranes of S clones. The protein had a molecular weight of 48 000 and consisted of four sub-units with at least two binding sites. Membrane preparations from R clones contained an immunologically identical but inactive protein (Strobel, 1973b), the two proteins differing by only four amino acid residues. Inactive protein from R leaves could be converted to an active, toxin-binding form by mild detergent suggesting that inability of the native TBP from R leaves to bind toxin resides in a slight difference in binding site conformation as a result of the amino acid differences. TBP is located at the plasma membrane surface since antisera to TBP cause agglutination of sugar-cane protoplasts. In addition toxin-binding activity of fractionated membranes co-sediments with plasma membrane markers on sucrose gradients and TBP can be labelled by procedures reputed to be specific for the cell surface (Strobel & Hess, 1974).

A physiological basis for the action of helminthosporoside in killing cells has been suggested. Toxin-treated protoplasts swell, cells show plasma membrane depolarization and treated tissues take up ^{86}Rb (an analogue of K^+) at a greater rate than controls, together suggesting that membranes exposed to toxin are less able to regulate ion movement (Strobel, 1974; Van Sambeck *et al.*, 1975). The basis of these effects appears to lie in the indirect activation by the toxin, of important membrane proteins, especially K^+-Mg^{2+}-stimulated ATPase. TBP itself does not have ATPase activity, presumably toxin binding to TBP results in a conformational change within the membrane, activating this, and other enzymes. Confirmation that TBP is the primary locus of toxin action lies in the demonstration that protoplasts of a resistant clone of tobacco, will adsorb or otherwise bind TBP to their surfaces, conferring toxin susceptibility (Strobel & Hapner, 1975). However, although specificity appears to reside in toxin-binding, the binding alone is insufficient to cause disease symptoms since TBPs have been detected in diverse toxin-insensitive plants such as mint and beet (Kenfield & Strobel, 1977). Further, binding of helminthosporoside to mint TBP does not activate K^+-Mg^{2+}-ATPase. An important general principle illustrated by this example therefore, is that the molecular event conferring specificity is not necessarily that which constitutes the crucial detrimental effect in the disease process.

Since TBP binds α-galactosides other than helminthosporoside it has been suggested that the real physiological role of TBP is in α-galactoside transport between cells: presumably there is a strong selective disadvantage to the possession of a protein, the only function of which is to recognize a parasite toxin. Evidence consistent with this view is that plasma membrane preparations of S and R clones bind α-galactosides such as raffinose equally well; specificity is only expressed in helminthosporoside binding (Kenfield & Strobel, 1977). Presumably alterations can occur in α-galactoside-binding-protein structure which affect helminthosporoside binding affinity but which do not interfere with apparently normal physiological function.

A number of authors have commented on the exciting and revolutionary nature of these studies since they provide a unique demonstration of the molecular basis of specific parasite virulence and host resistance mechanisms. A mechanism for specificity based on the binding of carbohydrate-containing molecules of parasite origin also has considerable implication for other instances of host-parasite recognition (Callow, 1977 and 17.4.3.3). In view of the highly significant nature of these findings therefore, reservations in experimental protocol and interpretation of data expressed by some authors (Daly, 1977; Wheeler, 1977) deserve serious attention and independent verification of a number of findings has been urged.

Helminthosporium maydis race T (Hmt) (Southern Corn Leaf Blight) produces a toxin (T-toxin) which only affects maize cultivars carrying the T-source of cytoplasmic male-sterility (cms-T). This non-Mendelian trait confers pollen grain abortion and was developed by breeders to facilitate F1 hybrid seed production. Normal (N) plants are unaffected by either Hmt fungus or T-toxin.

The genes for both male sterility and susceptibility to T-toxin are mitochondrial and maternally inherited (see also Chapter 11). Restriction endonuclease fragment analysis of mitochondrial DNAs from N and cms-T plants are distinctly different and there are specific differences in the pattern of mitochondrial polypeptides synthesized *in vitro* (Leaver, 1980). A 21 000 molecular weight polypeptide is replaced in cms-T mitochondria by a 13 000 molecular weight component. Nuclear genes that restore fertility (Trf restorer genes) reduce sensitivity to T-toxin and mitochondria from restored plants no longer synthesize the 13 000 component. Further, plants selected for toxin resistance as spontaneous revertants in tissue culture, are male-fertile, insensitive to T-toxin and do not synthesize the 13 000 polypeptide. Thus, suppression of 13 000 polypeptide synthesis, toxin sensitivity and male sterility appear to be different expressions of a single genetic defect under the control of the same nuclear restorer gene.

The significance of the 13000 polypeptide to toxin susceptibility is not known, indeed it is far from clear what the biochemical basis of T-toxin action is. One problem is that the chemical nature of T-toxin is in dispute, having been described as a mixture of five sterols or glycosidic sterol derivatives (Karr *et al.*, 1974), a mixture of polyketols (Kono & Daly, 1979) and an acetate ester of mannitol linked to two formyl-L-valine residues (Aranda *et al.*, 1978). The toxin described by Kono and Daly (1979) appears to have the highest and most specific activity (Payne *et al.*, 1980), but comparison is difficult in view of the different bioassays used by different authors.

Crude and semi-purified toxin preparations induce a whole variety of physiological effects but in view of the genetic evidence those of most interest are the effects on mitochondria, and several lines of evidence point to the mitochondrion as the primary site of toxin action (Gregory *et al.*, 1977). Thus, T-toxin-treated mitochondria from cms-T plants show inhibition of malate oxidation, stimulation of succinate oxidation, uncoupling of phosphorylation, swelling and distortion of cristae. Similar ultrastructural changes have been detected *in situ*. Watrud *et al.*, (1975) demonstrated an inhibition of mitochondrial malic dehydrogenase from cms-T plants but not from those with N cytoplasm. On rupture or removal of the outer mitochondrial membrane however, mitochondria from N plants became sensitive. Toxin binding studies showed that mitochondria from both N and cms-T plants bound terpenoid toxins I and II but greater activity was associated with the inner membrane of cms-T mitochondria. A role for both mitochondrial membranes is therefore indicated, the outer membrane perhaps restricting permeability to the toxin in N mitochondria, and the inner membrane with a higher toxin binding affinity in cms-T mitochondria. Gregory *et al.*, (1977) urge caution in the interpretation of toxin binding studies because of problems associated with shrinkage and swelling of mitochondria during isolation. Thus, the primary site of T-toxin action remains to be unequivocally determined. Indeed, it has been argued that the plasma membrane is the specific site of toxin action since low concentrations of toxin II induce electrolyte leakage, K^+ uptake is inhibited and a partial depolarization of the plasma membrane potential occurs (Mertz & Arntzen, 1977; 1978).

17.3.2.2 *Non host-specific toxins*

Seedling chlorosis of a number of plants caused by the pathogen *Alternaria tenuis* (*A. alternata*) can also be induced by culture filtrates of the fungus, from which the active toxin principle, 'tentoxin' a cyclic tetrapeptide may be isolated. The toxin appears to exert its effects by inhibiting photophosphorylation by binding to chloroplast coupling factor CF1 (Steele *et al.*, 1976). A difference in binding affinity of four orders of magnitude was detected between CF1 from sensitive and insensitive leaves. The inhibition of CF1-ATPase and therefore the lack of ATP synthesis presumably accounts for the ultimate inhibition of chloroplast development and consequent chlorosis. Rich (1977) has described structure/activity relationships.

Fusicoccum amygdali causes wilting and desiccation of almonds and peaches. Cultures of the fungus produce a number of toxins that will induce similar symptoms, the principle being fusicoccin (F) a glycosidic diterpene (Ballio *et al.*, 1968). F also causes a number of responses normally associated with auxin such as increased wall plasticity and increased rates of tissue elongation. Stimulation of cell enlargement both by auxin and F is accompanied by and probably mediated by secretion of protons from the cell and is dependent on oxidative and phosphorylative metabolism. Hyperpolarization of the plasma membrane following proton extrusion (the inside becomes more negative with respect to the outside) is partially balanced by uptake of K^+. It has been suggested that both F and auxin activate an energy-requiring proton extrusion mechanism at the cell surface based on an ion-exchange pump such as K^+-Mg^{2+}-ATPase (Marrè *et al.*, 1974; Fig. 17.2). The resulting decrease in

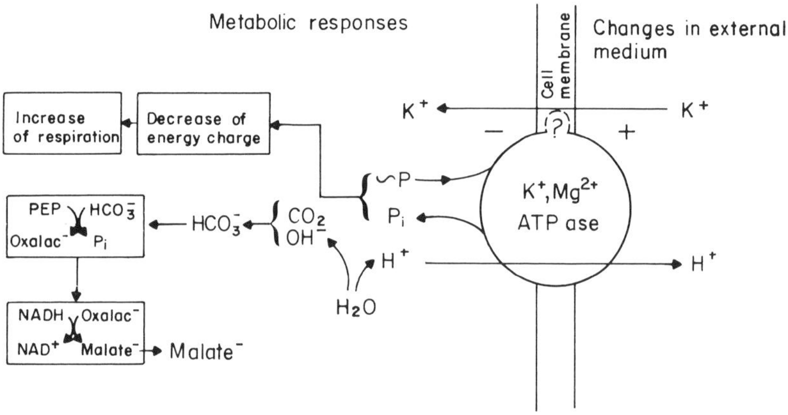

Fig. 17.2 Membrane model for the Fusicoccin-activated H^+/K^+ exchange system. Respiration rates increase due to decreased energy charge and alkalinization of the cytoplasm increases the rate of dark CO_2 fixation through PEP carboxylase (modified after Marrè, 1977).

energy charge (ATP/ADP + Pi becomes less) could account for the accompanying change in respiration rate. Other metabolic consequences have been discussed by Marrè (1977). The basis of the pathological effects of F lie, at least in part, on the stimulation of K^+ uptake by guard cells, both in the light and dark, (Turner, 1973). Increased guard cell turgor and the consequent loss of stomatal control of transpiration leads to desiccation.

Not all toxins are of low molecular weight. High molecular weight polysaccharides or glycopeptides are frequently implicated in wilting diseases although their modes of action are largely unknown. Van Alfen and Turner (1975) showed that uptake by elm seedlings of as little as 4μg of a crude high molecular weight glycopeptide toxin preparation from *Ceratocystis ulmi* (causal agent of Dutch Elm Disease), reduced stem water conductance by 79% and leaf water potential was decreased to a level at which the seedlings wilted, stomata closed and transpiration decreased. Synthetic high molecular weight dextrans had similar effects in direct proportion to their molecular weight. The effects on water economy are probably the basis of the wilting syndrome in Dutch Elm Disease. The mechanism of action of the glycopeptides appeared to be through a direct, physical blocking of the xylem by plugging the pores of pit membranes between vessels leading to cavitation. Van Alfen *et al.*, (cited in Strobel, 1976) showed that the *C. ulmi* glycopeptides were polydisperse with a mean molecular weight of 300 000 and consisted of 1,6-linked mannose and mannose-1,3-linked rhamnose linked to a peptide via serinyl and threonyl 0-glycosidic bonds.

Other mannose-containing glycopeptide phytotoxins have been isolated from *Phoma tracheophila* (Barash *et al.*, 1977) and *Cladosporium fulvum* (*Fulvia fulva*). The peptide moiety of *C. fulvum* is 0-glycosidically linked to a polysaccharide core containing galactose, mannose and glucose as major sugars, and to oligomannose side chains (Dow & Callow, 1979a). The core is phosphorylated in diester form and also bears galactofuranose side chains. In this context, a polydisperse mixture of such glycopeptides will bind reversibly to tomato leaf cells, inducing changes in membrane permeability and cell necrosis (Dow & Callow, 1979b). These activities are a function of the glycan moiety since periodate oxidation abolishes toxic activity. Digestion with exomannosidase also destroys activity suggesting the presence of mannose-binding receptors for this toxin on the host cell surface. Other glycosidases were without effect. The biological function of this glycopeptide 'toxin' is far from clear since it will also elicit phytoalexins suggesting a function in recognition and resistance-triggering.

Highly-branched β1,3 and 1,6-linked glucans from *Phytophthora* spp. exert toxicity through interference with plant water economy (Keen *et al.*, 1975, Woodward *et al.*, 1980). Although the mechanism(s) of action has yet to be described, direct blocking of the vascular system does not appear to be involved. An effect on plasma membrane function is suggested by the demonstration that potato protoplasts bear receptor sites for fungal β-glucans (Peters *et al.*, 1978).

17.4 THE MOLECULAR BASIS OF RESISTANCE AND SPECIFICITY

17.4.1 INTRODUCTION

Plants, like all living organisms, have evolved immune or resistance mechanisms of various types by which they are able to counter the advance of foreign organisms, the result being that a given parasite can usually only infect a distinct range of host plants. Despite the fact that plants in nature are surrounded by myriads of potential pathogens, disease, (i.e. susceptibility) is relatively rare. Plant resistance may be divided into 'general' or 'non-host' resistance where a plant is clearly outside the range of hosts a given parasite may infect, and 'specific' resistance, the result of genetic modification to the host, either naturally or through plant breeding, rendering it resistant to a pathogen that would otherwise infect it.

Every plant possesses a diverse array of mechanisms, operative at different levels, the sum total of which constitute resistance. Successful infection is dependent on the ability of the pathogen to avoid or overcome each in turn. Heath (1974) developed a conceptually useful 'switching-point' model, rather analogous to control mechanisms of metabolic pathways. The response of the pathogen at any one control point determines the subsequent progress of infection (Fig. 17.3). In functional terms, the mechanisms may be either 'constitutive' ('preformed') or 'induced' ('active' or 'facultative').

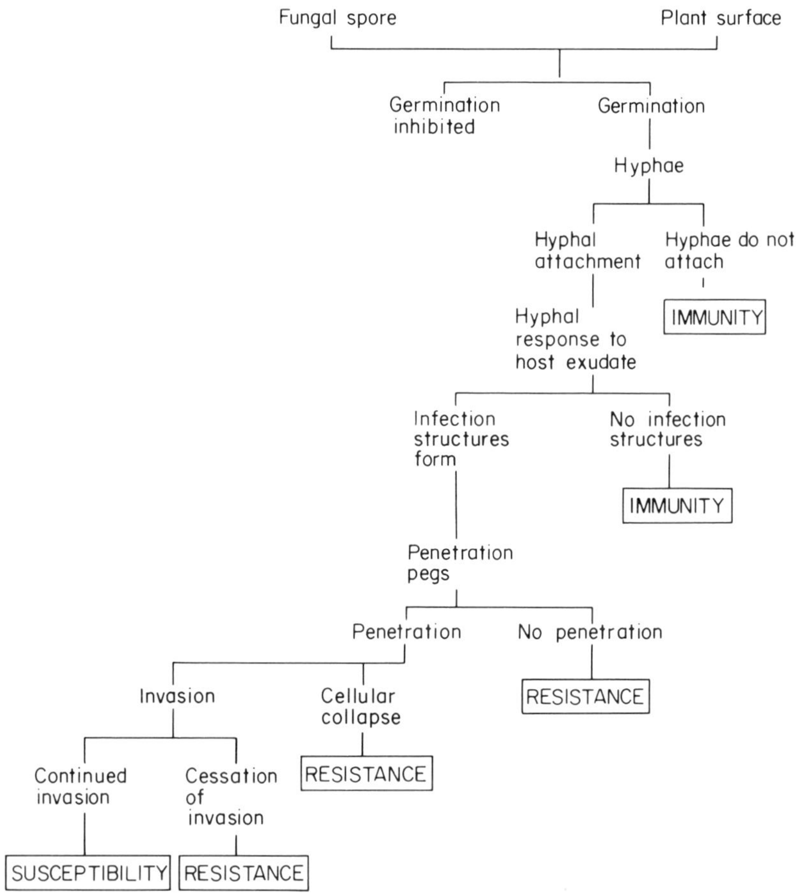

Fig. 17.3 Simplified version of the 'switching-point' concept of Heath (1974).

17.4.2 CONSTITUTIVE DEFENCE

Much of the general resistance exhibited by all plants to the majority of pathogens may be accounted for by the presence of mechanical barriers such as surface waxes, cuticle, suberized layers. Surface waxes create a hydrophobic surface which minimizes retention of water necessary for many fungal spores to germinate or which is unsuitable for infection structure differentiation. Alternatively the products of cutin hydrolysis, the hydroxy and epoxy fatty acids, are directly toxic to some fungi and infection success may lie in their detoxification through epoxy hydrase (Kolattukudy & Brown, 1975).

Cell walls of plants contain proteinaceous inhibitors of fungal polygalacturonases (PGs) (Anderson & Albersheim, 1972). These inhibitors are equally effective against endo-PGs of both virulent and avirulent races of pathogens and are present in equal concentration in resistant and susceptible hosts suggesting a role in general resistance. Plants also contain, and apparently secrete glycosidases such as β1,3-glucanase, chitinase and α-mannosidase, with the ability to degrade fungal cell walls. Nichols *et al.*, (1980) showed that a spectrum of glycosidases in pea which degrade *Fusarium* cell walls release molecules such as chitosan (a polymer of glucosamine) which is both a potent elicitor of pea phytoalexins and directly toxic to fungi (Hadwiger & Beckman, 1980). Fungi also secrete inhibitors of host glycosidases (Albersheim & Valent, 1974). Mirelman *et al.*, (1975) showed that wheat germ agglutinin (a plant lectin with binding specificity for N-acetyl-glucosamine-containing polymers such as fungal wall chitin) bound to hyphal walls of *Trichoderma viride* and inhibited growth. Since high concentrations of lectin were required it is likely that such effects are only significant in nature in relation to seed decay by soil microbes since seeds contain the highest concentrations of lectins (Callow, 1977).

Most plant tissues also contain high concentrations of secondary compounds that are toxic to micro-organisms, usually localized in the vacuole. Although many of these compounds, notably phenolics, phenol glycosides, tannins, saponins, and cyclic hydroxamates are generally considered to play an important role in general defence (Overeem, 1976; Schönbeck & Schlösser, 1976), the evidence is largely circumstantial. Here only a few compounds will be considered to illustrate general principles.

Tomatine (a steroid glycoalkaloid, Fig. 17.4) and the avenacins (glycosidic triterpenoids) are fungitoxic saponins present in tomato leaf and fruit, and oat root tissues respectively, at concentrations several orders of magnitude greater than those required to inhibit growth of many fungi. The distinct hydrophobic/hydrophilic asymmetry of the molecules permits binding to sterols in fungal plasma membranes leading to permeability changes and cell lysis. The efficacy of such molecules will depend on concentration around the infecting hyphae (unknown), pH (Dow & Callow, 1978) and the ability of the fungus to detoxify through secreted glycosidases, (Fig. 17.4). Thus, there is a correlation between the ability to degrade tomatine and the pathogenicity of *Septoria* spp. (Arneson & Durbin, 1968).

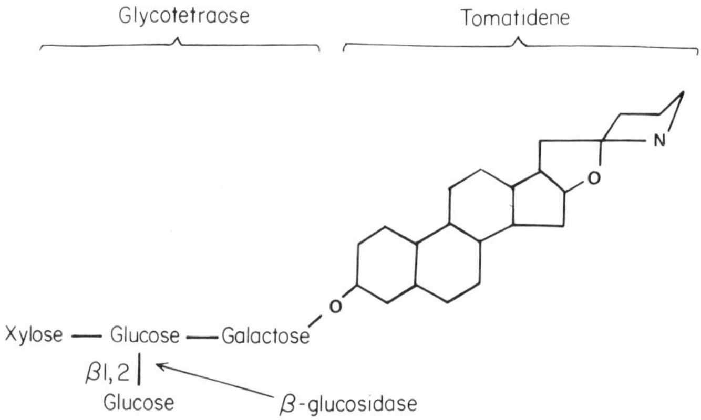

Fig. 17.4 Tomatine, the steroidal glycoalkaloid saponin. The molecule can be detoxified to β2 tomatine by β-glucosidase action as shown.

The vacuoles of plant cells contain substantial quantities of phenolics, as free phenols, phenol glycosides or polymerized tannins. Disruption of cell compartmentation by necrotrophic parasites in particular will bring such substrates into contact with phenol oxidase and peroxidase enzymes. The resulting o-quinone oxidation products are highly reactive and inhibitory to fungal growth. For example, on release from the apple leaf vacuole, phloridzin may be deglucosylated to phloretin by β-glucosidase (either of host or fungal origin). Neither compound is directly toxic to *Venturia inequalis* (apple scab) but the o-quinones formed by their oxidation are inhibitory to fungal growth. The o-quinones can inhibit enzymes such as PGs by binding to —SH or —NH_2 groups, or they may condense to form tannins which bind to, and precipitate enzymes (Overeem, 1976). Tannins may also condense with protein or polysaccharide substrates in the host, protecting them from fungal degradation.

The role of preformed secondary metabolites in defence may be clarified through the use of mutants. Induced mutants of a number of fungi with lower membrane sterol content were less sensitive to tomatine (Defago & Luescher, 1978). Recessive corn mutants with lower levels of cyclic hydroxamates showed reduced, but not totally abolished, resistance to corn pathogens (Day, 1974).

17.4.3 ACTIVE DEFENCE

Since defences are presumably costly in terms of energy required to synthesize the defensive agents, it would seem advantageous to the plant to possess mechanisms by which commitment to defence is increased when organisms are subject to attack, and decreased when the pathogen is absent (Callow, 1981). A considerable literature exists to show that invasion of plants results in the induction of active, energy-requiring responses superimposed on constitutive defence.

Active defence may be divided into two phases. There is now evidence to show that the compatible/incompatible 'decision' is made within the first few hours of an interaction and the initial determinative or '*recognition*' phase is viewed as one in which the host plant recognizes the potential invader as 'non-self', through components on the surface of the pathogen. The result is a triggering or *expression* of aspects of plant metabolism constituting resistance. Some of the mechanisms of active defence will now be considered.

17.4.3.1 *Active defence: formation of papillae and barriers*

A common response of many plant cells to both wounding and attempted fungal penetration is the formation of localized amorphous deposits ('papillae' or 'wall appositions') between the host plasma membrane and cell wall (Aist, 1976; Aist *et al.*, 1979). Such structures are generally assumed to play an important role in resistance especially to non-pathogens since they are often associated with the failure of penetration pegs to develop (e.g. Heath, 1972). In detailed time-course studies Aist *et al.*, (1979) showed that wall appositions formed early in interactions before actual penetration, were effective in inhibiting subsequent penetration, and appositions induced mechanically in *Brassica oleracea* root hairs were effective in limiting penetration by *Olpidium brassicae*. Whilst such observations and approaches clearly establish some degree of correlation between papilla formation and resistance, a direct causal relationship has yet to be established. It is possible that papilla formation has little to do with resistance but is a wound-healing response closely linked to a quite different resistance mechanism, or that papilla formation is just one of a number of resistance mechanisms triggered by the invading parasite (see the 'switching-point' model, Section 18.4.1). Heath (1979) and Vance and Sherwood (1977) have utilized inhibitors of protein and RNA synthesis in attempts to establish a causal role. In both cases inhibition of protein synthesis inhibited papilla formation and increased fungal invasion and growth. Caution is urged, however, in the interpretation of these experiments since inhibitors are rarely completely effective or specific. Allowing for this, at best these experiments suggest that papilla formation may require protein synthesis, but equally, resistant responses other than papilla formation might also require protein synthesis.

Papillae vary in chemical composition. Heath (1979) detected silica in dwarf bean appositions, phenolics and lignin were absent. In wheat, papillae and associated modifications of the upper epidermal cell wall ('haloes') contain lignin and polysaccharides, including callose-like material (Ride & Pearce, 1979). Lignification in wheat is an early response to infection and could control infection in a number of ways. Lignin is resistant to fungal cell wall degrading enzymes involved in primary penetration. The phenolic precursors are fungitoxic, either directly, or through oxidation to o-quinones and tannins, or the fungal wall may itself become lignified and inextensible.

Lignins are complex polymers of phenylpropanoid ($C_6 . C_3$) units derived

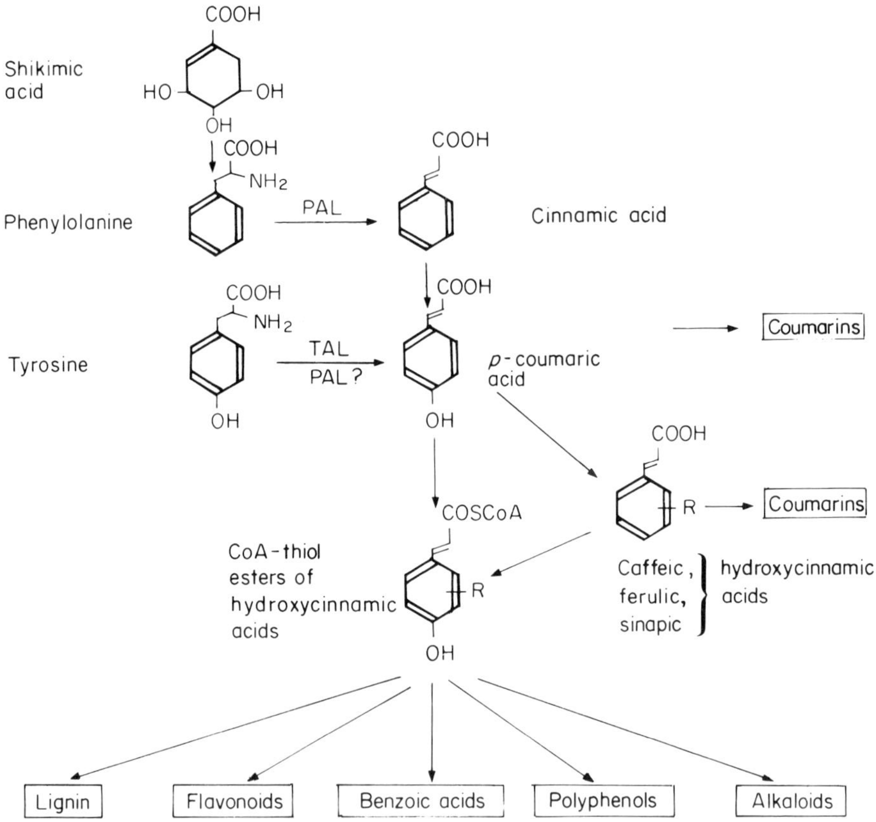

Fig. 17.5 The shikimate pathway of phenylpropanoid (C_6 . C_3) metabolism and phenolic biosynthesis (after Grisebach and Hahlbrock, 1974). R represents variously-linked —OH or —OMe groups.

from the amino acids phenylalanine and tyrosine via nonoxidative deamination to cinnamic and hydroxycinnamic acids (Fig. 17.5). The phenylpropanoid pool is important in disease resistance not only because it provides lignin precursors, but also a range of other secondary metabolites implicated in disease resistance, such as the isoflavanoid phytoalexins. The diversion of carbon skeletons of phenylalanine and tyrosine from protein synthesis to phenylpropane synthesis is controlled by the enzymes phenylalanine ammonia lyase (PAL) and tyrosine ammonia lyase (TAL, although in some species PAL and TAL appear to be activities of the same enzyme). Increases in PAL activity following infection (Friend, 1976) or application of elicitors of phytoalexin accumulation (17.4.3.3) are widely reported. Lignin deposition in canary reed grass papillae in response to non-pathogens was correlated with 75–100% increases in the activities of PAL/TAL, hydroxycinnamate-CoA ligase and peroxidase (Vance & Sherwood, 1976a,b). It has yet to be demonstrated that these increases are the result of *de novo* enzyme synthesis although cycloheximide did prevent both lignification and the increase in enzyme activity.

Reed canary grass papillae contain peroxidase (Vance & Sherwood, 1976b), presumably involved in lignin polymerization. The increased peroxidase activity was partly due to the appearance of new peroxidase isoenzymes which may have different specificities for the various cinnamoyl alcohol precursors of lignin. Similar changes in the spectrum of isoperoxidases in infected Japanese radish roots have been correlated with differences in the composition of lignin formed during infection compared with uninfected plants (Asada *et al.*, 1979).

17.4.3.2 *Active defence: hypersensitivity and phytoalexins*

Plant cells challenged by many incompatible or avirulent races of biotrophic or necrotrophic pathogens frequently undergo a 'hypersensitive response' (HR) in which fungal penetration of the first host cell is rapidly followed by cell disorganization, electrolyte leakage, increased synthesis of secondary metabolites, browning and necrosis. The conventional view of this response is that death of the host cell creates an unfavourable or toxic milieu for pathogen development. Biotrophic parasites rely on intact host cells for a continued supply of nutrients. The accumulation of antifungal compounds known as phytoalexins may directly inhibit the growth of both biotrophic and necrotrophic parasites. In some cases, e.g. the interactions between potato and *Phytophthora infestans* or lettuce and *Bremia lactucae* changes may be detected microscopically within minutes or 1–2 h of penetration (Kitizawa *et al.*, 1973; Maclean *et al.*, 1974; Maclean & Tommerup, 1979). In both of these cases host cell disorganization and necrosis clearly precede inhibition of fungal growth. In other cases, such as crown oat rust (Prusky *et al.*, 1980) necrotic haustoria may be detected within apparently healthy host cells, death of host cells occurring later. Király *et al.*, (1972) postulated that the HR was the consequence, not the cause of fungal cell death. However their conclusions were based on heat and antibiotic treatments designed to kill fungal hyphae selectively, and Maclean and Tommerup (1979) aptly point out that the course of events initiated by such treatments, even assuming they were specific, may be quite different to those initiated during genetically-controlled resistance. A further technical difficulty lies in defining cell 'death'. Ingram (1978) concludes that the term HR covers a variety of relatively ill-defined events which may, or may not have a causal role in resistance and he argues for abandonment of the term.

In most cases, hypersensitivity is accompanied by an expansion of secondary metabolism and the synthesis of phytoalexins. The HR and accumulation of phenolics in tobacco inoculated with TMV was inhibited by α-amino-oxyacetate, a competitive inhibitor of PAL (Massala *et al.*, 1980). The expansion of secondary metabolism is not necessarily restricted to the infected cell. Mansfield *et al.*, (1974) showed by microspectrofluorimetry that living host cells adjacent to necrotic cells of broad bean infected by *Botrytis cinerea* accumulated the phytoalexin wyerone acid (Fig. 17.6). It is possible that hypersensitive cell death is the result of the export of phytoalexins from living cells peripheral to the penetrated cell, since phytoalexins are also anti-plant as well as anti-fungal (see Hargreaves, 1980). Phytoalexin concentrations up to 10% by tissue dry weight have been detected in hypersensitive lesions (Keen, 1971).

The subject of phytoalexins has been reviewed many times (e.g. Deverall, 1976; Kuć, 1976; van Etten & Pueppke, 1976; Cruikshank, 1977). Phytoalexins may be defined as low molecular weight antibiotics that are both synthesized by, and accumulate in plants after exposure to micro-organisms. The 100 or so compounds that have been characterized fall into a limited number of chemical categories (Fig. 17.6). Their antimicrobial activity varies between 1–50 $\mu g\,cm^{-3}$ for ED_{50} values *in vitro*, and their effects on fungi are generally fungistatic rather than fungicidal. The ability to produce phytoalexins is widely distributed in the plant kingdom and is not usually specific to pathogen type, i.e. the same phytoalexins may be produced in response to different pathogens. Phytoalexins are also induced by a variety of biotic and abiotic agents, wounding, U.V. light, heavy metal ions and water stress. Their synthesis may be the result of active metabolism including *de novo* RNA (Yoshikawa *et al.*, 1977) and protein synthesis (Dixon & Lamb, 1979) or they may constitute products of cell necrosis. Although Yoshikawa (1978) claimed that biotic (β-glucan) and abiotic (Hg^{2+}) elicitors stimulate glyceollin accumulation in soybeans by different mechanisms (stimulated synthesis and reduced degradation in the presence of turnover respectively), Moesta and Grisebach (1980) have been unable to confirm this. Rathmell and Bendall (1971) showed that phaseollin accumulation in dwarf bean was the result of a specific stimulation of isoflavonoid metabolism, distinct from a general increase in phenolics associated with cell necrosis.

Phytoalexins have been implicated both in varietal-specific resistance and general resistance or immunity. Keen and Bruegger (1977) summarize evidence which strongly suggests a causal role for phytoalexins in specific resistance, at least in some systems:

(1) The ability to synthesize phytoalexins is not restricted to genetically resistant cultivars. Rather, it appears to be the rate and time at which phytoalexins accumulate which is important, as determined by the initial recognition phase of resistance (17.4.3.3). In some cases the correlation between inhibition of fungal growth and phytoalexin accumulation suggests a causal relationship (e.g. Yoshikawa *et al.*, 1979; Keen & Littlefield, 1979).

(2) All races of a given pathogenic species are generally equally sensitive to the phytoalexin.

(3) Where measured, localized levels of phytoalexins are sufficient to account for growth inhibition of the parasite.

(4) Metabolic inhibitors which block phytoalexin accumulation also reduce resistance.

(5) Transfer of phytoalexins to compatible associations promotes incompatibility.

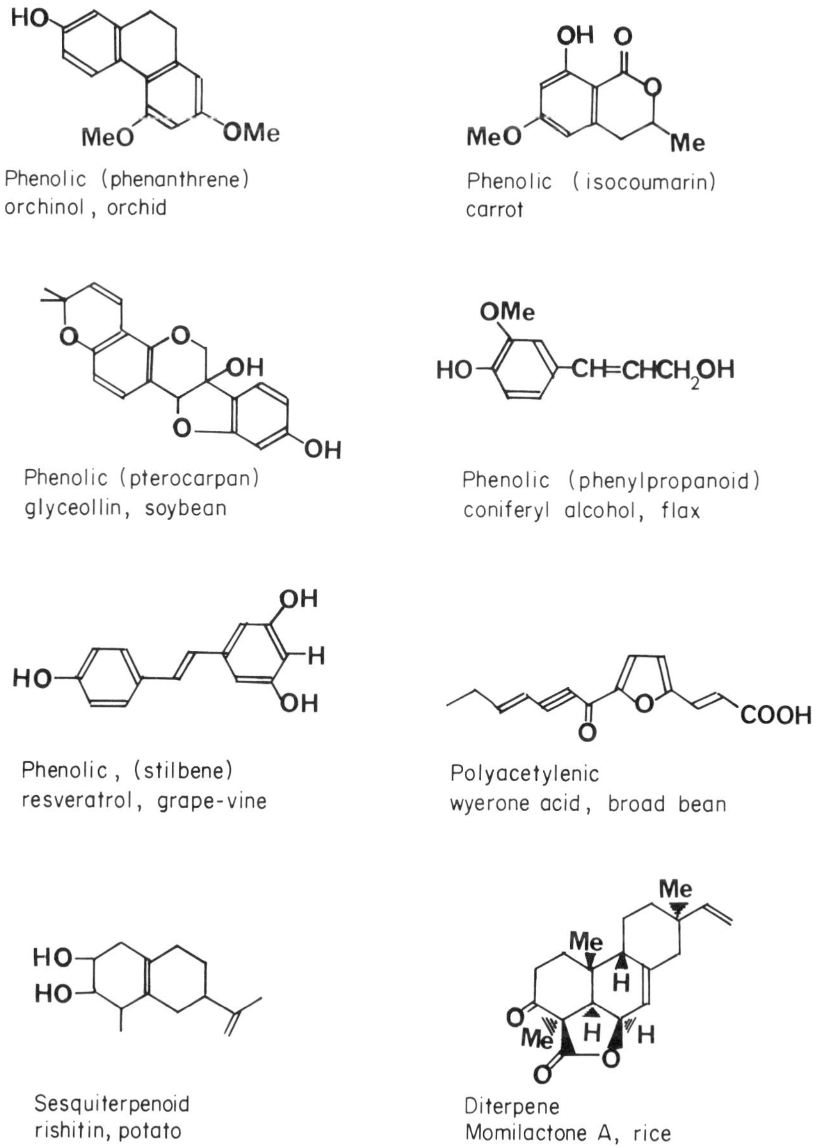

Fig. 17.6 Some common phytoalexin structures

A detailed chronology of the important events in the resistance of soybean to *Phytophthora megasperma sojae* is shown in Fig. 17.7.

In general resistance, a further consideration in addition to phytoalexin elicitation, is sensitivity of the parasite to the phytoalexin. Generally, fungi pathogenic to a given species are either less sensitive to the phytoalexins or have

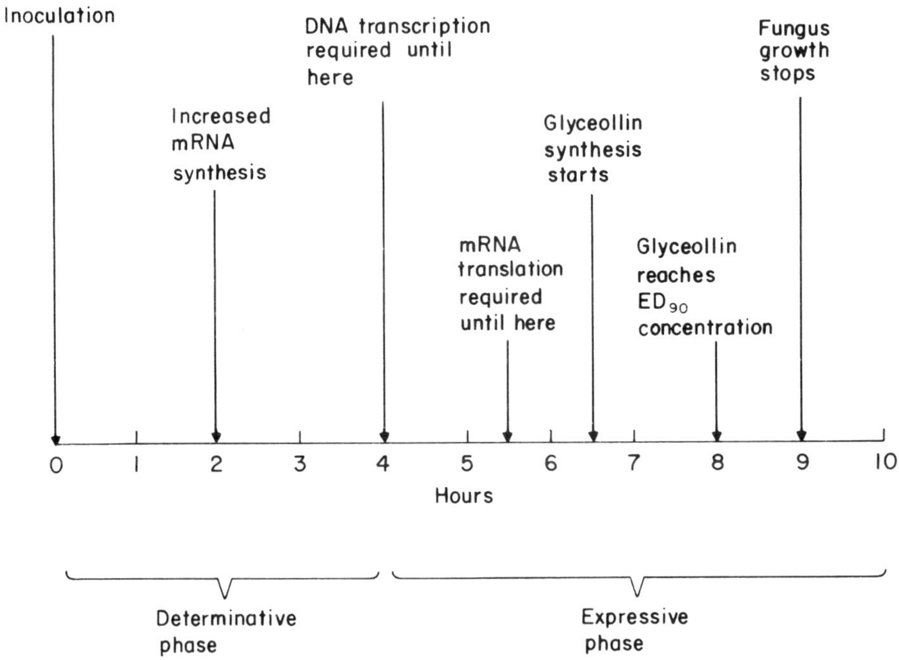

Fig. 17.7 Chronology of events in incompatible interactions between soybean and *Phytophthora megasperma sojae*. From Keen & Bruegger, 1977.

the ability to degrade them to a non-toxic form (van Etten, 1979). For example, three *forma speciales* of *Fusarium solani* viz. *F. solani pisi*, *F. solani phaseoli*, *F. solani cucurbitae* are characterized by their ability to parasitize peas, dwarf beans and cucurbits respectively. *F. solani pisi* was able to detoxify the pea phytoalexin pisatin, but not phaseollin, a dwarf bean phytoalexin. *F. solani phaseoli* could detoxify phaseollin but not pisatin, whilst *F. solani cucurbitae* were unable to degrade either pisatin or phaseollin.

Notwithstanding the evidence discussed here, evidence implicating phytoalexins in a causal role in resistance is largely circumstantial. Little is known of the concentrations of phytoalexins at the cellular level in relation to pathogen distribution and more direct approaches through the use of phytoalexin-deficient mutants of plants and phytoalexin-tolerant mutants of pathogens are urgently required.

17.4.3.3 *Active defence: surface recognition*

The ability to recognize and discriminate between 'self' and 'non-self' is a fundamental property of all organisms and, in general, proceeds via complementary molecular mechanisms located at the cell surface. Callow (1977), Clarke and Knox (1978) and Sequeira (1978; 1979) discuss a range of examples of cell recognition from plant, animal and bacterial literature, in which it has been established that mutual recognition in very diverse phenomena such as pollination, egg-sperm recognition, phagocytosis, mating reactions in fungi and algae, tissue differentiation and compatibility, allergenic and immune responses etc., are mediated by interactions between surface saccharide-containing moieties of polysaccharides and glycoproteins, and complementary carbohydrate-binding proteins, associating in a highly-specific, 'lock-key' manner (see also Fig. 7.4 in this volume).

Most systems of cultivar and race specific resistance are controlled by one, or at most a few genes of readily detectable effect (oligogenic or major gene systems), and where clearly analyzed conform to a so-called 'gene-for-gene relationship'. Genes for resistance and susceptibility in the host are associated with complementary genes for avirulence and virulence in the parasite as gene-pairs. Resistance or incompatibility is conferred only in the combination between host genes for resistance and parasite genes for avirulence, (Fig. 17.8). The apparent simplicity of this relationship and the fact that resistant cultivars can be obtained which are nearly isogenic to susceptible cultivars has resulted in the widespread use of such systems in biochemical pathology. In addition, the precise genetic relationships permit a critical testing of biochemical hypotheses. Cogent genetic arguments have been presented to suggest that the major implication of the gene-for-gene relationship is that specific recognition between host and parasite in systems of cultivar specificity, is for the incompatible

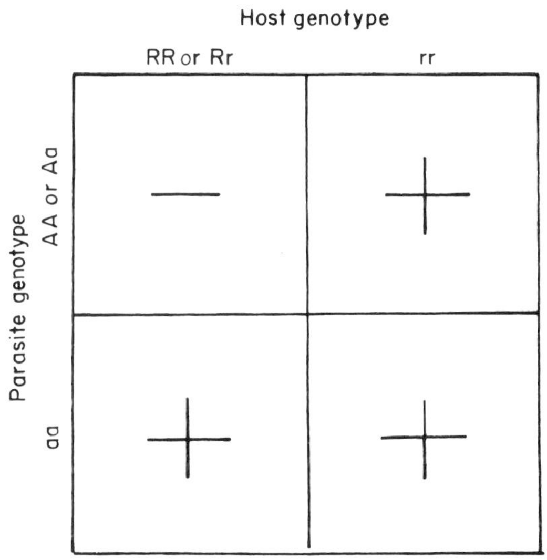

Fig. 17.8 The quadratic check in gene-for-gene interactions. R and r represent host genes for resistance (dominant) and susceptibility (recessive). A and a represent genes for avirulence (dominant) and virulence (recessive) in a diploid or dikaryotic parasite. Compatible and incompatible responses are denoted by + and − respectively.

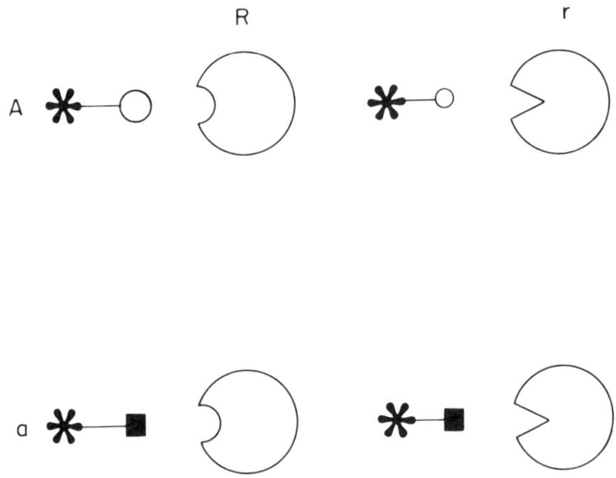

Fig. 17.9 Simple lock-key interpretation of the molecular implications of the gene-for-gene concept. Products of host R/r alleles are shown as surface-localized receptors with defined binding-site configurations. Products of parasite avirulence/virulence alleles are shown as molecules with a substantially common structure but specific recognized termini. Only in the case R/A can correct binding and recognition take place, triggering the resistant response (after Callow, 1981).

relationship (Ellingboe, 1979). The combination R : A, or rather between their gene products (Fig. 17.9), in effect constitutes a 'stop-signal' since it triggers incompatibility through a series of metabolic changes. The explosion of interest in the recognition of fungal polysaccharides and glycoproteins by plants lies in the belief that the association of such molecules with complementary plant receptors forms the molecular basis of recognition and resistance-triggering in cultivar-specific forms of resistance, the high informational content of polysaccharide structures conferring the necessary degree of specificity (Anderson-Prouty & Albersheim, 1975; Callow, 1977; Clarke & Knox, 1978). Lectins, specific carbohydrate-binding proteins or glycoproteins known to be present in plant cell walls and membranes represent an obvious candidate for the plant receptors, and have been implicated in some examples. Callow (1977) speculated that the products of host resistance genes could well be lectin-like surface receptors or 'sensors'.

Specificity in the infection of some legume roots by some species of *Rhizobium* appears to lie in the initial attachment of the bacterium to root hair surfaces via interactions between lectins on the root hair walls and carbohydrate receptors either in the lipopolysaccharide (LPS) portion of the cell wall of the bacterium (Wolpert & Albersheim, 1976) or the extracellular polysaccharide (EPS) capsule (Calvert *et al.*, 1978; Mort & Bauer, 1980). The purity of LPS fractions has been questioned (Schmidt, 1979) and Mort and Bauer (1980) demonstrated that the presence and composition of the EPS of *R. japonicum* varies with culture age, correlating with changes in the affinity of the cells for soybean lectin. In *R. trifolii* a surface heteropolysaccharide binds to clover lectin through 2-deoxyglucose residues (Dazzo & Hubbell, 1975). Antiserum raised to this material cross-reacts with clover root surfaces and it has been suggested that binding of *R. trifolii* to clover root surfaces may be mediated by multivalent clover lectin acting as a bridge between common surface antigens (see also Chapter 16).

Tobacco leaves infiltrated with incompatible strains of *Pseudomonas solanacearum* or the non-pathogen *P. pisi* undergo a form of the hypersensitive response and bacteria are immobilized by binding to host intercellular surfaces followed by encapsulation in a fibrillar wall pellicle (Goodman *et al.*, 1976; Sequeira *et al.*, 1977). Virulent bacteria neither attach nor do they elicit a hypersensitive response. Avirulent strains of *P. solanacearum* are agglutinable by potato lectin (specific for oligomers of N-acetyl-glucosamine) but virulent isolates are unable to bind this lectin (Sequeira & Graham, 1977). The lectin receptors appear to be LPS in the outer bacterial wall, and virulent strains escape wall binding and agglutination by potato lectin because the LPS receptors are masked by capsular EPS. A lectin with identical properties to potato lectin has been detected in tobacco leaf cell walls (Sequeira, 1979) and Anderson and Jasalavich (1979) and Goodman *et al.*, (1977) have demonstrated lectin-like bacterial agglutinins in bean cell walls and tobacco petioles respectively. The binding of LPS receptors to plant surface lectins also triggers resistance responses in addition to recognition and containment. Purified LPS from both avirulent and virulent strains of *P. solanacearum* and non-pathogenic bacteria bind to tobacco cell walls and induce disease resistance (Graham *et al.*, 1977; Mazzuchi *et al.*, 1979). A role for LPS in the recognition and attachment of virulent cells of *Agrobacterium tumefaciens* by plant cell walls has been suggested (Whatley *et al.*, 1976; Lippincott *et al.*, 1977). Clearly attachment of bacteria to plant cell walls can be the prelude to both compatibility and incompatibility.

Adhesion of zoospores of *Phytophthora cinnamomi* to maize roots is mediated by root surface carbohydrate-containing receptors in surface slimes (Hinch & Clarke, 1980). Removal or blockage of terminal fucose residues in the slime by enzymes and lectins abolishes zoospore binding. Zoospore surfaces bear complementary fucose-binding factors (Clarke, personal communication).

Plants have the ability to recognize fungal carbohydrate-containing polymers that potentiate both pathogenesis and resistance. Reference was made earlier (17.3.2.2) to the role of fungal β-glucans in the induction of wilting through effects on host cell membranes. This is consistent with the agglutination of potato protoplasts by the binding of β-glucans, possibly through β-glucan-specific lectins (Peters *et al.*, 1978; Marcan *et al.*, 1979; Marcan & Friend, 1979). Culture fluids and cell wall extracts of *Phytophthora megasperma sojae* (Pms) contain a number of components which will stimulate the accumulation of the phytoalexin glyceollin in soybean tissues (reviewed by Albersheim & Valent, 1978). Molecules of microbial origin which stimulate phytoalexin accumulation are termed 'elicitors' (Keen, 1975). The term is preferred to 'inducers' since it avoids implications of *de novo* protein synthesis, although this may be involved. Glyceollin elicitation is accompanied by an expansion in secondary metabolism as indicated by increases in the key enzyme, phenylalanine ammonia lyase (see also 17.4.3.1). The elicitors of Pms are highly active, less than 0.2μg (10^{-11}mol) per hypocotyl being sufficient to elicit enough glyceollin to inhibit Pms growth *in vitro*. However, the elicitors of Pms are non-specific, those from avirulent and virulent races being equally effective against R and S soybean cultivars. The elicitors are heat-stable, polydisperse (5–200 × 10^3) and appear to consist largely of β1,3 linked glucans with 6-linked branches, such molecules being common structural components of many fungal cell walls. Activity is dependent on the branched structure rather than the β1,3 link *per se* since linear β1,3 glucans such as laminaran are ineffective elicitors.

Similar glucan elicitors can be extracted from a wide variety of fungi including non-pathogens such as baker's yeast. Anderson (1978) isolated glucan elicitors from three species of *Colletotrichum* including species non-pathogenic towards *Phaseolus vulgaris*. All three preparations initiated browning and the accumulation of one of the dwarf bean phytoalexins, phaseollin. Many different plants also respond to the same glucan elicitor. It has been suggested therefore, that plants detect the presence of many non-pathogens by recognizing a structural component of the fungal cell wall. Since fungi are probably unable to survive without such vital components they are unable to escape detection by altering the structure of such components to a non-recognized form, (Albersheim & Valent, 1978). Such non-specific fungal elicitors probably constitute, therefore, the basis of at least one aspect of non-host or general resistance (Fig. 17.10).

Other types of elicitor have been described. The polydisperse glycopeptides from *Cladosporium fulvum*, described in 17.3.2.2 in the context of toxin action, also act as non-specific phytoalexin elicitors in tomato and other higher plants (de Wit & Roseboom, 1980). Again, the carbohydrate moieties, particularly the mannose residues, appear to constitute the elicitor-active fraction. Stekoll and West (1978) purified a heat-labile glycoprotein of 23 000 molecular weight from the saprotroph *Rhizopus stolonifer*, which at low concentration (10^{-7}M) elicited the castor bean phytoalexin, casbene. Not all elicitors are glycoproteins or polysaccharides. Monilicolin A, a polypeptide of 8000 molecular weight from the culture fluids of *Monilinia fruticola* stimulates phaseollin synthesis in *Phaseolus vulgaris* at low concentration, but is ineffective against peas and beans. This elicitor does not cause necrosis (Cruikshank & Perrin, 1968).

In contrast to the non-specific glucan elicitors from Pms described earlier, Keen and Legrand (1980) have partially characterized high molecular weight glycoproteins from Pms cell walls with race- and variety-specific activity in the elicitation of glyceollin. Wade and Albersheim (1979) have also isolated cell wall glycoproteins from Pms but these did not possess elicitor activity. However, when applied to soybean plants these glycoproteins were able to protect treated

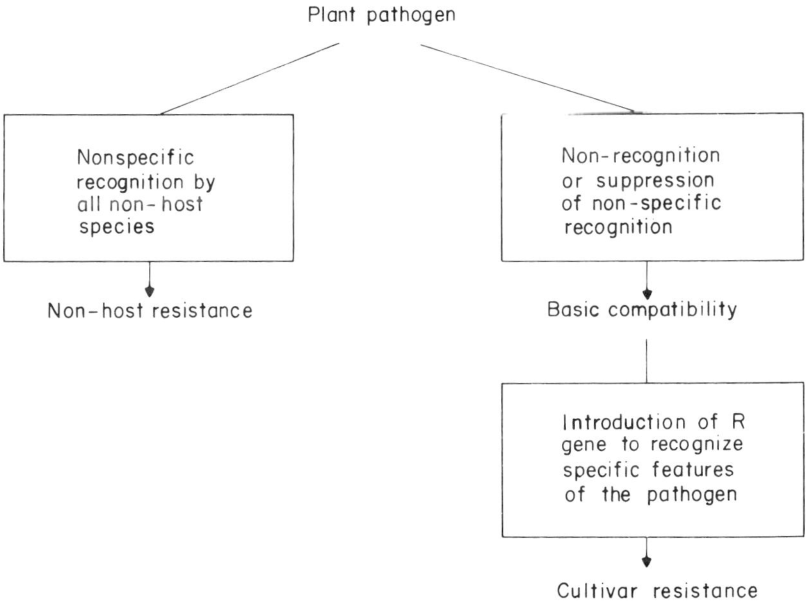

Fig. 17.10 Possible origins of compatibility and cultivar-specific resistance (after Heath, 1981).

plants against subsequent infection by Pms and protection was race-specific. Whether either of these glycoprotein fractions from Pms constitute the molecular determinants of gene-for-gene specificity remains to be seen and awaits more complete characterization of the molecules involved.

One of the criticisms often directed against the study of cell wall elicitors and other fractions is that their extraction frequently involves harsh procedures such as autoclaving or the use of alkali, thus potentially vitiating their significance in nature. However, as discussed earlier (Section 17.4.2) plant cells secrete glycosidases which can degrade fungal wall polymers. It seems possible that such enzymes might function in general defence by releasing soluble elicitors from the parasite cell walls. Thus, chitosan released from *Fusarium* cell walls by pea glycosidases, is a potent elicitor of the pea phytoalexin, pisatin (Nichols *et al.*, 1980).

An alternative and more specific role for fungal glucans has recently been suggested. A crude mixture of high molecular weight components from hyphal walls of *Phytophthora infestans* will elicit hypersensitivity and phytoalexin accumulation in genetically resistant and susceptible potato cultivars (Varns *et al.*, 1971). The hyphal wall components are bound by membrane fractions of potato (Doke *et al.*, 1979) or by protoplasts in which they induce cytoplasmic aggregation and cell lysis (Doke & Tomiyama, 1980a). The hypersensitive response and phytoalexin accumulation in potato tissue inoculated with an incompatible race or hyphal wall components, can be prevented or 'suppressed' by water-soluble glucans from compatible races of the fungus. Similar fractions from incompatible races are less effective (Garas *et al.*, 1979). The suppressive glucans appear to bind to protoplast surfaces (Doke & Tomiyama, 1980b), and have been partially characterized as β1,3 and 1,6-linked glucans of 17–23 glucose units, with both uncharged, and anionic, phosphorylated forms. It has been suggested therefore, that the elicitors from the hyphal walls of *P. infestans* may bind to receptors on the host cell surface and that the glucan suppressors or 'hypersensitivity-inhibiting factors' may block the binding sites in a competitive manner. It has been shown that the glucans are released into the germination fluids of *P. infestans* cystospores (Doke *et al.*, 1980). More information is required on the nature and mode of action of the *P. infestans* elicitor and suppressor fractions before the significance of this proposed mechanism for R-gene specificity can be assessed, but as proposed, it is consistent with the observation that prior inoculation of potato with a compatible race of *P. infestans* reduces the ability to respond to a compatible race through the hypersensitive response (Varns & Kúc, 1971).

Finally, mention must be made of the demonstration by Hargreaves and Bailey (1978) of 'constitutive' or 'endogenous' elicitors in *Phaseolus vulgaris*. Uncharacterized factors appear to be released from host cells on injury, infection, or following treatment with abiotic elicitors such as Hg^{2+} (Hargreaves, 1979) with the ability to diffuse to surrounding living cells where they stimulate phytoalexin accumulation. The high apparent affinity of dead or dying cells for phytoalexins then results in their sequestration in the original hypersensitive cell where they may inhibit pathogen growth. It may be speculated that the transmission to the rest of the cell of initial surface recognition stimuli could be achieved through the release of endogenous elicitors. This is in accordance with the recent demonstration that both abiotic (Hg^{2+}) and biotic (glucan) elicitors of glyceollin accumulation in soybean exert their action through stimulated phytoalexin synthesis (Moesta & Grisebach, 1980).

17.4.4 MOLECULES AND THE EVOLUTION OF RESISTANCE

The non-specific recognition of the majority of pathogens by all plants, through universally-present surface components of pathogens, logically constitutes the basis of at least some aspects of non host-resistance. If this is the case, then host susceptibility, i.e. susceptibility in those plants constituting 'host species' is best explained by the ability of successful, or compatible, pathogens to suppress or divert the general, non-specific recognition so that the fungus is recognized as 'self', resulting in disease (Fig. 17.10). Heath (1981) has argued a persuasive case supporting this concept of suppressed resistance or induced compatibility as the basis of compatibility at the species level. Race- and cultivar-specific forms of resistance have probably evolved from systems of more general resistance and induced susceptibility through plant breeding. Single resistance genes are incorporated into host species from wild relatives, permitting the new cultivar to recognize the pathogen as 'non-self' once more, resulting in resistance, either through the same, or a different set of resistance mechanisms to those involved in general resistance (Fig. 17.10). Anderson-Prouty and Albersheim (1975) and Callow (1977) have discussed how such a pattern of man-guided evolution has resulted in the gene-for-gene systems with multiple gene loci and alleles for resistance.

The studies on the molecular basis of resistance and recognition are consistent with this evolutionary scenario in a number of respects, although it must be emphasized that so few systems have been critically examined in depth that the dangers of over-generalization are obvious. Thus, at the level of non-host or general resistance a number of cell wall components such as some β glucans, present in a wide range of fungi, appear to act as non-specific elicitors of resistance reactions in a wide range of non-hosts. The evidence for suppressors in the induction of compatibility at the species level is perhaps less convincing. Although a number of suppressors have been implicated in induced susceptibility (Heath, 1981) to date only the glucan suppressors of *P. infestans* have been partially characterized with respect to structure and mode of action, and they appear to donate cultivar specificity rather than a basic compatibility at host species level. Heath (1981) suggests an interesting theoretical solution to this apparent difficulty. In addition the EPS of virulent Pseudomonads could be considered to be a form of suppressor since it effectively blocks recognition of the bacterium by masking LPS components.

If basic compatibility at the species level truly represents a form of induced susceptibility or suppressed resistance, as seems logical and for which there is good biological evidence, then the incorporation of major genes for resistance into otherwise compatible host species, in effect creates in this new cultivar the ability to recognize the pathogen as 'non-self' once more. In theory the recognition of any parasite surface component will suffice, providing it has sufficient potential for structural variation to confer specificity. Callow (1977) suggested that molecules important in parasite nutrition, such as surface glycoprotein invertases (Callow *et al.*, 1980) would be ideal candidates for recognition. In effect, genes coding for such molecules constitute the alleles for avirulence and their products, directly or indirectly become specific elicitors or specificity factors as claimed for *P. megasperma sojae*. Anderson-Prouty and Albersheim (1975) suggested that avirulence gene products could be specific glycosyl transferases with the capacity to create specific polysaccharide structures in the glycosyl moieties of glycoproteins, and Ziegler and Albersheim (1977) and West *et al.*, (1980) have shown that the glycosyl portions of *P. megasperma sojae* invertases are race-specific. These proposals are entirely consistent with one of the major implications of the gene-for-gene system, that recognition conditions resistance as a positive interaction between the products of host resistance and parasite avirulence genes (Fig. 17.9). It also explains the fact that in most cases avirulence is dominant and virulence recessive. In anthropomorphic terms, for the parasite to escape detection by the new cultivar containing a new R gene, it must mutate to virulence presumably by creating structural differences in the molecule recognized by the plant. If the recognized molecule is an important enzyme, this alteration, whilst promoting some aspects of parasite success, could also reduce fitness if the catalytic properties of the enzyme are affected. Hence the field observation that in the absence of appropriate conditioning R genes parasites tend to regain the avirulence forms of the genes concerned (van der Plank, 1968).

17.5 CONCLUDING REMARKS

In this brief review it is hoped that some of the atmosphere of the currently exciting prospects in biochemical plant pathology has been conveyed. The continued application of advanced techniques and thinking from fields such as molecular genetics and cell biology will facilitate our understanding of the mechanisms of pathogenicity and resistance. What of the practical implications?

Molecular studies on disease resistance may permit the identification of markers for plant breeders, thereby increasing the efficacy of crosses. More effective use of germ plasm banks as sources of resistance genes may be obtained by, for example, screening of germ plasm for the ability to synthesize phytoalexins rapidly in efforts to increase the level of general resistance in crops. Elicitors may facilitate such procedures by removing variability due to living pathogens.

The use of phytoalexins themselves as fungicides seems unpromising since their biological activity is often less than conventional synthetic agents, they are not systemic or therapeutic and would be difficult to synthesize chemically. In addition they are also anti-plant. The application of potent elicitors such as glucans would also presumably elevate the concentrations of phytoalexins in tissues, to the detriment of the plant and, potentially, any organism consuming it. A more promising approach would appear to be the use of agents such as the dichlorocyclopropanes which appear to potentiate the natural processes of resistance when required, i.e. when the plant is challenged by a potential pathogen (Cartwright *et al.*, 1977).

Finally, enzyme-targeted fungicides may be developed, i.e. compounds directed against unique metabolic systems of the parasite, or at least those which are sufficiently different in properties from the corresponding processes in the host. The identification of such systems requires continued basic research in the molecular processes constituting pathogenicity and resistance and the development of simpler systems in which molecular interactions can be more precisely studied.

Acknowledgement

The author wishes to thank Dr. M. C. Heath for permission to examine an unpublished paper.

FURTHER READING

ALBERSHEIM P. & VALENT B. S. (1978) Host-pathogen interactions in plants. *J. Cell Biol.*, **78**, 627–43.

CALLOW J.A. (Ed.) *Biochemical Plant Pathology*. John Wiley, London. (In press)

DAY P. R. (1974) *Genetics of Host–Pathogen Interactions*, Freeman, San Francisco.

DEVERALL B. (1977) *Defence Mechanisms of Plants*, Cambridge University Press, London.

HEDIN P. A. (1977) *Host Plant Resistance to Pests*. A.C.S. Symposium **62**, American Chemical Society, Washington.

HORSFALL J. G. & COWLING E. R. (1978) *Plant Pathology: An Advanced Treatise*. Academic Press, New York, San Francisco, London.

WHEELER H. (1975) *Plant Pathogenesis*, Elsevier-North Holland, Amsterdam.

CHAPTER 18

TUMOURS AS A RESULT OF GENE TRANSFER IN PLANTS[1]

18.1 INTRODUCTION

The study of pathological or aberrant states in living organisms has often turned out to be an efficient and rewarding way to study fundamental aspects of cell biology. Prime examples of this can be found in studies on bacterial and animal viruses. Plant molecular biology is, by comparison, relatively undeveloped. This is, to some extent, due to the absence of known DNA viruses suitable for experimental study. Only Cauliflower Mosaic Virus (CaMV) which contains a circular double-stranded DNA genome, has been studied in any detail. Recently, recombinant DNA techniques, with their potential for gene isolation and amplification, have opened new ways to study the relationship between structure and function of bacterial and animal genes. Again, the application of these new methods to plant genes has lagged behind although studies on plant chloroplast and mitochondrial genomes have recently made dramatic progress (see Chapter 9).

Essential to such plant studies would be the development of so-called plant gene vectors, or better still, of multiple host gene vectors. Ideally, such gene-vectors are DNA replicons that are easy to isolate, have single cleavage sites for a number of restriction enzymes and can readily be introduced and replicated in one or several host systems (e.g. *E. coli*/plant, *E. coli*/yeast/plant).

Apart from the fact that such gene-vectors can be used to isolate and amplify single genes or gene clusters, their main interest probably resides in their use for 'reversed genetics'. In order to probe for the function of a given DNA segment or nucleotide sequence, site-specific alterations can be introduced into the isolated genes and the effect of such alterations (e.g. deletions, insertions, base substitutions) can subsequently be determined by reintroducing the modified gene into a suitable host cell to study its expression.

Two gene-vector systems for plants are currently under study. One is the Cauliflower Mosaic Virus (CaMV). Howell *et al.*, (1980) have recently demonstrated that CaMV cloned in an *E. coli* vector is infectious in turnips. One can therefore expect rapid progress in the elucidation of the site and mechanism of replication of this viral DNA and thus establish whether or not this DNA can be used as a plant gene vector system.

The other system is the Ti-plasmid. Ti-plasmids are harboured by a group of gram-negative soil bacteria (*Agrobacteria*) and are responsible for the capacity

[1] Co-authors: M. van Montagu, M. Holsters, J. P. Hernalsteens, J. Leemans, H. de Greve, Ch. Shaw, P. Zambryski, H. M. Goodman, L. Willmitzer, L. Otten and G. Schröder.

of these bacteria to induce so-called 'crown gall' tumours on most dicotyledonous plants. (For recent reviews, see Gordon, 1981; Schilperoort *et al.*, 1979; 1980; Schell *et al.*, 1979; van Montagu & Schell, 1979; Schell & van Montagu, 1980; van Montagu *et al.*, 1980). Once established, crown gall tumours proliferate autonomously in tissue culture on simple media devoid of growth hormones, in the absence of the inciting bacteria. The tumorous state of crown gall tissue results from the transfer of Ti-plasmid DNA from *Agrobacterium tumefaciens* to the plant cells. This system is therefore uniquely suited for the study of the genetic determination and control of growth and morphogenesis in plant cells.

Most importantly this system represents a natural gene-vector for plant cells, evolved by and for the benefit of the bacteria that harbour Ti-plasmids (Schell *et al.*, 1979). Crown gall cells, as a direct result of genetic transformation by the Ti-plasmid, produce various substances called opines (i.e. derivatives of basic amino acids, such as arginine and lysine). Free-living *Agrobacteria* utilize these opines as sources of carbon and nitrogen. The genetic information both for the synthesis of opines in transformed plant cells and for their catabolism by free-living *Agrobacteria* is carried by Ti-plasmids. Because different Ti-plasmids induce and catabolize different opines, the Ti-plasmids can be classified according to the type of opine they determine (Guyon *et al.*, 1980).

Bacteria are known to be able to conquer an ecological niche by acquiring the capacity to catabolize certain organic compounds, not readily degradable by most other bacterial species. In several cases, the genes determining this degradative capacity have been found to be part of extrachromosomal plasmids. Several groups of soil bacteria, especially those living in and around the rhizosphere of plants, are able to decompose organic compounds released by plants. Clearly, with the advent of Ti-plasmids, *Agrobacteria* have carried this capacity one step further, by genetically forcing plant cells—via a gene transfer mechanism—to produce specific compounds (opines) which they are uniquely equipped to catabolize. This novel type of parasitism has therefore been called 'genetic colonization' (Schell *et al.*, 1979).

This chapter deals with recent developments in Ti-plasmid research with special emphasis on those aspects directly relevant to the use of the Ti-plasmid as a gene-vector system suited for the detailed analysis of plant gene structure and function.

18.2 Ti-PLASMIDS AS NATURAL GENE-VECTORS FOR PLANTS

18.2.1 GENETIC AND FUNCTIONAL ORGANIZATION OF Ti-PLASMIDS

Ti-plasmids are double-stranded covalently-closed DNA circles. There are many different isolates which vary in size but these can be allocated to one of two general classes specifying either octopine or nopaline synthesis. The

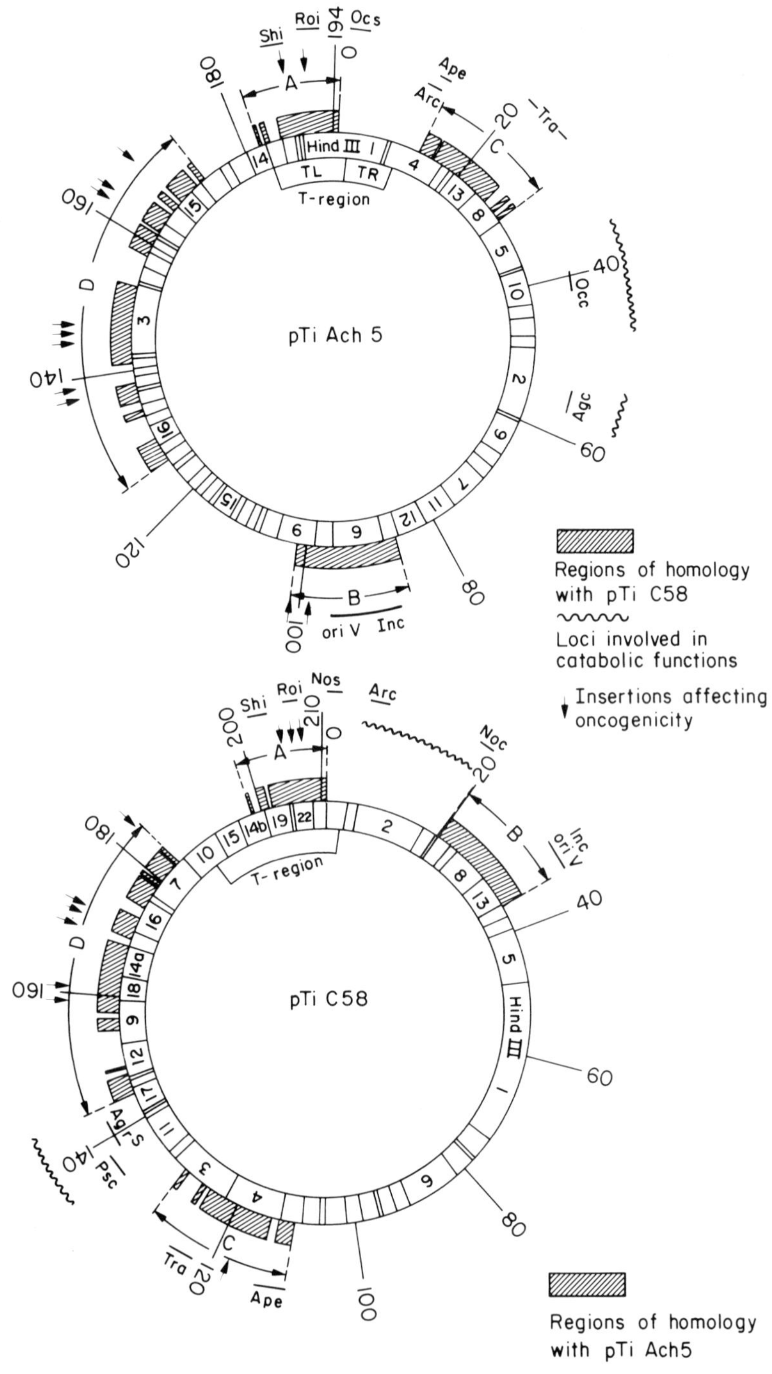

pTi Ach 5
Shi
Roi
Ocs
A
B
C
D
Arc
Ape
Tra
Occ
Agc
ori V
Inc
Hind III
TL
TR
T-region
Regions of homology with pTi C58
Loci involved in catabolic functions
Insertions affecting oncogenicity
pTi C58
Nos
Noc
Agr^S
Psc
Regions of homology with pTi Ach5

octopine plasmid Ach5 illustrated in Figure 18.1 consists of 194 000 base pairs and the nopaline plasmid C58 has 213 000 base pairs.

By comparing the properties of *Agrobacterium* strains with and without a given Ti-plasmid, it has been possible to define Ti-plasmid encoded functions. Ti-plasmids are responsible for the following properties of *Agrobacteria*: (i) crown gall tumour induction; (ii) specificity of opine synthesis in transformed plant cells; (iii) catabolism of specific opines; (iv) agrocin sensitivity; (v) conjugative transfer of Ti-plasmids; and (vi) catabolism of arginine and ornithine (Zaenen *et al.*, 1974; van Larebeke *et al.*, 1974; 1975; Schell, 1975; Watson *et al.*, 1975; Engler *et al.*, 1975; Bomhoff *et al.*, 1976; Genetello *et al.*, 1977; Kerr *et al.*, 1977; Petit *et al.*, 1978a; 1978b; Firmin & Fenwick, 1978; Klapwijk *et al.*, 1978; Ellis *et al.*, 1979; Guyon *et al.*, 1980).

In order to establish further whether these properties are determined by plasmid genes, mutant plasmids were isolated by transposon-insertion mutagenesis (Hernalsteens *et al.*, 1978; Holsters *et al.*, 1980; van Montagu & Schell, 1979; Ooms *et al.*, 1980; de Greve *et al.*, 1981) and by deletion (Koekman *et al.*, 1979; Holsters *et al.*, 1980). The different mutations were localized on the physical maps established for nopaline (Depicker *et al.*, 1980) and octopine (de Vos *et al.*, 1981) Ti-plasmids with restriction endonucleases, thus allowing the establishment of functional genetic maps for these plasmids. A summary of these results is shown in Fig. 18.1.

A number of general conclusions can be derived from this genetic analysis of Ti-plasmids. Mutants of all the expected Ti-determined phenotypes have been found and assigned to specific regions of the Ti-plasmid maps. Several regions of Ti-plasmids must be directly or indirectly involved in determining the different steps leading to neoplastic transformation. Indeed, mutants affecting so-called Onc (oncogenicity) functions are distributed over about half of the Ti-plasmid map; the other half does not appear to carry any genes involved in transformation. No Onc$^-$ mutants were observed in this region (right half of maps in Fig. 18.1) of various Ti-plasmids and large deletions, eliminating most of this region, can be obtained that do not affect crown gall formation by these plasmids.

One class of mutations localized within the so-called T-region of Ti-plasmids is of particular interest, since this region is transferred and maintained

---

Fig. 18.1 Functional map of a standard octopine (Ach5) and nopaline (C58) Ti-plasmid. The four areas of homology, a, b, c, d, between these plasmids are indicated by the hatched boxes. The arrow heads indicate the sites where non-tumorogenic (Onc$^-$) mutants were localized. The wavy lines indicate the areas where functions involved in opine catabolism were mapped. The location of some further mutants such as mutants in incompatibility (Inc), origin of replication (oriV), shoot induction (Shi), root induction (Roi), transfer function (Tra), phage exclusion (Ape) as well as biosynthesis of nopaline (Nos) and octopine (Ocs), catabolism of nopaline (Noc), catabolism of octopine (Occ), catabolism of arginine (Arc), catabolism of agropine (Agc) and catabolism of phosphorylated sugars (Psc) are indicated. The T-region is a segment of the plasmid DNA which is transferred to and maintained in transformed cells.

in transformed plant cells (see below). The mutations thus far localized in this region are of four general types:

(1) mutant plasmids inducing crown gall tumours in which no opine synthesis occurs. These mutations demonstrate that opine synthesis and tumour induction and maintenance are coded for by different genes. Furthermore, since these opine synthesis deficient Ti-plasmid mutants still allow *Agrobacteria* harbouring them to catabolize opines, it must be concluded that genes involved in opine synthesis (after transfer to the plant cells) are not in any way involved in opine catabolism.

(2) Mutants that totally or incompletely abolish tumour formation.

(3) Mutants that strikingly influence the organogenic activity of transformed plant cells. Three types of such mutants have been isolated thus far, each localized in a well-defined segment of the T-region: (i) shoot formers, (ii) root formers and (iii) shoot *and* root formers. From this last class of mutants normal plants can be regenerated which synthesize opines in all their tissues (stems, leaves and roots).

(4) Mutants that do not visibly affect any of the known phenotypes of crown galls.

On the basis of the localization of these various mutants, a functional map of the T-region can be drawn (Fig. 18.2). The most important conclusion from this genetic analysis is that the T-region must contain genes that play a direct, active and specific role after transfer into plant cells. Some of these genes specify opine synthesis, whereas other genes directly or indirectly control plant cellular growth and differentiation.

It is important to note that all the mutations which have thus far been found to affect tumour formation and tumour growth and differentiation are localized in Ti-plasmid sequences, both inside and outside the T-region, that are homologous between different types of Ti-plasmids (see Fig. 18.1 and Engler *et al.*, 1981). Interestingly, both the genes in the T-region involved in opine synthesis and the various genes outside the T-region, involved in opine and amino acid catabolism, have different (non-homologous) DNA sequences in different Ti-plasmid types. On the other hand, most of the Tra functions (involved in the conjugative transfer of Ti-plasmids) are localized in DNA sequences conserved amongst different types of Ti-plasmids. The Ti-plasmids studied thus far turn out to be repressed for auto-transferability, i.e. the conjugative Tra functions are normally repressed but are induced by the presence of specific opines, thus further stressing the central role of opine synthesis and catabolism in the evolution of Ti-plasmids (Petit *et al.*, 1978b; Klapwijk *et al.*, 1978).

18.2.2 THE TRANSFER OF THE T-REGION TO PLANT CELLS

To establish whether or not Ti-plasmid DNA is transferred and maintained in crown gall plant cells, DNA–DNA hybridizations were performed. The most detailed results were obtained by 'Southern blotting' hybridizations involving

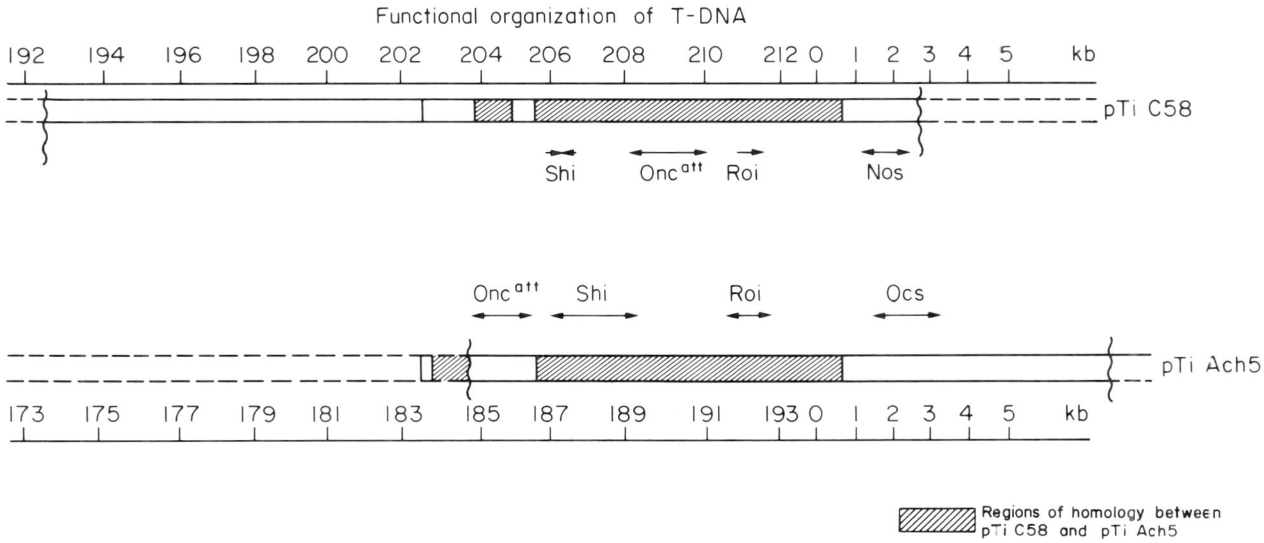

Fig. 18.2 The functional organization of the T-region of the nopaline plasmid, pTiC58, and the octopine plasmid, pTiAch5. The regions of homology between these plasmids are indicated by the hatched boxes. The coordinates are given in kilobases (kb). Horizontal arrows mark the extent of certain characteristic properties such as attenuated tumorogenicity (Onc^{att}), the capacity to synthesize nopaline (Nos) or octopine (Ocs), shoot induction (Shi) and root induction (Roi) are indicated. It is important to note that although the Tiregions of octopine and nopaline Ti-plasmids differ significantly in DNA sequence, they nevertheless have a similar organization. The central part is made up by a DNA stretch which is highly conserved in different Ti-plasmids (Depicker *et al.*, 1978; Chilton *et al.*, 1978). It is in this conserved DNA that most of the mutations affecting tumour maintenance or tumour growth and differentiation, are localized. To the right of this conserved region the genes involved in specifying opine synthesis are found. In nopaline tumours the T-DNA extends to the left of the conserved region. No phenotypes have thus far been associated with this region of the T-DNA.

DNA isolated from several independently isolated crown gall tissues. The DNAs were digested with various restriction enzymes and the fragments thus produced were separated by electrophoresis in agarose gels. The separated DNA fragments were then denatured and transferred to nitrocellulose sheets. Plasmid sequences present in the crown gall DNA were detected by DNA–DNA hybridization: the nitrocellulose sheets bearing the plant DNA were soaked in solutions of radioactive plasmid DNA and hybridization detected with X-ray film. Radioactively-labelled isolated DNA fragments covering the whole of the Ti-plasmid were used as probes in the Southern analysis (de Beuckeleer *et al.*, 1978; Schell *et al.*, 1979; Zambryski *et al.*, 1980; Thomashow *et al.*, 1980; Yadav *et al.*, 1980; Lemmers *et al.*, 1980).

The results show that all crown gall tumours contain a DNA segment, called the T-DNA, which is homologous to DNA sequences in the Ti-plasmid used to induce the tumour line. In all cases, this T-DNA corresponds to, and is colinear with, a continuous stretch of Ti-plasmid DNA, which has therefore been called the T-region (see above).

In different tumours induced on tobacco by the nopaline strains C58 and T37, the T-DNA segments were found to be identical, with a size of about 23 kb (kilobases). The T-DNA in octopine-producing tobacco crown gall lines was found to be smaller (about 14–15 kb) and somewhat variable in size, although always derived from the T-region. A detailed restriction endonuclease map of the T-DNA of nopaline strains has been published (Depicker *et al.*, 1980).

By preparing DNA from purified nuclei, chloroplasts and mitochondria isolated from crown gall tissues, it was shown that the T-DNA is located in the nucleus and not in chloroplast or mitochondria (Willmitzer *et al.*, 1980; Chilton *et al.*, 1980). When left- or right-hand border fragments of the T-region were used as probes in the Southern blot hybridization analysis, a small number of bands (usually 1–3) became apparent. The exact number and the sizes of these composite fragments, presumably consisting of T-DNA covalently linked to plant DNA, varied from tumour line to line (and sometimes within a given line) as a function of the restriction enzyme used. These observations indicated that more than one copy of T-DNA is integrated at more than one site in plant DNA.

In contrast, when internal fragments of the T-region were used as hybridization probes, only single bands were revealed, corresponding in each case to the full length of the T-region restriction fragment used as probe. The bands hybridizing to different internal fragments of the T-region demonstrated the same relative intensity.

Taken together, these data suggest that tumour cells contain a limited number (possibly up to five) of complete copies of T-DNA associated with different plant DNA sequences. Further demonstration of the integration of T-DNA in plant DNA and an understanding of the structure of T-DNA was obtained when a complete *Eco*RI digest of T37 tobacco crown gall DNA was cloned in a phage λ vector, thus allowing the isolation and detailed study of

T-DNA border sequences (Zambryski *et al.*, 1980). Two clones were studied in detail. One clone contains the left and right borders of the T-DNA linked together and the other clone contains the right end of the T-DNA linked to repetitive plant DNA sequences. These data therefore suggest that the T-DNA in nopaline tumours is organized as tandem repeats which can be inserted into repetitive sequences of plant DNA.

It should be noted, however, that it has not yet been shown whether or not the T-DNA is a stable part of a plant chromosome. It is also possible that there may be two states of the T-DNA, one integrated and one not.

The sum of the evidence derived from comparative studies involving different tumour lines induced with the same Ti-plasmid, tumour lines induced with Ti-plasmid mutants carrying transposon-insertions in the T-region (Lemmers *et al.*, 1980) and also from a detailed study of the nucleotide sequences bordering the T-DNA segment in cloned T-DNA fragments (Zambryski *et al.*, 1980) suggests that the 'ends' of the T-region are involved in the integration of the T-DNA. Either the T-region ends recombine before integration into the plant DNA or the entire T-region is excised and subsequently inserted.

The T-DNA therefore has one central property in common with bacterial transposons since it appears to be a discrete unit of DNA with the capacity to integrate into non-homologous DNA.

18.3 EXPRESSION OF T-DNA IN PLANT CELLS

18.3.1 TRANSCRIPTION OF T-DNA SEQUENCES

The first positive evidence that the T-DNA is transcribed within the plant cell was reported by Drummond *et al.*, (1977), who demonstrated that total RNA isolated from crown gall tumours hybridized to a specific fragment of the Ti-plasmid assumed to be present within the plant cell. A more extensive study was performed by Gurley *et al.*, (1979), who studied transcription in three different octopine tumours and found that the right part of the T-DNA was transcribed most actively, whereas little transcription was detected from the left part. A serious drawback in both studies, however, was that the extent of the T-DNA in the tumour lines studied was not known at the time. Therefore, these studies did not allow the construction of a detailed picture of the transcribed parts of the T-DNA. Since the transcription studies by Drummond *et al.*, (1977) and Gurley *et al.*, (1979) were performed with total RNA, it was important to study in detail the transcription of the T-DNA in two octopine tumours with special emphasis on comparison between nuclear and polysomal RNA. For these studies, tumour lines with accurately mapped T-DNAs were used.

The following approach was used, (i) nuclear RNA was isolated from crown gall cells and labelled *in vitro* by polynucleotide kinase. Furthermore, radioactive cDNA copies of nuclear RNA were synthesized using reverse transcriptase and oligo dT as primer. (ii) Polysomal RNA was isolated from crown gall cells; after separation into poly(A) containing and poly(A) minus fractions by oligo dT-cellulose chromatography, the RNAs were labelled by polynucleotide kinase. (iii) Highly purified and physiologically active nuclei were isolated from crown gall cells and incubated under conditions allowing RNA synthesis by endogenous RNA polymerase. Synthesized RNA was labelled by incorporation of radioactivity from α-^{32}P-UTP which was included in the incubation medium.

To map transcribed regions, the labelled RNA was hybridized to Southern blots containing cloned parts of the T-DNA digested by various restriction enzymes. To find out which RNA polymerase is responsible for the transcription of the T-DNA, transcription was studied both in the presence and in the absence of various concentrations of α-amanitin.

These studies have indicated that the T-DNA is transcribed over its entire length. However, well-defined regions of the T-DNA are actively transcribed, whereas other parts are only weakly transcribed. Surprisingly, no extensive differences were found between nuclear transcripts and polysomal transcripts. This might indicate that extensive processing steps do not occur in the formation of polysomal RNA. T-DNA derived transcripts were found both in the poly(A) containing and the poly(A) minus fractions of the polysomal RNA. Both fractions gave the same relative hybridization pattern of T-DNA fragments. By comparing the transcriptional pattern of the T-DNA (Fig. 18.3) with the functional organization of the T-region (Fig. 18.2), it can be seen that the areas that are most actively transcribed correspond to those that are genetically correlated with opine synthesis (right end of T-region) and with tumour growth (left part of conserved region). The results of the transcription studies are summarized in Fig. 18.3.

Finally, the transcription of the T-DNA in isolated nuclei was shown to be inhibited by low concentrations of α-amanitin (0.7μg/ml). This concentration is known to inhibit specifically RNA polymerase II from plants. Therefore the transcription of the prokaryotic derived T-DNA seems to be provided by the host RNA polymerase II.

18.3.2 TRANSLATION OF T-DNA DERIVED mRNA

To demonstrate that the transcription of T-DNA results in the production of functional mRNAs, T-DNA specific transcripts were isolated, translated *in vitro*, and the radioactive protein products identified.

Messenger RNA was isolated from transformed (B6, octopine) and from habituated tobacco tissue culture cells and tested for translational activity in two cell-free systems for protein synthesis. Polysomal RNA as well as polyA-containing RNA was efficiently translated in wheat germ extracts and in a rabbit reticulocyte lysate. Inhibition studies showed that 0.5mM pm^{7}G suppressed activity by about 90%, suggesting that the majority of mRNAs were capped with $5' \rightarrow 5'$ 7 methyl guanine as would be expected for eukaryotic mRNA. Gel electrophoretic analysis in the presence of dodecyl-sulphate, re-

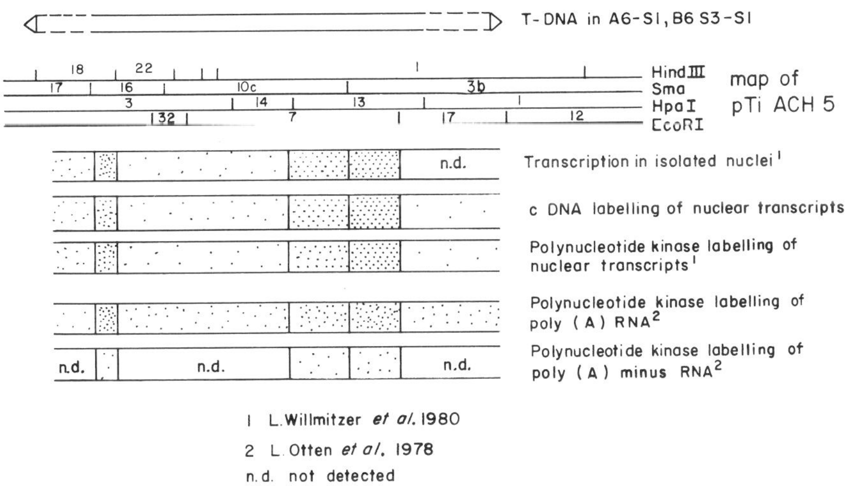

Fig. 18.3 Mapping of nuclear and polysomal transcripts of T-DNA in two octopine tobacco tumours (A6-S1,B6S3-S1) by polynucleotide kinase labelling, cDNA and transcription in isolated nuclei. For further explanation see text.

vealed a broad distribution of radioactive protein products of translation with molecular weights ranging from 10 to 90 × 10^3 Daltons. In all of these experiments mRNA from transformed and habituated tissues showed no significant differences.

To detect T-DNA specific mRNAs, poly(A)-containing RNA from both tissues was hybridized to two clones derived from the T-region, which covered the entire length of the T-DNA. After separation of the DNA–RNA hybrids from non-hybridized RNA by chromatography on agarose, the DNA–RNA hybrids were denatured, and the RNA was recovered and translated in the wheat germ extract. Figure 18.4 shows the autoradiographic pattern of the radioactive proteins after separation on polyacrylamide gels containing dodecyl-sulphate. Lane 3 (habituated cells) showed some background of radioactivity without distinct protein bands. Most likely this represents mRNA activity which has been carried through the hybridization procedure in a non-specific way. In contrast, lane 2 (transformed cells) showed four distinct protein bands appearing above the background. These proteins are good candidates for T-DNA specific proteins. These four proteins do not represent major proteins; they were never detected as distinct bands of radioactive protein after translation of total mRNA from transformed or from habituated cells.

Experiments in progress will show which part(s) of the T-DNA code for these proteins. Of special interest is the protein with $M_r = 41 \times 10^3$ Daltons, since its size is close to the published values for lysopine dehydrogenase (M_r = 38–39 × 10^3 Daltons). Experiments with specific antibodies will be necessary to demonstrate unequivocally lysopine dehydrogenase as a product of the T-DNA. (Lysopine dehydrogenase produces opines of the octopine series).

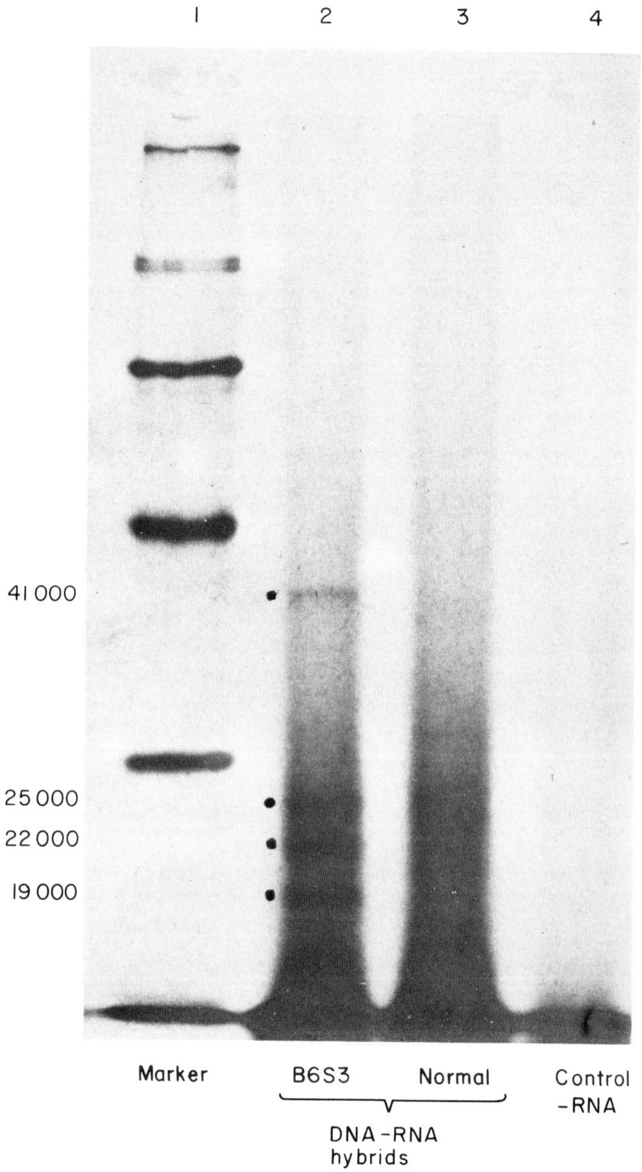

Fig. 18.4 Autoradiography of T-DNA derived proteins synthesized *in vitro*. Poly(A)-containing mRNAs from transformed (octopine crown gall on tobacco-line A6) or habituated tissues (tobacco) were separately hybridized to DNA clones covering the entire length of T-DNA. After removal of non-hybridized RNA by chromatography on agarose, the hybrids were melted and mRNAs were translated in a cell-free system for protein synthesis prepared from wheat germ. Radioactive proteins were separated by polyacrylamide gel electrophoresis in presence of dodecyl-sulphate and detected by autoradiography. Lane 1: marker proteins; lane 2: proteins from RNA of transformed cells; lane 3: proteins from RNA of habituated cells; lane 4: control without RNA.

T-DNA specific proteins have recently been described (McPherson *et al.*, 1980). In those experiments, a different procedure was employed for DNA–RNA hybridizations, and the tissue used was a different tobacco line which had, however, also been transformed with B6 (octopine plasmid). The size of the T-DNA specific proteins (M_r = 15, 16.5, <30, and 30×10^3 Daltons) was different from those found in the experiments described above, shown in Figure 18.4 (M_r = 19.5, 22, 25, and 41×10^3 Daltons). These discrepancies could be due to inaccurate molecular weight determinations, since one would expect that T-DNA encoded proteins would be identical in different tissues. Other explanations are possible, however, and more experiments with different tissues would seem necessary to clarify these points.

18.4 THE DEVELOPMENT OF THE Ti-PLASMID AS AN EXPERIMENTAL GENE VECTOR

In the previous section we have summarized the evidence showing that Ti-plasmids in *Agrobacterium* are natural gene vectors which can transfer a number of genes into the plant nucleus such that these genes are stably maintained and expressed. The obvious question therefore arose (Schell & van Montagu, 1977) whether or not the Ti-plasmid could be developed as a general plant gene vector. In order to answer this question the following points had to be settled.

18.4.1 ARE GENES, INSERTED IN THE T-REGION, CO-TRANSFERRED TO THE PLANT NUCLEUS?

In view of the observed involvement of the 'ends' of the T-region in the integration of T-DNA, one could expect that any DNA segment inserted between these 'ends' would be co-transferred, provided no function essential for T-DNA transfer and stable maintenance was inactivated by the insertion. The genetic analysis of the T-region by transposon-insertion provided us with Ti-plasmid mutants suited to test this hypothesis (see Section 18.2.1). A Tn*7* insertion in the nopaline synthase locus (see Fig. 18.2), produced a Ti-plasmid able to initiate T-DNA transfer and tumour formation. Analysis of the DNA extracted from these tumours showed that the T-region containing the Tn*7* segment had been transferred as a single 38×10^3 base pair segment without any major rearrangements (Hernalsteens *et al.*, 1980).

18.4.2 CAN GENES, INSERTED VIA T-DNA INTO PLANT NUCLEI, BE EXPRESSED?

The expression of the Tn*7* insertion in the nopaline synthase locus (see Section 2.1) was assayed as a model system for other genes inserted in this area. Tn*7* codes for a dihydrofolate reductase which is resistant to methotrexate

(Tennhammar-Ekman & Sköld, 1979). Suspension cultures of both untransformed and crown gall tobacco tissue were found to be completely inhibited by 2μg/ml of methotrexate. In contrast, cultures established with Tn*7* containing crown gall cells grew well on media containing 2μg/ml of methotrexate. In addition, nuclei isolated from these methotrexate tobacco lines were shown to synthesize Tn*7* transcripts. Further experiments are in progress to determine whether transcription of the inserted Tn*7* is initiated by a T-DNA promoter or by a Tn*7* promoter.

18.4.3 HOW CAN ONE INTRODUCE ISOLATED GENES IN THE T-REGION VIA IN-VITRO RECOMBINATION?

Because of the size of Ti-plasmids (M_r about 130×10^6) and because a large number of Ti-plasmid genes are involved in the transformation mechanism, it did not appear to be feasible to develop a 'mini-Ti' cloning vector with unique cloning sites at appropriate locations within the T-region and with all functions essential for T-DNA transfer and stable maintenance. An alternative way was therefore developed to introduce genes at specified sites in the T-region of a functional Ti-plasmid. The principle was to make an 'intermediate vector' consisting of a common *E. coli* cloning vehicle, for example the pBR322 plasmid, into which an appropriate fragment of the T-region of a Ti-plasmid is inserted. Single restriction sites in this T-region fragment can then be used to insert the chosen DNA sequence (see Fig. 18.5, stage 1). This 'intermediate vector' is subsequently introduced, via transformation or mobilization, into an *A. tumefaciens* strain carrying a Ti-plasmid which has been made constitutive for transfer (Holsters *et al.*, 1980) and which already carries antibiotic resistance markers (e.g. streptomycin, sulfonamide) cloned into the same T-region restriction site as that chosen to insert the DNA to be transferred. These resistance markers were introduced into this site by a procedure essentially identical to the one described here. Recombination *in vivo* will transfer the DNA of interest into the appropriate site of the Ti-plasmid as illustrated in Fig. 18.6 (stage 2). We have thus created a situation analogous to the one described under (stage 1) and therefore expected that the Ti-plasmid thus engineered to contain the desired DNA, would transfer this DNA into plant cells. This was recently shown to be the case with a model system in which a DNA fragment derived from the plasmid R702 and carrying streptomycin and sulfonamide resistance genes, was successfully introduced into tobacco crown gall cells. It should be noted that the introduction of inserts and substitutions in all parts of the T-region by these methods should allow us to probe further for the different T-DNA functions, this approach being equivalent to 'site specific' mutagenesis.

18.4.4 CAN NORMAL PLANTS BE REGENERATED FROM T-DNA CONTAINING PLANT CELLS?

The ultimate aim of many gene transfer attempts in plants is to produce fertile cultivars harbouring and transmitting new genetic properties. It was, therefore,

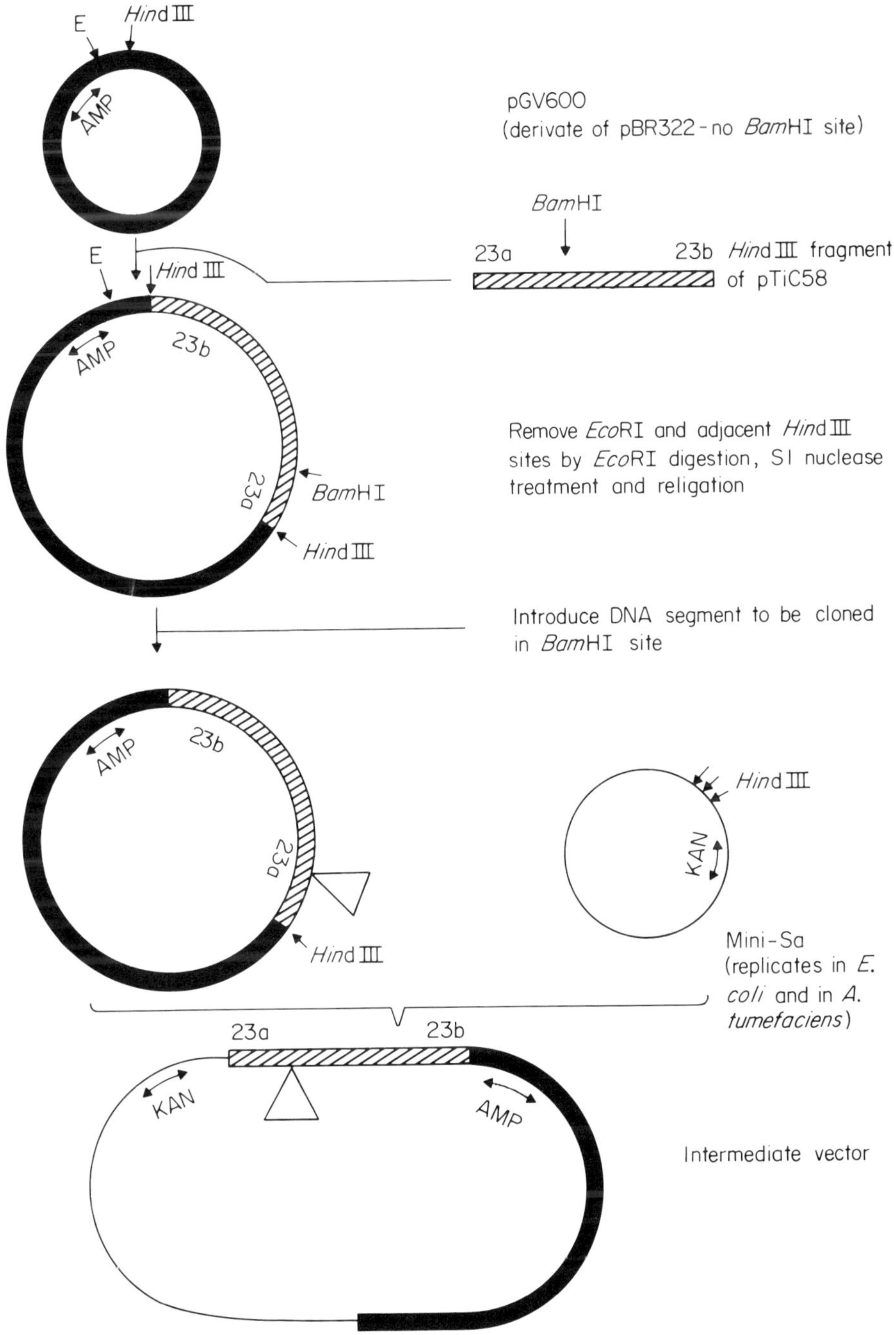

Fig. 18.5 Directed gene transfer into plant cells. (Stage 1).

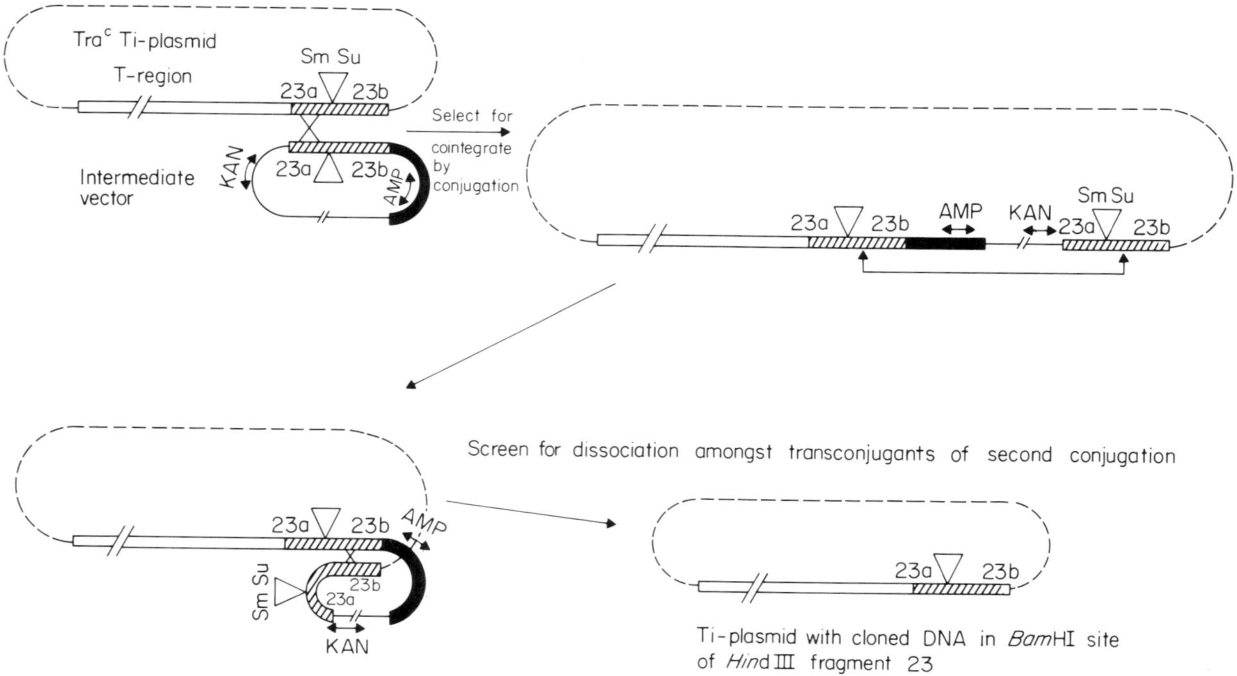

Fig. 18.6 Directed gene transfer into plant cells. (Stage 2).

essential to determine whether T-DNA transfer could be dissociated from neoplastic transformation and whether normal, fertile plants could be derived from T-DNA transformed plant cells. In order to answer this question, a large set of insertion mutants were obtained with an octopine TiB6S3 plasmid (de Greve *et al.*, 1981). One of the mutant plasmids (pGV2100) was clearly less oncogenic on tobacco and sunflower hypocotyls when compared to the wild-type plasmid. Tumours appeared only after prolonged incubation time. Furthermore, shoots proliferated from the greenish tumours, in contrast to the undifferentiating white tumours induced by strains harbouring the wild-type TiB6S3 plasmid. The Tn7 insertion in pGV2100 was mapped and found to be located in the left arm of the common DNA of the T-region (Fig. 18.2).

As a test for transformation, tumour tissue and the shoots on tobacco were assayed for the presence of lysopine dehydrogenase (Otten & Schilperoort, 1978). The tumour tissue was found to be positive. Most of the shoots were negative but some of the proliferating shoots were positive. One such shoot was grown further on growth-hormone-free media and found to develop roots and later to grow into a fully normal, flowering plant. Each part of this plant—leaves, stem and roots—was found to contain lysopine dehydrogenase activity, and polysomal RNA was found to contain T-DNA transcripts homologous to the opine synthesis locus. No transcripts of the conserved segment of the T-region (Figs 18.2 and 18.3) were observed.

These observations therefore demonstrate that normal plants can be obtained from plant cells transformed with Ti-plasmids genetically altered in specific segments of the T-region. Seeds obtained by self-fertilization of these plants produced new plants with active T-DNA linked genes, thus demonstrating that genes introduced into plant nuclei, via the Ti-plasmid, can be sexually inherited.

18.5 GENERAL CONCLUSIONS

The phenomenon of crown gall formation on plants is of fundamental importance in the study of the molecular biology of plant development. The abnormal growth pattern of crown gall plant cells has been shown to be the direct consequence of the presence and expression of specific 'tumour' genes. These genes have been transferred into the plant nuclei by gram-negative bacteria, which contain natural gene vectors in the form of large Ti-plasmids. These Ti-plasmids carry the plant tumour-inducing genes and can transfer them to the plant nucleus. Because these tumour genes and a series of mutations have been isolated which affect directly the growth and morphogenetic properties of the transformed plant cells, the crown gall system represents a model system uniquely suited to the study of the genes and gene products involved in the control of cellular growth and differentiation. The Ti-plasmid, because of its gene vector properties, is also suited to the analysis, via reversed genetics, of the relation between structure and function of plant genes. Finally, the Ti-plasmid can be used to introduce selected and isolated genes into plants and may therefore open new avenues for plant breeding.

FURTHER READING

LEDOUX L. (ed) (1975) *Genetic Manipulations with Plant Materials.* Plenum Press, New York.

LEAVER C. J. (ed) (1981) *Genome Organization and Expression in Plants.* Plenum Press, New York.

TIMMIS K. & PUHLER A. (eds) (1981) *Plasmids of Medical, Environmental and Commercial Importance.* Elsevier, Amsterdam.

CLETON F. J. & SIMONS J. W. (eds) (1980) *Genetic Origins of Tumour Cells.* Martinus Nijhoff, The Hague.

SOME PROBLEMS FOR THE FUTURE

CHAPTER 19

DETERMINATION AND RELATED ASPECTS OF PLANT DEVELOPMENT

19.1 INTRODUCTION

This chapter does not purport to present a comprehensive summary of the conclusions of the preceding chapters, but its aim is to draw attention to and discuss certain aspects of plant development about which there is, at present, very little information at the molecular level, but which are very important and constitute major gaps in our knowledge and understanding of development. Since the available 'hard' information on these topics is sparse and the discussion includes a large element of speculation and conjecture, the approach might be said to border on the philosophical, but if such an overview helps to provide a general framework for thinking about such a complex process as development, it may serve a useful purpose.

The achievement of the mature state requires, at all levels of organization, the attainment of a specific form and the establishment of orderly patterns of differentiated cells showing qualitative differences. Thus, development must involve processes which lead to a precise spatial organization. However, since development is also a sequential process, involving a highly ordered sequence of events, it also presents problems of temporal coordination. Thus, both *spatial* and *temporal control* of developmental processes is required.

Most of the orderly changes occurring during plant development will proceed even in a constant environment, as seen in leaf development, so that in such cases control is *autogenous*, but a number of major steps in development are regulated by environmental factors, as in the phenomena of photomorphogenesis when a seedling penetrates the soil surface and emerges from darkness into light, and in the photoperiodic control of flowering; in such cases, the internal control mechanisms must be operated by external factors.

The attainment of the normal shape of a plant organ requires controlled patterns of cell division, including the correct orientation of the mitotic spindles. Some interesting suggestions as to how polarized cell division may be regulated are made by Trewavas in Chapter 2.

Differentiation involves selective gene expression—the right genes must be expressed in the right cells at the right time. We are still ignorant as to how gene expression is regulated in eukaryotes, and until this problem is solved it will remain a major obstacle to a full understanding of cell differentiation. Some important findings have come from studies on the establishment of polarity and cell differences in developing *Fucus* eggs, (described in Chapter 7), and the importance of unequal division as one process whereby cell heterogeneity is established is becoming well recognized.

The potentiality of plant tissues to regenerate buds, roots or embryoids when grown under aseptic conditions has rightly captured the imagination of plant scientists, because of its great practical importance, and has led to a climate of opinion in which it is assumed that all living plant cells are totipotent and could be induced to regenerate whole plants if we could find the right treatment. A tacit corollary to this latter assumption is that there are no intrinsic differences in the developmental potential of meristematic cells and that cell differences are established and maintained by the internal cellular environment (i.e. by influences from surrounding cells and tissue) to which they are subject. Long-established evidence that relatively stable differences in metabolism and developmental potential may exist between plant cells and can persist even when they are cultured on the same nutrient medium, and other evidence for the existence of stable alternative states in shoot apical meristems, has attracted less attention and the implications have been largely ignored. Moreover, although it is generally recognized that plant hormones may evoke very varied responses in different tissues of the same plant, little attention has been paid to the occurrence of differences in the 'competence' of target cells. Hence, much of the following discussion is devoted to a summary of the evidence for 'determination' in plant cells and of its implications for plant development.

19.2 SPATIAL AND TEMPORAL COORDINATION IN DEVELOPMENT

There are several characteristic features of growth and differentiation in plants which have to be borne in mind:

(1) At the cellular level there is a rather sharp separation between the phases of cell division and cell enlargement and since the two phases tend to be synchronized in neighbouring cells they are, in the main, reflected in the development of larger units, so that zones of cell division and of cell enlargement can be recognized in the growing regions of shoots and roots. In the development of leaves, flowers and fruits, the separation of cell division and enlargement tends to be temporal rather than spatial.

(2) The indeterminate pattern of growth of higher plants by the continued activity of apical meristems leads to steady accretion of mature tissue in the older parts of the plant body, while meristematic activity is maintained at the shoot and root apices. However, the plant body is not as sharply divided into mature and meristematic regions as might appear since potential meristematic zones still remain in the older parts of the shoot in (i) the vascular cambium, (ii) the axillary buds and (iii) developing flowers and fruits.

(3) The overall structure and organization of the constituent units of the shoot (stem, internodes, leaves, flowers, etc.) are established during the meristematic phase of development, at which stage the primordia of these various

units are quite small—of the order of a few millimetres or less. Consequently the distances over which movement of nutrients and regulatory substances is necessary are quite short and diffusion and cell-to-cell (i.e. symplastic) transport are probably adequate.

(4) On the other hand, the distances over which the various growth centres of the whole plant need to be coordinated are very much greater than in the meristematic regions. Spatial coordination within the whole plant involves hormonal control, and it may be argued (Wareing, 1977) that IAA, the transport of which is strictly polar, is the hormone which is primarily concerned with the spatial coordination of growth centres in the shoot; this idea has been discussed in detail in Chapter 3.

(5) Development not only involves processes which require spatial coordination, but it is also an ordered sequential process calling for a high degree of temporal coordination. Again, it is necessary to make a distinction between temporal coordination during the meristematic phase, and that during the life-cycle of the whole plant. It has been suggested that the proper sequence of events during the meristematic phase may be achieved by establishing a kind of chain reaction in which one step initiates the next step in the overall process.

(6) Temporal coordination at the whole plant level ensures that the various phases of the life-cycle (germination, vegetative growth, flowering, fruiting, senescence and dormancy), occur in an orderly sequence and are appropriately synchronized with seasonal climatic changes.

(7) Although the shoot and root are exposed to very different conditions, the environmental conditions do not normally vary appreciably within the different parts of the shoot system, so that spatial coordination is, in general, determined by autogenous control systems, although the pattern of spatial organization may be modified by environmental factors.

(8) Temporal coordination may involve either autogenous or environmental control. Thus, in day-neutral species the transition from vegetative growth to flowering will occur even under constant daylength conditions, whereas in daylength-sensitive species the onset of flowering depends upon changing photoperiods. Winter chilling temperatures constitute the second major factor whereby the plant life-cycle is geared to the annual climatic cycle.

(9) As in spatial coordination within the whole plant, hormones appear to play a crucial role in temporal coordination. Although the nature of the flowering hormone still eludes us, all the evidence suggests that the photoperiodic induction of flowering involves the production of a hormonal stimulus in the leaves from whence it is transported to the shoot apical meristems, where it brings about a radical change in the structure of the meristem from one with an indeterminate growth pattern to that of a determinate organ. Similarly, the induction of tuberization in the potato (Gregory, 1956) and of bud dormancy in woody plants (Wareing, 1954) appear to be mediated through hormonal signals produced in the leaves under inductive photoperiods.

(10) The extent to which the major groups of growth hormone are involved

in temporal coordination is not fully established, but endogenous levels of gibberellins and cytokinins are strongly modulated by both photoperiod and chilling temperatures and there is much evidence suggesting that dormancy in buds and seeds may be regulated by variations in endogenous levels of gibberellins and possibly of cytokinins (Noodén & Weber, 1978).

(11) Where hormones are involved in temporal coordination they frequently act on specific target cells, as in the transformation of vegetative shoot apices in response to a flowering stimulus. The activation of the cells of the abscission layer by ethylene as the leaf senesces (p. 530) is another good example of hormonal action on target cells. On the other hand, it is not possible to identify specific target cells in the effects of winter chilling on dormant embryos.

19.3 DEVELOPMENT AND GENE EXPRESSION

That development is essentially a process involving controlled gene expression, whereby the right genes are expressed in the right cells in the right sequence, is effectively a truism, since many genes are apparently only expressed in specific parts of the plant at particular stages of development, as first pointed out by Haldane (1932). The real problems, therefore, concern the level at which such selective gene expression is controlled and how it is achieved. The brilliant achievements of Jacob and Monod in elucidating the regulation of enzyme synthesis in bacteria gave us a real insight into how the expression of certain genes in these procaryotes is achieved at the transcriptional level. The temptation to apply the same concepts to the regulation of gene expression in eukaryotes was irresistible and most of us assumed that the selective control of gene expression in eukaryotes was also exerted at the transcriptional level. However, as often happens in biology, the most obvious hypothesis has proved to be over-simplistic and early attempts to interpret the control of plant development by hormones and phytochrome in terms of the regulation of transcription have proved unrewarding in the main, with the result that there is some scepticism as to the validity of the selective gene expression concept in relation to development. However, it is important not to 'throw out the baby with the bath water', since in the final analysis it seems inescapable that development must involve the orderly regulation of gene expression.

The evidence for enzyme differentiation in plant tissues presented by Trewavas in Table 2.1 (p. 10), and of enzyme or protein changes which are accompanied by accumulation of the appropriate messenger RNA listed in Table 2.2 (p. 12), are consistent with the hypothesis that the differentiation of plant tissues involves selective gene expression. Moreover, there is good evidence from studies involving DNA–RNA hybridization that different plant tissues contain qualitatively different repetitive and single copy RNA transcripts. Thus, studies with the tobacco plant (Goldberg, 1980) have shown that there are about 25 000–30 000 different species of RNA in each of the main types of tissue studied (leaf, stem, root, petal, ovary, anther) and of these, about 8000 species of mRNA are common to all types of tissue. The remaining diverse species of mRNA are apparently characteristic of each type of tissue, so that each tissue possesses a unique set of mRNAs which can comprise the transcripts of thousands of diverse structural genes.

However, the occurrence of specific mRNAs does not, by any means, imply that there is selective DNA transcription in each type of tissue. It is now known that the immediate product of transcription is not mRNA but heterogeneous nuclear RNA (HnRNA), which undergoes a quite complex processing to yield mRNA. Thus, it is possible to envisage that the production of specific mRNAs is achieved at the stage of HnRNA processing. Indeed, this has recently been suggested by Davidson and Britten (1979), who point out that whereas large numbers of single copy sequences are present in HnRNA, only a small fraction of these are represented in the mRNA in the cytoplasm of a given cell type. Moreover, whereas the sets of mRNA in a given tissue are mainly absent from other types of tissue, nevertheless these sequences appear to be ubiquitously present in the HnRNA in all cell types. The implication of these observations is that each differentiated cell nucleus includes not only all of the genes ever utilized in the organism, but also transcripts of all or most of these genes.

Most HnRNAs turn over very rapidly, with a half-life, in the sea-urchin egg, for example, of about 20 minutes. Davidson and Britten suggest that the regulation of gene expression is achieved by determining which of the potential mRNA precursors survive, are processed and are transported to the cytoplasm. It is postulated that certain mRNA precursors of structural genes are protected from degradation by forming RNA–RNA duplexes with transcripts of non-structural gene sequences. These non-structural gene sequences apparently consist of interspersed repetitive and single copy sequences and are complementary to the repeated sequences in the transcripts of the structural genes. Thus, according to this model, the survival of some mRNAs in any given cell type depends upon the production of specific regulatory non-structural gene sequences, but how this latter specificity is achieved in any given cell type remains unknown.

Whether or not the hypothesis of Davidson and Britten will prove correct remains to be seen, but studies on the tobacco plant (Goldberg, 1980) are consistent with the hypothesis that control of gene expression is exercised at a post-transcriptional stage. Thus, while only 75% of leaf mRNA sequences are represented in the polysomes of the stem, all of the leaf mRNAs are represented in the stem HnRNA. Thus leaf structural genes, which are not utilized to produce proteins in the stem, are nevertheless transcribed into stem HnRNA, suggesting that 'post-transcriptional processing or selection mechanisms play a central role in the regulation of plant gene expression' (Goldberg, 1980).

It is also possible to envisage that gene expression is regulated at the translational or post-translational stages, and Trewavas (Chapter 2) has indicated the criteria which need to be met before control at these levels can be regarded as rigorously demonstrated. Nevertheless, the evidence for the occurrence of long-lived mRNA in *Acetabularia* (Chapter 6), *Blastocladiella* (Chapter 5) and seeds (Chapter 13), presented earlier, strongly suggests that control may also occur at the translational stages. However, in these cases we must assume that the specificity of the proteins produced is determined by the nature of the long-lived mRNAs and that it is the *time* of gene expression which is controlled at the translation stage.

It will be apparent that views regarding the regulation of gene expression in eukaryotes are very fluid at present, but the subject is advancing rapidly and it should not be too long before it is possible to resolve many of the special problems of gene expression in development.

19.4 THE ESTABLISHMENT OF CELL HETEROGENEITY

The establishment of cell differences during the process of cell maturation clearly constitutes one of the central problems in the study of development. As Vöchting (1878) stated, the fate of a cell is determined by its position, implying that the particular pattern of differentiation which a given cell pursues is determined by influences from outside the cell itself. Such influences may originate from the external environment or they may emanate from surrounding cells and be part of the 'internal' environment to which the cell is subject.

Cell differences are probably controlled by several different mechanisms. Undoubtedly one of the most important means whereby divergent patterns of differentiation are established is by unequal division, which has been discussed in Chapters 2 and 4. In unequal division cell differences arise as a result of influences, possibly electrical potential gradients, which bring about unequal distribution of cytoplasmic factors in the mother cell *before* the division occurs (Chapter 2). This process would seem to be a very effective means of establishing cell differences and there is no doubt that it plays an important role in cell differentiation in lower plants; it also appears to be important in the early development of the zygote of higher plants, the first division of which is unequal, and in the later stages of their development, as well as in the development of pollen grains and the stomata and root-hair cells of grasses.

In some instances the pattern of cell differentiation appears to be regulated by hormonal factors, as in the stimulation of xylem differentiation by IAA. However, since the same hormone can evoke very different responses in different parts of the plant, it is clear that the particular pattern of differentiation is determined by the 'competence' of the target tissue (p. 529), and the number of instances in which hormones themselves determine a specific pattern of cell differentiation is probably quite limited. Consequently, we need to consider whether or not there are other mechanisms whereby cell differences are induced by influences from neighbouring cells.

Induction of differentiation by one tissue in another is well-established in animal development and there is good evidence for the exchange of informational molecules between different tissues. There is little evidence for induction of this type in plant tissues, although it has been suggested that there may be 'homogenetic induction' of one differentiating xylem cell on another (Lang, 1965).

However, we have to bear in mind that development involves not only the establishment of spatial pattern differentiation, but also temporal control, so that the correct *sequence* of events is achieved. The regular sequence of changes seen in the development of an organ such as a leaf or a flower strongly suggests that once these particular pathways have been entered, then all the subsequent stages follow inexorably as a chain reaction in which the attainment of one stage triggers off the next one. A mechanism of this type has been suggested for flower development (Heslop-Harrison, 1964). It is postulated that once the transition to the flowering stage has begun, a gene complex A is activated in the first-formed primordia and these genes produce an inducer X which moves to the next set of primordia and there activates a gene complex B, and so on through the different classes of organ. This hypothesis postulates the occurrence of short-range intercellular inducer molecules; the existence of such molecules has yet to be demonstrated but we now know that short-range informational molecules are involved in 'cell-recognition', as in pollen-stigma interactions (Chapter 11), in which glycoproteins appear to play a crucial role. There is also evidence that cell recognition interactions occur between somatic cells. There is, as yet, no evidence that such interactions are important in cell differentiation, but the possible role of informational molecules, such as glycoproteins, in differentiation would seem worth bearing in mind.

Apart from the possible role of short-range inducer molecules, we also have to consider the possibility that differentiation may be regulated by properties of a meristem or an organ primordium as a whole, as in the well-known model of Turing (1952), in which it is postulated that leaf initiation may be controlled by chemical standing waves. Again, there is the possibility that electrical fields within a meristem or developing organ may be important in regulating patterns of histogenesis. Certainly, there is strong evidence that electrical potential differences, resulting in the electrophoresis of informational molecules, are important in polarity and unequal division in *Fucus* eggs and other cells, as discussed in Chapters 2 and 7. However, there is little evidence that unequal division is important in the vegetative shoot apical meristems in higher plants; on the other hand, the patterns of differentiation appear to be controlled by the properties of the meristem as a whole (p. 538), and electrical fields may well play a controlling role in some way other than through unequal division.

19.5 DETERMINATION IN PLANTS

In the developing animal embryo various regions become successively committed to specific pathways of development, so that the cells become increasingly specialized and show progressive restriction of 'developmental potential'. This progressive commitment to particular pathways is referred to as 'determination'. The determined state is remarkably stable and can be transmitted without reversal through many cell generations. One of the most remarkable examples of this phenomenon is shown in the development of insects, such as *Drosophila*, in which groups of cells, known as the imaginal discs, are laid down at a very early stage in the larva, but they are already determined to give rise to various parts of the adult fly and they can retain the state of determination through many cell divisions without loss (Gehring, 1972).

In plants the progressive commitment observed during animal development is less conspicuous in the plant as a whole because of its 'open' pattern of growth by apical meristems, but in organs such as leaves, flowers or fruits in which the pattern of development is 'closed', the analogies with animal development are closer. Thus, both surgical and aseptic methods involving excision of leaf primordia have shown that the fern leaf primordium is not irrevocably determined in its early stages, and can be made to develop into a bud, but beyond a certain stage it appears to become determined as a leaf (Wardlaw, 1949; Caponetti & Steeves, 1963). Thus, the leaf primordium as a whole shows determination, but the question arises as to whether this determination is a function of the cellular organization of the leaf primordium as a whole, or whether it involves changes in the intrinsic properties of its constituent cells.

Attempts to answer this question have given some evidence of determination as leaf cells, although the results were not entirely conclusive (Steeves, 1966). There are relatively few other clear examples of organ determination in plants, although there are a number of interesting observations suggestive of organ determination, including floral organs (Hicks, 1980). Hicks and McHughen (1974) found that isolated placental tissue of tobacco in aseptic culture gave rise to elongate, style-like structures with terminal stigmas. Thus, isolated placental tissue appears to be in a state of determination as 'female' tissue.

As well as determination shown during development of plant organs, it may be argued that shoot apical meristems may also exhibit properties analogous to determination (Wareing, 1971; 1978; 1979). There are a number of well-known examples in which a shoot apex can exist in two alternative stable states (Wareing, 1959; Brink, 1962). Thus, it is characteristic of species which show a vernalization response that, once the shoot apices have become fully vernalized, it is impossible to 'devernalize' them, except during a relatively short period after termination of the chilling treatment (Purvis, 1961). Moreover, once the shoot apices of winter rye (*Secale cereale*) are fully vernalized, the state appears to be transmitted without dilution through many successive cell divisions.

Again, the occurrence of free-living gametophyte and sporophyte generations in lower plants appears to indicate that their apical cells can exist in two alternative states. These two alternative states are apparently not dependent upon the state of ploidy, since haploid sporophytes and diploid gametophytes can be produced experimentally.

The occurrence of two different states within the shoot system is also demonstrated by the phenomenon of 'phase change' in woody plants. Many woody plants show a distinct 'juvenile' phase, during which they are unable to initiate flowers, but ultimately they attain the mature or 'adult' condition in which flowers are initiated if other conditions are favourable (Doorenbos, 1965; Zimmermann, 1976). Although the presence or absence of the ability to flower is the primary distinguishing characteristic of the two phases, a number of differences in vegetative characters may also be recognized, such as differences in leaf shape, phyllotaxis, presence or absence of spines, rooting ability etc.

The two alternative phases show great stability, as shown by the ability of rooted cuttings from the adult part of the vine of ivy (*Hedera helix*) to retain their adult characters for many years, and that of scions of juvenile beech (*Fagus sylvatica*) grafted into the adult part of a mature tree to retain their juvenile characters. However, reversion of adult cuttings to the juvenile state can be induced by relatively high temperatures or by treatment with gibberellic acid (Robbins, 1957).

It is clear from these various examples that the shoots of some species can exist in alternative states, and that the differences between juvenile and adult ivy, or between vernalized and unvernalized rye plants, which are manifest in the mature, differentiated tissues of the shoots must reflect differences in the states of the apical meristems in these various phases. The various examples we have considered, in which shoot apices can exist in alternative states, seem to be analogous to the phenomenon of determination and thus the shoot apex can be 'determined' as either juvenile or adult, vernalized or unvernalized, gametophytic or sporophytic (Wareing, 1971). However, the development of an organ, in both plants and animals, is a progressive process involving a series of changes which follow each other in regular succession, with no long time interval between them. By contrast, in phenomena such as phase change, vernalization and the alternation of generations, although there are short transitional stages between the alternative states, once these have completed there is no further progressive change; that is to say, the two alternative phases represent *stable* states, which preserve their identity through many cell generations.

The question arises as to whether or not the meristematic cells of the apices are themselves 'determined'; that is, are the meristematic cells of juvenile and adult apices of ivy, or of vernalized and unvernalized rye, already determined along these respective pathways of development? Isolated shoot apices of embryos of winter rye can be chilled, and they subsequently give rise to mature vernalized plants (Purvis, 1940). Again very small pieces of the adult shoot apex of *Citrus*, bearing only three small leaf primordia, can be successfully grafted on to seedling stocks and are then found to retain their adult character (Navarro *et al.*, 1975). Thus, the shoot apical regions themselves can exist in alternative states, but this is still not evidence that the meristem cells themselves are determined, since even these very small excised apices will have included some parts which have begun differentiation into specific organs (e.g. leaf primordia) or tissue (e.g. procambium). However, callus cultures derived from juvenile and adult stem tissues grown in the same nutrient medium show differences in their rates of growth and in the fact that 'juvenile' calluses show abundant regeneration of adventitious roots while 'adult' calluses show little capacity for root regeneration (Stoutemeyer & Britt, 1965). On the other hand, such calluses are normally established from mature stem or petiole tissue and may not necessarily reflect the state of determination of the meristematic cells in shoot apices.

19.6 CELL DETERMINATION AND HABITUATION IN TISSUE CULTURES

As Meins and Binns (1979) have pointed out, the phenomenon known as 'habituation' in plant tissue cultures appears to present an example of cell determination. Callus cultures derived from the pith parenchyma of tobacco require to be supplied with auxin and cytokinin as well as the other nutrients constituting the basal culture medium, for continued growth in culture. However, when such cultures are grown through a number of passages they sometimes lose the requirement for exogenous kinetin and can then be subcultured many times on a medium lacking kinetin. That the change involved in such 'habituation' is a property of the individual cells is shown by the fact that cultures derived from single cells of habituated tissue will continue to grow on a cytokinin-free medium (Lutz, 1971; Binns & Meins, 1973).

Experiments were carried out to determine whether the changes occurring in habituation are 'epigenetic' or due to gene mutation. Pith cultures of certain varieties of tobacco habituate rapidly when they are kept on a basal medium lacking kinetin at 35°C, instead of the normal 25°C used for such cultures. The rate of habituation was calculated to be greater than 10^{-3} conversions per cell generation (Meins, 1977), which is 100–1000 times faster than the rates of somatic mutation in tobacco. Meins concluded that habituation is a directed process which may be induced by high temperature and which occurs at a rate well above that expected for mutation.

That habituation is reversible was shown by experiments in which fertile tobacco plants were regenerated from both habituated and non-habituated tissue (Binns & Meins, 1973; Meins, 1977). Pith cultures from both types of plant were found to be cytokinin-requiring (i.e. non-habituated). Thus, the habituated state is evidently reversed at some point in the process of regeneration of whole plants, so that the pith tissue is always non-habituated. This result shows that habituation is reversible and that it does not affect the totipotency of the genome. Clearly, determination in plants differs from that in animals in being reversible and, as Meins and Binns (1979) point out, stability does not imply permanence.

Both habituated and non-habituated tissues are similar in their requirement for exogenous kinetin when grown under certain conditions, e.g. low auxin, incubation at 16°C or treatment with 5-bromodeoxyuridine, but the habituated tissue retains its capacity to grow on a kinetin-deficient medium under 'permissive' conditions (Meins, 1976). Thus, habituation resembles determination in that it involves changes in cell commitment which persist whether or not the committed state is expressed.

The molecular basis of habituation and non-habituation is not understood, but as pointed out below (p. 528), it is not difficult to postulate model circuits in which there is end-product feedback in enzyme-controlled processes involving the biosynthesis of a growth substance, which will account for the continued activation or repression of enzyme synthesis in biochemical differentiation. Meins and Binns (1978) have produced evidence that this situation may apply to cytokinin production in habituated tissues. They grew cloned habituated tissues for several transfers at 16°C, at which temperature they require cytokinin for cell division. Some cultures were supplied with optimal concentrations of kinetin, whereas others were supplied with low concentrations of kinetin. After several transfers at 16°C, the cultures were transferred to 25°C and tested for habituation. It was found that the tissues supplied with high kinetin concentrations remained habituated, whereas those that had been kept on low kinetin at 16°C had shifted to the non-habituated state. These results are consistent with the hypothesis that the habituated state is maintained by a positive feedback loop involving cytokinins.

Several lines of evidence indicate that tobacco cells in culture exhibit different degrees of cytokinin autotrophy and that the conversion of these cells to the fully autotrophic state is a gradual process (Meins & Binns, 1977). This finding suggests that the transition between the habituated and non-habituated states is not controlled by a simple 'on/off' mechanism.

19.7 THE DIFFERENTIATED STATE

Differentiation is the process by which cells become increasingly specialized in structure and/or function. However, in plants the situation is complicated by the fact that the attainment of the mature condition frequently involves a considerable increase in cell volume as a result of vacuolation, with corresponding increases in area of the cell wall. It is during this process of vacuolation and enlargement that the differences in structure and function also develop. Consequently, the term differentiation is frequently applied to include the process of vacuolation and enlargement, even where no differences are apparent between neighbouring cells during this phase, but this use of the same term for two different processes leads to confusion of thought and it is preferable to apply the term *maturation* to the process of cell enlargement and vacuolation, and to restrict the term differentiation to denote the origin of differences in structure and function.

The most obvious and striking differences in the states of differentiation between plant cells concern the plastids, the cell walls and the endo-membrane system. There are, no doubt, also quantitative differences between cells affecting mitochondria and other organelles, although it seems probable that the main metabolic pathways involving respiration, intermediary metabolism and protein and nucleic acid turnover will be active in all living cells. It is evident that there must be great differences between cells and tissues in other aspects of metabolism affecting photosynthesis and the synthesis of starch, storage proteins and secondary metabolites. There are also certain overall differences between shoots and root tissues, the latter being heterotrophic with respect to β-vitamins, such as thiamine; shoots and roots also show differences in their capacity for the synthesis of certain amino acids (Pate, 1966). Although it is taken for granted that cells and tissues differ markedly in their metabolic capacities and enzymic activities, there is remarkably little precise information on the subject, but the examples of differences between tissues with respect to specific enzymes given in Table 2.1 (p. 10) provide some evidence for such specialization. It is clear that in addition to the visible manifestations of differentiation involving the cell wall and plastids, there is also *biochemical differentiation* at the molecular level. Thus, differences in gene expression are manifest not only during the processes of cell differentiation and maturation, but are maintained also in mature cells and tissues.

It seems probable that the mechanism controlling the stable pattern of active enzymes characteristic of a given type of cell (i.e. 'selective gene expression') is the same in both mature, non-dividing cells in the intact plant and in dividing cells in aseptic culture or in a meristem; if dividing cells retain their special enzymic complement through a number of divisions we say that they are 'determined', but if it could be shown that non-dividing, mature tissues retain their enzymic characteristics when transferred to a different environment (e.g. by grafting) then it would seem equally legitimate to say that they were determined.

We do not know how the stable differences involved in biochemical differentiation are maintained, but it is not difficult to postulate feedback loops in metabolic pathways in which the product of an enzyme-controlled reaction induces its own synthesis. In the absence of exogenous inducer the enzyme will not be produced, but temporary contact with the inducer will cause the production of the enzyme and of its product, so that the loop becomes permanently 'locked' in the active condition. There is evidence suggesting that the biosynthesis of plant hormones may involve such types of loop (Wareing & Phillips, 1978). For example, exogenous gibberellin will stimulate the germination of certain dormant seeds, which give rise to seedlings able to synthesize gibberellin. The stimulation of endogenous auxin synthesis by a single application of exogenous IAA in fruit set may provide a similar example.

In addition to the evidence for tissue-specific enzymes referred to above, there is also evidence for the occurrence of organ-specific antigenic determinants in plants (Raff *et al.*, 1979). Stems, leaves, pistils and anthers of *Prunus avium* examined by immunological methods were found to share some antigenic determinants while others were unique to a particular organ. Of special interest to the present discussion was the finding that callus cells derived from different organs shared some determinants while others were specific. *The parental organ antigens were still expressed in callus cells after four subcultures.* This observation provides striking evidence not only for the occurrence of organ-specific proteins, but also for the continued transmission of certain tissue or organ-specific characteristics when cell division is induced in mature, differentiated cells. It would appear that the induction of cell-division in mature cells does not necessarily completely wipe out the previous 'biochemical differentiation' of the cells, so that the dividing cells retain some of the characteristics of the parent cells. The requirement for cytokinins and auxin of callus derived from tobacco pith (p. 531) would seem to afford another example of the carry-over from a previously differentiated state to newly-dividing cells.

19.8 COMPETENCE

The concept of 'competence' originated from studies of animal development, in which it is applied to the ability of cells to respond to 'induction' at certain stages of embryo development. In plant development the term is often used to refer to the ability of different cells or tissues to respond in a characteristic and specific response to the same hormonal signal. Indeed, the wide spectrum of effects of a growth substance, such as IAA, in different tissues of a plant can only be attributed to differences in the competence or 'programming' of the various target tissues. It is clear that 'competence' is closely allied to determination in the sense that in both phenomena a given tissue is committed to a specific pattern of differentiation in advance of its expression; competence can indeed be regarded as a special aspect of determination.

At present we have no idea how the specific competence of a given cell is controlled. It is generally assumed that a hormone binds with a receptor, and Raven and Rubery (Chapter 3) suggest that 'the differentiation of a particular cell may be seen as reflecting that cell's complement of receptors and receptor-mediated events' (p. 37). However, to account for the specificity of a given cell's competence, we would have to postulate that either there are specific receptors for the same hormone in different target cells or, alternatively, that there is a single type of receptor (for a given hormone) and that the specificity lies in the subsequent events initiated by the hormone-receptor complex. At present it is not possible to decide between these two alternative hypotheses, although the second would seem to be the more likely; in which case, the molecular basis for competence is probably the same as that for determination in general, although its nature remains totally unknown. Competence clearly represents an aspect of selective gene expression, and a decade ago we should not have hesitated to suggest that competence involves selective gene activation and expression, but speculation would be unprofitable in the present state of uncertainty regarding the control of gene expression in eukaryotes.

The 'target cell' concept has been discussed by Osborne and her co-workers in relation to the effects of ethylene on abscission in *Phaseolus vulgaris* (Wright & Osborne, 1974) and *Sambucus nigra* (Osborne & Sargent, 1976). Separation occurs only at well defined junctions between two classes of cells; (1) cells *distal* to the abscission zone, which expand in response to auxin, and which senesce rather readily, and (2) cells *proximal* to the abscission zone, which do not readily senesce and which enlarge in response to ethylene rather than auxin. A single row of the cells on the proximal side are formed at the pulvinus-petiolar junction during the development of the stem and leaf. The formation of a separation layer is stimulated by ethylene formed when the leaf becomes senescent. Ethylene stimulates the expansion of the proximal layer of cells, and this is accompanied by nucleic acid and protein synthesis, and increased activity of cellulase. The significant point for the present discussion is the formation, during shoot development, of a special layer of target cells which have the competence to respond to ethylene in a manner different from that of neighbouring cells. Osborne (personal communication) has recently shown that in *Sambucus* the target proximal cells of the abscission layer can be detected immunologically, by their different antigenic properties, in their quiescent state before the hormone-induced response.

Other examples of target cells would appear to be provided by (1) the pericycle, which in many species responds to high auxin levels by initiating adventitious roots, and (2) the aleurone layer of cereals, such as barley, which responds to gibberellin by secreting α-amylase and other enzymes.

These examples of target cells appear to represent special cases of differentiation in which the cell is determined with respect to certain hormone-induced processes, but in which the response is delayed for some unknown reason. In most types of cell, hormone-induced aspects of differentiation occur during cell maturation, but in target cells certain hormone-induced changes are delayed until long after the main phase of maturation has been completed.

It would appear, from the discussion in this and the two preceding sections, that determination may occur in: (1) cells which are undergoing repeated cell division; (2) cells which are about to undergo maturation and differentiation; (3) resting cells which have undergone differentiation; and (4) target cells which have undergone vacuolation, but which are capable of further developmental changes. These ideas are summarized in Fig. 19.1.

At first sight, the concept of determination might seem incompatible with that of totipotency of plant cells. However, it is not suggested that a given cell is at the *same time* both determined and totipotent, but that before a cell can exhibit totipotency in regeneration any previous state of determination must be reversed. Regeneration of buds, roots or embryoids does not normally take place from apical meristem cells, but from cells which had previously undergone vacuolation and maturation, but which undergo 'de-differentiation' with the resumption of meristematic activity, as in the formation of adventitious roots in pericyclic tissue, or adventitious buds by parenchymatous or callus tissue. It is

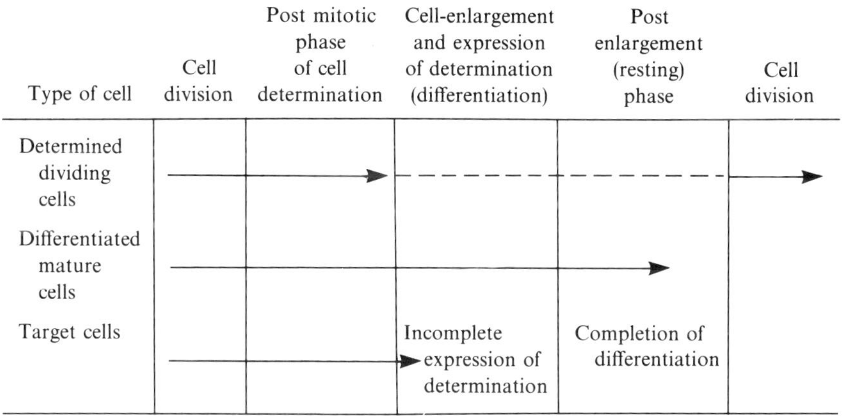

Fig. 19.1 Suggested scheme to illustrate differences in the time of expression of determination between various cell types.

suggested that in cells which have undergone maturation and subsequent dedifferentiation, any previous state of determination which existed in the cells of the meristem from which they were derived would be obliterated or reversed.

It might also be objected that determination is not consistent with the high degree of plasticity of plants, on which Trewavas lays emphasis (p. 15 *et seq.*). However, these two concepts are not incompatible since, for example, both juvenile and adult shoots of woody plants will be modified by environmental factors. Thus, although a mature beech tree is capable of initiating flowers, whether it does so in any one year depends upon the weather, whereas a juvenile tree is incapable of flowering even under favourable environmental conditions.

19.9 GROWTH HORMONES AND DIFFERENTIATION

It has been argued above that the various specific responses to a given hormone in different parts of the plant are determined by the competence of the target cells, so that in such instances the hormone itself does not control the pattern of differentiation. However, there are a few types of response where there does, indeed, appear to be evidence that the pattern of differentiation can be controlled by growth substances, notably in the induction of bud and root apical meristems in callus tissue by auxins and cytokinins (Skoog & Miller, 1957). In this frequently-cited example the pattern of regeneration can be regulated by the relative concentrations of auxin and cytokinin in the medium, high cytokinin/auxin ratios promoting bud development and high auxin/cytokinin ratios root regeneration. It may well be that the natural regeneration of bud and root meristems is controlled by endogenous auxin/cytokinin ratios, as suggested by Vardjan and Nitsch (1961).

Growth substances, particularly IAA, also appear to play an important role in the regeneration of vascular tissue (Jacobs, 1952; Sachs, 1973), in cambial activity and in the differentiation of xylem cells (Roberts, 1976). However, although IAA is undoubtedly essential for cambial activity, other factors, including cytokinins, also appear to be involved in xylem differentiation (Savidge & Wareing, 1981).

There is some evidence that cambial derivatives show a 'competence' to produce either xylem on the centripetal side or phloem on the centrifugal side. Thus, Siebers (1971) carried out experiments in which small pieces of young interfascicular tissue were cut out of the hypocotyls of *Ricinus communis* before a cambium had developed and replaced in a reversed orientation. Later examination showed that cambial activity in the reversed pieces had led to the production of xylem on the outside and phloem on the inside, suggesting that although auxin is essential for the differentiation of xylem cells, the normal production of xylem on the centripetal side is determined by the competence of the cambial derivatives there. Since a cambium had not yet developed at the time that the interfascicular tissue pieces were excised and reversed, Siebers postulated that this tissue had already acquired a radial polarity, which was determined at the time of formation of the primary meristem ring. It is possible to identify a ring of smaller parenchyma cells in the hypocotyl which is derived from the primary meristem ring and in which the cambium is later formed. Because of its determined state he suggests that this should be called the 'precambial layer'.

Although it is possible to induce xylem differentiation in callus tissue by the application of growth substances, it should be borne in mind that only a small proportion of cells in an unorganized mass of callus do, in fact, differentiate into vascular tissue in response to growth substances, suggesting that other factors, including possibly the competence of the cells, are important.

The finding that sex-expression in monoecious and dioecious plants is affected by application of growth substances is sometimes cited as an example of hormonal control of differentiation. The production of female flowers is promoted by auxin in *Cannabis sativa* (Heslop-Harrison, 1956) and in *Cucumis sativus* the development of female flowers is promoted by ethylene and of male flowers by gibberellin (Galun, 1959). However, there is no reason to believe that the growth substances are regulating the initiation of stamen or gynoecium primordia, but almost certainly they are regulating the further growth and development of such primordia, after initiation.

Thus, the only clear-cut example of the pattern of differentiation being directly controlled by growth hormones appears to be the initiation of bud and root meristems in callus tissue. The transformation of a vegetative into a flowering apex affords another example of a drastic change in gene expression which, in some species, appears to be under hormonal control, but unfortunately we do not know the nature of the flowering stimulus.

There are a number of examples in which appearance of a hormone results in the synthesis of a specific enzyme, as in the stimulation of α-amylase synthesis by gibberellic acid (Filner & Varner, 1967) or of cellulase synthesis by auxin (Fan & Maclachlan, 1967). The hormone stimulates the production of specific mRNAs for such enzymes (Higgins *et al.*, 1976; Verma *et al.*, 1975), but there is no evidence to suggest that the hormone directly determines which parts of the genome are expressed in a given cell or tissue. There are only five main classes of endogenous growth substances recognized at present, whereas during the life-cycle of the plant, differentiation involves the selective regulation of thousands of genes, and it seems unlikely that the same small number of growth substances could specifically regulate the expression of so many genes. However, it is possible that the establishment of the major developmental pathways, such as those involved in bud and root initiation may be regulated by certain master genes controlling the activities of a large number of subordinate genes (Britten & Davidson, 1969), which become expressed during the subsequent stages of differentiation, and that hormones regulate the expression of such master genes.

It is interesting to note that the conclusion that, in general, the pattern of hormone-induced responses is controlled by the competence of the responding tissue is paralleled by the conclusion of Johnson (p. 392) that the pattern of gene expression in photomorphogenesis is apparently not controlled by phytochrome directly but, here also, by the competence of the tissues.

19.9.1 THE BASIS OF STABILITY OF APICAL MERISTEMS

The division of the plant body into shoot and root is one of the most striking aspects of plant differentiation and one which, in higher plants, is normally maintained with great stability throughout the life of the plant. This division of the plant body originates in the embryo with the establishment of polarity and the initiation of shoot and root apical meristems, and once established the polarity of the plant body is extremely stable and difficult, if not impossible, to reverse.

The stability of root meristems is evidently intrinsic to the organs themselves, since isolated roots of tomato and other species can be maintained in aseptic culture for many years without the appearance of any buds. Apical meristems show strong homeostatic properties, as shown by the fact that if their structure is disturbed by surgical treatments, the normal structure is rapidly restored by compensatory changes. The normal structures of shoot and root apices thus appear to represent stable, steady-state configurations.

Not only do apical meristems, of both shoot and root, show a general stability, but they also preserve their specific identity as shoot or root apices, as shown by the fact that there appears to be only one well-authenticated case of conversion of a root apex into a shoot apex in higher plants (Ballade, 1970) and no valid report of the transformation of a shoot apex directly into a root apex (Peterson, 1975).

There are thus two aspects of the stability of apical meristems, *viz.* (i) the continuity of their structure and organization as meristems, and (ii) the maintenance of their specific identity as shoot or root apices.

Our present understanding of the properties of apical meristems is based upon various types of experimental approach, especially the results of surgical experiments (Wardlaw, 1965). The present consensus of opinion appears to be that meristematic cells of both shoots and root apices are *totally uncommitted* and do not differ from each other in their 'developmental potential'. It is also widely accepted that the apex is a self-organizing region and that the pattern of differentiation occurring in the shoot apical region is controlled by the properties of the apex itself, independently of influences from existing, older differentiated tissues (Wardlaw, 1965).

These two current concepts (non-commitment of the promeristem cells and independence of influence from older tissues) would seem to be incompatible (Wareing, 1979), since if we accept both these concepts, then the differing properties of the two types of apex must depend solely upon their different structure and organization, and it is then difficult to see how this difference could be maintained so tenaciously. There would seem to be only two possible alternative hypotheses to account for the stability of the shoot and root apices and the maintenance of the differences between them: namely, either (1) the meristematic cells of shoot and root apices *are* uncommitted, and development into either shoot or root pathways is controlled by influences from older cells which have undergone partial or complete differentiation; or (2) the meristematic cells are *inherently* programmed into either root or shoot channels and are not totally uncommitted, so that the different patterns of organization in shoot and root apices are a *consequence* of the inherent differences in their constituent cell. On this latter view, meristem cells are, in fact, specialized cells just as gametes are specialized cells.

These two alternative hypotheses will now be examined in more detail.

19.9.2 INFLUENCES FROM OLDER, MATURE TISSUES

There has long been dispute as to whether apical meristems are self-determining regions or whether their structure and properties are controlled by influences proceeding acropetally from the older, mature regions, in which the morphological and anatomical patterns are already established (for a full discussion of this subject see Halperin, 1978).

Careful study of the inception of vascular tissue in shoots has established that the procambial strands develop acropetally and, in a number of species, including several gymnosperms, procambial tissue can be recognized in the region of the apical zone, above the positions of the youngest leaf primordia i.e. in some species the development of procambial tissue precedes the appearance of the leaf primordia with which it will later be associated. A similar acropetal pattern of protophloem development is found in roots. On these grounds Esau (1954) has argued that the pattern of initiation of leaf primordia at the shoot apex is regulated by influences from the older tissues.

On the other hand, experimental approaches by various workers have led to opposite conclusions. From the results of various surgical experiments it was concluded that the apex is a self-determining region, in which the patterns of morphogenesis and tissue differentiation are not controlled by influences from adjacent, older tissues (Wardlaw, 1965; Torrey, 1965). More recently, however, Larson (1975) has carried out a very detailed study of the pattern of initiation of procambial strands in the shoot apex of *Populus* and has reaffirmed the observations of earlier workers regarding development of procambial strands before the appearance of the associated leaf primordia; he considers that his observations indicate that the development of procambial tissue controls the pattern of leaf initiation.

There is also evidence that hormonal factors produced in the older tissues may influence the activity of apical meristems. Thus, there is strong circumstantial evidence that in some species the transition of the shoot apex from the vegetative to the floral condition may be regulated by a stimulus produced in the mature leaves.

It is possible that endogenous growth hormones also play an important role in the homeostasis of apical meristems. It is well-known that root initiation is promoted by high auxin levels, not only in the formation of adventitious roots on stems but also in callus cultures derived from tobacco pith, in which high auxin/cytokinin ratios produce roots and high cytokinin/auxin ratios produce buds (Skoog & Miller, 1957). It seems possible that not only are root apices initiated under conditions of high auxin/cytokinin concentrations, but that a continued supply of auxin from neighbouring tissues is required also to *maintain* the structure and organization of a root apex, since it is now well-established that the transport of IAA in the root apex is acropetal i.e. from older tissue towards the meristem (Batra *et al.*, 1975). Thus, it would seem that the differentiating or mature tissues do, indeed, supply auxins to the root meristem.

Since high cytokinin/auxin ratios promote bud regeneration in certain callus cultures, there is the possibility that bud initiation and bud growth may depend upon the supply of cytokinins from other parts of the plant. There is, indeed, evidence that young, expanding leaves may supply both auxin and cytokinin to the apical dome of shoot apices (Smith & Murashige, 1970; Shabde & Murashige, 1977).

From the evidence presented above it seems possible that the stability and homeostatic properties of root and shoot meristems may be dependent upon and controlled by the continued supply of growth hormones from the neighbouring tissues, and that the stability of apical meristems resides in the neighbouring mature tissues which produce and supply the hormones. This hypothesis does not, of course, exclude the possibility of the other types of influence from older tissues referred to above.

19.9.3 ARE THE MERISTEMATIC CELLS OF SHOOT AND ROOT APICES DETERMINED?

The conclusion that the alternative states in which the shoot apex can exist in phenomena such as phase change and vernalization represent aspects of 'determination' suggests that the differences between shoot and root apices may also reflect differences in their states of determination. From an operational standpoint, shoot and root apices do exhibit determination to the extent that they retain their identity over long periods of time. But the question is whether such determination reflects inherent differences in their constituent cells.

There is little direct evidence bearing on this question. However, Feldman and Torrey (1976) were able to isolate the quiescent centre of roots of maize and to demonstrate that the quiescent centre itself was able to give rise to new roots. Very small explants of shoot apices, consisting of only the apical dome regenerated whole shoots, provided they were supplied with auxin and cytokinin (Shabde & Murashige, 1977), but they would no doubt have included cells and tissues which had begun to undergo vacuolation and maturation. The most direct way of testing whether the cells of meristems are intrinsically determined would be to establish aseptic cultures from the promeristems of root and shoot apices and to compare their behaviour and growth factor requirements.

Evidence which at first sight appears to conflict with the view that shoot and root apical cells are determined along 'shoot' and 'root' pathways of development, appears to be provided by the fact that adventitious roots can develop from stem tissue and adventitious buds from roots. This evidence indicates that some mature stem and root cells are not irrevocably determined and can give rise to root and shoot apices, respectively. However, the development of adventitious roots and buds from pericycle or cortical tissue entails a degree of 'dedifferentiation' of mature cells and the renewal of meristematic activity. Thus, it is possible that during the process of maturation and subsequent dedifferentiation, the state of determination of the apical meristem cells from which they were derived may be obliterated.

It should be borne in mind that even if the meristem cells of shoot and root apices are determined as suggested, differentiation into a variety of cells and tissue types takes place within each type of meristem indicating that the cells are still undetermined at these levels of organization. The commonly accepted view that the pattern of differentiation within shoot and root apices is determined by the overall structure and organization of the apices is probably valid at this level of organization. Thus, the cells in the apical meristems of *Hedera* can be determined with respect to their general pathway of development, i.e. as juvenile or adult, and yet remain undetermined with respect to their future differentiation at the tissue and cellular level. It is this general type of determination which is postulated as possibly being involved in the differences in organization and behaviour of root and shoot apices.

As stated above, in order to account for the stability of apical meristems it seems necessary to postulate that either the meristem cells are subject to influences, possibly of a hormonal nature, from older tissues, or that the meristem cells are themselves 'determined' as shoot apical cells or root apical cells. The present available evidence does not enable us to distinguish unequivocally between these alternative hypotheses, and indeed they could both be valid. If we can assume that the differences between shoot and root apices reflect differences in the selective expression of the genome in the two organs, then we may say that the first hypothesis postulates that the differences in programming of the genome reside in the adjacent mature tissues which produce hormonal factors affecting the apical meristems, while according to the second hypothesis the dividing cells of the promeristem are themselves selectively 'programmed'.

19.10 HISTOGENESIS IN APICAL MERISTEMS

The importance of unequal or asymmetric division in establishing cell differences during the development of stomata in grasses and in various other systems is now well recognized. The first cell division in the fertilized *Fucus* egg is also unequal (Chapter 7), as it is in the zygote of higher plants (Chapter 11). It is pertinent, therefore, to consider whether unequal division plays an important role in the establishment of the regular patterns of differentiation of tissues observed in the zones of cell maturation in shoots and roots, in both lower and higher plants.

In algae, such as *Chara*, which have apical cells, and in mosses and ferns, the single apical cell divides unequally, the outer daughter cell continuing as the apical cell and the other giving rise to differentiated thallus cells after further unequal divisions. In certain mosses and a few pteridophytes, such as certain species of *Selaginella*, the various tissues of the mature stem can apparently be traced back to precise divisions of the single apical cell and its immediate derivatives. This problem has recently been studied in greater detail in the root of the aquatic fern, *Azolla*, using the electron microscope to follow the process at the ultra-structural levels (Gunning *et al.*, 1978). There is a single tetrahedral apical cell, from the three proximal faces of which derivative cells are cut off in succession. Each of these derivative cells undergoes further divisions and the pattern of subsequent divisions is extremely regular, so that the precise cell lineage leading to each cell of the mature root can be traced (see also Chapter 4). Some of these divisions are asymmetric (unequal) and others are symmetric. However, differentiation apparently does not necessarily depend upon unequal division since in some instances the daughter cells derived from an unequal division are similar, whereas in other instances symmetric divisions produce sister cells which undergo very different patterns of differentiation (e.g. outer sieve element and pericycle cells). Thus, it would appear that cell differentiation in the *Azolla* root is not necessarily preceded by an unequal division.

It is difficult to identify the initial (promeristem) cells in roots of higher plants, but it is nevertheless possible, by studying the patterns of cell division, to trace the cell lineages back to the margins of the quiescent centre (Feldman, 1977). Thus, the cell lineages which ultimately form specific tissues in the mature root (central cylinder, cortex, epidermis and root cap) can be seen to be established very close to the promeristem itself. However, we do not know whether the cells of any given lineage are already *determined* to give rise to certain tissues; that is to say, we do not know whether (1) the developmental potentialities of the early derivatives of the initial cells of the promeristem are already different and that these differences are transmitted through further divisions and so give rise to the various tissue zones of the mature root, or (2) the cell lineages are, at first, still capable of giving rise to any of the main zones of the root, and only later is the pattern of differentiation imposed on them by some unknown process.

The fact that in some roots cell lineages can apparently be traced back to specific layers of cells in the meristem region, is consistent with the Histogen Theory put forward by Hanstein in 1868 according to which three meristematic zones or histogens can be recognized in both shoot and root apices, known as plerome, periblem and dermatogen, which were held to give rise to the vascular cylinder, cortex and epidermis, respectively. Thus, according to the Histogen Theory, initiating cells of the various tissue regions are separate, and destined to give rise to different regions of the mature root. It is possible, therefore, that the quiescent centre does, indeed, contain cells of different developmental potential.

Although the histogen concept is still a tenable hypothesis for root meristems, it does not appear to be applicable to shoot apices of higher plants since the vascular tissue of leaf traces is formed from cortical cells and the pattern of development of these traces appears to cut across any potential histogen boundaries that might be held to occur in the apical region of the shoot.

Evidence of differences in the states of determination or competence of various shoot tissues to regenerate buds or roots has come from studies with aseptic cultures (Tran Thanh Van & Trinh, 1978). Cultures of monolayers of epidermis from various species have been shown to regenerate buds directly, and Tran Thanh Van states that superficial cells, in some cases including cortical parenchyma of stems and leaves, tend to regenerate buds, whereas stem segments form adventitious roots, which arise from inner tissues in the region of the pericycle or cambium. It is tempting to raise the question whether or not these inner and outer regions differing in their regenerative capacity correspond to tunica and corpus.

There is little or no evidence in the shoot apices of higher plants of unequal divisions leading to clearly defined cell lineages, such as can be seen in some of the pteridophytes, and histogenesis may be controlled by certain properties of the shoot apex as a whole, as suggested above. Certainly the shoot apex has the properties of a self-organizing region, with the capacity for self-regulation and regeneration.

19.11 PHASES IN ORGAN DEVELOPMENT

The early stages of the development of an organ of determinate growth, such as a leaf, involve polarized divisions of non-vacuolated cells, leading first to the formation of the primordium of the mid-rib, on the flanks of which develop the marginal meristems which give rise to the lamina. The further development of the lamina of a mesophytic dicotyledon involves predominantly anticlinal divisions, so that a thin lamina is formed in which there are several distinct layers of dividing cells, forming 'plate meristems'. As a result of the spatially and temporally coordinated cell divisions the form of the adult leaf is effectively established while it is still quite small.

The cessation of the cell division phase is followed by cell vacuolation and enlargement, and at the same time differentiation occurs. However, it seems probable that some tissues are already 'determined' during the cell division phase, since procambial cells can be distinguished at a very early stage of leaf development. Moreover, in many species the 'target cells' of the future abscission layer are laid down during leaf development (p. 530), although there is normally a considerable time lag between the formation of these target cells and the ultimate expression of their state of determination.

The same sequence of cell division and cell enlargement phases applies to the development of other organs, such as stem internodes, and the various parts of the flower; that is to say, the shape of the mature organ is effectively established during the meristematic phase by polarized and coordinated cell divisions, although some modification of the shape may occur during the cell vacuolation phase, by differences in the degree of cell enlargement along certain axes. Although the expression of cell differentiation is normally delayed until the phase of cell enlargement and vacuolation, it seems very probable that the pattern of differentiation is already determined before vacuolation commences.

19.12 PROBLEMS FOR THE FUTURE

Major advances in our understanding of plant development have taken place over the past 50–60 years. There have been three main types of approach. Firstly, there have been studies in what might be broadly termed 'experimental morphology', including the use of surgical techniques and more recently the use of isolated tissues and organs grown in aseptic culture which have led to important advances, including the experimental production of embryoids. Secondly, at least five major classes of plant growth substance have been discovered and their essential roles, but not their mechanism of action, in many aspects of plant growth and differentiation have been established. Thirdly, a vast body of information has been acquired about the environmental control of plant development, starting with the discovery of photoperiodism and vernalization and followed later by the isolation of phytochrome and the demonstration of its all-pervasive role in photomorphogenesis and other phenomena, again, however, with little progress on its mechanism of action. In general, it has not yet proved possible to integrate fully the knowledge gained from these three different types of approach, although growth substances have proved an important tool in the development of aseptic techniques, and some responses to environmental factors can be related to hormonal changes, e.g. in the role of ABA in plant responses to water stress.

However impressive the advances in our knowledge have been, it has to be admitted that there are certain fundamental aspects of development, in both plants and animals, of which our understanding is almost zero at the present time. The following are the major aspects of development of which a greater understanding is urgently required:

(1) the manner in which selective gene expression is achieved, both spatially and temporally, during development;
(2) the molecular basis of cell determination and cell competence;
(3) the mode of hormone action; the isolation of hormone receptors and their interaction with other processes controlling selective gene expression in differentiation;
(4) the achievement of form in developing organs and other plant parts by the regulation of the planes of cell division;
(5) the environmental control of development, including photomorphogenesis and the mode of action of phytochrome; the biochemistry of photoperiodism and vernalization;
(6) the control of flowering.

The first three topics are closely connected and satisfactory progress is unlikely to be achieved until we know how gene expression is regulated in eukaryotes, which is, therefore, the problem of highest priority. Once this is achieved, then rapid progress on the other problems can be expected.

The importance of membrane-controlled processes and electrical phenomena is indicated in such apparently diverse phenomena as polarized cell division, the establishment of cell differences and phytochrome-controlled effects, so that this approach is rightly receiving increasing attention.

The control of flowering remains a problem which still eludes a solution after some 40 years of intensive study. The lack of progress seems to indicate that our approach has been too simplistic and its ultimate solution may well reveal new and unsuspected types of control.

FURTHER READING

Doorenbos J. (1965) Juvenile and adult phases in woody plants. In: *Encyclopaedia of Plant Physiology* (Ed. Ruhland W.) Vol 15/1, 1222–35. Springer-Verlag, New York.

Halperin W. (1978) Organogenesis at the shoot apex. *Ann. Rev. Plant Physiol.*, **29**, 239–62.

Heslop-Harrison J. (1967) Differentiation. *Ann. Rev. Plant Physiol.*, **18**, 325–48.

Leaver C. J. (Ed.) (1981) *Genome Organisation and Expression in Plants.* Plenum, New York.

MEINS F. & BINNS A. (1978) Epigenetic clonal variation in the requirement of plant cells for cytokinins. In *The Clonal Basis for Development* (Eds Subtelny S. & Sussex I. M.), pp. 185–201. Academic Press, New York.

MEINS F. & BINNS A. N. (1979) Cell determination in plant development. *Bioscience*, **29**, 221–5.

SAVIDGE R.A. & WAREING P.F. (1981) Plant growth regulators and differentiation of vascular elements. In *Xylem Cell Development* (Ed. Barnett J.R.) pp. 192–234. Castle House Publications, Tunbridge Wells.

WAREING P. F. (1977) Growth Substances and Integration in the Whole Plant. In *Society for Experimental Biology Symposium No XXXI* Integration of Activity in the Higher Plant, pp. 337–65.

WAREING P. F. (1978) Determination in plant development. *Bot. Mag. Tokyo*, Special Issue, **1**, 3–18.

WAREING P. F. (1979) What is the basis of the stability of apical meristems? BPGRG Monograph No. 3—*Control of Plant Development*, 1–9.

ZIMMERMAN R. H. (Ed.) (1976) Symposium on Juvenility in Woody Perennials. *Acta Horticulturae*, **56**.

REFERENCES

ABELES F.B. (1973) *Ethylene in Plant Biology*. 302 pp. Academic Press, New York.

VON ABRAMS G.J. (1974) An effect of ornithine on degredation of chlorophyll and protein in excised leaf tissue. *Z. Pflanzenphysiol.*, **72**, 410–21.

ABROL Y.P., UPRETY D.C., AHUJA V.P. & NAIK M.S. (1971) Soil fertility levels and protein quality of wheat grains. *Aust. J. Agr. Res.*, **22**, 197–202.

ACKOFF R.L. (1980) From information to control. In *The Human Side of Information Processing*. (Ed. Bjørn-Andersen N.) North Holland Publishing Co., Amsterdam.

ACTON G.J., DRUMM H. & MOHR H. (1974) Control of synthesis *de novo* of ascorbate oxidase in the mustard seedling (*Sinapis alba* L.) by phytochrome. *Planta*, **121**, 39–50.

ADAMS C.A., WATSON T.G. & NOVELLIE L. (1975) Lytic bodies from cereals hydrolysing maltose and starch. *Phytochemistry*, **14**, 953–6.

ADAMS D.O. & YANG S.F. (1977) Methionine metabolism in apple tissue. *Plant Physiol.*, **60**, 892–6.

ADAMS D.O. & YANG S.F. (1979) Ethylene biosynthesis: identification of 1-aminocyclopropane-1-carboxylic acid as an intermediate in the conversion of methionine to ethylene. *Proc. Nat. Acad. Sci. USA*, **76**, 170–4.

AHARONI N. & LIEBERMAN M. (1979) Ethylene as a regulator of senescence in tobacco leaf discs. *Plant Physiol.*, **64**, 801–4.

AHARONI N., LIEBERMAN M. & SISLER H.D. (1979a) Patterns of ethylene production in senescing leaves. *Plant Physiol.*, **64**, 796–800.

AHARONI N., ANDERSON J.D. & LIEBERMAN M. (1979b) Production and action of ethylene in senescing leaf discs. *Plant Physiol.*, **64**, 805–9.

AIST J.R. (1976) Cytology of penetration and infection—fungi. *Encyclopaedia of Plant Physiology, New Series, Physiological Plant Pathology*, pp. 197–221. (Eds. Heitefuss R. & Williams P.H.). Springer-Verlag, Berlin.

AIST J.R., WATERMAN M.A. & ISRAEL H.W. (1979) Papillae and penetration: some problems, procedures and perspectives. *Recognition and Specificity in Plant-Host Parasite Interactions*. (Eds. Daly J.M. & Uritani I.) pp. 85–95. Japan Scientific Societies Press, Tokyo.

ALBAUM H.G., KAISER S. & NESTLER H.A. (1937) The relation of hydrogen ion concentration to the penetration of 3-indole acetic acid into *Nitella* cells. *Am. J. Bot.*, **24**, 513–18.

ALBERSHEIM P. (1978) Concerning the structure and biosynthesis of the primary cell walls of plants. *Biochemistry of Carbohydrates. International Review of Biochemistry*. Vol. 16, pp. 127–50. (Ed. Manners D.J.) University Park Press, Baltimore.

ALBERSHEIM P. & VALENT B. (1974) Host-pathogen interactions VII. Plant pathogens secrete proteins which inhibit enzymes of the host capable of attacking the pathogen. *Plant Physiol.*, **53**, 684–7.

ALBERSHEIM P. & VALENT B.S. (1978) Host-pathogen interactions in plants. *J. Cell Biol.*, **78**, 627–43.

ALBERTINI D.F. & ANDERSON E. (1977) Microtubule and microfilament rearrangements during capping of con A receptors on ovarian cells. *J. Cell. Biol.*, **73**, 111–27.

VAN ALFEN N.K. & TURNER N.C. (1975) Influence of a *Ceratocystis ulmi* toxin on water relations of elm (*Ulmus americana*). *Plant Physiol.*, **55**, 312–16.

ALLEN J.R.F., GREENWAY A.A. & BAKER D.A. (1979) Determinations of indole-3-acetic acid in xylem sap of *Ricinus communis* using mass fragmentography. *Planta*, **144**, 299–303.

ALLEN R.D., JACOBSEN L., JOAQUIN J. & JAFFE L.F. (1972) Ionic concentrations in developing Pelvetia eggs. *Developmental Biology*, **27**, 538–45.

D'AMATO F. (1952) Polyploidy in the differentiation and function of tissues and cells in plants. *Caryologia*, **iv**, 311–58.

D'AMATO F. (1959) C-mitosis and experimental polyploidy in plants. *Genetica Agraria*, **11**, 17–26.

D'AMATO F. (1964) Nuclear changes and their relationships to histological differentiation. *Caryologia*, **17**, 317–25.

D'AMATO F. (1977) *Nuclear Cytology in Relation to Development*. Cambridge University Press, Cambridge.

AMRHEIN N. (1977) The current status of cyclic AMP in higher plants. *Ann. Rev. Plant Physiol.*, **28**, 123–32.

ANDERSON A.J. (1978) Isolation from three species of Colletotrichum of glucan-containing polysaccharides that elicit browning and phytoalexin production in bean. *Phytopathol.*, **68**, 189–94.

ANDERSON A.J. & ALBERSHEIM P. (1972) Host-pathogen interactions V. Comparison of the abilities of proteins isolated from three varieties of *Phaseolus vulgaris* to inhibit the endopolygalacturonases secreted by three races of *Colletotrichum lindemuthianum*. *Physiol. Plant Pathol.*, **2**, 339–46.

ANDERSON A.J. & JASALAVICH C. (1979) Agglutination of Pseudomonad cells by plant products. *Physiol. Plant Pathol.*, **15**, 149–54.

ANDERSON G. (1979) Self-electrophoresis of membrane molecules and chromosomes: unifying hypothesis of cell cycle controls. *J. theoret. Biol.*, **77**, 1–18.

ANDERSON J.M. & CORMIER M.J. (1978) Calcium-dependent regulation of NAD kinase in higher plants. *Biochem. Biophys. Res. Commun.*, **84**, 595–602.

ANDERSON J.M. & ROWAN K.S. (1965) Activity of peptidase in tobacco leaf tissue in relation to senescence. *Biochem. J.* **97**, 741–6.

ANDERSON K.S., BAIN J.M., BISHOP D.G. & SMILLIE R.M. (1972) PS11 activity in a granal bundle sheath chloroplasts from *Zea mays*. *Plant Physiol.*, **49**, 461–6.

ANDERSON-PROUTY A.J. & ALBERSHEIM P. (1975) Carbohydrates, proteins, cell surfaces and the biochemistry of pathogenesis. *Ann. Rev. Plant Physiol.*, **26**, 31–52.

ANDRES J. & SMITH H. (1976) Evidence for a rapid effect of abscissic acid on amino acid metabolism in *Lemna*. *Plant Sci. Lett.*, **6**, 315–18.

ANS B. (1979) Formation d'une structure de type vasculaire primaire végétale dans un système de réaction-diffusion. *Flora*, **168**, 358–78.

ANSTINE W., JACOBSEN J.V., SCANDALIOS J.G. & VARNER J.E. (1970) Deuterium oxide as a density label of peroxidases in germinating barley embryos. *Plant Physiol.*, **45**, 148–52.

APEL K. (1979) Phytochrome-induced appearance of mRNA activity for the apoprotein of the light harvesting chlorophyll *a/b* protein of barley (*Hordeum vulgare*). *Eur. J. Biochem.*, **97**, 183–8.

APEL K. & KLOPPSTECH K. (1978) The plastid membranes of barley. *Eur. J. Biochem.*, **15**, 581–8.

APPLEBY C.A. (1974) Leghemoglobin. In *The Biology of Nitrogen Fixation.* (Ed. Quispel A.) pp. 521–54. Elsevier Pub. Co. Inc., New York.

APPLEBY C.A., TURNER G.L. & MAANICOL P.K. (1975a) Involvement of oxyleghaemoglobin and cytochrome P-450 in an efficient oxidative phosphorylation pathway which support nitrogen fixation in *Rhizobium*. *Biochim. biophys. Acta*, **387**, 461–75.

APPLEBY C.A., NICOLA N.A., HURRELL J.G.R. & LEACH S.J. (1975b) Characterization and improved separation of soybean leghaemoglobin. *Biochemistry*, **14**, 444–50.

ARANDA G., BERVILLE A., CASSINE R., FETIZON M. & PONET B. (1978) Recherches sur les metabolites produits par les races T et O d'*Helminthosporium maydis*. *Ann. Phytopathol.*, **10**, 375–9.

ARCHER B.L. & COCKBAIN E.G. (1969) Rubber transferase from *Hevea brasiliensis* Latex. In *Methods in Enzymology* (Eds. Colowick S.P. & Kaplan N.O.) Vol. XV, pp. 476–80. Academic Press, New York & London.

ARNESON P.A. & DURBIN R.D. (1968) Studies on the mode of action of tomatine as a fungitoxic agent. *Plant Physiol.*, **43**, 683–6.

ASADA Y., OLIGUCHI T. & MATSUMOTO I. (1979) Induction of lignification in response to fungal infection. *Recognition and Specificity in Plant-Host Parasite Interactions.* (Eds. Daly J. M. & Uritani I.) pp. 99–112. Japan Scientific Societies Press, Tokyo.

ASH J.F., LOUVARD D. & SINGER S.J. (1977) Antibody induced linkages of plasma membrane proteins to intracellular actomyosin-containing filaments in cultured fibroblasts. *Proc. Nat. Acad. Sci. USA*, **74**, 5584–8.

ASHFORD A.E. & JACOBSEN J.V. (1974) Cytochemical localization of phosphate in barley aleurone cells: the pathway of gibberellic acid-induced enzyme release. *Planta*, **120**, 81–105.

ASHTON N.W., GRIMSLEY N.H. & COVE D.J. (1979) Analysis of gametophytic development in the moss, *Physcomitrella patens* using auxin and cytokinin resistant mutants. *Planta*, **144**, 427–35.

ASPART L., COOKE R., MICHAUX-FERRIÈRE N. & DELSENY M. (1978) Ribosomal RNA synthesis in imbibing radish (*Raphanus sativus*) embryo axes. *Planta*, **148**, 17–23.

ATKINSON A.W., GUNNING B.E.S. & JOHN P.C.L. (1972) Sporopollenin in the cell wall of Chlorella and other algae: ultrastructure, chemistry and incorporation of ^{14}C-acetate, studied in synchronous cultures. *Planta*, **107**, 1–32.

ATKINSON A.W. JR., JOHN P.C.L. & GUNNING

B.E.S. (1974) The growth and division of the mitochondrion and other organelles during the cell cycle of *Chlorella*, studied by quantitative stereology and three-dimensional reconstruction. *Protoplasma*, **81**, 77–109.

Attridge T.H. (1974) Phytochrome-mediated synthesis of ascorbic acid oxidase in mustard cotyledons. *Biochim. biophys. Acta*, **362**, 258–65.

Attridge T.H. & Smith H. (1973) Evidence for a pool of inactive phenylalanine ammonialyase in *Cucumis sativus* seedlings. *Phytochem.*, **12**, 1569–74.

Attridge T.H. & Smith H. (1974) Density-labelling evidence for the blue-light-mediated activation of phenylalanine ammonia-lyase in *Cucumis sativus* seedlings. *Biochim. biophys. Acta*, **343**, 452–64.

Audus L.J. (1962) The mechanism of perception of gravity by plants. *Symp. Soc. Exp. Biol.*, **16**, 197–226.

Auger S., Baulcombe D. & Verma D.P.S. (1979) Sequence complexities of the poly(A)-containing mRNA in uninfected soybean root and the nodule tissue developed due to the infection by *Rhizobium*. *Biochim. biophys. Acta*, **503**, 496–507.

Auger S. & Verma D.P.S. (1980) Induction and expression of nodule-specific host genes in effective and ineffective root nodules of soybean. *Biochemistry*, (in press).

Avanzi S., Maggini F. & Innocenti A.M. (1973) Amplification of ribosomal cistrons during the maturation of metaxylem in roots of *Allium cepa*. *Protoplasma*, **76**, 197–210.

Avissar V.J. & Nadler K.D. (1978) Stimulation of tetrapyrrole formation in *Rhizobium japonicum* by restricted aeration. *J. Bacteriol.*, **135**, 782–9.

Bach M.K., Magee W.F. & Burris R.H. (1958) Translocation of photosynthetic products of soybean nodules and their role in nitrogen fixation. *Plant Physiol.*, **33**, 118–24.

Bachrach U. (1973) *Function of Naturally Occurring Polyamines*. Academic Press, New York.

Backs-Husemann D. & Reinert H. (1970) Embryobildung durch isolierte Einzelzellen aus Gewebekulturen von *Daucus carota*. *Protoplasma*, **70**, 49–60.

Bailey C.J., Cobb A. & Boulter D. (1970) A cotyledon slice system for the electron autoradiographic study of the synthesis and intracellular transport of the seed storage protein of *Vicia faba*. *Planta*, **95**, 103–18.

Bailey I.W. (1920) The cambium and its derivative tissues. III. A reconnaissance of cytological phenomena in the cambium. *Am. J. Bot.*, **7**, 417–34.

Bain J.M. & Mercer F.V. (1966a) Subcellular organization of the developing cotyledons of *Pisum sativum* L. *Aust. J. biol. Sci.*, **19**, 49–68.

Bajer A. (1968) Fine structure studies on phragmoplast and cell-plate formation. *Chromosoma*, **24**, 383–417.

Bajer A. (1973) Interaction of microtubules and the mechanism of chromosome movement (zipper hypothesis). I. General principle. *Cytobios.*, **8**, 139–60.

Bajer A. & Jensen C. (1969) Detectability of mitobic spindle microtubules with the light and electron microscopes. *J. de Microscopie*, **8**, 343–54.

Baker B.S., Carpenter A.T.C., Esposito R.E. & Sandler L. (1976) The genetic control of meiosis. *Ann. Rev. Genet.*, **10**, 53–134.

Bal A.K., Shantharam S. & Ratnam S. (1978) Ultrastructure of *Rhizobium japonicum* in relation to its attachment to root hairs. *J. Bact.*, **133**, 1393–400.

Bal A.K., Shantharam S. & Verma D.P.S. (1980) Changes with outer cell wall of *Rhizobium* during development of root nodule symbiosis in soybean. *Can. J. Microbiol.*, (in press).

Bal A.K., Verma D.P.S., Byrne H. & McLachlan G.A. (1976) Subcellular localization of cellulases in auxin-treated pea. *J. Cell Biol.*, **69**, 97–105.

Balinsky B.I. (1970) *An Introduction to Embryology*. W.B. Saunders, London.

Ballade P. (1970) Précisions nouvelles sur la caulogenèse apicale des racines du cresson (*Nasturtium officinale* R.Br.) *Planta* (Berl.), **92**, 138–45.

Ballio A., Brufani M., Casinovi C.G., Cerrini S., Fedeli W., Pellicciari R., Santurbano B. & Vaciago A. (1968) The structure of fusicoccin A. *Experientia*, **24**, 631–5.

Barash A.N.I., Zolel Z. & Strobel G.A. (1977) Purification and characterization of a phytotoxin produced by *Phoma tracheophila*, the causal agent of mal secco disease of citrus. *Physiol. Plant Pathol.*, **10**, 147–57.

Bard J. & Lauder I. (1974) How well does Turing's theory of morphogenesis work? *J. theoret. Biol.*, **45**, 501–31.

Barlow K.K., Lee J.W. & Vesk M. (1974) Morphological development of protein bodies in wheat. In The Royal Society of New Zealand, Bulletin 12, pp. 793–7. (Eds. Bieleski R.L., Ferguson A.R. & Cresswell M.M.)

Barlow P.W. (1974) Recovery of geotropism after removal of the root cap. *J. exp. Bot.*, **25**, 1137–46.

BARLOW P.W. (1975) The root cap. *The Development and Function of Roots.* (Eds. Torrey J.G. & Clarkson D.T.) pp. 21–54. Academic Press, London, New York & San Francisco.

BARLOW P.W. (1976a) Towards an understanding of the behaviour of root meristems. *J. theoret. Biol.*, **57**, 433–51.

BARLOW P.W. (1976b) The effect of ethylene on root meristems of *Pisum sativum* and *Zea mays. Planta*, **131**, 235–43.

BARLOW P.W. (1977a) The dynamic aspect of cytodifferentiation of cells. *Naturwissenschaften*, **64**, 532–3.

BARLOW P.W. (1977b) An experimental study of cell and nuclear growth and their relation to cell diversification within a plant tissue. *Differentiation*, **8**, 153–7.

BARLOW P.W. (1977c) New patterns of vascular development in roots of *Pisum* recovering from colchicine treatment. *Experientia*, **33**, 1153–4.

BARLOW P.W. (1978a) The concept of the stem cell in the context of plant growth and development. *Stem Cells and Tissue Homeostasis.* (Eds. Lord B.I., Potten C.S. & Cole R.J.) pp. 87–113. Cambridge University Press, Cambridge.

BARLOW P.W. (1978b) RNA metabolism in the quiescent centre and neighbouring cells in the root meristem of *Zea mays. Z. Pflphysiol.*, **86**, 147–57.

BARLOW P.W. & GRUNDWAG M. (1974) The development of amyloplasts in cells of the quiescent centre of *Zea* roots in response to removal of the root cap. *Z. Pflphysiol.*, **73**, 56–64.

BARLOW P.W. & MACDONALD P.D.M. (1973) An analysis of the mitotic cell cycle in the root meristem of *Zea mays. Proc. Roy. Soc. Lond.* B **183**, 385–98.

BARLOW R.B. & DIXON R.O.D. (1973) Choline Acetyltransferase in the nettle *Urtica dioica. Biochem. J.*, **132**, 15–18.

BARSTOW W.E. & LOVETT J.S. (1978) Ultrastructure of a reduced developmental cycle (minicycle) in *Blastocladiella emersonii. Exp. Mycol.*, **2**, 145–55.

BARTON R. (1966) Fine structure of mesophyll cells in senescing leaves of *Phaseolus. Planta*, **71**, 314–25.

BASHA S.M.M. & BEEVERS L. (1976) Glycoprotein metabolism in the cotyledons of *Pisum sativum* during development and germination. *Plant Physiol.*, **57**, 93–7.

BASSETT B., GOODMAN R.N. & NOVACKY A. (1977a) Ultrastructure of soybean nodules. I. Release of Rhizobia from the infection thread. *Canad. J. Microbiol.*, **23**, 573–82.

BASSETT B., GOODMAN R.N. & NOVACKY A. (1977b) Ultrastructure of soybean nodules. II: deterioration of symbiosis in ineffective nodules. *Canad. J. Microbiol.*, **23**, 873–83.

BATEMAN D.F. (1976) Plant cell wall hydrolysis by pathogens. *Biochemical Aspects of Plant-Parasite Relationships.* (Eds. Friend J. & Threlfall D.R.) pp. 79–103. Academic Press, London.

BATEMAN D.F. & BASHAM H.G. (1976) Degradation of plant cell walls and membranes by microbial enzymes. *Encyclopaedia of Plant Physiology, New Series, Physiological Plant Pathology.* (Eds. Heitefuss R. & Williams P.H.) pp. 316–55. Springer-Verlag, Berlin.

BATRA M.W., EDWARDS K.L. & SCOTT T.K. (1975) Auxin transport in roots: Its characteristics and relationship to growth. In *The Development and Function of Roots*, pp. 299–326. Academic Press, London.

BATT T. & WOOLHOUSE H.W. (1975) Changing sites of activity during senescence and sites of synthesis of photosynthetic enzymes in leaves of the labiate, *Perilla frutescens* L. *Brit. J. exp. Bot.*, **26**, 569–79.

BAUER W.D. (1977) Plant cell walls. *The Molecular Biology of Plant Cells.* (Ed. Smith H.) pp. 6–23. Blackwell Scientific Publications, Oxford.

BAUER W.D., TALOMADGE K.W., KEEGSTRA K. & ALBERSHEIM P. (1973) The structure of plant cell walls II. The hemicellulose of the walls of suspension-cultured sycamore cells. *Plant Physiol.*, **51**, 174–87.

BAULCOMBE D. & VERMA D.P.S. (1978) Preparation of a complimentary DNA for leghaemoglobin and direct demonstration that leghaemoglobin is encoded by the soybean genome. *Nucl. Acid. Res.*, **5**, 4141–53.

BAUMGARTNER B. & CHRISPEELS M.J. (1977) Purification and characterisation of vicilin peptidohydrolase, the major endopeptidase in the cotyledons of mung bean seedlings. *Eur. J. Biochem.*, **77**, 223–33.

BAUMGARTNER B., TOKUYASU K.T. & CHRISPEELS M.J. (1978) Localization of vicilin peptidohydrolase in the cotyledons of mung bean seedlings by immunofluorescence microscopy. *J. Cell Biol.*, **79**, 10–19.

BAUMGARTNER B., TOKUYASU K.T. & CHRISPEELS M.J. (1980) Immunocytochemical localization of reserve proteins in the endoplasmic reticulum of developing bean (*Phaseolus vulgaris*) cotyledons. *Planta*, **150**, 419–25.

BAYLEY S.T., COLVIN J.R., COOPER F.P. &

MARTIN-SMITH C.A. (1957) The structure of the primary epidermal cell wall of the *Avena* coleoptile. *J. Biochem. Biophys. Cytol.*, **3**, 171–82.

BEACHY R.N., BARTON K.A., THOMPSON J.F. & MADISON J.T. (1980) *In vitro* synthesis of the α and α′ subunits of the 7S storage proteins (conglycinin) of soybean seeds. *Plant Physiol.*, (in press).

BEACHY R.N., THOMPSON J.F. & MADISON J.T. (1978) Isolation of polyribosomes and messenger RNA active in *in vitro* synthesis of soybean seed proteins. *Plant Physiol.*, **61**, 139–44.

BEASLEY C.A. & TING I.P. (1974) Effects of plant growth substances on *in vitro* fiber development from unfertilized cotton ovules. *Am. J. Bot.*, **61**, 188–94.

BEAVERS L., LOVEYS B., PEARSON J.A. & WAREING P.F. (1970) Phytochrome and hormonal control of expansion and greening of etiolated wheat leaves. *Planta*, **90**, 286–94.

BECKER W.M. (1979) RNA polymerases in plants. In *Nucleic Acids in Plants*, Vol. 1 (Eds. Hall T.C. & Davies J.W.) pp. 111–42. CRC Press Inc., Cleveland.

BEDBROOK J.R. & KOLODNER R. (1979) The structure of chloroplast DNA. *Ann. Rev. Plant Physiol.*, **30**, 593–620.

BEDBROOK J.R., KOLODNER R. & BOGORAD L. (1977) *Zea mays* chloroplast ribosomal RNA genes are part of a 22,000 base pair inverted repeat. *Cell*, **11**, 739–49.

BEDBROOK J.R., SMITH S.M. & ELLIS R.J. (1980) Molecular cloning and sequencing of cDNA encoding the precursor to the small subunit of chloroplast ribulose-1,5-bisphosphate carboxylase. *Nature, Lond.*, **287**, 692–7.

BEER S. (1975) *Platform for Change*. John Wiley and Sons, London.

BEEVERS H. (1979) Microbodies in higher plants. *Ann. Rev. Plant Physiol.*, **30**, 159–93.

BEEVERS L. (1968) Growth regulator control of senescence in leaf discs of nasturtium (*Tropaeolum majus*). In *Biochemistry and Physiology of Plant Growth Substances*. (Eds. Wightman F. & Setterfield G.) pp. 1417–34. Runge, Ottawa.

BEEVERS L. & POULSON R. (1972) Protein synthesis in cotyledons of *Pisum sativum* L. I. Changes in cell-free amino acid incorporating capacity during seed development and maturation. *Plant Physiol.*, **49**, 476–81.

BELLIARD G., PELLETIER G., VEDEL F. & QUETIER F. (1978) Morphological characteristics and chloroplast DNA distribution in different cytoplasmic parasexual hybrids of *Nicotiana tabacum*. *Molec. gen. Genet.*, **165**, 231–7.

BELLIARD G., VEDEL F. & PELLETIER G. (1979) Mitochondrial recombination in cytoplasmic hybrids of *Nicotiana tabacum* by protoplast fusion. *Nature, Lond.*, **281**, 401–3.

BENEŠ K. & HADAČOVÁ V. (1980) The localization and the isoenzymes of α- and β-galactosidases in root tips. *Biol. Plant.*, **22**, 210–7.

BENGOCHEA T., ACASTER M.A., DODDS J.H., EVANS D.E., JERIE P.H. & HALL M.A. (1980a) Studies on ethylene binding by cell-free preparations from cotyledons of *Phaseolus vulgaris* L. II. Effects of structural analogues of ethylene and of inhibitors. *Planta*, **148**, 407–11.

BENGOCHEA T., DODDS J.H., EVANS D.E., JERIE P.H., NIEPEL B., SHAARI A.R. & HALL M.A. (1980b) Studies on ethylene binding by cell-free preparations from cotyledons of *Phaseolus vulgaris* L. I. Separation and characterisation. *Planta*, **148**, 397–406.

BENNETT J. (1977) Phosphorylation of chloroplast membrane polypeptides. *Nature, Lond.*, **269**, 344–6.

BENNETT J. (1979a) Chloroplast phosphoproteins. Phosphorylation of polypeptides of the light-harvesting chlorophyll protein complex. *Eur. J. Biochem.*, **99**, 133–7.

BENNETT J. (1979b) Chloroplast phosphoproteins. The protein kinase of thylakoid membranes is light-dependent. *FEBS Lett.*, **103**, 342–4.

BENNETT P.A. & CHRISPEELS M.J. (1972) *De novo* synthesis of ribonuclease and β-1,3-glucanase by aleurone cells of barley. *Plant Physiol.*, **49**, 445–7.

BENTRUP F.W. (1977) Electrical events during apex regeneration in *Acetabularia mediterranea*. *Progress in Acetabularia Research*. (Ed. Woodcock C.L.F.) pp. 249–54. Academic Press, New York.

BERGER S. & SCHWEIGER H.G. (1975a) 80 S ribosomes in *Acetabularia major*. Redundancy of rRNA cistrons. *Protoplasma*, **83**, 41–50.

BERGERSEN F.J. (1957) The occurrence of a previously unobserved polysaccharide in immature infected cells of root nodule of *Trifolium ambiguum*. M. Biels and other members of the *Trifoliae*. *Aust. J. biol. Sci.*, **10**, 15–24.

BERGERSEN F.J. (1965) Ammonia—an early stable product of nitrogen fixation by soybean root nodules. *Aust. J. biol. Sci.*, **18**, 1–9.

BERGERSEN F.J. (1971) Biochemistry of symbiotic nitrogen fixation in legumes. *Ann. Rev. Plant Physiol.*, **22**, 121–40.

BERGERSEN F.J. (1977) Physiological chemistry of dinitrogen fixation by legumes: In *A Treatise in Dinitrogen Fixation*. (Eds. Hardy

R.F.W. & Silver W.S.) pp. 519–55. John Wiley & Sons, New York.

BERGERSEN F.J. & BRIGGS M.J. (1958) Studies on the bacterial component of soybean root nodules. Cytology and organization in the host tissue. *J. gen. Microbiol.*, **14**, 482–90.

BERGERSEN F.J. & GOODCHILD D.J. (1973) Cellular location and concentration of leghaemoglobin in soybean root. *Aust. J. biol. Sci.*, **26**, 741–56.

BERGMAN K., BURKE P.V., CERDÁ-OLMEDO E., DAVID C.N., DELBRÜCK M., FOSTER K.W., GOODELL E.W., HEISENBERG M., MEISSNER G., ZALOKAR M., DENNISON D.S. & SHROPSHIRE W. JR. (1969) Phycomyces. *Bacteriol Rev.*, **33**, 99–157.

BERINGER J.E. (1976) Plasmid transfer in *Rhizobium*. In *Proc. 1st int. Symp. on Nitrogen Fixation*. Vol. 2. (Eds. Newton W.E. & Nyman C.J.) pp. 358–66. Washington State University Press, Pullman.

BERJACK P. &. LAWTON J.R. (1973) Prostelar autolysis: a further example of a programmed senescence. *New Phytol.*, **72**, 625–37.

BERNS M.W., RATTNER J.B., BRENNER S. & MEREDITH S. (1977) The role of the centriolar region in animal cell mitosis. A laser microbeam study. *J. Cell Biol.*, **72**, 351–77.

BERRIDGE M.J. (1976) Calcium, cyclic nucleotides and cell division. In *Calcium in Biological Systems.* (Ed. Duncan C.J.) No. XXX, pp. 219–31. Cambridge University Press, New York and London.

DE BEUCKELEER M., DE BLOCK M., DE GREVE H., DEPICKER A., DE VOS R., DE VOS G., DE WILDE M., DHAESE P., DOBBELAERE M.R., ENGLER G., GENETELLO C., HERNALSTEENS J.P., HOLSTERS M., JACOBS A., SCHELL J., SEURINCK J., SILVA B., VAN HAUTE E., VAN MONATGU M., VAN VLIET F., VILLARROEL R. & ZAENEN I. (1978) The use of Ti-plasmid as a vector for the introduction of foreign DNA into plants. *Proc. 4th Int. Conference on Plant Pathogenic Bacteria*, pp. 115–26. Angers.

BEWLEY J.D. & BLACK M. (1978) Physiology and biochemistry of seeds in relation to germination. *Development, Germination and Growth.* Vol. 1. 306 pp. Springer-Verlag, Berlin.

BEYER E.M. (1976) A potent inhibitor of ethylene action in plants. *Plant Physiol.*, **58**, 268–71.

BHJOWANI S.S. & COCKING E.C. (1972) Isolation of protoplasts from pollen tetrads. *Nature, New Biol.*, **239**, 29.

BIETZ J.A. & WALL J.S. (1972) Wheat gluten subunits: molecular weights determined by sodium dodecylsulphate-polyacrylamide gel electrophoresis. *Cereal Chem.*, **50**, 416–30.

BIETZ J.A. & WALL J.S. (1973) Isolation and characterization of gliadin-like subunits from glutenin. *Cereal Chem.*, **50**, 537–47.

BIETZ J.A., HUEBNER F.R., SANDERSON J.E. & WALL J.S. (1977) Wheat gliadin homology revealed through N-terminal amino acid sequence analysis. *Cereal Chem.*, **54**, 1070–83.

BILLET E.E., WALLACE W. & SMITH H. (1978) A specific and reversible macromolecular inhibitor of phenylalanine ammonia-lyase and cinnamic acid-4-hydroxylase in gherkins. *Biochim. Biophys. Acta*, **524**, 219–30.

BINNS A. & MEINS F. (1973) Habituation of tobacco pith cells for factors promoting cell division is heritable and potentially reversible. *Proc. Natl. Acad. Sci. USA*, **70**, 2660–2.

BISCHOFF F., STAUFFER R. & GRAY C. (1954) Physiochemical state of circulating steroids. *Amer. J. Physiol.*, **177**, 65–8.

BISSELING T., VANDEN BOSS R.C., & VAN KAMMEN A. (1978) The effect of ammonium nitrate on the synthesis of nitrogenase and the concentration of leghaemoglobin in pea root nodule induced by *Rhizobium leguminosarum. Biochim. biophys. Acta*, **539**, 1–11.

BISSELING T., VANDEN BOS R.C., WESTSTRATE M.W., MAKKAART M.J.J. & VAN KAMMEN A. (1979) Development of the nitrogen-fixing and protein-synthesizing apparatus of bacteriods in pea root nodules. *Biochim. biophys. Acta*, **562**, 515–26.

BLAAUW O.H. & BLAAUW-JENSEN G. (1970) The phototropic responses of *Avena* coleoptiles. *Acta Bot. Neerl.*, **19**, 755–63.

BLACK M. (1969) Light-controlled germination of seeds. In *Dormancy and Survival* (Ed. Woolhouse H.W.) pp. 193–217. Symp. Soc. Exp. Biol. xxiii. Cambridge University Press.

BLAGROVE R.J., GILLESPIE J.M. & RANDALL P.J. (1976) Effect of sulphur supply on the seed globulin of *Lupinus angustifolius. Aust. J. Plant Physiol.*, **3**, 173–84.

BLAIR G.E. & ELLIS R.J. (1973) Protein synthesis in chloroplasts. I. Light-driven synthesis of the large subunit of Fraction I protein by isolated pea chloroplasts. *Biochim. biophys. Acta*, **319**, 223–34.

BLAKELY L.M., RODAWAY S.J., HOLLEN L.B. & CROKER S.G. (1972) Control and kinetics of branch root formation in cultured root segments of *Haplopappus ravenii. Plant Physiol.*, **50**, 35–42.

BLOCH R. (1965) Histological foundations of differentiation and development in plants. In *Encyclopedia of Plant Physiology* XV. (Ed.

Lang A.) (i) pp. 156–86. Springer-Verlag, Berlin.

Bogorad L. (1976) Chlorophyll biosynthesis. In *Chemistry and Biochemistry of Plant Pigments.* (Ed. Goodwin T.W.) Vol. 1. pp. 29–75. Academic Press, London.

Bogorad L., Jolly S.O., Kidd G., Link G. & McIntosh L. (1980) Organisation and transcription of maize chloroplast genes. In *Genome Organisation and Expression in Plants.* (Ed. Leaver C.J.) pp. 291–304. Plenum Press, New York.

Bohlool B.B. & Schmidt E.L. (1974) Lectins: possible basis for specificity in the *Rhizobium*-legume symbiosis. *Science* **185**, 269–71.

Bohlool B.B. & Schmidt E.L. (1976) Immunofluorescent polar tips of *Rhizobium japonicum*: Possible site of attachment or lectin binding. *Bacteriology*, **125**, 1188–94.

Böhmer H. (1958) Untersuchungen über das Wachstum und den Feinbau der Zellwände in der *Avena*-Koleoptile. *Planta*, **50**, 461–97.

Boller T. & Kemde H. (1979) Hydrolytic enzymes in the central vacuole of plant cells. *Plant Physiol.*, **63**, 1123–32.

Bollini R. & Chrispeels M.J. (1978) Characterization and subcellular localization of vicilin and phytohemogglutinin, the two major reserve proteins of *Phaseolus vulgaris* L. *Planta*, **142**, 291–8.

Bollini R. & Chrispeels M.J. (1979) The rough endoplasmic reticulum is the site of reserve-protein synthesis in developing *Phaseolus vulgaris* cotyledons. *Planta*, (in press).

Bolwell G.P., Callow J.A., Callow M.E. & Evans L.V. (1977) Cross-fertilization in fucoid seaweeds. *Nature, Lond.*, **268**, 626–7.

Bolwell G.P., Callow J.A., Callow M.E. & Evans L.V. (1979) Fertilization in brown algae II. Evidence for lectin-sensitive complementary recognition in *Fucus serratus. J. Cell Sci.*, **36**, 19–30.

Bolwell G.P., Callow J.A. & Evans L.V. (1980) Fertilization in brown algae. III. Preliminary characterization of putative gamete receptors from eggs and sperm of *Fucus serratus. J. Cell Sci.*, **43**, 209–24.

Bomhoff G., Klapwijk P.M., Kester H.C.M., Schilperoort R.A., Hernalsteens J.P. & Schell J. (1976) Octopine and nopaline synthesis and breakdown genetically controlled by a plasmid of *Agrobacterium tumefaciens. Mol. Gen. Genet.*, **145**, 177–81.

Bond L. (1948) Origin and developmental morphology of root nodules of *Pisum sativum. Bot. Gaz.*, **109**, 411–34.

Bonner J. (1934) The relation of hydrogen ions to the growth rate of the *Avena* coleoptile. *Protoplasma*, **21**, 406–23.

Bonner J.T. (1955) *Cells and Societies.* Princeton University Press, Princeton, N.J.

Bonner J. (1965) *The Molecular Biology of Plant Development.* Oxford University Press, Oxford.

Börner R. & Ramshorn K. (1968) Untersuchungen über die Veranderung der Basenzusammensetzung der RNS an unterschiedlich differenzierten Wurzelgewebe von *Vicia faba. Biol. Zbl.*, **87**, 411–8.

Borodin I. (1976) Physiologische Untersuchungen uber die der beblatterten Sprosse Atmung. *Bot. Jahrb.*, **4**, 919–37.

Borthwick H.A., Hendricks S.B., Parker M.W., Toole E.H. & Toole U.K. (1952) A reversible photoreaction controlling seed germination. *Proc. Nat. Acad. Sci. USA.*, **38**, 662–6.

Bose J.C. (1913) *Researches on Irritability in Plants.* Longmans, Green & Co., London.

Böttger M. (1978a) Levels of endogenous indole-3-acetic acid and abscisic acid during the course of the formation of lateral roots. *Z. Pflanzenphysiol.*, **86**, 283–6.

Böttger M. (1978b) The occurrence of cis, trans- and trans, trans-xanthoxin in pea roots. *Z. Pflanzenphysiol.*, **86**, 265–8.

Boulter D. (1973) Seed proteins. In *Phytochemistry*, Vol. II. (Ed. Miller L.P.) pp. 30–61. Van Nostrand Reinhold Co.,

Boulter D. (1977) Protein synthesis in the cytoplasm. *The Molecular Biology of Plant Cells.* (Ed. Smith H.) pp. 256–79. Blackwell Scientific Publications, Oxford.

Bourguignon L.Y.W. & Singer S.J. (1977) Trans-membrane interactions and the mechanism of capping of surface receptors by their specific ligards. *Proc. Nat. Acad. Sci. Wash.*, **74**, 5031–5.

Bowden L. & Lord J.M. (1977) Serological and developmental relationships between endoplasmic reticulum and glyoxysomal proteins of castor bean endosperm. *Planta*, **134**, 267–73.

Bowen M.R. & Wareing P.F. (1969) The interchange of ^{14}C-kinetin and ^{14}C-gibberellic acid between the bark and xylem of willow. *Planta*, **89**, 108–25.

Bradbeer J.W., Ireland H.M.M., Smith J.W., Rest T. & Edge H.J.W. (1974) Plastid development in primary leaves of *Phaseolus vulgaris. New Phytol.*, **73**, 263–70.

Brandt A. (1976) Endosperm protein formation during kernel development of wild-type and a high-lysine barley mutant. *Cereal Chem.*, **53**, 890–901.

Brandt A. & Ingversen J. (1978) Isolation and translation of hordein messenger RNA from wild type and mutant endosperms in barley.

Carlsberg Res. Commun., **43**, 451–69.

BRANTON D., BULLIVANT S., GILULA N.B., MOOR H., MÜHLETHALER K., NORTHCOTE D.H., PACKER L., SATIR B. & P., SPETH V., STAEHLIN L.A., STEERE R.L. & WEINSTEIN R.S. (1975) Freeze-etching nomenclature. *Science*. **190**, 54–6.

BRAWLEY S.H., WETHERBEE R. & QUATRANO R.S. (1976) Fine-structural studies of the gamete and embryo of *Fucus vesiculosus* L. (Phaeophyta) I. Fertilization and pronuclear fusion. *J. Cell Sci.*, **20**, 233–54.

BRENNER S., BRANCH A., MEREDITH S. & BERNS M.W. (1977) The absence of centrioles from spindle poles of rat kangaroo (PtK_2) cells undergoing meiotic-like reduction division *in vitro*. *J. Cell Biol.*, **72**, 368–79.

BRIARTY L.G., COULT D.A. & BOULTER D. (1969) Protein bodies of developing seeds of *Vicia faba*. *J. exp. Bot.*, **20**, 358–72.

BRIARTY L.G., HUGHES C.E. & EVERS A.D. (1979) The developing endosperm of wheat—a stereological analysis. *Ann. Bot.*, **44**, 641–58.

BRICKER T.M. & NEWMAN D.W. (1980) Quantitative changes in the chloroplast thylakoid polypeptide complement during senescence. *Z. Pflanzenphysiol. Bd.*, **98** S, 339–46.

BRIGGS W.R. (1963) Mediation of phototropic responses of corn coleoptiles by lateral transport of auxin. *Plant Physiol.*, **38**, 237–47.

BRIGGS W.R. (1964) Phototropism in higher plants. In *Photophysiology*. (Ed. Giese A.C.) pp. 223–71. Academic Press, New York.

BRIGGS W.R. (1976) The nature of the blue light photoreceptor in higher plants and fungi. In *Light and Plant Development*. (Ed. Smith, H.) pp. 7–18. Butterworths, London, Boston.

BRILL W.J. (1974) Genetics of N_2-fixing organisms. In *The Biology of Nitrogen-Fixation*. (Ed. Quispel A.) pp. 639–60. Elsevier Pub. Co. Inc., New York.

BRINK A. (1962) Phase change in higher plants and somatic cell heredity. *Q. Rev. Biol.*, **37**, 1–22.

BRISTOW J.M. (1975) The structure and function of roots in aquatic vascular plants. In *The Development and Function of Roots*. (Eds. Torrey J.G. & Clarkson D.T.). pp. 221–36. Academic Press, London.

BRITTEN R.J. & DAVIDSON E.H. (1969) Gene regulation for higher cells: a theory. *Science, N.Y.* **165**, 349–57.

BRITTEN R.J. & DAVIDSON E.H. (1979) Regulation of gene expression: possible role of repetitive sequences. *Science*, **204**, 1052–9.

BROOKER J.C., CHEUNG C.P. & MARCUS A. (1977) The physiology and biochemistry of seed dormancy and germination. In *Protein Synthesis and Seed Germination*. (Ed. Khan A.) pp. 347–56. Elsevier/North Holland Biomedical Press, New York.

BROOKER J.D., TOMASZEWSKI M. & MARCUS A. (1978) Preformed messenger RNAs and early wheat embryo germination. *Plant Physiol.*, **61**, 145–419.

BROWN C.M. & DILWORTH M.J. (1975) Ammonia assimilation by *Rhizobium* cultures and bacteroids. *J. gen. Microbiol.*, **86**, 39–48.

BROWN E.G., AL-NAJAFI T. & NEWTON R.P. (1979) Adenosine 3′:5′-cyclic monophosphate, adenylate cyclase and a cyclic AMP binding-protein in *Phaseolus vulgaris*. *Phytochemistry*, **18**, 9–14.

BROWN R. (1963) Cellular differentiation in the root. *Soc. exp. Biol. Symp.* **17**, pp. 1–18.

BROWN R. & BROADBENT D. (1950) The development of cells in the growing zones of the root. *J. Exp. Bot.*, **1**, 249–64.

BROWN R.M. & MONTEZINOS D. (1976) Cellulose microfibrils: visualization of biosynthetic and orienting complexes in association with the plasma membrane. *Proc. Nat. Acad. Sci. USA.*, **73**, 143–7.

BROWN R.M. & ROMANOVICZ D.K. (1976) Biogenesis and structure of golgi-derived cellulosic scales in *Pleurochrysis*. I. Role of the endomembrane system in scale assembly and exocytosis. *Applied Polymer Symposium*. **28**, 537–85. (1976) John Wiley & Sons, New York.

BROWN W.V. & JOHNSON C. (1962) The fine structure of the grass guard cell. *Amer. J. Bot.*, **49**, 110–15.

BRUINSMA J. (1977) Hormonal regulation of phototropism in dicotyledonous. In *Plant Growth Regulation*. (Ed. Pilet P.E.) pp. 218–25. Springer-Verlag, Berlin, New York.

VAN BRUNT J. & HAROLD F.M. (1980) Ionic control of germination in zoospores of *Blastocladiella emersonii*. *J. Bacteriol.*, **141**, 735–44.

BÜHLER B., DRUMM H. & MOHR H. (1978) Investigations on the role of ethylene in phytochrome-mediated photomorphogenesis. I. Anthocyanin synthesis. *Planta*, **142**, 109–17.

BUIATTI M. (1977) DNA amplification and tissue culture. In *Applied and Fundamental Aspects of Plant Cell, Tissue and Organ Culture*. (Eds. Reinert J. & Bajaj Y.P.S.) pp. 358–74. Springer-Verlag, Berlin, Heidelberg & New York.

BULOW J.F.W. & DOBEREINER J. (1975) Potential for nitrogen fixation in maize genotypes in Brazil. *Proc. Nat. Acad. Sci. USA.*, **72**, 2389–93.

BUNGER W. & FEIERABEND J. (1980) Capacity

for RNA synthesis in 70s ribosome—deficient plastids of heat-bleached rye leaves. *Planta*, **149**, 163–9.

Burg S.P. (1973) Ethylene in plant growth. *Proc. Nat. Acad. Sci. USA*, **70**, 591–7.

Burkard G., Canaday J., Crouse E., Guillemaut P., Imbault P., Keith G., Keller M., Mubumbila M., Osorio L., Sarantoglou V., Steinmetz A. & Weil J.H. (1980) Transfer RNAs and aminoacyl—tRNA synthesis in plant organelles. In *Genome Organisation and Expression in Plants*. (Ed. Leaver C.J.) pp. 313–20. Plenum Press, New York.

Burr B. & Burr F.A. (1976) Zein synthesis in maize endosperm of polyribosomes attached to protein bodies. *Proc. Nat. Acad. Sci. USA*, **73**, 515–9.

Burr B., Burr F.A., Rubenstein I. & Simon M.N. (1978) Purification and translation of zein messenger RNA from maize endosperm protein bodies. *Proc. Nat. Acad. Sci. USA*, **75**, 696–700.

Burström H.G. (1979) In search of plant growth paradigm. *Amer. J. Bot.*, **66**, 98–104.

Burton W.G., Grabowy C.T. & Sager R. (1979) Role of methylation in the modification and restriction of chloroplast DNA in *Chlamydomonas*. *Proc. Nat. Acad. Sci. USA*, **76**, 1390–4.

Busby C. & Gunning B.E.S. (1980) Observations on preprophase bands of microtubules in uniseriate hairs, stomatal complexes of sugarcane and *Cyperus* root meristems. *Europ. J. Cell Biol.*, (in press).

Butler R.D. & Simon E.W. (1970) Ultrastructural aspects of senescence in plants. *Adv. Gerontol. Res.*, **3**, 73–129.

Butterfass Th. (1963) Die Abhängigkeit der Plastidenvermehrung von der Reproduktion der Erbsubstanz im Kern. *Berichten der Deutschen Botanischen Gesellschaft* Vol. LXXVI (5) 123–34.

Butterfass Th. (1979) *Patterns of Chloroplast Reproduction*. A Developmental Approach to Protoplasmic Plant Anatomy. Cell Biology Monographs Vol. 6. 205 pp. Springer-Verlag, Vienna.

Byers M. & Boulton J. (1979) Effects of nitrogen and sulphur fertilizers and the yield, N and S content, and amino acid composition of the grain of spring wheat. *J. Sci. Food Agric.*, **30**, 251–63.

Byrde R.J.W. & Archer S.A. (1977) Host inhibition or modification of extracellular enzymes of pathogens. In *Cell Wall Biochemistry Related to Specificity in Host-Plant Pathogen Interactions*. (Eds. Solheim B. & Raa J.) pp. 213–38. Universitetsforlaget, Oslo.

Byrde R.J.W. & Fielding A.H. (1968) Pectin methyl*trans*-eliminase as the maceration factor of *Sclerotinia fructigena* and its significance in brown rot of apple. *J. Gen. Microbiol.*, **52**, 287–97.

Caboche M. & Bacchelerie J.P. (1977) RNA methylation and control of eucaryotic RNA biosynthesis. Effects of cycloleucine, a specific inhibitor of methylation on ribosomal RNA maturation. *Eur. J. Biochem.*, **74**, 19–29.

Caers L.I., Peumans W.J. & Carlier A.C. (1979) Preformed and newly synthesized messenger RNA in germinating wheat embryos. *Planta*, **144**, 491–6.

Cairns E., Doerfler W. & Schweiger H.G. (1978) Expression of a DNA animal virus genome in a plant cell. *FEBS Lett.*, **96**, 295–7.

Caldwell B.E. (1966) Inheritance of a strain-specific ineffective nodulation in soybean. *Crop. Sci.*, **6**, 426–8.

Caldwell B.E. & Vest G. (1968) Nodulation interactions between soybean genotypes and serogroups of *Rhizobium japonicum*. *Crop. Sci.*, **8**, 680–2.

Caldwell B.E. & Vest H.G. (1977) Genetic aspects of nodulation and dinitrogen fixation by legumes: the macrosymbiont. In *A Treatise on Dinitrogen Fixation*. Sec. III. (Eds. Hardy R.W.F. & Silver W.S.) Wiley–Interscience, London.

Callow J.A. (1974) Ribosomal RNA, fraction 1 protein synthesis, and ribulose diphosphate carboxylase activity in developing and senescing leaves of cucumber. *New Phytol.*, **73**, 13–20.

Callow J.A. (1977) Recognition, resistance and the role of plant lectins in host-parasite interactions. In *Advances in Botanical Research*. (Eds. Preston R.D. & Woolhouse H.W.) pp. 1–49. Academic Press, London, New York, San Francisco.

Callow J.A. (1981) Antifungal compounds and disease resistance in plants. In *The Use of Naturally Occurring Plant Products in Pest and Disease Control*. (Eds. Odhiambo T. & Bowers W.) (in press).

Callow J.A. (1981) Antifungal compounds and disease resistance in plants. *Curr. Adv. trop. Sci.*, **2**, (in press).

Callow J.A., Callow M.E. & Woolhouse H.W. (1972) *In vitro* protein synthesis, ribosomal RNA synthesis and polyribosomes in senescing leaves of *Perilla*. *Cell Differentiation*, **1**, 79–90.

Callow J.A., Long D.E. & Lithgow E.D. (1980) Multiple molecular forms of invertase in maize smut infections. *Physiol. Plant Pathol.*, **16**, 93–107.

Callow M.E., Evans L.V., Bolwell G.P. &

CALLOW J.A. (1978a) Fertilization in brown algae I. SEM and other observations on *Fucus serratus J. Cell Sci.*, **32**, 45–54.

CALLOW M.E., COUGHLAN S.J. & EVANS L.V. (1978b) The role of Golgi bodies in polysaccharide sulphation in *Fucus* zygotes. *J. Cell Sci.*, **32**, 337–56.

CALLOW M.E. & WOOLHOUSE H.W. (1973) Changes in nucleic acid metabolism in regreening leaves of *Perilla. J. exp. Bot.*, **24**, 285–94.

CALOW P. (1979) *Life Cycles.* Chapman and Hall, London.

CALVERT H.E., LALONDE M., BHUVANESWARI T.V. & BAUER W.D. (1978). Role of lectins in plant-microorganism interactions IV. Ultrastructural localization of soybean lectin binding sites on *Rhizobium japonicum. Canad. J. Microbiol.*, **24**, 785–93.

CAMERON-MILLS V. & INGVERSEN J. (1978) *In vitro* synthesis and transport of barley endosperm proteins: reconstitution of functional rough microsomes from polyribosomes and stripped microsomes. *Carlsberg Res. Commun.*, **43**, 471–89.

CANDE W.Z. & RAY P.M. (1976) Nature of cell-to-cell transfer of auxin in polar transport. *Planta*, **129**, 43–52.

CANTINO E.C. (1961) The relationship between biochemical and morphological differentiation in non-filamentous aquatic fungi. *Symp. Soc. gen. Microbiol.*, **11**, 243–71.

CAPLAN A.I. & ORDAHL C.P. (1978) Irreversible gene repression model for control of development. *Science*, **201**, 120–30.

CAPONETTI J.D. & STEEVES T.A. (1963) Morphogenetic studies on excised leaves of *Osmunda cinnamomea* L. Morphological studies on leaf development in sterile nutrient culture. *Canad. J. Bot.*, **41**, 545–56.

CARLIER A.R., MANICKAM A. & PEUMANS W.J. (1980) Characterization of a maturation-specific mRNA in dry mung bean embryonic axes. *Planta*, **149**, 227–33.

CARLSON P. & CHALEFF R.S. (1974) Forced association between higher plants and bacterial cells *in vitro. Nature, Lond.*, **252**, 393.

CARPITA N., SABULARSE O., MONTEZINOS D. & DELMER D.P. (1979) Determination of the pore size of cell walls of living plants. *Science*, **205**, 1144–7.

CARR D.J. (1976) Plasmodesmata in growth and development. In *Intercellular Communication in Plants.* (Eds. Gunning B.E.S. & Robards A.W.) Springer-Verlag, Berlin.

CARTWRIGHT D., LANGCAKE P., PRYCE R.J., LEWORTHY D.P. & RIDE J.P. (1977) Chemical activation of host defence mechanisms as a basis for crop protection. *Nature, Lond.*, **267**, 511–12.

CASE R.M. (1978) Synthesis, intracellular transport and discharge of exportable proteins in the pancratic acinar cell and other cells. *Biol. Rev.*, **53**, 211–354.

CASEY R. (1979a) Immunoaffinity chromatography as a means of purifying legumin from *Pisum* (Pea) seeds. *Biochem. J.*, **177**, 509–20.

CASEY R. (1979b) Genetic variability in the structure of the α-subunits of legumin from *Pisum*: a 2-dimensional gel electrophoresis study. *Heredity*, **43**, 265–72.

CASEY R. (1979) Genetic variability in the α-subunits of legumin from *Pisum*—a two-dimensional gel electrophoresis study. *Heredity*, **43**, 265–72.

CASEY R. & SANGER E. (1980) Purification and some properties of a 7S seed storage protein from *Pisum* (pea). *Biochem. Soc. Trans.*, **8**, 658.

CASHMORE A.R. (1979) Reiteration frequency of the gene coding for the small subunit of RDP carboxylase. *Cell*, **17**, 383–8.

CASHMORE A.R., BROADHURST M.K. & GRAY R.E. (1978) Cell free synthesis of leaf protein. *Proc. Nat. Acad. Sci. USA*, **75**, 655–7.

CASTLE E.S. (1933) The refractive indices of whole cells. The physical basis of the positive phototropism of *Phycomyces. J. gen. Physiol.*, **17**, 41–62.

CASTROVIEJO M., TARRAGO-LITVAK L. & LITVAK S. (1975) Partial purification and characterization of two cytoplasmic DNA polymerases from ungerminated wheat. *Nucl. Acids Res.*, **2**, 2077–90.

ČATSKÝ J., TICHÁ I. & SOLÁROVÁ J. (1976) Ontogenetic changes in the internal limitations to bean leaf photosynthesis. I. Carbon dioxide exchange and conductances for carbon dioxide transfer. *Photosynthetica*, **10**, 394–402.

CAVALIER-SMITH T. (1975) The origin of nuclei and of eukaryotic cells. *Nature, Lond.*, **256**, 463–8.

CHAFE S.C. & WARDROP A.B. (1970) Microfibril orientation in plant cell walls. *Planta*, **92**, 13–24.

CHALY N., POSSINGHAM J.V. & THOMSON W.W. (1980) Chloroplast division in spinach leaves examined by scanning electron microscopy and freeze-etching. *J. Cell Sci.*, (in press).

CHALY N.M. & SETTERFIELD G. (1975) Organization of the nucleus, nucleolus, and protein-synthesizing apparatus in relation to cell development in roots of *Pisum sativum. Canad. J. Bot.*, **53**, 200–18.

CHAMBERS R. & CHAMBERS E.L. (1961) *Explora-*

tions into the Nature of the Living Cell. Harvard University Press, Cambridge, USA.

CHEN J. & JONES M.E. (1979) Effect of 6-azuaridine on *de novo* pyrimidine biosynthesis in cultured Ehrlich ascites cells. *J. biol. Chem.*, **254**, 4908–14.

CHEN D. & OSBORNE D.J. (1970) Ribosomal genes and DNA replication in germinating wheat embryos. *Nature, Lond.*, **225**, 336–40.

CHEN K., WILDMAN S.G. & SMITH H.H. (1977) Chloroplast DNA distribution in parasexual hybrids as shown by polypeptide composition of Fraction I protein. *Proc. Nat. Acad. Sci. USA*, **74**, 5109–12.

CHENIAL G. & EVANS H.J. (1960) Physiological studies on nodule-nitrate reductase. *Plant Physiol.*, **35**, 454–62.

CHERRY J.H. (1977) Hormone Action. In *The Molecular Biology of Plant Cells.* (Ed. Smith H.) pp. 329–64. Blackwell Scientific Publications, Oxford.

CHEUNG C.P. & SUHADOLNIK R.J. (1978) Regulation of RNA synthesis in early germination of isolated wheat (*Triticum aestivum* L.) embryo. *Nature, Lond.*, **271**, 357–8.

CHIANG S-H. & GIFFORD E.M. (1971) Development of the root of *Ceratopteris thalictroides* with special reference to apical segmentation. *J. Indian bot. Soc.*, **50A**, 96–106.

CHIANG Y-L. (1970) Macro- and microscopic structure of the root of *Ceratopteris pteridoides* (Hook.) *Hieron. Taiwania*, **15**, 31–49.

CHIBNALL A.C. (1939) *Protein Metabolism in the Plant.* 306 pp. Yale University Press, New Haven.

CHILD J.J. (1975) Nitrogen fixation by a *Rhizobium* sp. in association with non-leguminous plant cell culture. *Nature, Lond.*, **253**, 350–1.

CHILTON M.D., DRUMMOND M.M., MERLO D.J., SCEALSY D., MONTOYA A.L., GORDON M.P. & NESTER E.W. (1977) Stable incorporation of plasmid DNA into higher plant cells: the molecular bases of crown gall tumerogenesis. *Cell*, **11**, 263–71.

CHILTON M.-D., DRUMMOND M.H., MERLO D.J. & SCIAKY D. (1978) Highly conserved DNA of Ti-plasmids overlaps T-DNA, maintained in plant tumors. *Nature, Lond.*, **275**, 147–9.

CHILTON M.-D., SAIKI R.K., YADAV N., GORDON M.P. & QUETIER F. (1980) T-DNA from *Agrobacterium* Ti plasmid is in the nuclear DNA fraction of crown gall tumor cells. *Proc. Nat. Acad. Sci. USA*, **77**, 4060–4.

CHOE H.T. & THIMAN K.V. (1974) The senescence of isolated chloroplasts. *Planta*, **121**, 201–3.

CHOE H.T. & THIMAN K.V. (1975) The metabolism of oat leaves during senescence. III. The senescence of isolated chloroplasts. *Plant Physiol.*, **55**, 828–34.

CHOE H.T. & THIMAN K.V. (1977) The retention of photosynthetic activity by senescing chloroplasts of oat leaves. *Planta*, **135**, 101–7.

CHOLODNY N.G. (1924) Uber die hormonale Wirkung der Organspitze bei der geotropischen Krummung. *Ber. Deut. Bot. Ges.*, **42**, 356–62.

CHOLODNY N.G. (1927) Wuchshormone und Tropismen bei den Pflanzen. *Biol. Zentrabl.*, **47**, 604–26.

CHRISPEELS M.J. (1976) Biosynthesis, intracellular transport and secretion of extracellular macromolecules. *Ann. Rev. Plant Physiol.*, **27**, 19–38.

CHRISPEELS M.J., BAUMGARTNER B. & HARRIS N. (1976) Regulation of reserve protein metabolism in the cotyledons of mung bean seedlings. *Proc. Nat. Acad. Sci.*, **73**, 3168–72.

CHRISPEELS M.J. & VARNER J.E. (1967) Gibberellic acid-enhanced synthesis and release of α-amylase and ribonuclease by isolated barley aleurone layers. *Plant Physiol.*, **42**, 398–406.

CHU N. & TEWARI K.K. (1979) Arrangement of the ribosomal RNA genes in the restriction endonuclease map of pea chloroplast DNA. Citation by Dyer T.A. & Bedbrook J.R. Genes coding for chloroplast ribosomal RNA. In *Genome Organisation and Expression in Plants.* (Ed. Leaver C.J.) pp. 305–11. Plenum Press, New York.

CHUA N.H. (1980) Synthesis of chloroplast proteins. *Hoppe-Seylers Z. für Physiol. Chemie*, **361**, 496–7.

CHUA N-H. & GILLHAM N.W. (1977) The sites of synthesis of the principal thylakoid membrane polypeptides in *Chlamydomonas reinhardii. J. Cell Biol.*, **74**, 441–52.

CHUNG S.R., DURAND R. & DURAND B. (1979) Differential cytokinin binding to diaecious plant ribosomes. *FEBS Lett.*, **102**, 211–5.

CIBA FOUNDATION (1975) The Structure and Function of Chromatin. *Ciba Foundation Symposium* **28**. Elsevier, Amsterdam.

CIHA A.J., BRENNER M.L. & BRUN W.A. (1977) Rapid separation and quantification of abscissic acid from plant tissues using high performance liquid chromatography. *Plant Physiol.*, **59**, 821–6.

CIONINI P.G., BENNICI A., ALPI A. & D'AMATO F. (1976) Suspensor, gibberellin and *in vitro* development of *Phaseolus coccineus* embryos. *Planta*, **131**, 115–7.

CLARKE A.E., ANDERSON R.L. & STONE B.A. (1979a) Form and function of arabinogalactans and arabinogalactan-proteins. *Phytochem.*, **18**, 521–40.

CLARKE A.E. & KNOX R.B. (1978) Cell recognition in flowering plants. *Quart. Rev. Biol.*, **53**, 3–28.

CLARKE K.J., MCCULLY M.E. & MIKI N.K. (1979b) A developmental study of the epidermis of young roots of *Zea mays* L. *Protoplasma*, **98**, 283–309.

CLAUSS H., LÜTTKE A., HELLMANN F. & REINERT J. (1970) Chloroplastenvermehrung in kernlosen Teilstücken von *Acetabularia mediterranea* und *Acetabularia cliftonii* und ihre Abhängigkeit von inneren Faktoren. *Protoplasma*, **69**, 313–29.

CLELAND R. & RAYLE D.L. (1977) Re-evaluation of the effect of calcium ions on auxin-induced elongation. *Plant Physiol.*, **60**, 709–12.

CLOWES F.A.L. (1958) Protein synthesis in root meristems. *J. exp. Bot.*, **9**, 229–38.

CLUTTER M., BRADY T., WALBOT V. & SUSSEX I. (1974) Macromolecular synthesis during plant embryogeny. Cellular rate of RNA synthesis in diploid and polytene cells in bean embryos. *J. Cell Biol.*, **63**, 1097–102.

COEN D.M., BEDBROOK J.R., BOGORAD L. & RICH A. (1977) Maize chloroplast DNA fragment encoding the large subunit of ribulose bisphosphate carboxylase. *Proc. Nat. Acad. Sci. USA.*, **74**, 5487–91.

COHEN R.J. (1974) Cyclic AMP levels in *Phycomyces* during a response to light. *Nature, Lond.*, **251**, 144–6.

COHEN R.J. (1978) Aberrant cyclic nucleotide regulation in a behavioural mutant of *Phycomyces*. *Plant Sci. Lett.*, **13**, 315–9.

COHEN R.J. & ATKINSON M.M. (1978) Activation of *Phycomyces* adenosine 3′,5′-monophosphate phosphodiesterase by blue light. *Biochem. Biophys. Res. Comm.*, **83**, 616–21.

COLEMAN A.W. (1978) Visualization of chloroplast DNA with two fluorochromes. *Exp. Cell Res.*, **114**, 95–100.

COLEMAN A.W. (1979) Use of the fluorochrome 4′6-diamidino-2-phenylindole in genetic and developmental studies of chloroplast DNA. *J. Cell Biol.*, **82**, 299–305.

COLLANDER R. (1949) Die Verteilung organischer Verbindungen Zwischen Äther und Wasser. *Acta. chim. Scand.*, **3**, 717–47.

COLLANDER R. (1959) Cell membranes: their resistance to penetration and their capacity for transport. *Plant Physiology: A treatise* Vol. II. (Ed. Steward F.C.) pp. 3–102. Academic Press, New York.

COLOHOUN A.J. & HILLMAN J.R. (1972) The effects of abscisic acid in senescence in leaf discs of radish, *Raphanus sativus* L. *Planta*, **105**, 213–24.

CONDE M.F., PRING D.R. & LEVINGS C.S. III. (1979) Maternal inheritance of organelle DNA's in *Zea mays-Zea perennis* reciprocal crosses. *J. Hered.*, **70**, 2–4.

COOK G.M.W. & STODDART R.W. (1973) *Surface Carbohydrates of the Eucaryotic Cell*. Academic Press, New York.

COOPER R.M., RANKIN B. & WOOD R.K.S. (1978) Cell wall degrading enzymes of vascular wilt fungi II. Properties and modes of action of polysaccharidases of *Verticillium albo-atrum* and *Fusarium oxysporum* f. sp. *lycopersici*. *Physiol. Plant Pathol.*, **13**, 101–34.

COOPER R.M. & WOODS R.K.S. (1975) Regulation of synthesis of cell wall degrading enzymes by *Verticillium albo-atrum* and *Fusarium oxysporum* f.sp. *lycopersici*. *Physiol. Plant Pathol.*, **5**, 135–56.

COTTE K.J. & DUFFAU M.L. (1977) Peptide hormone receptors. *Ann. Rev. Physiol.*, **39**, 529–57.

COVENTRY D.R. & DILWORTH M.J. (1976) Synthesis and turnover of leghaemoglobin in lupin root nodules. *Biochim. biophys. Acta*, **447**, 1–10.

CRAIG S., GOODCHILD D.J. & HARDHAM A.R. (1979a) Structural aspects of protein accumulation in developing pea cotyledons. I. Qualitative and quantitative changes in parenchyma vacuoles. *Aust. J. Plant Physiol*, **6**, 81–98.

CRAIG S., GOODCHILD D.J. & MILLERD A. (1979b) Immunofluorescent localization of pea storage proteins in glycol methacrylate embedded tissue. *J. Histochem. Cytochem.*, **27**, 1312–6.

CRAIG S., MILLERD A. & GOODCHILD D.J. (1980a) Structural aspects of protein accumulation in developing pea cotyledons. III. Immunocytochemical localization of legumin and vicilin using antibodies shown to be specific by the enzyme-linked immunosorbent assay (ELISA). *Aust. J. Plant Physiol.*, **7**, 339–51.

CRAIG S., GOODCHILD D.J. & MILLER C. (1980b) Structural aspects of protein accumulation in developing pea cotyledons. III. Three dimensional reconstructions of vacuoles and protein bodies from serial sections. *Aust. J. Plant Physiol.*, **7**, 329–37.

CRAIG S. & MILLERD A. (1981) Pea seed storage proteins—immunocytochemical localization with protein A-gold by electron microscopy. *Protoplasma*, **105**, 333–9.

CRAN D.G. & POSSINGHAM J.V. (1972a)

Variation of plastid types in spinach. *Protoplasma*, **74**, 345–56.

CRAN D.G. & POSSINGHAM J.V. (1972b) Two forms of division profile in spinach chloroplasts. *Nature New Biol.*, **235**, 142.

CRISWELL J.G., HAVELKA V.D., QUEBEDEAUX B. & HARDY R.W.F. (1976) Adaptation of nitrogen fixation of intact soybean nodules to altered rhizosphere pO_2. *Plant Physiol.*, **58**, 622–5.

CROSS B.E. (1954) Gibberellic acid, Part I. *J. chem. Soc.*, 4670–6.

CROSS J.W. & BRIGGS W.R. (1978) Properties of a solubilized microsomal auxin-binding protein from coleoptiles and primary leaves of *Zea mays*. *Plant Physiol.*, **62**, 152–7.

CROSS J.W., BRIGGS W.R., DOHRMANN U.C. & RAY P.M. (1978) Auxin receptors of maize coleoptile membranes do not have ATPase activity. *Plant Physiol.*, **61**, 581–4.

CROUCH M.L. (1979) Storage proteins as embryo-specific developmental markers in zygotic, microspore and somatic embryos of *Brassica napus*. PhD. Thesis, Yale University.

CROY R.R.D., GATEHOUSE J.A., EVANS I.M. & BOULTER D. (1980a) Characterization of the storage protein subunits synthesized *in vitro* by polyribosomes and RNA from developing pea (*Pisum sativum* L.) I. Legumin. *Planta*, **148**, 49–56.

CROY R.R.D., GATEHOUSE J.A., TYLER M. & BOULTER D. (1980b) The purification and characterization of a third storage protein (convicilin) from the seeds of pea (*Pisum sativum* L.). *Biochem. J.*, **191**, 509–16.

CRUIKSHANK I.A.M. (1977) A review of the role of phytoalexins in disease resistance mechanisms. *Pont. Acad. Scient. Scripta Varia*, **41**, 1–61.

CRUIKSHANK I.A.M. & PERRIN D. (1968) The isolation and partial characterisation of monilicolin A, a polypeptide with phaseollin-inducing activity from *Monilinia fructicola*. *Life Sci.*, **7**, 449–58.

CUATRECASAS P. & HOLLENBERG M.D. (1976) Membrane receptors and hormone action. *Adv. Protein Chem.*, **30**, 251–451.

CUATRECASAS P., HOLLENBERG M.D., CHANG K-J. & BENUET V. (1975) Hormone receptor complexes and their modulations of membrane functions. *Rec. Prog. Horm. Res.*, **31**, 37–9.

CUMING A.C. & LANE B.G. (1979) Protein synthesis in imbibing wheat embryos. *Eur. J. Biochem.*, **99**, 217–24.

CURRIER W.M. & STROBEL G.A. (1974) Chemotaxis of *Rhizobium* spp. to a glycoprotein produced by Birds foot—Trefoil roots. *Science*, **196**, 434–6.

CUTTING J.A. & SCHULMAN H.M. (1969) The site of heme synthesis in soybean root nodules. *Biochim. biophys. Acta*, **192**, 486–93.

DALLING M.J., BOLAND, G. & WILSON J.H. (1976) Relation between acid proteinase activity and redistribution of nitrogen during grain development in wheat. *Aust. J. Plant Physiol.*, **3**, 721–30.

DALY J.M. (1977) Summary of points from contribution and discussions. In *Specificity in Plant Diseases.* (Eds. Wood R.K.S. & Graniti A.) pp. 231–4. Plenum Press, New York & London.

DANIEL R.M. & APPLEBY C.A. (1972) Anaerobic-nitrate, symbiotic and aerobic growth of *Rhizobium japonicum*: Effect on cytochrome P450, other hemoproteins, nitrate and nitrite reductase. *Biochim. biophys. Acta*, **275**, 347–54.

DANIELSSON C.E. (1949) Seed globulins of the Gramineae and Leguminoseae. *Biochem. J.*, **44**, 387–400.

DANIELSSON C.E. (1952) A contribution to the study of the synthesis of the reserve proteins in ripening pea seeds. *Acta Chem. Scand.*, **6**, 149–59.

DANIELSSON C.E. (1956) Plant proteins. *Ann. Rev. Plant Physiol.*, **7**, 215–36.

DART P.J. (1974) The infection process. In *The Biology of Nitrogen Fixation.* (Ed. Quispel A.) pp. 381–429. Elsevier-North-Holland, Amsterdam.

DART P.J. (1977) Infection and development of leguminous nodules. In *A Treatise on Dinitrogen Fixation.* Section III. Biology. (Eds. Hardy R.W.F. & Silver W.S.) pp. 367–472. John Wiley & Sons, New York.

DARWIN C. (1880) *The Power of Movement in Plants.* John Murray, London.

DARWIN C. & DARWIN F. (1881) *The Power of Movement in Plants.* Appleton-Century, New York.

DASHEK W.V. & ROSEN W.G. (1966) Electron microscopical localization of chemical components in the growth zone of lily pollen tubes. *Protoplasma*, **61**, 192–204.

DASHET W.V. & CHRISPEELS M.J. (1977) Gibberellic acid enhanced synthesis and release of cell wall degrading endoxylanase by isolated aleurone layers of barley. *Plants*, **134**, 251–6.

DATTA P.C. (1971) Karyological anatomy of fibre development in the leaves of *Agave americana* L. var. *marginata alba* Teel. *Ann. Bot.*, **35**, 421–7.

DATTA & RODAWAY, Personal communication.

DAUWALDER M. & WHALEY W.G. (1974)

Patterns of incorporation of [^{3}H] galactose by cells of *Zea mays* root tips. *J. Cell Sci.*, **14**, 11–27.

DAUWALDER M., WHALEY W.G. & KEPHART J.E. (1969) Phosphatases and differentiation of the Golgi apparatus. *J. Cell Sci.*, **4**, 455–97.

DAVEY R.A. & DUDMAN W.F. (1979) The carbohydrate of storage glycoproteins from seeds of *Pisum sativum*: characterization and distribution of component polypeptides. *Aust. J. Plant Physiol.*, **6**, 435–47.

DAVEY R.A., DUDMAN W.F. & HIGGINS T.J.V. (1980) Characterization of the 12 000 and 14 000 molecular weight components of vicilin from *Pisum sativum*.

DAVIDSON D. & PERTENS E. (1978) Differences in volumes of sister nuclei in binucleate cells: evidence for asymmetry of mitosis. *Canad. J. Bot.*, **56**, 2363–9.

DAVIDSON E.H. (1976) *Gene Activity in Early Development*. 452 pp. Academic Press, New York.

DAVIDSON E.H. & BRITTEN R.J. (1979) Regulation of gene expression: possible role of repetitive sequencies. *Science, N.Y.* **204**, 1052–9.

DAVIES D.D. (1977) Regulation of enzyme levels and activity. In *The Molecular Biology of Plant Cells*. (Ed. Smith H.) pp. 306–29. Blackwell Scientific Publications, Oxford.

DAVIES J.W., KAESBERG P. & DIENER T.O. (1974) Potato spindle tuber viroid. An investigation of Viroid RNA as a messenger for protein synthesis. *Virology*, **61**, 281–6.

DAVIES P.J. & RUBERY P.H. (1978) Components of auxin transport in stem segments of *Pisum sativum* L. *Planta*, **142**, 211–19.

DAVIS B. (1979) Influence of cotyledons upon α-amylase activity in pea embryonic axes. *Plant Physiol.*, **63**, 1010–15.

DAY P.R. (1974) *Genetics of Host-Pathogen Interaction*. Freeman, San Francisco.

DAZZO F.B. (1980) Lectins and their saccharide receptors as determinants of specificity in the *Rhizobium*—legume symbiosis. In *The Cell Surface: Mediator of Developmental Processes*. (Eds. Subtelny S. & Wessells N.) pp. 279–303. Academic Press, New York.

DAZZO F.B. & BRILL W.J. (1977) Receptor sit on clover and alfalfa roots for *Rhizobium*. *Appl. environ. Microbiol.*, **33**, 132–6.

DAZZO F.B. & HUBBELL D.M. (1975) Cross-reactive antigens and lectins as determinants of symbiotic specificity in the *Rhizobium*-clover association. *Appl. Microbiol.*, **30**, 1017–23.

DAZZO F.B., NAPOLI C.A. & HUBBELL D. (1976) Adsorption of bacteria to root as related to host specificity in the *Rhizobium*–clover symbiosis. *App. environ. Microbiol.*, **32**, 166–71.

DAZZO F.B., YANKE W.E. & BRILL W.J. (1978) Trifollin: a *Rhizobium* recognition protein from white clover. *Biochim. biophys. Acta*, **539**, 276–86.

DEFAGO G. & LUESCHER D. (1978) Induced mutants for studying the role of saponins as resistance factors. *Proc. 3rd Int. Congress of Plant Pathology*. p. 218.

DELBRÜCK M. & SHROPSHIRE W. (1960) Action and transmission spectra of *Phycomyces*. *Plant Physiol.*, **35**, 194–204.

DELSENY M., ASPART L. & GUITTON Y. (1977) Disappearance of stored polyadenylic acid and mRNA during early germination of radish (*Raphanus sativus* L.) embryo axes. *Planta*, **135**, 125–8.

DEMCHENKO N.P. & IVANOV V.B. (1978) Dependence of the duration of the mitotic cycle and G_2 period in sister cells of wheat root meristem on the ratio of their lengths. *Ontogenez*, **9**, 278–87.

DEMMIG B. & GIMMLER H. (1979) Effect of divalent cations on cation fluxes across the chloroplast envelope and on photosynthesis of intact chloroplasts. *Z. Naturforsch.*, **34C**, 233–41.

DENNISON D.S. (1961) Tropic responses of *Phycomyces* sporangiophores to gravitational and centrifugal stimuli. *J. gen. Physiol.*, **45**, 23–8.

DENNISON D.S. (1979) Phototropism. *Encyclopaedia of Plant Physiology*. Physiology of Movements New Series. Vol. 7. pp. 506–66. Springer-Verlag, Berlin.

DEPICKER A., VAN MONTAGU M. & SCHELL J. (1978) Homologous DNA sequences in different Ti-plasmids are essential for oncogenicity. *Nature, Lond.*, **275**, 150–3.

DEPICKER A., DE WILDE M., DE VOS G., DE VOS R., VAN MONTAGU M. & SCHELL J. (1980) Molecular cloning of overlapping segments of the nopaline Ti-plasmid pTiC58 as a means to retriction endonuclease mapping. *Plasmid*, **3**, 193–211.

DERBYSHIRE E. & BOULTER D. (1976) Isolation of legumin-like protein from *Phaseolus aureus* and *Phaseolus vulgaris*. *Phytochemistry*, **15**, 411–4.

DERBYSHIRE E., WRIGHT D.J. & BOULTER D. (1976) Legumin and vicilin, storage proteins of legume seeds. *Phytochemistry*, **15**, 3–24.

DESHPANDE B.P. (1976a) Observations on the fine structure of plant cell walls. I. Use of permanganate staining. *Ann. Bot.*, **40**, 433–7.

DESHPANDE B.P. (1976b) Observations on the fine structure of plant cell walls. II. The micro-fibrillar framework of the parenchymatous cell wall in *Cucurbita*. *Ann. Bot.*, **40**, 439–42.

DEVERALL B.J. (1976) Current perspectives in research on phytoalexins. In *Biochemical Aspects of Plant-Parasite Relationships.* (Eds. Friend J. & Threlfall D.R.) pp. 208–23. Academic Press, London.

DEVRIES G.E. (1980) Transport of sugars and organic acids in *Rhizobium leguminosarum* and their role in symbiosis. Thesis, Leiden State University, Holland.

DICKINSON C.H. & LUCAS J.A. (1977) *Plant Pathology and Plant Pathogens.* Blackwell Scientific Publications, Oxford.

DIECKERT J.W. & DIECKERT M.C. (1976) The chemistry and cell biology of the vacuolar proteins of seeds. *J. Food Sci.*, **41**, 475–82.

DÍEZ J.L., DE LA TORRE C. & LÓPEZ-SÁEZ J.F. (1971) Auxin deficiency at the onset of root growth in *Allium cepa*. *Planta*, **97**, 364–6.

DÍEZ J.L., LÓPEZ-SÁEZ J.F. & GONZÁLEZ-BERNÁLDEZ F. (1970) Growth components in *Allium* roots. *Planta*, **91**, 87–95.

DIGBY J. & FIRN R.D. (1976) A critical assessment of the Cholodny-Went theory of shoot geotropism. *Curr. Adv. Plant Sci.*, **8**, 953–60.

DIGBY J. & FIRN R.D. (1979) An analysis of the changes in growth rate occurring during the initial stages of geocurvature in shoots. *Plant Cell Environ.* **2**, 145–8.

DILWORTH M.J. & KIDBY D.K. (1968) Localization of iron and legheamoglobin in the legume root nodules by electron microscope autoradiography. *Exp. Cell Res.*, **49**, 148–59.

DILWORTH M.J. (1969) The plant as the genetic determinant of leghemoglobin production in the legume root nodule. *Biochim. biophys. Acta*, **184**, 432–41.

DIXON R.A. & LAMB C.J. (1979) Stimulation of *de novo* synthesis of L-phenylalanine ammonia lyase in relation to phytoalexin accumulation in *Colletotrichum lindemuthianum*-elicitor treated cell suspension cultures of French bean (*Phaseolus vulgaris*). *Biochim. biophys. Acta.*, **586**, 453–63.

DIXON R.O.D. (1967) The origin of the membrane envelope surrounding the bacteria and bacteroids and the presence of glycogen in clover root nodules. *Arch. Mikrobiol.*, **56**, 156–66.

DIXON R.O.D. (1969) *Rhizobia* (with particular reference to the relationship with host plants). *Ann. Rev. Microbiol.*, **23**, 137–58.

DIXON R.O.D. (1972) Hydrogenase in legume root nodule bacteroids: occurrence and properties. *Arch. Mikrobiol.*, **85**, 193–201.

DOHRMAN V., HERTEL R. & KOWALIK H. (1978) Properties of auxin binding sites in different subcellular fractions from maize coleoptiles. *Planta*, **140**, 97–106.

DOKE N., GARAS N.A. & KUĆ J. (1979) Partial characterization and aspects of the mode of action of a hypersensitivity-inhibiting factor (HIF) isolated from *Phytophthora infestans*. *Physiol. Plant Pathol.*, **15**, 127–40.

DOKE N., GARAS N.A. & KUĆ J. (1980) Effect on host hypersensitivity of suppressors released during the germination of *Phytophthora infestans* cystospores. *Phytopathology*, **70**, 35–9.

DOKE N. & TOMIYAMA K. (1980a) Effect of hyphal wall components from *Phytophthora infestans* on protoplasts of potato tuber tissues. *Physiol. Plant Pathol.*, **16**, 169–76.

DOKE N. & TOMIYAMA K. (1980b) Suppression of the hypersensitive response of potato tuber protoplasts to hyphal wall components by water soluble glucans isolated from *Phytophthora infestans*. *Physiol. Plant Pathol.*, **16**, 177–86.

DONACHIE W.D. (1968) Relationship between cell size and time of initiation of DNA replication. *Nature, Lond.*, **219**, 1077–9.

DONOVAN G.R. & LEE J.W. (1978) Effect of the nitrogen source on grain development in detached wheat heads in liquid culture. *Aust. J. Plant Physiol.*, **5**, 81–7.

DOORENBOS J. (1953) Review of the literature on dormancy in buds of woody plants. *Mededelingen van de Landbouwhodgeschool te Waginengen*, **53**, 1–21.

DOORENBOS J. (1965) Juvenile and adult phases in woody plants. In *Encyclopaedia of Plant Physiology.* (Ed. Ruhland W.) 15/1, pp. 1222–35. Springer-Verlag, Berlin.

DORÉE M. (1973) Etude de l'absorption de l'adenine et des adenines-N6-substituées par les cellules d'*Acer pseudoplatanus* I. Caractères généraux de l'absorption. *Physiol. Veg.*, **11**, 267–90.

DORÉE M. & GUERN J. (1973) Short-term metabolism of some exogenous cytokinins in *Acer pseudoplatanus* cells. *Biochim. biophys. Acta*, **304**, 611–22.

DOSIER L.W. & RIOPEL J.L. (1977) Differential enzyme activity during trichoblast differentiation in *Elodea canadensis*. *Am. J. Bot.*, **64**, 1049–56.

DOSIER L.W. & RIOPEL J.L. (1978) Origin, development, and growth of differentiating trichoblasts in *Elodea canadensis*. *Am. J. Bot.*, **65**, 813–22.

DOW J.M. & CALLOW J.A. (1978) A possible role for α-tomatine in the varietal-specific resistance of tomato to *Cladosporium fulvum*. *Phytopath. Z.*, **92**, 211–16.

DOW J.M. & CALLOW J.A. (1979a) Partial characterization of glycopeptides from culture filtrates of *Fulvia fulva* (Cooke) Ciferri (syn. *Cladosporium fulvum*), the tomato leaf mould pathogen. *J. gen. Microbiol.*, **113**, 57–66.

DOW J.M. & CALLOW J.A. (1979b) Leakage of electrolytes from isolated leaf mesophyll cells of tomato induced by glycopeptides from culture filtrates of *Fulvia fulva* (Cooke) Ciferri (syn. *Cladosporium fulvum*). *Physiol. Plant Pathol.*, **15**, 27–34.

DRAKE G. (1979) Flux studies and compartmentation analysis of Gibberellin A_1 in oat coleoptiles. *J. exp. Bot.*, **30**.

DRAKE G. & CARR D.J. (1978) Plasmodesmata, tropisms and auxin transport. *J. exp. Bot.*, **29**, 1309–18.

DRAKE G. & CARR D.J. (1979) Symplastic transport of Gibberellin: evidence from inhibitor and flux studies. *J. exp. Bot.*, **30**.

DRAKE G.A., CARR D.J. & ANDERSON W.P. (1978) Plasmolysis, plasmodesmata and the electrical coupling of oat coleoptile cells. *J. exp. Bot.*, **29**, 1205–14.

DREW M.C., JACKSON M.B. & GIFFARD, S. (1979) Ethylene-promoted adventitious rooting and development of cortical air spaces (aerenchyma) in roots may be adaptive responses to flooding in *Zea mays* L. *Planta*, **147**, 83–8.

DREYER E.M. & WEISENSSEEL M.H. (1979) Phytochrome-mediated uptake of calcium in *Mougeotia* cells. *Planta*, **146**, 31–9.

VANDEN DRIESSCHE T. (1966) Circadian rhythms in *Acetabularia*: photosynthetic capacity and chloroplast shape. *Exp. Cell Res.*, **42**, 18–30.

DRIVDAHL R.H. & THIMAN K.V. (1977) The proteases of senescing oat leaves. I. Purification and general properties. *Plant Physiol.*, **59**, 1059–63.

DRIVDAHL R.H. & THIMAN K.V. (1978) Proteases of senescing oat leaves. II. Reactions to substrates and inhibitors. *Plant Physiol.*, **61**, 501–5.

DRUMMOND M.H., GORDON M.P., NESTER E.W. & CHILTON M.D. (1977) Foreign DNA of bacterial plasmid origin is transcribed in crown gall tumours. *Nature, Lond.*, **269**, 535–6.

DUDMAN W.F. & MILLERD A. (1975) Immunochemical behaviour of legumin and vicilin from *Vicia faba*: survey of related proteins in the Leguminosae subfamily Faboideae. *Biochem. Systematics & Ecol.*, **3**, 25–33.

DUHAMEL DU MONCEAU H.L. (1758) *La Physique des Arbres*. Paris.

DULLAART J. (1967) Quantitative estimation of indoleacetic acid and indolecarboxylic acid in root nodules and roots of *Lupinus luteus* L. *Acta Bot. Neerl.*, **16**, 222–30.

DULLAART J. (1970) The auxin contents of root nodules of *Alnus glutinos* L. vill. *J. exp. Bot.*, **21**, 975–84.

DURANTE M., CREMONINI R., BRUNORI A. AVANZI S. & INNOCENTI A.M. (1977) Differentiation of metaxylem cell line in the root of *Allium cepa* I. DNA heterogeneity and ribosomal cistrons of two different stages of differentiation. *Protoplasma*, **93**, 289–303.

DURE L.S. III (1979) role of stored messenger RNA in late embryo development and germination. In *The Plant Seed: Development, Preservation and Germination*. (Eds. Rubenstein I., Phillips R.L., Green C.E. & Gengenbach B.G.) pp. 113–27. Academic Press, New York.

DURE L. & WATERS L. (1965) Long lived messenger RNA: evidence from cotton seed germination. *Science*, **147**, 410–11.

DURLEY R.C. & PHARIS R.P. (1972) Partition coefficients of 27 Gibberellins. *Phytochem.*, **11**, 317–26.

DURLEY R.C., SASSA T. & PHARIS R.P. (1979) Metabolism of tritiated gibberellin A_{20} in immature seeds of dwarf pea, cv. Meteor. *Plant Physiol.*, **64**, 214–19.

DUVICK D.N. (1965) Cytoplasmic pollen sterility in corn. *Adv. Genet.*, **13**, 1–56.

DYAR M.T. (1950) Some observations on starch synthesis in pea root tips. *Am. J. Bot.*, **37**, 786–92.

EAST E.M. & MANGELSDORF A.J. (1925) A new interpretation of the hereditary behavior of self-sterile plants. *Proc. Natl. Acad. Sci. USA.*, **11**, 166–71.

EAST J.W., NAKAYAMA T.O. & PARKMAN S.B. (1972) Changes in stachyose, raffinose, sucrose, and monosaccharides during germination of soybeans. *Crop. Sci.*, **12**, 7–9.

EATON-MORDAS C.A. & MOORE K.G. (1978) Seed glycoproteins of *Lupinus angustifolius*. *Phytochemistry*, **17**, 619–21.

EBERHARDT F. (1955) Der Atmungsverlauf alternder Blätter und reifender Fruchte. *Planta*, **45**, 57–67.

EBERT J.D.E. & SUSSEX I.M. (1970) *Interacting Systems in Development*. Holt Reinhart and Winston, New York.

EDELMAN G.M. (1976) Surface modulation in cell recognition and cell growth. *Science*, **192**, 218–26.

EL-ANTABLY H.M.M. (1975) Redistribution of endogenous indoleacetic acid, abscisic acid and gibberellins in geotropically stimulated

Ribes nigrum roots. *Z. Pflanzenphysiol.*, **75**, 17–24.

EL-ANTABLY H.M.M. & LARSEN P. (1974) Redistribution of endogenous gibberellins in geotropically stimulated roots. *Nature, Lond.*, **250**, 76.

ELBEIN A.D. (1979) The role of lipid-linked saccharides in the biosynthesis of complex carbohydrates. *Ann. Rev. Plant Physiol.*, **30**, 239–72.

VAN ELDIK L.J., GROSSMAN A.R., IVERSON D.B. & WATTERSON D.M. (1980) Isolation and characterisation of calmodulin from Spinach leaves and *in vitro* translation mixtures. *Proc. Natl. Acad. Sci. USA*, **77**, 1912–16.

ELLFOLK N. & SIEVERS G. (1971) The primary structure of soybean leghaemoglobin. *Acta. Chim. Scand.* **25**, 3532–4.

ELLINGBOE A.H. (1979) Inheritance of Specificity: The Gene-for-Gene Hypothesis. In *Recognition and Specificity in Plant Host–Parasite Interactions.* (Eds. Daly J.M. & Uritani I.) pp. 3–15. Japan Scientific Societies Press, Tokyo and University Park Press, Baltimore.

ELLIOT M.C. (1977) Auxins and the regulation of root growth. In *Plant Growth Regulation*, (Ed. Pilet P.E.) pp. 100–8. Springer-Verlag, Berlin, Heidelberg & New York.

ELLIS J., KERR A., TEMPÉ J. & PETIT A. (1979) Arginine catabolism: a new function of both octopine and nopaline Ti-plasmids of *Agrobacterium. Molec. gen. Genet.*, **173**, 263–9.

ELLIS R.J. (1979) The most abundant protein in the world. *Trends in biochem. Sci.*, **4**, 241–4.

VAN DEN ENDE H. (1976) *Sexual Interactions in Plants:* The role of specific substances in sexual reproduction. Academic Press, London.

ENGLER G., HOLSTERS M., VAN MONTAGU M., SCHELL J., HERNALSTEENS J.P. & SCHILPEROORT R.A. (1975) Agrocin 84 sensitivity: a plasmid determined property in *Agrobacterium tumefaciens. Molec. gen. Genet.*, **138**, 345–9.

ENGLER G., DEPICKER A., MAENHAUT R., VILLARROEL-MANDIOLA R., VAN MONTAGU M., & SCHELL J. (1981) Physical mapping of DNA base sequence homologies between an octopine and a nopaline Ti-plasmid of *Agrobacterium tumefaciens. J. mol. Biol.*, (submitted).

ENSER U. & HEBER U. (1980) Metabolic regulation by pH gradients. Inhibition of photosynthesis by indirect proton transfer across the chloroplast envelope. *Biochim. biophys. Acta*, **592**, 587–91.

ERNAUX T., HIERNAUX J. & NICOLIS G. (1978) Turing's theory in morphogenesis. *Bull. math. Biol.*, **40**, 771–89.

ESAU K. (1954) Primary vascular differentiation in plants. *Biol. Rev.* **29**, 46–86.

ESAU K. (1977) *The Anatomy of Seed Plants.* 2e. John Wiley & Sons, New York.

ESAU K. & CHARVAT I. (1978) On vessel member differentiation in the bean (*Phaseolus vulgaris* L.). *Ann. Bot.*, **42**, 665–77.

ESHEL A. & WAISEL Y. (1973) Heterogeneity of ion uptake mechanisms along primary roots of maize seedlings. In *Ion Transport in Plants.* (Ed. Anderson W.P.) pp. 531–7. Academic Press, London, New York.

ETHERTON B. & DEDOLPH R.R. (1972) Gravity and intracellular differences in membrane potentials of plant cells. *Plant Physiol.*, **49**, 1019–920.

VAN ETTEN H.D. (1979) Relationship between tolerance to isoflavonoid phytoalexins and pathogenicity. In *Recognition and Specificity in Plant Host–Parasite Interactions.* (Eds. Daly J.M. & Uritani I.). pp. 301–16. Japan Scientific Societies Press, Tokyo and University Park Press, Baltimore.

VAN ETTEN H.D. & PUEPPKE S.G. (1976) Isoflavonoid phytoalexins. In *Biochemical Aspects of Plant–Parasite Relationships.* pp. 239–89. Academic Press, London.

EVANS A. & SMITH H. (1976) Localization of phytochrome in etioplasts and its regulation *in vitro* of gibberellin levels. *Proc. Nat. Acad. Sci. USA*, **73**, 138–42.

EVANS I.M., BOULTER D., FOX R.L. & KANG B.T. (1977) The effects of sulphur fertilizers on the content of sulpho-amino acids in seeds of cowpea (*Vigna unguiculata*). *J. Sci. Food Agric*, **28**, 161–6.

EVANS I.M., CROY R.R.D., HUTCHINSON P., BOULTER D., PAYNE P.I. & GORDON M.E. (1979) Cell free synthesis of some storage protein subunits by polyribosomes and RNA isolated from developing seeds of pea. *Planta*, **144**, 455–62.

EVANS I.M., CROY R.R.D., BROWN P. & BOULTER D. (1980) Synthesis of complementary DNAs to partially purified mRNAs coding for the storage proteins of *Pisum sativum* L. *Biochim. biophys. Acta*, **610**, 81–95.

EVANS J.H., RUIZ-ARGUESO T. & RUSSELL S.A. (1978) Relationship between hydrogen metabolism and nitrogen fixation in legumes. In *Limitations and Potentials for Biological Nitrogen Fixation.* (Eds Dobereiner J. *et al.*) Plenum Publishing Corp., New York.

EVANS L.S. & VAN'T HOF J. (1975) Is polyploidy necessary for tissue differentiation in higher plants. *Amer. J. Bot.*, **62**, 1060–4.

FALK H., LIEDVOGEL B. & SITTE P. (1974) Circular DNA in isolated chromoplasts. *Z. Naturforschung C.*, **29C**, 541–4.

FAN D. & MACLACHLAN G.A. (1967) Massive synthesis of ribonucleic acid and cellulase in the pea epicotyl in response to indoleacetic acid with and without concurrent cell division. *Plant Physiol.*, **42**, 1114–22.

FAWCETT D.W. (1966) *The Cell: its Organelles and Inclusions.* W.B. Saunders, Philadelphia.

FELDMAN L.J. (1975) Cytokinins and quiescent center activity in roots of *Zea.* In *The Development and function of Roots,* (Eds. Torrey J.G. & Clarkson D.T.) pp. 55–72. Academic Press, London.

FELDMAN L.J. (1977) The regeneration and elaboration of primary vascular tissue patterns in roots of *Zea. Bot. Gaz.*, **138**, 393–401.

FELDMAN L.J. (1979a) Cytokinin biosynthesis in roots of corn. *Planta*, **145**, 315–21.

FELDMAN L.J. (1979b) The proximal meristem in the root apex of *Zea mays* L. *Ann. Bot.*, **43**, 1–9.

FELDMAN L.J. & TORREY J.G. (1975) The quiescent center and primary vascular pattern formation in cultured roots of *Zea. Canad. J. Bot.*, **53**, 2796–803.

FELDMAN L.J. & TORREY J.G. (1976) The isolation and culture *in vitro* of the quiescent centre of *Zea mays. Am. J. Bot.*, **63**, 345–55.

FELLER U.K. (1978) Changes in nitrogen contents and in proteolytic activities in different parts of field grown wheat ears (*Triticum aesivum* L.) during maturation. *Plant Cell Physiol.*, **19**, 1487–95.

FELLER U.K. (1979) Effect of changed source sink relations in proteolytic activities and on nitrogen mobilisation in field-grown wheat (*Triticum aestivum* L.). *Plant Cell Physiol.*, **20**, 1577–83.

FELLER U.K., SOONG T-S.T. & HAGEMAN R.H. (1977) Proteolytic activities and leaf senescence during grain development of field-grown corn (*Zea mays* L.). *Plant Physiol.*, **59**, 290–4.

FERGUSON J.E., DICKINSON D.B. & RHODES A.M. (1979) Analysis of endosperm sugars in sweet corn inbred (Illinois 677A) which contains the sugary enhancer (SE) gene and comparison of SE with other corn genotypes. *Plant Physiol.*, **63**, 416–20.

FERRARI T.E., BRUNS D. & WALLACE D.H. (1981) Isolation of a plant glycoprotein involved with control of intercellular recognition. *Plant Physiol.*, (in press).

FERRARI T.E. & WALLACE D.H. (1977) A model for self-recognition and regulation of the incompatibility response of pollen. *Theor. appl. Genet.*, **50**, 211–25.

FERRER A., DELSENY M. & GUITTON Y. (1979) Isolation and characterization of subribosomal ribonucleoprotein particles from radish seeds and seedlings. *Plant Sci. Lett.*, **14**, 31–42.

FILNER B., HERTEL R., STEELE C. & FAN V. (1970) Some aspects of geotropism in coleoptiles. *Planta*, **94**, 333–54.

FILNER P. & VARNER J.E. (1967) A test for *de novo* synthesis of enzymes. Density labelling with H_2O^{18} of barley α-amylase induced by gibberellic acid. *Proc. Nat. Acad. Sci. USA*, **58**, 1520–6.

FIRMIN J.L & FENWICK G.R. (1978) Agropine—a major new plasmid-determined metabolite in crown gall tumours. *Nature, Lond.*, **276**, 842–4.

FIRN R. & DIGBY J. (1977) The role of the peripheral cell layers in the geotropic curvature of sunflower hypocotyles: a model of shoot geotropism. *Aust. J. Plant Physiol.*, **4**, 337–47.

FIRN R.D. & DIGBY J. (1980) The establishment of tropic curvatures in plants. *Ann. Rev. Plant Physiol.*, **31**, 131–48.

FIRN R.D. DIGBY J. & PINSENT C. (1977) Evidence against the involvement of gibberellins in the differential growth which causes geotropic curvature. *Z. Pflphysiol.*, **82**, 179–85.

FLAVELL R. (1974) A model for the mechanism of cytoplasmic male sterility in plants, with special reference to maize. *Plant Sci. Lett.*, **3**, 259–63.

FOARD D.E., HABER A.H. & FISHMAN T.N. (1965) Initiation of lateral root primordia without completion of mitosis and without cytokinesis in uniseriate pericycle. *Am. J. Bot.*, **52**, 580–90.

DI FONZO N., FORNASARI E., SALAMINI F. & SOAVE C. (1977) SDS-protein subunits in normal, opaque-2 and floury-2 maize endosperms. *Maydica*, **22**, 77–88.

FORD P.J. (1976) Control of gene expression during differentiation and development. In *The Developmental Biology of Plants and Animals.* (Eds. Graham C.F. & Wareing P.F.) pp. 302–48. Blackwell Scientific Publications, Oxford.

FORDE B.G. & LEAVER C.J. (1980) Nuclear and cytoplasmic genes controlling synthesis of variant mitochondrial polypeptides in male-sterile maize. *Proc. Nat. Acad. Sci. USA*, **77**, 418–22.

FORDE B.G., OLIVER R.J. & LEAVER C.J. (1978) Variation in mitochondrial translation products associated with male-sterile cytoplasms in maize. *Proc. Natl. Acad. Sci. USA.*, **75**, 3841–5.

Fowler M.W. & ap Rees T. (1970) Carbohydrate oxidation during differentiation in roots of *Pisum sativum*. *Biochim. biophys. Acta*, **201**, 33–44.

Fox J.E. & Erion J.L. (1975) A cytokinin binding protein associated with higher plant ribosomes. *Biochem. biophys. Res. Communs.*, **64**, 694–700.

Fox J.E., Pratt H.M., Shewry P. R. & Miflin B.J. (1977) Hordein synthesis. In *Acides Nucléiques et Synthèse des Protéines chez les Végétaux*. (Eds. Bogorad L. & Weil J.H.) pp. 502–9. Plenum Press, New York.

Françon M. (1961) *Progress in Microscopy*. Pergamon Press, London.

Frankel R., Scowcroft W.R. & Whitfeld P.R. (1979) Chloroplast DNA variation in isonuclear male-sterile lines of *Nicotiana*. *Molec. gen. Genet.*, **169**, 129–35.

Fraser D.E. & Bidwell R.G.S. (1974) Photosynthesis and photorespiration during the ontogeny of the bean plant. *Canad. J. Bot.*, **52**, 2561–70.

Frey-Wyssling A. (1953) *Submicroscopic Morphology of Protoplasm*. Elsevier Publishing Co., Amsterdam.

Friend J. (1976) Lignification in infected tissue. In *Biochemical Aspects of Plant–Parasite Relationships*. (Eds. Friend J. & Threlfall D.R.) pp. 291–303. Academic Press, London.

Fritsch H. & Beevers H. (1979) ATP citrate lyase from germinating castor bean endosperm. *Plant Physiol.*, **63**, 687–91.

Frosch S., Drumm H. & Mohr H. (1977) Regulation of enzyme levels of phytochrome in mustard cotyledons: multiple mechanisms. *Planta*, **136**, 181–6.

Frydman V.M., Gaskin P. & MacMillan J. (1974) Qualitative and quantitative analyses of gibberellins throughout seed maturation in *Pisum sativum* cv. Progress No. 9. *Plants*, **118**, 123–32.

Fuchsman W. & Appleby C.A. (1979a) Separation and determination of relative concentrations of the homogeneous components of soybean leghaemoglobins by isoelectric focusing. *Biochim. biophys. Acta*, **579**, 314–24.

Fuchsman W.H. & Appleby C.A. (1979b) CO and O_2 complexes of soybean leghaemoglobins: pH effects upon infrared and visible spectra. Comparisons with CO and O_2 complexes of myoglobin and hemoglobin. *Biochemistry*, **18**, 1309–21.

Fujimura T. & Komamine A. (1979) Synchronization of somatic embryogenesis in a carrot cell suspension culture. *Plant Physiol.*, **64**, 162–4.

Fukuda H. & Komamine A. (1980) Direct evidence for cytodifferentiation to tracheary elements without intervening mitosis in a culture of single cells isolated from the mesophyll of *Zinnia elegans*. *Plant Physiol.*, **65**, 61–4.

Fulcher R.G. McCully M.E., Setterfield G. & Sutherland J. (1976) β-1-3 glucans may be associated with cell plate formation during cytokinesis. *Canad. J. Bot.*, **54**, 539–42.

Furuya A. (1978) Photocontrol of developmental processes in fern gametophytes. *Bot. Mag.*, Tokyo. (Special Issue) 1,219–42.

Gabella M. & Pilet P-E. (1979) Effect of fusicoccin on the maize root elongation and on the pH of the medium. *Z. Pflanzenphysiol.*, **93**, 23–30.

Galston A.W. (1974) Plant photobiology in the last half century. *Plant Physiol.*, **54**, 427–36.

Galston A.W. & Satter R.L. (1976) Light, clocks and ion flux: an analysis of leaf movement. In *Light and Plant Development*. (Ed. Smith H.) pp. 159–83. Butterworth, London.

Galun E. (1959) Effects of gibberellic acid and napthalene acetic acid on sex expression and some morphological characters in the cucumber plant. *Phyton (Argentina)*, **13**, 1–8.

Gamow R.I. & Bottger B. (1977) Phycomyces: distribution of growth velocities in the growing zone. *Plant Physiol.*, **59**, 901–6.

Garas N.A., Doke N. & Kuc J. (1979) Suppression of the hypersensitive reaction in potato tubers by mycelial components from *Phytophthora infestans*. *Physiol. Plant Pathol.*, **15**, 117–26.

Gardiner M. & Chrispeels M.J. (1975) Involvement of the Golgi apparatus in the synthesis and secretion of hydroxyproline-rich cell wall glycoproteins. *Plant Physiol.*, **55**, 536–41.

Gardner, G., Sussman M.R. & Kende H. (1978) *In vitro* cytokinin binding to a particular fraction from protenemata of *Funaria hygrometrica*. *Planta*, **143**, 67–73.

Gatehouse J.A., Croy R.R.D. & Boulter D. (1980) Isoelectric-focusing properties and carbohydrate content of pea (*Pisum sativum* L.) legumin. *Biochem. J.*, **185**, 497–503.

Gehring W. (1972) The stability of the determined state in cultures of imaginal disks in *Drosophila*. In *The Biology of Imaginal Disks*. (Eds. Ursprung H. & Nöthinger R.) pp. 35–7. Springer-Verlag, Berlin.

Geiger D.B. (1975) Phloem loading and associated processes. In *Phloem Transport*. (Eds. Arnoff S., Dainty J., Gorham P.G., Srivastava L.M. & Swanson C.A.) pp. 251–81. Plenum Press, New York.

Genetello Ch., van Larebeke N., Holsters M., Depicker A., van Montagu M. & Schell J. (1977) Ti-plasmids of *Agrobac-*

terium as conjugative plasmids. *Nature, Lond.*, **265**, 561–3.

GENGENBACH B.G., GREEN C.E. & DONOVAN C.M. (1977) Inheritance of selected pathotoxin resistance in maize plants regenerated from cell cultures. *Proc. Nat. Acad. Sci. USA*, **74**, 5113–7.

GEPSTEIN S. & THIMAN K.V. (1980) Changes in the abscisic acid content of oat leaves during senescence. *Proc. Nat. Acad. Sci. USA.*, **77**, 2050–3.

GERSHON D. (1961) Genetic studies on effective nodulation in lotus spp. *Canad. J. Microbiol.*, **7**, 961–3.

GERSHON N.D. (1978) Model for capping of membrane receptors based on boundary surface effects. *Proc. Nat. Acad. Sci. USA.*, **75**, 1357–60.

GIANAZZA E., VIGLIENGHI V., RIGHETTI P.G., SALAMINI F. &. SOAVE C. (1977) Amino acid composition of zein molecular components. *Phytochemistry*, **16**, 315–7.

GIAQUINTA R.T. (1979) Phloem loading of sucrose: involvement of membrane ATPase and H^+ transport. *Plant. Physiol.*, **63**, 744–8.

GIBBONS G.S.B. & WILKINS M.B. (1970) Growth inhibitor production by root caps in relation to geotropic responses. *Nature, Lond.*, **226**, 558–9.

GIDDINGS T.H., BROWER D.L. & STAEHELIN L.A. (1980) Visualization of particle complexes in the plasmamembrane of *Micrasterias denticulata* associated with the formation of cellulose fibrils in primary and secondary walls. *J. Cell Biol.*, **84**, 327–39.

GILKES N.R. & CHRISPEELS M.J. (1980) Endoplasmic reticulum of mung bean cotyledons. *Plant Physiol.*, **65**, 600–4.

GILKEY J.C., JAFFE L.F., RIDGWAY E.B. & REYNOLDS G.T. (1978) A free calcium wave traverses the activating egg of the medaka. *Oryzias latipes. J. Cell Biol.*, **76**, 448–66.

GILROY J., WRIGHT D.J. & BOULTER D. (1979) Homology of basic subunits of legumin from *Glycine max* and *Vicia faba. Phytochemistry*, **18**, 315–6.

GIMÉNEX-MARTÍN G., DE LA TORRE C. & LÓPEZ-SÁEZ (1977) Cell division in higher plants. In *Mechanisms and Control of Cell Division.* (Eds. Rost T.L. & Gifford E.M.) pp. 261–307. Dowden, Hutchinson & Ross, Stroudsburg.

GLAUERT A.M. & GLAUERT R.H. (1958) Araldite as an embedding medium for electron microscopy. *J. Biophys. Biochem. Cytol.*, **4**, 191–4.

GLINKA Z. & REINHOLD L. (1971) Abscissic acid raises the permeability of plant cells to water. *Plant Physiol.*, **48**, 103–5.

GOLDBERG R.B. (1980) Structural gene expression in higher plants. In *Genome Organization and Expression in Plants.* (Ed. Leaver C.) pp. 117–26. NATO Advanced Study Institute. Plenum Press, New York and London.

GOLDBERG R.B. & TIMBERLAKE W.E. (1980) Hybridisation of scDNA does not lead to overestimates of mRNA complexity. *Nature, Lond.*, **283**, 601–2.

GOLDSMITH M.H.M. (1967) Movement of pulses of labelled auxin in corn coleoptiles. *Plant Physiol.*, **42**, 258–63.

GOLDSMITH M.H.M. (1977) The polar transport of auxin. *Ann. Rev. Plant Physiol.*, **28**, 439–78.

GOODMAN R.N., HUANG P.Y., HUANG J.S. & THAIPANICH V. (1977) Induced resistance to bacterial infection. In *Biochemistry and Cytology of Plant–Parasite Interaction.* (Eds. Tomiyama K., Daly J.M., Uritani I., Oku H. & Ouchi S.), pp. 35–42. Kodansha Ltd., Tokyo and Elsevier, New York.

GOODMAN R.N., HUANG P.Y. & WHITE J.A. (1976) Ultrastructural evidence for immobilization of an incompatible bacterium, *Pseudomonas pisi* in tobacco leaf tissue. *Phytopathol.*, **66**, 754–64.

GONZÁLEZ-FERNÁNDEZ A., GIMÉNEZ-MARTIN G., DÍEZ J.L., DE LA TORRE C. & LÓPEZ-SÁEZ J.F. (1971) Interphase development and beginning of mitosis in the different nuclei of polynucleate homokaryotic cells. *Chromosoma*, **36**, 100–11.

GOODWIN B.C. & PATEROMICHELAKIS S. (1979) The role of electrical fields, ions and the cortex in the morphogenesis of *Acetabularia. Planta*, **145**, 427–36.

GORDON M.P. (1979) Tumor formation in plants. In *Proteins and Nucleic Acids.* Vol. 6. (Ed. Marcus A.) Academic Press, New York (in press).

GORSKI J. & GANNON F. (1976) Current models of steroid hormone action: a critique. *Ann. Rev. Physiol.*, **38**, 425–50.

GOUG C.S. & LOVETT J.S. (1977) Regulation of protein synthesis in *Blastocladiella* zoospores: Factors for synthesis in nonsynthetic spores. *Expt. Mycol.*, **1**, 138–51.

GRAHAM T.A. & GUNNING B.E.S. (1970) Localization of legumin and vicilin in bean cotyledon cells using fluorescent antibodies. *Nature, Lond.*, **228**, 81–2.

GRAHAM T.L., SEQUIERA L. & HUANG T.R. (1977) Bacterial lipopolysaccharides as inducers of disease resistance in tobacco. *Appl. env. Microbiol.*, **34**, 424–32.

GRANT J.S. & BEWLEY J.D. (1979) A dual role for the endosperm and its galactomannan reserves in the germinative physiology of fenugreek (*Trigonella foenum-graecum* L.), an endospermic leguminous seed. *Planta*, **147**, 145–50.

Green J.R. & Northcote (1978) The structure and function of glycoproteins synthesized during slime-polysaccharide production by membranes of the root-cap cells of maize (*Zea mays*). *Biochem. J.*, **170**, 599–608.

Green P.B. (1969) Cell morphogenesis. *Ann. Rev. Plant Physiol.*, **20**, 365–94.

Green P.B. (1976) Growth and cell pattern formation on an axis: critique of concepts, terminology and modes of study. *Bot. Gaz.*, **37**, 187–202.

Green P.B. (1980) Organogenesis—a biophysical view. *Ann Rev. Plant Physiol.*, **31**, 51–82.

Gregory F.G. & Sen G.K. (1937) Physiological studies in plant nutrition. VI. The relation of respiration rate to the carbohydrate and nitrogen metabolism of the barley leaf as determined by nitrogen and potassium deficiency. *Ann. Bot.*, **1**, 521–61.

Gregory L.E. (1956) Some factors for tuberization in the potato plant. *Am. J. Bot.*, **43**, 281–8.

Gregory P., Earle E.D. & Gracen V.E. (1977) Biochemical and ultrastructural aspects of southern corn leaf blight disease. In *Host Plant Resistance to Pests*. Am. Chem. Soc. Symp., **62**. (Ed. Hedin P.A.) pp. 90–114. American Chemical Society, Washington.

Gregory W.C. (1940) Experimental studies on the cultivation of excised anthers in nutrient solution. *Am. J. Bot.*, **27**, 687–92.

Gressel J. (1979) Blue light photoreception. *Photochem. Photobiol.*, **30**, 749–54.

de Greve h., Decraemer H., Seurinck J., van Montagu M. & Schell J. The functional organization of the octopine *Agrobacterium tumefaciens* plasmid pTiB6S3. *Plasmid*, (submitted).

Grierson D. (1977) The nucleus and the organisation and transcription of nuclear DNA. In *The Molecular Biology of Plant Cells*. (Ed. Smith H.) Blackwell Scientific Publications, Oxford.

Grierson D. (1981) RNA processing and other post-transcriptional modifications. In *Encyclopaedia of Plant Physiology*. Vol. 16 (Ed. Parthier B.) (in press). Springer-Verlag, Berlin.

Grisebach H. & Hahlbrock K. (1974) Enzymology and regulation of flavonoid and lignin biosynthesis in plants and plant cell suspension cultures. In *Metabolic Regulation of Secondary Plant Products*. Recent Advances in Phytochemistry, **8**. (Eds. Runeckles V.C. & Conn E.E.) pp. 21–52. Academic Press, New York.

Grison R. & Pilet P-E. (1978) Analyse critique des péroxydases de la racine de maïs. *Plant Sci. Lett.*, **13**, 213–8.

Grossman A., Bartlett S. & Chua N.H. (1980a) Energy-dependent uptake of cytoplasmically synthesized polypeptides by chloroplasts. *Nature, Lond.*, **285**, 625–8.

Grossman A.R., Bartlett S.G., Schmidt G.W. & Chua N.H. (1980b) Post-translational uptake of cytoplasmically synthesized proteins by intact chloroplasts *in vitro*. *Ann. N.Y. Acad. Sci.*, **343**, 266–74.

Guha S. & Maheshwari S.C. (1964) *In vitro* production of embryos from anthers of *Datura*. *Nature, Lond.*, **204**, 497.

Guldager P. (1978) Immunoelectrophoretic analysis of seed proteins from *Pisum sativum* L. *Theor. Appl. Genet.*, **53**, 241–50.

Gunning B.E.S. (1977) Transfer cells and their roles in transport of solutes in plants. *Sci. Progr.*, **64**, 539–68.

Gunning B.E.S., Hardham A.R. & Hughes J.E. (1978b) Preprophase bands of microtubules in all categories of formative and proliferative cell division in *Azolla* roots. *Planta*, **143**, 145–60.

Gunning B.E.S., Hardham A.R. & Hughes J.E. (1978c) Evidence for initiation of microtubules in discrete regions of the cell cortex in *Azolla* root-tip cells, and an hypothesis on the development of cortical assays of microtubules. *Planta*, **143**, 161–79.

Gunning B.E.S., Hughes J.E. & Hardham A.R. (1978a) Formative and proliferative cell divisions, cell differentiation and developmental changes in the meristem of *Azolla* roots. *Planta*, **143**, 121–44.

Gunning B.E.S. & Pate J.S. (1974) Transfer cells. In *Dynamic Aspects of Plant Ultrastructure*. (Ed. Robards A.W.) McGraw-Hill, London.

Gunning B.E.S. & Robards A.W. (1976) *Intercellular Communication in Plants: Studies in Plasmodesmata*. pp. xiv & 387. Springer-Verlag, Berlin.

Gunning B.E.S. & Steer M.W. (1975) *Ultrastructure and the Biology of Plant Cells*. Chapter 9.5 Amyloplasts and Starches. pp. 116–19. Edward Arnold, London.

Gurley W.B., Kemp J.D., Alber M.J., Sutton D.W. & Callis J. (1979) Transcription of Ti plasmid-derived sequences in three octopine-type crown gall tumor lines. *Proc. Nat. Acad. Sci. USA.*, **76**, 2828–32.

Gutknecht J. & Walter A. (1980) Transport of auxin (indoleacetic acid) through lipid bilayer membranes. *J. Membrane Biol.*, (submitted).

GUYON P., CHILTON M-D., PETIT A. & TEMPÉ J. (1980) Agropine in 'null type' crown gall tumors: evidence for the generality of the opine concept. *Proc. Nat. Acad. Sci. USA.*, **77**, 2693–7.

HAAS D.L. & CAROTHERS Z.B. (1975) Some ultrastructural observations on endodermal cell development in *Zea mays* roots. *Am. J. Bot.*, **62**, 336–48.

HABER A.H. (1962) Nonessentiality of concurrent cell divisions for degree of polarization of leaf-growth. I. Studies with radiation-induced mitotic inhibition. *Am. J. Bot.*, **49**, 563–89.

HABERLANDT G. (1900) Über die Perception des geotropischen Reizes. *Ber. Deutsch. Bot. Ges.*, **18**, 261.

HABERLANDT G. (1909) *Physiologische Pflanzenanatomie.* W. Engelmann, Leipzig.

HABERLANDT G. (1965) *Physiological Plant Anatomy.* Reprint edition. Today and Tomorrow. Book Agency, New Delhi.

HADAČOVÁ V. (1974) Malate and lactate dehydrogenase isoenzymes in growth zones of *Vicia faba* L. *Structure and Function of Primary Root Tissue.* (Ed. Kolek J.) pp. 227–34. Veda, Bratislava.

HADWIGER L.A. & BECKMAN J.M. (1980) Chitosan as a component of pea-*Fusarium solani* interactions. *Plant Physiol.*, **66**, 199–204.

HAGUE D.R. (1975) Studies of storage proteins of higher plants. I. Concanavalin A from three species of the genus *Canavalia. Plant Physiol.*, **55**, 636–42.

HAHLBROCK K. & GRISEBACH H. (1979) Enzymic controls in the biosynthesis of lignin and flavonoids. *Ann. Rev. Plant Physiol.*, **30**, 105–30.

HAHN H., DEZACKS R. & KENDE H. (1974) Cytokinin formation in pea seeds. *Naturwissenschaften*, **61**, 170.

HALDANE J.B.S. (1932) The time of action of genes and its bearing on some evolutionary problems. *Am. Nat.*, **66**, 5–24.

HALE C.C. & ROUX S.J. (1980) Photoreversible calcium fluxes induced by phytochrome in oak coleoptile cells. *Plant Physiol.*, **65**, 658–62.

HALL D., MOLLENHAUER H.H. & MORRÉ D.J. (1965) Evidence for secretion of cell dispersing enzymes from maize root cap and epidermis. *Proc. Indiana Acad. Sci.*, **75**, 65.

HALL J.L. (1978) *Electron Microscopy and Cytochemistry of Plant Cells.* Elsevier/North Holland Biomedical Press, Amsterdam.

HALL N.P., KEYS A.J. & MERRETT M.S. (1978) Ribulose-1,5-diphosphate carboxylase protein during flag leaf senescence. *J. exp. Bot.*, **39**, 31–7.

HALL T.C., MA Y., BUCHBINDER B.U., PYNE J.W., SUN S.M. & BLISS F.A. (1978c). mRNA for G1 protein of French bean seeds. *Proc. Nat. Acad. Sci. USA*, **75**, 3196–200.

HALL T.C., WEPPRICH R.K., DAVIES J.W. WEATHERS L.G. & SEMANICK J.G. (1974) Functional distinctions between the ribonucleic acids from citrus exocortis viroid and plant viruses. *Virology*, **61**, 486–92.

HALLIER U.W., SCHMITT J.M., HEBER U., CHAIANOVA S.S. & VOLODARSKY A.D. (1978) Ribulose-1,5-bisphophate carboxylase-deficient plastome mutants of *Oenothera. Biochim. biophys. Acta*, **504**, 67–83.

HALPERIN W. (1978) Organogenesis at the shoot apex. *Ann. Rev. Plant Physiol.*, **29**, 239–62.

HAMBLIN J. & KENT S.P. (1973) Possible role of phytohaemoglutinin in *Phaseolus vulgaris* L. *Nature, New Biol.*, **245**, 28–30.

HAMMETT J.R. & KATTERMAN F.R. (1975) Storage and metabolism of poly (adenylic acid)-mRNA in germinating cotton seeds. *Biochemistry*, **14**, 4375–9.

HAMPP R. & WELLBURN A.R. (1976) Early changes in the envelope permeability of developed chloroplasts. *J. exp. Bot.*, **27**, 778–84.

HANSTEIN J. (1868) Die Scheitelzellgruppe im Vegetationspunkt der Phanerogamen. *Festschr. niederrhein. Gesellschaft nat. Heilk. zum 50-jahr. Jubilaüm Univ. Bonn*, pp. 109–43.

HARA I., OHMIYA M. & MATSURBARA H. (1978) Pumpkin (*Cucurbita* sp) seed globulin. III. Comparison of subunit structures among seed globulins of various Cucurbita species and characterization of peptide components. *Plant Cell Physiol.*, **19**, 237–43.

HARAGUCHI N. & SHIMIZU S. (1970) Photosynthetic activities in tobacco plants. II. Relationship between photosynthetic activity and chlorophyll content. *Bot. Mag.*, **83**, 411–18.

HARDHAM A.R. & GUNNING B.E.S. (1977) The length and deposition of cortical microtubules in plant cells fixed in glutaraldehyde-osmium tetroxide. *Planta*, **134**, 201–3.

HARDHAM A.R. & GUNNING B.E.S. (1978) The structure of cortical microtubule assays in plant cells. *J. Cell Biol.*, **77**, 14–34.

HARDHAM A.R. & GUNNING B.E.S. (1979). Interpolation of microtubules into cortical assays during cell elongation and differentiation in roots of *Azolla pinnata. J. Cell Sci.*, **37**, 411–42.

HARDIE D.G. (1975) Control of carbohydrase formation by gibberellic acid in barley endosperm. *Phytochemistry*, **14**, 1719–22.

HARDWICK K., WOOD M.E. & WOOLHOUSE H.W. (1968) Photosynthesis and respiration in relation to leaf age in *Perilla frutescens* L. *Britt. New Phytol.*, **67**, 79–86.

HARDY R.W.F., HOLSTEN R.D., JACKSON E.K. & BURNS R.C. (1968) The acetylene-ethylene assay for N_2-fixation: laboratory and field evaluation. *Plant Physiol.*, **43**, 1185–207.

HARGREAVES J.A. (1979) Investigations into the mechanism of mercuric chloride stimulated phytoalexin accumulation in *Phaseolus vulgaris* and *Pisum sativum*. *Physiol. Plant Pathol.*, **15**, 279–87.

HARGREAVES J.A. (1980) A possible mechanism for the phytotoxicity of the phytoalexin phaseollin. *Physiol. Plant Pathol.*, **16**, 351–7.

HARGREAVES J.A. & BAILEY J.A. (1978) Phytoalexin production by hypocotyls of *Phaseolus vulgaris* in response to constitutive metabolites released by damaged bean cells. *Physiol. Plant Pathol.*, **13**, 89–100.

HARMPP R. & DE FILLIPIS L.F. (1980) Plastid protease activity and prolamellar body transformation during greening. *Plant Physiol.*, **65**, 663–8.

HARRIS B. & DURE L. III (1978) Developmental regulation in cotton seed germination: polyadenylation of stored messenger RNA. *Biochem.*, **17**, 3250–6.

HARRIS N. (1979) Endoplasmic reticulum in developing seeds of *Vicia faba*. A high voltage electron microscope study. *Planta*, **146**, 63–9.

HARRIS N. & BOULTER D. (1976) Protein body formation in cotyledons of developing cowpea (*Vigna unguiculata*) seeds. *Ann. Bot.*, **40**, 739–44.

HARRIS N. & CHRISPEELS M.J. (1980) The endoplasmic reticulum of mung bean cotyledons: quantitative morphology of cisternal and tubular ER during seedling growth. *Planta*, **148**, 293–303.

HARRIS P.J. & NORTHCOTE D.N. (1970) Patterns of polysaccharide biosynthesis in differentiating cells of maize root tips. *Biochem. J.*, **120**, 479–91.

HARRIS P., OSBORN M. & WEBER K. (1980) Distribution of tubulin-containing structures in the egg of the sea-urchin *Strongylocentrotus purpuratus* from fertilization through first cleavage. *J. Cell Biol.*, **84**, 668–79.

HART J.W. & SABNIS D.D. (1977) Microtubules. In *The Molecular Biology of Plant Cells*. (Ed. Smith H.) pp. 160–82. Blackwell Scientific Publications, Oxford.

HARTMANN E.V. & SCHMID K. (1980) Effects of blue light on biopotential changes in etiolated hypocotyl hooks of beans. In *The Blue Light Syndrome*. (Ed. Senger H.) pp. 221–31. Springer-Verlag, Berlin (in press).

HARTMANN K.M. (1966) A general hypothesis to interpret 'high energy phenomena' of photomorphogenesis on the basis of phytochrome. *Photochem. Photobiol.*, **5**, 349–66.

HARTMANN K.M. (1967) Photoreceptor problems in photomorphogenic responses under high-energy conditions (UV–blue–far-red). In *Book of Abstracts European Photobiology Symposium Hvar (Yugoslavia)*. pp. 29–32.

HARTUNG W., GIMMLER H., HEILMANN B. & KAISER G. (1980) The site of ABA metabolism in mesophyll cells of *Spinacia oleracea*. *Plant Sci. Lett.* **18**, 359–64.

HARTWELL L.H. (1978) Cell division from a genetic perspective. *J. Cell. Biol.*, **77**, 627–37.

HATCH M.D., KAGAWA T. & CRAIG S. (1975) Subdivision of C_4-pathway species based on differing C_4 acid decarboxylating systems and ultrastructural features. *Aust. J. Plant Physiol.*, **2**, 111–28.

HATCH M.D. & SLACK C.R. (1968) A new enzyme for the interconversion of pyruvate and phosphopyruvate. *Biochem. J.*, **106**, 141–6.

HAUPT W. (1972) Localization of phytochrome within the cell. In *Phytochrome*. (Eds. Mitrakos K., Shropshire W. Jr.) Academic Press, New York, London.

HAUPT W. & FEINLEIB M.E. (Eds) (1979) *Physiology of Movements, Encyclopaedia of Plant Physiology* Vol. 7, NS. 725 pp. Springer-Verlag, Berlin.

HAUPT W. & WEISENSEEL M.H. (1976) Physiological evidence and some thoughts on localized responses, intracellular localization and action of phytochrome. In *Light and Plant Growth*. (Ed. Smith, H.) pp. 63–74. Butterworths, London.

HAWKER L. (1932) A quantitative study on the geotropism of seedlings with special reference to the nature and development of their statolith apparatus. *Ann. Bot.*, **46**, 121–57.

HEALE J.B. & GUPTA D.P. (1972) Mechanism of vascular wilting induced by *Verticillium alboatrum*. *Trans. Brit. mycol. Soc.*, **58**, 19–28.

HEATH I.B. (1974) A unified hypothesis for the role of membrane bound enzyme complexes and microtubules in plant cell wall synthesis. *J. theoret. Biol.*, **48**, 445–9.

HEATH I.B. (1980) Behaviour of kinetochores during mitosis in the fungus *Saprolegnia ferax*. *J. Cell Biol.*, **84**, 531–46.

HEATH M.C. (1972) Ultrastructure of host and

non-host reaction to cowpea rust. *Phytopathol.*, **62**, 27–38.

HEATH M.C. (1974) Light and electron microscope studies of the interactions of host and non-host plants with cowpea rust *Uromyces phaseoli* var. *vignae*. *Physiol. Plant Pathol.*, **4**, 403–14.

HEATH M.C. (1979) Partial characterization of the electron-opaque deposits formed in the non-host plant, French bean, after cowpea rust infection. *Physiol. Plant Pathol.*, **15**, 141–8.

HEATH M.C. (1981) The absence of active defence mechanisms in compatible host–pathogen interactions. In *Active Defence Mechanisms in Plants*. (Ed. Wood R.K.S.) Plenum Press, New York.

HEBER U.W. (1974) Metabolite exchange between chloroplasts and cytoplasm. *Ann. Rev. Plant Physiol.*, **25**, 393–421.

HEDDEN P., MACMILLAN J. & PHINNEY B.O. (1978) The metabolism of the gibberellins. *Ann. Rev. Plant Physiol.*, **29**, 149–92.

HEIDEMANN S.R. & KIRSCHNER M.W. (1975) Aster formation in eggs of *Xenopus laevis*. Induction by isolated basal bodies. *J. Cell Biol.*, **67**, 105–17.

HEILMAN B., HARTUNG W. & GIMMLER H. (1980). The distribution of abscisic acid between chloroplasts and cytoplasm of leaf cells and the permeability of the chloroplast envelope for abscisic acid. *Z. Pflphysiol.*, **97**, 67–78.

HEITEFUSS R. & WILLIAMS P.H. (1976) (Eds) *Physiological Plant Pathology. Encyclopaedia of Plant Physiology*, New Series. Springer-Verlag, Berlin.

HELDT H.W., CHOU C.J., LILLEY R.M.C. & PORTIS A. (1978) The role of fructose and sedoheptulose bisphosphatase in the control of CO_2 fixation. Evidence from the effects of Mg^{2+} concentration, pH and H_2O_2. In *Proc. 4th Int. Congr. Photosynthesis*, (Eds. Hall D.O., Coombs J. & Goodwin T.W.) pp. 469–78. The Biochemical Society, London.

HELDT H.W., WERDAN K., MILOVANCEV M. & GELLER G. (1973) Alkalinisation of the chloroplast stroma caused by light-dependent proton flux into the thylakoid space. *Biochim. biophys. Acta*, **314**, 224–41.

HEMBERG T. (1965) The significance of inhibitors and other chemical factors of plant origin in the induction and breaking of rest periods. In *Encyclopedia of Plant Physiology* Vol. XV/2. (Ed. Lang, A.) pp. 669–98. Springer-Verlag, Berlin.

HENDRICKS S.B. & BORTHWICK H.A. (1967) The function of phytochrome in regulation of plant growth. *Proc. Nat. Acad. Sci. USA*, **58**, 2125–30.

HENNESSY S.W. & CANTINO E.C. (1972) Lag-phase sporogenesis in *Blastocladiella emersonii*: Induced formation of unispored plantlets. *Mycologia*, **64**, 1066–87.

HENSON I.E. & WHEELER C.T. (1977) Hormones in plant-bearing nitrogen-fixing root nodules. Cytokinin levels in roots and root nodules of non-leguminous plants. *Z. Pflphysiol.*, **84**, 179–82.

HEPLER P.K. (1976) The blepharoplast of *Marsilea*: its *de novo* formation and spindle association. *J. Cell Sci.*, **21**, 361–90.

HEPLER P.K. (1980) Membranes in the mitotic apparatus of barley cells. *J. Cell Biol.*, **86**, 490–9.

HEPLER P.K. & JACKSON W.T. (1968) Microtubules and early stages of cell-plate formation in the endosperm of *Haemanthus katherinae* Baker. *J. Cell Biol.*, **38**, 437–46.

HEPLER P.K. & NEWCOMB E.H. (1967) Fine structure of cell plate formation in the apical meristem of *Phaseolus* roots. *J. ultrastruct. Res.*, **19**, 498–513.

HEPLER P.K. & PALEVITZ B.A. (1974) Microtubules and microfilaments. *Ann. Rev. Plant Physiol.*, **25**, 309–62.

HERNALSTEENS J.P., DE GREEVE H., VAN MONTAGU M. & SCHELL J. (1978) Mutagenesis by insertion of the drug resistance transposon Tn7 applied to the Ti plasmid of *Agrobacterium tumefaciens*. *Plasmid*, **1**, 218–25.

HERNALSTEENS J.P., VAN VLIET F., DE BEUCKELEER M., DEPICKER A., ENGLER G., LEMMERS M. HOLSTERS M., VAN MONTAGU M. & SCHELL J. (1980) The *Agrobacterium tumefaciens* Ti-plasmid as a host vector system for introducing foreign DNA in plant cells. *Nature, Lond.*, **287**, 654–6.

HERRMANN R.G., KOWALLIK K.V. & BOHNERT H.J. (1976) Structural and functional aspects of the plastome. I—The organization of the plastome. *Portugali. Acta Biol.*, **14** A; 91–110.

HERMANN R.G. & POSSINGHAM J.V. (1979) Plastid DNA—The plastome. In *The Plastids*. (Ed. Reinert J.) Springer-Verlag, Vienna.

HERTEL R. & LEOPOLD A.C. (1963) Versuche zur analyse des Auxintransports in der Koleoptile von *Zea mays* L. *Planta*, **59**, 535–62.

HESLOP-HARRISON J. (1956) Auxin and sexuality in *Cannabis sativa*. *Physiol. Plant.*, **9**, 588–97.

HESLOP-HARRISON J. (1964a)

HESLOP-HARRISON J. (1964b) Cell walls, cell membranes, and protoplasmic connections during meiosis and pollen development. In *Pollen Physiology and Fertilization*. (Ed.

Linskens H.F.) pp. 39–47. Elsevier/North-Holland, Amsterdam.

Heslop-Harrison J. (1975) Incompatibility and the pollen-stigma interaction. *Ann. Rev. Plant Physiol.*, **26**, 403–25.

Heslop-Harrison J. (1978a) Genetics and physiology of angiosperm incompatibility systems. *Proc. Roy. Soc. Lond.*, B **202**, 73–92.

Heslop-Harrison J. (1978b) Recognition and response in the pollen-stigma interaction. In *Cell-Cell Recognition.* (Ed. Curtis A.S.G.) pp. 121–38. Symp. Soc. Exp. Biol. No. 32, Cambridge University Press.

Heslop-Harrison J. (1978c) *Cellular Recognition Systems in Plants.* Studies in Biology, Vol. 100. Edward Arnold, London.

Heslop-Harrison J. (1979) An interpretation of the hydrodynamics of pollen. *Am. J. Bot.*, **66**, 737–43.

Heslop-Harrison J. (1980) Aspects of the structure, cytochemistry and germination of rye pollen (*Secale cereale* L.). *Ann. Bot.*, **44**, (Suppl. 1) 1–47.

Heslop-Harrison J., Heslop-Harrison Y., Knox R.B. & Howlett B. (1973) Pollen wall proteins: gametophytic and sporophytic fractions in the pollen walls of the Malvaceae. *Ann. Bot.*, **37**, 403–12.

Heslop-Harrison Y. (1975) Enzyme release in carnivorous plants. In *Lysosomes in Biology and Pathology.* (Eds. Dingle J.T. & Dean R.T.) pp. 525–71. North-Holland, Amsterdam.

Hewitt J.A. (1978) Diffusion gradients, membrane receptors and the acquisition of orientational information by cells. *J. theoret. Biol.*, **74**, 297–306.

Heyes J.K. (1960) Nucleic acid changes during cell expansion in the root. *Proc. Roy Soc. Lond., B* **152**, 218–30.

Heyes J.K. (1977) Metabolic aspects of cell growth and development in the root. In *Physiology of the Garden Pea.* (Eds. Sutcliffe J.F. & Pate J.S.) pp. 153–81. Academic Press, London, New York & San Francisco.

Heyes J.K. & Dale J.E. (1971) A virescens mutant of *Phaseolus vulgaris*: photosynthesis and metabolic changes during leaf development. *New Phytol.*, **70**, 415–26.

Heyn A.N.J. (1933) Further investigations on the mechanism of cell elongation and the properties of the cell wall in connection with elongation. I. The load extension relationship. *Protoplasma*, **19**, 78–96.

Heyn A.N.J. (1966) The microcrystalline structure of cellulose in cell walls of cotton, ramie and jute fibres as revealed by negative staining of sections. *J. Cell Biol.*, **29**, 181–97.

Hicks G.S. (1980) Patterns of organ development in plant tissue culture and the problem of organ determination. *Bot. Rev.*, **46**, 1–23.

Hicks G.A. & McHughen A. (1974) Altered morphogenesis of placental tissues of tobacco *in vitro.* Stigmatoid and carpelloid outgrowths. *Planta*, **121**, 193–6.

Higashi S. (1967) Transfer of clover infectivity of *Rhizobium trifolii* to *Rhizobium phaseoli* as mediated by an episomic factor. *J. gen. app. Microbiol.*, **13**, 391–403.

Higgins T.J.V. & Jacobsen J.V. (1978) Phytohormones and subcellular structural modification. In *Phytohormones and Related Compounds—A Comprehensive Treatise, Vol. 1.* (Eds. Letham D.S., Goodwin P.B. & Higgins T.J.V.) pp. 419–56. Elsevier/North-Holland, Amsterdam.

Higgins T.J.V. & Spencer D. (1977) Cell-free synthesis of pea seed proteins. *Plant Physiol.*, **60**, 655–61.

Higgins T.J.V. & Spencer D. (1981) Precursor forms of pea storage proteins: modification by microsomal membranes during cell-free translation. *Plant Physiol.*, **67**, 205–11.

Higgins T.J.V., Zwar J.A. & Jacobsen J.V. (1976) Gibberellic acid enhances the level of translatable RNA for amylase in barley aleurone layers. *Nature, Lond.*, **260**, 166–9.

Highfield P.E. & Ellis R.J. (1978) Synthesis and transport of the small subunit of chloroplast ribulose bisphosphate carboxylase. *Nature, Lond.*, **271**, 420–4.

Higinbotham N. (1973). Electropotentials of plant cells. *Ann. Rev. Plant Physiol.*, **24**, 25–46.

Hill J.E. & Breidenbach R.W. (1974) Proteins of soybean seeds. II. Accumulation of the major protein components during seed development and maturation. *Plant Physiol.*, **53**, 747–51.

Hilton J.R. & Smith H. (1980) The presence of phytochrome in purified barley etioplasts and its *in vitro* regulation of biologically-active gibberellin levels in etioplasts. *Planta*, **148**, 312–18.

Hinch J.M. & Clarke A.E. (1980) Adhesion of fungal zoospores is mediated by carbohydrate determinants of the root slime. *Physiol. Plant Pathol.*, **16**, 303–7.

Hinegardner R.T. (1971) An improved fluorometric assay for DNA. *Anal. Biochem.*, **39**, 197–201.

Ho T-H.D. & Varner J.E. (1978) Density labeling of proteins and ^{13}C-labeled amino acids. No accumulation of an inactive α-amylase precursor in barley aleurone cells.

Arch. Biochem. Biophys., **187**, 441–6.

HOAD G.V. (1978) Effect of water stress on abscissic acid levels in white lupin (*Lupinus albus* L.) fruit, leaves and phloem exudate. *Planta*, **142**, 287 90.

HOCK B. (1970) Die zeitliche Dauer der Isocitrat Lyase Synthese im Kotyledonen von Wassermelonenkeimlingen. *Planta*, **93**, 26–38.

HOD Y. & HERSHKO A. (1976) Relationship of the pool of intracellular valine to protein synthesis and degradation in cultured cells. *J. biol. Chem.*, **251**, 4458–67.

HODGKINSON K.C. (1974) Influence of partial defoliation on photosynthesis, photorespiration and transpiration by lucerne leaves of different ages. *Aust. J. Plant Physiol.*, **1**, 561–78.

HOEKSTRA F.A. & BRUINSMA J. (1979) Protein synthesis of binucleate and trinucleate pollen and its relationship to tube emergence and growth. *Planta*, **146**, 559–66.

HOFFMANN P. (1962) Untersuchungen über Photosynthese und Atmung von Laubblättern verschedenen alters. *Flora*, (Jena), **152**, 622–54.

HOGETSU T. & SHIBAOKA H. (1978a) The change of pattern in microfibril arrangement on the inner surface of the cell wall of *Closterium acerosum* during cell growth. *Planta*, **140**, 7–14.

HOGETSU T. & SHIBAOKA H. (1978b) Effects of colchicine on cell shape and on microfibril arrangement in the cell wall of *Closterium acerosum*. *Planta*, **140**, 15–18.

HOLL F.B. (1973) Host-determined genetic control of nitrogen-fixation in the *Pisum-Rhizobium* symbiosis. *Canad. J. Genetics Cytol.*, **15**, 659.

HOLL F.B. & LARUE T.A. (1976) Genetics of legume host plants. In *Proc. 1st Int. Symp. on Nitrogen Fixation*. Vol. II. (Eds. Newton W.E. & Nyman C.J.) pp. 391–9. Washington State University Press. Pullman, Washington.

HOLLAND M.J. & HOLLAND J.P. (1978) Isolation and identification of yeast messenger ribonucleic acids coding for enolase, glyceraldehyde-3-phosphate dehydrogenase and phosphoglycerate kinase. *Biochemistry*, **17**, 4900–7.

HOLMES M.G. & MCCARTNEY A. (1976) Spectral energy distribution in the natural environment and its implications for phytochrome function. In *Light and Plant Development*. (Ed. Smith H.) pp. 467–76. Butterworths, London, Boston.

HOLMES M.G. & SMITH H. (1975) The function of plants growing in the natural environment. *Nature, Lond.*, **254**, 512–4.

HOLMES M.G. & SMITH H. (1977a) The function of phytochrome in the natural environment I. Characterization of daylight for studies in photomorphogenesis and photoperiodism. *Photochem. Photobiol.*, **25**, 533–8.

HOLMES M.G. & SMITH H. (1977b) The function of phytochrome in the natural environment. The influence of vegetation canopies on the spectral energy distribution of natural daylight. *Photochem. Photobiol.*, **25**, 539–45.

HOLMES M.G. & WAGNER E. (1980) A re-evaluation of phytochrome involvement in time measurement in plants. *J. theoret. Biol.*, **83**, 255–65.

HOLSTEN R.D., BURNS R.C., HARDY R.W.F. & HEBERT R.R. (1971) Establishment of symbiosis between *Rhizobium* and Plant cell *in vitro*. *Nature, Lond.*, **232**, 113–76.

HOLSTERS M., SILVA B., VAN VLIET F., GENETELLO C., DE BLOCK M., DHAESE P., DEPICKER A., INZÉ D., ENGLER G., VILLARROEL R., VAN MONTAGU M., & SCHELL J. (1980) The functional organization of the nopaline *A. tumefaciens* plasmid pTiC58. *Plasmid*, **3**, 212–30.

HOPP H.E., ROMERO D.A. & PONT-LEZICA R. (1978) On the inhibition of cellulose synthesis by coumarin. *FEBS Lett.*, **86**, 259–62.

HOPPER J.E., ASCHER P.D. & PELOQUIN S.J. (1967) In activation of self-incompatibility following temperature pre-treatment of styles in *Lilium longiflorum*. *Euphytica*, **16**, 215–20.

HORNER M. & MOTT R.L. (1980) The frequency of embryogenic pollen grains is not increased by *in vitro* anther culture in *Nicotiana tabacum*. *Planta*, **147**, 156–8.

HORSFALL J.G. & COWLING E.B. (1978) *Plant Pathology: An Advanced Treatise*. Vols. I–IV. Academic Press, New York.

HORTON J.C. & KEEN N.T. (1966) Sugar repression of endopolygalacturonase and cellulase synthesis during pathogenesis by *Pyrenochaeta terrestris* as a resistance mechanism in onion pink root rot. *Phytopathology*, **56**, 908–16.

HOVER J.M. & GUSTAVSON F.G. (1926) Rates of respiration as related to age. *J. gen. Physiol.*, **10**, 33–9.

HOWELL C.R. (1976) Use of enzyme-deficient mutants of *Verticillium dahliae* to assess the importance of pectolytic enzymes in symptom expression of *Verticillium* wilt of cotton. *Physiol. Plant Pathol.*, **9**, 279–83.

HOWELL S.H., WALKER L.L. & DUDLEY R.K. (1980) Cloned cauliflower mosaic virus DNA infects turnips (*Brassica rapa*). *Science*, **208**, 1265–7.

Hsu F.C. (1979) Abscisic acid accumulation in developing seeds of *Phaseolus vulgaris* L. *Plant Physiol.*, **63**, 552–6.

Huang B., Rodaway S., Wood A. & Marcus A. (1980) RNA synthesis in germinating embryos of soybean and wheat. *Plant Physiol.*, **65**, 1155–9.

Hubbell D.M. & Elkan G.H. (1967) Correlation of physiological characteristics with nodulating ability in *Rhizobium japonicum*. *Canad. J. Microbiol.*, **13**, 235–41.

Hubbell D.M., Norales V.M. & Umali-Garcia M. (1978) Pectolytic enzyme in *Rhizobium*. *App. environ. Microbiol.*, **35**, 210–13.

Huber W., de Fekete M.A.R. & Ziegler U. (1973) Enzyme des Starkeumsatzes in den Wurzelhaube von *Zea mays* L. *Planta*, **112**, 343–56.

Huffaker R.C. & Miller B.L. (1978) Reutilisation of ribulose bisphosphate carboxylase. In *Photosynthetic Carbon Assimilation*. (Eds. Siegelman H.W. & Hind G.) pp. 139–52. Plenum Press, New York.

Hughes J.E. & Gunning B.E.S. (1980) Glutaraldehyde-induced deposition of callose. *Canad. J. Bot.*, **58**, 250–8.

Hughes R.C. (1976) *Membrane Glycoproteins*. Academic Press, London.

Hughes T.A., Lecce J.G. & Elkan G.H. (1979) Modified-fluorescent technique using rhodomine for studies of *Rhizobium japonicum* soybean symbiosis. *App. environ. Microbiol.*, **37**, 1243–4.

Hunt R.E. & Pratt L.H. (1980) Radioimmunoassay of phytochrome content in green light-grown oats. *Plant Cell and Environment*, **3**, 91–5.

Hunter W.J. & Elkan G.H. (1975) Role of pectic and cellulolytic enzyme in the invasion of the soybean by *Rhizobium japonicum*. *Canad. J. Microbiol.*, **21**, 1254–8.

Hurrell J.G.R. & Leach S.J. (1977) The amino acid sequence of soybean leghaemoglobin C_2. *FEBS Lett.*, **80**, 23–6.

Hyams J.S. & Borisy G.G. (1978) Nucleation of microtubules *in vitro* by isolated spindle pole bodies of the yeast *Saccharomyces cerevisiae*. *J. Cell Biol.*, **78**, 401–14.

Ihl M. (1976) Indole-acetic acid binding proteins in soybean cotyledon. *Planta*, **131**, 223–8.

Ihle J.N. & Dure L.S. (1972) The developmental biochemistry of cotton seed embrogenesis and germination. III. Regulation of the biosynthesis of enzymes utilized in germination. *J. Biol. Chem.*, **247**, 5048–55.

Ingle J., Bietz D. & Hageman R.H. (1965) Changes in composition during development and maturation of maize seeds. *Plant Physiol.*, **40**, 835–9.

Ingle J. & Sinclair J. (1972) Ribosomal RNA genes and plant development. *Nature, Lond.*, **235**, 30–32.

Ingram D.S. (1978) Cell death and resistance to biotrophs. *Ann. appl. Biol.*, **89**, 291–5.

Inoué S. & Ritter H. (1978) Mitosis in *Barbulanympha*. II. Dynamics of a two-stage anaphase, nuclear morphogenesis and cytokinesis. *J. Cell Biol.*, **77**, 655–84.

Insel P.A. (1978) Membrane-active hormones: receptors and receptor regulation. *Int. Rev. Biochem.*, **20**, 1–44.

Ivanov V.B. (1968) The growth of corn seedling root cells irradiated with high X-ray doses. II. Cell growth with a complete blocking of mitosis. *Tsitologiya*, **10**, 1105–17.

Ivanov V.B. (1971) Critical size of the cell and its transition to division. I. Sequence of transition to mitosis for sister cells in the corn seedling root tip. *Ontogenez*, **2**, 524–35.

Ivanov V.N., Misharin S.I., Reimers F.E. & Khavkin E.E. (1974) Dynamics of the mitochondrial and cytoplasmic antigens in growing and mature cells of corn roots. *Dokl. Akad. Nauk. SSSR*, **218**, 1229–32.

Ivanov V.B. & Filipenko V.N. (1976) Role of asymmetric mitosis in differentiation of rhizoderm cells of festucoid grasses. *Dokl. Akad. Nauk SSSR*, **230**, 712–15.

Iversen T-H. & Larsen P. (1973) Movement of amyloplasts in the statocytes of geotropically stimulated roots. The preinversion effect. *Physiol. Plant*, **28**, 172–81.

Iversen T.H., Pedersen K. & Larsen P. (1968) Movement of amyloplasts in the root cap cells of geotropically sensitive roots. *Physiol. Plant.*, **21**, 811–19.

Iversen T.H. & Rommelhoff A. (1978) The starch statolith hypothesis and the interaction of amyloplasts and endoplasmic reticulum in root geotropism. *J. exp. Bot.*, **113**, 1319–28.

Izhar J. & Frankel R. (1971) Mechanism of male sterility in *Petunia*: the relationship between pH, callase activity in the anthers, and the breakdown of microsporogenesis. *Theor. appl. Genet.*, **41**, 104–8.

Jackson W.A., Kwik K.D., Volk R.J. & Butz R.G. (1976) Nitrate influx and efflux by intact wheat seedlings: effects of prior nitrate nutrition. *Planta*, **132**, 149–56.

Jacob F. (1964) Uber die Funktion eines Karotin—Lichtschirmes Chromosporer. Pilobolus—Arten. *Flora*, **155**, 309–33.

Jacob F. & Monod J. (1961) Genetic regulatory

mechanisms in the synthesis of proteins. *J. molec. Biol.*, **3**, 318–56.

Jacobs M. & Hertel R. (1978) Auxin binding to subcellular fractions from *Cucurbita* hypocotyls: *in vitro* evidence for an auxin transport carrier. *Planta*, **142**, 1–10.

Jacobs W.P. (1952) The role of auxin in differentiation of xylem around a wound. *Am. J. Bot.*, **39**, 301–9.

Jacobs W.P. (1979) *Plant Hormones and Plant Development*, p. 21. Cambridge University Press.

Jacobsen L.V. (1977) Regulation of ribonucleic acid metabolism by plant hormones. *Ann. Rev. Plant Physiol.*, **28**, 537–64.

Jacobsen J.V. & Varner J.E. (1967) Gibberellic acid induced synthesis of protease by isolated aleurone layers of barley. *Plant Physiol.*, **42**, 1596–600.

Jaenicke L. & Seferiadis K. (1975) Die Stereochemie von Fucoserraten, dem Gameten-Lockstoff der Braunalge *Fucus serratus* L. *Chem. Ber.*, **108**, 225–32.

Jaffe L.F. (1968) Localization in the developing *Fucus* egg and the general role of localizing currents. *Adv. Morphogen.*, **7**, 295–328.

Jaffe L.F. (1969a) Localisation in the developing *Fucus* egg and the general role of localising currents. *Adv. Morphogen.*, **7**, 295–328.

Jaffe L.F. (1969b) On the centripetal course of development the *Fucus* egg and self electrophoresis. In *Communication in Cell Development*. (Ed. Lang A.) Developmental Biology suppl. 3, pp. 83–111.

Jaffe L.F. (1980) Control of plant development by steady ionic currents. Membrane Transport in Plants. (Eds. Spanswick R.M., Lucas W.J. & Dainty J.). Elsevier, Amsterdam.

Jaffe L.F. & Nuccitelli R. (1977) Electrical controls of development. *Ann. Rev. Biophys. Bioeng.*, **6**, 445–76.

Jaffe L.F., Robinson K.R. & Nuccitelli R. (1974) Local cation entry and self-electrophoresis as an intracellular localisation mechanism. In *Membrane Transport in Plants.* (Eds. Zimmerman U. & Dainty J.) pp. 226–34. Springer-Verlag, Berlin.

James T.W. & Spencer M.S. (1979) Cyanide-insensitive respiration in pea cotyledons. *Plant Physiol.*, **64**, 431–4.

James W.O. (1953) *Plant Respiration.* Clarendon Press, Oxford.

Janczewski E. de (1874) Recherches sur l'accroissement terminal des racines dans les phanérogames. *Ann. Sci. nat. Bot. 5ᵉ sér.* **20**, 162–201.

Jaworski A.J., Lu W.B.L. & Bateman O.M. (1976) Metabolism of polyadenylic acid sequences during germination of *Blastocladiella emersonii* zoospores. *Arch. Biochem. Biophys.*, **117**, 516–24.

Jayaram M., Leutwiler L. & Delbruck M. (1980) Light-induced carotene synthesis in mutants of *Phycomyces* with abnormal phototropism, *Photochem. Photobiol.*, **32**, 241–5.

Jayaram M., Presti D. & Delbruck M. (1979) Light induced carotene synthesis in *Phycomyces. Exp. Mycol.*, **3**, 42–52.

Jeffs R.A. & Northcote D.H. (1967) The influence of indol-3yl-acetic acid and sugar on the pattern of induced differentiation in plant tissue culture. *J. Cell Sci.*, **2**, 77–88.

Jenkins G.I. & Woolhouse H.W. (1981) Photosynthetic electron transport during senescence of the primary leaves of *Phaseolus vulgaris* L. I. Non-cyclic electron transport. *J. exp. Bot.*, **32**, 467–78.

Jenni B. & Stutz E. (1978) Physical mapping of the ribosomal DNA region in *Euglena gracilis. Eur. J. Biochem.*, **88**, 127–34.

Jennings A.C. & Morton R.K. (1963a) Changes in carbohydrate, protein and non-protein nitrogenous compounds of developing wheat grain. *Aust. J. biol. Sci.*, **16**, 318–31.

Jennings A.C. & Morton R.K. (1963b) Changes in nucleic acids and other phosphorous-containing compounds of developing wheat grain. *Aust. J. biol. Sci.*, **16**, 323–41.

Jennings A.C. & Morton R.K. (1963c) Amino acids and protein synthesis in developing wheat endosperm. *Aust. J. biol. Sci.*, **16**, 384–94.

Jennings A.C., Morton R.K. & Palk B.A. (1963) Cytological studies of protein bodies of developing wheat endosperm. *Aust. J. biol. Sci.*, **16**, 366–74.

Jensen R.G. & Bahr R.T. (1977) Ribulose 1,5-bisphosphate carboxylase-oxygenase. *Ann. Rev. Plant Physiol.*, **28**, 379–400.

Jensen R.G. & Bassham J.A. (1968) Photosynthesis by isolated chloroplasts. III. Light activation of the carboxylation reaction. *Biochim. biophys. Acta*, **153**, 227–34.

Jensen W.A. (1961) Relation of primary cell wall formation to cell development in plants. In *Synthesis of Molecular and Cellular Structure.* (Ed. Rudnick D.) pp. 89 110. Ronald Press Co., New York.

Jensen W.A. (1972) The embryo sac and fertilization in angiosperms. Univ. of Hawaii, Harold L. Lyon Arboretum Lecture No. 3, 1–31.

JENSEN W.A., SCHULTZ P. & ASHTON M.E. (1977) An ultrastructural study of early endosperm development and synergid changes in unfertilized cotton ovules. *Planta*, **133**, 179–89.

JERIE P.H., SHAARI A.R. & HALL M.A. (1979) The compartmentation of ethylene in developing cotyledons of *Phaseolus vulgaris* L. *Planta*, **144**, 403–7.

JERIE P.H., SHAARI A.R., ZERONI M. & HALL M.A. (1978) The partition coefficient of $^{14}C_2H_4$ in plant tissue as a screening test for metabolism of compartmentation of ethylene. *New Phytol.*, **81**, 499.

JESAITAS A.J. (1974) Linear dichroism and orientation of the *Phycomyces* photopigment. *J. gen. Physiol.*, **63**, 1–21.

JESAITIS A.J., HENERS P.R., HERTEL R. & BRIGGS W.R. (1977) Characterisation of a membrane fraction containing a b-type cytochrome. *Plant Physiol.*, **59**, 941–7.

JOHNSON C.B. (1976) Rapid activation by phytochrome of nitrate reductase in the cotyledons of *Sinapis alba*. *Planta*, **128**, 127–31.

JOHNSON C.B. (1980) The effect of red light in the high irradiance reaction of phytochrome. *Plant, Cell Environ.*, **3**, 45–51.

JOHNSON C.B., HOLLOWAY B.R., SMITH H. & GRIERSON D. (1973) Isoenzymes of acid phosphatase in germinating peas. *Planta*, **115**, 1–10.

JOHNSON C.F., ATTRIDGE T. & SMITH H. (1975) Regulation of phenylalanine ammonia-lyase synthesis by cinnamic acid. Its implications for the light mediated regulation of the enzyme. *Biochem. biophys. Acta*, **385**, 11–19.

JOHNSON E.M. & ALLFREY V.G. (1978) Post-synthetic modifications of histone primary structure: phosphorylation and acetylation as related to chromatin conformation and function. In *Biochemical Actions of Hormones*, Vol. 5, (Ed. Litwack G.) pp. 1–51. Academic Press, New York, San Francisco & London.

JOHNSON S.A., LOVETT J.S. & WILT F.H. (1977) The polyadenylated RNA of zoospores and growth phase cells of the aquatic fungus, *Blastocladiella*. *Dev. Biol.*, **56**, 329–42.

JOHNSSON A. & PICKARD B.G. (1979) The threshold stimulus for geotropism. *Physiol. Plant.*, **45**, 315–19.

JOHNSTON A.W.B., BEYNON J.L., BUCHANAN-WOLLASTON A.V., SETCHELL S.M., HIRSCH P.R. & BERINGER J.E. (1978) High frequency transfer of nodulating ability between strains and species of *Rhizobium*. *Nature, Lond.*, **276**, 634–6.

JOHNSTON A.W.B., BERINGER J.E., BEYNON J.L., BREWIN N., BUCHANAN-WOLLASTON A.V. & HIRSCH P.R. (1979) Plasmid and the *Rhizobium*-legume symbiosis. Proc. 3rd Int. Symp. on Nitrogen Fixation, Madison, (in press)

JOHNSTON A.W.B., BREWSTER V. & DAVIES D.R. (1977) Seed proteins of peas in relation to nitrogen fixation. *Ann. Bot.*, **41**, 381–5.

JONES D.D. & MORRÉ D.J. (1967) Golgi apparatus mediated polysaccharide secretion by outer root cap cells of *Zea mays*: II. Isolation and characterization of the secretory product. *Zsch. Pflphysiol.*, **56**, 166–9.

JONES D.D. & MORRÉ D.J. (1973) Golgi apparatus mediated polysaccharide secretion by outer root cap cells of *Zea mays*. III. Control by exogenous sugars. *Physiol. Plant.*, **29**, 68–75.

JONES P.C.T. (1972) Central role for ATP in determining some aspects of animal and plant behaviour. *J. theoret. Biol.*, **34**, 1–13.

JONES R.A., LARKINS B.A. & TSAI C.Y. (1976) Reduced synthesis of zein *in vitro* by a high lysine mutant of maize. *Biochem. Biophys. Res. Commun.*, **69**, 404–10.

JONES R.A., LARKINS B.A. & TSAI C.Y. (1977) Storage protein synthesis in maize. *Plant Physiol.*, **59**, 733–7.

JONES R.L. & CHEN R.F. (1976) Immunohistochemical localization of α-amylase in barley aleurone cells. *J. Cell Sci.*, **20**, 183–98.

JONES R.L. & PRICE J.M. (1970) Gibberellic acid and the fine structure of barley aleurone cells. III. Vacuolation of the aleurone cell during the phase of ribonuclease release. *Planta*, **94**, 191–202.

JONES R.L. & STODDART J.L. (1977) Gibberellins and seed germination. In *The Physiology and Biochemistry of Seed Dormancy and Germination.* (Ed. Khan A.A.) pp. 77–109. Elsevier/North-Holland Biomedical Press, Amsterdam.

JONKERS G.M., TIMMERHANS H.A.T., DE JONG F.M., LAMBERTS S.W.J. & BIRKENHÄGER J.C. (1980) An energy-dependent corticosterone uptake system in the rat liver cell. *FEBS Lett.*, **111**, 403–6.

JORDAN G., TIMMIS J.N. & TREWAVAS A.J. (1981) The plant nucleus. In *Comprehensive Treatise of Plant Biochemistry*. (Eds. Conn E.C., Stumpf P.K. & Talbot E.) Vol. 1, Academic Press, New York.

JULIANO B.O. & BOULTER D. (1976) Extraction and composition of rice endosperm glutelin. *Phytochemistry*, **15**, 1601–6.

KAFATOS F.C. (1972) mRNA stability and cellular differentiation. *Karoliaska Symposium on Research Methods in Reproductive Endocrinology*, **5**, 319–41.

KAHL G. (1973) Genetic and metabolic regulation in differentiating plant storage cells. *Bot. Rev.*, **39**, 274–99.

KAHN F.R. & MCFADDEN B.A. (1979) Enzyme profiles in seedling development and the effect of itaconate, an isocitrate lyase-directed reagent. *Plant Physiol.*, **64**, 228–31.

KANNANGARA C.G. & WOOLHOUSE H.W. (1968) Changes in the enzyme activity of soluble protein fraction in the course of foliar senescence in *Perilla frutescens* L. Britt. *New Phytol.*, **67**, 533–42.

KAPIL R.N. & TIWARI R.N. (1978) Plant embryological investigations and fluorescence microscopy: an assessment of integration. *Int. Rev. Cytol.*, **53**, 291–331.

KAPOOR A.C. & GUPTA Y.P. (1977) Changes in proteins and amino acids in developing soybean seed and the effect of phosphorus nutrition. *J. Sci. Food Agric.*, **28**, 113–20.

KARR A.L., KARR D.B. & STROBEL G.A. (1974) Isolation and partial characterization of four host-specific toxins of *Helminthosporium maydis* (race T). *Plant Physiol.*, **53**, 250–7.

KASARDA D.D., BERNARDIN J.E. & NIMMO C.C. (1976) Wheat proteins. In *Advances in Cereal Science and Technology*. (Ed. Pomeranz Y.) pp. 158–236. American Association of Cereal Chemists Inc., St. Paul.

KATZENELLENBOGEN B.S. & GORSKI J. (1972) Estrogen action *in vitro*. Induction of the synthesis of a specific uterine protein. *J. biol. Chem.*, **247**, 1299–305.

KAUR-SAWHNEY R. & GALSTON A.W. (1979) Interactions of polyamines and light on biochemical processes involved in leaf senescence. *Plant, Cell Environ.*, **2**, 189–96.

KAWASHIMA N. & MITAKE T. (1969) Studies on protein metabolism in higher plants. VI. Changes in ribulose diphosphate carboxylase activity and fraction I protein content in tobacco leaves with age. *Agr. biol. Chem.*, **33**, 539–43.

KAWASHIMA N. & WILDMAN S.G. (1972) Studies on fraction I protein. IV. Mode of inheritance of primary structure in relation to whether chloroplast or nuclear DNA contains the code for a chloroplast protein. *Biochim. biophys. Acta*, **262**, 42–50.

KAWATA S. & MATSUI S. (1977) On the cortex formation in the apical meristems of the elongating crown roots of rice plants. *Jap. J. Crop Sci.*, **46**, 403–13.

KEEN N.T. (1971) Hydroxyphaseollin production by soybeans resistant and susceptible to *Phytophthora megasperma* var. *sojae*. *Physiol. Plant Pathol.*, **1**, 265–76.

KEEN N.T. (1975) Specific elicitors of plant phytoalexin production: determinants of race specificity in pathogens? *Science*, **187**, 74–5.

KEEN N.T. & BRUEGGER B.B. (1977) Phytoalexins and chemicals that elicit their production in plants. In *Host Plant Resistance to Pests*, ACS Symposium Series **62**. (Ed. Hedin P.A.) pp. 1–26. American Chemical Society, Washington D.C.

KEEN N.T. & LEGRAND M. (1980) Surface glycoproteins: evidence that they may function as the race-specific phytoalexin elicitors of *Phytophthora megasperma* f.sp. *glycinea*. *Physiol. Plant Pathol.*, **17**, 175–92.

KEEN N.T. & LITTLEFIELD L. (1979) The possible association of phytoalexins with resistance gene expression in flax to *Melampsora lini*. *Physiol. Plant Pathol.*, **14**, 265–80.

KEEN N.T., WANG M.C., BARTNICKI-GARCIA S. & ZENTMEYER G.A. (1975) Phytotoxicity of mycolaminarans, β 1,3 glucans from *Phytophthora* spp. *Physiol. Plant Pathol.*, **7**, 91–7.

KEFFORD N.P., BROCKWELL J. & ZWAR J.A. (1960) The symbiotic synthesis of auxin by legumes and nodule bacteria and its role in nodule development. *Aust. J. biol. Sci.*, **13**, 456–67.

KELLER W.A. & STRINGHAM G.R. (1978) Production and utilization of microspore-derived haploid plants. In *Frontiers of Plant Tissue Culture*. (Ed. Thorpe T.E.) pp. 113–22. Int. Assoc. Plant Tissue Culture, University of Calgary, Alberta.

KEMBLE R.J. (1980) A rapid single leaf assay for detecting the presence of 'S'-male-sterile cytoplasm in maize. *Theor. appl. Genet.*, **57**, 97–100.

KEMBLE R.J. & BEDBROOK J.R. (1980) Low molecular weight circular and linear DNA in mitochondria from normal and male-sterile *Zea mays* cytoplasm. *Nature, Lond.*, **284**, 565–6.

KEMBLE R.J., GONN R.E. & FLAVELL R.B. (1980) Classification of normal and male-sterile cytoplasms in maize. II. Electrophoretic analysis of DNA species in mitochondria. *Genetics*, **95**, 451–8.

KEITH B. & SRIVASTAVA L.M. (1978) Effects of colchicine and lumicolchicine on hypocotyl elongation, respiration rates and microtubules in gibberellic-acid-treated lettuce seedlings. *Planta*, **139**, 301–3.

KENDE H. & GARDNER G. (1976). Hormone

binding in plants. *Ann. Rev. Plant Physiol.*, **27**, 267–90.

Kenfield D.S. & Strobel G.A. (1977) Biochemical aspects of plant disease resistance and susceptibility. In *Host Plant Resistance to Pests.* (Ed. Hedin P.A.) pp. 35–46. Am. Chem. Soc. Symp., **62**, 35–46.

Kerr A., Manigault P. & Tempé J. (1977) Transfer of virulence *in vivo* and *in vitro* in *Agrobacterium*. *Nature, Lond.*, **265**, 560–1.

Keusch L. (1968) Die Mobilisierung des Reservemannans im kiemenden Dattelsemen. *Planta*, **78**, 321–50.

Khavkin E.E. & Perelyaeva A.I. (1970) Phenylalanine ammonia-lyase and accumulation of soluble phenolic compounds in mature cells of corn roots and coleoptiles. *Dokl. Akad. Nauk SSSR*, **193**, 227–30.

Khavkin E.E., Varakina N.N. & Perelyaeva A.I. (1971) Connection between mitochondrial and cytoplasmic intensification of respiration in elongating maize root cells. *Fiziol. Rastenii*, **18**, 42–9.

Khoo U. & Wolf M.J. (1970) Origin and development of protein granules in maize endosperm. *Am. J. Bot.*, **57**, 1042–50.

Kidby D.K. & Goodchild D.J. (1966) Host influence in the ultrastructure of root nodules of *Lupinus luteus* and *Ornithopus sativus*. *J. gen. Microbiol.*, **45**, 147–52.

King P.J. & Street H.E. (1977) Growth patterns in cell cultures. In *Plant Tissue and Cell Culture.* (Ed. Street H.E.) pp. 307–87. Blackwell Scientific Publications, Oxford.

Kipps A.E. & Boulter D. (1974) Origins of the amino acids in pods and seeds of *Vicia faba* L. *New Phytol.*, **73**, 675–84.

Király Z., Barna B. & Ersek T. (1972) Hypersensitivity as a consequence, not the cause of plant resistance to infection. *Nature, Lond.*, **239**, 456–8.

Kirby E.G. & Vasil I.K. (1979) Effect of pollen protein diffusates on germination of eluted pollen samples of *Petunia hybrida in vitro*. *Ann. Bot.*, **44**, 361–7.

Kitazawa K., Inagaki H. & Tomiyama K. (1973) Cinephotomicrographic observations on the dynamic responses of protoplasm of a potato plant cell to infection by *Phytophthora infestans*. *Phytopath. Z.*, **76**, 80–6.

Klapwijk P.M., Scheuldermon T. & Schilperoort R.A. (1978) Coordinated regulation of octopine degradation and conjugative transfer of Ti-plasmids in *Agrobacterium tumefaciens*: evidence for a common regulatory gene and separate operons. *J. Bacteriol.*, **136**, 775–85.

Klein S., Baranholz H. & Budnik A. (1971) The initiation of growth in isolated lima bean axes. Physiological and fine structural effects of actinomycin D, cycloheximide and chloramphenicol. *Plant, Cell Physiol.*, **12**, 41–60.

Klemm E. & Ninnemann, H. (1979) Nitrate reductase: a key enzyme in blue light-promoted conidiation and absorbance change of *Neurospora*. *Photochem. Photobiol.*, **29**, 629–32.

Klucas R.V. & Evans H.J. (1968) An electron donor system for nitrogenase-dependent acetylene reduction by extracts of soybean nodules. *Plant Physiol.*, **43**, 1458–60.

Klucas R.V., Koch B., Russell S. & Evans H.J. (1968). Purification and some properties of nitrogenase from soybean (*Glycine max.* Merr.) nodules. *Plant Physiol.*, **43**, 1906–12.

Knapf G. & Jacobi G. (1975) Dark starvation and plant metabolism. II. CO_2 fixation in isolated chloroplasts. *Planta*, **128**, 135–43.

Knapp E. (1931) Entwicklungsphysiologische Untersuchungen an Fucaceen-eiern. I. Zur Kenntnis der Polarität der Eier von *Cystosira barbata*. *Planta*, **14**, 731–51.

Knee M. (1977) Cell wall degradation in senescent fruit tissue in relation to pathogen attack. In *Cell Wall Biochemistry Related to Host–Plant Pathogen Interactions.* (Eds. Solheim B. & Raa J.) pp. 259–62. Universitetsforlaget, Oslo.

Knoth R. & Hagemann R. (1977) Struktur und Funktion der genetischen Information in den Plastiden. XVI. Die Feinstruktur der Plastiden und der elektronenmikroskopische Nachweis echter Mischzellen in Blättern der Plastommutationen auslösenden Genmutante albostrians von *Hordeum vulgare* L. *Biol. Zentralbl.*, **96**, 141–50.

Knoth R. von, Herrmann F.H., Böttger H. & Börner T. (1974) Struktur und Funktion der genetischen Information in den Plastiden. XI. DNA in normalen und mutierten Plastiden der Sorte 'Mrs. Parker' von *Pelargonium zonale*. *Biochem. Physiol. Pfl.*, **166**, 129–48.

Knox R.B., Heslop-Harrison J. & Heslop-Harrison Y. (1975). Pollen-wall proteins: localization and characterization of gametophytic and sporophytic fractions. In *Suppl. 1 biological Journal Linn. Soc.*, **7**, 177–87.

Knypl J.S. & Mazurczyk W. (1972) Arrest of yellowing in senescing leaf discs of maize by growth retardants, coumarin and inhibitors of RNA and protein synthesis. *Biol. Plant.*, **12**, 199–207.

Koekman B.T., Ooms G., Klapwijk P.M. & Schilperoort R.A. (1979) Genetic map of an octopine Ti-plasmid. *Plasmid*, **2**, 347–57.

KOIE B., INGVERSEN J., ANDERSON A.J., DOLL H. & EGGUM B.O. (1976) Composition and nutritional quality of barley proteins. In *Evaluation of Seed Protein Alterations by Mutation Breeding*, pp. 55–61. International Atomic Energy Agency, Vienna.

KOLATTUKUDY P. & BROWN L. (1975) Fate of naturally-occurring epoxy acids: a soluble epoxide hydrase which catalyses *cis* hydration, from *Fusarium solani pisi*. *Arch. biochem. Biophys.*, **166**, 599–607.

KOLDERUP F. (1975) Effects of temperature, photoperiod and light quantity on protein production in wheat grains. *J. Sci. Food Agric.*, **26**, 583–92.

KONINGS H. (1967) On the mechanism of the transverse distribution of auxin in geotropically exposed pea roots. *Acta Bot. Neerl.*, **16**, 161–76.

KONO Y. & DALY J.M. (1979) Characterisation of the host-specific pathotoxin produced by *Helminthosporium maydis* race T, affecting corn with Texas male-sterile cytoplasm. *Bioorganic Chem.*, **8**, 391–7.

KONONOWICZ A.K. (1978) RNA synthesis and changes in template-DNA activity during the growth and differentiation of parenchyma cells of the primary cortex of roots of *Zea mays* and *Tulipa Kaufmaniana*. *Folia histochem. Cytochem.*, **16**, 123–38.

KONONOWICZ A.K., KURAN H. & OLSZEWSKA M.J. (1976) Cytochemical and ultrastructural studies of changes in the nucleus and nucleolus during cell differentiation in the root cortex of *Haemanthus Katharinae*. *Folia histochem. Cytochem.*, **14**, 151–64.

KOOP H.U. (1980) The life cycle of *Acetabularia* (Dasycladales, Chlorophyceae): A compilation of evidence for meiosis in the primary nucleus. *Protoplasma*, (in press).

KOOP H.U. & KIERMAYER O. (1980) Protoplasmic streaming in the giant unicellular green alga *Acetabularia mediterranea* (Dasycladales). II. Differential sensitivity of movement systems to substances acting on microfilaments and microtubuli. *Protoplasma*, (in press).

KOOP H.U., SCHMID R., HEUNERT H.H. & MILTHALER B. (1978) Chloroplast migration: A new circadian rhythm in *Acetabularia*. *Protoplasma*, **97**, 301–10.

KOOP H.U., SCHMID R., HEUNERT H.H. & SPRING H. (1979) Spindle formation and division of the giant primary nucleus of *Acetabularia* (Chlorophyta, Dasycladales). *Differentiation*, **14**, 135–46.

KORNBERG R.D. (1977) Structure of chromatin. *Ann. Rev. Biochem.*, **46**, 931–54.

KOVALEV A.G. & OBRUCHEVA N.V. (1976) Characteristics of the initial stage of the sigmoid curve of root growth. *Fiziol. Rastenii*, **23**, 340–6.

KOVALEV A.G. & OBRUCHEVA N.V. (1977) Cellular analysis of the S curve of root growth. I. Peculiarities of division and elongation of cells in the horse chestnut root. *Ontogenez*, **8**, 397–405.

KOWALLIK K.V. & HERRMANN R.G. (1972) Do chromoplasts contain DNA? I. Electron-microscopic investigation of *Narcissus* chromoplasts. *Protoplasma*, **74**, 1–6.

KRAMER D., ANDERSON W.P. & PRESTON J. (1978) Transfer cells in the root epidermis of *Atriplex hastata* L. as a response to salinity: a comparative cytological and X-ray microbe investigation. *Aust. J. Plant Physiol.*, **5**, 739–47.

KRAMER D., LÄUCHLI A., YEO A.R. & GULLASCH J. (1977) Transfer cells in roots of *Phaseolus coccineus*: Ultrastructure and possible function in exclusion of sodium from the shoot. *Ann. Bot.*, **41**, 1031–40.

KRAPF G. & JACOBI G. (1975) Dark starvation and plant metabolism. II. CO_2 fixation in isolated chloroplasts. *Planta*, **128**, 135–43.

KREDICH N.M. (1980) Inhibition of nucleic acid methylation by cordycepin. *J. biol. Chem.*, **255**, 7380–5.

KUBAI D.F. (1975) The evolution of the mitotic spindle. *Int. Rev. Cytol.*, **43**, 167–227.

KUBO H. (1939) Über das Hämoprotein aus den Wurzellknöllchen von Leguminosen. *Acta. Phytochim.*, **11**, 195–200.

KUĆ J. (1976) Phytoalexins. In *Physiological Plant Pathology, Encyclopaedia of Plant Physiology*, New Series. (Eds. Heitefuss R. & Williams P.H.) pp. 632–52. Springer-Verlag, Berlin.

KUNDU K. K. & AUDUS L.J. (1974) Root growth inhibitors from root cap and root meristem of *Zea mays*. L. *J. exp. Bot.*, **25**, 479–85.

KUNG S.D., GRAY J.C., WILDMAN S.G. & CARLSON P.S. (1975) Polypeptide composition of Fraction I protein from parasexual hybrid plants in the genus *Nicotiana*. *Science*, **187**, 353–5.

KURZ W.G.W. & LARUE T.A. (1975) Nitrogenase activity in *Rhizobia* in absence of plant host. *Nature, Lond.*, **256**, 407–8.

LAIDMAN D.L., COLBORNE A.J., DOIG R.I. & VARTY K. (1974) The multiplicity of induction systems in wheat aleurone tissue. In *Mechanisms of Regulation of Plant Growth*. (Eds. Bieleski R.L., Ferguson A.R. & Cresswell M.M.) pp. 581–590. The Royal Society of

New Zealand, Wellington.

LAMPORT D.T.A. (1970) Cell wall metabolism. *Ann. Rev. Plant Physiol.*, **21**, 235–70.

LAMPORT D.T.A. (1977) Structure, biosynthesis and significance of cell wall glycoproteins. *Rec. Adv. Phytochem.*, **11**, 79–115.

LAMPPA G.K. & BENDICH A.J. (1979a) Chloroplast DNA sequence homologies among vascular plants. *Plant Physiol.*, **63**, 660–8.

LAMPPA G.K. & BENDICH A.J. (1979b) Changes in chloroplast DNA levels during development of pea (*Pisum sativum*). *Plant Physiol.*, **64**, 126–30.

LAMPPA G.K., ELLIOT L.V. & BENDICH A.J. (1980) Changes in chloroplast number during pea leaf development: An analysis of a protoplast population. *Planta*, **148**, 437–43.

LANCE A. (1958) Infrastructure des cellules du méristème apical et des jeunes ébauches foliaires de *Chrysanthemum segetum* L. (Composées). *Ann. Sci. Nat. Bot.*, 11ᵉ série **19**, 165–202.

LANG A. (1965) Progressiveness and contagiousness in plant differentiation and development. In *Encyclopaedia of Plant Physiology* (Ed. Ruhland W.) Vol. 15/1, pp. 409–23. Springer-Verlag, Berlin.

VAN LAREBEKE N., ENGLER G., HOLSTERS M., VAN DEN ELSACKER S., ZAENEN I., SCHILPEROORT R.A., & SCHELL J. (1974) Large plasmid in *Agrobacterium tumefaciens* essential for crown gall-inducing ability. *Nature, Lond.*, **252**, 169–70.

VAN LAREBEKE N., GENETELLO C., SCHELL J., SCHILPEROORT R.A., HERMANS A.K., HERNALSTEENS J.P. & VAN MONTAGU M. (1975) Acquisition of tumour-inducing ability by non-oncogenic agrobacteria as a result of plasmid transfer. *Nature, Lond.*, **255**, 742–3.

LARKINS B.A., BRACKER C.E. & TSAI C. (1976a) Storage protein synthesis in maize. *Plant Physiol.*, **57**, 740–5.

LARKINS B.A., JONES R.A. & TSAI C.Y. (1976b) Isolation and *in vitro* translation of zein messenger RNA. *Biochem.*, **15**, 5506–11.

LARKINS B.A. & HURKMAN W.J. (1978) Synthesis and deposition of zein in protein bodies of maize endosperm. *Plant Physiol.*, **62**, 256–63.

LARKINS B.A., PEDERSEN K., HANDA A.K., HURKMAN W.J. & SMITH L.D. (1979) Synthesis and processing of maize storage proteins in *Xenopus laevis* oocytes. *Proc. Nat. Acad. Sci. USA*, **76**, 6448–52.

LARSON R.R. (1975) Development and organisation of the primary vascular system in *Populus deltoides* according to phyllotaxy. *Am. J. Bot.*, **62**, 1084–99.

LATIES G. (1959). Active transport of salt into plant tissues. *Ann. Rev. Plant Physiol.*, **10**, 87–112.

LAUGHNAN J.R. & GABAY S.J. (1975) An episomal basis for instability of S male sterility in maize and some implications for plant breeding. In *Genetics and Biogenesis of Mitochondria and Chloroplasts*. (Eds. Birky C.W., Perlman P.S. & Byers T.J.) pp. 330–49. Ohio State University Press, Columbus.

LEA E.J.A. & COLLINS J.C. (1979) The effect of the plant hormone abscissic acid on lipid bilayer membranes. *New Phytol.*, **82**, 11–18.

LEADBEATER B. & DODGE J.D. (1967) An electron microscopic study of nuclear and cell division in a dinoflagellate. *Arch. Mikrobiol.*, **57**, 239–54.

LEAVER C.J. (1980) Mitochondrial genes and male sterility in plants. *TIBS*, **5**, 248–52.

LEAVER C.J. & LOVETT J.S. (1974) An analysis of protein and RNA synthesis during encystment and outgrowth (germination) of *Blastocladiella* zoospores. *Cell Diff.*, **3**, 165–92.

LEDBETTER M.C. & PORTER K.R. (1963) A 'microtubule' in plant cell fine structure. *J. Cell Biol.*, **19**, 239–50.

LEBLANC H.J. JR. (1971) The effect of uncouplers of oxidative phosphorylation on lipid bilayer membranes—CCCP. *J. Membrane Biol.*, **4**, 227–51.

LEE J.W. (1978) Influence of nitrogen source on nitrogen metabolism in detached wheat heads. *Aust. J. Plant Physiol.*, **5**, 779–85.

LEE S.L.J., GRACEN V.E. & EARLE E.D. (1979) The cytology of pollen abortion in C-cytoplasmic male-sterile corn anthers. *Am. J. Bot.*, **66**, 656–67.

LEE S.L.J., EARLE E.D. & GRACEN V.E. (1980) The cytology of pollen abortion in S cytoplasmic male-sterile corn anthers. *Am. J. Bot.*, **67**, 237–45.

LEECH J.H., MOLLENHAUER H.H. & WHALEY W.G. (1963) Ultrastructural changes in the root apex. *Symp. Soc. exp. Biol.*, **17**, 74–84.

LEECH R., RUMSBY M.G. & THOMSON W.W. (1973) Plastid differentiation, acyl lipid and fatty acid changes in developing green maize leaves. *Plant Physiol.*, **52**, 240–5.

LEGOCKI R.P. & VERMA D.P.S. (1979) A nodule-specific plant protein (Nodulin-35) from soybean. *Science*, **205**, 190–3.

LEGOCKI R.P. & VERMA D.P.S. (1980) Identification of 'nodule-specific' host proteins involved in the development of *Rhizobium*-legume symbiosis. *Cell*, **20**, 153–63.

LEMMERS M., DE BEUCKELEER M., HOLSTERS M., ZAMBRYSKI P., DEPICKER A., HERNALSTEENS

J.P., van Montagu M. & Schell J. (1980) Internal organization, boundaries and integration of Ti-plasmid DNA in nopaline crown gall tumours. *J. molec. Biol.*, **144**, 355–78.

Lenci F. & Colombetti G. (1978). Photobehaviour of microorganisms. A biophysical approach. *Ann. Rev. Biophys. Bioeng.*, **7**, 341–61.

de Leo P. & Sacher J.A. (1970) Control of ribonuclease and acid phosphatase by auxin and abscisic acid during senescence of *Rhoeo* leaf sections. *Plant Physiol.*, **46**, 806–11.

Leong T.Y. (1978) The role of protein synthesis in the circadian clock. Purification and properties of a polypeptide which tentatively is involved in the clock. Thesis, 98 pp. Freie Universität, Berlin.

Leong T.Y. & Schweiger H.G. (1979) The role of chloroplast-membrane-protein synthesis in the circadian clock. *Eur. J. Biochem.*, **98**, 187–94.

Letham D.S. (1974) Regulators of cell division in plant tissues. Distribution coefficients for cytokinins. *Planta*, **118**, 361–4.

Letham D.S. (1978) Naturally-occurring plant growth regulators other than the principal hormones of higher plants. In *Phytohormones and Related Compounds—a Comprehensive Treatise, Vol. 1*, Eds. Letham D.S., Goodwin P.B. & Higgins T.J.V.) pp. 349–417. Elsevier/North-Holland, Amsterdam.

Letham D.S., Goodwin P.B. & Higgins T.J.V. (Eds) (1978a). *Phytohormones and Related Compounds—a Comprehensive Treatise, Vol. 1. The Biochemistry of Phytohormones and Related Compounds.* pp. xxxvi + 641. Elsevier/North-Holland, Amsterdam.

Letham D.S., Goodwin P.B. & Higgins T.J.V. (Eds) (1978b) *Phytohormones and Related Compounds—a Comprehensive Treatise. Vol. 2. Phytohormones and the Development of Higher Plants.* pp. xxvi + 448. Elsevier/North-Holland, Amsterdam.

Letham D.S., Higgins T.J.V., Goodwin P.B. & Jacobsen J.V. (Eds.) (1978c). Phytohormones in retrospect. In *Phytohormones and Related Compounds: a Comprehensive Treatise. Vol. 1. The Biochemistry of Phytohormones and related compounds.* pp. 1–27. Elsevier/North-Holland, Amsterdam.

Letham D.S., Shannon J.S. & McDonald I.R.C. (1967) Regulators of cell division in plant tissues. III. *Tetrahedron*, **23**, 479–86.

Letvenuk L.J. & Peterson R.L. (1976) Occurrence of transfer cells in vascular parenchyma of *Hieracium florentinum* roots. *Canad. J. Bot.*, **54**, 1458–71.

Levings C.S. III, Kim B.D., Pring D.R., Conde M.F., Mans R.J., Laughnan J.R. & Gabay-Laughnan S.J. (1980) Cytoplasmic reversion of cms-S in maize: association with a transpositional event. *Science*, **209**, 1021–3.

Levings C.S. III. & Pring D.R. (1977) Diversity of mitochondrial genomes among normal cytoplasms of maize. *J. Hered.*, **68**, 350–4.

Lewis D. (1952) Serological reactions of pollen incompatibility substances. *Proc. roy. Soc. Lond. B*, **140**, 127–35.

Lewis D.H. (1973) Concepts in fungal nutrition and the origin of biotrophy. *Biol. Rev.*, **48**, 261–78.

Lewis O.A.M. & Pate J.S. (1973) The significance of transpirationally derived nitrogen in protein synthesis in fruiting plants of pea (*Pisum sativum* L.). *J. exp. Bot.*, **24**, 596–606.

Lezzi M. (1970) Differential gene activation in isolated chromosomes. *Int. Rev. Cytol.*, **29**, 127–68.

Li H.J. & Eckhardt R. (1977) *Chromatin and Chromosome Structure.* Academic Press, New York.

Libbenga K.R. & Harkes P.A.A. (1973) Initial proliferation of cortical cells in the formation of root nodules in *Pisum sativum* L. *Planta*, **114**, 17–28.

Libbenga K.R., van Iren F., Bogers, R.J. & Schraag-Lamers M.F. (1973) The role of hormones and gradients in the initiation of cortex proliferation and nodule formation in *Pisum sativum* L. *Planta*, **114**, 29–39.

Lieberman M. (1979) Biosynthesis and action of ethylene. *Ann. Rev. Plant Physiol.*, **30**, 533–91.

Lin T.S. & Kolattukudy P.E. (1980) Isolation and characterization of a cuticular polyester (cutin) hydrolyzing enzyme from phytopathogenic fungi. *Physiol. Plant Pathol.*, **17**, 1–15.

Link G., Coen D.M. & Bogorad L. (1979) Differential expression of the gene for the large subunit of RDP carboxylase in maize leaf cell types. *Cell*, **15**, 725–31.

Linschitz H., Kasche V., Butler W.L. & Siegelman H.W. (1966) The kinetics of phytochrome conversion. *J. biol. Chem.*, **241**, 3395–403.

Lippincott B.B., Whatley M.H. & Lippincott J.A. (1977) Tumor induction by *Agrobacterium* involves attachment of the bacteria to a site on the host plant cell wall. *Plant Physiol.*, **59**, 388–90.

Ljunggren H. (1969) Mechanism and pattern of *Rhizobium* invasion into leguminous root hairs. *Physiol. Plant.* Supp. **V**, 1–82.

Lloyd C. (1979) The shapely cell's cycle. *Trends in biochem. Sci.* **4**, 187–9.

Locy R. & Kende H. (1978) The mode of secretion of α-amylase in barley aleurone layers. *Planta*, **143**, 89–99.

LOEWENSTEIN W.R. & ROSE B. (1978). Calcium in (junctional) intercellular communication and a thought on its behaviour in intracellular communication. *Ann. N.Y. Acad. Sci.*, **307**, 285–307.

LONGO C.P. (1968) Evidence for *de novo* synthesis of isocitratase and malate synthetase in germinating peanut cotyledons. *Plant Physiol.*, **43**, 660–4.

LÓPEZ-SÁEZ J.F., GIMÉNEZ-MARTÍN G. & GONZÁLEZ-FERNÁNDEZ A. (1966) Relative velocity of the division cycle in sister cells. *Experimentia*, **22**, 501–2.

LORD J.M. (1978) Evidence that a proliferation of endoplasmic reticulum precedes the formation of glyoxysomes and mitochondria in germinating castor bean endosperm. *J. exp. Bot.*, **29**, 13–23.

LOVETT J.S. (1963) Chemical and physical characterization of 'nuclear caps' isolated from *Blastocladiella* zoospores. *J. Bacteriol.*, **85**, 1235–46.

LOVETT J.S. (1968) Reactivation of ribonucleic acid and protein synthesis during germination of *Blastocladiella* zoospores and the role of the ribosomal nuclear cap. *J. Bacteriol.*, **96**, 962–9.

LOVETT J.S. (1975) Growth and differentiation of the water mold *Blastocladiella emersonii*: Cytodifferentiation and the role of ribonucleic acid and protein synthesis. *Bacteriol. Rev.*, **39**, 345–404.

LOVETT J.S. (1976) Regulation of protein metabolism during spore germination. In *The Fungal Spore: Form and Function.* (Eds. Weber D.J. & Hess W.M.) pp. 189–240. John Wiley and Sons, New York.

LOVETT J.S., GONG C.S. & JOHNSON S.A. (1977) Zoospore germination and early development of *Blastocladiella emersonii*. In *Eucaryotic Microbes as Model Developmental Systems.* (Eds. O'Day D.H. & Horgen P.A.) pp. 402–24. Marcel Dekker, Inc., New York.

LU B.C. (1978) Meiosis in *Coprinus*. VIII. A time course study of the fusion and division of the spindle pole body during meiosis. *J. Cell Biol.*, **76**, 761–6.

LUDLOW M.M. & WILSON G.L. (1971) Photosynthesis of tropical pasture plants. III. Leaf age. *Aust. J. biol. Sci.*, **24**, 1077–87.

LUND E.J. (1947) *Bioelectric Fields and Growth.* Texas University Press, Austin.

LUTHE D.S. & PETERSON D.M. (1977). Cell-free synthesis of globulin by developing oat (*Avena sativa*) seeds. *Plant Physiol.*, **59**, 836–41.

LUTZ A. (1971) Aptitudes morphogénétiques des cultures des tissus d'origine unicellulaire. In *Les Cultures des Tissus des Plantes*, pp. 163–8. Editions du CNRS, Paris.

LYNE R.L. & AP REES T. (1971) Invertase and sugar content during differentiation of roots of *Pisum sativum*. *Phytochem.*, **10**, 2593–9.

MCCOMB J.A., ELLIOT J. & DILWORTH M.J. (1975) Acetylene reduction by *Rhizobium* in pure culture. *Nature, Lond.*, **256**, 408–9.

MCCREA J.M. & HERSCHBERGER C.L. (1978) Chloroplast DNA codes for tRNA from cytoplasmic polyribosomes. *Nature, Lond.*, **274**, 717–9.

MCINTOSH L., POULSEN C. & BOGORAD L. (1980) Chloroplast gene sequence for the large subunit of ribulose bisphosphate carboxylase of maize. *Nature, Lond.*, **288**, 556–60.

MACKENZIE C.R. & JORDAN D.C. (1974) Ultrastructure of root nodules formed by ineffective strains of *Rhizobium meliloti*. *Canad. J. Microbiol.*, **20**, 755–8.

MCLAREN A.C., TURNER, R.G., BOLAND N.N. & HOBBS B.E. (1970) Dislocation structure of the deformation lamellae in synthetic quartz: a study by electron and optical microscopy. *Contr. Mineral. Petrol.*, **29**, 104–15.

MACLEAN D.J., SARGENT J.A., TOMMERUP I.C. & INGRAM D.S. (1974) Hypersensitivity as the primary event in resistance to fungal parasites. *Nature, Lond.*, **249**, 186–7.

MACLEAN D.J. & TOMMERUP I.C. (1979) Histology and physiology of compatibility and incompatibility between lettuce and the downy mildew fungus, *Bremia lactucae*, Regel. *Physiol. Plant Pathol.*, **14**, 291–312.

MCNEIL M., ALBERSHEIM P., TAIZ L. & JONES R.L. (1975) The structure of plant cell walls. *Plant Physiol.*, **55**, 64–8.

MACNICOL P.K. (1977) Synthesis and interconversion of amino acids in developing cotyledons of pea (*Pisum sativum* L.). *Plant Physiol.*, **60**, 344–8.

MCNITT R. (1974) Centriole ultrastructure and its possible role in microtubule formation in an aquatic fungus. *Protoplasma*, **80**, 91–108.

MCNITT R.E. & SHEN-MILLER J. (1978) Quantitative assessment of ultrastructural changes in primary roots of corn (*Zea mays* L.) after geotropic stimulation. *Plant Physiol.*, **61**, 644–8.

MCPHERSON D.C. (1939) Cortical air spaces in the roots of *Zea mays* L. *New Phytol.*, **38**, 190–202.

MCPHERSON J.C., NESTER E.W. & GORDON M.P. (1980) Proteins encoded by *Agrobacterium tumefaciens* Ti plasmid DNA (T-DNA) in crown gall tumors. *Proc. Nat.*

Acad. Sci. USA, **77**, 2666–70.

MADISON J.T., THOMPSON J.F. & MUENSTER A.E. (1976) Deoxyribonucleic acid, ribonucleic acid, protein and uncombined amino acid content of legume seeds during embryogeny. *Ann. Bot.*, **40**, 745–56.

MAHESHWARI P. (Ed.) (1963) *Recent Advances in the Embryology of Angiosperms*. International Society of Plant Morphologists, Delhi.

MAIER R.J. & BRILL W.J. (1976) Ineffective and non-nodulating mutant strains of *Rhizobium japonicum*. *J. Bacteriol.*, **127**, 763–9.

MAIER R.J. & BRILL W.J. (1978) Involvement of *Rhizobium japonicum* O-antigen in soybean nodulation. *J. Bacteriol.*, **133**, 1295–9.

MAIER R.J., BISHOP P.E. & BRILL R.J. (1978a) Transfer from *Rhizobium japonicum* to *Azotobacter vinelandii* of genes required for nodulation. *J. Bacteriol.*, **134**, 1199–201.

MAIER R.J., CAMPBELL H., HANUS F.J., SIMPSON F.B., RUSSELL S.A. & EVANS H.J. (1978b) Expression of hydrogenase activity in free-living *Rhizobium japonicum*. *Proc. Nat. Acad. Sci.*, USA **75**, 3258–62.

MAITI I.B. & KOLATTUKUDY P.E. (1979) Prevention of fungal infection in plants by specific inhibitors of cutinase. *Science*, **205**, 507–8.

MAKLON A.E.S. (1976) An examination of the development of sodium and chloride absorption capacities in beetroot discs. *J. exp. Bot.*, **27**, 651–7.

MAKOVETZKI S.N. & GOLDSCHMIDT E.E. (1976) A requirement for cytoplasmic protein synthesis during chloroplast senescence in the aquatic plant *Anacharis canadensis*. *Plant Cell Physiol.*, **17**, 859–62.

MALNOE P., ROCHAIX J-D., CHUA N.H. & SPAHR P-F. (1979) Characterization of the gene and messenger RNA of the large subunit of ribulose 1,5-diphosphate carboxylase in *Chlamydomonas reinhardii*. *J. molec. Biol.*, **133**, 417–34.

MANIKAM A & CARLIER A.R. (1980) Isolation and function of a low molecular weight protein of mung bean embryonic axes. *Planta*, **149**, 234–40.

MANSFIELD J.W., HARGREAVES J.A. & BOYLE F.C. (1974) Phytoalexin production by live cells in broad bean leaves infected with *Botrytis cinerea*. *Nature, Lond.* **252**, 316–7.

MANTON I. (1966) Observations on scale production in *Pyramimonas amylifera*. *J. Cell Sci.*, **1**, 429–38.

MANTON I. & CLARKE B. (1955) Observations with the electron microscope on the internal structure of the spermatozoid of *Fucus*. *J. exp. Bot.*, **7**, 416–32.

MARCAN H. & FRIEND J. (1979) Lectins in potato tuber tissues and relevance to host-parasite interactions. *Arabinogalactan-protein News*, **2**, 61–3.

MARCAN H., JARVIS M.C. & FRIEND J. (1979) Effect of methyl glucosides and oligosaccharides on cell death and browning of potato tuber discs induced by mycelial components of *Phytophthora infestans*. *Physiol. Plant Pathol.*, **14**, 1–9.

MARCUM J.M., DEDMAN J.R., BRINKLEY B.R. & MEANS A.R. (1978) Control of microtubule assembly-disassembly by calcium-dependent regulator protein. *Proc. Nat. Acad. Sci. USA.*, **75**, 3771–5.

MARCUS A. & FEELEY J. (1964) Activation of protein synthesis in the imbibition of seed germination. *Proc. Nat. Acad. Sci.*, USA **51**, 1075–9.

MARCUS A., FEELEY J. & VOLCANI T. (1966) Protein synthesis in imbibed seeds. III. Kinetics of amino acid incorporation, ribosome activation and polysome formation. *Plant Physiol.*, **41**, 1167–72.

MARES D.J., KNARSTOG K. & STONE B.A. (1975) Early stages in the development of the wheat endosperm. I. The changes from free nuclear to cellular endosperm. *Aust. J. Bot.*, **23**, 311–26.

MARES D.J., STONE B.A., JEFFREY C. & KNARSTOG K. (1977) Early stages in the development of wheat endosperm. II. Ultrastructural observations on cell wall formation. *Aust. J. Bot.*, **25**, 599–613.

MARMÉ D. (1974) Binding properties of the plant photoreceptor phytochrome to membranes. *J. supramol. Struct.*, **20**, 751–68.

MARRÈ E. (1977) Effects of fusicoccin and hormones on plant cell membrane activities; observations and hypotheses. In *Regulation of Cell Membrane Activities in Plants*. (Eds. Marrè E. & Ciferri O.) pp. 185–201. Elsevier/North-Holland Biomedical Press, Amsterdam.

MARRÈ E. (1979a) Fusicoccin: a tool in plant physiology. *Ann. Rev. Plant Physiol.*, **30**, 273–88.

MARRÈ E. (1979b) Integration of solute transport in cereals. In *Recent Advances in the Biochemistry of Cereals*. Phytochem. Soc. Symp. **16**, 3–23. (Eds. Laidman D.L. & Wyn Jones R.G.). Academic Press, London.

MARRÈ E., LADO P., ROSSI-CALDOGNO F., COLOMBO R. & DE MICHELIS M.I. (1974) Evidence for the coupling of proton extrusion and K^+ uptake in pea internode segments treated with fusicoccin or auxin. *Plant Sci. Lett.*, **3**, 365–74.

MARRIOTT K.M. & TANNER W. (1979) Dolichylphosphate-dependent glycosyl transfer reactions in the endoplasmic reticulum of castor bean endosperm. *Plant Physiol.*, **64**, 445–9.

MARTIN C. & THIMAN K.V. (1972) The role of protein synthesis in the senescence of leaves. I. The formation of protease. *Plant Physiol.*, **49**, 64–71.

MARTINEK K. & BEREZIN I.V. (1979) Artificial light-sensitive enzymatic systems as chemical amplifiers of weak light signals. *Photochem. Photobiol.*, **29**, 637–49.

MARTY F. (1978) Cytochemical studies on GERL, provacuoles, and vacuoles in root meristematic cells of *Euphorbia*. *Proc. Nat. Acad. Sci. USA.*, **75**, 852–6.

MASCARENHAS J.P. (1975) The biochemistry of angiosperm pollen development. *Bot. Rev.*, **41**, 260–314.

MASER M.D., O'BRIEN T.P. & MCCULLY M.E. (1967) Shadowing of ultrathin sections of epoxy embedded tissues as an aid in three-dimensional interpretation. *J. de Microscopie*, **6**, 305–12.

MASSALA R., LEGRAND M. & FRITIG B. (1980) Effect of α-aminooxyacetate, a competitive inhibitor of PAL and the hypersensitive resistance of tobacco to TMV. *Physiol. Plant Pathol.*, **16**, 213–26.

MASSART J. (1922) Sur la polarité des organes végétaux. Recueil de l'Institut Botanique. *Leo Errera*, **10**, 107–15.

MATHILE P.M. & WINKENBACH F. (1971) Function of lysosomes and lysosomal enzymes in the senescing corolla of Morning Glory. *J. exp. Bot.*, **22**, 759–71.

MATILE P. (1978) Biochemistry and function of vacuoles. *Ann. Rev. Plant Physiol.*, **29**, 193–213.

MATTHEWS J.A. & MIFLIN B.J. (1980) *In vitro* synthesis of barley storage proteins. *Planta*, **149**, 262–8.

MATTHEWS R.A., EVANS L.V. & CALLOW M.E. (1976) Isolation and characterization of Golgi apparatus from eggs of *Fucus* (Pheophyceae). *J. Phycol.*, **12**, 435–8.

MAYER A.M. & POLJAKOFF-MAYBER A.J. (1975) *The Germination of Seeds*. Pergamon Press, Oxford.

MAYER A.M. & SHAIN Y. (1968) Zymogen granules in enzyme liberation and activation in pea seeds. *Science*, **162**, 1283–4.

MAZZUCHI U., BAZZI C. & PUPILLO P. (1979) The inhibition of susceptible and hypersensitive reactions by protein-lipopolysaccharide complexes from phytopathogenic Pseudomonads: relation ship to polysaccharide antigenic determinants. *Physiol. Plant Pathol.*, **14**, 19–30.

MECHAM D.K., KASARDA D.D. & QUALSET C.O. (1978) Genetic aspects of wheat gliadin proteins. *Biochem. Genet.*, **16**, 831–53.

MEINS F. (1976) 5-Bromodeoxyuridine: a specific inhibitor of cytokinin-habituation in tobacco cell culture. *Planta*, **129**, 239–44.

MEINS F. (1977) Reversal of the neoplastic state in plants. *Am. J. Pathol.*, **89**, 687–702.

MEINS F. & BINNS A. (1977) Epigenetic variation of cultured somatic cells: evidence for gradual changes in the requirement for factors promoting cell division. *Proc. Nat. Acad. Sci. USA*, **74**, 2928–32.

MEINS F. & BINNS A. (1978) Epigenetic clonal variation in the requirement of plant cells for cytokinins. In *The Clonal Basis for Development*. (Eds. Subtelny S. & Sussex I.M.) pp. 185–201. Academic Press, New York.

MEINS F. & BINNS A.N. (1979) Cell determination in plant development. *BioScience*, **29**, 221–5.

MEISTRICH M.L., FORK R.L. & MATRICON J. (1970). Phototropism in *Phycomyces* as investigated by focused laser radiation. *Science*, **169**, 370–1.

MELCHER U. (1979) *In vitro* synthesis of a precursor to the methionine-rich polypeptide of the zein fraction of corn. *Plant Physiol.*, **63**, 354–8.

MELCHER U. & VARNER J.E. (1971) Protein release by barley aleurone layers. *J. Inst. Brew.*, **77**, 456–61.

MENDIOLA-MORGENTHALER L.R., MORGENTHALER J-J. & PRICE C.A. (1976) Synthesis of coupling factor CF_1 protein by isolated spinach chloroplasts. *FEBS Lett.*, **62**, 96–100.

MEPHAM R.H. & LANE G.R. (1969) Formation and development of the tapetal periplasmodium in *Tradescantia bracteata*. *Protoplasma*, **68**, 175–92.

MERGENHAGEN D. & SCHWEIGER H.G. (1973) Recording the oxygen production of a single *Acetabularia* cell for a prolonged period. *Exp. Cell Res.*, **81**, 360–4.

MERRICK W.C. & DURE L. (1972) The developmental biochemistry of cotton seed embryogenesis and germination. IV. Levels of cytoplasmic and chloroplastic transfer ribonucleic acid species. *J. biol. Chem.*, **247**, 7988–99.

MERSEY B. & MCCULLY M.E. (1978) Monitoring of the course of fixation of plant cells. *J. Micros.*, **114**, 49–76.

MERTZ S.M., JR. & ARNTZEN C.J. (1977) Selective inhibition of K^+, Na^+, Cl^- and PO_4^{3-} uptake in *Zea mays* L. by *Bipolaris*

(*Helminthosporium*) *maydis* race T pathotoxin. *Plant Physiol.*, **60**, 363–9.

MERTZ S.M. JR. & ARNTZEN C.J. (1978) Depolarisation of the electrogenic transmembrane potential of *Zea mays* L. by *Bipolaris* (*Helminthosporium*) *maydis* race T toxin, azide, cyanide, dodecyl succinic acid or cold temperature. *Plant Physiol.*, **62**, 781–3.

MEYER A.M. (1969a) Versuche zur 1. positiven und zur negative phototrophischen Krümmung des *Avena* koleoptile I. Lichtperception und Absorptiongradent. *Z. Pflphysiol.*, **60**, 418–33.

MEYER A.M. (1969b) Versuche zur 1. positiven und zur negativen phototropischen Krümmung der *Avena* koleoptile: II. Die Inversion durch Paraffinol. *Z. Pflphysiol.*, **61**, 129–34.

MEYER B. & STUBBE W. (1974) Das Zahlenverhaltnis von mu7tterlichen und väterlichen Plastiden in den Zygoten von *Oenothera erythrosepala* Borbas (Syn. *Oe. lamarckiana*). *Ber. Dtsch. Bot. Ges.*, **87**, 29–38.

MIERNYK J.A., TRELEASE R.N. & CHAINSKI J.S. (1979) Malate synthase activity in cotton and other ungerminated oil seeds. *Plant Physiol.*, **63**, 1068–71.

MIFLIN B.J., MATTHEWS J.A., BURGESS S.R. & SHEWRY P.R. (1979). The synthesis of barley storage proteins. In *Genome Organisation and Expression in Plants.* (Ed. Leaver C.J.). Plenum Press, New York.

MIFLIN B.J. & SHEWRY P.R. (1979) The biology and biochemistry of cereal seed prolamins. In *Seed Protein Improvement in Cereals and Grain Legumes.* pp. 137–58. SM-230/79. International Atomic Energy Agency, Vienna.

MIKOLA L. & MIKOLA J. (1980) Mobilization of proline in the starchy endosperm of germinating barley grain. *Planta*, **149**, 149–54.

MILBORROW B.V. (1978) Abscissic acid. In *Phytohormones and Related Compounds—a Comprehensive Treatise. Vol. 1. The Biochemistry of Phytohormones and Related Compounds.* (Eds. Letham D.S., Goodwin P.B. & Higgins T.J.V.) pp. 295–347. Elsevier/North-Holland, Amsterdam.

MILBORROW B.V. (1979) Antitranspirants and the regulation of abscissic acid content. *Aust. J. Plant Physiol.*, **6**, 249–54.

MILLER B.L. & HUFFAKER R.C. (1979) Changes in endopeptidases, protein and chlorophyll in senescing barley leaves. In *Agronomy Abstracts, American Society of Agronomy 71st Annual Meeting*, p. 91. Fort Collins, Colorado.

MILLER C.O. (1979) Cytokinin inhibition of respiration by cells and mitochondria of soybean, *Glycine max* (L.) Merrill. *Planta*, **146**, 503–11.

MILLER C.O., SKOOG F., OKUMURA F.S., VON SALLZA M.H. & STRONG F.M. (1956) Isolation, structure and synthesis of kinetin, a substance promoting cell division. *J. Am. Chem. Soc.*, **78**, 1375–80.

MILLER K.R. & MILLER G.J. (1978). Organization of the cell membrane in *Euglena. Protoplasma*, **95**, 11–24.

MILLER R.J. & KOEPPE D.E. (1971) Southern corn leaf blight: susceptible and resistant mitochondria. *Science*, **173**, 67–9.

MILLERD A., RANDALL P.J. & THOMSON J.A. (1979) Heterogeneity of sulphur content in the storage proteins of pea cotyledons. *Planta*, **146**, 463–6.

MILLERD A. & SPENCER D. (1974) Changes in RNA-synthesizing activity and template activity in nuclei from cotyledons of developing pea seeds. *Aust. J. Plant Physiol.*, **1**, 331–41.

MILLERD A., THOMSON J.A. & SCHROEDER H.E. (1978) Cotyledonary storage proteins in *Pisum sativum*. III. Patterns of accumulation during development. *Aust. J. Plant Physiol.*, **5**, 519–34.

MILLERD A. & WHITFELD P.R. (1973) Deoxyribonucleic acid and ribonucleic acid synthesis during the cell expansion phase of cotyledon development in *Vicia faba* L. *Plant Physiol.*, **51**, 1005–10.

MILLS G.M. & CANTINO E.C. (1978) The lipid composition of the *Blastocladiella emersonii* γ and the function of γ-particle lipid in chitin formation. *Exp. Mycol.*, **2**, 99–109.

MINCHIN F.R. & PATE J.S. (1973) The carbon balance of a legume and the functional economy of its root nodules. *J. exp. Bot.*, **24**, 259–71.

MIRELMAN D., GALUN E., SHARON N. & LOTAN R. (1975) Inhibition of fungal growth by wheat germ agglutinin. *Nature, Lond.*, **256**, 414–5.

MISHARIN S.I., ANTIPINA A.I., REIMERS F.E. & KHAVKIN E.E. (1974) Phase, tissue, and organ specific proteins of corn seedlings. *Dokl. Akad. Nauk SSSR*, **219**, 473–6.

MITCHELL P. (1970). Membranes of cells and organelles: morphology, transport and metabolism. *Symp. Soc. gen. Microbiol.*, **20**, 121–66.

MITCHISON G.J. (1980) The dynamics of auxin transport. *Proc. roy. Soc. B.*, **209**, 489–511.

MIURA G.A. & MILLER C.O. (1969) 6-(γ,γ-Dimethyl allylamino) purine as a precursor of zeatin. *Plant Physiol.*, **44**, 372–6.

MOESTA P. & GRISEBACH H. (1980) Effects of biotic and abiotic elicitors on phytoalexin metabolism in soyabean. *Nature, Lond.*, **286**, 710–1.

MOHR H. (1966) Differential gene activation as a mode of action of phytochrome. *Photochem. Photobiol.*, **5**, 469–83.

MOHR H. (1978) Pattern specification and realization in photomorphogenesis. *Bot. Mag. Tokyo Special Issue*, **1**, 199–217.

MOHR H. (1979) *Lectures on Structure and Significance of Science.* Springer-Verlag, Berlin.

MOHR H., DRUMM H., SCHMIDT R. & STEINITZ B. (1979) The effect of light pretreatments on phytochrome-mediated induction of anthocyanin and phenylalanine ammonia-lyase. *Planta*, **146**, 369–76.

MOLLENHAUER H.H. & MOLLENHAUER B.A. (1978) Changes in the secretory activity of the Golgi apparatus during the cell cycle in root tips of maize (*Zea mays* L.). *Planta*, **138**, 113–8.

MONNIER M. (1978) Culture of zygotic embryos. In *Frontiers of Plant Tissue Culture.* (Ed. Thorpe T.A.) pp. 277–286. International Association of Plant Tissue Culture. Calgary.

MONTAGNOLI G. (1977) Biological effects of light on proteins: enzyme activity modulation. *Photochem. Photobiol.*, **26**, 679–83.

VAN MONTAGU M., HOLSTERS M., ZAMBRYSKI P., HERNALSTEENS J.P., DEPICKER A., DE BEUCKELEER M., ENGLER G., LEMMERS M., WILLMITZER M. & SCHELL J. (1980) The interaction of *Agrobacterium* Ti-plasmid and plant cells. *Proc. roy. Soc. B*, **210**, 351–65.

VAN MONTAGU M. & SCHELL J. (1979) The plasmids of *Agrobacterium tumefaciens.* In *Plasmids of Medical, Environmental and Commercial Importance.* (Eds. Timmis K. & Pühler A.) pp. 71–96. Elsevier/North-Holland, Amsterdam.

MOORE A.E. & STONE B.A. (1972). *β* 1-3 glucanase in *Nicotiana* leaves. *Planta*, **104**, 93–109.

MOREIRA M.A., HERMODSON M.A., LARKINS B.A. & NIELSEN N.C. (1979) Partial characterization of the acidic and basic subunits of glycinin. *J. biol. Chem.*, **254**, 9921–6.

MORI T., WAKABAYASHI Y. & TAKAGI S. (1978) Occurrence of mRNA for storage protein in dry soybean seeds. *J. Biochem.*, **84**, 1103–11.

MORGAN, D.C., O'BRIEN T. & SMITH H. (1980) Rapid photomodulation of stem extension in light-grown *Sinapis alba* L. Studies in kinetics, site of perception and photoreceptor. *Planta*, (in press).

MORGAN D.G. & SMITH H. (1978) The relationship between phytochrome photoequilibrium and development in light grown *Chenopodium album* L. *Planta*, **142**, 187–93.

MOROHASHI Y. & BEWLEY J.D. (1980) Development of mitochondrial activities in pea cotyledons during and following germination of the axis. *Plant Physiol.*, **66**, 70–3.

MORRIS D.A., KADIR G.O. & BARRY J. (1973) Auxin transport in intact pea seedlings (*Pisum sativum* L.): the inhibition of transport by 2,3,5 tri-iodobenzoic acid. *Planta*, **110**, 173–82.

MORRIS D.A. & THOMAS A.G. (1974). Specificity of auxin transport in intact pea seedlings (*Pisum sativum* L.). *Planta*, **118**, 225–34.

MORRIS D.A. & THOMAS A.G. (1978) A microautoradiographic study of auxin transport in the stem of intact pea seedlings (*Pisum sativum* L.). *J. exp. Bot.*, **29**, 147–57.

MORRISON I.N. & O'Brien T.P. (1976) Cytokinesis in the developing wheat grain; division with and without a phragmoplast. *Planta*, **130**, 57–67.

MORT A.J. & BAUER W.D. (1980) Composition of the capsular and extracellular polysaccharides of *Rhizobium japonicum* changes with culture age and correlation with binding of soyabean seed lectin to the bacteria. *Plant Physiol.*, **66**, 158–64.

MORTON R.K., PALK B.A. & RAISON J.K. (1964) Intracellular components associated with protein synthesis in developing wheat endosperm. *Biochem. J.*, **91**, 522–8.

MORY Y.Y., CHEN D. & SARID S. (1972) Onset of DNA synthesis in germinating wheat embryos. *Plant Physiol.*, **49**, 20–23.

MOSSE B. (1964) Electron-microscopic studies of nodule development in some clover species. *J. gen. Microbiol.*, **36**, 49–66.

MOTHES K. (1933) Die Vakuminfiltration in Ernährungsversuch. (Dargestellt an Untersuchungen über die Assimilation des Ammoniaks). *Planta*, **19**, 117–38.

MOTT R.L. & STEWARD F.C. (1972) Solute accumulation in plant cells. I. Reciprocal relations between electrolytes and non-electrolytes. *Ann. Bot.*, **36**, 621–39.

MOZER T.J. (1980a) Partial purification and characterization of the mRNA for α-amylase from barley aleurone layers. *Plant Physiol.*, **65**, 834–7.

MOZER T.J. (1980b) Control of protein synthesis in barley aleurone layers by the plant hormones gibberellic acid and abscisic acid. *Cell*, **20**, 479–85.

MUELLER S.C. & BROWN R.M. JR. (1980) Evidence for an intramembrane component associated with a cellulose microfibril-synthesizing complex in higher plants. *J. Cell Biol.*, **84**, 315–26.

MUELLER W.C. & BECKMAN C.H. (1976) Ultrastructure and development of phenolic-storing cells in cotton roots. *Canad. J. Bot.*, **54**, 2074–82.

MUELLER W.C. & BECKMAN C.H. (1978) Ultrastructural localization of polyphenoloxidase and peroxidase in roots and hypocotyls of cotton seedlings. *Canad. J. Bot.*, **56**, 1579–87.

MULAY B.N. & SALUJA S.K. (1957) Histology of some desert grasses. *J. Ind. bot. Soc.*, **36**, 106–11.

MÜLLER D.G. & JAENICKE L. (1973) Fucoserraten, the female sex attractant of *Fucus serratus* L. (Phaeophyta). *FEBS Lett.*, **30**, 137–9.

MÜLLER D.G. & SEFERIADIS K. (1977) Specificity of sexual chemotaxis in *Fucus serratus* and *Fucus vesiculosus* (Phaeophyceae). *Z. Pflphysiol.*, **84**, 85–94.

MURPHY G.J.P. (1979) Plant hormone receptors: comparison of naphthalene acetic acid binding by maize extracts and by a non-plant protein. *Plant Sci. Lett.*, **15**, 183–91.

MURPHY M.N. SR. & LOVETT J.S. (1966) RNA and protein synthesis during zoospore differentiation in synchronized cultures of *Blastocladiella*. *Dev. Biol.*, **14**, 68–95.

MURRAY D.R. (1979) Proteolysis in the axis of the germinating pea seed. II. Changes in polypeptide composition. *Planta*, **147**, 117–21.

MURRAY D.R., PEOPLES M.B. & WATERS S.P. (1979) Proteolysis in the axis of the germinating pea seed. Changes in protein degrading enzyme activities of the radicle and primary root. *Planta*, **147**, 111–6.

MURRAY M.G., CUELLAR R.E. & THOMPSON W.F. (1978) DNA sequence organization in the pea genome. *Biochem.*, **17**, 5781–90.

MUSSELL H. & STRAND L.L. (1976) Pectic enzymes: involvement in pathogenesis and possible relevance to tolerance and specificity. In *Cell Wall Biochemistry Related to Specificity in Host–Plant Pathogen Interactions.* (Eds. Solheim B. & Raa J.) pp. 31–61. Universitetsforlaget, Oslo.

MUTO S. & BEEVERS H. (1974) Lipase activities in castor bean endosperm during germination. *Plant Physiol.*, **54**, 23–8.

NADEAU R. & RAPPAPORT L. (1972) Metabolism of gibberellin A_1 in germinating bean seeds. *Phytochem.*, **11**, 1611–6.

NAGAHASHI J. & BEEVERS L. (1978) Subcellular localization of glycosyl transferases involved in glycoprotein biosynthesis in the cotyledons of *Pisum sativum* L. *Plant Physiol.*, **61**, 451–9.

NAGL W. (1974) The *Phaseolus* suspensor and its polytene chromosomes. *Z. Pflphysiol.*, **73**, 1–44.

NAGL W. (1976a) Nuclear organization. *Ann. Rev. Plant Physiol.*, **27**, 39–69.

NAGL W. (1976b) Early embryogenesis in *Tropaeolum majus* L. evolution of DNA content and polyteny in the suspensor. *Plant Sci. Lett.*, **7**, 1–6.

NAGL W. (1979) Replication of the eukaryotic chromosome. *Prog. Bot.*, **41**, 161–72.

NAIR T.V.R., GROVER H.L. & ABROL Y.P. (1978) Nitrogen metabolism of the upper three leaf blades of wheat at different soil nitrogen levels. II. Protease activity and mobilisation of reduced nitrogen to the developing grains. *Physiol. Plant.*, **42**, 293–300.

NAKAYAMA N., SUGIMORO I. & ASAHI T. (1980) Presence in dry pea cotyledons of soluble succinate dehydrogenase that is assembled into the mitochondrial inner membrane during seed imbibition. *Plant Physiol.*, **65**, 229–33.

NASRALLAH M.E., WALLACE D.H. & SAVO R.M. (1972) Genotype-protein-phenotype relationships in self-incompatibility in higher plants. *Genet. Res.*, **20**, 151–60.

NAVARRETE M.H. & BERNABEU C. (1978) Soluble polypeptides from meristematic and mature cells of *Allium cepa* L. roots. *Planta*, **142**, 147–51.

NAVARRO L., ROISTACHER C.N. & MURASHIGE T. (1975) Improvement of shoot-tip grafting *in vitro* for virus-free citrus. *J. Am. Soc. Hort. Sci.*, **100**, 471–9.

NEALES T.F., ANDERSON M.J. & WARDLAW I.F. (1963) The role of leaves in the accumulation of nitrogen by wheat during ear development. *Aust. J. agric. Res.*, **14**, 725–36.

NĚMEC B. (1900) Über die Art der Wahrnehmung des Schwerkraftreizes bei den Pflanzen. *Jb. wiss. Bot.*, **36**, 80.

NESS P.J. & WOOLHOUSE H.W. (1980a) RNA synthesis in *Phaseolus* chloroplasts. I. Ribonucleic acid synthesis in chloroplast preparations from *Phaseolus vulgaris* L. leaves and solubilisation of the RNA polymerase. *J. exp. Bot.*, **31**, 223–33.

NESS P.J. & WOOLHOUSE H.W. (1980b) RNA synthesis in *Phaseolus* chloroplasts. II. Ribonucleic acid synthesis in chloroplasts from developing and senescing leaves. *J. exp. Bot.*, **31**, 235–45.

NETTANCOURT D. DE (1977) *Incompatibility in Angiosperms.* Springer-Verlag, Berlin.

Neumann D. & Weber E. (1978) Formation of protein bodies in ripening seeds of *Vicia faba*. *Biochem. Physiol. Pflanzen.*, **173**, 167–80.

Newbury H.J. & Smith H. (1981) Immunochemical evidence for phytochrome regulation of the specific activity of ascorbate oxidase in mustard seedlings. *Eur. J. Biochem.*, **117**, 575–80.

Newcomb E.H. (1963) Cytoplasm–cell-wall relationships. *Ann. Rev. Plant Physiol.*, **14**, 43–64.

Newcomb E.H. (1969) Plant microtubules. *Ann. Rev. Plant Physiol.*, **20**, 253–88.

Newcomb E.H. & Bonnett H.T. jr. (1965) Cytoplasmic microtubule and wall microfibril orientation in root hairs of radish. *J. Cell Biol.*, **27**, 575–89.

Newcomb W. (1976) A correlated light and electron microscope study of symbiotic growth and differentiation in *Pisum sativum* root nodules. *Canad. J. Bot.*, **54**, 2163–86.

Newcomb W. (1978) The development of cells in the coenocytic endosperm of the African blood lily *Haemanthus katherinae*. *Canad. J. Bot.*, **56**, 483–501.

Newcomb W., Syono K. & Toorrey J.G. (1977) Development of an ineffective pea root nodule: morphogenesis, fine structures and cytokinin biosynthesis. *Canad. J. Bot.*, **55**, 1891–1907.

Newman I.A. & Briggs W.R. (1972) Photochrome-mediated electric potential changes in oat seedlings. *Plant Physiol.*, **50**, 687–93.

Nichols E.J., Beckman J.M. & Hadwiger L.A. (1980) Glycosidic enzyme activity in pea tissue and pea-*Fusarium solani* interactions. *Plant Physiol.*, **66**, 199–204.

Nishio T. & Hinata K. (1979) Purification of an S specific glycoprotein in self-incompatible *Brassica campestris* L. *Jap. J. Genet.*, **54**, 307–11.

Nitsch C. (1974) La culture de pollen isolé sur milieu synthétique. C. R. *Acad. Sci. Paris*, **278**, 1031–4.

Nomura T. & Akazawa T.A. (1973) Enzymic mechanism of starch breakdown in germinating rice seeds. *Plant Physiol.*, **51**, 979–81.

Noodén L.D. & Leopold A.C. (1978) Phytohormones and the endogenous regulation of senescence and abscission. In *Polyhormones and Related Compounds. A Comprehensive Treatise*, Vol. II. (Eds. Letham D.S., Goodwin P.B. & Higgins T.J.V.). Elsevier/North-Holland, Amsterdam.

Noodén L.D. & Weber J.A. (1978) Environmental and hormonal control of dormancy in terminal buds of plants. In *Dormancy and Developmental Arrest*. (Ed. Clutter M.E.) Academic Press, New York.

Norris R.D., Lea P.J. & Fowden L. (1973) Aminoacyl-tRNA synthetases in *Triticum aestivum* L. during seed development and germination. *J. exp. Bot.*, **24**, 615–25.

Norris R.D., Lea P.J. & Fowden L. (1975) tRNA species in the developing grain of *Triticum aestivum*. *Phytochem.*, **14**, 1683–6.

Northcote D.H. (1972) The chemistry of the plant cell wall. *Ann. Rev. Plant Physiol.*, **23**, 113–32.

Northcote D.H. (1979a) The involvement of the Golgi apparatus in the biosynthesis and secretion of glycoproteins and polysaccharides. *Biomembranes*, **10**, 51–76.

Northcote D.H. (1979b) Biochemical mechanisms involved in plant morphogenesis. In *Differentiation and the Control of Development in Plants—Potential for Chemical Modification. British Plant Growth Regulator Group Monograph No. 3.* (Ed. George E.C.) pp. 11–20. Agricultural Research Council, Wantage.

Nougarède A. & Rembur J. (1980) Extension cellulaire, division cellulaire et niveaux de ploïdie des cellules corticales de l'épicotyle du pois en cours d'élongation. *Canad. J. Bot.*, **58**, 486–501.

Novák B. & Bentrup F.W. (1972) An electrophysiological study of regeneration in *Acetabularia mediterranea. Planta*, **108**, 227–44.

Novák B. & Sironval C. (1975) Inhibition of regeneration of *Acetabularia mediterranea* enucleated posterior stalk segments by electrical isolation. *Plant. Sci. Lett.*, **5**, 183–8.

Novák B. & Sironval C. (1976) Circadian rhythm of the transcellular current in regenerating enucleated posterior stalk segments of *Acetabularia mediterranea. Plant Sci. Lett.*, **6**, 273–83.

Novotný A.M. & Forman M. (1975) The composition and development of cell walls of *Fucus* embryos. *Planta*, **122**, 67–78.

Nuti M.P., Lepidi A.A., Prakash R.K., Scehilperoort R.A. & Cannon F.C. (1979) Evidence for nitrogen fixtion (nif) genes on indigenous *Rhizobium* plasmids. *Nature, Lond.*, **282**, 533–5.

Nutman P.S. (1956) The influence of the legume in root nodule symbiosis. A comparative study of host determinants and functions. *Biol. Rev. Camb. philos. Soc.*, **31**, 109–51.

Nutman P.S. (1957) Symbiotic effectiveness in nodulated red clover. III. Further studies on inheritance of ineffectiveness in the host.

Heredity, **ii**, 157–73.

Nutman P.S. (1969) Genetics of symbiosis and nitrogen fixation in legumes. *Proc. roy. Soc. B*, **172**, 417–37.

Oakley B.R. & Dodge J.D. (1974) Kinetochores associated with the nuclear envelope in the mitosis of a dinoflagellate. *J. Cell Biol.*, **63**, 322–5.

O'Brien T.P. (1970) Further observations on hydrolysis of the cell wall in the xylem. *Protoplasma*, **69**, 1–14.

O'Brien T.P. (1972) The cytology of cell-wall formation in some eukaryotic cells. *Bot. Rev.*, **38**, 87–118.

O'Brien T.P. & McCully M.E. (1969) *Plant Structure and Development*. Collier Macmillan, New York.

O'Brien T.P. & Thimann K.V. (1967) Observations on the fine structure of the oat coleoptile. III. Correlated light and electron microscopy of the vascular tissues. *Protoplasma*, **63**, 443–78.

Obrucheva N.V. (1969) Primary stages of lignification in roots of corn seedlings. *Dokl. Akad. Nauk SSSR*, **185**, 713–6.

Ochiai-Yamagi S., Fukazawa C. & Harada K. (1978) Formation of storage protein components during soybean seed development. *Agric. biol. Chem.*, **42**, 697–702.

Oertli J.J. & Richardson W.F. (1970). The mechanism of boron immobility. *Physiol. Plantarum*, **23**, 108–16.

O'Farrell P.H. (1975) High resolution two-dimensional electrophoresis of proteins. *J. biol. Chem.*, **250**, 4007–21.

O'Gara F. & Shanmugam K.T. (1976) Regulating of nitrogen fixation by *Rhizobia*; Export of fixed N_2 as NH_4^+. *Biochim. biophys. Acta*, **437**, 313–21.

Ohlrogge J.B., García-Martínez J.L., Adams D. and Rappaport L. (1980) Uptake and subcellular compartmentation of Gibberellin A_1 applied to leaves of barley and cowpea. *Plant Physiol.*, **66**, 422–7.

Ohmori K. & Gottlieb D. (1965) Development of respiratory enzyme activities during spore germination. *Phytopath.*, **55**, 1238–336.

Ojha M. (1978) Characterization of spinach DNA. *Cytobiologie*, **15**, 185–98.

Okamoto K. & Akazawa T. (1980) Enzymic mechanism of starch breakdown in germinating rice seeds. *Plant Physiol.*, **65**, 81–4.

Oksala T. & Therman E. (1977) Endomitosis in tapetal cells of *Eremurus* (Liliaceae). *Am. J. Bot.*, **64**, 866–72.

Olesen P. (1978) Studies on the physiological sheaths in roots I. Ultrastructure of the exodermis in *Hoya carnosa* L. *Protoplasma*, **94**, 325–40.

Olszewska M.J. (1976) Autoradiographic and ultrastructural study of *Cucurbita pepo* root cells during their growth and differentiation. *Histochemistry*, **49**, 157–75.

Olszewska M.J. & Kononowicz A.K. (1979) Activities of DNA polymerases and RNA polymerases detected *in situ* in growing and differentiating cells of root cortex. *Histochemistry*, **59**, 311–23.

Ooms G., Klapwijk P.M., Poulis J.A. & Schilperoort R.A. (1980) Characterization of Tn*904* insertions in octopine Ti plasmid mutants of *Agrobacterium tumefaciens*. *J. Bacteriol.*, **144**, 82–91.

Oostrom H., Loopik-Detmers M.A. van & Libbenga K.R. (1975) A high affinity receptor for indoleacetic acid in cultured tobacco pith cells. *FEBS Lett.*, **59**, 194–7.

Öpik H. (1968) Development of cotyledon cell structure in ripening *Phaseolus vulgaris* seeds. *J. exp. Bot.*, **19**, 64–76.

Oram R.N., Doll H. & Køie B. (1975) Genetics of two storage protein variants in barley. *Hereditas*, **80**, 53–8.

Osborne D.J. (1967) Hormonal regulation of leaf senescence. *Symp. Soc. exp. Biol.*, **21**, 305–22.

Osborne D.J. & Sargent J.A. (1976) The positional differentiation of ethylene-responsive cells in rachis abscission zones in leaves of *Sambucus nigra* and their growth and ultrastructural changes at senescence and separation. *Planta*, **130**, 203–10.

Osborne T.B. (1924) The vegetable proteins. In *Monographs on Biochemistry*. (Eds. Plummer R.H. & Hopkins F.G.) Longman, Green & Co., London.

Otten L.A. & Schilperoort R.A. (1978) A rapid microscale method for the detection of lysopine and nopaline dehydrogenase activities. *Biochim. biophys. Acta*, **527**, 497–500.

Overeem J.C. (1976) Pre-existing antimicrobial substances in plants and their role in disease resistance. In *Biochemical Aspects of Plant–Parasite Relationships*. (Eds. Friend J. & Threlfall D.R.) pp. 195–206. Academic Press, London.

Ownby J.D., Ross C.W. & Key J.L. (1975) Studies on the presence of adenosine cyclic 3′:5′ monophosphate in oat coleoptiles. *Plant Physiol.*, **55**, 346–51.

Padmanabhan U. & Green B. (1978) The kinetic complexity of *Acetabularia* chloroplast DNA. *Biochim. biophys. Acta*, **521**, 67–73.

PAGAN J.D., CHILD J.J., SCOWCROFT W.R. & GIBSON A.H. (1975) Nitrogen fixation by *Rhizobium* cultured on a defined medium. *Nature, Lond.*, **256**, 406–7.

PAI H.S., DEHM P., SCHWEIGER M., RAHMSDORF H.J., PONTA H., HIRSCH-KAUFFMANN M. & SCHWEIGER H.G. (1975) Protein kinase of *Acetabularia. Protoplasma*, **85**, 209–18.

PALEVITZ B.A. (1978) Cortical microtubules in plant cells: a high voltage electron microscope study. *J. Cell Biol.*, **79**, 278a.

PALMITER R.D. (1972) Regulation of protein synthesis in chick oviduct. II. Modulation of polypeptide elongation and initiation rates by estrogen and progesterone. *J. biol. Chem.*, **247**, 6770–80.

PALMITER R.D. & SCHIMKE R.T. (1973) Regulation of protein synthesis in chick oviduct. *J. biol. Chem.*, **248**, 1502–13.

PANJIPAN B. (1979) Protonation scheme in acid-induced DNA strand separation. *Trends in biochemical Science*, **4**, N210–11.

PARADIES H.H., GÖKE L. & WERZ G. (1977) On the spatial structure of a plant cell wall protein. Secondary structure of a cell wall protein from *Acetabularia. Protoplasma*, **93**, 249–65.

PARANJOTHY K. & WAREING P.F. (1971) The effect of abscisic acid, kinetins and 5-fluorouracil on ribonucleic acid and protein synthesis in senescing radish leaf discs. *Planta*, **99**, 112–19.

PARK W.D., LEWIS E.D. & RUBENSTEIN I. (1980) Heterogeneity of zein mRNA and protein in maize. *Plant Physiol.*, **65**, 98–106.

PARKER M.L. (1979) Morphology and ultrastructure of the gravity sensitive leaf sheath base of the grass *Echinochloa colonum* L. *Planta*, **145**, 471–7.

PARKER M.W., HENDRICKS S.B., BORTHWICK H.A. & SCULLY N.J. (1946) Action spectrum for the photoperiodic control of floral initiation of short day plants. *Botan. Gaz.*, **108**, 1–26.

PASCOE M.J. & INGLE J. (1978) Distinction between nuclear satellite DNAs and chloroplast DNA in higher plants. *Plant Physiol.*, **62**, 975–7.

PATE J.S. (1958) Abscisic acid inhibition of root nodule initiated in *Pisum sativum. Planta*, **100**, 181–90.

PATE J.S. (1966) Photosynthesizing leaves and nodulated roots as donors of carbon to protein of the shoot of the field pea (*Pisum arvense* L.). *Ann. Bot.*, **30**, 93–109.

PATE J.S. (1977) Functional biology of dinitrogen fixation by legumes. In *A Treatise on Dinitrogen Fixation Sec. II. Biology*. (Eds. Hardy Q.W.F. & Silver W.S.) pp. 473–517. Wiley-Interscience, New York.

PATE J.S. (1980) Transport and partitioning of nitrogenous solutes. *Ann. Rev. Plant Physiol.*, **31**, 313–40.

PATE J.S. & FLINN A.M. (1973) Carbon and nitrogen transfer from vegetative organs to ripening seeds of field pea (*Pisum arvense*) L. *J. exp. Bot.*, **24**, 1090–9.

PAULL R.E., JOHNSON C.M. & JONES R.L. (1975) Studies on the secretion of maize root cap slime. I. Some properties of the secreted polymer. *Plant Physiol.*, **56**, 300–6.

PAULL R.E. & JONES R.L. (1976a) Studies on the secretion of maize root-cap slime. *Plant Physiol.*, **57**, 249–56.

PAULL R.E. & JONES R.L. (1976b) Studies on the secretion of maize root cap slime V. The cell wall as a barrier to secretion. *Z. Pflphysiol.*, **79**, 154–64.

PAULSON R.E. & SRIVASTAVA L.M. (1968) The fine structure of the embryo of *Lactuca sativa*. I. Dry embryo. *Canad. J. Bot.*, **46**, 1437–45 + 5 plates.

PAYNE E.S., BROWNRIGG A., YARWOOD A. & BOULTER D. (1971) Changing protein synthetic machinery during development of seeds of *Vicia faba. Phytochem.*, **10**, 2299–303.

PAYNE G., KNOCHE H.W., KONO Y. & DALY J.M. (1980) Biological activity of purified host-specific pathotoxins produced by *Bipolaris* (*Helminthosporium*) *maydis* race T. *Physiol. Plant Pathol.*, **16**, 227–39.

PAYNE P.I. (1976) The long-lived messenger ribonucleic acid of flowering plant seeds. *Biol. Rev.*, **51**, 329–63.

PAYNE P.I. (1977) Synthesis of poly A-rich RNA in embryos of rye during imbibition and early germination. *Phytochem.*, **16**, 431–4.

PEABODY D.C. & MOTTA J.L. (1979) The ultrastructure of nuclear division in *Armillaria mellea*: meiosis I. *Canad. J. Bot.*, **57**, 1860–72.

PEARSON J.A., THOMAS K. & THOMAS H. (1978) Nucleic acids from leaves of a yellowing and a non-yellowing variety of *Festuca pratensis* Huds. *Planta*, **144**, 85–7.

PECK E.J., JR., BURGMER J. & CLARK J.H. (1973) Estrophilic binding sites of the uterus. Relation to uptake and retention of estradiol *in vivo*. *Biochem.*, **12**, 4596–603.

PEDERSEN K., BLOOM K.S., ANDERSON J.N., GLOVER D.V. & LARKINS B.A. (1980) Analysis of the complexity and frequency of zein genes in the maize genome. *Biochem.*, **19**, 1644–50.

PENG H.B. & JAFFE L.F. (1976) Cell-wall formation in *Pelvetia* embryos. A freeze-fracture study. *Planta*, **133**, 57–71.

PENNY P. (1971) Growth-limiting protein in relation to auxin-induced elongation in lupin cotyledons. *Plant Physiol.*, **48**, 720–3.

PENNY P. & PENNY D. (1978). Rapid responses to phytohormones. In *Phytohormones and Related Compounds—a Comprehensive Treatise. Vol. 2. Phytohormones and the Development of Higher Plants* (Eds. Letham D.S., Goodwin P.B. & Higgins T.J.V.) pp. 537–97. Elsevier/North-Holland, Amsterdam.

PEOPLES M.B. & DALLING M.J. (1978) Degradation of ribulose 1,5-bisphosphate carboxylase by proteolytic enzymes from crude extracts of wheat leaves. *Planta*, **138**, 153–60.

PEOPLES M.B. & DALLING M.J. (1979) Intra-cellular localisation of acid peptide hydrolases in wheat leaves. *Plant Physiol.*, **63**, Suppl. 159.

PEOPLES M.B., FRITH G.J.T. & DALLING M.J. (1979) Proteolytic enzymes in green wheat leaves. IV. Degradation of ribulose-1,5-bisphosphate carboxylase by acid proteases isolated on DEAE cellulose. *Plant & Cell Physiol.*, **20**, 253–8.

PERBAL G. (1978) The mechanism of geoperception in lentil roots. *J. exp. Bot.*, **29**, 611–38.

PERCIVAL E. & MIAN A.J. (1973) Fucose-containing polysaccharides in brown seaweeds. *Proc. VIIth Int. Seaweed Symposium.* pp. 443–6, Sapporo, Japan 1971.

PEREZ C.M., PERSON A.A., RESURRECCION A.R., VILLAREAL R.M. & JULIANO B.O. (1975) Enzymes of carbohydrate metabolism in the developing rice grain. *Plant Physiol.*, **56**, 579–83.

PERNOLLET J-C. (1978) Protein bodies of seeds: ultrastructure, biochemistry, biosynthesis and degradation. *Phytochem.*, **17**, 1473–80.

PETERS B.M., CRIBBS D.H. & STELZIG D.A. (1978) Agglutination of plant protoplasts by fungal cell wall glucans. *Science*, **201**, 364–5.

PETERSON D.M. & SMITH D. (1976) Changes in nitrogen and carbohydrate fractions in developing oats groats. *Crop Sci.*, **16**, 67–71.

PETERSON L.W. & HUFFAKER R.C. (1975) Loss of ribulose 1,5-diphosphate carboxylase and increase in proteolytic activity during senescence of detached primary barley leaves. *Plant Physiol.*, **55**, 1009–15.

PETERSON L.W., KLEINKOPF G.E. & HUFFAKER R.C. (1973) Evidence for lack of turnover of ribulose 1,5-diphosphate carboxylase in barley leaves. *Plant Physiol.*, **51**, 1042–5.

PETERSON R.L. (1975) The initiation and development of root buds. In *The Development and Function of Roots.* (Eds. Torrey J.G. & Clarkson D.T.) pp. 125–62. Academic Press, London.

PETIT A., DESSAUX Y. & TEMPÉ J. (1978a) The biological significance of opines. I. A study of opine catabolism by *Agrobacterium tumefaciens. Proc. IVth Int. Conf. Plant Path. Bact.* Angers, pp. 143–52.

PETIT A., TEMPÉ J., KERR A., HOLSTERS M., VAN MONTAGU M. & SCHELL J. (1978b) Substrate induction of conjugative activity of *Agrobacterium tumefaciens. Nature, Lond.*, **271**, 570–2.

PEUMANS W.J. & CARLIER A.R. (1977) Messenger ribonucleoprotein particles in dry wheat and rye embryos. *In vitro* translation and size distribution. *Planta*, **136**, 195–201.

PHILLIPS D.A. (1971) Abscissic acid inhibition of root nodule initiation in *Pisum sativum. Planta*, **100**, 181–90.

PHILLIPS D.A. (1974) Factors affecting the reduction of acetylene by *Rhizobium* soybean cell association *in vitro. Plant Physiol.*, **53**, 67–72.

PHILLIPS D.A. & TORREY J.G. (1972) Studies on cytokinin production of *Rhizobium. Plant Physiol.*, **49**, 11–15.

PHILLIPS D.R. & FLETCHER R.A. (1969) Ribonuclease in leaves of *Phaseolus vulgaris* during maturation and senescence. *Physiol. Plant.*, **22**, 1050–4.

PHILLIPS I.D.J. (1969). Apical dominance. In *Physiology of Plant Growth and Development.* (Ed. Wilkins M.J.) pp. 164–202. McGraw-Hill, New York.

PHILLIPS I.D.J. (1972) Diffusible gibberellins and phototropism in *Helianthus annuus. Planta*, **106**, 363–7.

PHILLIPS I.D.J. & HARTUNG W. (1976) Longitudinal and lateral transport of (3, 4-^{3}H) gibberellin A, and $_3$-indolyl(acetic acid-2-^{14}C) in upright and geotropically responding green internode segments from *Helianthus annuus. New Phytol.*, **76**, 1–9.

PICKARD B.G. & THIMANN K.V. (1964) Transport and distribution of auxin during tropistic response. II. The lateral migration of auxin in phototropic coleoptiles. *Plant Physiol.*, **39**, 341–50.

PICKARD B.G. & THIMANN K.V. (1966) Geotropic response of wheat coleoptiles in absence of amyloplast starch. *J. gen. Physiol.*, **49**, 1065–86.

PICKETT-HEAPS J.D. (1966) Incorporation of radioactivity into wheat xylem cells. *Planta*, **71**, 1–14.

PICKETT-HEAPS J.D. (1967a) Preliminary attempts at ultrastructural polysaccharide localization in root tip cells. *J. Histochem. Cytochem.*, **15**, 442–55.

Pickett-Heaps J.D. (1967b) Further observations on the Golgi apparatus and its functions in the cells of the wheat seedling. *J. ultrastruct. Res.*, **18**, 287–303.

Pickett-Heaps J.D. (1967c) Ultrastructure and differentiation in *Chara* sp. II. Mitosis. *Aust. J. biol. Sci.*, **20**, 883–94.

Pickett-Heaps J.D. (1968) Further ultrastructural observations on polysaccharide localization in plant cells. *J. Cell Sci.*, **3**, 55–64.

Pickett-Heaps J.D. (1969) The evolution of the mitotic apparatus: an attempt at comparative ultrastructural cytology in dividing plant cells. *Cytobios.*, **1**, 257–80.

Pickett-Heaps J.D. (1971) The autonomy of the centriole: fact or fallacy? *Cytobios.*, **3**, 205–14.

Pickett-Heaps J.D. (1972) Cell division in *Tetraedon. Ann. Bot.*, **36**, 693–701.

Pickett-Heaps J.D. (1974) Plant microtubules. In *Dynamic Aspects of Plant Ultrastructure.* (Ed. Robards A.W.). McGraw-Hill, London.

Pickett-Heaps J.D. (1975a) Aspects of spindle evolution. *Ann. N.Y. Acad. Sci.*, **253**, 352–61.

Pickett-Heaps J.D. (1975b) *Green Algae.* Sinauer Associates Inc., Mass., USA.

Pickett-Heaps J.D. & Fowke L.C. (1969) Cell division in *Oedogonium.* I. Mitosis, cytokinesis and cell elongation. *Aust. J. biol. Sci.*, **22**, 857–94.

Pilet P-E. (1961) Auxins and the process of aging in root cells. In *Plant Growth Regulation.* (Eds. Wightman F. & Setterfield G.) pp. 167–78. Iowa State University Press, Ames, Iowa.

Pilet P-E. (1971) *Les Parois Cellulaires.* Doin, Paris.

Pilet P-E. (1973) Growth inhibitor from the root cap of *Zea mays. Planta*, **111**, 275–8.

Pilet P-E. (1977) Growth inhibitors in growing and geostimulated maize roots. In *Plant Growth Regulation.* pp. 115–28. Springer-Verlag, Berlin.

Pinto da Silva P. (1972) Translational mobility of the membrane intercalated particles of human erythrocyte membranes. pH dependent reversible aggregation. *J. Cell. Biol.*, **53**, 777–87.

van der Plank J.E. (1968) *Disease Resistance in Plants.* Academic Press, New York.

Poff K.L. & Butler W.L. (1974) Absorbance changes induced by blue light in *Phycomyces blakesleeanus* and *Dictyostelium discoideum. Nature, Lond.*, **248**, 799–801.

Polevoy V.V. & Salmatova T.S. (1977) Auxin, proton pump and cell trophics. In *Regulation of Cell Membrane Activities in Plants.* (Eds. Marrè E. & Ciferri O.) pp. 209–216. Elsevier/North-Holland, Amsterdam.

Pollock C.J. & Lloyd E.J. (1978) Acid invertase activity during senescence of excised leaf tissue of *Lolium temulentum. Z. Pflphysiol.*, **90**, 79–84.

Poo M-M., Law J.W., Onida N. & Choa A.W. (1979) Electrophoresis and diffusion in the plane of the membrane. *Biophys. J.*, **26**, 1–22.

Poo M.M. & Robinson K.R. (1977) Electrophoresis of con A receptors along embryonic cell membranes. *Nature, Lond.*, **265**, 602–5.

Pooviah B.W. & Leopold A.C. (1973) Deferral of leaf senescence with calcium. *Plant Physiol.*, **52**, 236–9.

Possingham J.V. (1980) Plastid replication and development in the life cycle of higher plants. *Ann. Rev. Plant Physiol.*, **31**, 113–29.

Possingham J.V. & Rose R.J. (1976) Chloroplast replication and chloroplast DNA synthesis in spinach leaves. *Proc. Roy. Soc. Lond. B.*, **193**, 295–305.

Possingham J.V. & Saurer W. (1969) Changes in chloroplast number per cell during leaf development in spinach. *Planta*, **86**, 186–94.

Prakash R.K., Hookyas P.J.J., Ledboer A.M., Kijne J.W., Schilperoot R.A., Nuti M.P., Lepidi A.A., Casse F., Boucher C., Julliot J.S. & Dénarié J. (1980) Detection, isolation, and characterization of large plasmids in *Rhizobium.* In *Nitrogen Fixation Vol. II.* (Eds. Newton W.E. & Orme-Johnson W.H.) pp. 139–63. University Park Press, Baltimore.

Pratt L.H. (1978) Molecular properties of phytochrome. *Photochem. Photobiol.*, **27**, 81–105.

Pratt L.H. & Marmé D. (1976) Red light-enhanced phytochrome pelletability. Re-examination and further characterization. *Plant Physiol.*, **58**, 686–92.

Presti D. & Delbrück M. (1978) Photoreceptors for biosynthesis, energy storage and vision. *Plant, Cell & Environ.*, **1**, 81–100.

Preston K.R. & Kruger J.E. (1979) Physiological control of exo- and endoproteolytic activities in germinating wheat and their relationship to storage protein hydrolysis. *Plant Physiol.*, **64**, 450–4.

Preston R.D. (1979) Polysaccharide conformation and cell wall function. *Ann. Rev. Plant Physiol.*, **30**, 55–78.

Price H.J., Sparrow A.H. & Nauman A.F. (1974) Evolutionary and developmental considerations of the variability of nuclear parameters in higher plants. I. Genome volume, interphase chromosome volume, and estimated DNA content of 236 gymnosperms. *Brookhaven Symp. Biol.*, **25**, 390–421.

PRING D.R., CONDE M.F. & LEVINGS C.S. III. (1980) DNA heterogeneity within the C group of maize male-sterile cytoplasms. *Crop Sci.*, **20**, 159–62.

PRING D.R. & LEVINGS C.S. III. (1978) Heterogeneity of maize cytoplasmic genomes among male-sterile cytoplasms. *Genetics*, **89**, 121–36.

PRING D.R., LEVINGS C.S. III, HU W.W.L. & TIMOTHY P.H. (1977) Unique DNA associated with mitochondria in the 'S' type cytoplasm of male-sterile maize. *Proc. Nat. Acad. Sci. USA*, **74**, 2904–8.

PRUSKY D., DINOOR A. & JACOBY B. (1980) The sequence of death of haustoria and host cells during the hypersensitive response of oat to crown rust. *Physiol. Plant Pathol.*, **17**, 33–40.

PRYKE J.A. & AP REES T. (1976) Activity of the pentose phosphate pathway during lignification. *Planta*, **132**, 279–84.

PUCHEL M., MUNTZ K., PARTHIER B., AURICH O., BASSUNER R., MANTEUFFEL R. & SCHMIDT P. (1979) RNA metabolism and membrane-bound polysomes in relation to globulin biosynthesis in cotyledons of developing field beans (*Vicia faba* L.) *Eur. J. Biochem.*, **96**, 321–9.

PUEPPKE G., BAUER W.D., KEOGSTRA K. & FERGUSON A.L. (1978) Role of lectins in plant-microorganisms interactions. II. Distribution of soybean lectins in tissues of *Glycine max* L. Merr. *Plant Physiol.*, **61**, 779–84.

PUISEUX-DAO S. (1970) *Acetabularia and Cell Biology*, 162 pp. Logos Press, London.

PULL S.P., PUEPPKE S.G., HYMOWITZ T. & ORF J.H. (1978) Soybean lines lacking 120 000-dalton seed lectin. *Science*, **200**, 1277–9.

PURDY R.E. & KOLATTUKUDY P.E. (1975) Hydrolysis of plant cuticle by plant pathogens. Purification, amino acid composition and molecular weight of two isoenzymes of cutinase and a non-specific esterase from *Fusarium solani* f. *pisi*. *Biochem.*, **14**, 2824–31.

PURVIS O. (1940) Vernalisation of fragments of embryo tissue. *Nature, Lond.*, **145**, 462.

PURVIS O. (1961) The physiological analysis of 'vernalisation'. In *Encyclopaedia Plant Physiology*. (Ed. Ruhland W.) Vol. 16, 76–122.

QUAIL P.H. (1978) Irradiation-enhanced phytochrome pelletability in *Avena: in vivo* development of a potential to pellet and the role of Mg^{++} in its expression *in vitro*. *Photochem. Photobiol.*, **27**, 147–53.

QUAIL P.H. (1979) Plant cell fractionation. *Ann. Rev. Plant Physiol.*, **30**, 425–84.

QUATRANO R.S. (1968) Rhizoid formation in *Fucus* zygotes: dependence on protein and ribonucleic acid syntheses. *Science*, **162**, 468–70.

QUATRANO R.S. (1974) Developmental biology: development in marine organisms. In *Experimental Marine Biology*. (Ed. Marisca R.) pp. 303–46. Academic Press, New York.

QUATRANO R.S. (1978) Development of cell polarity. *Ann. Rev. Plant Physiol.*, **29**, 489–510.

QUATRANO R.S. HOGSETT W.S. & ROBERTS M. (1978) Localization of a sulfated polysaccharide in the rhizoid wall of *Fucus distichus* (Phaeophyceae) zygotes. In *Proc. IXth International Seaweed Symposium*. (Ed. Jensen A. & Stein J.R.) pp. 113–123. Santa Barbara 1977.

QUATRANO R.S. & STEVENS P.T. (1976) Cell wall assembly in *Fucus* zygotes. I. Characterization of the polysaccharide components. *Plant Physiol.*, **58**, 224–31.

QUETIER F. & VEDEL F. (1979) Physicochemical characteristics and restriction endonuclease analysis of mitochondrial DNA from higher plants. In *Proc. NATO/FEBS Conf. Genome Organisation and Expression in Plants*, 178 pp. Plenum Press, New York.

QUINN P.J. (1976) *The Molecular Biology of Cell Membranes*. Macmillan, London.

RACUSEN D. & FOOTE M. (1971) The major glycoprotein in germinating bean seeds. *Canad. J. Bot.*, **49**, 2107–11.

RACUSEN R. (1976) Phytochrome control of electrical potentials and intercellular coupling in oat-coleoptile tissue. *Planta*, **132**, 25–9.

RACUSEN R. & MILLER K. (1972) Phytochrome-induced adhesion of mung bean root tips to platinum electrodes in a direct current field. *Plant Physiol.*, **49**, 654–5.

RACUSEN R. & SATTER R.L. (1975) Rhythmic and phytochrome-regulated changes in transmembrane potential in *Samanea pulvini*. *Nature, Lond.*, **255**, 408–10.

RADLEY M. (1967) Site of production of gibberellin-like substances in germinating barley embryos. *Planta*, **75**, 164–71.

RAFF J.W., HUTCHINSON J.F., KNOX R.B. & CLARKE A.E. (1979) Cell recognition: antigenic determinants of plant organs and their cultured callus cells. *Differentiation*, **12**, 179–86.

RAGHAVEN V. (1976) *Experimental Embryogenesis in Vascular Plants*. Experimental Botany Vol. 10. Academic Press, London.

RAGHAVEN V. (1979) Totipotency of plant cells:

evidence for the totipotency of male germ cells from anther and pollen culture studies. In *Plant Cell and Tissue Culture.* (Ed. Sharp W.R., Larsen P.O., Paddock E.F. & Raghaven V.) pp. 155–78. Ohio State University Press, Columbus.

Raghaven V. & Torrey J.G. (1963) Growth and morphogenesis of globular and older embryos of *Capsella* in culture. *Am. J. Bot.*, **50**, 540–51.

Raghaven V. & Torrey J.G. (1964) Inorganic nitrogen nutrition of the seedlings of the orchid *Cattleya. Am. J. Bot.*, **50**, 540–51.

Railton I.D. & Reid D.M. (1974) Studies on gibberellins in shoots of light grown peas. I. A re-evaluation of the data. *Plant Sci. Lett.*, **2**, 157–63.

Rake A.V. (1972) Lack of DNA homology between the legume *Glycine max* and its symbiotic *Rhizobium* bacteria. *Genetics*, **71**, 19–24.

Randall P.J., Thomson J.A. & Schroeder H.E. (1979) Cotyledonary storage proteins in *Pisum sativum.* IV. Effects of sulphur, phosphorous, potassium and magnesium deficiencies. *Aust. J. Plant Physiol.*, **6**, 11–24.

Rangaswamy N.S. & Shivanna K.R. (1967) Induction of gamete compatibility and seed formation in axenic cultures of a diploid self-incompatible species of *Petunia. Nature, Lond.*, **216**, 937–9.

Rapp B.J. & Randall D.D. (1980) Pyruvate dehydrogenase complex from germinating castor bean endosperm. *Plant Physiol.*, **65**, 314–8.

Rappaport L. & Wolf N. (1969). The problem of dormancy in potato tubers and related structures. In *Dormancy and Survival.* Symp. Soc. exp. Biol. XXIII. (Ed. Woolhouse H.W.) p. 219–41. Cambridge University Press, Cambridge.

Rathmell W. & Bendall D. (1971) Phenolic components in relation to phytoalexin biosynthesis in hypocotyls of *Phaseolus vulgaris. Physiol. Plant Pathol.*, **1**, 351–62.

Rau W. (1979) Photoregulation of carotenoid biosyntheses: an example of photomorphogenesis. *Pigments in Plants.* (Ed. Czygan H.Ch.) Fischer Verlag, Jena (in press).

Raven J.A. (1975) Transport of indoleacetic acid in plant cells in relation to pH and electrical potential gradients, and its significance for polar IAA transport. *New Phytol.*, **74**, 163–72.

Raven J.A. (1977a) H^+ and Ca^{2+} in phloem and symplast: relation of relative immobility of the ions to the cytoplasmic nature of the transport paths. *New Phytol.*, **79**, 465–80.

Raven J.A. (1977b) the evolution of vascular land plants in relation to supracellular transport processes. *Adv. bot. Res.*, **5**, 153–217.

Raven J.A. (1979). The possible role of membrane electropheresis in the polar transport of IAA and other solutes in plant tissues. *New Phytol.*, **82**, 285–91.

Raven J.A. (1980a) Nutrient transport in microalgae. *Adv. Microb. Physiol.*, **21**, 47–226.

Raven J.A. (1980b) Short- and long-distance transport of boric acid in plants. *New Phytol.*, **84**, (2), (in press).

Raven J.A. (1981) Introduction to metabolic control. In *Mathematics and Plant Physiology.* (Eds. Rose D.A. & Charles-Edwards D.A.). Academic Press, London.

Raven J.A. & Smith F.A. (1976) Nitrogen assimilation and transport in vascular land plants in relation to intracellular pH regulation. *New Phytol.*, **76**, 415–31.

Raven J.A. & Smith F.A. (1977) Characteristics, functions and regulation of active proton extrusion. In *Regulation of Cell Membrane Activities in Plants.* (Eds. Marrè E. & Ciferri O.). North-Holland, Amsterdam.

Ray P.M. (1963) Sugar composition of oat-coleoptile cell walls. *Biochem. J.*, **89**, 144–50.

Ray P.M. (1977) Auxin-binding sites of maize coleoptiles are localized on the membranes of the endoplasmic reticulum. *Plant Physiol.*, **59**, 594–9.

Rayle D.L. & Cleland R. (1977) Control of plant cell enlargement by hydrogen ions. *Curr. Topics devel. Biol.*, **11**, 187–214.

Rees D.A. (1969) Structure, conformation and mechanism in the formation of polysaccharide gels and networks. *Advan. Carbohydr. Chem. Biochem.*, **24**, 267–332.

Reid D.M., Tuing M.S., Durley R.C. & Railton I.D. (1972) Red-light-enhanced conversion of tritiated gibberellin A9 into other gibberellin-like substances in homogenates of etiolated barley leaves. *Planta*, **108**, 67–75.

Reid J.S.G. (1971) Reserve carbohydrate metabolism in germinating seeds. *Planta*, **100**, 131–42.

Reid J.S.G. & Bewley J.D. (1979) A dual role for the endosperm and its galactomannan reserves in the germinative physiology of fenugreek (*Trigonella foenum-graecum* L.) an endospermic leguminous seed. *Planta*, **147**, 145–50.

Reiss R-D. & Herth W. (1978) Visualization of the Ca^{2+}-gradient in growing pollen tubes of *Lilium longiflorum* with chlorotetracycline fluorescence. *Protoplasma*, **97**, 373–7.

REISSIG J.L. (1977) The divalent cation ionophore A23187 induces branching in *Neurospora. J. Cell Biol.*, **75**, CD340.

REPORTER M. & HERMINA N. (1975) Acetylene reduction by transfilter suspension cultures of *Rhizobium japonicum. Biochem. biophys. Res. Comm.*, **64**, 1126–33.

REST J.A. & VAUGHAN J.G. (1972) The development of protein and oil bodies in seeds of *Sinapis alba* L. *Planta*, **105**, 245–62.

RICARDO C.C.P. & AP REES T. (1970) Invertase activity during the development of carrot roots. *Phytochem.*, **9**, 239–47.

RICH D.B. (1977) Effects of structure of active compounds on biological activity and specificity. In *Specificity in Plant Diseases.* (Eds. Wood, R.K.S. & Graniti A.) pp. 169–81, Plenum Press, New York.

RICHARDSON S.D. (1957) The effect of leaf age on the rate of photosynthesis in detached leaves of tree seedlings. *Acta Bot. Neerl.*, **6**, 445–57.

RICHMOND A.E. & LANG A. (1957) Effect of kinetin on protein content and survival of detached *Xanthium* leaves. *Science*, **125**, 650–1.

RIDE J.P. & PEARCE R.B. (1979) Lignification and papilla formation at sites of attempted penetration of wheat leaves by non-pathogenic fungi. *Physiol. Plant Pathol.*, **15**, 79–92.

REIZMAN H., WEIR E., LEAVER C.J., TITUS D.E. & BECKER W.M. (1980) Regulation of glyoxysomal enzymes during germination of cucumber. *Plant Physiol.*, **65**, 40–46.

RIS H. (1975) Primitive mitotic mechanisms. *Biosystems*, **7**, 298–304.

RIS H. & PLAUT W. (1962) The ultrastructure of DNA-containing areas in the chloroplast of *Chlamydomonas. J. Cell Biol.*, **13**, 383–91.

RITTER H., INOUÉ S. & KUBAI D. (1978) Mitosis in *Barbulanympha.* I. Spindle structure, formation, and kinetochore arrangement. *J. Cell Biol.*, **77**, 638–54.

RIVIER L. & PILET P-E. (1974) Indolyl-3-acetic acid in cap and apex of maize roots: identification and quantification by mass fragmentography. *Planta*, **120**, 107–12.

ROBARDS A.W. (1976) Plasmodesmata in higher plants. In *Intercellular Communication in Plants.* (Eds. Gunning B.E.S. & Robards A.W.) pp. 15–59. Springer-Verlag, Berlin.

ROBARDS A.W., JACKSON S.M., CLARKSON D.T. & SANDERSON J. (1973) The structure of barley roots in relation to the transport of ions into the stele. *Protoplasma*, **77**, 291–311.

ROBARDS A.W. & KIDWAI P. (1972) Microtubules and microfibrils in xylem fibres during secondary cell wall formation. *Cytobiologie*, **6**, 1–21.

ROBBINS W.J. (1957) Gibberellic acid and the reversal of adult *Hedera* to a juvenile state. *Am. J. Bot.*, **44**, 743–6.

ROBENEK H. & PEVELING E. (1977) Ultrastructure of the cell wall regeneration of isolated protoplasts of *Skimmia japonica* Thunb. *Planta*, **136**, 135–45.

ROBERTS K. (1974) Crystalline glycoprotein cell walls of algae: their structure, composition and assembly. *Phil. Trans. R. Soc. Lond. B.*, **268**, 129–46.

ROBERTS K. & HYAMS J.S. (1979) *Microtubules.* Academic Press, London.

ROBERTS L.M. & LORD J.M. (1979) Developmental changes in the activity of the messenger RNA isolated from germinating castor bean endosperm. *Plant Physiol.*, **64**, 630–4.

ROBERTS L.W. (1976) *Cytodifferentiation in Plants.* Cambridge University Press, London.

ROBERTSON J.G., LYTTLETON P., BULLIVANT S. & GRAYSTON G.F. (1978a) Membranes in Lupin root nodules. I. The role of golgi bodies in the biogenesis of infection threads and peribacteroid membranes. *J. Cell Sci.*, **30**, 129–49.

ROBERTSON J.G., WARBURTON M.P., LYTTLETON P., FORDYCE A.M. & BULLIVANT S. (1978b) Membranes in lupin root nodules. II. Preparation and properties of peribacteroid membranes and bacteroid envelope inner membranes from developing lupin nodules. *J. Cell Sci.*, **30**, 151–74.

ROBINS R.J. (1976) The nature of the stimuli causing digestive juice secretion in Venus flytrap. *Planta*, **128**, 263–5.

ROBINOW C.F. (1963) Observations on cell growth, mitosis and division in the fungus *Basidiobolus ranarum. J. Cell Biol.*, **17**, 123–52.

ROBINSON D.G. (1977) Plant cell wall synthesis. *Adv. Bot. Res.*, **5**, 89–151.

ROBINSON K.R. & JAFFE L.F. (1975) Polarizing fucoid egg drive and calcium current through themselves. *Nature, Lond.*, **187**, 70–72.

ROBINSON N.E. & BRYANT J.A. (1975) Onset of nucleic acid synthesis during germination of *Pisum sativum. Planta*, **127**, 63–8.

ROCHAIX J.D., MALNOE, P., CH N.H. & SPARH P.F. (1979) Gene localization and transcription of the chloroplast DNA of *Chlamydomonas reinhardii.* In *Proc. NATO/FEBS Conf. Genome Organisation and Expression in Plants.* 178 pp.

RODAWAY S. & MARCUS A. (1980) *In vivo* synthesis of 6-azauridine 5-triphosphate and incorporation of 6-azauridine into RNA of germinating wheat embryonic axes. *J. biol. Chem.*, **255**, 8402–4.

Roelofsen P.A. (1965) Ultrastructure of the wall in growing cells and its relation to the direction of growth. *Adv. bot. Res.*, **2**, 69–145.

Roland J-C. (1978) Cell wall differentiation and stages involved with intercellular gas space opening. *J. Cell Sci.*, **32**, 325–36.

Roland J-C., Lembi C.A. & Morré D.J. (1972) Phosphotungstic acid. Chromic acid as a selective electron-dense stain for plasma membrane of plant cells. *Stain Technol.*, **47**, 195–200.

Roland J-C. & Vian B. (1979) The wall of the growing plant cell: its three-dimensional organization. *Int. Rev. Cytol.*, **61**, 129–66.

Roldán J.M. & Butler W.L. (1980) Photoactivation of nitrate reductase from *Neurospora erassa. Photochem. Photobiol.*, **32**, 375–81.

Romanovicz D.K. & Brown R.M. (1976) Biogenesis and structure of golgi-derived cellulosic scales in *Pleurochrysis*. II. Scale composition and supramolecular structure. *Applied Polymer Symposium No. 28*, 587–610. John Wiley & Sons Inc., New York.

Rose R.J., Cran D.G. & Possingham J.V. (1974) Distribution of DNA in dividing spinach chloroplasts. *Nature, Lond.*, **251**, 641–2.

Rose R.J. & Possingham J.V. (1976) The localization of ^{3}H-thymidine incorporation in the DNA of replicating spinach chloroplasts by electron microscope autoradiography. *J. Cell Sci.*, **20**, 341–55.

Rosen W.G., Gawlik S.R., Dashek W.V. & Siegesmund K.A. (1964) Fine structure and cytochemistry of *Lilium* pollen tubes. *Am. J. Bot.*, **51**, 61–71.

Rosene H.F. & Lund E.J. (1953) Bioelectric fields and correlation in plants. In *Growth and Differentiation in Plants.* (Ed. Loomis W.) pp. 219–52. University of Iowa Press.

Rosnitschek I. & Theimer R.R. (1980) Properties of a membrane triglyceride lipase of rapeseed (*Brassica napus* L.) cotyledons. *Planta*, **148**, 193–8.

Rowsell E.V. & Goad L.G. (1962) Latent β-amylase of wheat: its mode of attachment to glutenin and its release. *Biochem. J.*, **84**, 73–4P.

Royer H-D. & Sager R. (1979) Methylation of chloroplast DNAs in the life cycle of *Chlamydomonas. Proc. Nat. Acad. Sci. USA*, **76**, 5794–8.

Rubery P.H. (1977) The specificity of carrier-mediated auxin transport by suspension-cultured crown gall cells. *Planta*, **135**, 275–83.

Rubery P.H. (1978) Hydrogen ion dependence of carrier-mediated auxin uptake by suspension-cultured crown gall cells. *Planta*, **142**, 203–6.

Rubery P.H. (1979a) The effects of 2,4-dinitrophenol and chemical modifying reagents on auxin transport by suspension-cultured crown gall cells. *Planta*, **144**, 173–8.

Rubery P.H. (1979b) The effects of [3-phenyl-1,2,3 thiadiazol-(5-yl)thio] acetic acid, a synthetic growth regulator, on auxin uptake by tobacco stem segments and crown gall suspension culture cells. *Plant Sci. Lett.*, **14**, 365–71.

Rubery P.H. (1980) The mechanism of transmembrane auxin transport and its relation to the chemiosmotic hypothesis of the polar transport of auxin. *Proceedings of the Madison Wisconsin Meeting on Plant Growth Substance.* July, 1979.

Rubery P.H. (1981) Auxin receptors. *Ann. Rev. Plant Physiol.*, **32**, 569–96.

Rubery P.H. & Northcote D.H. (1968) Site of phenylalanine ammonia-lyase activity and synthesis of lignin during xylem differentiation. *Nature, Lond.*, **219**, 1230–4.

Rubery P.H. & Sheldrake A.R. (1973) Effect of pH and surface change on cell uptake of auxin. *Nature, New Biol.*, **244**, 285–8.

Rubery P.H. & Sheldrake A.R. (1974) Carrier-mediated auxin transport. *Planta*, **118**, 101–21.

Rubin P.M. & Randall D.D. (1977) Regulation of plant pyruvate dehydrogenase complex by phosphorylation. *Plant Physiol.*, **60**, 34–9.

Rudolph K. (1976) Non-specific toxins. In *Physiological Plant Pathology, Encyclopaedia of Plant Physiology.* New Series, **4**. (Eds. Heitefuss R. & Williams P.H.) pp. 270–315. Springer-Verlag, Berlin.

Russell S.D. (1980) Participation of male cytoplasm during gamete fusion in an angiosperm. *Science*, **210**, 200–1.

Sabater B. & Rodriguez M.T. (1978) Control of chlorophyll degradation in detached leaves of barley and oat through effect of kinetin on chlorophyllase levels. *Physiol. Plant.*, **43**, 274–6.

Sabatini D.D., Bensch K. & Barnett R.J. (1963) Cytochemistry and electron microscopy. The preservation of cellular ultrastructure and enzymatic activity by aldehyde fixation. *J. Cell Biol.*, **17**, 19–58.

Sacher J.A. & Davies D.D. (1974) Demonstration of *de novo* synthesis of RNAase in *Rhoeo* leaf sections by deuterium oxide labelling. *Plant Cell Physiol.*, **15**, 157–62.

SACHS J. (1873) Über das Wachsthum der Haupt- und Nebenwurzeln. *Arb. bot. Inst. Würzburg*, **1**, 385–474.

SACHS T. (1973) The induction of vessel differentiation by auxin. In *Plant Growth Substances*, pp. 900–906. Proc. 8th Int. Conf. Tokyo. Hirokawa Publ. Co.

SACHS T. (1978) Patterned differentiation in plants. *Differentiation*, **11**, 65–73.

SACRISTÁN-GÁRATE A., NAVARETTE M.H. & DE LA TORRE C. (1974) Mitochondrial biogenesis and turnover during a single cycle in meristematic cells. *Cytobios.*, **11**, 21–31.

SAGER R. (1972) *Cytoplasmic Genes and Organelles*. Academic Press, New York.

SAHLMAN K. & FAHRAEUS G. (1963) An electron microscope study of root hair infection by *Rhizobium*. *J. gen. Microbiol.*, **33**, 425–7.

SAKAI B. (1951) Caryotype analysis of Leguminous plants. I. *Senshokutai (Chromosome)/ Kromosomo*, **11**, 425–9 (translation from the original Japanese title of article).

VAN SAMBECK J.W., NOVACKY A. & KARR A.L. (1975) Effect of helminthosporoside on transmembrane potentials in leaf mesophyll cells of *Saccharum*. *Plant Physiol. Suppl.*, **56**, 53.

SAMISH R.M. (1954) Dormancy in woody plants. *Ann. Rev. Plant Physiol.*, **5**, 183–204.

SANDAKHCHIEV L.S., PUCHKOVA L.I., PIKALOV A.V., KHRISTOLUBOVA N.B. & KISELEVA E.V. (1972) Subcellular localization of morphogenetic factors in anucleate *Acetabularia* at the stages of genetic information transfer and expression. In *Biology and Radiobiology of Anucleate Systems*. II. *Plant Cells*. (Eds. Bonotto S., Goutier R., Kirchmann R. & Maisin J.R.) pp. 297–320. Academic Press, New York.

SANDERS R.F., CARLSON R.W. & ALBERSHEIM P. (1978) A *Rhizobium* mutant incapable of nodulation and normal polysaccharide secretion. *Nature, Lond.*, **271**, 240–2.

SANDS H. & RICKENBERG H.V. (1978) Assessment of the role of cyclic nucleotides as hormonal mediators. *Int. Rev. Biochem.*, **20**, 45–80.

SATO S. & ASAHI T. (1975) Biochemical properties of mitochondrial membrane from dry pea seeds and changes in the properties during imbibition. *Plant Physiol.*, **56**, 816–20.

SATTER R.L., MARIONEF P. & GALSTON A.W. (1972) Phytochrome-controlled nyctinasty in *Albizzia julibrissia* IV. Auxin effects on leaf movement and K^+ flux. *Plant Physiol.*, **50**, 235–41.

SAUNDERS P.F. & COOKE R.J. (1975) Phytochrome-mediated changes in extractable gibberellin activity in a cell-free system isolated from etiolated wheat leaves. *Planta*, **124**, 319–28.

SAURER W. & POSSINGHAM J.V. (1970) Studies on the growth of spinach leaves (*Spinacea oleracea*). *J. exp. Bot.*, **21**, 151–8.

SAVELBERGH R. & VAN PARIJS R. (1971) Ultrastructural study of protein body formation in ripening seeds of *Pisum sativum* L. *Arch. int. Physiol. Biochim.*, **79**, 1040–1.

SAVIDGE R.A. & WAREING P.F. (1981) Plant growth regulators and the differentiation of vascular elements. In *Aspects of Xylem Cell Differentiation*. (Ed. Barnett J.R.) Castle House Publications, London.

SAWHNEY V.K. & SRIVASTAVA L.M. (1975) Wall fibrils and microtubules in normal and gibberellic-acid-induced growth of lettuce hypocotyls. *Canad. J. Bot.*, **53**, 824–35.

SCANDALIOS J.G. (1977) Isozymes: genetic and biochemical regulation of alcohol dehydrogenase. In *Regulation of Enzyme Synthesis and Activity in Higher Plants*. (Ed. Smith H.) pp. 129–53. Academic Press, London.

SCHARPÉ A. & VAN PARIJS R. (1973) The formation of polyploid cells in ripening cotyledons of *Pisum sativum* L. in relation to ribosome and protein synthesis. *J. exp. Bot.*, **24**, 216–22.

SCHEFFER R.P. (1976) Host-specific toxins in relation to pathogenesis and disease resistance. In *Physiological Plant Pathology*, New Series, **4**. (Eds. Heitefuss R. & Williams P.H.) pp. 247–69. Springer-Verlag, Berlin.

SCHELL J. (1975) The role of plasmids in crown gall formation by *A. tumefaciens*. In *Genetic Manipulations with Plant Materials*. (Ed. Ledoux L.) pp. 163–81. Plenum Press, New York.

SCHELL J. & VAN MONTAGU M. (1977) The Ti plasmid of *Agrobacterium tumefaciens*, a natural vector for the introduction of *NIF* genes in plants? *Genetic Engineering for Nitrogen Fixation*, (Ed. Hollaender A.) pp. 159–79. Plenum Press, New York.

SCHELL J. & VAN MONTAGU M. (1980) The Ti-plasmids of *Agrobacterium tumefaciens* and their role in crown gall formation. In *Genome Organization and Expression in Plants*. (Ed. Leaver C.J.) pp. 453–70. Plenum Press, New York.

SCHELL J., VAN MONTAGU M., DE BEUCKELEER M., DE BLOCK M., DEPICKER A., DE WILDE M., ENGLER G., GENETELLO C., HERNALSTEENS J.P., HOLSTERS M., SEURINCK J., SILVA B., VAN VLIET F. & VILLARROEL R. (1979) Interactions and DNA transfer between *Agro-*

bacterium tumefaciens, the Ti-plasmid and the plant host. *Proc. roy. Soc. Lond. B*, **204**, 251–66.

Schilperoort R.A., Hooykaas P.J.J., Klapwijk P.M., Koekman B.P., Nuti M.P., Ooms G. & Prakash R.K. (1979) Characters on large plasmids in Rhizobiaceae involved in the interaction with plant cells. In *Plasmids of Medical, Environmental and Commercial Importance*. (Eds. Timmis K. & Pühler A.) pp. 339–52. Elsevier, Amsterdam.

Schilperoort R.A., Klapwijk P.M., Ooms G. & Wullems G.J. (1980) Plant tumours caused by bacterial plasmids: crown gall. In *Genetic Origins of Tumour Cells*, (Eds. Cleton F.J. & Simons J.W.) pp. 87–108. Martinus Nijhoff, The Hague.

Schimper A.F.W. (1885) Untersuchungen über die Chlorophyllkörper und die ihnen homologen Gebilde. *Jahrb. Wiss. Bot.*, **16**, 1–247.

Schletz K. (1979) Phototaxis bei Volvox-Pigmentsysteme des Lichtrichtungsperzeption. *Z. Pflphysiol.*, **77**, 189–211.

Schmidt E.L. (1979) Initiation of plant–root microbe interactions. *Ann. Rev. Microbiol.*, **33**, 355–76.

Schmidt W., Hart J., Filner P. & Poff K.L. (1977) Specific inhibition of phototropism in corn seedlings. *Plant Physiol.*, **60**, 736–8.

Schmoyer I.R. & Lovett J.S. (1969) Regulation of protein synthesis in zoospores of *Blastocladiella*. *J. Bacteriol.*, **100**, 854–64.

Schneider E.A. & Wightman F. (1974) Metabolism of auxin in higher plants. *Ann. Rev. Plant Physiol.*, **25**, 487–513.

Schnepf E., Diechgräber G. & Ljubešić N. (1976) The effects of colchicine, ethionine and deuterium oxide on microtubules in young *Sphagnum* leaflets. A quantitative study. *Cytobiologie*, **13**, 341–53.

Schopfer P.H. (1977) Phytochrome control of enzymes. *Ann. Rev. Plant Physiol.*, **28**, 223–52.

Schönbeck F. & Schlösser E. (1976) Preformed substances as potential protectants. In *Physiological Plant Pathology, Encyclopaedia of Plant Physiology*, New Series. (Eds. Heitefuss R. & Williams P.H.) pp. 653–78. Springer-Verlag, Berlin.

Schroder J., Kreuzaler F., Schafter E. & Hahlbrock K. (1979) Concomitant induction of phenylanine ammonia lyase and flavone synthase mRNAs in plant cells. *J. biol. Chem.*, **254**, 57–65.

Schröter K., Läuchli A. & Sievers A. (1975) Mikroanalytische Identifikation von Bariumsulfat-Kristallen in den Statolithen der Rhizoide von *Chara fragilis* Desv. *Planta*, **122**, 213–25.

Schubert K.R. & Evans H.J. (1976) Hydrogen evolution: a major factor affecting the efficiency of nitrogen fixation of nodulated symbionts. *Proc. Nat. Acad. Sci.* USA, **73**, 1207–11.

Schulz D. & Jarosch R. (1980) Rotating microtubules as a basis for anaphase spindle elongation in diatoms. *Europ. J. Cell Biol.*, **20**, 249–53.

Schulz S.P. & Jensen W.A. (1969) *Capsella* embryogenesis: the suspensor and the basal cell. *Photoplasma*, **67**, 139–63.

Schulze E. (1878) Über Zersetzung und Neubildung von Eiweisstoffen in Lupinenkeimlingen. *Landw.Jb.*, **7**, 411–35.

Schürmann P. & Buchanan B.B. (1975) Role of ferredoxin in the activation of sedoheptulose diphosphatase in isolated chloroplasts. *Biochim. biophys. Acta*, **376**, 189–92.

Schwabe W.W. (1970) The control of leaf senescence in *Kleinia articulata* by photoperiod. *Ann. Bot.*, **34**, 43–55.

Schwartz A. & Koller D. (1978) Phototropic response to vectorial light in leaves of *Lavatera cretica* L. *Plant Physiol.*, **61**, 924–8.

Schweiger H.G. (1976) Nucleocytoplasmic interaction in *Acetabularia*. *Handbook of Genetics*, **5**, 451–75.

Schweiger H.G. & Schweiger M. (1977) Circadian rhythms in unicellular organisms: an endeavour to explain the molecular mechanism. *Intern. Rev. Cytol.*, **51**, 315–42.

Schwinghamer E.A. (1970) Requirement for riboflavin for effective symbiosis on clover by an auxotrophic mutant strain of *Rhizobium trifolii*. *Aust. J. biol. Sci.*, **23**, 1187–96.

Schwinghamer M.W. & Symons R.H. (1977) Translation of the four major RNA species of Cucumber mosaic virus in plant and animal cell free systems and in toad oocytes. *Virology*, **79**, 83–108.

Scott B.I.H. (1967) Electric fields in plants. *Ann. Rev. Plant Physiol.*, **18**, 409–18.

Scott B.I.H., Gulline H. & Pallaghy C.K. (1968) The electrochemical state of cells of broad bean roots. I. Investigation of elongating roots of young seedlings. *Aust. J. biol. Sci.*, **21**, 185–200.

Scott M.G. & Peterson R.L. (1979a) The root endodermis in *Ranunculus acris*. I. Structure and ontogeny. *Canad. J. Bot.*, **57**, 1040–62.

Scott M.G. & Peterson R.L. (1979b) The root endodermis in *Ranunculus acris*. II. Histochemistry of the endodermis and the synthesis of phenolic compounds in roots. *Canad. J. Bot.*, **57**, 1063–77.

Scott N.S. & Possingham J.V. (1980) Chloroplast DNA in expanding spinach leaves. *J. exp. Bot.*, (in press).

Scott N.S. & Smillie R.M. (1967) Evidence for the direction of chloroplast ribosomal RNA synthesis by chloroplast DNA. *Biochem. Biophys. Res. Comm.*, **28**(4), 598–603.

Seago J.L. & Wolniak S.M. (1976) Cortical ontogeny in roots. I. *Zea mays*. *Am. J. Bot.*, **63**, 220–5.

Seagull R.W. (1978) Arrangement of microtubules and microfilaments during oriented secondary wall growth. *Proc. 9th Int. Con. on E.M.* Toronto, **2**, 262–3.

Selitrennikoff C.P., Allin D. & Sonneborn D.R. (1976) Chitin biosynthesis during *Blastocladiella* zoospore germination: Evidence that the hexosamine biosynthetic pathway is post-translationally activated during cell differentiation. *Proc. Nat. Acad. Sci. USA*, **73**, 534–8.

Sen S., Payne P.I. & Osborne D.J. (1975) Early ribonucleic acid synthesis during the germination of rye (*Secale cereale*) embryos and the relationship to early protein synthesis. *Biochem. J.*, **148**, 381–7.

Senger H. (Ed.) (1980) *The Blue Light Syndrome*. Springer-Verlag, Berlin.

Sengupta C., deLuca V., Bailey D.S. & Verma D.P.S. (1981) Post-translational processing of 7S and 11S components of soybean storage proteins. *Plant molec. Biol.*, (in press).

Senser F. & Schotz F. (1964) Untersuchungen über die Chloroplastenentwicklung bei *Oenothera*. II. Der Albivelutina-typ. *Planta*, **62**, 171–90.

Sequeira L. (1978) Lectins and their role in host–pathogen specificity. *Ann. Rev. Phytopath.*, **16**, 453–81.

Sequeira L. (1979) Recognition between plant hosts and parasites. In *Host–Parasite Interfaces*. (Ed. Nickol B.B.) pp. 71–84. Academic Press, New York.

Sequeira L., Gaard G. & de Zoeten G.A. (1977) Interaction of bacteria and host cell walls: its relation to mechanisms of induced resistance. *Physiol. Plant Pathol.*, **10**, 43–50.

Sequeira L. & Graham T.L. (1977) Agglutination of avirulent strains of *Pseudomonas solanacearum* by potato lectin. *Physiol. Plant Pathol.*, **11**, 43–54.

Šesták Z. (1977) Photosynthetic characteristics during ontogenesis of leaves. 2. Photosystems, components of electron transport chains, and photophosphorylations. *Photo synthetica*, **11**, 449–74.

Setter T.L., Brun W.L. & Brenner M.L. (1980a) Stomatal closure and photosynthetic inhibition in soybean leaves induced by petiole girdling and pod removal. *Plant Physiol.*, **65**, 884–7.

Setter T.L., Brun W.A. & Brenner M.L. (1980b). Effect of obstructed translocation on leaf ABA, and associated stomatal closure and photosynthetic decline. *Plant Physiol.*, **65**, 1111–5.

Shabde M. & Murashige T. (1977) Hormonal requirements of excised *Dianthus caryophullus* L. shoot apical meristem *in vitro*. *Am. J. Bot.*, **64**, 443–8.

Sharp L.W. (1926) *Introduction to Cytology*. 2e. McGraw-Hill, New York.

Shaw S. & Wilkins M.B. (1973) The source and latent transport of growth inhibitors in geotropically stimulated roots of *Zea mays* and *Pisum sativum*. *Planta*, **109**, 11–26.

Shaykh M., Soliday C. & Kolattukudy P.E. (1977a) Proof for the production of cutinase by *Fusarium solani* f. *pisi* during penetration into its host, *Pisum sativum*. *Plant Physiol.*, **60**, 170–2.

Shaykh M., Kolattukudy P.E. & Davis R. (1977b) Production of novel extracellular cutinase by the pollen and the chemical composition and ultrastructure of the stigma cuticle of nasturtium (*Tropaeolum majus*). *Plant Physiol.*, **60**, 907–15.

Sheldrake A.R. (1973) The production of hormones in higher plants. *Biol. Rev.*, **48**, 509–59.

Sheldrake A.R. (1979) Effects of osmotic stress on polar auxin transport in *Avena* mesocotyl segments. *Planta*, **145**, 113–7.

Sheldrake A.R. & Northcote D.H. (1968). The production of auxin by tobacco internode tissue. *New Phytol.*, **67**, 1–13.

Shewry P.R., Faulks A.J., Pickering R.A., Jones I.T., Finch R.A. & Miflin B.J. (1980) The genetic analysis of barley storage proteins. *Heredity*, **44**, 383–9.

Shewry P.R., Autran J-C., Nimmo C.C., Lew E.J-L. & Kasarda D.D. (1980a) N-terminal amino acid sequence homology of storage protein components from barley and a diploid wheat. *Nature, Lond.*, **286**, 520–2.

Shewry P.R., Faulks A.J., Pickering R.A., Jones I.T., Finch R.A. & Miflin B.J. (1980b) The genetic analysis of barley storage proteins. *Heredity*, **44**, 383–90.

Shropshire W. (1971) Phototropic bending rate in *Phycomyces* as a function of average growth rate and cell radius. In *First European Biophysics Congress*. (Eds. Broder E., Locker A. & Springer-Lederer H.) pp. 111–4. Verlag Weiner Med. Akad., Vienna.

Shropshire W. jr. (1962) Phototropism of *Phycomyces*. *J. gen. Physiol.*, **45**, 949–58.

Shropshire W. jr. (1974) Phototropism. Introductory lecture No. 24, 6 pp. *Progress in Photobiology*. (Ed. Schenck G.O.) Deutsche Gess. F. Lichtforsch., Frankfurt.

SHROPSHIRE W. JR. (1980) Carotenoids acting as primary photoreceptors in blue light phenomena. In *The Blue Light Syndrome*, pp. 172–86. Springer-Verlag, Berlin.

SHUTTLEWORTH J.E. & BLACK M. (1977) The role of cotyledons in phototropism of de-etiolated seedlings. *Planta*, **135**, 51–55.

SIDDELL S.G. & ELLIS E.J. (1975) Protein synthesis in chloroplasts. VI. Characteristics and products of protein synthesis *in vitro* in etioplasts and developing chloroplasts from pea leaves. *Biochem. J.*, **146**, 675–85.

SIEBERS A.M. (1971) Initiation of radial polarity in the interfascicular cambium of *Ricinus communis* L. *Acta Bot. Neerl.*, **20**, 211–20.

SIEVERS A. & SCHRÖTER K. (1971) Versuch einer Kansanalyse der geotropischen Reaktionskette in *Chara*-Rhizoid. *Planta*, **96**, 339–53.

SIEVERS A. & VOLKMANN D. (1979) Gravitropism in single cells. In *Encyclopaedia Plant Physiol.* New Series **7**. pp. 567–72. Springer-Verlag, Berlin.

SILVERMAN P.M. & EPSTEIN P.M. (1975) Cyclic nucleotide metabolism coupled to cytodifferentiation of *Blastocladiella emersonii*. *Proc. Nat. Acad. Sci. USA*, **72**, 442–6.

SILVERTHORNE J. & ELLIS R.J. (1980) Protein synthesis in chloroplasts. VIII. Differential synthesis of chloroplast proteins during spinach leaf development. *Biochim. biophys. Acta*, **607**, 319–30.

SILVIUS J.E. & SNYDER F.W. (1979) Comparative enzymic studies of sucrose metabolism in taproots and fibrous roots of *Beta vulgaris* L. *Plant Physiol.*, **64**, 1070–3.

SIMCHEN G. (1978) Cell cycle mutants. *Ann. Rev. Genet.*, **12**, 161–91.

SIMON E.W. (1967) Types of leaf senescence. *Symp. Soc. exp. Biol.*, **21**, 215–30.

SIMON E.W. (1978) The symptoms of calcium deficiency in plants. *New Phytol.*, **80**, 1–15.

SIMPSON E. & VARNER J. (1980) Determination of the degradation rate of corn leaf RuBP carboxylase *in vivo* with ^{3}H-acetic anhydride. (in press).

SINNOTT E.W. & BLOCH R. (1941) The relative position of cell walls in developing plant tissues. *Am. J. Bot.*, **28**, 607–17.

SIOUFI A., PERCHERON F. & COURTORS J.E. (1970) Nucléoside-diphosphate-ases et métabolisme glucidique au cours de la germination chez le fénugrec. *Phytochem.*, **9**, 991–9.

SKAAR H. & JOHNSSON A. (1978) Rapid, blue-light induced transpiration in *Avena*. *Physiol. Plant.*, **43**, 390–6.

SKENE K.G.M. (1970) The gibberellins of developing bean seeds. *J. exp. Bot.*, **21**, 236–46.

SKOOG F. & MILLER C. (1957) Chemical regulation of growth and organ formation in plant tissues cultured *in vitro*. *Symp. Soc. exp. Biol.*, **11**, 118–30.

SMART M.G. & O'BRIEN T.P. (1979) Observations on the scutellum. III. Ferulic acid as a component of the cell wall in wheat and barley. *Aust. J. Plant Physiol.*, **6**, 485–91.

SMILLIE R.M. (1962) Photosynthetic and respiratory activities of growing pea leaves. *Plant Physiol.*, **37**, 716–21.

SMITH D.L. (1973) Nucleic acid, protein, and starch synthesis in developing cotyledons of *Pisum arvense* L. *Ann. Bot.*, **37**, 795–804.

SMITH F.A. & RAVEN J.A. (1979) Intracellular pH and its regulation. *Ann. Rev. Plant Physiol.*, **30**, 289–312.

SMITH H. (1972) Light quality and germination: ecological implications. In *Seed Ecology*. (Ed. Heydecker W.) pp. 219–31. Butterworth, London.

SMITH H. (1975) *Phytochrome and Photomorphogenesis*. McGraw-Hill, Maidenhead.

SMITH H. (1976) Phytochrome-mediated assembly of polyribosomes in etiolated bean leaves: Evidence for post-transcriptional regulation of development. *Eur. J. Biochem.*, **65**, 161–70.

SMITH H., BILLETT E.E. & GILES A.B. (1977) The photocontrol of gene expression in plants. In *Regulation of Enzyme Synthesis and Activity in Higher Plants*. (Ed. Smith H.) pp. 93–127. Academic Press, London.

SMITH J.D. (1949) The concentration and distribution of haemoglobin in the root nodules of leguminous plants. *Biochem.*, **44**, 585–91.

SMITH J.D. & SILVERMAN P.M. (1973) Lipid turnover during morphogenesis in the water mold *Blastocladiella emersonii*. *Biochem. biophys. Res. Comm.*, **54**, 1191–7.

SMITH J.M. & ELLIS R.J. (1979) Processing of small sub-unit precursor of ribulose bisphosphate carboxylase and its assembly into whole enzyme are stromal events. *Nature, Lond.*, **278**, 662–4.

SMITH M.M. & O'BRIEN T.P. (1979) Distribution of autofluorescence and esterase and peroxidase activities in the epidermis of wheat roots. *Aust. J. Plant Physiol.*, **6**, 201–19.

SMITH R.H. & MURASHIGE T. (1970) *In vitro* development of isolated shoot apical meristems of angiosperms. *Am. J. Bot.*, **57**, 562–8.

SMITH R.L., BOUTON J.M., SCHANK S.C.,

QUESENBERRY K.M., TAYLER M.E., MILANS J.R., GASKIN M.H. & LITTLE R.C. (1976) Nitrogen fixation in grasses inoculated with *Spirillum lipoferum*. *Science*, **193**, 1003–5.

SOAVE C., SUMAN N., VIOTTI A. & SALAMINI F. (1978) Linkage relationships between regulatory and structural gene loci involved in zein synthesis in maize. *Theor. Appl. Genet.*, **52**, 263–7.

SOLIDAY C.L. & KOLATTUKUDY P.E. (1976) Isolation and characterisation of a cutinase from *Fusarium roseum culmorum* and its immunological comparison with cutinases from *F. solani pisi*. *Arch. biochem. Biophys.*, **176**, 334–43.

SOLIDAY C.L. & KOLATTUKUDY P.E. (1977) Biosynthesis of cutin. *Plant Physiol.*, **59**, 1116–21.

SOLL D.R. & SONNEBORN D.R. (1971) Zoospore germination in *Blastocladiella emersonii*: cell differentiation without protein synthesis? *Proc. Nat. Acad. Sci. USA*, **68**, 459–63.

SOLL D.R. & SONNEBORN D.R. (1972) Zoospore germination in *Blastocladiella emersonii*. IV. Ion control over cell differentiation. *J. Cell Sci.*, **10**, 315–33.

SONG P.S. & MOORE T.A. (1974) On the photoreceptor pigment for phototropism and phototaxis: is a carotenoid the most likely candidate? *Photochem. Photobiol.*, **19**, 435–41.

SONNENBICHLER J. (1979) Advances in chromatin research. *Naturwissenschaften*, **66**, 244–50.

SOPANEN T. (1979) Development of peptide transport activity in barley scutellum during germination. *Plant Physiol.*, **64**, 570–4.

SOPANEN T., UUSKALLIO M. & MIKOLA S.N.T. (1980) Characteristics and development of leucine transport activity in the scutellum of germinating barley grain. *Plant Physiol.*, **65**, 249–53.

SPENCER D. & HIGGINS T.J.V. (1980) The biosynthesis of legumin in maturing pea seeds. *Biochem. Int.*, **1**, 502–9.

SPENCER D. & HIGGINS T.J.V. (1980) Protein synthesis in developing pea seeds: short-term *in vivo* labelling studies. *Plant Physiol.*, (submitted).

SPENCER D., HIGGINS T.J.V., BUTTON S.C. & DAVEY R.A. (1980) Pulse-labeling studies on protein synthesis in developing pea seeds and evidence of a precursor form of legumin small subunit. *Plant Physiol.*, **66**, 510–15.

SPENCER T. & HARRIS H. (1964) Regulation of enzyme synthesis in an enucleate cell. *Biochem. J.*, **91**, 282–6.

SPERTEL R.B. (1978) On the decay of lateral phase separations in biological membranes. *J. theoret. Biol.*, **71**, 1–9.

SPIEGEL S. & MARCUS A. (1975) Polyribosome formation in early wheat embryo germination independent of either transcription or polyadenylation. *Nature, Lond.*, **256**, 228–30.

SPIEGEL S., OBENDORF R. & MARCUS A. (1975) Transcription of ribosomal and messenger RNAs in early wheat embryo germination. *Plant Physiol.*, **56**, 502–7.

SPRING H., GRIERSON D., HEMLEBEN V., STÖHR M., KROHNE G., STADLER J., & FRANKE W.W. (1978) DNA contents and numbers of nucleoli and pre-rRNA-genes in nuclei of gametes and vegetative cells in *Acetabularia mediterranea*. *Exp. Cell Res.*, **114**, 203–15.

SPURR A.R. (1969) A low-viscosity epoxy resin embedding medium for electron microscopy. *J. ultrastruct. Res.*, **26**, 31–43.

SRIVASTAVA B.I.S. (1972) RNA-DNA hybridisation competition studies on senescing barley leaves. *New Phytol.*, **71**, 93–7.

SRIVASTAVA L.M. & PAULSON R.E. (1968) The fine structure of the embryo of *Lactuca sativa*. II. Changes during germination. *Canad. J. Bot.*, **46**, 1447–53 + 7 plates.

SRIVASTAVA L.M., SAWHNEY V.K. & BONETTEMAKER M. (1977) Cell growth, wall deposition and correlated fine-structure of colchicine-treated lettuce hypocotyl cells. *Canad. J. Bot.*, **55**, 902–17.

VAN STADEN J. & BUTTON J. (1978) The cytokinin content of aseptically cultured pea fruits. *Z. Pflphysiol.*, **87**, 129–35.

VAN STADEN J. & DAVEY J.B. (1979) The synthesis, transport and metabolism of endogenous cytokinins. *Plant, Cell and Environment*, **2**, 93–106.

STANIER R.Y. (1970) Some aspects of the biology of cells and their possible evolutionary significance. In *Organization and Control in Prokaryotic and Eukaryotic Cells.* (Eds. Charles H.P. & Knight B.C.J.G.) 20th Symp. Soc. Gen. Microbiol. pp. 1–38. Cambridge University Press, London.

STANLEY R.G. & LINSKENS H.F. (1974) *Pollen.* Springer-Verlag, Berlin.

STEBBINGS H. & BENNETT C.E. (1975) The sleeve element in microtubules. In *Microtubules and Microtubule Inhibitors.* (Eds. Borgers M. & de Brabander M.) North-Holland, Amsterdam.

STEELE J.A., UCHYTIL T.F., DURBIN R.D., BHATNAGAR P. & RICH D.H. (1976) Chloroplast coupling factor 1, a species-specific receptor for tentoxin. *Proc. Nat. Acad. Sci. USA*, **73**, 2245–8.

STEEVES T.A. (1966) On the determination of leaf primordia in ferns. In *Trends in Plant*

Morphogenesis. (Ed. Cutter E.G.) pp. 200–19. Longman, London.

STEINMETZ A., MUBUMBILA M., KELLER M., BURKARD G., WEIL J.H., DRIESEL A.J., CROUSE E.J., GORDON K., BOHNERT H.J. & HERRMANN R.G. (1978) Mapping of the tRNA genes on the circular DNA molecule of *Spinacia oleracea*. In *Chloroplast Development*. (Ed. Akoyunoglou G. *et al.*) Elsevier/North-Holland Biomedical Press, Amsterdam.

STEKOLL M. & WEST C.A. (1978) Purification and properties of an elicitor of castor bean phytoalexin from culture filtrates of the fungus *Rhizopus stolonifer*. *Plant Physiol.*, **61**, 38–45.

STERN H. & HOTTA Y. (1973) Biochemical control of meiosis. *Ann. Rev. Genet.*, **7**, 37–66.

VAN STEVENNINCK R.F.M. (1964) A comparison of chloride and potassium fluxes in red beet tissue. *Physiologia Plantarum*, **17**, 757–70.

STODDART J.L., THOMAS H. & ROBERTSON A. (1973) Protein synthesis patterns in barley embryos during germination. *Planta*, **112**, 309–21.

STOREY R. & BEEVERS L. (1977) Proteolytic activity in relationship to senescence and cotyledonary development in *Pisum sativum* L. *Planta*, **137**, 37–44.

STOUTEMEYER V.T. & BRITT O.K. (1965) The behaviour of tissue cultures from English and Algerian ivy in different growth phases. *Am. J. Bot.*, **52**, 805–10.

STREET H.E. (1977) Plant cell culture. In *The Molecular Biology of Plant Cells*. (Ed. Smith H.) pp. 393–417. Blackwell Scientific Publications, Oxford.

STROBEL G.A. (1973a) The helminthosporoside-binding protein of sugar cane. Its properties and relationship to susceptibility to the eye spot disease. *J. biol. Chem.*, **248**, 1321–8.

STROBEL G.A. (1973b) Biochemical basis of the resistance of sugar cane to eyespot disease. *Proc. Nat. Acad. Sci. USA*, **70**, 1693–6.

STROBEL G.A. (1974) The toxin-binding protein of sugar-cane, its role in the plant and disease development. *Proc. Nat. Acad. Sci. USA*, **71**, 4232–6.

STROBEL G.A. (1976) Toxins of plant pathogenic bacteria and fungi. In *Biochemical Aspects of Plant–Parasite Relationships*. (Eds. Friend J. & Threlfall D.R.) pp. 135–59. Academic Press, London.

STROBEL G.A. & HAPNER K. (1975) Transfer of toxin susceptibility to plant protoplasts via the helminthosporoside-binding protein of sugarcane. *Biochem. biophys. Res. Comm.*, **63**, 1151–6.

STROBEL G.A. & HESS W.M. (1974) Evidence for the presence of the toxin-binding protein in the plasma membrane of sugar-cane cells. *Proc. Nat. Acad. Sci. USA*, **71**, 1413–7.

SUBERKROPP K.F. & CANTINO E.C. (1972) Environmental control of motility and of encystment in *Blastocladiella emersonii* zoospores at high population densities. *Trans. Brit. mycol. Soc.*, **59**, 463–75.

SUGITA M., YOSHIDA K. & SASAKI K. (1979) Germination-induced changes in chromosomal proteins of spring and winter wheat embryos. *Plant Physiol.*, **64**, 780–5.

SULLIVAN D., BRISSON N., GOODCHILD B., VERMA D.P.S. & THOMAS D.Y. (1981) Molecular cloning and organization of two leghaemoglobin genomic sequences of soybean. *Nature, Lond.*, **289**, 516–8.

SUN S.M., BUCHBINDER B.U. & HALL T.C. (1975) Cell-free synthesis of the major storage protein of the bean, *Phaseolus vulgaris* L. *Plant Physiol.*, **56**, 780–5.

SUN S.M., MOORE T.A. & SONG P.S. (1972) Molecular luminescence studies of flavins. I. The excited states of flavins. *J. Amer. Chem. Soc.*, **94**, 1730–40.

SUN S.M., MUTSCHLER M.A., BLISS F.A. & HALL T.C. (1978). Protein synthesis and accumulation in bean cotyledons during growth. *Plant Physiol.*, **61**, 918–23.

SUN S.M., SIGHTON J.L. & HALL T.C. (1981) Intervening sequences in a plant gene-comparison of the partial sequence of cDNA and genomic DNA of French bean phaseolin. *Nature, Lond.*, **289**, 37–41.

SUTCLIFFE J.F. & SEXTON R.F. (1974) Enzymatic changes during the differentiation of tissues in young pea roots. In *Structure and Function of Primary Root Tissue*. (Ed. Kolek J.) pp. 203–19. Veda, Bratislava.

SUTTER E. (1944) Die osmotische Bestimmung des Heteroauxins und Versuche über seine Aufrahme durch die Pflanze. *Ber. Schweiz. botan. Ges.*, **54**, 197–244.

SUZUKI R., KONDO N. & FUJII T. (1979) Distribution of growth regulators in relation to the light-induced geotropic responsiveness in *Zea* roots. *Planta*, **145**, 323–9.

SWEENEY B. & HAXO F.T. (1961) Persistence of a photosynthetic rhythm in enucleated *Acetabularia*. *Science*, **134**, 1361–3.

SYONO K., NEWCOMB W. & TORREY J.G. (1976) Cytokinin production in relation to the development of pea root nodules. *Canad. J. Bot.*, **54**, 2155–62.

SYONO K. & TORREY J.G. (1976) Identification of cytokinins in root nodules of garden pea, *Pisum sativum* L. *Plant Physiol.*, **57**, 602–6.

TAIZ L. & HONIGMAN W.A. (1976) Production

of cell wall hydrolyzing enzymes by barley aleurone layers in response to gibberellic acid. *Plant Physiol.*, **58**, 380–6.

TAIZ L. & JONES R.L. (1970) Gibberellic acid, β-1,3-glucanase and the cell walls of barley aleurone layers. *Planta*, **92**, 73–84.

TAKEGAMI T. & YOSHIDA K. (1975) Remarkable retardation of the senescence of tobacco leaf discs by cordycepin, an inhibitor of RNA polyadenylation. *Plant Cell Physiol.*, **16**, 1163–6.

TAKIMOTO A. & IKEDA K. (1961) Effect of twilight on photoperiodic induction in some short day plants. *Plant Cell Physiol.*, **2**, 213–20.

TANADA T. (1968) A rapid photoreversible response of barley root tips in the presence of 3-indoleacetic acid. *Proc. Nat. Acad. Sci. USA*, **59**, 376–80.

TAYLOR F.J.R. (1979) Symbionticism revisited: a discussion of the evolutionary impact of intracellular symbioses. *Proc. roy. Soc. Lond. B.*, **204**, 267–86.

TELZER B.R. & ROSENBAUM J.L. (1979) Cell cycle-dependent, in vitro assembly of microtubules onto the pericentriolar material of HeLa cells. *J. Cell Biol.*, **81**, 484–97.

TENNHAMMER-EKMAN B. & SKÖLD O. (1979) Trimethoprim resistance plasmids of different origin encode different drug-resistant dihydrofolate reductases. *Plasmid*, **2**, 334–46.

TETLEY R.M. & THIMAN K.V. (1974) The metabolism of oat leaves during senescence. I. Respiration, carbohydrate metabolism and the action of cytokinins. *Plant Physiol.*, **54**, 294–303.

TEWARI K.K. & WILDMAN S.G. (1966) Chloroplast DNA from tobacco leaves. *Science*, **153**, 1269–71.

THANH V.H. & SHIBASAKI K. (1978) Major proteins of soybean seeds. Subunit structure of β-conglycinin. *J. Agric. Food Chem.*, **26**, 692–5.

THIMANN K.V. (1936) On the physiology of formation of nodules in legume roots. *Proc. Nat. Acad. Sci. USA*, **22**, 511–4.

THIMANN K.V. (1952) *The Action of Hormones in Plants and Invertebrates*, p. 51. Academic Press, New York.

THIMANN K.V. (1978) Senescence. In *Controlling Factors in Plant Development*. (Eds. Shibaoka H., Funiya M., Katsumi M. & Takivots A.) pp. 19–43. Botanical Magazine, Tokyo. Special issue 1.

THIMANN K.V. & CURRY G.M. (1960) Phototropism and phototaxis. *Comp. Biochem.*, **1**, 243–309.

THIMANN K.V. & CURRY G.L. (1961) Phototropism. In *Light and Life*. (Eds. McElroy W.D. & Glass B.) p. 651. Johns Hopkins Press, Baltimore.

THOMAS B. & DICKINSON H.G. (1979) Evidence for two photoreceptors controlling growth in de-etiolated seedlings. *Planta*, **146**, 545–50.

THOMAS H. (1975) Leaf growth and senescence in grasses. *Report of Welsh Plant Breeding Station*, Aberystwyth, 133–8.

THOMAS H. (1976) Delayed senescence in leaves treated with the protein synthesis inhibitor MDMP. *Plant Sci. Lett.*, **6**, 369–77.

THOMAS H. (1977) Ultrastructure, polypeptide composition and photochemical activity of chloroplasts during foliar senescence of a non-yellowing mutant genotype of *Festuca pratensis* Huds. *Planta*, **137**, 53–60.

THOMAS H. (1978) Enzymes of nitrogen mobilisation in detached leaves of *Lolium temulentum* during senescence. *Planta*, **142**, 161–9.

THOMAS H. & BINGHAM M.J. (1977) Isoenzymes of esterase naphthyl-acetate in senescing leaves of *Festuca pratensis*. *Phytochem.*, **16**, 1887–9.

THOMAS H. & STODDART J.L. (1975) Separation of chlorophyll degradation from other senescence processes in leaves of a mutant genotype of meadow fescue (*Festuca pratensis*). *Plant Physiol.*, **56**, 438–41.

THOMAS H. & STODDART J.L. (1977) Biochemistry of leaf senescence in grasses. *Ann. appl. Biol.*, **85**, 461–3.

THOMAS H. & STODDART J.L. (1980) Leaf senescence. *Ann. Rev. Plant Physiol.*, **31**, 83–111.

THOMAS S.M., HALL N.P. & MERRETT M.J. (1978) Ribulose 1,5-bisphosphate carboxylase-oxygenase activity and photorespiration during the ageing of flag leaves of wheat. *J. exp. Bot.*, **29**, 1161–8.

THOMASHOW M.F., NUTTER R., MONTOYA A.L., GORDON M.P. & NESTER E.W. (1980) Integration and organisation of Ti-plasmid sequences in crown gall tumors. *Cell*, **19**, 729–39.

THOMPSON E.W. & LANE B.G. (1980) Relation of protein synthesis in imbibing wheat embryos to the cell-free translational capacities of bulk mRNA from dry and imbibing embryos. *J. biol. Chem.*, **255**, 5965–70.

THOMSON J.A. & DOLL H. (1979) Genetics and evolution of seed storage proteins. In *Seed Protein Improvement in Cereals and Grain Legumes*, pp. 109–24. International Atomic Energy Agency, Vienna.

THOMSON J.A. & SCHROEDER H.E. (1978) Cotyledonary storage proteins in *Pisum sativum*. II. Hereditary variations in components of the legumin and vicilin fractions. *Aust. J. Plant Physiol.*, **5**, 281–94.

THOMSON J.A., MILLERD A. & SCHROEDER H.E. (1979) Genotype-dependent patterns of accumulation of seed storage proteins in *Pisum*. In *Seed Protein Improvement in Cereals and Grain Legumes*, pp. 231–40. International Atomic Energy Agency, Vienna.

THOMSON J.A., SCHROEDER H.E. & DUDMAN W.F. (1978) Cotyledonary storage proteins in *Pisum sativum*. I. Molecular heterogeneity. *Aust. J. Plant Physiol.*, **5**, 263–79.

THOMSON J.A., SCHROEDER H.E. & TASSIE A.M. (1980) Cotyledonary storage proteins in *Pisum sativum*. V. Further studies on molecular heterogeneity in the vicilin series of holoproteins. *Aust. J. Plant Physiol.*, **7**, 271–82.

THORNLEY J.H.M. (1972) A balanced quantitative model for root : shoot ratios in vegetative plants. *Ann. Bot.*, **36**, 431–41.

THORNLEY J.H.M. (1977) Root : shoot interactions. *Symp. Soc. exp. Biol.*, **31**, 367–89.

TIDD B.K. (1964) Dissociation constants of the gibberellins. *J. chem. Soc.*, 1521–2.

TIMOTHY D.H., LEVINGS C.S., PRING D.R., CONDE M.F. & KERMICLE J.L. (1979) Organelle DNA variation and systematic relationships in the genus *Zea:* Teosinte. *Proc. Nat. Acad. Sci. USA*, **76**, 4420–4.

TIPPETT J.T., HOLLAND A.A., MARKS G.C. & O'BRIEN T.P. (1976) Penetration of *Phytophthora cinnamomi* into disease tolerant and susceptible eucalypts. *Arch. Microbiol.*, **108**, 231–42.

TIPPIT D.H. & PICKETT-HEAPS J.D. (1974) Experimental investigations into morphogenesis in *Micrasterias*. *Protoplasma*, **81**, 271–96.

TIPPIT D.H., SCHULZ D. & PICKETT-HEAPS J.D. (1978) Analysis of the distribution of spindle microtubules in the diatom *Fragilaria*. *J. Cell Biol.*, **79**, 737–63.

TISSIERES A. (1974) Ribosome research: historical background. In *Ribosomes*. (Eds. Nomura M., Tissieres A. & Lengyel P.) pp. 3–12. Cold Spring Harbour Laboratory, New York.

TJEPKEMA J.D. (1971) Oxygen transport in the soybean nodule and the function of leghaemoglobin. Ph.D. Thesis. University of Michigan, Ann Arbor, USA.

TOBIN E.M. (1978) Light regulation of specific mRNA species in *Lemna gibba* L. G-3. *Proc. Nat. Acad. Sci. USA*, **75**(10), 4749–53.

DE LA TORRE C., SACRISTAN-GARATE A. & NAVARETTE M.H. (1979) Dynamics of the nuclear envelope during cell cycle in plants. *Cytobios*, **24**, 25–31.

DE LA TORRE C. & GONZÁLEZ-FERNÁNDEZ A. (1979) Cell cycle mapping by irradiating cells with bromosubstituted DNA segments. *Photochem. Photobiol.*, **29**, 977–81.

TORREY J. G. (1955) On the determination of vascular patterns during tissue differentiation in excised pea roots. *Am. J. Bot.*, **42**, 183–98.

TORREY J.G. (1957) Auxin control of vascular formation in regenerating pea root meristems grown *in vitro*. *Am. J. Bot.*, **44**, 859–70.

TORREY J.G. (1961) Kinetin as trigger for mitosis in mature endomitotic plant cells. *Exp. Cell Res.*, **23**, 281–99.

TORREY J.G. (1965) Physiological bases of organisation and development in the root. In *Encyclopaedia of Plant Physiology*. (Ed. Ruhland W.) Vol. 15/1, pp. 1256–327. Springer-Verlag, Berlin.

TORREY J.G. (1975) Tracheary element formation from single isolated cells in culture. *Physiologia Plantarum*, **35**, 158–65.

TORREY J.G. (1976) Root hormones and plant growth. *Ann. Rev. Plant Physiol.*, **27**, 435–59.

TORRES-PERRERA J., MEHLHORN R., KEITH A.J. & PACKER L. (1974) Changes in membrane lipid structure of illuminated chloroplasts. *Arch. biochem. Biophys.*, **160**, 90–9.

TRAN THANH VAN K. (1977) Regulation of morphogenesis. In *Plant Tissue Culture and its Bio-technological Application*. (Eds. Barz W., Reinhard E. & Zenk M.H.) pp. 367–85. Springer-Verlag, Berlin.

TRAN THANH VAN K. & TRINH H. (1978) Plant propagation: non-identical and identical copies. In *Propagation of Higher Plants Through Tissue Culture*. (Eds. Hughes K.W., Henke R. & Constantin M.) pp. 134–58. University of Tennessee Symposium Proceedings. U.S. Dept. of Energy.

TRAN THANH VAN M. & CHLYAH A. (1976) Différentiation de boutons floraux, de bourgeons végétatifs, de racines et de cal à partir de l'assise sous-épidermique des ramifications florales de *Nicotiana tabacum* Wisc. 38. Etude infra-structurale. *Canad. J. Bot.*, **54**, 1979–96.

TRAN THANH VAN M. & TRINH H. (1978) Morphogenesis in thin layers: concept, methodology and results. In *Frontiers of Plant Tissue Culture*. (Ed. Thorpe T.A.) pp. 37–48. Proc. 4th Internat. Con. Plant Tissue & Cell Culture.

TRAVIS R.L., GENG S. & BERKOWITZ R.L. (1979) Analysis of the distribution of potassium-stimulated adenosine triphosphatase activity in soybean root. *Plant Physiol.*, **63**, 1187–90.

TREBST A. (1974) Energy conservation in photosynthetic electron transport of chloroplasts. *Ann. Rev. Plant Physiol.*, **25**, 423–58.

TRENDELENBURG M.F., SPRING H., SCHEER U. & FRANKE W.W. (1974) Morphology of nucleolar cistrons in a plant cell, *Acetabularia mediterranea*. *Proc. Nat. Acad. Sci. USA*, **71**, 3626–30.

TREWAVAS A.J. (1976a) Plant growth substances. In *Molecular Aspects of Gene Expression in Plants*. (Ed. Bryant J.A.) pp. 249–99. Academic Press, London.

TREWAVAS A.J. (1976b) Post-translational modification of protein by phosphorylation. *Ann. Rev. Plant Physiol.*, **27**, 349–74.

TREWAVAS A.J. (1979a) The dynamics of meristem control by growth substances. In *British Plant Growth Regulator Group, Monograph 3. Control of Plant Development* (Ed. George E.) pp. 39–56. Wessex Press, Wantage.

TREWAVAS A.J. (1979b) Plant Growth substances: What is the molecular basis of their action. *What's New in Plant Physiol.*, **10**, 33–6.

TREWAVAS A.J. (1979c) What is the molecular basis of plant hormone action? *Trends in Biochem. Sci.*, **4**N, 199–202.

TREWAVAS A.J. (1979d) Nuclear phosphoproteins in germinating cereal embryos and their relationship to the control of mRNA synthesis and the onset of cell division. In *Recent Advances in the Biochemistry of Cereals*. Phytochemical Society Symposium. **16**, pp. 175–208. (Eds. Laidman D.L. & Wyn Jones R.G.) Academic Press, London.

TREWAVAS A.J. & STRATTON B.R. (1977) Control of plant growth by protein kinase. In *Nucleic Acids and Protein Synthesis in Plants*. (Eds. Weil J. & Bogorad C.) pp. 309–19. Plenum Press, New York.

TRUESDELL L.C. & CANTINO E.C. (1971) The induction and early events of germination in the zoospore of *Blastocladiella emersonii*. *Curr. Topics devel. Biol.*, **6**, 1–44.

TU J.C. (1975a) Structural similarity of the membrane envelopes of Rhizobial bacteroids and the host plasma membrane as revealed by freeze-fracturing. *J. Bacteriol.*, **122**, 691–4.

TU J.C. (1975b) Rhizobial root nodules of soybean as revealed by scanning and transmission electron microscopy. *Phytopathol.*, **65**, 447–54.

TUCKER A., ROBERTSON N.G. & GRIERSON D. (1980) Changes in polygalcturonase isoenzymes during 'ripening' of normal and mutant tomato fruit. *Europ. J. Biochem.*, **112**, 119–24.

TURING A.N. (1952) The chemical basis of morphogenesis. *Phil. Trans. Roy. Soc. B.*, **237**, 37–72.

TURNER N.C. (1973) Action of fusicoccin on the potassium balance of guard cells of *Phaseolus vulgaris*. *Am. J. Bot.*, **60**, 717–25.

TYREE M.T., PETERSON C.A. & EDGINTON L.V. (1979) A simple theory regarding ambimobility of xenobiotics with special reference to the mematocide, oxamyl. *Plant Physiol.*, **63**, 367–74.

TZAGOLOFF A., MACINO G. & SEBALD W. (1979) Mitochondrial genes and translation products. *Ann. Rev. Biochem.*, **48**, 419–41.

UDVARDY J., FARKAS G.L., MAURRE E. & FORTI G. (1967) The effects of sucrose and light on the level of soluble and particle-bound ribonuclease activities in excised *Avena* leaves. *Physiol. Plant.*, **20**, 781–8.

UNGER M.W. & HARTWELL L.H. (1976) Control of cell division in *Saccharomyces cerevisiae* by methionyl-tRNA. *Proc. Nat. Acad. Sci.* USA, **73**, 1664–8.

USCIATI M., CODACCIONI M. & GUERN J. (1972) Early cytological and biochemical events induced by a 6 benzylaminopurine application on inhibited axillary buds of *Acer arietinum* plants. *J. exp. Bot.*, **23**, 1009–20.

VANCE C.P. & SHERWOOD R.T. (1976a) Regulation of lignin formation in reed canary grass in relation to disease resistance. *Plant Physiol.*, **57**, 915–19.

VANCE C.P. & SHERWOOD R.T. (1976b) Soluble and cell wall peroxidases in reed canary grass in relation to disease resistance and localised lignin formation. *Plant Physiol.*, **57**, 920–2.

VANCE C.P. & SHERWOOD R.T. (1977) Lignified papilla formation as a mechanism for protection in reed canary grass. *Physiol Plant. Pathol.*, **10**, 247–56.

VAN'T HOF J. (1966) Experimental control of DNA synthesizing and dividing cells in excised root tips of *Pisum*. *Am. J. Bot.*, **53**, 970–6.

VAN'T HOF J. (1973) Two principal points of control in the mitotic cycle of pea meristem cells: energy considerations, characterization, and radiosensitivity. In *Advances in Radiation Research—Biology and Medicine*, Vol. 2. (Eds. Duplan J.F. & Chapiro A.) pp. 881–94. Gordon & Breach Science Publishers, New York.

VARDJAN M. & NITSCH J.P. (1969) La régénération chez *Cichorium endiva* L. Études des auxines et des 'kinines' endogenèse. *Bull. Soc. bot. Fr.*, **108**, 363–74.

VARNER J. (1975) Hormone mediated integration of seedling physiology. *Adv. exp. Med. Biol.*, **62**, 65–78.

VARNER J.E. (1978) Mode of action of plant hormones. *Int. Rev. Biochem.*, **20**, 189–218.

Varner J.E. & Chandra G.R. (1964) Hormonal control of enzyme synthesis in barley endosperm. *Proc. Nat. Acad. Sci. USA.*, **52**, 100–6.

Varns J.L., Currier W.W. & Kuć J. (1971) Specificity of rishitin and phytuberin accumulation by potato. *Phytopathol.*, **61**, 968–71.

Varns J.L. & Kuć J. (1971) Suppression of rishitin and phytuberin accumulation and hypersensitive response in potato by compatible races of *Phytophthora infestans*. *Phytopathol.*, **61**, 178–81.

Vasil I.K. (1967) Physiology and cytology of anther development. *Biol. Rev.*, **42**, 327–73.

Vasil I.K. (1973) The new biology of pollen. *Naturwissenschaften*, **60**, 247–53.

Vedel F. & Quétier F. (1978) Hydrolyse spécifique de l'ADN chloroplastique et de l'ADN mitochondrial des végétaux supérieurs par les enzymes de restriction. *Physiol. Vég.*, **16**(3), 411–25.

Vedel F., Quétier F. & Bayen M. (1976) Specific cleavage of chloroplast DNA from higher plants by EcoR I restriction nuclease. *Nature, Lond.*, **263**, 440–2.

Venis M.A. (1977) Receptors for plant hormones. *Adv. bot. Res.*, **5**, 53–89.

de Veris G.E. (1980) Transport of sugars and organic acids in *Rhizobium leguminosarum* and their role in symbiosis. Thesis. Leiden State University, Holland.

Verma D.P.S. (1980) Expression of host genes during symbiotic nitrogen fixation. In *Genome Organization and Expression in Plants.* (Ed. Leaver C.J.) pp. 439–52. Plenum Press, New York.

Verma D.P.S. & Bal A.K. (1976) Intracellular site of synthesis and localization of leghaemoglobin in soybean root nodules. *Proc. Nat. Acad. Sci. USA*, **73**, 3843–7.

Verma D.P.S., MacLachlan G.A., Byrne H. & Ewing D. (1975). Regulation and *in vitro* translation of mRNA for cellulase from auxin-treated pea epicotyls. *J. biol. Chem.*, **250**, 1019–26.

Verma D.P.S., Nash D.T. & Schulman H.M. (1974) Isolation and *in vitro* translation of soybean leghaemoglobin mRNA. *Nature, Lond.*, **251**, 74–7.

Verma D.P.S., Zogbi V. & Bal A.K. (1978a) A cooperative action of plant and *Rhizobium* in dissolving host cell wall during development of symbiosis. *Plant Sci. Lett.*, **13**, 137–42.

Verma D.P.S., Kazazian V., Zogbi V. & Bal A.K. (1978b) Isolation and characterization of the membrane envelope enclosing the bacteriods in soybean root nodules. *J. Cell Biol.*, **78**, 919–36.

Verma D.P.S., Hunter N. & Bal A.K. (1978c) Asymbiotic association of *Rhizobium* with pea epicotyls treated with plant hormone. *Planta*, **138**, 107–10.

Verma D.P.S., Ball S., Guérin C. & Wanamaker L. (1979) Leghemoglobin biosynthesis in soybean root nodules. Characterization of nascent and released peptides and the relative rate of synthesis of major leghaemoglobins. *Biochem.*, **18**, 476–83.

Verma D.P.S., Haugland R., Auger S., Brisson N., Legocki R.P. & Lacroix L. (1981) Expression of leghemoglobin and other host genes in effective and ineffective nodules of soybean. *Biochim. biophys. Acta*, (in press).

Vest G. (1970) Rj_3—A gene conditioning ineffective nodulation in soybean. *Crop Sci.*, **10**, 34–5.

Vest G. & Caldwell B.F. (1972) Rj_4—A gene conditioning ineffective nodulation in soybean. *Crop. Sci.*, **12**, 692–3.

Vian B. & Roland J-C. (1972) Différentiation des cytomembranes et renouvellement du plasmalemme dans les phénomènes de sécrétions végétales. *J. Microscop.*, **13**, 119–36.

Viell-Maek A. (1978) Genetische Einflüsse auf Wachstum und Differenzierung des Wurzelspitzen meristems von *Antirrhinum majus* L. und einigen Mutanten. *Biol. Zbl.*, **97**, 701–14.

Vieweg G.H. (1974) Enzyme des Saccharosestoffwechsels in Wurzeln. *Planta*, **116**, 349–59.

Villiers T.A. (1972) Seed dormancy. In *Seed Biology*, vol. II. (Ed. Kozlowski T.T.) pp. 219–81. Academic Press, New York.

Viotti A., Sala E., Alberi P. & Soave C. (1975) RNA metabolism and polysome profiles during seed development in normal and opaque-2 maize endosperms. *Maydica*, **20**, 111–24.

Viotti A., Balducci C. & Weil J.H. (1978a) Adaptation of the tRNA population of maize endosperm for zein synthesis. *Biochim. biophys. Acta*, **517**, 125–32.

Viotti A., Sala E., Alberi P. & Soave C. (1978b) Heterogeneity of zein synthesized *in vitro*. *Plant Sci. Lett.*, **13**, 365–75.

Viotti A., Sala E., Marotta R., Alberi P., Balducci C. & Soave C. (1979) Genes and mRNAs coding for zein polypeptides in *Zea mays*. *Eur. J. Biochem.*, **102**, 211–22.

Virgin H.I. (1962) Light-induced unfolding of the grass leaf. *Physiol. Plant.*, **15**, 380–9.

Virtanen A.I., Jorma J., Lincola H. & Linnaslmi A. (1947) On the relation between nitrogen fixation and leghaemoglobin. *Acta Chemica. Scandinavica*, **1**, 90–111.

Vöchting H. (1878) *Über Organbildung im Pflanzen Reich.* Max Cohen, Bonn.

VOLKMANN D. & SIEVERS A. (1979) Graviperception in multicellular organisms. In *Encyclopaedia of Plant Physiology*. New Series, **71**, 573–626.

DE VOS G., DE BEUCKELEER M., VAN MONTAGU M. & SCHELL J. Restriction endonuclease mapping of the octopine tumor inducing pTiAch5 of *Agrobacterium tumefaciens*. *Plasmid*, submitted.

WACHTEL A.W., GETTNER M.E. & ORNSTEIN L. (1966) Microtomy. In *Physical Techniques in Biological Research* 2e. IIIA. (Ed. Pollister A.W.) Academic Press, New York.

WADA M. & O'BRIEN T.P. (1975) Observations on the structure of the protonema of *Adiantum capillusveneris* L. undergoing cell division following white-light irradiation. *Planta*, **126**, 213–27.

WADDINGTON C.H. (1956) *Principles of embryology*. Allen & Unwin, London.

WADDINGTON C.H. (1957) *The Strategy of the Genes*. Allen & Unwin, London.

WADE M. & ALBERSHEIM P. (1979) Race-specific molecules that protect soyabeans from *Phytophthora megasperma* var *sojae*. *Proc. Nat. Acad. Sci. USA*, **76**, 4433–7.

WAGNER G. & BELLINI E. (1976) Light-dependent fluxes and compartmentation of calcium in the green alga Mougeotia. *Z. Pflphysiol.*, **79**, 283–91.

WAINWRIGHT C.M. (1977) Sun-tracking and related leaf movements in a desert lupine (*Lupinus arizonicus*). *Am. J. Bot.*, **64**, 1032–41.

WAISMAN D.M., STEVENS F.C. & WANG J.H. (1978) Purification and characterisation of a Ca^{2+}-binding protein in *Lumbricus terrestris*. *J. biol. Chem.*, **253**, 1106–13.

WALBOT V. (1977) The dimorphic chloroplasts of the C_4 plant *Panicum maximum* contain identical genomes. *Cell*, **11**, 729–37.

WALBOT V. & COE E.H. (1979) Nuclear gene *iojap* conditions a programmed change to ribosome-less plastids in *Zea mays*. *Proc. Nat. Acad. Sci. USA*, **76**, 2760–4.

WALBOT V. & DURE L.S. (1976) Developmental biochemistry of cotton seed embryogenesis and germination VII. Characterization of the cotton genome. *J. molec. Biol.*, **101**, 503–36.

WALD G. & DUBUY H.G. (1936) Pigments of the oat coleoptile. *Science*, **84**, 247.

WALKER D.A. (1976) Regulatory mechanisms in photosynthetic carbon metabolism. *Curr. Topics in Cell. Regulation*, **11**, 203–41.

WALLACE D.H. (1979) Interactions of S alleles in sporophytically controlled self incompatibility of *Brassica*. *Theor. Appl. Genet.*, **54**, 193–201.

WALTON D.C. (1980) Biochemistry and physiology of abscisic acid. *Ann. Rev. Plant Physiol.*, **31**, 453–89.

WALTON J.D., EARLE E.D., YODER O.C. & SPANSWICK R.M. (1979) Reduction of steady state ATP levels in susceptible maize mesophyll protoplasts by *Helminthosporium maydis* race T toxin. *Plant Physiol.*, **63**, 806–10.

WALTON D.C. & SOOFI G. (1969) Germination of *Phaseolus vulgaris* III. The role of nucleic acid and protein synthesis in the initiation of axis elongation. *Plant Cell Physiol.*, **10**, 307–15.

WANG J.H. & WAISMAN D.M. (1979) Calmodulin and its role in the second messenger system. *Curr. Topics in Cell. Regulation*, **15**, 47–107.

WANGENHEIM K-H. VON (1975) A major component of the radiation effect: interference with endocellular control of cell proliferation and differentiation. *Int. J. radiat. Biol.*, **27**, 7–30.

WANGENHEIM K-H. VON (1976) A mechanism for the endocellular control of cell differentiation and cell proliferation. *J. theoret. Biol.*, **59**, 205–22.

WANGERMANN E. (1974) The pathway of transport of applied idolyl-acetic acid through internode segments. *New Phytol.*, **73**, 623–36.

WARDLAW C.W. (1949) Experiments in organogenesis in ferns growth (Suppl.), **13**, 93–131.

WARDLAW C.W. (1955) *Embryogenesis in Plants*. John Wiley & Sons, New York.

WARDLAW C.W. (1965) The organisation of the shoot apex. In *Encyclopaedia of Plant Physiology*. (Ed. Ruhlard W.) Vol. 15/1, 966–1076. Springer-Verlag, Berlin.

WARDLAW I.F. (1968) The control and pattern of movement of carbohydrates in plants. *Bot. Rev.*, **34**, 79–105.

WARDROP A.B. (1955) The mechanism of surface growth in parenchyma of *Avena coleoptiles*. *Aust. J. Bot.*, **3**, 137–48.

WARDROP A.B. (1964) The structure and formation of the cell wall in xylem. In *The Formation of Wood in Forest Trees*. (Ed. Zimmerman M.H.) Academic Press, New York.

WARDROP A.B., WOLTERS-ART M. & SASSEN M.M.A. (1979) Changes in microfibril orientation in the walls of elongating plant cells. *Acta Bot. Neerl.*, **28**, 313–33.

WARDROP A.J. & POLYA G.M. (1977) Properties of a soluble auxin-binding protein from dwarf bean seedlings. *Plant Sci. Lett.*, **8**, 155–63.

WARDROP A.J. & POLYA G.M. (1980a) The ligand specificity of bean leaf soluble auxin-binding protein. *Plant Physiol.*, **66**, 105–11.

WARDROP A.J. & POLYA G.M. (1980b) Co-purification of pea and bean leaf soluble auxin-binding proteins with ribulose-1,5-bisphosphate carboxylase. *Plant Physiol.*, **66**, 112–18.

Wareing P.F. (1954) Growth studies in woody species. VI. The locus of photoperiodic perception in relation to dormancy. *Physiol. Plant*, **7**, 261–77.

Wareing P.F. (1959) Problems of juvenility and flowering in trees. *J. Linn. Soc. Lond.*, **56**, 282–9.

Wareing P.F. (1971) Some aspects of differentiation in plants. *Symp. Soc. exp. Biol.*, **25**, 323–44.

Wareing P.F. (1977) Growth substances and integration in the whole plant. *Symp. Soc. exp. Biol.*, **31**, 337–65.

Wareing P.F. (1978) Determination in plant development. *Bot. Mag. Tokyo*, Special Issue **1**, 3–18.

Wareing P.F. (1979) What is the basis of the stability of apical meristems? BPGRG Monograph No. 3—*Control of Plant Development*, 1–9.

Wareing P.F. & Phillips I.D.J. (1978) *The Control of Growth and Differentiation in Plants*. 2e. p. 210. Pergamon Press, Oxford.

Wareing P.H. & Thompson A.G. (1976) Rapid effects of red light on hormone levels. In *Light and Plant Development*. (Ed. Smith H.) pp. 285–94. Butterworth, London.

Warmke H.E. & Lee S-L.J. (1977) Mitochondrial degeneration in Texas cytoplasmic male-sterile corn anthers. *J. Hered.*, **68**, 213–22.

Warmke H.E. & Lee S-L.J. (1978) Pollen abortion in T-cytoplasmic male-sterile corn (*Zea mays*): a suggested mechanism. *Science*, **200**, 561–3.

Warner T.J. & Ross J.D. (1981) Phytochrome control of maize coleoptile section elongation: the role of cell wall extensibility. *Plant Physiol.*, (in press).

Warner T.J., Ross J.D. & Coombs J. (1981) Phytochrome control of maize coleoptile section elongation: a rapid loss of photoreversibility. *Plant Physiol.*, **67**, 355–7.

Waterkeyn L. & Bienfait A. (1970) On a possible function of the callosic special wall in *Ipomoea purpurea* (L.) Roth. *Grana*, **10**, 13–20.

Waters S.P., Peoples M.B., Simpson R.J. & Dalling M.J. (1980) Nitrogen redistribution during grain growth in wheat (*Triticum aestivum* L.) I. Peptide hydrolase activity and protein breakdown in the flag leaf, glumes and stem. *Planta*, **148**, 422–8.

Watrud L.S., Baldwin J.K., Miller R.J. & Koeppe D.E. (1975) Production of a sensitive response to *Helminthosporium maydis* race T toxin in resistant mitochondria of corn (*Zea mays*, L.) by removal of the outer membrane. *Plant Physiol.*, **56**, 216–21.

Watson B., Currier T.C., Gordon M.P., Chilton M-D. & Nester E.W. (1975) Plasmid required for virulence of *Agrobacterium tumefaciens*. *J. Bacteriol.*, **123**, 255–64.

Webster P.L. & van't Hof J. (1970) DNA synthesis and mitosis in meristems: requirements for RNA and protein synthesis. *Am. J. Bot.*, **57**, 130–9.

Weir E.M., Riezman H., Grienenberger J.M., Becker W.M. & Leaver C.J. (1980) Regulation of glyoxysomal enzymes during germination of cucumber. *Eur. J. Biochem.*, **112**, 469–77.

Weisenseel M.H. (1979) Growth movements directed by light: induction of polarity. In *Encyclopaedia of Plant Physiology, New Series* Vol. 7. (Eds. Haupt W. & Feinleib M.E.) pp. 485–505. Springer-Verlag, Berlin.

Weisenseel M.H., Dorn A. & Jaffe L.F. (1979) Natural H^+ ion currents traverse growing roots and hairs of barley. (*Hordeum vulgare* L.) *Plant Physiol.*, **64**, 512–18.

Weisenseel M.H., Nuccitelli R. & Jaffe L.F. (1975) Large electrical currents traverse growing pollen tubes. *J. Cell Biol.*, **66**, 556–67.

Weisenseel M.H. & Ruppert H.K. (1977) Phytochrome and calcium ions are involved in light induced depolarization in *Nitella*. *Planta*, **137**, 225–9.

Weisenseel M.H. & Smeibidl E. (1973) Phytochrome controls the water permeability in *Mougeotia*. *Z. Pflphysiol.*, **70**, 420–31.

Weiss R.L., Goodenough D.A. & Goodenough U.W. (1977) Membrane particle arrays associated with the basal body and with contractile vacuole secretion in *Chlamydomonas*. *J. Cell Biol.*, **72**, 133–43.

Wellburn A.R. & Hampp R. (1976) Fluxes of gibberellic and abscissic acids, together with that of adenosine 3′,5′ cyclic phosphate, across plastid envelopes during development. *Planta*, **131**, 95–6.

Wellburn F.A.M. & Wellburn A.R. (1979) Conjoined mitochondria and plastids in the barley mutant *Albostrians*. *Planta*, **147**, 178–9.

Went F.W. (1925) Over het verschil in gevogliheid von top en basis der colesptielen van *Avena* voor licht. *Proc. Kon. Ned. Akad. Wetensch.*, **34**, 1021–7.

Went F.W. (1929) Wuchstoff und Wachstum. *Rec. Trav. Bot. Neerl.*, **25**, 1–116.

Went F.W. & Thimann K.V. (1937) *Phytohormones*. Macmillan, New York.

WERKER E. & KISLEV M. (1978) Mucilage on the root surface and root hairs of *Sorghum*: heterogeneity in structure, manner of production and site of accumulation. *Ann. Bot.*, **42**, 809–16.

WERNER D., MORSCHEL E., STRIF R. & WINCHBACH B. (1980) Development of nodules of *Glycine max* infected with an ineffective strain of *Rhizobium japonicum*. *Planta*, **147**, 320–9.

WEST C., WADE M., MCMILLAN C., III. & ALBERSHEIM P. (1980) Purification and properties of invertases extractable from *Phytophthora megasperma* var. *sojae mycelia*. *Arch. biochem. Biophys.*, **201**, 25–35.

WETMORE R.H. & RIER J.P. (1963) Experimental induction of vascular tissues in callus of angiosperms. *Am. J. Bot.*, **50**, 418–30.

WEYERS J.B.D. (1978) The hormonal control of stomatal aperture in *Commelina communis* L. by abscissic acid. Ph.D. Thesis, University of Glasgow, pp. 172–3.

WHATLEY J.M. (1974) Chloroplast development in primary leaves of *Phaseolus vulgaris*. *New Phytol.*, **73**, 1097–110.

WHATLEY J.M. (1978) A suggested cycle of plastid developmental interrelationships. *New Physiol.*, **80**, 489–502.

WHATLEY J.M., JOHN P. & WHATLEY F.R. (1979) From extracellular to intracellular: the establishment of mitochondria and chloroplasts. *Proc. roy Soc. Lond. B.*, **204**, 165–87.

WHATLEY M.H., BODWIN J.S., LIPPINCOTT B.B. & LIPPINCOTT J.A. (1976) Role for *Agrobacterium* cell envelope lipopolysaccharides in infection site attachment. *Infect. Immun.*, **13**, 1080–3.

WHEELER H. (1975) *Plant Pathogenesis*. Elsevier/North-Holland, Amsterdam.

WHEELER H. (1977) The role of phytotoxins in specificity. In *Specificity in Plant Disease*. (Eds. Wood R.K.S. & Graniti A.) pp. 217–30. Plenum Press, New York.

WHEELER H. (1978) Disease alterations in permeability and membranes. In *Plant Disease: An Advanced Treatise*, Vol. 3. (Eds. Horsfall J.G. & Cowling E.B.) pp. 327–47. Academic Press, New York.

WHITELAM G.C., JOHNSON C.B. & SMITH H. (1979) The control by phytochrome of nitrate reductase in the curd of light-grown cauliflower *Photochem. Photobiol.*, **30**, 589–94.

WHYTE P. & LUCKWILL L.C. (1966) Sensitive bioassays for gibberellins based upon retardation of leaf senescence in *Rumex obtusifolius*. *Nature, Lond.*, **210**, 1360.

WICK S.M. & HEPLER P.K. (1980) Localization of calcium containing antimonate precipitates during mitosis. (in press).

WIENAND U., BRÜSCHKE C. & FEIX G. (1979) Cloning of double stranded DNAs derived from polysomal mRNA of maize endosperm: isolation and characterization of zein clones. *Nucleic Acids Research.*, **6**, 2707–15.

WIENAND U. & FEIX G. (1978) Electrophoretic fractionation and translation *in vitro* of poly(rA)-containing RNA from maize endosperm. *Eur. J. Biochem.*, **92**, 605–11.

VAN DER WILDEN W., HERMAN E.M. & CHRISPEELS M.J. (1980) Protein bodies of mung bean cotyledons as autophagic organelles. *Proc. Nat. Acad. Sci.* USA, **77**, 428–32.

WILDMAN S.G., JOPE C. & ATCHISON B.A. (1974) Role of mitochondria in the origin of chloroplast starch grains. Description of the phenomenon. *Plant Physiol.*, **54**, 231–7.

WILLIAMS L.F. & LYNCH D.L. (1954) Inheritance of a non-nodulating character in soybean. *Agronomy*, **46**, 28–9.

WILLIAMS R.F. & RIJVEN A.H.G.C. (1965) The physiology of growth in the wheat plant. II. The dynamics of leaf growth. *Aust. J. biol. Sci.*, **18**, 721–43.

WILLIAMS W.T. (1967) Numbers, taxonomy and judgement. *Bot. Rev.*, **33**, 379–86.

WILLMITZER L., DE BEUCKELEER M., LEMMERS M., VAN MONTAGU M. & SCHELL J. (1980) The Ti-plasmid derived T-DNA is present in the nucleus and absent from plastids of crown gall cells. *Nature, Lond.*, **287**, 359–61.

WILSON D.W. (1940) *The Biochemistry of Symbiotic Nitrogen Fixation*. University of Wisconsin Press, Madison.

WILSON K. (1957) Extension growth in primary cell walls with special reference to *Elodea canadensis*. *Ann. Bot.*, **21**, 1–12.

WIMPEE C.F. & RAWSON J.R.Y. (1979) Characterization of the nuclear genome of pearl millet. *Biochim. biophys. Acta*, **562**, 192–206.

WINFREE A.T. (1973) Scroll-shaped waves of chemical activity in three dimensions. *Science*, **181**, 937–9.

WIPF L. & COOPER D.C. (1940) Somatic doubling of chromosomes and nodular infection in certain leguminosae. *Am. J. Bot.*, **27**, 821–4.

DE WIT P.J.G.M. & ROSEBOOM P.H.M. (1980) Isolation, partial characterization and specificity of glycoprotein elicitors from culture filtrates, mycelium and cell walls of *Cladosporium fulvum* (syn. *Fulvia fulva*). *Physiol. Plant Pathol.*, **16**, 391–408.

WITTENBACH V.A. (1978) Breakdown of

ribulose bisphosphate carboxylase and change in proteolytic activity during dark-induced senescence of wheat seedlings. *Plant Physiol.*, **62**, 604–8.

Wittenberg J.B., Bergerson F.J., Appleby C.A. & Turner G.L. (1974) Facilitated oxygen diffusion: the role of leghaemoglobin in nitrogen fixation by bacteroids isolated from soybean root nodules. *J. biol. Chem.*, **249**, 4057–66.

Woledge J. (1972) The effect of shading on the photosynthetic rate and longevity of grass leaves. *Ann. Bot.*, **36**, 551–61.

Wolpert J.S. & Albersheim P. (1976) Host–symbiont interactions I. The lectins of legumes interact with the O-antigen containing lipopolysaccharides of their symbiont *Rhizobia*. *Biochem. biophys. Res. Comm.* **70**, 729–37.

Wong S.C., Cowan I.R. & Farquhar G.D. (1979) Stomatal conductance correlates with photosynthetic capacity. *Nature, Lond.*, **282**, 424–6.

Wong W.-J.L. & ap Rees T. (1971) Carbohydrate oxidation in stele and cortex isolated from roots of *Pisum sativum*. *Biochim. biophys. Acta*, **252**, 296–304.

Wood R.K.S. (1976) Killing of protoplasts. In *Biochemical Aspects of Plant–Parasite Relationships.* (Eds. Friend J. & Threlfall D.R.) pp. 105–16. Academic Press, London.

Woodcock A.E.R. & Wilkins M.B. (1969) The geoelectric effect in plant shoots. I. The characteristics of the effect. *J. exp. Bot.*, **20**, 156–69.

Woodcock C.L.F. (1971) The anchoring of nuclei by cytoplasmic microtubules in *Acetabularia*. *J. Cell Sci.*, **8**, 611–21.

Woodcock C.L.F. & Bogorad L. (1970) Evidence for variation in the quality of DNA among plastids of *Acetabularia mediterranea*. *J. Cell Biol.*, **44**, 361–75.

Woodcock C.L.F. & Fernandez-Moran H. (1968) Electron microscopy of DNA conformations in spinach chloroplasts. *J. molec. Biol.*, **31**, 627–31.

Woodward J.R., Keane P.J. & Stone B.A. (1980) Structures and properties of wilt-inducing polysaccharides from *Phytophthora* spp. *Physiol. Plant Pathol.*, **16**, 439–54.

Woodward R.G. & Rawson H.M. (1976) Photosynthesis and transpiration in dicotyledonous plants. II. Expanding and senescing leaves of soybean. *Aust. J. Plant Physiol.*, **3**, 257–67.

Woolhouse H.W. (1967) *The Nature of Senescence in Plants.* Symp. Soc. exp. Biol., **21**, 179–214.

Woolhouse H.W. (1968) Leaf age and mesophyll resistance as factors in the rate of photosynthesis. *Hilger J.*, **11**, 15–20.

Woolhouse H.W. (1974) Longevity and senescence in plants. *Sci. Progr. Oxford*, **61**, 123–47.

Woolhouse H.W. (1978a) Senescence processes in the life cycle of flowering plants. *Bioscience*, **28**, 25–31.

Woolhouse H.W. (1978b) Light gathering and carbon assimilation processes in photosynthesis, their adaptive modifications and significance for agriculture. *Endeavour, New Series*, **2**, 35–40.

Woolhouse H.W. (1978c) Cellular and metabolic aspects of senescence in higher plants. In *The Biology of Aging.* (Eds. Behnke J.A., Finch C.E. & Moment G.B.) pp. 83–9. Plenum Press, New York.

Woolhouse H.W. (1980) Plant physiology in relation to crop production—the genetic link. In *Physiological Processes Limiting Plant Productivity.* 30th Easter School in Agricultural Science, University of Nottingham, (Ed. Johnson C.B.) Butterworth, Sevenoaks.

van der Woude W.J. & Morré D.J. (1968) Endoplasmic-reticulum-dictyosome-secretory vesicle associations in pollen tubes of *Lilium longiflorum* Thunb. *Proc. Indiana Acad. Sci.*, **77**, 164–70.

van der Woude W.J., Morré D.J. & Bracker C.E. (1971) Isolation and characterization of secretory vesicles in germinated pollen of *Lilium longiflorum*. *J. Cell Sci.*, **8**, 331–51.

Wright D.J. & Boulter D. (1972) The characterization of vicilin during seed development in *Vicia faba* L. *Planta*, **105**, 60–5.

Wright K. & Northcote D.H. (1976) Identification of β 1 $\rightarrow$ 4 glucan chains as part of a fraction of slime synthesised within the dictyosomes of maize root caps. *Protoplasma*, **88**, 225–39.

Wright M., Mousdale D.M.A. & Osborne D.J. (1978) Evidence for a gravity regulated level of endogenous auxin controlling cell elongation and ethylene production during geotropic bending in grass nodes. *Biochem. Physiol. Pflanzen*, **172**, 581–96.

Wright M. & Osborne D.J. (1974) Abscission in *Phaseolus vulgaris*. The positional differentiation and ethylene-induced expansion growth of specialised cells. *Planta*, **120**, 163–70.

Wright M. & Osborne D.J. (1977) Gravity-regulation of cell elongation in nodes of the grass *Echinochloa colonum*. *Biochem. Physiol. Pflanzen*, **171**, 379–92.

Wright S.T.C. (1961) Growth and cellular differentiation in the wheat coleoptile. *J. exp. Bot.*, **12**, 303–18.

WRIGHT S.T.C. (1963) Cellular differentiation at the molecular level with special references to proteins. In *Cell Differentiation*. Symp. Soc. Exp. Biol. Med. **XVII**. (Ed. Fogg G.E.) pp. 18–40. Cambridge University Press.

WRIGLEY C.W., DUCROS D.L., ARCHER M.J., DOWNIE P.G. & ROXBURGH C.M. (1980) The sulfur content of wheat endosperm proteins and its relevance to grain quality. *Aust. J. Plant Physiol.*, **7**, 755–66.

WRIGLEY C.W. & SHEPHERD K.W. (1973) Electro-focusing of grain proteins from wheat genotypes. *Ann. New York Acad. Sci.*, **209**, 154–62.

WUTTKE H-G. (1976) Circular DNA in chromoplasts of *Tulipa gesneriana*. *Planta*, **132**, 317–19.

YADAV N.S., POSTLE K., SAIKI R.K., THOMASHOW M.F. & CHILTON M-D. (1980) T-DNA of a crown gall teratoma is covalently joined to host plant DNA. *Nature, Lond.*, **287**, 458–61.

YAMANE H., MUROFUSHI N. & TAKAHASHI N. (1975) Metabolism of gibberellins in maturing and germinating bean seeds. *Phytochem.*, **14**, 1195–200.

YEMM E.W. (1956) The metabolism of senescent leaves. In *CIBA Coloquia on Ageing.*, **2**, 202–14 (Ed. Wolstenholme G.E.W.).

YOCUM C.S. (1964) Recent studies of symbiotic nitrogen fixation. *Science*, **146**, 432.

YOMO H. & SRINIVASAN K. (1973) Protein breakdown and formation of protease in attached and detached cotyledons of *Phaseolus vulgaris* L. *Plant Physiol.*, **52**, 671–3.

YOMO H. & VARNER J.E. (1971) Hormonal control of a secretory tissue. In *Current Topics in Developmental Biology*. Vol. 6, pp. 111–44 (Eds. Moscona A.A. & Monroy A.). Academic Press, London.

YOO B.Y. & CHRISPEELS M.J. (1980) The origin of protein bodies in developing soybean cotyledons: a proposal. *Protoplasma*, **103**, 201–4.

YORK D.W., EARLE E.D. & GRACEN V.E. (1980) Ultrastructural effects of *Helminthosporium maydis* race T on isolated corn mitochondria and mitochondria within corn protoplasts. *Canad. J. Bot.*, **58**, 1562–70.

YOSHIDA Y. (1961) Nuclear control of chloroplast activity in *Elodea* leaf cells. *Protoplasma*, **54**, 476–92.

YOSHIDA Y., LAULHÈRE J-P., ROZIER C. & MACHE R. (1978) Visualization of folded chloroplast DNA from spinach. *Biol. Cellulaire*, **32**, 187–90.

YOSHIKAWA M. (1978) Diverse modes of action of biotic and abiotic phytoalexin elicitors. *Nature, Lond.*, **275**, 546–7.

YOSHIKAWA M., MASAGO H. & KEEN N.T. (1977) Activated synthesis of poly (A)-containing messenger RNA in soyabean hypocotyls infected with *Phytophthora megasperma* var. *sojae*. *Physiol. Plant Pathol.*, **10**, 125–38.

YOSHIKAWA M., YAMAUCHI K., & MASAGO H. (1979) Biosynthesis and biodegradation of glyceollin by soyabean hypocotyls infected with *Phytophthora megasperma* var. *sojae*. *Physiol. Plant Pathol.*, **14**, 157–69.

YOULE R.J. & HUANG A.H.C. (1978) Albumin storage proteins in the protein bodies of castor bean. *Plant Physiol.*, **61**, 13–16.

YOUNG E.M. & RAFF R.A. (1979) Messenger ribonucleoprotein particles in developing sea urchin embryos. *Dev. Biol.*, **72**, 24–40.

ZAENEN I., VAN LAREBEKE N., TEUCHY H., VAN MONTAGU M. & SCHELL J. (1974). Supercoiled circular DNA in crown gall inducing *Agrobacterium* strains. *J. molec. Biol.*, **86**, 109–27.

ZAMBRYSKI P., HOLSTERS M., KRUGER K., DEPICKER A., SCHELL J., VAN MONTAGU M. & GOODMAN H.M. (1980) Tumor DNA structure in plant cells transformed by *A. tumefaciens*. *Science*, **209**, 1385–91.

ZAMIR D., JONES R.A. & KEDAR N. (1980) Anther culture of male-sterile tomato (*Lycopersicon esculentum* Mill.) mutants. *Plant Sci. Lett.*, **17**, 353–61.

ZERBAN H., WEHNER M. & WERZ G. (1973) Über die Feinstruktur des Zellkernes von *Acetabularia* nach Gefrierätzung. *Planta*, **114**, 239–50.

ZETSCHE K. (1968) Regulation der UDPG-Pyrophosphorylase-aktivität in *Acetabularia*. I. Morphogenese und UDPG-Pyrophosphorylase-Synthese in kernhaltigen und kernlosen Zellen. *Z. Naturforschung*, **23**b, 369–76.

ZIEGLER E. & ALBERSHEIM P. (1977) Host–pathogen interactions XIII. Extracellular invertases secreted by three races of a plant pathogen are glycoproteins which possess different carbohydrate structures. *Plant Physiol.*, **59**, 1104–10.

ZIMMERMAN B.K. & BRIGGS W.R. (1963a) Phototropic dosage-response curves for oat coleoptiles. *Plant Physiol.*, **38**, 248–53.

ZIMMERMAN B.K. & BRIGGS W.R. (1963b) A kinetic model for phototropic responses of oat coleoptiles. *Plant Physiol.*, **38**, 253–61.

ZIMMERMAN J.L. & GOLDBERG R.B. (1977) DNA

Sequence organization in the genome of *Nicotiana tabacum*. *Chromosoma*, **59**, 227–52.

ZIMMERMAN R.H. (Ed.) (1976) Symposium on juvenility in woody perennials *Acta Horticulturae*, **56**.

ZIMMERMANN U. (1978) Physics of turgor and osmoregulation. *Ann. Rev. Plant Physiol.*, **29**, 121–92.

INDEX